Can Başkent · Thomas Macaulay Ferguson
Editors

Graham Priest on Dialetheism and Paraconsistency

Editors
Can Başkent
Department of Computer Science
University of Bath
Bath, UK

Thomas Macaulay Ferguson
Cycorp
Austin, TX, USA

Saul Kripke Center
New York, NY, USA

ISSN 2211-2758   ISSN 2211-2766 (electronic)
Outstanding Contributions to Logic
ISBN 978-3-030-25364-6   ISBN 978-3-030-25365-3 (eBook)
https://doi.org/10.1007/978-3-030-25365-3

© Springer Nature Switzerland AG 2019
This work is subject to copyright. All rights are reserved by the Publisher, whether the whole or part of the material is concerned, specifically the rights of translation, reprinting, reuse of illustrations, recitation, broadcasting, reproduction on microfilms or in any other physical way, and transmission or information storage and retrieval, electronic adaptation, computer software, or by similar or dissimilar methodology now known or hereafter developed.
The use of general descriptive names, registered names, trademarks, service marks, etc. in this publication does not imply, even in the absence of a specific statement, that such names are exempt from the relevant protective laws and regulations and therefore free for general use.
The publisher, the authors and the editors are safe to assume that the advice and information in this book are believed to be true and accurate at the date of publication. Neither the publisher nor the authors or the editors give a warranty, expressed or implied, with respect to the material contained herein or for any errors or omissions that may have been made. The publisher remains neutral with regard to jurisdictional claims in published maps and institutional affiliations.

Cover photo credit: © Steve Pyke

This Springer imprint is published by the registered company Springer Nature Switzerland AG
The registered company address is: Gewerbestrasse 11, 6330 Cham, Switzerland

Contents

1. Introduction to Graham Priest on Dialetheism and Paraconsistency 1
   Thomas Macaulay Ferguson and Can Başkent

2. Modal Meinongianism: Conceiving the Impossible 3
   Franz Berto

3. The Number of Logical Values 21
   Ross T. Brady

4. Respects for Contradictions 39
   Paul Égré

5. Hegel and Priest on Revising Logic 59
   Elena Ficara

6. Paraconsistent or Paracomplete? 73
   Hartry Field

7. Priest's Anti-Exceptionalism, Candrakīrti and Paraconsistency 127
   Koji Tanaka

8. Looting Liars Masking Models 139
   Diderik Batens

9. Inferential Semantics, Paraconsistency, and Preservation of Evidence 165
   Walter Carnielli and Abilio Rodrigues

10. A Model-Theoretic Analysis of Fidel-Structures for mbC 189
    Marcelo E. Coniglio and Aldo Figallo-Orellano

11. Unity, Identity, and Topology: How to Make Donuts and Cut Things in Half 217
    A. J. Cotnoir

12	Contradictory Information: Better Than Nothing? The Paradox of the Two Firefighters 231
	J. Michael Dunn and Nicholas M. Kiefer

13 Variations on the Collapsing Lemma 249
   Thomas Macaulay Ferguson

14 Dialetheic Conditional Modal Logic 271
   Patrick Girard

15 Priest on Negation... 285
   Lloyd Humberstone

16 From Iff to Is: Some New Thoughts on Identity in Relevant Logics 343
   Edwin Mares

17 The Difficulties in Using Weak Relevant Logics for Naive Set Theory 365
   Erik Istre and Maarten McKubre-Jordens

18 ST, LP and Tolerant Metainferences 383
   Bogdan Dicher and Francesco Paoli

19 Annotated Natural Deduction for Adaptive Reasoning 409
   Patrick Allo and Giuseppe Primiero

20 Denotation, Paradox and Multiple Meanings 439
   Stephen Read

21 Two Negations Are More than One 455
   Greg Restall

22 Inconsistency and Incompleteness, Revisited 469
   Stewart Shapiro

23 GP's LP 481
   Neil Tennant

24 Expanding the Logic of Paradox with a Difference-Making Relevant Implication 507
   Peter Verdée

25 On Non-transitive "Identity" 535
   Heinrich Wansing and Daniel Skurt

26 At the Limits of Thought 555
   Zach Weber

27 Some Comments and Replies 575
   Graham Priest

| 28 | Crossing Boundaries | 677 |

Graham Priest

Graham Priest: Publications................................. 691

Editors and Contributors

About the Editors

Can Başkent received his Ph.D. from the City University of New York in 2012. He subsequently worked as a postdoctoral researcher at IHPST, Paris, and INRIA and the University of Bath. Currently, he is a postdoctoral researcher in Ethical AI at Oxford Brookes University. His research interests include modal and non-classical logics and foundations of game theory.

Thomas Macaulay Ferguson works as an ontologist for the Cyc artificial intelligence project and is an affiliate research scholar at the Saul Kripke Center at the City University of New York. He is the author of *Meaning and Proscription in Formal Logic* (Springer 2017) and, with Graham Priest, is coauthor of the supplemental *Dictionary of Logic* (Oxford University Press 2016). His research includes work on philosophical logic, metaphysics, and the philosophy of mathematics.

Contributors

Patrick Allo Centre for Logic and Philosophy of Science, Vrije Universiteit Brussel, Oxford Internet Institute, University of Oxford, Oxford, UK

Can Başkent Department of Computer Science, University of Bath, Bath, UK

Diderik Batens Centre for Logic and Philosophy of Science, Ghent University, Ghent, Belgium

Franz Berto Department of Philosophy, University of St Andrews, St Andrews, UK;
Institute for Logic, Language and Computation (ILLC), University of Amsterdam, Amsterdam, The Netherlands

Ross T. Brady La Trobe University, Melbourne, VIC, Australia

Walter Carnielli State University of Campinas, Campinas, Brazil

Marcelo E. Coniglio Institute of Philosophy and the Humanities (IFCH), Centre for Logic, Epistemology and the History of Science (CLE), University of Campinas (UNICAMP), Campinas, Brazil

A. J. Cotnoir University of St Andrews, St Andrews, Scotland

Bogdan Dicher Centre for Philosophy of the University of Lisbon, Lisbon, Portugal

J. Michael Dunn Indiana University, Bloomington, USA

Paul Égré CNRS, ENS, EHESS, PSL University, Paris, France

Thomas Macaulay Ferguson City University of New York, New York City, USA;
Saul Kripke Center, New York, NY, USA;
Cycorp, Austin, TX, USA

Elena Ficara University of Paderborn, Paderborn, Germany

Hartry Field Philosophy Department, New York University, New York, NY, USA

Aldo Figallo-Orellano Centre for Logic, Epistemology and the History of Science (CLE), University of Campinas (UNICAMP), Campinas, Brazil;
Department of Mathematics, National University of the South (UNS), Bahía Blanca, Argentina

Patrick Girard University of Auckland, Auckland, New Zealand

Lloyd Humberstone Department of Philosophy and Bioethics, Monash University, Melbourne, VIC, Australia

Erik Istre Queensbury, USA

Nicholas M. Kiefer Cornell University, Ithaca, USA

Edwin Mares Victoria University of Wellington, Wellington, New Zealand

Maarten McKubre-Jordens Christchurch, New Zealand

Francesco Paoli Department of Pedagogy, Psychology, Philosophy, University of Cagliari, Cagliari, Italy

Graham Priest Department of Philosophy, The Graduate Center, City University of New York, New York, USA;
Department of Philosophy, The University of Melbourne, Melbourne, Australia

Giuseppe Primiero Department of Philosophy, University of Milan, Milan, Italy

Stephen Read Arché Research Centre, University of St Andrews, St Andrews, Scotland

Greg Restall Philosophy Department, The University of Melbourne, Melbourne, Australia

Abilio Rodrigues Federal University of Minas Gerais, Belo Horizonte, Brazil

Stewart Shapiro The Ohio State University, Columbus, USA

Daniel Skurt Ruhr University Bochum, Bochum, Germany

Koji Tanaka School of Philosophy, Research School of Social Sciences, Australian National University, Canberra, Australia

Neil Tennant Department of Philosophy, The Ohio State University, Columbus, OH, USA

Peter Verdée University Catholique de Louvain, Louvain-la-Neuve, Belgium

Heinrich Wansing Ruhr University Bochum, Bochum, Germany

Zach Weber University of Otago, Dunedin, New Zealand

Chapter 1
Introduction to Graham Priest on Dialetheism and Paraconsistency

Thomas Macaulay Ferguson and Can Başkent

Abstract We provide a short introduction to the volume "Graham Priest on Dialetheism and Paraconsistency."

Keywords Graham Priest · Paraconsistency · Dialetheism

During a time in which the humanities and sciences have progressed steadily toward hyperspecialization, Graham Priest's philosophical output over nearly half a century is exceptional in its breadth (to say nothing of its depth). Across his papers, books, and lectures, Priest has brought an outsider's eye to philosophy—Priest was trained as a mathematician and his knowledge of philosophy is proudly self-taught—which has been a characterizing feature of his work. This outsider's eye is a catalyst for both Priest's willingness to serve as an iconoclast to the idols of Western analytic philosophy and his talent for fostering the kind of synthetic dialogue necessary for alternatives to these idols.

As Priest's short intellectual autobiography in this volume illustrates, his research has touched on seemingly everything; his mark can be found in myriad fields, including political and legal philosophy, Eastern philosophy, game theory, artificial intelligence, and continental philosophy. Of all the fronts in Priest's insurgent career, he is arguably best known for his role as an logician. Although we have little doubt that similar volumes could be devoted to Priest qua metaphysician or philosopher of language, it is this role on which this volume focuses.

T. M. Ferguson (✉)
Saul Kripke Center, New York, NY, USA
e-mail: tferguson@gradcenter.cuny.edu

Cycorp, Austin, TX, USA

C. Başkent
Department of Computer Science, University of Bath, Bath, UK
e-mail: can@canbaskent.net

© Springer Nature Switzerland AG 2019
C. Başkent and T. M. Ferguson (eds.), *Graham Priest on Dialetheism and Paraconsistency*, Outstanding Contributions to Logic 18,
https://doi.org/10.1007/978-3-030-25365-3_1

In his role as an logician, Priest is known for championing the position of dialetheism—the thesis that true contradictions are a part of the fabric of the world—and the techniques of paraconsistency—the property of a consequence relation according to which the hypothesis of a contradiction does not entail everything.

Paraconsistency had lain implicitly in the fabric of several projects in philosophical logic before Priest took up its development. For example, in the communities of relevance (or relevant) logic, the core motivational thesis is that valid entailments require that a hypothesis must be relevant to its consequences. Paraconsistency is a necessary companion of this thesis; the irrelevance between, say, a contradiction in the language of mathematics and an arbitrary statement in the language of biology means that no entailment relation holds between "$0=1$ and not-$0=1$" and "frogs have wings." Paraconsistency likewise accompanies other projects, including connexive logic or discussive logic.

Priest further radicalized the notion of paraconsistency by arguing for the position of dialetheism. Moving from the model-theoretic vantage point from which there exist inconsistent but nontrivial models to the thesis that reality itself includes inconsistencies was indeed a radical move (the preface to the second edition of Priest's *In Contradiction* includes an involved discussion of the trials Priest faced in publishing this landmark work). In the Western tradition, the consistency of the world had been taken nearly as an axiom since Aristotle. To take a relatively inoffensive tool of the nonclassical logician and lend it the gravity of a metaphysical thesis running counter to philosophy's first principles was a risky move; to this day, the central claim of dialetheism is as likely as not to be at the receiving end of the "incredulous stare." But throughout it all, Priest has continued the work of producing calm, clear, and compelling argumentation in its favor.

We believe that the papers of this volume serve as a further demonstration of the breadth and reach of Priest's endeavors in logic. The dimensions along which their contents range—from sympathetic to critical, from philosophical to technical, from analytic to continental—are wide ranging. Each piece seizes on some facet of Priest's work in logic and offers new contributions to his legacy; while there are undoubtedly gaps—it would take multiple volumes to touch on everything that Priest has worked on—it is our opinion that the work included in this volume provides a great representation of the arc of Priest's work and shows that the debates ignited by Priest's work are as compelling today as they've ever been.

Chapter 2
Modal Meinongianism: Conceiving the Impossible

Franz Berto

> *It is impossible to construct a regular polygon of nineteen sides with ruler and compass; it is possible but very complicated to construct one of seventeen sides. In whatever sense I can imagine the possible construction, I can imagine the impossible construction just as well*
>
> David Lewis, *On the Plurality of Worlds*

Abstract Modal Meinongianism—the version of Meinongianism invented by Graham Priest—presupposes that we can think about absolute impossibilities. I defend the view that we can, by tidying up a couple of loose ends in Priest's arguments to this effect from his book *Towards Non-Being*.

Keywords Conceivability and possibility · Modal Meinongianism · Impossible worlds

2.1 Comprehension and Intentionality

In 2005, Graham Priest reinvented Meinongianism: the view that some objects do not exist, thus existence, *pace* Quine (1948), is not captured by the quantifier. He did it by publishing a slim book, *Towards Non-Being*, which included a new approach to one of the Meinongianism's core problems: which conditions characterize objects? Let me explain.

Any Meinongian theory needs some principle stating which objects are admitted by the theory, and which properties they can have. Principles of this kind have been

---

F. Berto (✉)
Department of Philosophy, University of St Andrews, St Andrews, UK
e-mail: fb96@st-andrews.ac.uk; F.Berto@uva.nl

Institute for Logic, Language and Computation (ILLC), University of Amsterdam, Amsterdam, The Netherlands

called *Characterization* or (in analogy with set theory) *Comprehension Principles*, and the problem of finding a good one has been called the Characterization Problem.

Why *problem*? In its naive version (of which it is unclear whether it was ever endorsed by anyone), Meinongianism subscribes to what Parsons (1980) called an 'Unrestricted Comprehension Principle' for objects:

(UCP) For any condition $A[x]$, with free x, some object satisfies $A[x]$.

The principle looks intuitive. Take such features as *x is a detective, x lives in Victorian London at 221b Baker Street, x is Moriarty's archenemy, x has amazing powers of observation and deduction, x always wears a deerstalker...*, etc. If $A[x]$ stands for the conjunction of the corresponding predicates, then according to the (UCP) an object is characterized by $A[x]$. Call it "Sherlock Holmes", h. Then Holmes really has the relevant properties, $A[h]$.

This cannot work, however. As remarked by Priest (2005), p. xix, via the (UCP) one can prove anything whatsoever. Let $A[x]$ be $x = x \land B$, with B an arbitrary formula. By the (UCP), something, b, is such that $b = b \land B$, from which B follows by Conjunction Elimination. The Naive Comprehension Principle of set theory, granting a set for any condition $A[x]$, also produced notorious problems. Mathematicians who did not want to abandon Cantor's paradise had to work around the Principle in order to fix it. So did philosophers reluctant to abandon Meinong's paradise (?) work around the (UCP) in order to fix it.

Nuclear Meinongians, Parsons (1980), Routley (1980), Jacquette (1996), limited the Principle to a restricted vocabulary. They distinguished between two kinds of predicates (with the corresponding properties), called nuclear and extranuclear, and only conditions $A[x]$ including just nuclear predicates were allowed to deliver objects. It was essential that existence be extranuclear.

Dual copula Meinongians, Zalta (1983, 1997), made a distinction between two ways in which things can be ascribed properties: ordinary predication expressing property instantiation or exemplification, and *encoding*. Encoding did not in general entail exemplification. The relevant nonexistent objects could then encode features of any kind—provided $A[x]$ did not mention encoding itself, otherwise a kind of self-referential paradox would ensue (see Rapaport 1978).

Priest (2005), pp. xix and 84, came up with (what Berto 2012 later on called) a Qualified Comprehension Principle:

- (QCP) For any condition $A[x]$, with free x, some object satisfies $A[x]$ at some world.

Reference to worlds is embedded in the Principle; thus, since Berto (2008) the view has come to be called "modal Meinongianism". There is no restriction at all on $A[x]$ in the (QCP); and "satisfying" is not encoding: it expresses ordinary property instantiation. However, when object o is characterized as $A[x]$, $A[o]$ may not hold at the actual world (though it may). It holds at some world or other, that is, at those worlds that realize the situation *envisaged by the person* who uses the characterizing condition.

Speaking of "envisaging" takes us to the core of the issue I want to discuss. The subtitle of Priest's book was: *The Logic and Metaphysics of Intentionality*. On the one hand, the book aimed to give a general treatment of the logic and semantics of intentional states: representational states of the mind which are directed to objects, scenarios, or circumstances.[1] In particular, the semantics invalidated various intuitively undesirable forms of logical omniscience, that is, of logical closure properties of the relevant mental states.

On the other hand, intentionality itself was taken, as it often is by Meinongians, as a main motivation for accepting nonexistent objects. Meinongians have conjectured that (current, actual) nonexistents may be admitted by considering past and future times, or unrealized possibilities. But it is fair to say that the most promising candidates for nonexistence come from the realm of intentionality. This is also how Priest motivated the (QCP):

> Cognitive agents represent the world to themselves in certain ways. These may not, in fact, be accurate representations of this world, but they may, none the less, be accurate representations of a different world. For example, if I imagine Sherlock Holmes, I represent the situation much as Victorian London (so, in particular, for example, there are no aeroplanes); but where there is a detective that lives in Baker St, and so on. The way I represent the world to be is not an accurate representation of our world, but our world could have been like that; there *is* a world that is like that. (Priest 2005, p. 84)

Although here Priest speaks of a way of representing the world such that our world could have been like that, in the (QCP) above "world" does not stand just for *possible* world, that is, way things could be or have been. The worlds semantics of *Towards Non-Being* included *impossible* worlds: ways things could not be or have been (Berto 2013; Kiourti 2010; Nolan 2013).

Priest expanded on a technique used by Rantala (1982) in epistemic logic in order to get rid of logical omniscience phenomena. He had in the language of his theory intentional operators of the kind "$x\Psi A$" (x Ψ's that A: hopes that A, fears that A, believes that A), interpreted as restricted quantifiers over possible and impossible worlds. He admitted anarchic impossible worlds not closed under any nontrivial relation of logical consequence (he called them, thus, "open worlds"). By having truth conditions allowing access to such worlds, the relevant operators easily defied closure under nearly any nontrivial consequence relation in their turn (it being clear that, if one wanted, conditions on accessibility could be added to give to specific Ψ's more logical backbone).

Accessibility relations in the semantics were interpreted, thus, in an intentional sense: '$wR^x_\Psi w_1$' meant that world w_1 is cognitively accessed by intentional agent x, who, at world w, Ψ's that something is the case. It is true at w that $x\Psi A$ iff A is true at all the accessed w_1—which may be possible, or impossible.

The first edition of Priest's book largely took for granted that impossible worlds *are* cognitively accessible. That is, we can think about the absolutely impossible:

---

[1] Priest dealt both with intentional states directed toward objects, such as *fearing John, dreaming of Obama, imagining a tree in the garden*, and with so-called propositional states, such as *fearing that John comes along, dreaming that Obama wins the elections again, imagining that the tree starts talking to me*. In the following, we will deal only with the latter kind of states.

that which holds at no possible world whatsoever.[2] But the greatly expanded and revised second edition of the book, published in mid-2016, has more to say on this matter—and rightly so, because a venerable philosophical tradition denies that we can think the impossible.

The most quoted authority here is Hume, who formulated what I will call, stealing throughout this paper a label that has been used for something else, Hume's Principle (HP):

> 'Tis an establish'd maxim in metaphysics, that whatever the mind clearly conceives includes the idea of possible existence, or in other words, that [HP:] nothing we imagine is absolutely impossible. (*Treatise*, I, ii, 2)[3]

Hume took the absolutely impossible to coincide with the logically impossible, but we need not follow him on this. What matters is that it not be what is at times called the nomologically impossible: the impossible relative to some body of natural laws, like the laws of physics or biology. All parties in the debate around (HP) agree that we can conceive the biologically or physically impossible, e.g., my jumping 1 mile up in the air, or (if Einstein was right) a starship's moving faster than the speed of light. So, for instance, in *Positivismus und Realismus* Schlick maintained that, while the merely practically impossible is conceivable, it is the logically impossible, such as a contradiction, which is not.[4]

Although the issue pops up in various parts of the 2016 edition of his book, Priest addresses it mainly in Chapter 9, called *Possibility, Impossibility and Conceivability*. The purpose of *this* paper is to expand on Priest's arguments against (HP). Drawing largely on Berto and Schoonen (2017),[5] I will defend the view that we can conceive the impossible. But I will try to tidy up a couple of loose ends in Priest's line of argumentation, and to develop in greater detail a plausible view of how such conceiving is to be understood. I will describe two different conceptions of conceiving, which, borrowing terminology from Kung (2014), I will call the *telescopic* and the *stipulative* (if that reminds you of Kripke, you are on the right track). I will argue that the modal Meinongian should subscribe to the latter, and that the latter is independently more plausible than the former anyway.

---

[2] Philosophers debate on the nature of absolute necessity, and thus impossibility, but it is fair to say that the three main kinds of absolute necessities/impossibilities are usually taken to be the logical, the mathematical and the metaphysical. I will not get into the issue of whether one of them is reducible to another (e.g., the mathematical to the logical, as it is for logicists).

[3] As Yablo (1993), p. 4, has remarked, in spite of that "in other words" it is doubtful that here Hume is really giving the same maxim twice. It is one thing to say that, when we (clearly) conceive something, what is conceived comes with the idea that it could exist embedded in by default. It is another thing to say that we can only imagine the possible. It is the latter claim that will be on stage in the following, as the target (HP).

[4] Contradictions are often invoked as a paradigmatic case of absolute impossibility, and will come handy later on, too. The example may not sound good in the context of a discussion of Priest's work, given that he is (in)famous for believing that some contradictions are true. However, modal Meinongianism can be formulated as a consistent theory: one can be a modal Meinongian without thereby being a dialetheist.

[5] I am very grateful to the Editors of *Synthese* for allowing me to reuse that material.

A defense of the view that we can conceive the impossible is crucial to the modal Meinongian research program. If we cannot, the whole apparatus of impossible worlds becomes pointless as a way to give a semantics of intentionality. The modal Meinongian need not dispute the weaker claim that representing a situation in our mind in a certain way may often provide good, albeit defeasible, evidence to the effect that the situation is possible. As Yablo says, "in slogan form: *conceiving involves the appearance of possibility*" (Yablo 1993, p. 5). What needs to be disputed is the stronger (HP), the claim that we cannot conceive the impossible.[6]

I will not say much on the (way stronger) claim that *any* impossibility is conceivable—aside from the following remark[7]: it may seem that, in a sense, the modal Meinongian view is committed to the way stronger claim as well (and Priest is sympathetic to that in the aforementioned Chapter 9, esp. p. 194). Otherwise, why have (in principle, accessible) impossible worlds of all sorts? But one may retort that having logically anarchic worlds of any kind in the semantics is just a practical choice, motivated by the vague boundaries of what we can, in general, conceive. It may be that some impossibilities are just inconceivable, for the same reason that some possibilities are inconceivable, namely that the logical, cognitive or computational complexity of the relevant scenario is just too large for our finite human minds. Where to put the complexity boundary is a difficult issue better left to empirical psychology, and it may be that a fuzzy answer is the best we can hope for. In a logical and semantic modeling like the one aimed for in Priest's book, one may thus safely bracket the issue by being extremely generous with the supply of anarchic worlds one works with.

2.2 Minimal Conceiving

Of the two notions involved in (HP), the possibility is nowadays reasonably under control after the Twentieth Century development of possible worlds semantics. Conceivability is in a messy state. In *Towards Non-Being II*, p. 192, Priest starts by understanding conceivability in a rather minimal sense. He draws on the Oxford English Dictionary, according to which to conceive is "to take or admit into the mind, to form in the mind, to grasp with the mind". Given this characterization, he then claims that he can conceive of "anything that can be described in terms that I understand" (p. 194).

The weaker the relevant notion of conceivability, the harder it is to argue that we cannot conceive impossibilities. And conceivability, in such a minimal sense connected to "grasping with the mind" or "understanding a description of something", seems to me *clearly* to allow us cognitive access to the impossible. To deny this, one would seem to be forced to make one of two moves: (1) claim that linguistic

---

[6] Arguments for the claim that we can conceive impossible situations can be found in Byrne (2007), Fiocco (2007), Jago (2014).

[7] Triggered by a nice suggestion by two anonymous reviewers.

representations allegedly describing impossibilities, such as logical falsities, actually are meaningless so that there is nothing for us to grasp; or (2) claim that although these are meaningful, we cannot understand them.

As a general thesis, the latter is simply incredible in the face of the compositionality of learnable languages. Let P be any simple, meaningful, intelligible sentence, such as "This table is round". Surely P cannot become unintelligible because we stick a negation in front of it; so $\neg P$ must be intelligible, too. And surely two such sentences cannot deliver an unintelligibility once we conjoin them, $P \wedge \neg P$. So the latter must be intelligible, too.

Someone who came *close* to making the first claim is Wittgenstein (1922). I say "came close", because for Wittgenstein's *Tractatus* tautologies, logical truths, and their negations, logical falsities, are notoriously *sinnlos* (4.461). They "say nothing"(Ibid.). However, even for Wittgenstein they "are, however, not senseless [*unsinnig*]" but "part of the symbolism in the same way that '0' is part of the symbolism of arithmetic" (4.462). There is a debate among Wittensteinians, on what the difference between *sinnlos* and *unsinnig* amounts to, but we need not enter into this. One straightforward interpretation of the Wittgensteinian view, phrased in the contemporary terminology of possible worlds, is that the informative job of a sentence is to split into two the totality of possible worlds: those in which the sentence is true and those in which it is false. The former group is taken as giving the proposition expressed by the sentence in standard possible worlds semantics. But then tautologies and their negations, being true everywhere and nowhere in the modal space respectively, don't split, and turn out to be uninformative: "I know, e.g., nothing about the weather, when I know that it rains or it does not rain" (4.461).

Even if one buys the view that logical truths and falsities are uninformative,[8] though, that does not make them *meaningless*. Even if the distinction between saying and showing at the core of the *Tractatus* is right (and some, including perhaps the later Wittgenstein, may doubt it), that $P \vee \neg P$ and $P \wedge \neg P$ show something about the logical form of reality rather than informing us of what obtains in it, does not make them meaningless strings, provided that P is meaningful to begin with.

This is one of the few issues on which Priest and Quine, who otherwise disagree on lots of things in logic and ontology, may come to an agreement. Quine makes the point in *On What There Is*, as a response to fictional philosopher Wyman, sometimes taken as representing Meinong's view. Wyman believes that things like Pegasus ought to be admitted in our ontological catalogue, as *possibilia*, for otherwise, it would make no sense to even say that Pegasus is not. By parity of reasoning, objects Quine, we ought to admit the round square cupola on Berkeley College; otherwise, it would make no sense to even say that *it* is not. But accepting this, claims Quine,

---

[8] I do not buy this view either. Take a cognitive (as opposed to merely environmental) conception of information, and consider what can be learned by a rational, finite and fallible agent—one of us. We can learn that a complex formula, whose truth value we were ignorant of until we computed its long truth table, is a tautology. For all we knew before carrying out the computation, the formula's being false was a way things could be. In this sense, *pace* Wittgenstein (6.1251), there *are* surprises in logic. A beautiful book defending this view is Jago (2014).

brings inconsistency. Wyman reacts by declaring that inconsistent conditions are just meaningless. I find Quine's reply spotless:

> Certainly the doctrine [of the meaninglessness of contradictions] has no intrinsic appeal; and it has led its devotees to such quixotic extremes as that of challenging the method of proof by *reductio ad absurdum* – a challenge in which I sense a *reductio ad absurdum* of the doctrine itself.
> Moreover, the doctrine of meaninglessness of contradictions has the severe methodological drawback that it makes it impossible, in principle, ever to devise an effective test of what is meaningful and what is not. It would be forever impossible for us to devise systematic ways of deciding whether a string of signs made sense – even to us individually, let alone other people – or not. For it follows from a discovery in mathematical logic, due to Church [1936], that there can be no generally applicable test of contradictoriness. (Quine 1948, pp. 34–5)

In *What Is So Bad About Contradictions*, Priest is on the same page:

> If contradictions had no content, there would be nothing to disagree with when someone uttered one, which there (usually) is. Contradictions do, after all, have meaning. If they did not, we could not even understand someone who asserted a contradiction, and so evaluate what they say as false (or maybe true). We might not understand what could have brought a person to assert such a thing, but that is a different matter and the same is equally true of someone who, in broad daylight, asserts the clearly meaningful 'It is night'. (Priest 1998, p. 417)

2.3 Conceiving as Imagining

There appears to be a more substantive sense of "conceiving"—one that could be taken as lending some support to (HP). Right after resorting to the aforementioned minimal sense of "conceiving", Priest adds:

> I intend to use *conceive* here as roughly synonymous with *imagine*: the sort of imagination employed by scientists, mathematicians, philosophers, novelists, political reformers, theologians, visionaries, and so on [Fn: OED, *to imagine*: 'to form a mental image of, to represent to oneself in imagination, to create as a mental conception, to conceive']. *In imagination, a state of affairs or an object is brought before the mind*, and may be considered, enjoyed, its consequences thought through, and so on. (Priest 2005, p. 192, last italic mine)

Now, this sense of "conceiving" as "imagining a state of affairs" seems to be more substantive than merely grasping the meaning of a sentence. It is close enough to a notion one can find in Yablo (1993), Chalmers (2002), and dubbed by the latter "positive conceivability". Positively conceiving that S is understood as a mental operation different from merely supposing or assuming that S, as when we make an assumption in a mathematical proof. Instead, we represent a *situation*, or a state of affairs, in our mind, a configuration of objects and properties of which S is a truthful description:

> Positive notions of conceivability require that one can form some sort of positive conception of a situation in which S is the case. One can place the varieties of positive conceivability under the broad rubric of *imagination*: to positively conceive of a situation is to imagine (in

some sense) a specific configuration of objects and properties. [...] Overall, we can say that S is positively conceivable when one can imagine that S: that is, when one can imagine a situation that verifies S. (Chalmers 2002, p. 150)

Similarly, Yablo (1993) has it that the conceivability of S amounts to the imaginability of a world verifying S (he grants that we do not imagine the relevant world in all detail; we will come to issues of *detail* later on). And it seems to me that something like this is the notion typically at issue in debates around (HP) (see e.g., Hill 1997; Gendler and Hawthorne 2002; Stoljar 2007; Kung 2010; Balcerak Jackson 2016).

Characterizing (the relevant) *imagination* is in its turn difficult. The best that can be done, I think, is to point at some features that make conceivability as imagination differ from alternative intentional states.[9] Thus, imagining that P is distinct from believing that P in that one who imagines a situation making P true does not thereby commit to the actuality of that situation. Another difference (see e.g., Nichols and Stich 2003; Wansing 2015) is that, although there can be involuntary exercises of it, imagining can be voluntary in ways in which believing cannot: the agent just sets out to represent a certain scenario. So I can imagine that New York is in Canada but I cannot make myself believe it, for I have overwhelming evidence of the contrary.

So understood, that is, as a mental representation of a situation verifying some claims, carried out largely on a voluntary basis, imagination is an everyday business. We simulate alternatives to reality in our mind, in order to explore what would and would not happen if they were realized. This can often help us to cope with reality itself, by improving future performance, allowing us to make contingency plans, etc. (see e.g., the works in Markman et al. 2009).

Imagination is also distinguished, obviously, from (veridical) perception in that the target situations need not be real. To use a metaphor from Williamson (2007), in imagination our perceptions are left "offline". However, imagination is at times taken as involving some surrogate of sensory perception (typically, but not only, of the visual kind). For want of a better term for something that is close to perception but is not quite the real thing, it is common to use the "quasi-" prefix: people speak of quasi-visual or quasi-auditory imaginings (see e.g., Gendler 2011). Metaphors such as that of the mind's eye have been around for centuries.

Now whatever one makes of such loose characterizations, it seems to me that one should not take them as implying that the only imaginable scenarios are those that involve exclusively perceptual qualities (or, quasi-perceptual, whatever that means exactly). Otherwise, we could never imagine situations involving abstract objects or abstract features of concrete objects. But whether the imaginability of scenarios of these kinds entails their absolute possibility is precisely what is discussed in various debates on (HP). Thus, in such debates "imagination" seems to be normally understood broadly enough. This is a point stressed also by Williamson (2007), who

---

[9]I agree with Yablo on this: "Almost never in philosophy are we able to analyze an intentional notion outright, in genuinely independent terms: so that a novice could learn, say, what memory and perception *were* just by consulting their analyses. About all one can normally hope for is to locate the target phenomenon relative to salient alternatives, and to find the kind of internal structure in it that would explain some of its characteristic behavior." (Yablo 1993, pp. 25–6).

makes of imagination the key notion in a full-fledged counterfactual epistemology of metaphysical modality.

If such a rough characterization of the phenomenon at issue is sufficient, we can move on to the next act. That's where Kripke enters the stage.

2.4 Kripkean Error Theory

The post-Kripkean acceptance that, contrary to what much philosophical tradition believed, there are a posteriori necessities, may seem to hit (HP) hard. For identities such as those between Hesperus and Phosphorus or between water and H_2O, are empirical discoveries. Could we then not conceive of things as being otherwise, and so conceive the impossible? It seems easily imaginable that water may have turned out to have a different chemical constitution. At the time of *The Meaning of Meaning*, Putnam was clear:

> We can perfectly well imagine having experiences that would convince us (and that would make it rational to believe that) water isn't H_2O. In that sense, it is conceivable that water isn't H_2O. It is conceivable but it isn't logically possible! Conceivability is no proof of logical possibility [...] Human intuition has no privileged access to metaphysical necessity. (Putnam 1975, p. 133)

However, things may be not so simple for the (HP)-denier. In *Naming and Necessity*, Kripke proposed a different diagnosis of the phenomenon, which according to Kung (2014) amounts to an attempt to explain the appearances away via a kind of error theory.

The key idea is that some imaginings are compatible with their authors' making errors in appreciating the represented content. Specifically, they may involve misidentifications. A posteriori necessary truths often give us an "illusion of contingency": it may have turned out on empirical investigation, one thinks, that Hesperus is not Phosphorus or that water is not H_2O. Then these matters must be contingent. Kripke explains the illusion by resorting to intentional doppelgangers. We can think we are imagining a scenario in which water is not H_2O. What we actually intend, though, is a situation qualitatively identical to, or indiscernible from, one we may find ourselves in, and in which we face some fluid that has the same phenomenal features of water (say, a colorless, odorless, tasteless liquid, etc.), without being H_2O. We can also imagine having cherished that watery stuff with the name "water". But such an imagining is not the representation of an impossibility, that is, of (what we actually refer to as) water not being what it necessarily has to be. The illusion comes from misjudging our own representation, misidentifying that doppelganger of water with water.

To generalize: when we seem to imagine a situation S falsifying an a posteriori necessity P, (a) we actually imagine a qualitatively indiscernible scenario $S_1 \neq S$, such that (b) S_1 is possible and (thus) no falsifier of P, and (c) we confuse S_1 with S.[10]

Error theories do not have a great track record in philosophy, and I think this one is no exception. In the following Section, I will argue that the strategy does not generalize seamlessly—as it should, if it is the case that, as required by (HP), we can *never* conceive the impossible. In the Section after that, I will argue that the strategy is based on a dubious view of conceivability as imagination.

2.5 Does the Strategy Generalize?

The strategy of redescribing represented wannabe-impossibilities as represented possibilities + misidentification, it seems to me, just won't work in all cases. One example, proposed in Wright (2002), is that of first-person counterpossible conjectures.

If Kripke is right, Wright claims, I am essentially a human being, and necessarily tied to my actual biological originators. But I can imagine myself as having been born from different parents. I can also imagine myself, say by putting myself at center stage in a fantasy story, as being an elf, an alien, a monkey. Can my imagining these scenarios, which essentialists usually consider metaphysically impossible, be explained away as my imagining possible situations involving an intentional doppelganger of mine, which I mistakenly identify with myself? It seems not, says Wright. For I do not individuate myself *qua* thinking subject by means of phenomenal, surface appearances, as I individuate water by its external appearances of colorless, tasteless liquid. When I imagine myself in a clearly possible counterfactual situation, such as my being in the Grand Canyon instead of Europe, "no mode of presentation of the self need feature in the exercise before it can count as presenting a scenario in which I am in the Grand Canyon" (Wright 2002, p. 436). The same holds for my counterpossible imagining myself as a monkey: this is not easily redescribable as my imagining a doppelganger which is a monkey, and mistakenly taking the substitute to be me. I imagine *myself* in this case as well.

Another area in which Kripkean redescription doesn't appear to be available has to do with mathematical conjectures and impossibilities. First, it seems that we can conceive necessary truths of mathematics whose truth value we ignore as false, or vice versa. A mathematician may genuinely conceive that Goldbach's Conjecture (Every even integer larger than 2 is the sum of two primes) is wrong: she may also

---

[10]Here is a passage of *Naming and Necessity*, in which Kripke appears to endorse such an error theory. It is the famous example of the table: "But whatever we imagine counterfactually having happened to [the table] other than what actually did, one thing we cannot imagine happening to this thing is that it, given that it is composed of molecules, should still have existed and not have been composed of molecules. We can imagine having discovered that it wasn't composed of molecules. But once we know that this is a thing composed of molecules—that this is the very nature of the substance of which it is made—we can't then, at least if the way I see it is correct, imagine that this thing might have failed to have been composed of molecules." (Kripke 1980, pp. 126–7).

try to see what would follow from this. Suppose that the conjecture is indeed true. If mathematical necessity is unrestricted, then it is unrestrictedly impossible for some even number (larger than two) not to be the sum of two primes. Still, we cannot easily redescribe the mathematician's representation of the relevant impossibility as the conceiving of a false doppelganger of the conjecture. What could such a doppelganger be? As Priest claims:

> Take Goldbach's conjecture again. I have no difficulty in conceiving this, and no trouble conceiving its negation, though one of these is mathematically impossible. Indeed, mathematicians must be able to conceive these things, so that they understand what it is of which they are looking for a proof, or so that they can infer things from them, in an attempted *reductio* proof. (Priest 2005, p. 193)

Proven conjectures, such as Fermat's Last Theorem, make the case more vivid. Take a competent, but skeptic mathematician, who imagines she can find some mistake in Andrew Wiles' proof, or even direct counterexamples to the Theorem. The person understands the content of the Theorem pretty well: it's a simple claim on Diophantine equations. It is implausible to redescribe the situation as the mathematician's imagining counterexamples to an intentional duplicate of Fermat's Theorem. There appears to be no content misidentification going on here. Wright also concludes from similar cases that "for a large class of impossibilities, there are still determinate ways things would seem if they obtained" (Wright 2002, p. 437).

2.6 The Telescopic and Stipulative Views of Imagination

It seems to me that the error theory comes with what Kung (2014) called a "telescopic view" of imagination. In this view, "seeing in imagination" is interpreted as an activity very close to physical, perceptual seeing (recall the aforementioned "quasi-" jargon, often used in accounts of mental imagery). This appears to support (HP) by analogy: just as (veridical) visual perception only shows what is actual, so imaginative vision or quasi-vision only shows what is possible. When we imagine a scenario where P, we look with the metaphorical eye of the mind at a situation making P true. What cannot happen is that such mental telescope has us look at the impossible: if the scenario *shows up*, it is *there* to be seen. What can happen is that we fail to appreciate exactly what scenario we are looking at. Imagining is like looking at a photograph: if we see a snapshot of a girl (leaving photoshop tricks aside), the girl must exist or have existed. But who's that girl? Valery or Laura?

Talk of "telescopic view" by Kung is meant to remind us of Kripke arguing, notoriously, *against* a telescopic view of our access to worlds in *Naming and Necessity*. Kung thinks that Kripke's line of argumentation there applies against Kripke's own error theory for the imagining of impossibilities.

The relevant context is the problem of transworld identification in the philosophy of possible worlds—a problem which, as far as I know, is originally due to Kaplan (1969). This is an epistemic issue, not to be confused with the problem of transworld identity (see Paul and Jago 2013; Mackie 2006), which as far as I know is due to Chisholm (1967). The latter can itself be phrased in different ways (Divers 2002, Chap. 16, makes the relevant distinctions), but it is, in any case, an issue of (modal) metaphysics. The former has to do with how can we *know* whether we have a case of transworld identity in some sense or other.

In Kaplanian terms: which of the individuals in a possible world w is the "transworld heir" of an individual in a different possible world (say, the actual one, @)? Given our own Saul Kripke at @, we are supposed to carry out some investigation among the individuals in w, with the aim of locating the Kripke representative there. The problem seems intractable, insofar as w may include several individuals who resemble Kripke in various respects and can compete for the role. Here is one individual whose fingerprints and facial expression are indiscernible from those of our own beloved Kripke, but who never did philosophy and had a career as a drug dealer. Here's another one who does not quite look like Kripke, but who has written a book called *Naming and Necessity*, where he defends the view that there are necessary a posteriori truths, etc.

Scholars tend to consider transworld identity as a real issue (unless one is a counterpart theorist), and transworld identification as a pseudo-problem, precisely under the influence of Kripke. This pseudo-problem comes, for Kripke, from a purely qualitative conception of how worlds represent possibilities. Other worlds, says Kripke, are not something we glance at via the famous telescope. We need not represent alternative situations in purely qualitative terms: "generally, things aren't 'found out' about a counterfactual situation, they are stipulated." Kripke (1980), p. 49, et cetera: the story is so well known that it hardly needs rehearsing (see also Plantinga 1974, p. 95; Chihara 1998).[11]

I suggest, following Kung (2014) again, that imagination may work more like Kripkean stipulation than like a Kripkean telescope. It has, that is, an arbitrary *labeling* component. One need not deny that imagination, in general, has a qualitative or phenomenological component as well, but the presence of the former component, the labeling or stipulation, seems to be enough to defeat (HP). In particular, the identity of the represented objects in an exercise of imagination can, in general, be stipulated—it does not need to be *discovered*.

I imagine Valery swimming in the Atlantic Ocean, and the phenomenology of the mental imagery can be such that the represented girl is relevantly similar to Valery: hair color, eyes, body. But what makes my imagining count as a representation of a scenario in which *Valery* swims in the Ocean is that I label that represented woman as *Valery*. Now, as easily as I can imagine Valery as swimming in the Ocean—a possible scenario—I can represent Valery as having been born from different parents

---

[11] "There is something amiss when a claim of the type 'Suppose that Socrates had never gone into philosophy...' is met with the challenge to demonstrate how you know that it is Socrates that is the object of your supposition. The same might be said of the question how you know that the subject of the proposition that Socrates is a philosopher is the same as the subject of the proposition that Socrates was married [...]. The theme that unites the last two deflationary thoughts is that one can 'give' a possible world or a representational content in a non-qualitative way by relying on stipulation." (Divers 2002, p. 272).

from the ones she actually has, or as a cleverly disguised robot. But the two latter scenarios, if Kripke is right, are metaphysically impossible.

Kung (2014) conjectures that such stipulative component is what gives to imagine its power to access the impossible. And in *Towards Non-Being II*, Priest is rather explicit in acknowledging this stipulative feature of imagination:

> [W]hen I imagine that water is not H_2O I am imagining something about *water*. The imagination is *de re*. In the same way, when I imagine that Sarah Palin was the US Vice President after the 2012 US election, I am imagining something about *Palin*. When I imagine that Routley found a box that was empty and not empty, it is *him* that I imagine. And when I imagine that 361 is a prime number (it isn't) I am imagining something about that very number. (Priest 2005, p. 195)

Now, let us see how this stipulative view of conceivability as imagination is to be applied to (the interpretation of) the modal Meinongian theory. Given a condition $A[x]$, some object, o, can in principle be conceived by some cognitive agent c as satisfying it. Then if $\Vdash$ is making true, Ψ the relevant intentional state, and @ the actual world, we have @ $\Vdash c\Psi A[o]$. Even when $A[x]$ is the inconsistent "x is a round square", c is really conceiving that the very object o is round and square. And (QCP) guarantees that, at some world w, o is a round square: $w \Vdash A[o]$.

What is not in our powers to stipulate is that w be a *possible* world and, a fortiori, that $w = @$. In general, we cannot stipulate at which worlds objects have the properties they are characterized as having—whether these are possible worlds, and whether they include the actual one. This is true also when we embed an explicit reference to worlds in the characterizing $A[x]$, e.g., via an "actuality operator" that works as a world pointer, pointing at @ (see Berto 2012, pp. 174–5 for the details). What one can fantasize about and what is or could be the case is at times severely different things.[12]

2.7 Issues of Granularity

The stipulative view can help us to address another objection to the imaginability of impossibilities. In order for a certain kind of intentional state to count as a representation of a situation verifying P, the objection goes, the situation at which P must be represented in *some* relevant detail. To elucidate what this means with an example, here are two acts of imagination. First, I imagine a situation in which a bunch of mathematicians issue a press conference and declare that they have refuted

---

[12]"When I use the word 'Socrates' inside an explicitly worldly context, 'at w', or inside an implicitly worldly (modal or counterfactual) context [...] I do not thereby make it the case, nor do I come to know, that such a world is a *possible* world. It is this crucial point that underlies the complaint against [the claim that stipulative conceivability entails possibility]. It is one question how we know which objects are the objects of our *de re* modal thought and talk, and perhaps there stipulation has a legitimate role. It is another question altogether how we know what is modally true of those objects, and there stipulation has no legitimate role to play." (Divers 2002, p. 273).

Goldbach's Conjecture, thereby triggering admiration from the whole world. Second, I imagine building step by step a perfeclty detailed, valid proof, starting with Peano's axioms and ending with the negation of Goldach's Conjecture. Bracket the problem whether the exercise is mentally feasible (for the proof may be too complex for a human to mentally go through it with no external aid from paper, computers, or else).

In the latter case, the mental imagery is, in a sense, too fine-grained to count as a *merely* imagined, nonactual scenario: for if I actually go step by step through a sound refutation of the Conjecture, representing each step in full detail, then I have *actually* refuted the conjecture, if only in the private of my mind. On the other hand, the former case errs on the side of defect, by being too generic and relevantly disconnected a mental representation for it to count as my actually imagining that Goldach's conjecture is false. One could as well describe the envisaged scenario as one in which a bunch of folks makes a press statement.

Thus, the objection goes, merely imagining that Goldbach's Conjecture is false without actually refuting it must be something in the middle between these two extremes. Which scale, or bunch of scales, must it be in the middle *of*, can be a matter of debate: detail of the mental imagery, topicality, relevance. How we measure and locate imaginings across the scales may be an open issue. The scales may even be orthogonal to each other, so that a unique score for a certain act of imagination may be unfeasible. But this is of limited importance for our purposes. What matters is that one cannot just generally stipulate that one has imagined a certain scenario, and be guaranteed to *succeed* independently from concerns of fine-grainedness. One can properly claim to have imagined a situation such that P only when a sufficient level of structural detail in the mental imagery is reached, and that level may be generally unreachable for impossible P's.

I take it that Peter van Inwagen, a subscriber (it seems to me) to the telescopic view of imagination, has something like this objection in mind in the following passage:

> In my view, we cannot imagine worlds in which there are naturally purple cows, time machines, transparent iron, a moon made of green cheese, or pure phenomenal colors in addition to those we know. Anyone who attempts to do so will either fail to imagine a world or else will imagine a world that only seems to have the property of being a world in which the thing in question exists. Can we imagine a world in which there is transparent iron? Not unless our imaginings take place at a level of structural detail comparable to that of the imaginings of condensed matter physicists who are trying to explain, say, the phenomenon of superconductivity. (Van Inwagen 1998, p. 79)

The proper answer, I think, consists in distinguishing between (1) succeeding in thinking about a certain scenario and (2) succeeding in gaining evidence that the scenario is possible. That it can be hard to succeed in the first sense due to granularity problems which are relevant for the issue of whether we can imagine the impossible, it seems to me, just *presupposes* the telescopic view of imagination. If imagination worked as a telescope, then indeed we may have reasons to doubt that one succeeds in imagining one situation rather than another, unless the imagining comes with a fine-grained enough level of structural detail in the relevant mental imagery. But that imagination works thus cannot simply be assumed, on pain of begging the question against the subscriber to the stipulative view.[13]

What we may not succeed in, when our imaginative exercise does not come with the right level of structural detail, is getting evidence that *that* is a possible scenario. The lack of detail is one of the things that can mislead us on the modal status of the envisaged situation. Again, the modal Meinongian who subscribes to the stipulative view of imagination need not deny that, when one's imaginative exercise is carried out at some appropriate level of structural detail, that gives defeasible evidence that the scenario is possible. But this is not what is required to save (HP), the stronger view that we just have no cognitive access to the impossible via our imagination.

References

Balcerak Jackson, M. (2016). On imagining, supposing and conceiving. In A. Kind & P. Kung (Eds.), *Knowledge through imagination*. Oxford: Oxford University Press. forthcoming.
Berto, F. (2008). Modal meinoniganism for fictional objects. *Metaphysics*, 9, 205–18.
Berto, F. (2012). *Existence as a real property*. Synthse Library. Dordrecht and New York: Springer.
Berto, F. (2013). Impossible worlds. *The Stanford Encyclopedia of Philosophy*.
Berto, F., & Schoonen, T. (2017). Conceivability and possibility: Some dilemmas for humeans. *Synthese*. Online first.
Byrne, A. (2007). Possibility and imagination. *Philosophical Perspectives*, 21, 125–44.
Chalmers, D. (2002). Does conceivability entail possibility? In T. Gendler & J. Hawthorne (Eds.), *Conceivability and possibility* (pp. 145–99). Oxford: Oxford University Press.
Chihara, C. (1998). *The worlds of possibility*. Oxford: Oxford University Press.
Chisholm, R. (1967). Identity through possible worlds: Some questions. *Noûs*, 1, 1–8.
Divers, J. (2002). *Possible worlds*. London and New York: Routledge.
Fiocco, M. (2007). Conceivability, imagination and modal knowledge. *Philosophy and Phenomenological Research*, 74, 364–80.
Gendler, T. (2011). Imagination. *The Stanford Encyclopedia of Philosophy*.
Gendler, T., & Hawthorne, J. (Eds.). (2002). *Conceivability and possibility*. Oxford: Oxford University Press.
Hill, C. (1997). Imagininability, conceivability, possibility and the mind-body problem. *Philosophical Studies*, 87, 61–85.
Jacquette, D. (1996). *Meinongian logic: The Semantics of existence and nonexistence*. DeGruyter: Berlin and New York.
Jago, M. (2014). *The impossible. An essay on hyperintensionality*. Oxford: Oxford University Press.

---

[13] There is a different issue concerning cases in which one supposedly fails to conceive what one meant to conceive, nicely pointed out to me by an anonymous referee. The issue has to do with one's lacking certain information which is plainly required for the stipulation to succeed. I'll just pick the referee's example: one thinks one imagines that Goldbach's Conjecture has been refuted, but one has misunderstood the content of Goldbach's Conjecture. In fact, what the person labels thus is the claim that there is a greatest prime number. It seems to me that in this case one actually misdescribes what one is conceiving, but that this failure is just orthogonal to the distinction between the stipulative and telescopic conceptions. In particular, it has little to do with "insufficient granularity" in the sense relevant for van Inwagen's point: one can imagine in the greatest detail the standard proof that there is no largest prime while mistakenly labeling the proved claim as 'the negation of Goldbach's Conjecture'.

Kaplan, D. (1969). Transworld heirlines. In M. Loux (Ed.), *The possible and the actual* (pp. 88–109). Ithaca: Cornell University Press.
Kiourti, I. (2010). *Real impossible worlds: The bounds of possibility.* PhD thesis, University of St Andrews.
Kripke, S. (1980). *Naming and necessity.* Oxford: Blackwell.
Kung, P. (2010). Imagining as a guide to possibility. *Philosophy and Phenomenological Research, 81,* 620–63.
Kung, P. (2014). You really do imagine it: Against error theories of imagination. *Nous, 50,* 90–120.
Mackie, P. (2006). *How things might have been: Individuals, kinds, and essential properties.* Oxford: Oxford University Press.
Markman, K., Klein, W., & Surh, J. (Eds.). (2009). *Handbook of imagination and mental simulation.* New York: Taylor and Francis.
Nichols, S., & Stich, S. (2003). *Mindreading: An integrated account of pretence, self-awareness, and understanding other minds.* Oxford: Oxford University Press.
Nolan, D. (2013). Impossible worlds. *Philosophy Compass, 8,* 360–372.
Parsons, T. (1980). *Nonexistent objects.* New Haven, CT: Yale University Press.
Paul, L., & Jago, M. (2013). Transworld identity. *Stanford Encyclopedia of Philosophy.* https://plato.stanford.edu/entries/identity-transworld/.
Plantinga, A. (1974). *The nature of necessity.* Oxford: Clarendon Press.
Priest, G. (1998). What is so bad about contradictions? *Journal of Philosophy, 95,* 410–26.
Priest, G. (2005). *Towards non-being: The logic and metaphysics of intentionality.* Oxford: Oxford University Press. 2nd expanded ed. 2016.
Putnam, H. (1975). *Mind, language and reality* (Philosophical Papers). Cambridge: Cambridge University Press.
Quine, W. (1948). On what there is. *Review of Metaphysics, 48,* 21–38.
Rantala, V. (1982). Impossible world semantics and logical omniscience. *Acta Philosophica Fennica, 35,* 106–115.
Rapaport, W. (1978). Meinongian theories and a russellian paradox. *Noûs, 12,* 153–180.
Routley, R. (1980). *Exploring Meinong's jungle and beyond.* Canberra: RSSS Australian National University.
Stoljar, D. (2007). Two conceivability arguments compared. *Proceedings of the Aristotelian Society, 107,* 27–44.
Van Inwagen, P. (1998). Modal epistemology. *Philosophical Studies, 92,* 67–84.
Wansing, H. (2015). Remarks on the logic of imagination: A step towards understanding doxastic control through imagination. *Synthese.* Online first.
Williamson, T. (2007). *The philosophy of philosophy.* Oxford: Blackwell.
Wittgenstein, L. (1922). *Tractatus Logico-Philosophicus.* London: Routledge Kegan Paul.
Wright, C. (2002). The conceivability of naturalism. In T. Gendler & J. Hawthorne (Eds.), *Conceivability and Possibility* (pp. 401–39). Oxford: Oxford University Press.
Yablo, S. (1993). Is conceivability a guide to possibility? *Philosophy and Phenomenological Research, 53,* 1–42.
Zalta, E. (1983). *Abstract objects: An introduction to axiomatic metaphysics.* Dordrecht: Reidel.
Zalta, E. (1997). A classically-based theory of impossible worlds. *Notre Dame Journal of Formal Logic, 38,* 640–660.

Franz Berto works on nonmainstream ontology, nonclassical logic, nonstandard computation, and other deviant views in theoretical philosophy.

He is Professor of Logic and Metaphysics at the Department of Philosophy, University of St Andrews, and also works at the Institute for Logic, Language and Computation (ILLC), University of Amsterdam. He has also worked at the University of Aberdeen UK, at the Institute for Advanced Study, University of Notre Dame USA, at the Sorbonne-Ecole Normale Supérieure in

Paris, at the University of Italian Switzerland, at the Universities of Padua, Venice, and Milan-San Raffaele in Italy. He has written books for Oxford University Press, Blackwell, Bloomsbury, Springer, College Publications, and papers in the *Review of Symbolic Logic, Mind, Journal of Philosophical Logic, The Philosophical Quarterly, Philosophical Studies, Synthese, Erkenntnis, the Australasian Journal of Philosophy, the European Journal of Philosophy, Dialectica, Philosophia Mathematica*. He coauthors four entries of the *Stanford Encyclopedia of Philosophy*.

Chapter 3
The Number of Logical Values

Ross T. Brady

Abstract We argue that formal logical systems are four-valued, these four values being determined by the four deductive outcomes: A without ~A, ~A without A, neither A nor ~A, and both A and ~A. We further argue that such systems ought to be three-valued, as any contradiction, A and ~A, should be removed by reconceptualisation of the concepts captured by the system. We follow by considering suitable conditions for the removal of the third value, neither A nor ~A, yielding a classically valued system. We then consider what values are appropriate for the meta-theory, arguing that it should be three-valued, but reducible to the two classical values upon the decidability of the object system.

Keywords Deductive outcomes · Conceptualisation · Meta-theory · Finite matrices · Infinite matrices

3.1 Introduction

I have known Graham Priest for 40 years, having met him soon after his arrival in Australia from St Andrews in the United Kingdom in 1976. He presented a paper to the Australasian Association for Logic at the A.N.U. in that year introducing his three-valued logic of paradox LP. This started him on a journey into a study of contradiction, paraconsistency, dialetheism and beyond. It is an honour to present a paper to this volume on some aspects of this work as our researches have taken us over many overlapping topics even though our positions have differed at times over the years.

At the World Congress in Universal Logic (WCUL5), held at the University of Istanbul during June 2015, Graham Priest presented a paper on the number of logical values, taking into account the principles of Buddhism. So, it is an appropriate topic for this volume and it gives me an opportunity to present my own account of the

---

R. T. Brady (✉)
La Trobe University, Melbourne, VIC 3086, Australia
e-mail: ross.brady@latrobe.edu.au

matter, which will fit into my other general philosophical works on logic that I have been writing since the 'blueprint' paper (Brady 2007).

In his WCUL5 paper, Priest argued for the four familiar logical values that are derived by taking truth and falsity as independent of each other, but he also considered the addition of a fifth value in line with Buddhist principles. Such a fifth value was based on a 'garbage in–garbage out' principle, meaning that if any component of a sentence is regarded as 'garbage' then the whole sentence is also 'garbage'. Such a value would suit sentences such as 'Sunday is having a bath', where Sunday is understood as a day of the week, so that if this sentence was immersed in a compound sentence then that compound sentence would be a case of 'garbage out' as a result of 'Sunday is having a bath' being 'garbage in'.

We will argue in Sect. 3.2 for the removal of this fifth value on the grounds that some modicum of sentential meaning is needed for deductive logical arguments to take place. This leaves us with the four familiar values representing the four possible proof outcomes available for an independent negation, viz.

(i) A without ~A,
(ii) ~A without A,
(iii) neither A nor ~A and
(iv) both A and ~A.

Since we are interested in values, each of the formulae A will need to represent a single sentence and so we will subsequently use the term 'sentence' instead of 'formula' in such contexts. It should be noted too that these represent deductive outcomes and not semantic values as one would find in matrix-generated logics. We will compare these two types of values in Sects. 3.2 and 3.7 below.

In Sect. 3.3, we will make a case for dropping the contradictory value, (iv) in the above list. This will depend on ideal formal systems that represent conceivable concepts and would involve re-conceptualizing any concept or concepts that lead to contradiction or indeed making appropriate changes to the logic itself. We will say that there ought to be three values due to this idealization of systems. There is usually some subpart of such a system that is classical and hence two-valued, and this can apply to the system as a whole in some cases. In Sect. 3.4, this is seen as the ultimate idealization, creating a classical system, but this classicality requires extra special circumstances rather than being taken for granted.

In Sect. 3.5, we take a formal system itself as the central concept that is formalized in a meta-theory and apply the above considerations to a formalized meta-theory. We argue for three values on the grounds that formal systems are seen to be consistent in that both proof and non-proof cannot apply to the one formula, where the methodology for proving non-theoremhood is sound. However, there is still further idealization of formal systems yielding a meta-theory with a two-valued metalogic. In Sect. 3.6, we argue that a decidable logic suffices for this purpose. Again, the classicality is not taken for granted, but requires proof for it to happen.

We use Dugundji's work in his (1940) to reject finite many-valued matrix logics in Sect. 3.7 as unintuitive. We then go on to explore infinite-valued logics. In conclusion, we say that formal systems in general are four-valued, but ought to be three-valued.

3 The Number of Logical Values

Two-valuedness requires extra special effort. Meta-theory is three-valued, but two-valuedness requires decidability.

3.2 The Argument for Four Values

Goddard and Routley in their opus (1973) introduced the term 'non-significant' for sentences under the 'garbage in–garbage out' umbrella and developed a major body of theory about them. (See the three-valued matrix logic on p. 261, representing the 'garbage in–garbage out' approach.) In their general treatment of non-significance, there were other matrix logics (see p. 342 for example) and also a two-sorted logic, with significant sentences forming a sub-sort of the full sentential sort. However, the 'garbage in–garbage out' account is the favoured one.

The question now is whether there should be such a non-significant value in a logic, despite Goddard and Routley's use of a three-valued matrix logic to capture it. Is there a need for meaninglessness or nonsense, as it is sometimes called, in logic? What one needs to do is to go back to what logic is all about. First, we are focusing on deductive rather than inductive arguments here, in accordance with the above literature. Deductive arguments are valid when their conclusions are certain, given the premises. This is opposed to good inductive arguments where the conclusions are probably true, given the premises. Moreover, inductive logic is itself not clearly defined as a logical system. Now, it is a question of how conclusions can be certain, given the premises. One would expect that this certainty is brought about by some meaningful relationship between the premises and the conclusion. This is indeed the basis for relevant logics. Further, how can any sentence be certain without analysing its meaning, either by itself or in relation to other sentences? So, some level of meaningfulness is central to deductive reasoning.

On a more basic level, as argued in Brady (2015a), logic is about proof of conclusions from premises. As further discussed in Brady (2019), this is as opposed to logic being about propositions or determined by its truth-theoretic semantics, which are the main alternatives in the literature. Briefly, the problem with logic being about propositions is that propositions are true or false, and nothing else, which essentially constricts logic to that of classical logic and also unduly constricts the application of logic to sets of sentences that are consistent and complete. Logic can indeed be applied to situations with incomplete information, based on the meanings that are given, i.e. meaning does not need to be full meaning.

The problem with truth-theoretic semantics is the interpretation of disjunction and existential quantification. Because of formula induction being used to set up the semantics, each disjunction $A \vee B$ is established through one of its disjuncts A or B in this process, and similarly each existential statement $\exists xA$ is established through one of its instantiations A^a/x, for some instance a. (These disjunctive and existential instantiations are called 'witnesses', often used in the context of completeness proofs but also more widely.) However, in Hilbert-style proof theory or natural deduction, which is what is being referred to here, it is quite possible to have a disjunctive step in a proof without one of its disjuncts and an existential step without one of its instantiations. This is exemplified by the form of the respective elimination rules ∨E and ∃E of natural deduction, where no disjunctive or existential witness need occur in the main proof or any subproof. Moreover, when comparing these two situations with the meanings of disjunction and existential quantification, it is the proof theory that correctly captures their respective meanings. A disjunction A ∨ B means 'at least one of A or B', but it does not tell you which one it is, this being in conformity with the role the rule ∨E plays in proof theory. However, in truth-theoretic semantics, the formula induction requires a disjunctive or existential witness, which conforms to Henkin's proof of the completeness of predicate calculus in his (1949) and the Routley–Meyer proof of completeness for relevant logics in (1982). Such witnesses do not form part of the meanings of disjunction and existential quantification, but are required due to the formula-inductive process of setting up a world. (See Brady 2017 for a fuller presentation of these issues.) Thus, it is proof theory that captures the meanings of '∨' and '∃', and can capture the meanings of the other connectives and quantifiers, if put together with due care.

What we can also say here is that each of these approaches does involve meanings in different ways, whether it is to justify the derivation of a conclusion from premises, to capture the logical concepts in an axiomatic proof theory or in natural deduction, or to determine the truth and falsity of propositions, or to capture the meaning in a truth-theoretic semantics. So, meaninglessness does not fit into these ways of understanding logic. As stated above, meaning does not have to be full, and a disjunction without one of its disjuncts is such an example, as is an existential statement without an instantiation. So, 'meaningful' in this context does not imply negation-completeness as reasoning can take place based on incomplete information. (Note that negation-completeness requires that either A or ~A is a theorem, for each sentence A.) Indeed, such reasoning is ubiquitous. Further, non-significance, treated using the favoured 'garbage in–garbage out' approach, does not play any role in the determination of a valid argument, which is determined totally using the other 'significant' values.

Thus, we concern ourselves with the four deductive outcomes, for sentences A:

(i)   A without ~A,
(ii) ~A without A,
(iii) neither A nor ~A and
(iv) both A and ~A.

These are our four values, which are proof-theoretic values, as opposed to the more familiar semantic values that occur in truth tables and more generally in finite matrix logics. (More on this in Sect. 3.7 below.) However, we have argued here and elsewhere in Brady (2017) against such semantic approaches to logic.

These values apply for the applications of both classical and non-classical logics, as indicated in Brady (2007), since case (iii) applies for the Gödel sentence G in Peano Arithmetic and case (iv) applies to membership of the Russell set R in itself in naïve set theory, both based on classical logic. Cases (i) and (ii) precisely specify the classical Boolean negation, whilst case (iii) is a case of underspecification of concepts and case (iv) is a case of overspecification. (See Brady 2015a, b for further discussion of this.)

Given the above, we examine the relation between the Law of Excluded Middle (LEM) and the Disjunctive Syllogism (DS) and the respective values (iii) and (iv). What one can say is that the presence of the LEM, $A \vee \sim A$, though seemingly stating that A does not take the value (iii), can, when applied to the Gödel sentence G in classical Peano Arithmetic, take the value (iii). (See Brady 2019 for further discussion of the LEM.) Again, the DS ($\sim A, A \vee B \Rightarrow B$) is not sufficient to guarantee that the value (iv) cannot be taken by A, as (iv) can be taken by the membership of the Russell set R in itself, in classical naïve set theory, where it trivializes the theory. The essential point here is that the LEM and the DS are proof-theoretic, whilst negation consistency and completeness are meta-theoretic.

As initially discussed in Brady (2007) and more recently in (2015a) and (2019), negation is not completely determined in that the negative determinations, $\sim A$ in (i), A in (ii), A in (iii) and $\sim A$ in (iii) are meta-theoretic, whilst the remainder are proof-theoretically determined. The point here is that non-proof may not be recursive, unlike positive proof. This contrasts with the other connectives, &, $\vee$ and $\rightarrow$, which are determined by the respective relations between their components. Note that the connective '$\rightarrow$' in our preferred logic MC is interpreted as meaning containment, whilst the rule '$\Rightarrow$' preserves truth and is meta-theoretic as it concerns derivations in a system. (See the Appendix for the axiomatization of MC.) Here, we take truth and meaning as the key semantic concepts for the determination of a logic, with meaning containment having been subject to a refining process, starting with the system DJ^d of Brady (1996) and (2006) and then moving to the logic MC by weakening distribution and strengthening the metarule, as we argued for in Brady and Meinander (2013). Further refinement is possible, but the author thinks it to be unlikely, given the roundedness of the concepts captured by MC. (See Brady 2015a for further discussion on the connectives.) The quantifiers are added in MCQ as a natural extension of the concepts of conjunction and disjunction.

3.3 The Argument for Three Values

In arguing for three values, we first consider what formal systems are meant to achieve. Hilbert's formal systems, as originally introduced in accordance with his philosophy of mathematics, called 'formalism', were sets of axioms and rules, expressed using recursively generated symbolism. Derived theorems were also recursively generated by applying the rules, initially to the axioms and then to further theorems. Such formal systems were initially seen as systems of symbol manipulation that, at this general level, would clearly support the four values given in Sect. 3.2 above. However, Hilbert added simple consistency as a key constraint, among others, to put onto such systems. (We note too that Hilbert proposed that such consistency should be proved using finitary methods.) This would then restrict the values to the three: (i),

(ii) and (iii), and these are the three values that we are arguing for in this section. (See Körner 1960 for an account of formalism, together with its pros and cons.)

However, getting back to the general symbol manipulation systems, without the consistency constraint, these are seen to be too broad as they stand when trying to capture logical systems, both pure and applied. When setting up a logical system, we need to capture the associated logical concepts, in particular, negation, conjunction, disjunction and implication/entailment, with logical equivalence defined in the usual way. Further, we may add universal and existential quantification, and maybe identity or modal operators, etc. These logical concepts should be axiomatically captured in accordance with their meanings, with appropriate axioms and rules. This provides a conceptual constraint on formalization, such formal systems being idealized to achieve a reasonable level of conceptual clarity. This also extends to applied logical systems with non-logical axioms and rules capturing non-logical concepts. This is what formal systems are generally set up to achieve.

Certainly, conceptual clarity is an ideal property for formal systems to achieve, especially for non-logical concepts, but usually attempts at axiomatizing such concepts fall short. As discussed in Brady (2015a) and (2019), concepts can be underdetermined or overdetermined by their respective axiomatizations. As argued in (2019), underdetermination is ubiquitous in that reasoning can take place with incomplete information being assumed, as was stated above in Sect. 3.2. Concepts are often somewhat vague and thus understated in an axiomatization and people do not always go to the trouble of capturing all concepts in fullness, such as to achieve negation-completeness for the whole system. Overdetermination, on the other hand, involves overstating a concept or concepts to the point of contradiction. In such a situation, in order to avoid inconsistency, people would be inclined to re-examine the concepts to see if such a system can be made consistent by fixing up the axiomatization. It would be thought that a conceptual clash between concepts would have taken place or a particular concept would have been overdetermined. (See Brady 2015a, b for more on this point.) Once this re-examination of concepts has been achieved in such a way as to produce consistency, the three values (i), (ii) and (iii) remain, as occurs in Hilbert's simple consistency requirement. Thus, it can be said that these three values are an ideal worth aiming for.

Let us delve further into how consistency might be achieved in practice. A yardstick that can be used to achieve a consistent formal system is that of the capture of conceivable concepts. A simple way of creating conceivable concepts is to obtain a mental grasp of them, in some manner or other. Such concepts would have to be finite or, if infinite, expressible through recursion, which involves two steps. These steps then make such infinite concepts mentally graspable. On the other hand, inconceivable concepts do not relate to anything we can think about. Finite conceivable concepts are consistent, at the very least, in that a mental picture has to be consistent to be able to be grasped. If the concept is abstract, however, then there should be some concrete examples, any one of which could then be used to determine the consistency of the abstract concept. It is then hoped that the above processes would enable consistency to be proved.[1]

3   The Number of Logical Values

A problem with conceivability is that infinite totalities cannot be grasped in their entirety but nevertheless form a common and essential part of mathematical practice. This too puts them beyond the scope of Hilbert's finitary methods, standardly used to prove consistency. Given the importance of consistency to this exercise, we will need to examine the issue of infinite totalities and their use in proving consistency. We look into two examples. In Brady (2012), Peano arithmetic is proved consistent using finitary methods using metavaluations, which are proof-theoretic valuations, which in turn enables the consistency proof to be kept finitary. (Note that this uses a logic slightly weaker than our favoured logic MCQ.) However, in Brady (2014), metavaluations are also used to prove the consistency of naïve set theory, but denumerable ordinals are used to structure them and transfinite induction on these ordinals is used to prove consistency. (These ordinals are all denumerable because the fixed points at the end of the transfinite sequences are all determined by the denumerability of the set of all formulae.) Because this method does not involve the construction of models, unlike the proof in Brady (2006), there is no set theory used, other than the usage of the ordinals. Further, transfinite induction is a three-step process, which is an extension of the two-step processes of recursion and mathematical induction. So, on this basis, we believe there is a case to extend the consistency proving techniques from the finitary to include the transfinite denumerable ordinals, especially as the ordinals used can be put into one-to-one correspondence with the natural numbers. Further, every limit ordinal is approached by a denumerable sequence of ordinals below it and so the essential difference in this method is that every limit ordinal represents an infinite totality that is not conceivable as a totality. In some cases, this may not be sufficient to prove consistency outright, in which case the matter would be unresolved. For such cases, the inconsistent value (iv) would be left open as a possibility. This would apply too to systems for which only a relative consistency is established. An example is the set theory NBG that is relatively consistent with respect to ZF set theory.

Nevertheless, it is possible to axiomatize concepts that are inconceivable, either directly or indirectly. A direct example would be that of the round square, which is both round and square, which is explicitly contradictory. Clearly, this is inconceivable and can be made conceivable by removal of its roundness or its squareness. An indirect example would be the Liar sentence, 'This very sentence is false'. Self-reference is conceivable as it is clear what the reference is, even though it refers to itself. However, using classical logic, it leads to contradiction, that is, it is both true and false, taking our value (iv) above. In this case, as shown in Brady (2015a, b), it is the logic that is at fault in that it includes the LEM, which is not justified when applied to the Liar sentence in the context. This is a case of overreach or overdetermination within the logic. As also shown in (2015a, b), similar overdetermination applies to the other paradoxes, both set-theoretic and semantic. (We will examine extended semantic paradoxes in Sect. 3.6 below when dealing with metalogic.) When set up with an appropriate logic, such as the logic MCQ of meaning containment (see the Appendix below), we would not be able to derive these paradoxes due to consistency being proved in Brady (2006) (but using the slightly stronger logic DJ^dQ). We should note that this proof applies to naïve set theory as well as truth theory, with a very similar consistency proof.

3.4 The Argument for Two Values

The case for the two values, (i) and (ii), simply requires negation-completeness, in addition to simple consistency, by deleting the value (iii) from the three-valued account of Sect. 3.3. This would require precise determination of concepts in an applied logic, with no over- or underdeterminations, and would require much care in ensuring that either A or ~A occurs, for each sentence A. This would then yield a classical (possible) world, in the Kripke sense. This would represent a further idealization, over and above what is required for the three-valued case. As we argued in Sect. 3.2, the addition of the LEM is not adequate to guarantee negation-completeness.

Support can be given for these two values by recognizing that Boolean negation is the intended negation, as it is the only realistic full negation concept. (See discussion in Brady 2015a and 2019.) However, this intention cannot always be realized, due to the lack of recursion in the infinite case, which leaves incompleteness possibilities open. Logical formal systems themselves are intrinsically infinite and, since negation-completeness is meta-theoretically defined, Boolean negation can easily fall short. There can also be problems with (large) finite systems, where there is insufficient specification of all the details, for whatever reason.

Let us look into how such a two-valued system might be obtained. In Brady (2019), we discuss classical recapture and a key example that might help show us the way. This classical recapture, however, introduces the LEM and the DS into the logical system, but, as mentioned in Sect. 3.2, these are not sufficient on their own to establish the two values (i) and (ii). The example given is that of Peano arithmetic, based on a slight weakening of MCQ, and shown to be simply consistent in Brady (2012) using finitary methods. In this logic, all instances of the LEM must be proved through one of its disjuncts, which yields negation-completeness for the sub-theory generated by instances of the LEM. Further, we add the DS to the logic on the grounds that it is admissible due to the simple consistency. This completes the classical recapture and produces a two-valued sub-theory.

However, to achieve a two-valued system, we would still need negation-completeness, for the whole system. What we need is for the LEM to hold for all formulae of the system, in conjunction with the Priming Property (If $A \vee B$ is a theorem then either A or B is a theorem), whereby either A or ~A will hold, yielding negation-completeness. However, for classical logic and some strong relevant logics such as R, E and T, the LEM holds all right, but Priming fails. This is in contrast with metacomplete logics, where Priming holds but the LEM fails. (See the papers, Meyer 1976 and Slaney 1984, 1987 for the basics on metacomplete logics.) What we can do is to consider classical systems which are also metacomplete, so as to combine the LEM with Priming. This has been pursued in Brady (2010), where the sentential variables are treated as constants and introduced in accordance with a recursive specification. A two-sorted approach is introduced with classical sentential constants in one sort, shown to be negation-complete, and general sentential constants in the other sort. A one-sorted approach is also introduced where classicality is derived within the sentential system using rejection (thus making the system decidable). Predicate systems for both approaches are introduced, but with a finite domain of individual constants, with predicate constants over such a domain, and with no free variables. Thus, considerable constraints are applied in enabling negation-completeness to be established.

3.5 Extending the Arguments to the Meta-theory

We further examine the meta-theory of a formal system (or systems), where the formal system itself is an object of a special kind of logical theory, to which the preceding arguments can then apply. That is, whatever logic is used in the object theory is also applied to the meta-theory. The object theory can have two, three or four values, but we will argue that the meta-theory of such an object theory will have at most the values (i), (ii) and (iii). That is, the value (iv) is to be removed, on the grounds that the formal system is simply consistent within itself.

What would such an inconsistency involve? It would mean that a formula A is both provable and not provable. (Note that we are dealing with formulae here, as the statement of its proof is a sentence.) A proof of the formula A is obtained by a recursive process that applies rules to axioms and to further theorems. The non-proof of a formula A would use some meta-theoretic method that relies on this recursive process of proof to establish its methodology. This usually involves some modelling, either truth-theoretic or algebraic, or some alternative proof-theoretic method such as metavaluations, rejection, a cut-free Gentzen system or a normalized natural deduction system. (It is important too that such methodologies are sufficiently determinate to ensure that a formula outside its scope is indeed a non-theorem. This is achieved through recursive processes of their own.) In each case, the recursive proof procedure is used to establish that the method or modelling works. Thus, any non-proof of a formula A, which is establishable by such methods, can only be for such a formula that is outside the scope of the recursive proof process. Thus, non-proof cannot overlap with proof and the metalogic of proof is three-valued as indicated above. (We call these 'sound' methodologies as a soundness proof, which applies the recursive proof process and is needed to ensure that the methodology encompasses all the proofs, enables non-proof not to overlap with proof.) Furthermore, formal systems are conceivable concepts in that the recursion is conceivable by involving just a two-step process using axioms and rules. However, this would be extended to a three-step process with the use of metarules, as occurs in the axiomatization of MCQ in the Appendix below. And, conceivable concepts are generally consistent, as we discussed in Sect. 3.3. Indeed, this would put formal systems on a par with primitive recursive arithmetic, as far as conceivability goes, and such arithmetic is simply consistent. (See Brady 2012.)

3.6 The Case for a Two-Valued Meta-theory

Meta-theory is currently assumed to be two-valued, even when applied to logics that are many-valued. This, we believe, is due to meta-theory being regarded as a branch of pure mathematics and, since mathematics has been developed in the medium of classical logic for the last 100 years or so, this is just projected upon meta-theory. And, this stands to reason as most pure mathematics itself is conducted in the meta-theory, due to its abstract nature. Further, saying things in a classical meta-theory enhances what might only be stated non-classically in the object theory. Whatever we say about meta-theory will also project back to the whole of mathematics itself. Moreover, that includes what has been said in Sect. 3.5 above. So, we should continue with what we have been arguing for in Sect. 3.5 and also in Sect. 3.4, realizing its enormity.

Let us continue with the application of what has been said about logic in general in Sects. 3.2–3.4 to the study of formal systems in particular. In Sect. 3.5, we argued for three-valuedness on the grounds that the ways of establishing non-proof essentially involve the recursive proof process in determining the 'sound' methodologies that are used. What we need to show here is that there is a meta-theoretic proof process which enables all formulae A, that are not provable, to be shown to be so by such a process. That is, putting these two together, we need a meta-theoretic proof process, or processes, to establish, for each formula A, that A is either provable or not. That is, the formal system needs to be decidable. So, decidability ensures that the meta-theory of a formal system is two-valued, that is, with just the two classical values, (i) and (ii). For the classical predicate calculus, this is not possible, but for a weaker logic such as the logic MC of meaning containment, and its quantificational extension MCQ, decidability is a real possibility, with the latter still to be worked upon. The decidability of the sentential logic MC has been studied but the work is not yet published. A sketch of such work appears in section (x) of Brady (2017).

We note the difference between the criteria for the use of the LEM, in particular, in formal logical systems and in their meta-theory. The LEM in formal systems attempts to ensure negation-completeness and certainly holds if indeed the system is negation-complete. The LEM in meta-theory attempts to ensure that the formal system is decidable, and indeed holds when the system is decidable. So, we cannot fudge these two uses, as occurs in the derivation of the Extended Liar Paradox and in other extended paradoxes. (The Extended Liar sentence is 'This very sentence is not true', with the sense of 'not true' taken from the meta-theory). The derivation of this paradox licenses the use of the LEM on account of its being included in the metalogic, in which case a contradiction can be derived when the sentence is put in the object theory, regardless of whether the LEM is in the object-logic or not. This transfer from the meta-theory to the object theory is not warranted on account of the above differences in criteria for the LEM in the respective theories. Thus, the Extended Liar paradox is not derivable in the standard way, and we can put our account of logical values to some use in the solution of certain paradoxes. This also has ramifications for the idea of semantic closure, as discussed in Brady (2015a, b).

3.7 The Argument Against Other-Valued Logics

Logical values have traditionally been understood in terms of matrix logics, which have a certain number of specific values, divided up into designated and undesignated, with each of the designated values representing a kind of truth, which is then used to define validity using formula induction. Though we have argued against the semantic account of logical values in Sect. 3.2 and matrix logics are an example of such a semantic account, it is instructive to look into what specifically might go wrong for matrix logics. Of course, our proof-theoretic values work quite differently.

Using a method due to Gödel, Dugundji in (1940), by way of showing that there is no characteristic matrix for any of the Lewis systems S1–S5, introduced a technique that provides unintuitive valid formulae in a very wide range of matrix logics, including possible candidate matrices for S1–S5. Such a valid formula D for an n-valued matrix logic with connectives $\vee$ and $\leftrightarrow$ is

$$(p_1 \leftrightarrow p_2) \vee (p_1 \leftrightarrow p_3) \vee (p_2 \leftrightarrow p_3) \vee (p_1 \leftrightarrow p_4) \vee (p_2 \leftrightarrow p_4)$$
$$\vee (p_3 \leftrightarrow p_4) \vee (p_1 \leftrightarrow p_5) \vee \ldots \ldots, \vee (p_n \leftrightarrow p_{n+1}),$$

where the p_i are sentential variables in the matrix logic.

Simply put, D is a disjunction of all the formulae of the form: $(p_i \leftrightarrow p_j)$, where i and j range over all the natural numbers from 1 to n + 1, such as to satisfy i < j.

The point here is that there is one more variable than the total number of values in the matrix logic. So, at least one of these sentential variable pairs must take the same value in any valuation in the matrix logic, making the equivalence between them an instance of $p \leftrightarrow p$, which must be included as a disjunct in the above disjunction, for each valuation. However, $p \leftrightarrow p$ is valid in a wide range of matrix logics, indeed with any reasonable concept of logical equivalence, and this would then be reflected in the designated values down the main diagonal of the $\leftrightarrow$-matrix. This is sufficient to show that one of the disjuncts of D is designated, for each valuation of the matrix logic. The formula D can then be made valid by requiring that the rules $A \Rightarrow A \vee B$ and $B \Rightarrow A \vee B$ preserve designated values, which again would hold with any reasonable concept of disjunction. This would be ensured by putting a designated value in the matrix for $A \vee B$ whenever either A or B takes a designated value. So, for any reasonable matrix logic, the formula D would be valid. However, the formula is unintuitive in that there should be a sufficient number of non-equivalent sentence pairs, that is, n(n + 1)/2 such pairs based on an n-valued matrix logic, that can be taken into account in general argumentation. There is no general reason to restrict the number of non-equivalent sentences, as logic is widely applicable. This argument then applies to two-valued classical logic, the many-valued logics of Lukasiewicz, and many others in the literature.

A particular many-valued approach can be singled out, due to Shramko and Wansing (see their book 2011), where they move from Frege's classical two-valued logic to Dunn's four-valued logic, and then to a 16-valued logic, by following the same algebraic process of taking successive power sets over the sets of values. That is, moving from $\{T, F\}$ to $\{\emptyset, \{T\}, \{F\}, \{T, F\}\}$, and then on to the 16 elements:

$\emptyset, \{\emptyset\}, \{\{T\}\}, \{\{F\}\}, \{\{T, F\}\}, \{\emptyset, \{T\}\}, \{\emptyset, \{F\}\}, \{\emptyset, \{T, F\}\}, \{\{T\}, \{F\}\}, \{\{T\}, \{T, F\}\},$
$\{\{F\}, \{T, F\}\}, \{\emptyset, \{T\}, \{F\}\}, \{\emptyset, \{T\}, \{T, F\}\}, \{\emptyset, \{F\}, \{T, F\}\}, \{\{T\}, \{F\}, \{T, F\}\},$
$\{\{\emptyset, \{T\}, \{F\}, \{T, F\}\}\}$.

These movements start with a set of separate and complete values but then, by taking their power set, allow each of these values to be completely independent of each other, creating all the possible combinations. As we have argued, the above four-valued logic is appropriate as the sentences A and ~A are independent of each other in a general proof-theoretic setting. The question now is: Why cannot these four values be regarded as independent, in a similar way to that of the two values? The answer lies in the fact that the four values are obtained by a meta-theoretic examination of the possibilities occurring in the object language. So, the four values are not represented by sentences in the object language, for them to be meta-theoretically examined, as this would illicitly import meta-theory into the object language. By restricting the object language to include A and ~A, we are only able to examine its combinations meta-theoretically. Moreover, we do, quite appropriately, examine object-proof meta-theoretically and apply the same object-logic to the meta-theory, as we did in Sect. 3.5.

To conclude this section, let us look into infinitely valued matrix logics, as Dugundji's argument does not apply to them. We start by looking into two very general examples that one can create for almost all logics. The first is a trivial formula-based matrix logic, where the values are just the individual formulae and the matrices for the connectives are just filled out in accordance with the formation rules. That is, for each valuation v, $v(A) = A'$, where A' is a some uniform substitution instance of a formula A, obtained by substituting formulae (uniformly) for each of the sentential variables of A. For example, for the connective &, we put $v(A\&B) = (A\&B)' = A'\&B' = v(A)\&v(B)$, for all valuations v and formulae A and B, and similarly for the other connectives. Note that $A'\&B' = (A\&B)'$, because of the uniformity of the substitution into the respective sentential variables. We just designate the theorems, and we have created an infinite characteristic matrix model for the logic. This can be easily seen since all substitution instances A' of a theorem A are also theorems, that is, all valuations of A are designated, in which case soundness follows. Conversely, if all instances of a formula A are theorems then trivially A is a theorem, yielding completeness. Though not useful in itself, this infinite matrix model shows that there are always infinite matrix characterizations for any logic with a uniform substitution rule. However, it does show that validity in this case is defined in terms of theorem-hood and thus induction on proof steps is needed to determine validity rather than formula induction which is of little or no use.

The second example of an infinite matrix logic is more useful, but mainly as a technique for proving completeness for algebraic-style semantics. This matrix logic is proof-based rather than formula-based and is essentially a Lindenbaum algebra. As such, it accords with our proof-theoretic approach, in comparison with truth-theoretic approaches, which do not precisely capture some logical concepts due to their formula induction requirements. (See Brady 2017 for discussion of this.) This matrix logic is available for any logic with a substitution of equivalents rule for an equivalence connective $\leftrightarrow$. The values here are equivalence classes of formulae, such that v(A) = [A], where [A] is the equivalence class of A, for each formula A. (The equivalence class of A is the set of all formulae B such that A$\leftrightarrow$B is provable in the logic concerned.) To evaluate the connective &, we put v(A&B) = [A&B] = [A]&[B] = v(A)&v(B), where [A]&[B] is defined as [A&B]. This definition is possible since if A$\leftrightarrow$C and B$\leftrightarrow$D are provable then so is A&B $\leftrightarrow$ C&D, by the substitution of equivalents rule. Similar valuations are made for the other connectives, for which substitution of equivalence equally applies. We then designate the equivalence classes of each of the theorems, thus creating an infinite characteristic model for such a logic. Soundness follows by noting that, by uniformly substituting equivalence classes for the sentential variables of a theorem A, we obtain the equivalence class instance [A'], which is designated since A' is a theorem. Completeness follows by noting that if [A'] is designated, for all substitution instances A' of a formula A, then trivially A itself is a theorem. Again, validity is defined in terms of theoremhood with its associated induction on proof steps.

Also proof-based is the modelling used in content semantics, which is a slight variation on the above, where contents are used instead of equivalence classes, the content of a formula A, c(A), being defined as the set of all formulae B such that A $\rightarrow$ B is provable in the logic. Thus, contents can be understood as analytic closures when the logic is taken to be the logic MC of meaning containment. (See the Appendix for the axiomatization of MC.) However, there are more complex definitions for c(A $\rightarrow$ B), c(A&B) and c($\sim$A) in terms of their component contents. (See Brady 2006 for a full account, and more briefly in Brady 2017.)

We could indeed propose that these contents are the logical values, in that they play a key role in interpreting our preferred logical system MC, as set out in Brady (2006). Whilst there is a case to be made for this, contents are not formula-inductive like standard values and so we do need to differentiate our values from standard values, in any case. Furthermore, we do wish to relate our values, at least as close as we are able, with classical truth and falsity, which also aligns with the major discussion in the literature. Also, if we consider the $\{\sim, \&, \vee\}$-fragment of MC, which consists of the tautological entailments of Anderson and Belnap (1975), expressed as rules instead of entailments, then this basic fragment is four-valued in a standard formula-inductive sense, given that the rule form of distribution is validated in MC.

However, we still need to distinguish between infinite matrix logics that are more related to truth-theoretic semantics than to proof theory. The essential difference of concern is the use of disjunctive witnesses in truth-theoretic semantics, which causes a difference in the interpretation of disjunction between the semantics and the proof theory, as argued above in Sect. 3.2. (In logics with the LEM, such disjunctive witnesses apply to the LEM, which then ensures negation-completeness, as in truth-theoretic semantics of classically based systems.) So, we need to differentiate matrix logics that use witnesses in an essential way in determining validity from those that do not. Clearly, the finite logics and logics with truth-theoretic semantics do use witnesses due to the formula-inductive process. Also, infinite matrix logics that are set up in a conventional way using numbers as values, will also use witnesses in the process of determining the values of formulae and hence their validity. In such cases, disjunctions A ∨ B are usually evaluated in terms of their disjuncts, as the maximum value of the two, and thus the witnesses are called in. In the case of existentials, the supremum of all the numerical values is usually used, again defined by using witnesses. As a case in point, the Lukasiewicz three-valued logic can be recast as a Routley–Meyer model structure with two worlds, which is truth-theoretic semantics. (See Brady (1982) for the details.) It is quite straightforward to extend this relationship to that for the Lukasiewicz infinitely valued logic simply by taking its prime filters as worlds in a related infinite Routley–Meyer model structure.

In the two general examples given above, this simple valuation process breaks down in that provability is required to determine the designated values and in the case of equivalence classes (and also contents), deletions of the same equivalence classes (or contents) generated by different formulae would require provability again. Thus, these examples require algebraic-style operations to determine validity and their associated algebraic properties do follow the proof-theoretic properties very closely, as occurs in algebraic semantics in general. Unlike truth-theoretic semantics, which requires witnesses, algebraic semantics does not, as their completeness proofs use Lindenbaum algebras or such like, which are proof-theoretically defined without finding witnesses for each disjunction.[2] Again, the logical values we espouse are proof-theoretic and these do quite directly relate to algebraic semantics, which can then be used to construct an infinite matrix logic of a sort that does not require witnesses in the determination of validity. However, this does remove us from the familiar infinite matrix logics.

3.8 Conclusion

To briefly recap the main conclusions of the paper, we have argued that formal systems are four-valued, since some modicum of meaning is needed to drive logical deductions, and thus meaninglessness does not play a role in logic. These four values are determined by relative independence of the formulae A and ~A, giving rise to the four possibilities: A without ~A, ~A without A, neither A nor ~A and both A and ~A. We argued that values are determined from the proof theory rather than from the truth-theoretic semantics. However, logic ought to be three-valued as any contradiction, A and ~A, should be removed by re-conceptualization of the concepts captured by the formal system. This is in accord with what formal systems are generally set up to achieve, especially when the concepts they capture are conceivable. Two-valuedness requires extra special conditions to ensure negation-completeness, leaving the two classical values. Indeed, this work needs to be done to achieve what the classical logician takes for granted. This can be done for simple finite concepts and some that are recursively infinite, but this is not so easily achieved. (See 2010.) When applying this to meta-theory, we consider formal systems as special recursively defined objects of the theory, which are such as to ensure that meta-theory is three-valued. We then argue that two-valuedness requires decidability. So, we can say logic is four-valued, but ought to be at most three-valued, whilst meta-theory is three-valued and two-valued for decidable systems.

Whilst we used Dugundji's (and hence Gödel's) work to provide unintuitive valid formulae in finite matrix logics, this does leave open the question of values in infinite matrix logics. We tried to determine infinitely valued matrices with proof-theoretically determined values, and this led us to algebraic semantics, whose elements were presented as values in the infinite matrix format. Such values were equivalence classes or contents and as such are very general.

As advocated above, values are based on the relative independence of negation in a proof-theoretic context. As a further idea, this use of negation might be extended to involve other connectives, especially the '$\rightarrow$', to create other values, such as A $\rightarrow$ ~A and ~A$\rightarrow$A, for example. Indeed, if these two are added to A and ~ A, then there are 12 values instead of 16, as a matter of interest, since Modus Ponens would rule out the pair, A, A $\rightarrow$ ~A, in favour of the trio, A, A $\rightarrow$ ~A, ~A, and the pair, ~A, ~A$\rightarrow$A, in favour of the trio, ~ A, ~ A$\rightarrow$A, A. However, there would not be much point to this, as the criteria for satisfying the various combinations would take us away from mainstream properties of formal systems. Further, there is a point in singling out negation here since it is an incomplete concept, unlike the other connectives, as mentioned at the end of Sect. 3.2 and discussed in Brady (2015a) and (2019). Indeed, the four values are proof-theoretic values produced by their proof-theoretic independence rather than the traditional values that are semantically based and require formula induction. Also, the four values are based on classical truth and falsity, at least as close as can be achieved.

Notes

1. Though 'conceptual clarity' is somewhat imprecise in itself, concepts could still be precisified and consistency achieved for any such precisification. The conceivability of concepts may also be imprecise, but nevertheless the act of conceiving does require consistency in order for the conceiving to take place. And, it is consistency that really matters here.
2. In many algebraic semantics, the Priming Property holds, that is, if a $\vee$ b is true then either a is true or b is true. Though this creates disjunctive witnesses, this translates into a property for designated values only and may not apply to all valuations of disjunction. A similar point applies for existential quantification.

Acknowledgments I would like to thank the referee for a number of incisive comments on this paper. I would also like to thank the members of the University of Melbourne Logic Seminar for their interesting and pertinent discussion of this paper and, in particular, for pointing out a glaring error of reference. I would also like to thank Thomas Ferguson for encouraging me to continue with this paper, even after I felt that I could not complete it in a reasonable time.

Appendix

The required axiomatization of MCQ is set out below:

MC.
Primitives: ~, &, ∨, → .
Axioms.

1. A → A.
2. A&B → A.
3. A&B → B.
4. (A → B)&(A → C) → .A → B&C.
5. A → A∨B.
6. B → A∨B.
7. (A → C)&(B → C) → .A∨B → C.
8. ~~A → A.
9. A → ~B → .B → ~A.
10. (A → B)&(B → C) → .A → C.

Rules.

1. A, A → B ⇒ B.
2. A, B ⇒ A&B.
3. A → B, C → D ⇒ B → C → .A → D.

Metarule.

1. If A, B ⇒ C then D∨A, D∨B ⇒ D∨C.

We now add the quantifiers to yield MCQ. As in earlier presentations, we separate free and bound variables to simplify the conditions on the axioms.

MCQ.
Primitives: ∀, ∃,
a, b, c, ... (free variables)
x, y, z, ... (bound variables)

Axioms.

1. ∀xA → A^a/x.
2. ∀x(A → B) → .A → ∀xB.
3. A^a/x → ∃xA.
4. ∀x(A → B) → . ∃xA → B.

Rule.

1. A^a/x ⇒ ∀xA, where a does not occur in A.

Metarule.

1. If A, B ⇒ C then A, ∃xB ⇒ ∃xC,

where QR1 does not generalize on any free variable in A or in B. The same applies to the premises A and B of the metarule MR1 of MC.

References

Anderson, A. R., & Belnap, N. D. Jr. (1975). *Entailment: The logic of relevance and necessity* (Vol. 1, Princeton U.P.).
Brady, R. T. (1982). Completeness proofs for the systems RM3 and BN4. *Logique et Analyse, 25,* 9–32.
Brady, R. T. (1996). Relevant implication and the case for a weaker logic. *Journal of Philosophical Logic* (Vol. 25, pp. 151–183).
Brady, R. T. (2006). *Universal Logic*, CSLI Publs, Stanford.
Brady, R. T. (2007). Entailment: A Blueprint. In J.-Y. Beziau, W. Carnielli and D. Gabbay (eds.), *Handbook of Paraconsistent Logic*, College Publications, King's College, London (pp. 127–151).
Brady, R. T. (2010). Extending metacompleteness to systems with classical formulae. *Australasian Journal of Logic, 8,* 9–30.
Brady, R. T. (2012). The consistency of arithmetic, based on a logic of meaning containment. *Logique et Analyse, 55,* 353–383.
Brady, R. T. (2014). The simple consistency of Naïve set theory using meta valuations. *Journal of Philosophical Logic, 43,* 261–281.
Brady, R. T. (2015a). The use of definitions and their logical representation in paradox derivation. In *Presented to the 5th World Congress on Universal Logic, University of Istanbul, Istanbul, Turkey,* June 25–30, forthcoming in Synthese.
Brady, R. T. (2015b). Logic—the big picture. in J.-Y. Beziau, M. Chakraborty and S. Dutta (eds), *New Directions in Paraconsistent Logic*, Springer, New Delhi, pp. 353–373.
Brady, R. T. (2017). Some concerns regarding ternary-relation semantics and truth-theoretic semantics in general. *The IfCoLog Journal of Logics and their Applications, 4,* 755–781.
Brady, R. T. (2019). On the law of excluded middle. In *Ultralogic as Universal?, The Sylvan Jungle*, ed. by Zach Weber, Springer, Switzerland, (Vol. 4, pp. 161-183).
Brady, R. T. & Meinander, A. (2013). Distribution in the logic of meaning containment and in quantum mechanics. In *Paraconsistency: Logic and Applications*, ed. by Koji Tanaka, Francesco Berto, Edwin Mares and Francesco Paoli, Springer Publishing, Dordrecht (pp. 223–255).
Dugundji, J. (1940). Note on a property of matrices for Lewis and Langford's calculi of propositions. *The Journal of Symbolic Logic, 5,* 150–151.
Goddard, L., & Routley, R. (1973). *The logic of significance and context*. Edinburgh: Scottish Academic Press.
Henkin, L. (1949). The completeness of first-order functional calculus. *The Journal of Symbolic Logic, 14,* 159–166.
Körner, S. (1960). *The philosophy of mathematics*. London: Hutchinson.
Meyer, R. K. (1976). Metacompleteness. *Notre Dame Journal of Formal Logic, 17,* 501–516.
Routley, R., Meyer, R. K., Plumwood, V., & Brady, R. T. (1982). *Relevant logics and their rivals* (Vol. 1, Ridgeview, Atascadero, California).
Shramko, Y., & Wansing, H. (2011). *Truth and falsity: An inquiry into generalized logical values*. Dordrecht: Springer.
Slaney, J. K. (1984). A metacompleteness theorem for contraction-free relevant logics. *Studia Logica, 43,* 159–168.
Slaney, J. K. (1987). Reduced models for relevant logics without WI. *Notre Dame Journal of Formal Logic, 28,* 395–407.

Ross T. Brady is Adjunct Associate Professor of Philosophy at LaTrobe University. He has worked extensively on the consistency of naive set theory and its extension to higher order predicate theory, and the proof theory of relevant logics, especially Gentzen systems and natural deduction. He is the author of *Universal Logic*, published in the CSLI Lecture Notes series.

Chapter 4
Respects for Contradictions

Paul Égré

Abstract I discuss the problem of whether true contradictions of the form "x is P and not P" might be the expression of an implicit relativization to distinct respects of application of one and the same predicate P. Priest rightly claims that one should not mistake true contradictions for an expression of lexical ambiguity. However, he primarily targets cases of homophony for which lexical meanings do not overlap. There exist more subtle forms of equivocation, such as the relation of *privative opposition* singled out by Zwicky and Sadock in their study of ambiguity. I argue that this relation, which is basically a relation of general to more specific, underlies the logical form of true contradictions. The generalization appears to be that all true contradictions really mean "x is P in some respects/to some extent, but not in all respects/not to all extent". I relate this to the strict-tolerant account of vague predicates and outline a variant of the account to cover one-dimensional and multidimensional predicates.

Keywords Dialetheism · Borderline contradictions · Vagueness · Ambiguity · Gradability · Strict-tolerant logic · Truth

Dialetheism is the view according to which some sentences are both true and false: that is, false and true are not mutually exclusive properties, sometimes they cooccur. This is centrally the case of the Liar sentence: on the dialetheist view, the Liar is both true and false, and as a result, it is also true and not true, false and not false.

Over the last decade, several papers have documented the fact that the use of contradictory sentences of the form "x is P and not P" it not confined to dialetheist theorizing, nor limited to the predicates "true" and "false" in relation to Liar-like sentences, but that it is common in ordinary speakers in order to describe the borderline applicability of vague predicates. In a seminal study, Ripley (2011a) found that a square and a circle whose distance is intermediate between two more extremes cases are readily described by participants as "near and not near"; similarly, Alxatib and Pelletier (2011) found substantial assent to descriptions of the form "tall and not tall"

P. Égré (✉)
CNRS, ENS, EHESS, PSL University, Paris, France
e-mail: paul.egre@ens.fr

© Springer Nature Switzerland AG 2019
C. Başkent and T. M. Ferguson (eds.), *Graham Priest on Dialetheism and Paraconsistency*, Outstanding Contributions to Logic 18, https://doi.org/10.1007/978-3-030-25365-3_4

to apply to a man of middling height; and further studies have evidenced that these "borderline contradictions" (in the words of D. Ripley) are used in a wide range of gradable adjectives, including color adjectives (viz., Égré et al. 2013 on "yellow and not yellow") and relative gradable adjectives more generally (Égré and Zehr 2018).

The commonality of those contradictory expressions lends prima facie support to the dialetheist view. Opponents of dialetheism, on the other hand, need not take these expressions at face value. An epistemicist about vagueness might contend that an utterance of the form "*a* is tall and not tall" is a mere *façon de parler*, expressing a form of metalinguistic hesitance regarding the actual status of *a*. For the epistemicist, the real intent of such expressions should be to report epistemic ambivalence, viz.: "*I am tempted to say "a is tall", and I am also tempted to say "a is not tall", I just don't know which one is in fact correct*".

The problem with that objection, as forcefully put to me by Sam Alxatib (in a private communication), is that one does not see then why it appears illegitimate to use the same locution in cases of factual uncertainty. Suppose I am uncertain as to whether the binary Goldbach conjecture is true (as I should be in 2018, given the current state of mathematical knowledge). I have reasons to think it's true (it has not been disproved, moreover the ternary version was recently proved), and reasons to think it is not true (similar conjectures have been disproved for large numbers). It would be no good though to use the sentence: "Goldbach's conjecture is true and not true" to mean that I am tempted to say either, and just don't know which one holds as a matter of fact. There is more to the relation between vague predicates and borderline contradictions than the mere expression of factual uncertainty.[1]

A distinct and more delicate objection for dialetheism is to say that a sentence of the form "*a* is tall and not tall" really is shorthand for: "*a is tall in some sense, and not tall in some other sense*". This view is for example sketched by Kamp and Partee (1995), and it can be traced to Lewis's contention that acceptable contradictions in fact trade on some form of ambiguity or equivocation (see Lewis 1982; Priest 2006; Ripley 2011b; Kooi and Tamminga 2013; Cobreros et al. 2015b; Égré and Zehr 2018). That view, unlike the epistemicist view, does not rule out the simultaneous truth of both conjuncts. For "*a* is tall" and "*a* is not tall" can then express propositions that are true together, only true under different disambiguations of "tall".

(Priest 2006, 286–287) grants the objection:

> this may indeed be true; this is true; but this goes nowhere toward meeting the arguments for dialetheism. In any case, there seems to be little reason to believe, for most of the contradictions to which *In Contradiction* points, that they arise because of ambiguity. "This sentence is false", for example, is hardly ambiguous in the way that "He was hit by a bat" is. But even if it were, if there is no hope of disambiguating in practice, which would seem to be the case for such contradictions, there is effectively very little difference between such a view and dialetheism.

Priest makes two different points here. The first is a rejection of the idea that lexical ambiguity is in principle the right model to explain the acceptability of true

---

[1] See Schiffer (2003), MacFarlane (2010) on further arguments against the epistemicist attempt to reduce the ambivalence felt in vagueness to a form of factual uncertainty.

contradictions. The second is a concession that it might be the right model, but the denial that this perspective should alter the dialetheist perspective on contradictions. In this paper, I propose to clarify both points in relation to the semantic treatment of contradictory sentences involving vague predicates.

The issue I propose to examine in this paper is whether acceptable contradictions might be the expression of a phenomenon of multidimensionality rather than ambiguity, as suggested by Kamp and Partee, and drawing on recent work by Galit Sassoon on multidimensionality in adjectives and nouns. One claim of this paper is that the acceptability of contradictions involving adjectives in particular (including "true") might indeed be grounded in the availability of multiple respects of application, but provided those respects of comparison are closely related to each other in a way that is constitutive of the vagueness of the expression in question.

To show this, I start with a review of various examples of acceptable contradictions, and then go on to clarify the relation between those and the phenomenon of lexical ambiguity. My analysis, based on the classic work of Zwicky and Sadock (1975), draws particular attention to a distinction one can make between two forms of equivocation: equivocation based on homophony for lexemes that do not overlap in meaning, and equivocation based on lexemes that do overlap in meaning. Zwicky and Sadock mention in particular a relation of meaning overlap which they call "privative opposition". In Sect. 4.2, I highlight the importance of that notion with regard to acceptable contradictions. In Sect. 4.4, I show that the relation in question is congruent with the relation between strict and tolerant meaning used in our joint work with Cobreros, Ripley and van Rooij to model the acceptability of borderline contradictions. I outline a different version of the semantics for one-dimensional and multidimensional predicates, basically such that "x is P and not P" means "x is P to some extent, but not to all extent". I conclude, finally, with some remarks on whether or not this perspective should affect the ontology of dialetheism.

4.1 Acceptable Contradictions

The notion of a contradiction can be defined in various nonequivalent ways. In what follows, I call a contradiction a sentence of the form "x is P and not P", or one that is equivalent in virtue of classical logical rules. The Liar, for example, is not of that form, but it entails a sentence of that form, and is entailed by it, under minimal assumptions. And I say that a sentence of that form is an acceptable contradiction if the sentence can be used informatively. Acceptable contradictions in that sense come in at least two distinct linguistic forms, which may be called "and"-descriptions versus "neither"-descriptions (Égré and Zehr 2018). If you consider the Liar, the dialetheist accepts both:

(1)   a.   The Liar is true and not true.
       b.   The Liar is neither true nor not true.

From a logical point of view, both types of sentences appear to be equivalent, admitting de Morgan's laws, and the law of double negation to convert them into each other (see Ripley 2011a, b).

(2) a. $Pa \land \neg Pa$.
    b. $\neg(Pa \lor \neg Pa)$

From a linguistic point of view, however, the two types are not always used interchangeably (see Égré and Zehr 2018). In this paper, I focus on the "and"-descriptions, for one can find instances of "neither"-descriptions that seem not to have to do with the existence of true contradictions. This concerns, in particular, cases of presupposition failure (Spector 2012; Égré and Zehr 2018). Spector points out that in a situation in which John never smoked, there is a clear contrast between:

(3) a. *John stopped smoking and did not stop smoking.
    b. John neither stopped smoking nor did not stop smoking.

The latter is acceptable in response to someone asking: "when did John stop smoking?", but the former is clearly ruled out to convey the same thing. (3)-a may be acceptable to convey that John is a borderline case of someone having stopped smoking, but only if John was indeed a smoker. Setting vagueness aside, (3)-a and (3)-b mean different things.

Which examples of sentences of the form "x is P and not P" can we identify as common in ordinary language? The rest of this section lists some representative examples.

Multidimensional adjectives The first and probably the most common kind involves multidimensional adjectives like "intelligent", "beautiful", "good", and so on (see Klein 1980; Sassoon 2013). It is easy to imagine a context for the acceptability of:

(4) John is intelligent and not intelligent.

I could use the sentence to mean that John is intelligent because he is a very good mathematician, but not intelligent because he fails to have empathy for others, and sometimes takes wrong decisions because of that. I therefore intend to convey that John is intelligent in the respect that concerns mathematics, but not intelligent in the respect that concerns empathy and the understanding of other fellows. Similar occurrences can be found for a host of gradable adjectives, as soon as multiple respects are relevant. "John is rich and not rich" can be used to mean that John is rich in the respect of coming from a family with a large estate, but not rich with respect to his income; "this is good and not good" is often heard to mean that the situation referred to has both an upside and a downside, and so on.

One-dimensional adjectives

(5) John is tall and not tall.

As we know from the experimental studies of Ripley (2011a) and of Alxatib and Pelletier (2011), a sentence like (5) is acceptable to depict to a man whose size is intermediate between more extreme heights, to mean that the person is borderline tall. Égré and Zehr (2018) replicated the effect over a larger sample of similar adjectives, asking participants to imagine cases intermediate between more extreme ones.

A reason to distinguish "tall" from "intelligent" or "rich" is that "tall" fundamentally means "tall with respect to height". Note, however, that (5) could be uttered by reference to distinct points of comparison along the height dimension. Suppose child Mary describes her friend John as "tall" to her mother because John is significantly taller than her. The mother could respond "well, he is tall and not tall", thinking: "true, John is tall compared to you, but you know, he is not (so) tall compared to other teenagers of his age" (see Klein 1980; van Rooij 2011a; Burnett 2017 for the relativity of adjectives to a comparison class argument).

Nouns Kamp and Partee give the example of the following sentence as an acceptable contradiction:

(6)   Bob is a man and not a man.

They don't give much context for their example, but the sentence would, for example, be acceptable if "man" is taken to mean "male" in the first occurrence, and "brave man" or "aggressive man" in the second. It would mean that Bob is a man with respect to gender, but not a man with respect to the stereotypical respect of being sufficiently brave or sufficiently aggressive for a man (depending on what the speaker intends).

Many similar examples can be found. For instance, Pluto is considered a "dwarf planet" according to Resolution B6 of the International Astronomical Union, but also not a "planet" in the stricter sense of "planet" defined by the IAU in its Resolution B5(1) (see Égré 2013). Given the two resolutions, it would be perfectly sensible to say that according to the IAU,

(7)   Pluto is and isn't a planet.

to mean that Pluto is a planet with respect to being in the category of "dwarf planets", but not a planet with respect to the more restricted category "planet" defined by IAU Resolution B5 (it fails, in particular, to have "cleared the neighborhood around its orbit", IAU's discriminating criterion to rule out Pluto).

Verbs The availability of multiple respects is not restricted to adjectives and nouns, but also occurs in verbs. Consider the following example:

(8)   Q. Do you like John?
      A. I like him and I don't like him.

The sentence is perfectly natural to convey that I like John in some respects (for instance because he is charming, and generous, and funny), and that I don't like him in other respects (because he is so self-centered, never calls you unless you call him first, etc).

Summary The examples just surveyed are not meant to form an exhaustive list of acceptable contradictions. What appears from the list, however, is that each time contradictions can be paraphrased by means of an explicit specification of distinct respects of application. Our next task is twofold: first, to examine whether this availability of multiple respects is adequately viewed as a form of ambiguity. And second, whether it provides us with an adequate basis to semantically analyze contradictions.

4.2 Forms of Ambiguity

4.2.1 *Homophony*

Priest's first remark is that there is little reason to think that the acceptance of contradictions is akin to ambiguity. Priest is right, but it matters, in order to substantiate his claim, to define ambiguity more precisely. Typical cases of lexical ambiguity involve homophonous expressions that can be paired with meanings that are *disjoint* or *disconnected* in conceptual space (see Pinkal 1995; Bromberger 2012; Dautriche and Chemla 2016).

Consider Priest's example of the homophone "bat": a member of the Chiroptera species and a baseball bat are objects with very distinct properties, both functionally and in respects of perceptual similarity (Dautriche and Chemla 2016). We can say:

(9)   John was hit by a bat and John was not hit by a bat.

But in practice, one will forestall misunderstanding by putting different stress or using different gestures to mark the difference: "he was hit by a bat [gesture indicating a baseball hitting], not by a BAT [with gestures indicating a flying animal]". Also, as pointed by Ripley (2011a), it appears infelicitous in this case to use ellipsis to express the same thing (see Zwicky and Sadock 1975 on conjunction reduction):

(10)   #John was and wasn't hit by a bat.

Ripley notes that for a borderline case of nearness, on the other hand, it is fine to say:

(11)   The circle is and isn't near the square.

I think the same is in principle possible for the "planet" case we reviewed:

(12)   Pluto is and isn't a planet.

(12) can be used in my opinion to report that Pluto is both a planet in the sense of "dwarf planet", and not a planet in the sense of "having cleared the neighbordhood around its orbit". What is the difference with this case and the case of "bat" here?

4.2.2 *Polysemy*

The answer, it seems to me, again has to do with Pinkal's suggestion that for a typical case of an ambiguous expression, the distinct meanings are in fact completely disconnected (they lack a common more inclusive meaning). For a vague expression, on the other hand, Pinkal suggests that the various more precise meanings are connected and close to one another, with a common part (a form of polysemy, what Dautriche and Chemla 2016 call "motivated homophony").

In the case of "planet", the two meanings do indeed overlap in a sentence like (12). A dwarf planet, like a planet in the strict sense, needs to *orbit around the sun*, must *not be a satellite* of some other object orbiting the sun, and must have a *nearly round shape*. The difference between a dwarf planet and a planet in the strict sense is only that it lacks one characteristic feature that planets in the strict sense have. Hence the word "planet", as used by the IAU in 2006, is equivocal between a broad sense (inclusive of dwarf planets and planets in the strict sense) and a narrow sense (exclusive of dwarf planets), but the two senses are closely related, they overlap and are closer to each other than the two senses of "bat" are.

We may wonder if the same applies to other cases reviewed in the previous section. Consider:

(13)   Bob is and isn't a man.

Can the sentence be used to mean that Bob is male, but that Bob is not an aggressive male? I think the answer is positive. The case is similar to the previous one. "Aggressive man" implies or even presupposes "male" in this case, and thus the more restricted meaning overlaps with the more inclusive one.

What about:

(14)   John is and isn't rich.

Again, I think this will be fine to mean that John is rich with regard to his estate, but not rich with regard to his income, because both more precise meanings can in this case be subsumed under a common more inclusive meaning (such as: "having possessions ensuring material security") (see again Pinkal 1995). The sentence thus conveys that John is rich to some extent, but not as rich as one might expect.

The same is arguably the case with Ripley's "near and not near", or with Alxatib and Pelletier's "tall and not tall" examples: the more precise meanings that "near" or "tall" might have when negated and when not negated must be sufficiently close then. This will happen, and the sentence will be consistent, if the meanings overlap in such a way that the unnegated occurrence of "near" or "tall" is entailed by the negated occurrence (for example, if "tall" means "taller than 180 cm" in the first occurrence, and "taller than 186 cm" in the second).

Likewise, consider:

(15)   I like John and I don't.

Several enrichments are compatible with the sentence. Typically as we saw it means that I like John in some respects, but don't in others, and therefore that I fail to like him in all relevant respects. Overall, it conveys that even though I like John *to some extent*, I don't like him *very much*, compatible with the verb "like" maintaining a constant meaning.

Based on the previous examples, we therefore see that in all acceptable contradictions of the form "x is P and not P", each occurrence of P is open to adverbial modification. As a general template for this enrichment, we can consider the following schema:

(16) x is P [in some respects], and x is not P [in some respects].

The whole sentence is informative if the respects relevant to the second conjunct are distinct from the respects relevant to the first. An equivalent way of stating (16) is as follows:

(17) x is P [in some respects], and x is not [in all respects] P.

Although logically equivalent, (17) strikes me as a more significant way of representing the intended meaning of acceptable contradictions, in particular when quantification is done over extents instead of respects, as we shall see below. For adjectives in particular, this is a way of representing that P is instantiated to some extent, but not to all extent, and thereby of characterizing the borderline status of x relative to P.

4.2.3   Privative Opposites

The relation between the two understandings of the main predicate in (17) can be linked to an observation that Zwicky and Sadock (1975) made concerning what they called *privative opposites* as opposed to *polar opposites*. They define privative opposites as follows:

> U_1 and U_2 are *privative opposites* with respect to F if U_1 can be represented as identical to U_2, except that U_2 includes some specification for F that is lacking in U_1

As examples of privative opposites, they give the case of the non-homophonous expressions "parent" and "mother" relative to the feature gender, and the examples of the homophonous expressions "dog" versus "dog" relative to the same feature. "Dog" can mean a dog in the generic sense, or a male dog in the more specific sense. Importantly, it is possible to say of a female dog:

(18)   It is a dog and not a dog.

On the other hand, it is not possible to say of a father,

(19)   *It is a parent and not a parent.

in order to mean that the father is a parent but not a female parent: "parent", unlike "dog", does not have a more specific conventional meaning.[2] Zwicky and Sadock surmise in their paper that "ambiguities involving privative opposites are extremely

---

[2] Compare with Zwicky and Sadock's minimal pair: "This dog is not a dog", versus "*This parent is not a parent".

difficult to argue for with *any* syntactic test" (p. 24, emphasis theirs). However, I am not sure if it is entirely felicitous to use conjunction reduction to say of a female dog:

(20)   (?) It is and it isn't a dog.

Already for (18), contrastive stress on the second occurrence may be needed to make the sentence meaningful. The issue here is whether the two senses of "dog" are, like the two senses of "bat", stored as separate mental representations. My intuition on this case is that, despite the closer relationship in meaning between the general and the specific meaning, the case remains more similar to "this is a bat and not a bat" than it is to "this man is tall and not tall" to talk of a borderline tall man. However, I also think "tall and not tall" is closer in structure to "dog and not a dog" than to "bat and not a bat", that is, it can be viewed as a specific form of privative opposition. In agreement with Pinkal's remarks on the difference between vagueness and ambiguity, one may argue that the relevant meanings for "tall" are mentally much closer to each other than even the two meanings for "dog" are.

4.3 Polar Opposites

Is the proposed form for acceptable contradictions stated in (17) correct? Does it apply to all acceptable contradictions of interest to the dialetheist? To answer, we must consider whether some contradictions may not be acceptable even with predicates that stand in stronger opposition than privative opposites do, as do the antonymous adjectives "true" and "false". Those are cases of so-called "polar opposites" in the sense of Zwicky and Sadock, namely expressions that do appear to exclude each other completely. Zwicky and Sadock's definition is as follows (in their definition U_1 and U_2 stand for "understandings" of one or several terms, and F is a feature):

> U_1 and U_2 are *polar opposites* with respect to some semantic feature F if they are identical except that U_1 can be represented as having $+F$ where U_2 has $-F$ or the reverse.

Zwicky and Sadock assume that the features $+F$ and $-F$ are mutually exclusive. An example given by Zwicky and Sadock concerns the pair "father" and "mother", which are supposed to exclude each other relative to the feature "male".

The definition of polar opposites should in principle exclude their co-occurrence in a single object, but as with "True" and "False", it is easy to find counterexamples. First of all, contradictions involving polar antonyms are in fact acceptable in some contexts. Consider the polar opposite adjectives "rich" and "poor" (relative to the feature "having wealth"), and the following description of John's status:

(21)   John is rich and poor.

The sentence can perfectly be used to convey that John is rich in some or most respects, but also not rich, and therefore poor, in some distinct respects.[3]

A second problem for the exclusion idea is that even polar opposites can overlap in some cases. "Male" and "female" are polar opposites according to Zwicky and Sadock's definition (relative to the feature sex), but as we know from biology, the definition of those terms is problematic, and there are cases of overlap (see Fausto-Sterling 2000). This includes in particular cases of hermaphroditism. Snails, for example, are described as follows:

(22) Different snails reproduce differently, but most snails are "hermaphrodites." Being a hermaphrodite means that any given snail can be both male and female at the same time.[4]

Hermaphroditism is a case where the same individual has two functional capacities that are exclusive of each other in most other species, but that coexist in it. Consider now the following two sentences:

(23) Snails are male and female.

(24) (?) Snails are male and not male.

The first is attested. The second seems not to be attested (according to a Google search of occurrences), and to me a more natural way to express (23) by means of negation would be to say: "snails are male and non-male", using predicate negation. Even if it were attested, however, (24) ought not to mean that snails have some but not all of the features that are constitutive of male reproductive character. Instead, what is intended is that snails are *fully* male but *in part of their constitution*, and *fully* something other than male (namely female) *in part of their constitution*. In other words, the description in (24), supposing it was attested, has more to do with lack of *homogeneity* than with vagueness proper (or lack of clarity).[5]

An analogy that may help see the difference is by comparing a chessboard that is constituted of clear red squares and clear blue squares alternating, versus the same chessboard where all squares are of the same orange–reddish color. The former may be described as "red and blue", or as "red and non-red", to mean some of its parts are clearly red, and others clearly not red. The second may be described as "red and not red", but this time to refer to the fact that its homogeneous hue is neither clearly red nor clearly not red.[6]

---

[3] J. Zehr and I have empirical evidence indicating that polar contradictions of the form "x is P and $ant(P)$" (where $ant(P)$ is the lexical antonym of P) are less accepted than syntactic contradictions of the form "x is P and not P", but they are not systematically rejected. See Égré and Zehr (2018) for a brief mention of the finding, and Zehr (2014) for a discussion of antonymous adjectives.

[4] From http://scienceline.ucsb.edu/getkey.php?key=2578.

[5] On homogeneity and ways of characterizing the phenomenon, see Križ and Chemla (2015).

[6] I am assuming that each of the squares' color in the alternating blue–red chessboard is clearly perceptible as such. Brentano in his writings considers that one might see a homogeneous purple out of inhomogeneous red and blue tiles if the tiles are small enough. This is not a relevant coun-

We find a corresponding difference between hermaphroditism and cases of intersex: intersex is a case of vagueness, in my view, rather than inhomogeneity. The Intersex Society of North America defines intersex as follows[7]:

> "Intersex" is a general term used for a variety of conditions in which a person is born with a reproductive or sexual anatomy that doesn't seem to fit the typical definitions of female or male. For example, a person might be born appearing to be female on the outside, but having mostly male-typical anatomy on the inside. Or a person may be born with genitals that seem to be in-between the usual male and female types—for example, a girl may be born with a noticeably large clitoris, or lacking a vaginal opening, or a boy may be born with a notably small penis, or with a scrotum that is divided so that it has formed more like labia. Or a person may be born with mosaic genetics, so that some of her cells have XX chromosomes and some of them have XY.

Because intersex people are persons who, being borderline between typically male and typically female humans, share features of both of the more polar cases, they may be described as follows:

(25)   Intersex people are male and not male.

But this does not mean that intersex people have each of the male and the female reproductive capacities *fully* as respective parts of their constitution. As a matter of fact, the ISNA is explicitly opposed to the description of intersex as "hermaphrodites" precisely on the grounds that intersex is *partially* male, and *partially* female, but are neither fully. They have some male-defining features, and some female-defining features, which is very different from having both dispositions fully but "side by side" as it were (the way snails do in relation to reproduction: hermaphrodite snails, for example, can even self-fertilize, not a feature generally shared by intersex people).[8]

What should we conclude from this discussion of polar opposites? First, polar opposites can also be conjoined and predicated relative to different respects (as the

---

terexample to the distinction I am drawing, since I am assuming the tiles to be big enough to the eye. See Massin and Hämmerli (2017) for a discussion of Brentano's views on mixtures.

[7] http://www.isna.org/faq/what_is_intersex.

[8] See Fausto-Sterling (1993)'s typology, and her remarks about human intersex that "Although in theory it might be possible for a true hermaphrodite to become both father and mother to a child, in practice, the appropriate ducts and tubes are not configured so that egg and sperm can meet." Let me add that intersex people who militate for the recognition of a third category distinct from "male" and "female" may possibly be displeased with the description "intersex people are male and not male", and may prefer the description "intersex people are neither male nor not male", because they might prefer a description of their status in terms of exclusion of polar opposites ("neither male nor female") rather than a description in terms of the inclusion of both ("male and female"). If so, this preference could be explained along the pragmatic lines of Égré and Zehr (2018) to account for the overall preference for "neither"-descriptions over "and"-descriptions in relative gradable adjectives (that is by assuming that the default is to interpret "male" and "not male" in the strongest possible sense, see Alxatib and Pelletier 2011; Cobreros et al. 2015a). But this issue lies beyond the scope of the present paper. Incidentally, I note that hermaphrodites too are occasionally described in terms of a "neither"-description. A website on the internet even uses both types of descriptions with polar antonyms in the course of the same paragraph: "*Snails are neither male nor female. They are hermaphrodites, which means they are both male and female at the same time*". http://www.educationquizzes.com/nature-matters/2017/04/facts-about-the-slow-paced-and-shelled-snail/.

example of "rich and poor" indicates). Since "poor", on some theories at least (see Krifka 2007), is considered equivalent to "not rich", and conversely, the acceptability of conjunctions of polar antonyms is not by itself a challenge to the idea that acceptable contradictions might be a specific form of privative opposition. Secondly, we have seen that a conjunction like "male and female" can be used to express either borderline status (as in the case of intersex) between two polar properties, or inhomogeneity (as in the case of snails) in the instantiation of those properties. We may wonder which of those two cases is most relevant for an expression like "true and not true" as applied to the Liar. Below, I will argue that the "True and False" description of the Liar is better seen as a case of borderline status.

4.4 Quantifying over Extents

The generalization I stated is that "x is P and not P" ought to mean that "x is P [in some respects] and x is not [in all respects] P". In this section, I propose to clarify the semantic analysis we can give of such quantificational paraphrases for one-dimensional adjectives and for multidimensional predicates. The basic idea is that relevant respects determine different *extents* to which a property can be satisfied, and those extents can be quantified over. This analysis can be made to account for both one-dimensional adjectives and multidimensional predicates.

One-dimensional adjectives A one-dimensional adjectives like "tall" may not appear to involve any quantification over respects in a sentence like "John is tall and not tall", in particular, because expressions like "tall in some/all respects" are only marginally acceptable to native speakers (according to Sassoon and Fadlon 2016).

In van Rooij (2011b) and Cobreros et al. (2012), however, my coauthors and I proposed a treatment of borderline contradictions that comes very close to quantifying over respects. The leading idea there, originally due to van Rooij, is that vague predicates can be taken in two distinct meanings, a *strict* meaning and a *tolerant* meaning, and similarly for whole sentences. Both meanings stand in a relation of inclusion: the strict meaning is more restrictive than the tolerant meaning. In effect, the relation between the strict and the tolerant meaning is a relation of privative opposition.

The semantics as originally stated there did not involve respects, however, but it did involve quantification over reference points. That is, "John is tall" is true tolerantly in the relevant model provided there is an object a that is sufficiently similar to John and that counts as tall classically in the model. "John is not tall" is true tolerantly (or not true strictly), on the other hand, provided not every object a that is sufficiently similar to John is tall classically. The conjunction "John is tall and not tall" can therefore be true tolerantly if John is sufficiently near the boundary for tallness to be similar to distinct reference points on either side of that boundary.

Instead of stating the semantics in terms of similarity, it is possible to state it directly in terms of the availability of distinct reference points (or extents) along the dimension of height, conceiving those distinct reference points or extents to vary as a function of the various respects that may be relevant to judge of tallness in the background. Shifts of respects may correspond to implicit shifts in the relevant comparison class, as explained above in Sect. 4.1.[9]

In standard approaches to the meaning of gradable adjectives, "tall" denotes the function $\lambda x.height(x) \geq \theta$ for some contextually given threshold θ (see Kennedy 2007; Fara 2000). Because "tall" is vague, however, we may associate it instead with a set of admissible thresholds varying within an interval I that is lower- and upper bounded. Intuitively, the lower bound may be viewed as the smallest plausible threshold for "tall", and the upper bound as the highest plausible threshold for "tall" in the conversational context. "John is tall [to some extent]" would then be true if there is a threshold $\theta \in I$ such that $height(j) \geq \theta$; and "John is not [to all extent] tall" will be true if it is not the case that every threshold $\theta \in I$ is such that $height(j) \geq \theta$. As a result, "John is tall and not tall" will mean that John is taller than the lower bound, but smaller than the upper bound of the interval based on relevant respects.[10]

Various options can be considered to derive this logical form. One option is to piggyback on the strict-tolerant semantics but to define the tolerant and strict meaning of "tall" directly in terms of this existential versus universal quantification over admissible thresholds, and then rely on extant pragmatic accounts of the selection of tolerant versus strict meaning (see Cobreros et al. 2015a; Égré and Zehr 2018). Another option is to assume that the adverbials "to some extent", "to all extent" can freely enrich the semantic content of "tall".[11] This is not the place to discuss the choice between these options. The aspect of interest to us in relation to Priest's remarks is the observation that "tall" need not be *semantically* ambiguous to make sense of borderline contradictions then: implicit quantification over plausible thresholds can basically account for them.

Multidimensional predicates Sassoon (2013) offers an account of multidimensional predicates in language, including nouns and adjectives, which is helpful for us to analyze contradictions on the model of the previous paragraph. First of all, every multidimensional predicate P come with a finite set of respects or dimensions $R(P) = \{R_1, \ldots, R_n\}$. Sassoon distinguishes different ways in which respects can be manipulated for the ascription of a multidimensional predicate. She distinguishes the ascription of an MD predicate based on operations such as dimension counting, but also weighted averaging over the extents to which the dimensions are, and further possible operations.

---

[9]See Klein (1980), van Rooij (2011a), and Grinsell (2012) on the idea that judgments along a linear dimension of comparison might be a function of the comparison to distinct comparison classes.

[10]We need not suppose that the interval I is the same for every speaker. Similarly, the strict-tolerant account of vague predicates is not committed to the idea that strict and tolerant meaning is the same for all speakers.

[11]Multiple occurrences of the same predicate are not needed, in particular ellipsis is accommodated here: "x is and isn't tall" can directly mean "x is (to some extent) and is not (to all extent) tall", or "x is (to some extent) tall and is (to some extent) not tall", depending on how negation is handled. This coheres with the discussion given by Ripley (2011b) of inconstant content versus inconstant character in relation to vagueness: "tall" on the present account maintains a constant character.

For a multidimensional predicate P and a set of respects $R(P) = \{R_1, ..., R_n\}$, we may represent the meaning of P by the expression: $\lambda x. f(R_1(x), ..., R_n(x)) \geq \theta$, where $\lambda x. R_i(x)$ returns a degree corresponding to the extent to which R_i is satisfied, and f is a function from $\mathbb{R}^n$ to $\mathbb{R}$ that projects the separate extents to a single numerical dimension.[12] Consider an intersex person X with male and female characteristics, and the statement:

(26)   X is male and not male.

Assume that $R(male) = \{A, H, G\}$, that is the respects relevant for maleness consist of anatomical, hormonal, and genetic features. As a first approximation, let us assume the respects to be non-vague each, and so to map out objects to binary values (1 or 0). Let f be the addition function, counting the number of features that are instantiated. To say that "X is male and non male" may be represented by picking again an interval I of admissible values over the number of dimensions that need to be satisfied. Here it would suffice to set $I = [1, 3]$. "being male [to some extent/in some respects]" will be true of a if there is $\theta \in I$ such that $f(A(a), H(a), G(a)) \geq \theta$, and "not being male [to all extent/in all respects]" will be true if not for every $\theta \in I$ do we have $f(A(a), H(a), G(a)) \geq \theta$.

As pointed out by Sassoon, in general, the various features for a multidimensional predicate need not be binary, but can be graded. The previous analysis can accommodate this possibility, letting the extents to which respects can be satisfied be more fine-grained. Similarly, the function f can vary depending on how the features are integrated. In many cases, dimension counting will not suffice, but f will need to be a more complex function of the specific respects satisfied, and of the extent to which they are satisfied. Irrespective of the details, we can handle true contradictions based on multidimensional adjectives along the same lines as one-dimensional adjectives: "x is P and not P" is pragmatically enriched to mean that "x is P to some extent, but not to all extent", and the key ingredient is to have an interval of admissible threshold values over which to quantify along the combined dimension.

One remark worth adding is that for a multidimensional expression P with a set of features $R(P)$, it will generally not suffice for an object to satisfy a single P-relevant feature R_i in order to say that the object is "P and not P". The Moon, for example, is like a planet with respect to shape (it is round), but it is not a planet with respect to being a satellite. This is arguably not sufficient to say:

(27)   The Moon is and isn't a planet.

---

[12] Note that the meaning of the one-dimensional adjective "tall" can be subsumed under that definition. Note also that the above is my rendering of Sassoon's approach, not literally her approach. I am assuming, in particular, that for all multidimensional predicates, an integration and projection along a single dimension of magnitude is operative. This does not mean that respects won't need to be accessed separately in the semantics, but I leave a discussion of that aspect for another occasion. See Sassoon (2017) for more on respect accessing.

On the other hand, at the time Eris was discovered by astronomer Mike Brown and his team, it was a borderline case of planethood, and it would have been appropriate to say:

(28)   Eris is and isn't a planet.

The difference is that Eris met all of the *minimally relevant* respects for planethood, even though it failed to meet further criteria already deemed relevant.[13]

4.5 Contradictions Resolved?

Let us take stock. We have a clearer idea of the relation between true contradictions and the phenomenon of ambiguity. In support of Priest's remark, true contradictions are indeed not on a par with cases of pure homophony ("bat" vs. "bat"). They bear a closer relation to homophony based on privative opposition ("dog" vs. "dog"), but ultimately I have argued that quantification over respects, or better, over extents (of the combination of relevant respects), is, in fact, the right articulation of that idea. The one-dimensional and multidimensional predicates that support true contradictions are vague. In the present approach, this has the consequence that such predicates come with a set of variable extents, allowing us to say that the corresponding property applies to some but not to all extent.

To conclude this note, I propose to briefly discuss two issues. The first is whether the same analysis is applicable to the Liar. The second is whether "there is effectively very little difference" between the present view and dialetheism, to use the words of Priest.

4.5.1   The Liar

The Liar says of itself that it is not true. The dialetheist verdict is that the sentence is true and not true. Can this be restated within the terms of the present account? I think the answer is positive. We may handle "true" as a vague one-dimensional gradable adjective, satisfying the disquotation principle ($\overline{\phi}$ is true iff ϕ), and denoting the metalinguistic property $\lambda x.[\![x]\!] \geq \theta$ with x ranging over sentences, and the function $[\![\cdot]\!]$ taking values in the truth value set $\{0, 1/2, 1\}$. We can represent that "true" is vague in so far as it comes with the interval of admissible values $I = [1/2, 1]$. To say that "the Liar λ is true and not true", in agreement with the dialetheist view, can again be taken to mean that there is some θ in I such that $[\![\lambda]\!] \geq \theta$, but that not every $\theta \in I$ is such that $[\![\lambda]\!] \geq \theta$. This, obviously, is a way of stating that the Liar is partly true, but not fully or determinately true. In our joint work with Cobreros, Ripley and van Rooij, we refrain from making that move: "true" is a predicate only constrained

---

[13] See Brown (2010), Chap. 6: "Planet or not", and Égré (2013) for discussion.

by the disquotation principle, but we let assertability come in two degrees, strict and tolerant (see Cobreros et al. 2013, 2015b). The present, alternative view, on which "true" is vague and comes with variable admissible extents, is worth putting on the table, however, because it ties in with the analysis sketched above of gradable adjectives more generally and with the fact that "true" is indeed gradable in natural language.[14]

4.5.2 "True" and "False" as Vague

What about the second issue? Can we conclude that the idea of a true contradiction is done away with once we admit this implicit relativization to respects and extents? I am tempted to respond: Yes and No.

At one level of semantic analysis Yes, because each tokening of a sentence of the form "x is P and not P", for P a vague predicate, in fact relativizes P to distinct extents depending on its occurrence. Fundamentally therefore, the present account treats the acceptability of such contradictions as a specific form of context shift. As such, it may be subsumed under the general equivalence stated by Ripley (2011b) between so-called inconsistent (glutty) versus inconstant (context-sensitive) accounts of contradictions. Ripley, however, left as an open issue the nature of the empirical evidence susceptible to favor one kind of account over the other. What I have argued here is that adverbial modification (over respects and extents) is a general and productive mechanism, worthy of further empirical investigation, in terms of which all contradictions of the form "x is P and not P" can be paraphrased and classically interpreted.

At a more constitutive level, however, the answer is No, in agreement with Priest's citation at the beginning of this chapter, because the admission of an interval of variable extents for every vague property is viewed here as inherent to the phenomenon of vagueness. I have not said much about the sources of that variability here, but they may, in turn, be argued to originate in a conflicting multiplicity of anchoring or paradigmatic values for our vague concepts.[15] For an absolute gradable predicate like "true", for example, paradigmatic cases would include sentences that are determinately true, whose semantic value is therefore maximal in the range of admissible values. Likewise for "false", paradigmatic cases should include sentences whose semantic value is determinately false and minimal in the range of admissible values.

---

[14]See already (McGee 1990, 7–8) on the idea of treating "true" as a vague predicate. McGee, however, does not put emphasis on the gradability of "true" in support of that connection. Note that if "true" is gradable, it has all the features of an *absolute* gradable adjective, by some of the usual tests. In particular, one can say *completely true/perfectly true*, but not **completely tall/*perfectly tall* (see Kennedy 2007; Burnett 2017). Furthermore, McGee explicitly denies that ""true" is a vague predicate like ordinary vague predicates", because he sees "true" as overdetermined by conflicting rules, rather than underdetermined. But in my own understanding of ordinary vagueness, vagueness has at least as much to do with overdetermination as it has to do with undertermination.

[15]See Douven et al. (2013) for an outline of that view.

But to say that "true" and "false" are vague here is to make room for the possibility of borderline cases between those more typical cases. To represent the notion of a borderline case, we do need, just as Priest argues, and as evidenced by the Liar indeed, a range of values on which the properties "false" and "true", interpreted in the broadest sense, overlap (as encoded here by the value $1/2$). The present account, like Priest's, therefore accepts that some sentences are both true and false, to mean that some sentences are partially true, and partially false at the same time, even though no sentence can at the same time be fully true and fully false.

Acknowledgements I thank Can Başkent and Thomas Ferguson for their invitation and patience in awaiting this manuscript. I am particularly grateful to Olivier Massin, Galit Sassoon, Jeremy Zehr, and to two anonymous reviewers for detailed and helpful comments. Thanks also to Nick Asher, Heather Burnett, and Sam Alxatib for various exchanges, and to my accomplices Pablo, Dave, and Robert, for almost ten years thinking together about true contradictions. Thanks finally to Graham Priest for his inspiration and for enabling us to think so much more broadly and deeply about contradictions. This research was supported by the ANR program "Trivalence and Natural Language Meaning" (ANR-14-CE30-0010), and by the program FrontCog (ANR-17-EURE-0017).

References

Alxatib, S., & Pelletier, F. J. (2011). The psychology of vagueness: Borderline cases and contradictions. *Mind & Language, 26*(3), 287–326.

Bromberger, S. (2012). Vagueness, ambiguity, and the sound of meaning. In M. Frappier, D. Brown, R. DiSalle (Eds.), *Analysis and interpretation in the exact sciences*. The Western Ontario Series in Philosophy of Science (Vol. 78) (pp. 75–93). Springer.

Brown, M. (2010). *How I killed Pluto and why it had it coming*. Spiegel & Grau.

Burnett, H. (2017). *Gradability in natural language: Logical and grammatical foundations*. Oxford University Press.

Cobreros, P., Égré, P., Ripley, D., & van Rooij, R. (2012). Tolerant, classical, strict. *The Journal of Philosophical Logic, 41*(2), 347–385.

Cobreros, P., Égré, P., Ripley, D., & Van Rooij, R. (2013). Reaching transparent truth. *Mind, 122*(488), 841–866.

Cobreros, P., Égré, P., Ripley, D., & van Rooij, R. (2015a). Pragmatic interpretations of vague expressions: Strongest meaning and nonmonotonic consequence. *Journal of Philosophical Logic, 44*(4), 375–393.

Cobreros, P., Égré, P., Ripley, D., & van Rooij, R. (2015b). Vagueness, truth and permissive consequence. In D. Achouriotti, H. Galinon (Eds.), *Unifying the Philosophy of Truth* (pp. 409–430). Springer.

Dautriche, I., & Chemla, E. (2016). What homophones say about words. *PLoS ONE, 11*(9), e0162176.

Douven, I., Decock, L., Dietz, R., & Égré, P. (2013). Vagueness: A conceptual spaces approach. *The Journal of Philosophical Logic, 42*, 137–160.

Égré, P. (2013). What's in a planet? In M. Aloni & F. Roelofsen (Eds.), *The Dynamic, Inquisitive and Visionary Life of ϕ, ?ϕ, and $\Diamond\phi$, A Festtschift for J. M. Stokhof & F. Veltman* (pp. 74–82). Amsterdam: Groenendijk.

Égré, P., De Gardelle, V., & Ripley, D. (2013). Vagueness and order effects in color categorization. *Journal of Logic, Language and Information, 22*(4), 391–420.

Égré, P., & Zehr, J. (2018). Are gaps preferred to gluts? A closer look at borderline contradictions. In E. Castroviejo, G. W. Sassoon, & L. McNally (Eds.), *The semantics of gradability, vagueness, and scale structure–experimental perspectives* (pp. 25–58). Springer.

Fara, D. (2000). Shifting sands: An interest-relative theory of vagueness. *Philosophical topics*, 28(1):45–81. Originally published under the name "Delia Graff".

Fausto-Sterling, A. (1993). The five sexes: Why male and female are not enough. *The Sciences*, 33(2), 20–24.

Fausto-Sterling, A. (2000). *Sexing the body: Gender politics and the construction of sexuality*. Basic Books.

Grinsell, T. W. (2012). Avoiding predicate whiplash: Social choice theory and linguistic vagueness. *Semantics and Linguistic Theory*, 22, 424–440.

Kamp, H., & Partee, B. (1995). Prototype theory and compositionality. *Cognition*, 57(2), 129–191.

Kennedy, C. (2007). Vagueness and grammar: The semantics of relative and absolute gradable adjectives. *Linguistics and Philosophy*, 30(1), 1–45.

Klein, E. (1980). A semantics for positive and comparative adjectives. *Linguistics and Philosophy*, 4(1), 1–45.

Kooi, B., & Tamminga, A. (2013). Three-valued logics in modal logic. *Studia Logica*, 101(5), 1061–1072.

Krifka, M. (2007). Negated antonyms: Creating and filling the gap. In U. Sauerland, U. *Presupposition and implicature in compositional semantics* (pp. 163–177). Springer.

Križ, M., & Chemla, E. (2015). Two methods to find truth-value gaps and their application to the projection problem of homogeneity. *Natural Language Semantics*, 23(3), 205–248.

Lewis, D. (1982). Logic for equivocators. *Noûs*, 16, 431–441.

MacFarlane, J. (2010). Fuzzy epistemicism. In R. Dietz, R. & S. Moruzzi, S. (Eds.), *Cuts and clouds: Vagueness, its nature and its logic* (pp. 438–463). Oxford University Press.

Massin, O., & Hämmerli, M. (2017). Is purple a red and blue chessboard? Brentano on colour mixtures. *The Monist*, 100(1), 37–63.

McGee, V. (1990). *Truth, vagueness, and paradox: An essay on the logic of truth*. Hackett Publishing.

Pinkal, M. (1995). *Logic and lexicon: The semantics of the indefinite*. Kluwer Academics Publishers.

Priest, G. (2006). *In contradiction*. Oxford University Press.

Ripley, D. (2011a). Contradictions at the border. In R. Nouwen, H.C. Schmitz, & R. van Rooij (Eds.), *Vagueness in communication* (pp. 169–188). Springer.

Ripley, D. (2011b). Inconstancy and inconsistency. In *Reasoning Under Vagueness: Logical, Philosophical, and Linguistic Perspectives*, pages 41–58. College Publications.

van Rooij, R. (2011a). Implicit vs. explicit comparatives. In P. Égré, N. Klinedinst (Eds.), *Vagueness and language use* (pp. 51–72). Palgrave Macmillan.

van Rooij, R. (2011b). Vagueness, tolerance, and non-transitive entailment. In Cintrula, P., Fermüller, C., Godo, L., and Hàjek, P., (Eds.), *Understanding vagueness: Logical, philosophical and linguistic perspectives* (pp. 205–222). College Publications.

Sassoon, G. (2013). A typology of multidimensional adjectives. *Journal of Semantics*, 30, 335–380.

Sassoon, G., & Fadlon, J. (2016). *The role of dimensions in classification under predicates predicts their status in degree constructions*. Ms under review and revisions: Bar Ilan University.

Sassoon, G. W. (2017). Dimension accessibility as a predictor of morphological gradability. In *Compositionality and concepts in linguistics and psychology* (pp. 291–325). Springer.

Schiffer, S. (2003). *The things we mean*. Clarendon Oxford.

Spector, B. (2012). *Vagueness, (local) accommodation, presupposition and restrictors*. Course Notes on Trivalent Semantics for Vagueness and Presupposition, Vienna, May 2012, Class 6. https://www.dropbox.com/sh/pi9xwzj77fh0a7z/AACy5WYBqCvFu-leRcuMkG8Ha/HO6.pdf?dl=0.

Zehr, J. (2014). *Vagueness, presupposition and truth-value judgments*. PhD Thesis, ENS, Paris.

Zwicky, A., & Sadock, J. (1975). Ambiguity tests and how to fail them. *Syntax and Semantics*, 4(1), 1–36.

Paul Égré is Directeur de recherche at Institut Jean-Nicod (CNRS) and Professor in the Philosophy Department of Ecole normale supérieure in Paris. A large part of Égré's work over the last decade has been on the topic of vagueness, dealing with semantic, logical, and psychological aspects of the phenomenon, and fostering his interest in nonclassical and many-valued logics. Since 2011, paul Égré is also the Editor-in-Chief of the Review of Philosophy and Psychology.

Chapter 5
Hegel and Priest on Revising Logic

Elena Ficara

Abstract In this paper, I examine Hegel's and Priest's answers to three questions: "What is logic?" "Can logic be revised?" "If so, how is this done?" Considering that the two views are separated by more than 150 years in which logic has experienced its perhaps most radical revolution, the two accounts are surprisingly close. Both authors admit different meanings of "logic". In both accounts what could be called a metaphysical concept of logic plays an important role (Hegel calls it *das Logische*, Priest *logica ens*). Moreover, both authors hold that logic as theory can be revised. Yet the two conceptions also differ in some relevant respect.

Keywords Logic and metaphysics · Logic revision · Dialectics

5.1 Introduction

In this paper, I examine Hegel's and Priest's answers to three questions:
 "What is logic?"
 "Can logic be revised?"
 "If so, how is this done?"[1]

Considering that the two views are separated by more than 150 years in which logic has perhaps experienced its most radical revolution, the two accounts are surprisingly close. Both the authors admit different meanings of "logic". In both accounts what could be called a metaphysical concept of logic plays an important role (Hegel calls it *das Logische*, Priest *logica ens*). Moreover, both the authors hold that logic as theory can be revised. Yet the two conceptions also differ in some relevant respects.

In what follows, I first address the different meanings of "logic" in Hegel, i.e. "logic as theory", "natural logic", *das Logische, Verstandeslogik* and *Vernunft-*

---

[1] I derive this clarifying structure from Priest himself in 2014.

E. Ficara (✉)
University of Paderborn, Paderborn, Germany
e-mail: elena.ficara@upb.de

© Springer Nature Switzerland AG 2019
C. Başkent and T. M. Ferguson (eds.), *Graham Priest on Dialetheism and Paraconsistency*, Outstanding Contributions to Logic 18, https://doi.org/10.1007/978-3-030-25365-3_5

logik, as well as Hegel's theses on the necessity to subject logic as theory (as *Verstandeslogik*) to a critical analysis. In the second part, I present the three meanings of "logic" admitted by Priest in 2014 (but also with some reference to Priest (2006, 2016), i.e. *logica docens*, *logica utens* and *logica ens*, as well as Priest's theses concerning logic's revision.[2] In the last part, I compare the two accounts, hinting at one major difference between them. While Hegel postulates the idea of a rational logic, which contains the principle of its own revision and critique, Priest conceives revision as an external operation, which does not belong to logic as theory. More specifically, he intends the process of revision according to the standard mechanisms of rational theory choice, i.e. as following the usual criteria: adequacy to the data, simplicity, unifying power, etc. I conclude presenting a problem concerning the criterion that, in my view, is the most fundamental, but also the most enigmatic one: the adequacy to the data. If the data logic deals with are, as Priest holds, our intuitions about validity, and if our intuitions can be wrong, how can the data be a criterion of rationality? What is the relation between logical data and logical *entia*? Hegel's theory about *das Logische* could fruitfully enter the picture at this point, contributing to spelling out the link between *logica utens*, *logica docens* and *logica ens*, and, in particular, between logical data and logical *entia*.

5.2 Hegel[3]

5.2.1 What Is Logic?

The literature on Hegel's logic is fairly rich. The subject is dealt within different perspectives and for different aims, in a philosophical or historical approach, with exegetic or theoretical concern, and using formal or informal methods. The attention to contemporary philosophical logic places the present research closer to those works interested in the link between Hegel's thought and analytical philosophy.[4] More specifically, in what follows I focus on the semantics of some words in Hegel's

---

[2] I focus here only on Priest's own view on logical revision. For a detailed analysis of this view's relation to Quine's "change of logic, change of subject" see Priest (2006, chap. 10, 2016, 29–57). For a discussion of Priest's conception of logical revision in relation to Batens' and MacFarlane's view see Allo (2016, 3–31).

[3]*Note on citation: If the English translation of the works quoted is not available, I only provide the page number of the original German text and the translation is mine (Hegel's works are, as usual, quoted as Hegel Werke, followed by the indication of the volume and page). If the translation is available, I also provide the page number of the translation.

[4] In (1976, 75) Gadamer stressed that the works traditionally engaged in reading Hegel from a contemporary point of view normally consider the *Phenomenology of Spirit*, and are interested in examining the meaning of Hegel for epistemology and philosophy of mind, while assessing the relevance of Hegel's logic and metaphysics for contemporary philosophy is traditionally held as thorny. This judgement can now be partially revised. Today there are different works engaged in reading Hegel's logic and metaphysics from an analytical perspective. Among the most recent ones see Stekeler-Weithofer (1992), Bencivenga (2000), Burbidge (2004, 131–176), Berto (2005),

texts, namely: "logic [*die Logik*]", "the logical element [*das Logische*]", "intellectual logic [*Verstandeslogik*]", "rational logic [*Vernunftlogik*]" among others. In my view, presenting the meaning of these terms is crucial in order to give a preliminary answer to the still controversial question: "What does Hegel mean by 'logic'?".

As Hans-Georg Gadamer pointed out (1976, 78), Hegel coins a new expression, which cannot be found before him: "the logical" (*das Logische*). Gadamer suggests that Hegel uses it in the same way that the Greek philosophers used the word *logos*, that is, to denote the realm of *concepts* or *forms*, the universal and pure entities constituting and ruling human language and reasoning and expressing, at the same time, the structure of reality.[5]

In the Preface to the second edition of the *Science of Logic* Hegel writes:

> The forms of thought are, in the first instance, displayed and stored in human language [...] the Logical [*das Logische*] [is the] natural element [in which] human beings [live], indeed [their] own peculiar nature.[6]

Logical forms ("the forms of thought") are for Hegel, who follows the Ancient Greek tradition, forms of *true* thought, expression of what *really* follows from what. Not only that, they are objective entities, considerable as "facts" in every respect. These facts occur in a specific field, the field of human language and thought. Moreover, they are the distinctive feature (the "natural element") of human beings, and the field in which we live, act and interact. Hegel also writes that:

> The activity of thought which is at work in all our ideas, purposes, interests and actions is [...] unconsciously busy (natural logic) [...] To focus attention on this logical nature [...] this is the task.[7]

Whenever we think or speak or even simply live (act, have aims and interests), we use logical forms. They rule our thoughts and beliefs, and dominate our actions and interactions. Hegel uses here the expressions "logical nature" and "natural logic": the former denotes the natural field in which logical forms emerge; the latter expresses the natural and "unconscious" activity of using these forms. Now Hegel says that the task of logic as a discipline is to focus attention upon the forms of thought, making them the object of inquiry: they are used unconsciously, and we have to bring them into consciousness. This enterprise is what Hegel calls "*die Logik*", the theory or discipline that isolates and fixes the forms of valid inferences, "extracting them" from human language and life.

Forms are hence for Hegel objective occurrences, belonging to the domain of *das Logische,* and "logic" is the theory or the discipline that isolates and fixes them. Yet,

---

Rockmore (2005), Ameriks/Stolzenberg (2005, ed.), Redding (2007), Hammer (2007), Nuzzo (2010, ed.), Butler (2012), Brandom (2014, 1–15), Bordignon (2014), Pippin (2016, Chap. II).

[5] See Gadamer (1976, 78). In English translations, the term *das Logische* is often rendered with "logic" (see for instance Hegel 1969b, 36–37), but this could be misleading, as it risks overlooking important philosophical implications.

[6] Hegel Werke 5 26–27, Hegel (1969a, 36–37).

[7] Hegel Werke 5 26–27, Hegel (1969a, 36–37).

the connection between *das Logische* and *die Logik* is not so immediate and uncontroversial. In this respect, the further distinction between *Verstandeslogik* (intellectual logic) and *Vernunftlogik* (rational logic) is useful to introduce Hegel's answer to the second question.

5.2.2 Can Logic Be Revised?

In the Lectures *on Logic and Metaphysics* (1817) Hegel writes:

> Natural logic does not always follow the rules which are established in the logic as theory; these rules often tread down natural logic.[8]

Natural logic is the natural activity of thought, our everyday following rules without even knowing it. According to Hegel, natural logic does not follow all and only the rules established by *die Logik*. Logical rules as fixed by *die Logik* sometimes, indeed often, "tread down" the naturalness of thought. "Natural logic" so parts company with "logic as theory".[9] That our natural way of thinking is, in many senses, not strictly "logical", and that our reasoning is often ruled by "cognitive illusions" is for us quite obvious.[10] But Hegel has an opposite view: for Hegel what is wrong is, most frequently, logic, rather than the "natural" way of thinking. He stresses that the logical rules established by *die Logik* might be, and in fact often are, wrong, with respect to natural thought. This is precisely the dialectically relevant situation, which introduces Hegel's particular criticism of "logic" as an apparatus, a fixed institutional discipline as it was practiced in his times.

There are many examples of the "failures" of traditional logical rules in Hegel's texts.[11] One of the most simple ones, presented in the short article "Who thinks abstractly?" (1807), is the anecdote of a prosecution. Common people, when a lady claims that a murderer who is brought to the place of execution "is handsome", are shocked and remark: "how can one think so wickedly and call a murderer handsome?". Since we are normally committed to the (logico-metaphysical) view that the same subject cannot have opposite properties, we can conclude that those who call a criminal (a bad person) "good" (intelligent, handsome) "think wickedly". Common people, in the example, represent the normative instance of logic as theory, and the lady's remark its violation, yet a violation that is, as Hegel shows, absolutely legitimate.

---

[8] Hegel (1992, 8).

[9] In the *Jäsche Logik* (Kant Werke 2 439), Kant recalls that the distinction between natural logic (*natürliche Logik*) and scientific logic (*wissenschaftliche Logik*) is a common one in his times and is also known as the distinction between *logica naturalis* and *logica scholastica*. However, the true logic is for Kant only the scientific one. While in Kant natural logic is excluded from the scientific consideration, in Hegel it is an essential component of logic as a science.

[10] See on the errors of natural reasoning the classical account of Wason and Johnston-Laird (1972).

[11] See, for instance, Hegel Werke 2, 575–581, Hegel Werke 18, 526–538.

In this spirit, Hegel writes in the Preface to the second edition of the *Science of Logic* that the inference rules

> quite as well serve impartially error and sophistry and [...] however truth may be defined [...] they concern only correctness [*Richtigkeit*] [...] and not truth [*Wahrheit*].[12]

This means that, for Hegel, valid inferential forms are not always forms of truth, but can be, and often are, forms of falsity and error. Sometimes, in logic as theory there aren't any forms for expressing a true thought. Hence, what is true is ignored, or considered false.

In this regard, it is useful to mention Hegel's insights on how to react to this aberration of logic as theory. More specifically, another well-known Hegelian distinction deserves to be mentioned, the one between *Verstandeslogik* and *Vernunftlogik*. It is fairly evident that there are different ways of dealing with inconsistencies between "logic as theory" and "natural logic". If we stick to the validity of the rules, we have what Hegel calls "intellectual" or "finite logic" (*Verstandeslogik*); if we question the validity of the rules, we have instead what Hegel calls "speculative" or "rational logic" (*Vernunftlogik*).

With the expression "*Verstandeslogik*", Hegel generally refers to the traditional Aristotelian logic of his times, the theory of judgements, concepts and syllogisms, presented, among others, in Kant's *Jäsche Logik*. "*Vernunftlogik*", the logic of reason, is used instead as a synonym for *speculative* or also *dialectical* logic, and defines the logical enterprise in the specifically Hegelian sense. It involves a critical consideration of *Verstandeslogik*, of its basic logical concepts and forms, among others: the concept of the sentence, of concept, of contradiction, as well as the principle of excluded middle, the principle of identity and the principle of non-contradiction. In this respect, the fault of traditional logic according to Hegel is that it tends to be developed only in terms of intellectual logic, so failing in its duties towards *das Logische*.

We see then that Hegel postulates the idea of a rational logic (*Vernunftlogik*), which contains the critique of the forms established by logic itself (as *Verstandeslogik*). Hegel also stresses that it is one and the same discipline (namely, logic as a science, the science of logic) that individuates the forms and conceptual determinations and at the same time criticises, or revises, them. At this point, the final question can be addressed.

5.2.3 How Can Logic Be Revised?

In sum, we have:

Das Logische: the forms of true thought that are "deposited" in human language, thought, life, actions.

Natural logic: our use of these forms, a use that can also be unconscious.

---

[12]Hegel Werke 5, 29.

Die Logik: logic as theory, logic as a discipline that fixes forms of valid inference and can be taught.

Intellectual logic [*Verstandeslogik*]: again, logic as a discipline that limits itself to fix the forms of valid inference without asking about their capability to convey truth, and which excludes the possibility of revision.

Rational logic [*Vernunftlogik*]: logic as a discipline that analyses the forms fixed by *Verstandeslogik*, as well as the ones at the basis of our everyday thinking, living, acting.

In this light, the question about logic revision becomes, in Hegel, the question about the revision of intellectual logic in view of *das Logische*, the forms of true thought that are deposited in natural language and vernacular reasoning. To the question "How can logic be revised?" Hegel's answer would then be: "by analysing *das Logische*", that is, by extracting the forms of truth from their concrete integuments in natural language and thought.[13] Evidently, this descriptive, empirical inquiry is at the same time normative, since *das Logische* is the realm of forms and possibility conditions of true thought. As such, it contains the revision of the forms fixed by logic as theory whenever they are, as Hegel writes, "at the service of error and sophistry", and inadequate to express truth.

5.3 Priest

5.3.1 Revising Logic

In 2014, 212 Priest distinguishes between three meanings of the word "logic":

> *logica docens* (the logic that is taught) is what logicians claim about logic. It is what one finds in logic texts used for teaching. *Logica utens* (the logic which is used) is how people actually reason. The first two phrases are familiar from medieval logic. The third, *logica ens* (logic itself) is not. (I have had to make the phrase up). This is what *is* actually valid: what really follows from what.

and discusses these three senses of logic asking each of the following three target questions about them:

> Can logic be revised?
> If so, can this be done rationally?
> If so, how is this done?

Logica docens has been revised, hence it can be revised. Priest claims that the history of logic is not cumulative: in the history of logic, it is not the case that once something has been accepted it was never rejected. Rather, the historical development shows that something has to be given up. For example, syllogistic logic and classical logic give

---

[13] The expression "extracting the forms from their concrete integuments" is used by Russell (2009, 35). For an interpretation of Hegel's logic as philosophical logic in Russell's terms see Ficara 20+.

incompatible verdicts concerning the validity of some inferential forms: *Darapti* and *Camestros* are valid in syllogistic logic, invalid in classical logic. As Priest claims in 2006, 167:

> However one interprets traditional logic in classical logic, something has to been given up [...] classical logic is not (just) a more generous framework subsuming traditional logic [...] classical logic has given the thumbs down to *Darapti* and its ilk.

Hence, the same history of logic tells us that *logica docens* can be revised. The problem is rather to assess the reasons why one logic as theory is preferable to another.[14] Rivalry between logics arises when we apply logical systems to our practices of inferring:

> As pure logics, no logic is rival of any other [...] it is only when we apply them that a question of rivalry occurs [...] in the case of logic, we need to relate the theory to our practices of inferring.[15]

In order to apply a logic, we must have some way to identify structures in the formal language with claims in the vernacular. The procedure is

> largely tacit. It is a skill that good logicians acquire, but no one has spelled out the details in general. It is simple-minded, for example, to suppose that every sentence with a 'not' in is adequately represented in the language of a logical theory by whatever represents the sentence with 'not' deleted, and prefixed with '¬'.[16]

In other words, applying classical logic means seeing our vernacular reasoning in its perspective, asking, for example, whether the many kinds of negative judgements in natural language can be adequately grasped by the univocal form of classical negation.

Hence, the rivalry occurs because different logics give different verdicts concerning the validity of some inferences, verdicts that are based on the different meanings each logic attributes to the connectives. For example, $p, \neg p \vdash q$ is a classically valid inference, based on the classical interpretation of the meaning of negation. Paraconsistent logics give negation a different meaning, and question the validity of this inference.

On this basis, the way to assess the major or minor rationality of one logical account over the other follows, for Priest, the usual criteria of rational theory choice. Priest (2014, 7) claims that

> given any theory, in science, metaphysics, ethics, logic, or anything else, we choose the theory which best meets those criteria which determine a good theory.

The criteria are, among others, adequacy to the data for which the theory is meant to account, simplicity, non-ad hocness, unifying power, fruitfulness. I will return later to the criterion that, in my view, is the most fundamental, but also the most enigmatic

---

[14]See Priest (2014, 214).
[15]Priest (2006, 169).
[16]Priest (2006, 170).

one, the adequacy to the data. Now I shortly sum up Priest's view on the other two meanings of "logic" and on the possibility of their revision.

Logica utens is "the way that people actually reason" Priest (2014, 218). As the Wason Card Test shows, people reason invalidly in a systematic way. Subjects in the Wason study are shown four cards:

E K 4 7

and told that each of these cards has a letter on one of its sides and a number on its other side. They are then presented with the following rule:

If a card has a vowel on one side, then it has an even number on the other side.

The task is to name only those cards that need to be turned over in order to determine whether the rule is true or false. The vast majority say "E and 4", or "only E". Both answers are wrong. The correct answer is "E and 7".[17]

Priest (2014, 9) claims that *logica utens* is not a merely descriptive notion, as "it is constituted by the norms of an inferential practice. Subjects in the Wason Card Test can see, when it is pointed out to them, that they have violated appropriate norms".

Since *logica utens* is simply the way that people actually reason, and reasoning implies following norms, we can change the way we reason insofar as we start following different norms: As Priest (2014, 219) writes

> I was trained as a classical mathematician, and have no difficulty in reasoning in this way. But I have also studied intuitionistic logic, and can reason in this way too [...] clearly then it is possible to move from one *logica utens* to another.

The change can be rational (i.e. the revision of *logica utens* can be rational) insofar as some practices are better than others and we can change our practices accordingly. Evidently, one can be a relativist about logical practices, claiming that all (logical) practices are good in their own way, and no practice is better than others. However, Priest (2014, 220) states that

> a relativism about these practices [...] entails a relativism about truth [...] To take an extreme example: suppose that reasoning in one way, we establish that the theory of evolution is correct, but that reasoning in another way, we establish that creationism is true and the theory of evolution false. Something, surely, must be wrong with one of these forms of reasoning.

This claim is important, insofar as it reveals Priest's anti-relativistic attitude, and his (implicit) view on the relation between logic and metaphysics. Metaphysical views are grounded in reasoning practices and norms, as reasoning in a certain way forces us to admit that reality is made in a certain way. Conversely, one can/should also admit that reasoning practices and norms are grounded in metaphysical views about what there is and its nature. Priest implicitly assumes this insofar as he states that inferential norms are based on our views about the meanings of the connectives.

The revision of *logica utens* hence is possible, and we can revise *logica utens* insofar as we determine what *logica docens* is the best theory of reasoning (adopting

---

[17] Wason/Johnson-Laird (1972, 172–173).

the standard criteria of rational theory choice) and bring our practice into line with it.

Among the three concepts of logic, *logica ens* is perhaps the most controversial one. It is not admitted in medieval or modern logic. As Priest specifies, he had to make it up. Hegel, as I have shown, introduced a similar notion, *das Logische*, explicitly deriving it from the Greek word *lógos*.

Logica ens concerns the facts of validity, what really follows from what. Priest claims that in both current accounts of validity, in model-theoretic and proof-theoretic terms, validity can be spelled out as a relation between objects (in the first case we postulate models, i.e. sets as abstract objects, in the second a proof structure, and structures are, again, abstract objects). In this light, claims about validity are claims about special kinds of facts. Moreover, as we have seen both views are based on an account of the meaning of the logical connectives. Priest (2014, 222) stresses that the facts about validity are expressed through language, and one may suppose that the words involved may change their meanings. He writes

> if we change our theory, then our understanding of these meanings will change. This does not mean that the meanings of the vernacular words corresponding to their formal counterparts change.

Hence, our view about the meanings of the connectives may change, but this does not mean that the meaning of the connectives changes.

In sum, logic as theory as well as our inferential practices (the logic we use) can be revised. This can be done following the basic criteria of rational theory choice, i.e. adequacy to the data, fruitfulness, non-ad hocness, etc. Priest also endorses, as we have seen, a non-relativistic view on our inferential practices and logics as theories, insofar as he admits, beside logical theories and practices, logical *entia*. And whether logical *entia* can change is doubtful.

The question is at this point: Can our theories and practices be anchored to logical *entia*? If so, how? In what follows, I suggest to further analyse what is, in my view, the most important, but perhaps also the most enigmatic criterion, the criterion of logic's adequacy to the data.

5.3.2 What Counts as Data, in Logic?

In 2016, 41 Priest writes

> It is clear enough what provides the data in the case of an empirical science: observation and experiment. What plays this role in logic? The answer [...] is our intuitions about the validity or otherwise of vernacular inferences.

Thus some inferences, such as "if Juliette is French, she is European, she is French hence she is European" strike us as valid, others, such as "If Juliette is French she is European, she is not French hence she is not European" as invalid, and "any attempt that gets things the other way around is not adequate to the data". Yet logical "data" are "soft", which means that our intuitions could be in open contrast with received logical doctrine. For example "a is taller than b, b is taller than c, hence a is taller than c" strikes most of us as correct, yet according to classical logic it is not. As Priest (2016, 43) clarifies

> We can explain our initial reaction as follows. There is an evident suppressed premise, the transitivity of 'taller than': for all people x, y, and z, if x is taller than y and y is taller than z, then x is taller than z. It is the inference with this premise added that is valid. The premise is so obvious that we confuse the two inferences.

In this example, we see that the apparent incongruence between logic as theory and our intuitions about validity for Priest does not (always) mean that the intuitions are wrong. Good intuitions about validity that, from the point of view of logic as theory, are invalid, turn out to be valid as soon as we individuate the precise "norm" or "form" corresponding to them (insofar as we complete them enthymematically).

By the "data" of logic we can mean, more specifically, not particular inferences, but the forms of inference, for example, MP

A
If A then B
―――
B

But, Priest (2016, 43) continues

the pattern needs careful articulation. Neither of the following strikes us as valid:

If I may say so, that is a nice coat.
I may say so.
―――

That is a nice coat.

If he were here, he would be hopping mad.
He were here.
―――

He would be hopping mad.

[...] If theorization is to take account of such data, they are certainly much softer than those concerning individual inferences.

Data are the intuitions about what inferences (and what inferential forms) are valid and what are not. The views about validity, as we have seen, are based on views about the meanings of the connectives in vernacular reasoning.

Now there is here a duplicity in what counts as logical data, since by data we could mean both: what we (subjectively) think about the meanings of the connectives and inferential validity, whereby we could be wrong, and the objective meanings the connectives have, which cannot be wrong, since they express the way connectives are, and the way validity is (*logica ens*). How are the two aspects, i.e. data as the vernacular intuitions about validity and logical *entia* related? How can we say that our subjective intuitions about validity and the meanings of the connectives correspond to the objective meaning of the connectives?

5.4 Conclusion

Both Hegel and Priest admit different meanings of "logic". Hegel distinguishes between *die Logik*, i.e. institutionally established discipline, natural logic (our natural following logical rules without awareness about it), *das Logische* (the forms of true thought as facts that permeates our thinking, reasoning, action and life). These three meanings largely correspond to the three meanings individuated by Priest: *logica docens* (the logic that is taught), *logica utens* (the logic we use) and *logica ens* (the facts about validity). Both thinkers admit that there can be incongruences between the norms fixed by logic as theory and the way we normally reason in natural language. Both admit that, sometimes, logic as theory is wrong, as it gives an inadequate account of the inferences in natural reasoning, considering as valid inferences that, in vernacular reasoning, are not, and vice versa.

Yet the two accounts differ in some major respect. While Hegel postulates the idea of a rational logic, i.e. embeds his view of logical revision in a conception of logic as conceptual and philosophical (i.e. self-revising and truth-oriented) analysis of natural language, Priest sticks to the idea of logical revision as an external operation, which follows the model of rational theory choice among rival logical theories. Logic revision is for Hegel, as we have seen, the result of Hegel's very idea of logic as the analysis of das *Logische*, i.e. of logic as analysis and individuation of the forms of truth.[18] Revision, intended as the procedure of adjustment between theories and data, is an operation actuated by logic itself. In this sense, Hegel's idea of logic does not admit the distinction between pure and applied logic. In non-Hegelian terms, Hegel's logic as rational logic would then involve both the construction of a model and the reflection on the adequacy of the model.[19]

I suggest to conclude the comparison by hinting at the question concerning the criteria according to which the revision should take place. They are, for Priest, the standard ones of rational theory choice: adequacy to the data, fruitfulness, non-ad hocness, etc. For Hegel, the only criterion that orients the critique of logic as theory is *das Logische* as conceptual truth, as the way the connectives *are* and validity *is*. It is a realistic meaning of truth, the correspondence of the logical theories with the logical fact. Interestingly, a correspondence that is already given in the empirical data logic deals with (our intuitions about validity).

The adequacy to the data criterion shows that, also for Priest, truth is, up to a certain extent, a criterion of rationality and logical revision. As we have seen, Priest defines the logical data as our intuitions about what inferences are valid. In my view, in this regard the question:

> Why should the data be a criterion of rationality if they are our intuitions about validity and our intuitions can be wrong?

---

[18] For an interpretation of dialectical logic as an inquiry into the forms of truth see Ficara 20+.

[19] In the first attempts at formalising dialectics there are different interpretations of the formal counterpart of the "self-revising" moment in Hegel's logic. See among others Apostel (1979, 90) and Routley/Meyer (1979, 328).

arises. In principle, if we admit, as Priest does, that there is a *logica ens*, i.e. a way things stand concerning validity, and that there are logical data, i.e. an experiential field for logic, then spelling out the relation between logical *entia* and logical data becomes urgent.

The best (most rational) logic as theory should give an account of the logical *entia*, of how things stand concerning validity. But how can this happen, since the data the logics as theories have to stick to are given in vernacular reasoning, and vernacular reasoning often misleads us? In other words, Priest's theory about logic revision needs to be completed by a conception about the nature of the data, and their relation to truth, or *logica ens*.

At this point, Hegel's view on *das Logische* could enter the picture, contributing to complete the vision highlighted by Priest. Hegel's conception of *das Logische*, as we have seen, is precisely a theory about the relationship between *logica ens* and logical data.[20] It focuses on the interplay between data and *entia*, experience and possibility conditions of experience. It shows how *logica ens*, the forms or possibility conditions of valid and true thought, are *within* the experiential realm, the data, and at the same time make experience possible.[21]

References

Allo, P. (2016). Logic, reasoning and revision. *Theoria, 82*(1), 3–31.
Ameriks, K., & Stolzenberg, J. (2005) (ed.), *Internationales Jahrbuch des Deutschen Idealismus. International Yearbook of German Idealism: German Idealism and Contemporary Analytic Philosophy* (Vol. 3). Berlin New York: De Gruyter.
Apostel, L. (1979). Logica e dialettica in Hegel. In D. Marconi (Ed.), *La formalizzazione della dialettica* (pp. 85–113). Torino: Rosenberg & Sellier.
Bencivenga, E. (2000). *Hegel's dialectical logic*. Oxford: Oxford University Press.
Berto, F. (2005). *Che cosa è la dialettica Hegeliana?*. Padova: Il Poligrafo.
Bordignon, M. (2014). *Ai limiti della verità*. Il problema della contraddizione nella logica di Hegel, Pisa: ETS.
Brandom, R. (2014). Some Hegelian ideas of note for contemporary analytic philosophy. *Hegel Bulletin, 35*(1), 1–15.
Burbidge, J. (2004). Hegel's logic. In E. Gabbay, & J. Woods (ed.), *Handbook of the history of logic* (Vol. 3, pp. 131–176). Bonn: Elsevier.
Butler, C. (2012). *The dialectical method: A treatise Hegel never wrote*, New York: Prometheus Books.

---

[20]The approach to the justification of rules of inductive logic proposed by Goodman (1954) could be fruitful to grasp some aspects of Hegel's idea, and could be assumed to be at the basis of Priest's own view on logic revision. In Goodman's account, which is labelled now *reflective equilibrium* (see Daniels 2016), the intuitions count as data, and they can be wrong, but this is not a problem since theories and intuitions are amended via each other. In this way the mismatch between our intuitions and the theory is progressively adjusted. However, the idea of reflective equilibrium does not seem to admit the metaphysical meaning of logic, which is an essential part of Hegel's and Priest's account.

[21]I am grateful to two anonymous referees for their useful comments.

Daniels, N. (2016). Reflective equilibrium. In Edward N. Zalta (Ed.) *The stanford encyclopedia of philosophy* (Winter 2016 Edition). https://plato.stanford.edu/archives/win2016/entries/reflective-equilibrium/.
Ficara, E. (20+). *The form of truth. Hegel's Philosophical Logic*, Habilitationsschrift.
Gadamer, H.-G. (1976). *Hegel's dialectics: Five hermeneutical studies*. New Haven and London: Yale University Press.
Goodman, N. (1954). *Fact, fiction, and forecast*. University of London: Athlone Press.
Hammer, E. (2007). *German idealism*. Contemporary Perspectives, London: Routledge.
Hegel, G. W. F. (1969a, ff.). *Werke in zwanzig Bänden. Theorie Werkausgabe. New edition on the basis of the Works of 1832–1845 edited by Eva Moldenhauer and Karl Markus Michel*. Frankfurt a. M.: Suhrkamp (quoted as Hegel Werke, followed by the indication of the volume and page).
Hegel, G. W. F. (1969b). *Hegel's science of logic, English translation by A*. Miller, New York: Humanity Books.
Hegel, G. W. F. (1992). *Vorlesungen über Logik und Metaphysik (Heidelberg 1817)*. In K. Gloy, Hamburg: Meiner.
Kant, (1996). *Theorie-Werkausgabe Immanuel Kant. Werke in 12 Bänden hrsg. von W. Weischedel*, Frankfurt: Suhrkamp (quoted as Kant Werke followed by the indication of volume and page).
Nuzzo, A. (Ed.) (2010). *Hegel and the analytic tradition*, New York: Continuum.
Pippin, R. (2016). *Die Aktualität des Deutschen Idealismus*. Berlin: Suhrkamp.
Priest, G. (2006). *Doubt truth to be a Liar*. Oxford: Oxford University Press.
Priest, G. (2014). Revising logic, ch. 12 of P. Rush (Ed.), *The metaphysics of logic*, Cambridge: Cambridge University Press.
Priest, G. (2016). Logical disputes and the *a priori. Principios: Rivista de Filosofia 23*, 29–57.
Redding, P. (2007). *Analytic philosophy and the return of Hegelian thought*. Cambridge: Cambridge University Press.
Rockmore, T. (2005). *Hegel, idealism, and analytic philosophy*. New Haven: Yale University Press.
Routley, R., & Meyer, R. K. (1979). Logica dialettica, logica classica e non-contraddittorietà del mondo. In D. Marconi (Ed.), *La formalizzazione della dialettica* (pp. 324–353). Torino: Rosenberg & Sellier.
Russell, B. (2009). *Our Knowledge of the External World*, London: Routledge (first published in 1914).
Stekeler-Weithofer, P. (1992). *Hegels Analytische Philosophie. Die Wissenschaft der Logik als kritische Theorie der Bedeutung*, Paderborn: Schöningh.
Wason, P. C., & Johnston-Laird, P. N. (1972). *Psychology of reasoning: Structure and content*. Cambridge, Mass.: Harvard University Press.

Elena Ficara is Junior professor at the University of Paderborn. Her works include: Die Ontologie in der Kritik der reinen Vernunft (2006), Heidegger e il problema della metafisica (2010), Contradictions. Logic, History, Actuality (editor, 2014), and various papers on the importance of the history of philosophy (especially Hegel) for the history and philosophy of logic, among them 'Dialectic and Dialetheism' in History and Philosophy of Logic, 'Hegel's Glutty Negation' in History and Philosophy of Logic, 'Contrariety and Contradiction'. Hegel and the Berliner Aristotelismus' in Hegel-Studien.

Chapter 6
Paraconsistent or Paracomplete?

Hartry Field

Abstract This paper attempts a comprehensive account of the comparative merits of paracomplete and dialetheic approaches to the semantic paradoxes. It argues that aside from issues about conditionals, there can be no strong case for paracomplete approaches over dialetheic, or dialetheic over paracomplete, and indeed that in absence of conditionals, the two approaches are plausibly seen as notational variants. Graham Priest disagrees: many of his arguments favoring dialetheic solutions over paracomplete do not turn on issues about conditionals. The paper discusses his arguments on these points in some detail. On the matter of conditionals, it argues that extant paracomplete approaches are far better than extant dialetheic approaches, a fact that traces in part to dialetheists having focused too heavily on "relevant logics". It holds out hope for better dialetheic treatments of conditionals (including of how the distinct kinds of conditionals interact), but it also suggests that this is the one area where paracomplete approaches are inevitably better.

Keywords Conditionals · Informal provability · Paradox · Restricted quantification · Revenge

As Graham Priest has long argued, the paradoxes involving semantic notions such as truth, satisfaction, and denotation provide strong motivation for restricting classical logic in the presence of those notions. Not, I'd say, motivations that are obviously decisive: there are various ways in which one might keep classical logic by restricting the natural principles of truth, satisfaction, and denotation. But the classical options have been well explored, and all have a high cost. Priest pushed more forcefully than anyone before him the view that a non-classical option is better. His work has been a major inspiration for my own.

Priest advocates a particular non-classical approach, *paraconsistent dialetheism*. It is *dialetheic* because he takes the paradoxical arguments to show that for certain sentences, both they and their negations are true. No non-classicality yet, since one

H. Field (✉)
Philosophy Department, New York University, 5 Washington Pl,
New York, NY 10003, USA
e-mail: hartry.field@nyu.edu

could reject the inference rule ("T-Elim") that takes you from $True(\langle B\rangle)$ to B (and hence $True(\langle\neg B\rangle)$ to $\neg B$): "classical glut theorists" do that, to accept dialetheias in classical logic. But Priest accepts T-Elim, so he thinks that the paradoxes prove certain sentences B and also prove their negations. Also, unlike "subvaluationists" who accept such contradictory pairs but don't accept the rule of $\wedge$-Introduction and so don't infer the contradiction $B \wedge \neg B$, Priest accepts that rule: so he takes the paradoxical arguments to be arguments for certain contradictions. In classical logic, that would be a disaster, as would the subvaluationist alternative: a theory that entails contradictions or contradictory pairs is Post-inconsistent, i.e. it entails everything no matter how absurd. Priest's response is that classical logic needs to be weakened to a "paraconsistent" logic, one in which contradictions don't entail everything and hence can be accepted without accepting everything.[1] Hence my term 'paraconsistent dialetheism'. But that term is rather a mouthful, and from now on I'll usually simply say either 'paraconsistent' or 'dialetheic' and for the most part use those terms interchangeably to describe the view outlined in this paragraph. (But really neither term suffices, and indeed the two together aren't quite sufficient either.[2])

An alternative non-classical approach to resolving the paradoxes is to use "paracomplete" logics which block the paradoxical arguments to contradiction. Those arguments rest on the law of excluded middle or related principles (among which I include the metarules of $\supset$-Introduction and $\neg$-Introduction). The paracomplete logics in question restrict these laws enough to block the paradoxical arguments, without the need for dialetheias or paraconsistency.

In Field (2008), after surveying reasons for preferring a non-classical approach to the paradoxes, I compared the paracomplete and paraconsistent dialetheic alternatives and presented some considerations that seemed to me to favor the former over the latter (or at least, extant versions of the former over extant versions of the latter). None of these considerations was intended as anything like a knockdown argument favoring the paracomplete solutions over the paraconsistent ones. But more recent developments have, I think, considerably strengthened the case for paracomplete

---

[1] So if 'consistent' means Post-consistent, the paraconsistent view is that it is perfectly consistent to accept some contradictions. Priest prefers to use 'consistent' to mean negation-consistent (i.e., implies no contradictions); in that terminology, he thinks that the correct treatment of the paradoxes is inconsistent. This choice of terminology strikes me as unnecessarily provocative, but of course the issue is merely verbal.

[2] 'Dialetheic', meaning that the view accepts truth-value gluts (sentences such that both they and their negations are true), is insufficient, mainly because of the aforementioned glut theories in classical logic.

'Paraconsistent' is also insufficient to describe the view, since some people advocate paraconsistent logic not because they want to accept contradictions (or contradictory pairs) or even think that it can be rationally acceptable to do so, but because they think that logic should respect certain kinds of relevance requirements in addition to the preservation of rational acceptability.

Indeed, even the conjunction 'dialetheically paraconsistent' is insufficient to describe the view: the classical glut theorist's position that accepts *the truth of* the Liar sentence but not the Liar sentence itself could be combined with the idea that logical consequence involves relevance requirements over and above the preservation of rational acceptability. The correct description is "view that accepts contradictions in a paraconsistent logic".

over paraconsistent dialetheist. I'm under no illusion that these developments will persuade the committed dialetheist, but they do strengthen my own preferences and perhaps they will persuade the undecided.

The purpose of this paper is to give an updated account of the comparative merits of paracomplete and dialetheic approaches, as I see them. In Sect. 6.1, I suggest that any strong case for paracomplete over dialetheic, or dialetheic over paracomplete, must rest on their treatment of conditionals. In Sects. 6.2–6.4 I turn to the aforementioned recent developments, which involve conditionals: indeed, which involve two quite distinct kinds of conditionals and how they interact. I then move on to a consideration of several matters that Priest has raised in favor of dialetheic treatments over paracomplete ones (e.g. in his critical notice Priest 2010 of Field 2008). I won't discuss everything he says for lack of space, time and energy, but Sects. 6.7–6.12 respond to what I take to be the most important of his points not covered in Sects. 6.2–6.4. (These will include, in Sects. 6.11 and 6.12, a rather extensive treatment of the complex issues involving revenge.) I will also include some discussion (especially in Sect. 6.5) of issues on which I take us to be largely in agreement.

6.1 Logics Without Conditionals

If our only interest is in the logic of $\neg$, $\vee$, $\wedge$, $\exists$ and $\forall$, we can adequately handle the paradoxes using any of three very simple 3-valued logics. They all have the same model-theoretic semantics (Strong Kleene). In it, any model assigns to each k-place predicate a k-place function from the domain to the value space $\{0, \frac{1}{2}, 1\}$. Let a valuation be a model plus an assignment of objects in the domain to the variables. Then each valuation determines values for all atomic formulas, and then values for complex formulas by these rules:

- $|\neg A|_v = 1 - |A|_v$.
- $|A \vee B|_v = max\{|A|_v, |B|_v\}$.
- $|A \wedge B|_v = min\{|A|_v, |B|_v\}$.
- $|\exists x A|_v = max\{|A|_{v(o/x)} : o \text{ in the domain}\}$, where $v(o/x)$ is the valuation just like v except that the assignment to variable x is o.
- $|\forall x A|_v = min\{|A|_{v(o/x)} : o \text{ in the domain}\}$.

But the logics differ in their understanding of consequence. In Kleene logic (K_3) consequence is the preservation of value 1 in all valuations; this is paracomplete in that $A \vee \neg A$ needn't follow from everything (since A and hence $A \vee \neg A$ might have value $\frac{1}{2}$). In Priest's "Logic of Paradox" LP, consequence is the preservation in all valuations of the property of having a value in $\{1, \frac{1}{2}\}$; this is paraconsistent in that B needn't follow from $A \wedge \neg A$ (since A and hence $A \wedge \neg A$ might have value $\frac{1}{2}$ while B has value 0). In S_3 ("symmetric 3-valued"), B is a consequence of Γ iff there is no valuation in which B has a value lower than all the values of members of Γ; it is in effect the intersection of K_3 and LP, so is both paracomplete and paraconsistent.[3]

Once we consider semantic notions like truth and satisfaction, we will of course want to restrict the allowable models, to exclude ones that don't respect the logic of truth. My preference is to restrict to "Kripkean" models where (for any A) $True(\langle A\rangle)$ always gets the same value as A; that's due to my preference for what's called a "naive theory of truth". As we'll see in Sect. 6.6, Priest prefers a weaker restriction that I'll call "quasi-naivety".[4] But in any case, there will be a restriction on the allowable models. (I don't regard the issue of naivety vs. quasi-naivety as very central; and Priest could if he wished easily modify his account to one where truth and satisfaction are fully naive.)

The Liar paradox, involving a sentence Q that asserts its own untruth (or more accurately, a sentence that is *equivalent given uncontroversial background assumptions* to the claim that it is not true), is resolved in all these logics in basically the same way: in every allowed model satisfying the background assumptions, Q gets value $\frac{1}{2}$.

So in LP, Q and $\neg Q$ are each essentially theorems: they follow from the uncontroversial background assumptions. And that's fine because even together, they don't imply absurdities (unless you beg the question by saying that contradictions are in and of themselves absurd even if they don't imply everything).

In K_3, Q and $\neg Q$ are each essentially anti-theorems: given the uncontroversial background assumptions, acceptance of either would lead to absurdities, so we must reject both (in a sense of rejection that at the very least involves a commitment to nonacceptance). On this approach, the disjunction $Q \vee \neg Q$ is likewise an antitheorem: it too must be rejected. With S_3 it's less clear how to talk of acceptance and rejection. In it, perhaps what should be said is that one should neither accept nor reject Q, and the same for $\neg Q$, but that this isn't a matter of ignorance in any normal sense, since one also shouldn't accept $Q \vee \neg Q$ (though one shouldn't reject it either). In any case, I'll focus mostly on K_3 and LP.

From my point of view, the nice thing about K_3 and LP (and, for that matter, S_3) is that they allow for a naive theory of truth, one in which there is a very rich class of models (the "Kripkean" ones) in which

For any sentence A, $|True(\langle A\rangle)| = |A|$.

Given that the semantics is compositional, it follows that in these models

$|X| = |Y|$ whenever Y results from X by truth-substitutions (substitutions of $True(\langle A\rangle)$ for A or vice versa, for any A).

---

[3] Actually one can use Kleene semantics for other logics too, e.g. the non-transitive logic on which validity requires only that the drop in values from premises to conclusion not exceed $\frac{1}{2}$.

[4] He requires only that $|True(\langle A\rangle)|$ be at least $|A|$, and 0 when $|A|$ is 0; so that $|True(\langle A\rangle)|$ can be 1 when $|A|$ is $\frac{1}{2}$. In LP, this does not undermine the inference from $True(\langle A\rangle)$ to A (since the value $\frac{1}{2}$ is designated), but the inference from $\neg A$ to $\neg True(\langle A\rangle)$ does not come out valid. (Of course when A is a sentence asserting its own untruth, $|True(\langle A\rangle)|$ as well as $|A|$ must be $\frac{1}{2}$.)

6 Paraconsistent or Paracomplete? 77

Hence (given any of the above accounts of validity in terms of semantic values, provided that we restrict the models to Kripkean ones),

$X \models Y$ whenever Y results from X by truth-substitutions.

The possibility of such a naive approach to truth (and more generally, satisfaction) with this 3-valued semantics was shown by Kripke in his famous 1975 paper.

More fully, Kripke showed that for any classical 2-valued ω-model of a ground language without 'True' but adequate to syntax, you can expand it to a 3-valued ω-model for the corresponding language with 'True' (and more generally, 'Satisfies'), without altering the domain or the extensions of the ground-language predicates, in such a way that for any sentence A, $|True(\langle A \rangle)| = |A|$. (Analogously for 'Satisfies'. The restriction to ω-models is to ensure that the induction schema extends to formulas with 'True' and 'Satisfies'. The usual composition principles are also guaranteed.) We could call this "the ω-conservativeness" of truth.[5] That we get not merely consistency but ω-conservativeness is extremely important: among other things, it shows that the addition of 'True' or 'Satisfies' doesn't induce ω-inconsistency. All the naive truth results to be discussed in this paper will have these features.

Although naive truth (and satisfaction) is possible in K_3 and LP (and S_3),[6] neither of these logics provide a good resolution of the paradoxes: they fail because of severe expressive limitations, as we'll see. But suppose that we were to ignore their expressive limitations and ask which of K_3 and LP is better? I don't think a strong argument could be given either way. Indeed, there's a lot to be said for the position that the logics are just notational variants. Part of this position is that the advocates of the two logics simply mean different things by 'valid', and each can reinterpret the other's use of 'valid'.

Focus first on the case where the advocates of both K_3 and LP advocate the Kripkean restriction that the only models to be considered for the logic of truth are ones where for all A, $|True(\langle A \rangle)| = |A|$. In that case, the advocate of either logic can translate the other's 'valid' as 'valid*', where to call the inference from Γ to B valid* is to say that the inference from $\neg B$ to some disjunction of negations of members of Γ is valid.[7] The advocate of K_3 believes that the LP inferences are valid*, and similarly with K_3 and LP interchanged; so why isn't this just a dispute about which of two relations deserves the honorific 'valid'?

Well, you might point out, 'valid' isn't just an empty word, it has connections with acceptance: to regard an inference as valid is to suppose that acceptance of the premises commits you to acceptance of the conclusion.[8] Fair enough, but we can play the same game with acceptance: the advocate of each logic can take the other's

---

[5] Because of the restriction to ω-models, there is no conflict with standard results dating to Gödel and Tarski on how adding even a restricted truth predicate to arithmetic yields new arithmetic consequences when the usual composition rules plus extended induction are assumed.

[6] That is, although naive truth can be added to these logics in an ω-conservative fashion.

[7] The symmetry between 'valid' and 'valid*' would be neater in a multiple conclusion framework.

[8] Either take 'accept' to mean 'accept with certainty' or take 'accept the premises' to mean 'accept their conjunction'. (There is a more general formulation of the connection of validity to acceptance and rejection that doesn't require either course, but these will do for present purposes.)

apparent acceptance as acceptance*', where to accept* a sentence is to reject its negation. (Similarly, the other's apparent rejection is rejection*, the acceptance of the negation.) So the advocate of LP could argue:

> When the advocate of K_3 says "I think that the inference from $A \wedge \neg A$ to B is valid and that acceptance of the premise requires acceptance of the conclusion, but that the inference from an arbitrary B to $A \vee \neg A$ isn't valid and its conclusion needn't always be accepted", all she really means is that the inference from $\neg B$ to $\neg(A \wedge \neg A)$ is valid and the rejection of $\neg(A \wedge \neg A)$ should require the rejection of $\neg B$, but that the inference from $\neg(A \vee \neg A)$ to $\neg B$ isn't valid and $\neg(A \vee \neg A)$ needn't always be rejected. And under that interpretation, the advocate of K_3 is speaking what the advocate of LP regards as the truth.

By the same translation manual, the advocate of K_3 can argue that the advocate of LP speaks what the advocate of K_3 regards as the truth. This looks like a classic case of a verbal dispute.

For the dispute between K_3 and LP to be verbal, the reinterpretation of acceptance and rejection is required even independent of validity claims. One might wonder how the dispute can be verbal given that the advocate of LP accepts both the Liar sentence and its negation rather than rejects both, and that there is no obvious way to reinterpret 'True' or '$\neg$' or any other component of these sentences that would explain this. The answer is that each must reinterpret the other's speech acts and belief states: apparent acceptance is really rejection of the negation, and apparent acceptance is rejection of the negation.[9]

In arguing that the difference between truth theory in K_3 and in LP seems verbal, I've focused on the case where the advocates of both K_3 and LP advocate the Kripkean restriction that the only models to be considered for the logic of truth are ones where for all A, $|True(\langle A \rangle)| = |A|$. As I've mentioned, Priest does not advocate that restriction: in particular, he thinks that in some cases where A has value $\frac{1}{2}$ (though not in the case of the Liar sentence Q specified above), $True(\langle A \rangle)$ has value 1. Leaving aside why he thinks that for now, does it break the argument for the dispute being verbal? It doesn't if the K_3 advocate adopts the dual non-Kripkean view, that in some cases where A has value $\frac{1}{2}$, $True(\langle A \rangle)$ has value 0, so that the inference from $\neg True(\langle A \rangle)$ to $\neg A$ is invalid. The case for the verbal nature of the dispute then remains.

I'm inclined to think that before conditionals are introduced, the dispute between LP and K_3 really is verbal; but I do not insist on this. Even without claiming the dispute verbal, the duality makes it hard to see how there can be a strong case for one of LP and K_3 over the other, unless we broaden our horizons by including in the discussion conditionals or some other operators not present in LP and K_3 proper.

---

[9] Once one adds a conditional to the language, one can explain the verbal nature of the dispute in the conditional-free fragment slightly differently: we could use the conditional to define a weakening operator W (WB is $Q \to B$, where Q is a Liar sentence) and a strengthening operator S (SB is $\neg(Q \to \neg B)$); the advocate of K_3 interprets the acceptances and rejections of the LP-theorist as acceptances and rejections of the result of prefixing with a W, and analogously in the reverse direction using an S.

Priest evidently disagrees with this: he gives arguments, not based on conditionals, for the superiority of the dialetheic approach over the paracomplete. I'll consider these in Sects. 6.7–6.12.

6.2 Conditionals, in Classical Logic and Beyond

In classical logic, it's common to introduce a conditional $\supset$ by defining $A \supset B$ as $\neg A \vee B$. As is well known, this doesn't seem to be a very good representation of the ordinary English 'if ... then': consider

- If I run for President in 2020 I will be elected.
- If Sarah Palin is elected in 2020 she won't have run for that election.

Both seem false, but the first comes out true on the '$\supset$' reading since I won't run, and the second also comes out true on the assumption that Palin isn't elected. (The contrapositive of the second conditional is presumably true, so the ordinary indicative, unlike $\supset$, does not contrapose.)

However, there are uses of the conditional that the $\supset$ is well suited for. Consider the usual definition of restricted universal quantification in terms of unrestricted: "All A are B" is defined as "For anything at all, if it's A then it's B". The ordinary English conditional wouldn't work at all well in this definition. (For instance, "All the presidential nominees of the major parties in 2020 will be female" might be true. But "For every x, if x will be a major party presidential nominee in 2020 then x will be female" entails "If Ted Cruz will be a major party nominee then he will be female", which presumably isn't true for the ordinary English 'if ... then'.) By contrast, in classical logic, the $\supset$ seems like what we need for restricting universal quantification.

Upshot: In classical logic, the $\supset$ may not be a good account of the most common ordinary English conditional, but it is at least good for something.

When we come to non-classical logics, the $\supset$ account is much worse: it isn't any better at capturing the ordinary English conditional than it was in the classical case, and it doesn't work for restricting quantification either.

In the case of K_3, this is because the absence of excluded middle means that such basic laws as $A \supset A$, $A \supset A \vee B$, etc. will all be invalid. So on the usual account of restricted quantification, "All A are A", "All A are either A or B", etc. will fail. In the case of LP, it's because the absence of explosion will make modus ponens fail badly: if A has value $\frac{1}{2}$ and B has value 0, $A \supset B$ will have value $\frac{1}{2}$ and we'll have a counterexample to MP. Some (McGee 1985) have argued that MP ought to fail for embeddings of the ordinary English conditional, though the argument depends on the doubtful assumption that "If A, then if B then C" ought to be equivalent to "If A and B, then C". But even if McGee is right, that's irrelevant to the issue of conditionals in LP. One reason is that the McGee argument concerns only embedded conditionals, whereas in LP, MP for $\supset$ fails even with unembedded conditionals. But the main reason is that what is in consideration in this paragraph isn't the ordinary English conditional but the conditional used to restrict quantification, and in this context failure of MP means failure of the inference from 'All A are B' and 'This is A' to 'This is B'. That seems pretty clearly unacceptable.

So $\supset$ is unacceptable as a quantifier-restricting conditional in both K_3 and in LP, though for different reasons. (In S_3 it would have both the problems it has in K_3 and the problems it has in LP.)

One might think it inappropriate to define universal restricted quantification in terms of a conditional: perhaps universal restricted quantification should be primitive. But even if so, the conclusion is unchanged: K_3 and LP still don't have the resources for adequately defining universal restricted quantification, and no 3-valued valuation procedure for restricted quantification that obeys the monotonicity condition required for the (Kripke 1975) construction is remotely adequate. For instance, if for every assignment function in a given model, Ax gets value $\frac{1}{2}$, then the only remotely adequate *monotonic* procedure for evaluating 'All A are B' gives it value 1 when Bx gets value 1 for all assignments, and value $\frac{1}{2}$ otherwise. But that is the feature of the $\forall x(Ax \supset Bx)$ account of restricted quantification that led to the problems (for each of K_3, LP and S_3).

So it seems clear that if we are to pursue a non-classical account of truth for a language that has at least minimal expressive adequacy, we need it to have at least two new conditionals. One, which I'll denote ▷, is to be an analog of the ordinary English conditional; the other, which I'll denote →, is to be an analog of the quantifier-restricting ⊃. (Or alternatively, we need it to have both ▷ and a primitive restricted universal quantifier. But that possibility doesn't really need to be considered separately: it's easy to see that if we start out with a primitive restricted quantifier "All A are B", we can define a conditional → in terms of it, in such a way that "All A are B" can then be shown equivalent to $\forall x(Ax \to Bx)$.[10] The problems discussed below about getting an adequate → arise equally for getting an adequate account of universal restricted quantification if that is taken as primitive.)

When I say that → is to be an analog of ⊃, I mean that since in classical contexts quantifier restriction works by ⊃, → had better reduce to ⊃ in classical contexts. Of course, it must differ from ⊃ in being well behaved in non-classical contexts.

Analogously, when I say that ▷ is to be an analog of the ordinary English conditional, I mean that it is to reduce to "the" ordinary English 'if ... then' in classical contexts, but be well behaved in non-classical. (I put "the" in scare quotes because it may be that there is more than one ordinary English conditional; if there is more than one, we need an account of *each* that works in a non-classical context. But I'll content myself with getting one.)

The goal as regards '▷' is more elusive than as regards '→', because it is more controversial how the 'if ... then' of ordinary English works. A common framework is the semantics of variably strict conditionals in the general style of Lewis and Stalnaker, utilizing worlds and a ternary "comparative closeness" relation $x \leq_w y$ among them. ($x \leq_w y$ is read "the change from w to x is no greater than the change

---

[10]If we formalize "All A are B" as $\forall x_{Ax} Bx$, and v is a variable not free in either A or B, define $A \to B$ as $\forall v_{v=v \wedge A}(v = v \wedge B)$.

6 Paraconsistent or Paracomplete?

from w to y".) Both Lewis (1974) and Stalnaker (1968) assume that for any w, the resulting two-place relation is a weak order on a certain domain W_w (the set of worlds accessible from w): not only transitive and reflexive (on this domain) but also connected in the sense that for any x and y, either $x \leq_w y$ or $y \leq_w x$. A more general version that avoids the connectivity assumption was given in Lewis (1981) and Burgess (1981): for generality I follow that, though one can add connectivity and other assumptions if one likes.

Actually I want to generalize the Burgess framework slightly further, to allow for "non-normal worlds" of various stripes. It is by no means essential to my view that any form of non-normal worlds be allowed, but Priest thinks it important to allow them, so I want to stress that I can do so.

In the generalized framework, a *2-valued model M* for a first-order language enriched by ▷ consists of:

(i) A non-empty set W of worlds.
(ii) A division of these worlds into normal, moderately non-normal, and anarchic; either or both of the last two categories can be empty, but there must be at least one normal world.
(iii) For each $w \in W$, a subset W_w of W and a preorder (reflexive and translation relation) $\leq_w$ on W_w.
(iv) For each $w \in W$, a non-empty set U_w (the universe of w). Let U be the union of the U_w.
(v) For each $w \in W$ and k-place predicate p, a function p_w from U^k to $\{0,1\}$. (The set of k-tuples that get assigned value 1 at w is the *extension of p at w* in the model.) We require that the function $=_w$ (associated with '=') assigns 1 to $<o,o>$ for each $o \in U_w$ and assigns 0 to all other pairs.

In order to ensure Modus Ponens for ▷ I will impose one condition:

Weak centering, at least at normal worlds if w is normal then (a) $w \in W_w$, and (b) for any x in W_w, $w \leq_w x$.

And I allow the possibility of adding further conditions (e.g. connectivity of $\leq_w$, or Lewis's Limit Condition, or a stronger centering condition), either at all normal worlds or more broadly.

At all non-anarchic worlds, valuation rules for the first-order connectives proceed world by world, as in the semantics of classical modal logic. For the conditional we use the following rule, for such w:

General 2-valued ▷ $|A \triangleright B|_w = 1$ iff $(\forall x \in W_w)[|A|_x = 1 \supset (\exists y \leq_w x)[|A|_y = 1 \wedge (\forall z \leq_w y)(|A|_z = 1 \supset |B|_z = 1)]]$; 0 otherwise.

If like Lewis and Stalnaker one imposes the connectivity condition on $\leq_w$, this can be simplified:

Special 2-valued ▷ $|A \triangleright B|_w = 1$ iff either $(\forall y \in W_w)[|A|_y = 0]$ or $(\exists y \in W_w)[|A|_y = 1 \wedge (\forall z \leq_w y)(|A|_z = 1 \supset |B|_z = 1)]]$; 0 otherwise.

Validity is defined as the preservation of value 1 *at normal worlds*, in every model.

I'm not insistent that this framework provides the ultimately best treatment of English conditionals, but I do think it a reasonable shot. One thing I should stress is that I have called the members of W "worlds", not "possible worlds". Even for normal worlds, calling them "possible" seems unduly restrictive. For I don't insist that so-called "metaphysical necessities" like "water is H_2O" hold at all worlds, or even all normal ones; or that mathematical truths do. Also, I'm allowing anarchic worlds, at which even the valuation rules for $\neg$, $\wedge$, $\vee$ and the quantifiers can go awry.

Since "metaphysical necessities" and mathematics can fail even at normal worlds, why allow for moderately non-normal ones? The reason is that certain constraints on $\leq_w$ that are needed for validities we want needn't be imposed on non-normal w. For instance, we might allow Weak Centering to fail at non-normal worlds. In that case, "modus ponens might fail at such worlds", in the sense that B and $B \triangleright C$ might hold at them while C fails. That's compatible with the validity of modus ponens since validity is defined in terms only of normal worlds, but it affects the validity of the "embedded modus ponens" from $A \triangleright B$ and $A \triangleright (B \triangleright C)$ to $A \triangleright C$. Similarly, other structural requirements might be imposed on normal worlds only, giving other examples of validities without the corresponding "embedded validities".

Priest has argued for anarchic worlds, on the ground that we can give nontrivial sense to counterfactuals with logically impossible antecedents. I agree that we can make sense of such counterfactuals *in a context where an alternative logic has been clearly spelled out*, so that we can put constraints *appropriate to that context* on how the worlds behave. What this suggests to me is that we need different frameworks of worlds for different contexts. Priest apparently isn't happy with such a contextualist approach, he seems to want a context-independent framework where counter-logicals with antecedents like "if $\wedge$-Elimination weren't valid" can be handled even in absence of a hint as to what the logic without $\wedge$-Elimination is to be like. To simultaneously handle those and all other conditionals with different logically impossible antecedents, he needs anarchic worlds, with no constraints.

I'm not persuaded that this is a useful way to go, but there's no need to settle that here. If one wants to allow for worlds where $\wedge$-Elimination fails (even in absence of any indication how), I've allowed for it, by allowing anarchic worlds not governed by the usual rules for $\wedge$ and the other connectives. The cost of allowing them isn't that great: validity is defined in terms of preservation of truth (in all models) *at normal worlds*, none of which are anarchic; so the logic can be fairly standard except as regards conditionals whose antecedent is "far-fetched enough" to require the consideration of anarchic worlds.

Priest also tends, after allowing for anarchic worlds, to ignore them, since allowing for them would invalidate almost every law of conditionals.[11] Ignoring them for most purposes is perfectly sensible, since in most contexts of dealing with counterfactuals,

---

[11] See, for instance, Priest (2008, p. 167). Paragraph 9.4.6 advocates totally anarchic worlds, at which, for instance, the valuation rules for $\wedge$ fail. As a result, the inference from $A \to B \wedge C$ to $A \to B$ will come out invalid, by virtue of normal worlds from which such anarchic worlds are accessible. But in paragraph 9.4.7 and the subsequent development in the book, there is a shift to only moderately non-normal worlds in which the basic laws of FDE are held inviolate.

radical alteration of the logic isn't under consideration, and we can introduce a *special logic appropriate to such contexts* whose semantics will not involve anarchic worlds. I will follow Priest along this path: I will largely confine my attention to normal and moderately non-normal worlds. (And as I've said, nothing in my approach requires even the moderately non-normal worlds.)

The assumption of the generalized Burgess framework I've outlined here is made more in the spirit of a working assumption than out of a firm conviction that it is how ordinary indicative conditionals should be treated in a classical context. But given that working assumption, what an advocate of naive truth requires is that it can be generalized to the non-classical contexts required if 'true' is introduced and assumed to behave naively (at least at non-anarchic worlds). Indeed, what's required is that we can simultaneously generalize this conditional and a quantifier-restricting conditional, in a satisfactory way.

6.3 Target Laws, and One Framework for Achieving Them

6.3.1 Target Laws

There are various laws involving the two conditionals that one would expect to hold not just in classical contexts but in non-classical too; if they don't, it will mean that it is very hard to carry out "sustained ordinary reasoning" in the non-classical context. Of particular interest are certain laws involving the two conditionals together. I take it that each of the following is quite compelling:

(1) $[\forall x(Ax \to Bx) \wedge Ay] \triangleright By$ "If all A are B, and y is A, then y is B."
(1_c) $[\forall x(Cx \to Dx) \wedge \neg Dy] \triangleright \neg Cy$
   "If all C are D, and y is not D, then y is not C."
(2) $\forall x Bx \triangleright \forall x(Ax \to Bx)$ "If everything is B, then all A are B."
(2_c) $\forall x \neg Cx \triangleright \forall x(Cx \to Dx)$ "If nothing is C, then all C are D."
(2*) $\neg \forall x(Ax \to Bx) \triangleright \neg \forall x Bx$ "If not all A are B, then not everything is B."
(3) $\forall x(Ax \to Bx) \wedge \forall x(Bx \to Cx) \triangleright \forall x(Ax \to Cx)$
   "If all A are B and all B are C then all A are C."
(4a) $\forall x(Ax \to Bx) \wedge \forall x(Ax \to Cx) \triangleright \forall x(Ax \to Bx \wedge Cx)$
   "If all A are B and all A are C then all A are both B and C."
(4b) $\forall x(Ax \to Bx) \wedge \neg \forall x(Ax \to Bx \wedge Cx) \triangleright \neg \forall x(Ax \to Cx)$
   "If all A are B and not all A are both B and C then not all A are C."
(5) $\neg \forall x(Ax \to Bx) \triangleright \exists x(Ax \wedge \neg Bx)$
   "If not all A are B, then something is both A and not B."
(5*) $\neg \exists x(Ax \wedge \neg Bx) \triangleright \forall x(Ax \to Bx)$
   "If nothing is both A and not-B, then all A are B."
(6) $\exists x(Ax \wedge \neg Bx) \triangleright \neg \forall x(Ax \to Bx)$
   "If something is both A and not B, then not all A are B."

(1_c) and (2_c) follow from (1) and (2) respectively on the supposition that the $\rightarrow$ contraposes; I take that supposition to be plausible, but not entirely obvious. (2*) and (5*) would result from (2) and (5) on the assumption that $\triangleright$ contraposes, but as noted at the start of Sect. 6.2 (the Sarah Palin example), it doesn't. Despite not following, (2*) and (5*) seem as compelling as the unstarred versions. (There is some redundancy in the list: (2*) follows from (5), and both (2) and (2_c) follow from (5*).)

Of course in addition to laws like these, we'd expect a suitably rich supply of laws for $\rightarrow$ without $\triangleright$ and for $\triangleright$ without $\rightarrow$. For $\triangleright$ without $\rightarrow$, I'd assume that the basic axioms and rules of Burgess semantics are a reasonable target: these are

$A \triangleright A$
$[A \triangleright (B \wedge C)] \triangleleft \triangleright [(A \triangleright B) \wedge (A \triangleright C)]$
$[(A \triangleright C) \wedge (B \triangleright C)] \triangleright [(A \vee B) \triangleright C]$
$[A \triangleright (B \wedge C)] \triangleright [(A \wedge B) \triangleright C]$

together with Modus Ponens for $\triangleright$ and the metarule that if Y is a logical truth then so is $X \triangleright Y$ for any X.[12]

For $\rightarrow$ without $\triangleright$, it would be good to have as strong an approximation to $\supset$ as possible, compatibly with retaining the laws $A \rightarrow A$ and Modus Ponens and preserving naive truth. It isn't presently known what is the best that can be achieved, either in a paracomplete context or a dialetheic.

In the case of paracomplete logics, one can use the Brouwer fixed point theorem to show that the sentential fragment of the Łukasiewicz continuum logic is consistent with naive truth. (Indeed, it has the ω-conservativeness feature mentioned earlier in connection with Kripke, which among other things guarantees ω-consistency, the usual compositional principles, and the extendability of induction to formulas involving 'true' and 'satisfies'.)[13] Unfortunately, this striking result no longer holds when quantifiers are added: there seems to be no reasonable way to add quantifiers to Łukasiewicz logic without validating the "∃-exportation" rule $A \rightarrow \exists x Bx \models \exists x (A \rightarrow Bx)$ (when x isn't free in A), but that rule with the rest of the logic leads to paradox.[14] ∃-exportation is the infinitary generalization of

(?) $\quad A \rightarrow (B \vee C) \models (A \rightarrow B) \vee (A \rightarrow C)$,

a law which holds in Łukasiewicz because of the more fundamental law

(??) $\quad \models (B \rightarrow C) \vee (C \rightarrow B)$,

which reflects the fact that the semantic values in its standard semantics are linearly ordered. For naive truth in the context of quantification theory, we probably need a

---

[12] Some of these would not be acceptable if we allowed for anarchic worlds, but as indicated above, I will ignore those. Priest does too in the logics of paradox he discusses.

[13] A precisification and proof of the ω-conservativeness of naive truth in sentential Łukasiewicz continuum-valued logic can be found in Field (2008, Chap. 4). (I claim no originality: the result seems to have been known among the *cognoscenti* long before, as is indicated by the frequent allusions to the Brouwer theorem in related contexts.)

[14] See Restall (1992), Hajek et al. (2000), and Bacon (2013).

logic where (?) and (??) don't hold.[15] There are logics not that much weaker than Łukasiewicz that it's natural to try, such as the logic CK (*aka* RWK). (See Priest 2008, Sect. 11.5. Łukasiewicz logic is the result of adding to CK the unobvious axiom $((A \to B) \to B) \to A \vee B$; without it, (?) and (??) can't be proved in CK.) But it is not currently known whether naive truth can be added ω-conservatively to a quantified version of CK.

And the strongest logics of $\to$ plus quantification for which an ω-conservativeness result for naive truth and satisfaction *is* currently known are quite a bit weaker than CK. The earliest results in this general direction were provided by Ross Brady. His approach had several versions, including ones in paracomplete logics that are not paraconsistent: see Brady (1983). Unfortunately, the Brady conditionals are clearly unsuited as the conditional for restricted quantification, among other things because of their failure to obey the law $B \models A \to B$ and hence the restricted quantifier law

Everything is $B \models$ All A are B.

Brady himself agrees that his conditional is unsuited for this purpose, and for just this reason: see Beall et al. (2006), which calls for a different conditional for restricted quantification.[16] But using methods with some similarities to Brady's, I obtained a conservativeness result for more suitable logics. (See Field 2008.) One such logic strengthens the quantified version of the paraconsistent system **B** (described in Priest 2008, Chap. 10) by adding the explosion rule $A, \neg A \models B$ together with, among other things,

$\models (A \to \neg B) \to (B \to \neg A)$
$B \models A \to B$ (hence using the previous, $A \supset B \models A \to B$)
$A \wedge \neg B \models \neg(A \to B)$ (i.e. $\neg(A \supset B) \models \neg(A \to B)$)
$\neg(A \to B) \models A \wedge \neg B$ (i.e. $\neg(A \to B) \models \neg(A \supset B)$).[17]

(The last three are rules; their strengthenings to conditionals are not valid.)

Can we do better in our treatment of the restricted quantifier conditional? I don't know of any construction that will yield more of the laws that go into typical axiomatizations of CK. But I've come to realize that we can still do better, by expanding the range of sentences for which nice laws hold. Of course, the *really* nice laws are the classical laws, and I don't know how to greatly expand the range of sentences for which those hold. But the Łukasiewicz laws are also quite nice; and I now see

---

[15] I won't discuss here why I find it less promising to look for a theory that admits (?) but not the ∃-Exportation rule that is its infinitary generalization.

[16] I don't think the Brady conditional is suitable as an "ordinary" conditional either, for the same reasons to be offered in the next section in connection with Priest's conditionals. I'll come back to Brady's conditionals in Sect. 6.8, in connection with the naive theory of classes.

[17] The official system of Field (2008) didn't include this last axiom. Nor did it include the full axiom $\models (A \to B) \wedge (A \to C) \to [A \to (B \wedge C)]$, but only its rule weakening. However, the "first variation" described in 17.5 does yield the full system described in the text. I should have used that as the official system. (It also avoids a problem about truth-preservation pointed out in Standefer 2015.)

that one can modify the account in Field (2008) so as to get these to hold for any sentences where the use of quantifiers is limited in a certain way.

To explain that more fully, I need to say a bit about how the proof of consistency/ω-conservativeness proceeded. It was by a revision-theoretic construction, which can be viewed as assigning to each sentence (and world) a value in an infinite deMorgan algebra V that is only partially ordered. The V I used was a product (or subproduct) of the 3-valued lattice used in K_3, LP and S_3: that is, V was a set of functions from a set X to the set of "mini-values" 0, $\frac{1}{2}$ and 1.[18] It turns out that one can get substantially improved results by allowing the whole unit interval [0, 1] as "mini-values", so that now V is taken to be a set of functions from a set X to the unit interval [0, 1]: see Field (2018a). This lets one mimic Łukasiewicz semantics for the "typical" paradoxical sentences where that semantics works. Somewhat more precisely: in the modified framework, a "typical" paradoxical sentence A will get as value *a constant function*, one that assigns to each member of X the value that A would get (or one of the values it might get) in an attempt to handle A within Łukasiewicz semantics. The nonconstant functions in the new V would be used only to handle the cases that can't be handled within Łukasiewicz semantics. This has the advantage of allowing the richer laws of Łukasiewicz logic over a wide range.

I think this is an improvement over (Field 2008), but it isn't the recent progress I mostly want to talk about.

6.3.2 Achieving the Target Laws

My main focus in the last few years hasn't been on improving the system for $\rightarrow$, but on achieving a system for $\rightarrow$ and $\triangleright$ together that yields not only desirable laws for them individually, but also desirable laws for how they interact, like the ones with which I began Sect. 6.3.1. In Field (2016) I gave a proof of the consistency, indeed ω-conservativeness, of a logic that included all the above interaction laws, the basic Burgess laws, and the aforementioned $\rightarrow$ laws from Field (2008). It involved a generalization of the Burgess framework for variably strict conditionals from a space of 2-valued worlds to a space of V-valued worlds, where as above V is a subspace of $\{0, \frac{1}{2}, 1\}^X$ for some set X. One can easily adapt this to the case where V is a subspace of $[0, 1]^X$, to accommodate the improvement in the treatment of

---

[18]In Field (2014) I introduced as an alternative to (or generalization of) the revision construction "higher order fixed point construction" whose main laws are similar, but which avoids certain odd laws noted in Standefer (2015). But the construction that follows also avoids those odd laws, but in a revision-theoretic context, thus reducing the motivation for the more complicated "higher order fixed point" approach. In any case, the revision-theoretic and higher order fixed point approaches have a great deal in common, and what I say here about the revision-theoretic holds with little change for the higher order fixed point.

the restricted quantifier conditional just noted.[19] (On this too, the details are in Field 2018a.) At each world, each sentence (and each formula relative to any assignment of objects to the free variables) gets a value in this function-algebra V. (So the full evaluation space is a space $W \times V$ where W is the set of worlds.)

The rules for semantically evaluating ▷ obviously need generalizing from a 2-valued semantics to the V-valued semantics, but it is a straightforward generalization in that

(1) The space V contains obvious analogs of the two values 0 and 1, viz. the constant functions **0** and **1** which assign 0 and 1 respectively to every member of X; and
(2) The ω-conservativeness proof takes you from any standard 2-valued "ground model" for the language without 'True' to a corresponding V-valued model for the language with it, in which
   (a) the worlds, the accessibility relations and nearness relations among them, and their domains, are all the same, and the worlds are normal in the new model iff normal in the old; and
   (b) any formula that gets value 0 in the ground model at any world (relative to any assignment of values to the variables) gets **0** in the V-valued model at the same world (relative to the same assignment); and any formula that gets value 1 in the ground model at a world gets **1** in the V-valued model at that world.

Also, the rules for → generalize those for ⊃ in the same sense: among formulas with values in {**0**, **1**} (relative to an assignment of values to the variables), → behaves just like the classical ⊃.

In the case of the Kleene semantics, we saw that it is possible to use it for (at least) three different logics: the paracomplete logic K_3, the paraconsistent logic LP, and the symmetric logic S_3 that is both paraconsistent and paracomplete. With the value space enlarged to the function space V there are many more choices. The one I've advocated is the analog of K_3: validity is the preservation of value **1** at all normal worlds in all appropriate models. (I'd take the "appropriate" ones to be the standard models at which truth behaves naively, though one might allow for variations; e.g. as I've mentioned, Priest doesn't want to impose a requirement of full naivety.) An alternative which some might see as advantageous is an analog of S_3: for the inference from Γ to B to be valid, it must be the case that in all appropriate models and at all normal worlds in them, the value of B is at least the greatest lower bound (in V) of the values of members of Γ. There are also analogs of LP. For instance, one could require of a valid inference that in all appropriate models and all normal worlds in them, if each premise has value at least **1/2** then so does the conclusion, where **1/2** is the constant function with value $\frac{1}{2}$. Alternatively one could go for duality with the proposed generalization of K_3: take the inference from Γ to B as valid iff in all appropriate models and at all normal worlds in them, the value of B isn't **0** unless the greatest lower bound of the values of members of Γ is **0**.

---

[19] X can be viewed as a set of pairs of ordinals (one for ▷-conditionals and the other for →-conditionals), though a slightly more abstract "fiber bundle" representation can be used to exhibit the value-dependencies more faithfully.

I remarked in Sect. 6.2 that in adjudicating between K_3 and LP, no obvious grounds for choice is really compelling: indeed, there's a good case to be made that those apparently competing logics are notational variants. In the case of the V-valued logics sketched here, I think the situation quite different: even on the second version of the generalization of LP to the larger value space, *attractive laws for the conditionals $\rightarrow$ and $\triangleright$ break the symmetry*. That is, if one were to define validity in a V-valued logic in the second LP-like way, without altering the valuation rules for $\rightarrow$ and $\triangleright$, a great many attractive laws that hold on the K_3-like definition of validity would no longer hold (for instance, modus ponens, and the inference from $A \rightarrow B$ and $B \rightarrow C$ to $A \rightarrow C$); and while some new laws would hold, I think it would be hard to argue that those have anything like the appeal of the ones that have to be abandoned.

I doubt that Priest would dispute this; he doesn't advocate a dialetheic logic based on the above semantics, he favors an entirely different semantics, based in part on a standard semantics for "relevant" logics. I'll consider his favored conditionals in the next section.

6.4 Priest's Conditionals

One of the things that influenced me in developing the two-conditional account of the previous sections was Priest's joint paper (Beall et al. 2006). That paper proposed a type of theory with an "ordinary" conditional with some resemblance to the English indicative conditional, and a separate conditional to handle universal restricted quantification. Their "ordinary" conditional is some kind of "relevant" conditional; they leave open the details. Their restricted quantifier conditional is an "irrelevant" one, again with some details unspecified. An alternative restricted quantifier conditional is suggested in Chap. 18 of Priest (2006, pp. 254–5), in the context of a discussion of naive set theory; the differences between this and what is in the joint paper won't matter for what follows. In both cases, the systems are designed to allow for dialetheia. Neither Beall et al. nor Priest (2006) offers any proof that naive (or quasi-naive) assumptions about truth can be retained without triviality in their systems, but they clearly believe this to be the case, and I have no particular reason to doubt it.

In discussing these systems I will depart from the notation of Beall et al. (2006) and Priest (2006): they use the symbol '$\rightarrow$' for the "ordinary" conditional, whereas to facilitate comparison with the previous section I will use it here for their restricted quantifier conditional. (Their restricted quantifier conditional is not the same as mine, but using the same symbol seems natural since it was proposed for exactly the same role.)

For convenience here I will also use '$\triangleright$' for their "ordinary" conditional. However, it will turn out to be important that their version of the "ordinary" conditional isn't a variably strict one in the ballpark of the Lewis–Stalnaker–Burgess conditionals, but a "relevant" conditional whose logic is somewhat in the vicinity of the relevant logic

B. (See Priest 2008, Chap. 10 for a discussion of **B** and related relevant logics, and Priest 2002, Sect. 8.2 for the particular such conditional that Priest himself endorses, which extends **B** by adding the law of excluded middle.)[20] Technically the difference between variably strict conditionals and standard relevant conditionals is a big one; and it's a big difference philosophically too I think, for reasons that will emerge.

The appeal of "relevant" conditionals for Priest seems to be based on the view that they well approximate the ordinary English 'if ... then'. One piece of evidence that this is his rationale is that he repeatedly says in Priest (2008) that relevant logics like **B** avoid "the standard paradoxes of material implication": see for instance p. 205 and p. 208.

Another piece of evidence is a criticism he made in Priest (2010) (and earlier in Priest 2005) of my 2008 account of how conditionals can be added to truth theory. Priest's criticism was that the conditional $\rightarrow$ employed in my 2008 theory reduces to $\supset$ in classical contexts; as a result, he says, it fails to respect ordinary English usage, and he illustrates this with an example involving electric circuits. This involves a circuit of a battery, a light, and two switches, all in series. The disjunction of the conditionals "If Switch A is closed but Switch B isn't then the light is on" and "If Switch B is closed but Switch A isn't then the light is on" would be a logical truth if the conditionals were read as $\supset$, but intuitively both disjuncts are false. Priest is right about this, and it was in part due to his discussion there that I was led to expand my logic of naive truth to include a Burgess-like conditional, which handles the example well. (Both disjuncts will be false on any natural way of filling out the "nearness" relation for this example.) But what I want to focus on now is the fact that Priest uses the example to argue for the treatment of the paradoxes in a logic with a relevant conditional in the vicinity of System **B**. This makes pretty clear that he thinks that that logic is at least a pretty good approximation to the ordinary 'if...then'. (I'll come back to this.)

The reason that Priest (along with Beall et al.) wants an "irrelevant" conditional '$\rightarrow$' to restrict quantification, in addition to his "relevant" one, is that the latter does not obey the law $B \models A \triangleright B$; so if universal restricted quantification is defined in terms of it, "Everything is G" wouldn't imply "All F are G", which seems clearly unacceptable. So he and I agree on the need for the law $B \models A \rightarrow B$. Indeed I assume that we have a need for a conditional form $\models B \triangleright (A \rightarrow B)$, to get the restricted quantifier law (2) from Sect. 6.3.1.

---

[20] In some places he also contemplates weakening **B** by giving up even the rule form of contraposition: see the discussion of "quasi-naivety" below.

Priest advocates treating negation with the Routley * operator, which "interchanges gaps and gluts". Endorsing excluded middle involves ruling "gap worlds" out of the normal worlds, so evidently he takes the Routley * operator to yield a non-normal world whenever applied to a normal world.

6.4.1 Priest's Laws of Interaction

But it seems to me that the kind of theory proposed in Beall et al. (2006) and in Chap. 18 of Priest (2006) falls far short of adequacy as a logic for combining ordinary conditionals with restricted quantification. If this is the best that a dialetheist can do then the dialetheist approach is in very bad shape. After arguing this, I'll suggest that there may be prospects for somewhat better dialetheic approaches. But moving to those would require giving up the focus on standard relevant conditionals that has been characteristic of dialetheists' work on the paradoxes. And even so, there are in-principle limits on what kind of logic of restricted quantification can be attained.

Let's start with the in-principle limits. Two of the restricted quantifier laws that I myself find totally compelling are (1) and (5*) from Sect. 6.3.1, but let's just focus on their rule forms:

(1_r)   All F are G, c is $F \models c$ is G
(5^*_r)   Nothing is both F and not $G \models$ All F are G.

(These rule forms are weaker, assuming modus ponens for 'if...then' plus $\wedge$-introduction.) No dialetheist can accept both (1_r) and (5^*_r), if he or she accepts reasonable views of disjunction and unrestricted quantification (and the transitivity of inference, i.e. Cut). For by conjunction rules and Cut they together entail

Nothing is both F and not G, c is $F \models c$ is G;

that is,

Everything is either G or is not F, c is $F \models c$ is G.

And no dialetheist can accept that: taking G to be $x = x \wedge B$ where B is an absurdity, and F to be $x = x \wedge A$ where A is a dialetheia, this yields

$\neg A, A \models B$,

which the dialetheist rejects. So the dialetheist is forced to reject at least one of the rules (1_r) and (5^*_r). (Note that these principles don't involve the ordinary conditional, only the conditional implicit in the restricted quantifier.) The dialetheists I've asked accept (1_r) (which is in effect Modus Ponens for the quantifier-restricting conditional) and give up (5^*_r). But I see no independent motivation for giving up (5^*_r) that doesn't rely on ascribing to "All F are G" a modal character not present in the ordinary notion. (I'd have thought it more natural for a dialetheist, in the case where A is a dialetheia and $\rightarrow$ the restricted quantifier conditional, to regard $\forall x(x = x \wedge A \rightarrow x = x \wedge \bot)$ as itself a dialetheia: as true given that A is false, as well as false given that A is true. In that case, the conclusion of (5^*_r) as well as its premise would be correct for the F and G above; instead of rejecting (5^*_r), the dialetheist would impose a restriction on (1_r). Not that that's entirely comfortable either.)

By a similar argument, the dialetheist is forced to reject at least one of the rules (1_r) and

($2_{c,r}$)   Nothing is $F \models$ All F are G.

6 Paraconsistent or Paracomplete?

Again I see no independent motivation for giving up $(2_{c,r})$ that doesn't rely on inappropriately ascribing to restricted quantification a modal character.

Of course, I recognize that someone committed to alternative philosophical views may differently evaluate which laws are desirable, so I certainly don't claim that this is a knockdown objection to dialetheism. I can only report that I myself was much more inclined to consider dialetheism seriously before I realized (what probably should have been obvious to me all along, but wasn't) that it had this consequence for restricted quantification. Others can decide for themselves.

Now let's look at some other natural laws that we don't get in Beall et al. (2006) or Priest (2006, Chap. 18). Consider the full (1) from Sect. 6.3.1 (as opposed to the rule (1_r)):

(1) If all F are G, and c is F, then c is G.

Not only do the systems in Beall et al. (2006) and Priest (2006) not deliver this as a law, they rule it out by accepting the principle

(#) $A \triangleright B \models A \rightarrow B$

together with (1_r). For by defining predicates F and G in terms of sentences A and B as above, we see that the lawhood of (1) requires the lawhood of

$[(A \rightarrow B) \wedge A] \triangleright B$;

and (#) would then require the lawhood of

$[(A \rightarrow B) \wedge A] \rightarrow B$.

But that alleged law ("pseudo-modus ponens" for $\rightarrow$) is well known to be inconsistent with the genuine modus ponens for $\rightarrow$ (the rule $A \rightarrow B, A \models B$) that is required for (1_r), given anything close to naive truth. (That's a moral of the Curry paradox.) Of course, we could just insist that (1) shouldn't be a law, but I see no reason why a dialetheist should take this route: there seems to be little prior motivation for (#), despite it having been built into the Beall et al. and Priest (2006) systems.

I also have worries about their versions of the quantifier-restrictor $\rightarrow$ alone, even apart from its interaction with 'if...then'. I'm not clear enough about the details of their logics to pursue this, but the general worry is that despite its "irrelevance", (i) it isn't clear that their $A \rightarrow B$ reduces to $\neg A \vee B$ in classical contexts (i.e. where contradictions are rejected), and (ii) that if it doesn't then even in such contexts it isn't suitable as a device for restricting quantification.

Let's get back to the laws of how the quantifier-restrictor interacts with the ordinary conditional. The Beall et al. and Priest (2006) systems deliver very few of the laws mentioned in the previous section that are delivered by the paracomplete systems. They do deliver (4a), but except on versions where their relevant conditional (unlike the ordinary English conditional) is contraposable[21] they don't deliver some natural variants of it that hold in mine, including (4b) and

---

[21] Interestingly, the joint paper seems to suggest that its relevant conditional *should* be contraposable, though somewhat equivocally: see p. 595 middle.

(4a*) If not all A are both B and C then either not all A are B or not all A are C.

Their systems also deliver (2) (though not of course (2_c)), but only because of their acceptance of (#); if one modified the systems to drop (#), as is required to have a chance at (1), then their case for (2) collapses.[22] But as well as not delivering (1), their system doesn't deliver (3), (5) or (6) either.

I think these failures are due to basic features of the relevant conditional ▷ that they employ:

(i) the conditional is strict as opposed to variably strict: that is, $X \triangleright Y$ is true at a normal world w only if at *all* worlds x accessible from w, even non-normal ones, X is true only if Y is true;
(ii) for every normal world there are non-normal worlds accessible from it; and
(iii) at non-normal worlds, conditionals (including restricted quantifier conditionals) are very badly behaved.

Together, these make it very hard for claims of form $X \triangleright Y$ where X and/or Y contain a restricted quantifier to come out true.[23] (The system in the previous section, by contrast, didn't have feature (i) or require feature (ii); indeed it didn't commit to the existence of any non-normal worlds.)

6.4.2 The Irrelevance of Relevance

So far I have deferred consideration of a basic oddity in Priest's work on systems of conditionals that can be used for the paradoxes. That work has mostly been in terms of a "relevant" conditional in the vicinity of the relevant logic **B** (combined with an "irrelevant" one to restrict quantification); and as we've seen, the rationale for this seems to be that such a logic is at least a pretty good approximation to the ordinary 'if...then'.

The oddity is that quite independent of the paradoxes, that logic is clearly *not* a good candidate for the ordinary 'if...then'. For instance, in the relevant logic **B** and all similar systems, one of the validities is Antecedent Strengthening:

(AS) $A \triangleright C \models (A \wedge B) \triangleright C$.

And yet one of the most well-known features of the ordinary conditional is that this fails: from "If I get a reservation I'll eat there tonight" it doesn't follow that "If

---

[22] Without relying on (#), there's a way to modify the account of → to get (2) by brute force: see Beall (2009, pp. 125–6). But it has the serious disadvantage of invalidating (4a), and indeed even the rule form of that.

[23] Restricted quantifier conditionals are slightly better behaved than ordinary ones at non-normal worlds, which is what allows for (2) and (4).

The use of the "Routley *" to handle negation in the Beall et al. paper creates further difficulties for achieving acceptable laws.

I get a reservation and die immediately after doing so I'll eat there tonight".[24] So Priest's critique of Field (2008), that its conditional isn't the ordinary one, applies equally well the conditionals in the vicinity of **B** that he has used in his work on the paradoxes.

In his general work on conditionals Priest has proposed a variably strict semantics to model the ordinary English conditional; it isn't based on Burgess semantics (even with impossible worlds), but is in the same ballpark. I think Priest would do much better to make this conditional rather than relevant conditionals in the vicinity of **B** the focus of work on the paradoxes. And also, he'd do better to adopt a quantifier-restricting conditional that reduces to the material '⊃' in classical contexts, since any failure to so reduce would seem to show that restricted quantification doesn't behave as it should.[25]

I haven't explored the possibility of carrying out such a proposal in a dialetheic context. As I remarked at the end of Sect. 6.3, one couldn't get an acceptable dialetheic logic by simply taking over the model-theoretic construction I've given, but perhaps there are alternative approaches in more or less the same spirit that would be suitable for the dialetheist. I don't see any obvious reason why something like this couldn't be done, and in a way that yields laws far more satisfactory than those in Priest (2006) or the joint paper with Beall et al. Of course, this suggestion couldn't overcome the in-principle limitation for dialetheic treatments of restricted quantification that I noted above (the inability to have both (1_r) and $(5*_r)$), but I have acknowledged that not everyone finds that limitation devastating.

6.5 Modus Ponens, Conditional Introduction, and the Curry Paradox

Except in *radically* non-classical logics that restrict the classical structural rules or the substitutivity of identity, Curry's Paradox shows that no operator ⋙ (primitive or defined) can obey both the first and second of the following laws in the presence of a truth predicate that obeys the third and fourth:

(Modus Ponens for ⋙) $A, A \ggg B \models B$
(Conditional Introduction for ⋙)   If $\Gamma, A \models B$ then $\Gamma \models A \ggg B$,

---

[24] Another law validated in **B** and similar systems, but that fails for the ordinary conditional, is Transitivity: $A \triangleright B, B \triangleright C \models A \triangleright C$. That failure is almost an immediate consequence of the failure of (AS): Transitivity gives $(A \wedge B) \triangleright B, B \triangleright C \models (A \wedge B) \triangleright C$, and since $\models (A \wedge B) \triangleright B$ we get (AS).

[25] I'm not necessarily claiming that relevant conditionals are of no interest in connection with work on the paradoxes, only that they are of no use either as "ordinary" conditionals or for restricted quantification. Weber (2018) grants that they are of no use for these purposes, but proposes that they may nonetheless be of use in a third role, that of providing identity conditions for properties. I have a mixed reaction to this (see Field 2018b), but need take no stand here.

(T-Elim) $True(\langle A \rangle) \models A$
(Weak T-Introd)   If $\models A$ then $\models True(\langle A \rangle)$.

(Curry's paradox for $\ggg$ involves a sentence K equivalent to $True(\langle K \rangle) \ggg \bot$ where $\bot$ is some absurdity.) The two truth rules here are very weak, far weaker than demanded by naive truth or even the quasi-naivety that Priest advocates. For instance, many theories of truth that are broadly classical, in that they accept all classically valid inferences, accept both truth rules. Such theories avoid the Curry paradox (for the material $\supset$) by rejecting the classical *metarule* of $\supset$-Introduction.[26]

Moreover, if we interpret $\models$ broadly, so that $\models A$ when A follows from arithmetic, then even the special form of Conditional Introduction where Γ is empty is enough for triviality in the presence of Modus Ponens and the two truth rules. (The arithmetic you need is simply an instance of the diagonalization lemma.)

Since Priest and I accept the truth rules, we're forced to restrict either Modus Ponens or (even the special form of) Conditional Introduction, for each conditional, primitive or defined. (Of course, we can restrict Modus Ponens for some and Conditional Introduction for others.) And he and I tend to make the same choice for the conditionals we take as primitive: for both $\to$ and $\triangleright$ I accept Modus Ponens without restriction, but impose restrictions on Conditional Introduction; and Priest does the same for his analogs of these.[27]

I should emphasize that *restricting* Conditional Introduction means only that: for both Priest and me, there are restricted forms of Conditional Introduction that we count as valid. For instance, I accept

If $\Gamma, A \models C$ then $\Gamma, A \vee \neg A \models A \to C$.

Also, there are stronger forms $\models_s$ of validity for which I accept Conditional Introduction even without an excluded middle assumption:

If $\Gamma, A \models_s C$ then $\Gamma \models_s A \to C$ and hence $\Gamma \models A \to C$.

---

[26]These broadly classical theories are forced to reject other classical metarules too, like reasoning by cases, the rule that if $\Gamma, A \models C$ and $\Gamma, B \models C$ then $\Gamma, A \vee B \models C$. This is forced on them because they accept excluded middle and explosion: *given excluded middle*, reasoning by cases leads to the rule

(*) If $\Gamma, A \models C$ and $\Gamma, \neg A \models C$ then $\Gamma \models C$;

and with even the minimal truth rules and Explosion, (*) cannot be accepted in face of the Liar Paradox. Indeed, independent of any conditional, there is an obvious conflict simply between (*), (T-Elim), and

(Alternative Weak T-Introd) $A, \neg True(\langle A \rangle) \models C$.

[27]Recently he has shown some sympathy with keeping both Conditional Introduction and a version of Modus Ponens, by going for a radically non-classical logic that restricts the structural contraction rule. (I say that he keeps *a version of* Modus Ponens because he treats $A, A \to B \models B$ not as equivalent to $A \wedge (A \to B) \models B$, which he rejects, but to the weaker $A \circ (A \to B) \models B$ where $\circ$ is a new connective ("fusion") that is substantially stronger than $\wedge$.) I will ignore this more recent turn in his thought.

(While many valid forms of inference are valid in the stronger senses, Modus Ponens isn't; that's how this last restriction on Conditional Introduction evades the Curry paradox.) The situation for Priest is rather similar, though there are differences in detail about which restrictive forms he and I can accept.

It's also worth noting that for both of us, the decision to restrict Conditional Introduction rather than Modus Ponens isn't obviously mandatory. Indeed, in my own case (and I imagine that something similar holds for Priest's), an obvious way to get a logic with Conditional Introduction *at least for empty* Γ (or more generally, for Γ that include excluded middle for other members of Γ) is to keep the basic semantics precisely the same, but choose the "S_3-like" definition of validity in terms of the partial ordering on V rather than the "K_3-like" definition in terms of the designated value **1**. Obviously, this would require restrictions on Modus Ponens. I'm inclined to think this a less satisfactory logic overall, but this is a judgment call; and there might be alterations that would lead to something better.

I'd also like to say something about a charge often leveled against both Priest and me: that there's an unappealing disunity between our treatments of the Curry Paradox and the Liar Paradox.

I think the claim of disunity is overstated, especially as applied to paracomplete solutions. A key point to remember is that the Curry argument can be applied to *any* conditional (or indeed any operator, whether or not it's deemed a "real conditional"). In particular, it applies to $\supset$, so for a naive theorist, we must give up either Modus Ponens for $\supset$ or $\supset$-Introduction. But $\supset$-Introduction is equivalent to excluded middle, given reasonable disjunction laws plus transitivity of entailment; similarly, Modus Ponens for $\supset$ is equivalent to explosion given reasonable disjunction laws plus strong transitivity of entailment, i.e. Cut. So for this conditional, there just is no difference between the solution to the Curry and the solution to the Liar.

This argument can be adapted to suggest that at least for the paracomplete theorist, the Curry Paradox is fundamental and the Liar derivative. Call a conditional $\ggg$ "weak" if $\neg A \lor B \models A \ggg B$, and "somewhat weak" if whenever $\neg A \lor B$ is valid, $A \ggg B$ is also valid. There's a persuasive argument, from the assumption of excluded middle, that weak conditionals should obey full Conditional Introduction and that somewhat weak ones should obey the special form of it with no side formulas Γ. [For suppose $\Gamma, A \models B$. Then reasonable disjunction laws yield $\Gamma, A \lor \neg A \models \neg A \lor B$, hence with excluded middle, $\Gamma \models \neg A \lor B$; hence by the weakness assumption, $\Gamma \models A \ggg B$. When Γ is $\emptyset$, the special weakness assumption suffices.] So if one avoids the Curry Paradox for a conditional one regards as weak [alternatively somewhat weak] by rejecting full Conditional Introduction [alternatively the special form of it] for that conditional, then one should go for a paracomplete logic.[28] And once one has a paracomplete logic, the Liar sentence and its negation can both be rejected, thus avoiding the Liar paradox.

This isn't an argument for a paracomplete account over a dialetheic, because dialetheists won't reject Conditional Introduction for conditionals they regard as weak:

---

[28] This assumes very minimal laws for disjunction, and that consequence is transitive.

(i) Trivially, ⊃ is weak, but it would beg the question against any theory built around LP to assume that ⊃-Introduction is to be rejected.

(ii) In my view, any adequate restricted quantifier conditional → is also weak; there is an easy argument to this from the assumption that $\forall x(\neg Ax \vee Bx)$ entails "All A are B".[29] But as we've seen, the dialetheist typically rejects that assumption. And even if the dialetheist were to keep the assumption, we've seen that he would need to reject modus ponens for that conditional; that would suffice for blocking the Curry paradox, so there would be no need to reject the standard introduction rule for that conditional.

(iii) In my view, the ordinary indicative is somewhat weak, if anarchic worlds can be ruled out of consideration; and in my view also, Conditional Introduction for it is to be rejected. But here too a typical dialetheist will probably reject the first, and if not could reject the second.

So there's no argument here against the dialetheist, but that wasn't the point.

The point rather was to rebut the charge that for a paracomplete theorist, there is a disunity between the resolutions of the Curry Paradox and of the Liar. My argument is that the resolution of the Curry leads, by natural assumptions, to the resolution of the Liar.

Can a dialetheist make a similar connection between the two paradoxes? Perhaps not directly, when the Curry is for a conditional that obeys Modus Ponens: for then it's essential to reject the corresponding Curry sentence, in contrast to how the dialetheist treats the Liar.[30] However, there is a variant form of Curry sentence: a sentence κ that "says of itself" that if it's true then it isn't true: κ is equivalent to $True(\langle\kappa\rangle) \ggg \neg True(\langle\kappa\rangle)$. It's natural to think that the resolution of the variant Curry should have much in common with the resolution of the standard Curry. And for the variant Curry, there's no obvious reason why the dialetheist can't treat it, like the Liar, as a dialetheia.

Indeed, we can use the variant Curry to argue from the assumption that Modus Ponens and the truth rules are to be preserved to the conclusion that *either excluded middle or explosion* needs restriction. Actually, we need an additional assumption, the Weakening rule $B \models A \ggg B$; because of this, the argument to follow can't be used for the ordinary indicative, but it can for any plausible versions of the restricted quantifier conditional. Here's a sketch of the argument:

By T-Elim, $True(\langle\kappa\rangle) \models \kappa$, so by choice of κ, $True(\langle\kappa\rangle) \models True(\langle\kappa\rangle) \ggg \neg True(\langle\kappa\rangle)$.
So by Modus Ponens (and structural contraction), $True(\langle\kappa\rangle) \models \neg True(\langle\kappa\rangle)$.

---

[29](a) The entailment is hard to explain unless the quantifier-restricting conditional is weak; indeed, (b) it can actually be shown to entail that it's weak, using the definition of → in terms of restricted quantification suggested in note 10.

[30]By contrast, if we restrict Modus Ponens for a conditional $\ggg$ that is strong in the sense that $A \ggg B \models \neg A \vee B$, then we must restrict Explosion. [For Explosion gives $A, \neg A \models B$, which easily yields $A, \neg A \vee B \models B$, which by strongness assumption yields $A, A \ggg B \models B$, i.e. Modus Ponens.] So resolving the $\ggg$-Curry for a strong conditional by restricting Modus Ponens would make the standard paraconsistent resolution of the Liar very natural. This however doesn't provide much of an argument for the unity of the two paradoxes for a typical dialetheist: the only conditional that the typical dialetheist holds to be both strong and to fail Modus Ponens is the ⊃, and for it the resolution of the Curry Paradox *just is* the resolution of the Liar.

With structural reflexivity and reasoning by cases, this yields $True(\langle \kappa \rangle) \vee \neg True(\langle \kappa \rangle) \models \neg True(\langle \kappa \rangle)$. So by $\ggg$-Weakening,

(a) $True(\langle \kappa \rangle) \vee \neg True(\langle \kappa \rangle) \models True(\kappa) \ggg \neg True(\langle \kappa \rangle)$.

That is, $True(\langle \kappa \rangle) \vee \neg True(\langle \kappa \rangle) \models \kappa$. So by a slightly stronger version of T-Introd than before,

(b) $True(\langle \kappa \rangle) \vee \neg True(\langle \kappa \rangle) \models True(\kappa)$.

So by (a) and (b) together with Modus Ponens, $True(\langle \kappa \rangle) \vee \neg True(\langle \kappa \rangle) \models True(\kappa) \wedge \neg True(\langle \kappa \rangle)$;

so we must restrict either Excluded Middle or Explosion to avoid Post-inconsistency.

Again, a Curry-like argument pretty much forces *either* a dialetheic *or* a paracomplete logic, and either one resolves the Liar Paradox.

There are more connections that could be shown, but I think this is enough to argue that the charge sometimes leveled both against Priest and me that there is a disconnect between our treatments of the Curry Paradox and the Liar Paradox is dubious. We can be viewed as going our separate ways on the Liar Paradox because of different choices on how to resolve certain versions of the Curry or variants of it.

6.6 Naivety Versus Quasi-naivety

A small point on which Priest and I have diverged is that I prefer a fully naive theory of truth while he prefers one that is only quasi-naive: his disallows the inferential rule

($\neg$**T-Introd**) $\neg A \models \neg True(\langle A \rangle)$,

and so of course it also disallows the validity of the conditionals $\neg A \triangleright \neg True(\langle A \rangle)$ and $\neg A \rightarrow \neg True(\langle A \rangle)$. (It does allow $A \leftrightarrow True(\langle A \rangle)$ and $\neg True(\langle A \rangle) \rightarrow \neg A$, and the analogs using $\triangleright$.) So while taking a dialetheic stance to the "falsity-Liar" $Q*$ that asserts its own falsity (i.e. the truth of its own negation), he is not forced to simultaneously take a dialetheic stance to the question of its *truth*: he needn't say that $Q*$ is not true as well as being true. Exactly why he should want to resist saying that it is both true and not true is unclear to me. And any apparent advantage would seem to be undercut by two facts. First, he still needs to take a dialetheic attitude to the *falsity* of $Q*$ (so he must say that it is true and not false, as well as true and false). Second, he still must take a dialetheic attitude to the truth of the "untruth Liar" Q (and to its falsity as well), where being untrue is just not being true. From a naive point of view, taking a stance on $Q*$ that differs from the one that he's required to take on Q is excess complication: there's no reason to distinguish between being false (having a true negation) and not being true.[31]

In addition to my doubts about the motivation for quasi-naivety, I initially wondered how Priest squares it with his proofs of the nontriviality of his logic of truth:

---

[31] The dual line for the paracomplete theorist would be to reject the rule

($\neg$**T-Elim**) $\neg True(\langle A \rangle) \models \neg A$

these all work by constructions on which $True(\langle A \rangle)$ gets the same value as A in every world in every model, guaranteeing full naivety. An unsatisfactory response is that while these proofs directly show the nontriviality of an extended logic with ($\neg$T-Introd) (and indeed with $\neg True(\langle A \rangle) \leftrightarrow \neg A$), it follows that the weaker logic without that is also nontrivial. The reason that that's unsatisfactory is that Priest wants to assert more than that ($\neg$T-Introd) isn't valid: he wants to assert that it isn't valid *because there are (or might well be) cases where it's reasonable to accept* $\neg A$ *but reject* $\neg True(\langle A \rangle)$. And my puzzle was that the nontriviality proofs go by a construction that doesn't allow for treating $\neg A$ and $\neg True(\langle A \rangle)$ differently.

But note 37 of Priest (2010) points out that it is easy to modify the construction so as to allow failures of ($\neg$T-Introd) or its conditional strengthening *for sentences in the ground language*; all the construction requires is that we get all instances of $\neg A \rightarrow \neg True(\langle A \rangle)$ *for which A contains 'True'*. He's right, and as a result, the construction allows that we might not have full naivety for dialetheia *that don't arise from the notion of truth or related notions like satisfaction*. That wouldn't cover the case of $Q*$, which I took to be central to his motivation, but perhaps I was wrong to so take it. If it is Priest's position that naivety fails only for certain sentences in the ground language, then I accept his footnote as an answer to my puzzlement.

I still see three advantages of full naivety over the quasi-naivety that Priest prefers, though I don't want to make a big deal of any of them. The first is simplicity: I've already noted that the failure of ($\neg$T-Introd) would complicate the logic of truth without obvious benefit, but I should add that it raises further issues on which I'm not completely sure how Priest would answer. If ($\neg$T-Introd) fails, then calling a sentence untrue is in effect negating it in a stronger than usual way. One question is whether this "strengthening of negation" stops at the first step, i.e. whether $\neg True(\langle A \rangle) \models \neg True(\langle True(\langle A \rangle) \rangle)$; and if it doesn't stop there, whether it nontrivially iterates indefinitely.

A second advantage of full naivety is that without it there are limits on the use of 'True' as a device of generalization. "If not everything Joe said was true then something Sara said was true" seems as if it should be equivalent, modulo the assumption that Joe said $A_1, \ldots, A_n$ and nothing else and that Sara said $B_1, \ldots, B_m$ and nothing else, to "If $\neg A_1 \vee \cdots \vee \neg A_n$ then $B_1 \vee \cdots \vee B_m$". But on Priest's view that equivalence will fail, since the $\neg A_i$ won't imply the corresponding $\neg True(\langle A_i \rangle)$. Priest has however pointed out (2010, p. 130), that on his view, while 'True' by itself doesn't work for such generalizations, an artful combination of 'True' and 'False' does. I won't pursue the issue further here.

A smaller advantage of full naivety is that it offers an easy answer to the frequently heard charge that non-classical theories of truth are pointless since they save the truth schema in name only. The charge is that the connective '$\leftrightarrow$' in the non-classical logician's preferred version of "$True(\langle A \rangle) \leftrightarrow A$" is some contrived connective,

---

while accepting $A \leftrightarrow True(\langle A \rangle)$ plus $\neg A \rightarrow \neg True(\langle A \rangle)$. This would allow us to accept the untruth of the falsity-Liar $Q*$, though we'd still have to reject both its falsity and its non-falsity (so it would still be inappropriate to call it a "truth-value gap"). And for the untruth Liar, we'd still need to take a paracomplete attitude to its truth as well as to its falsity. Again this seems like excess complication.

far from what motivates the idea that $True(\langle A\rangle)$ should be equivalent to A. The naive theorist has an easy answer: the equivalence between $True(\langle A\rangle)$ and A is best spelled out by the claim that intersubstitution of one for the other (except inside quotation marks, attitude reports, and so on) should *preserve inferential validity*. Such a preservation of inferential validity is guaranteed by the schema $True(\langle A\rangle) \leftrightarrow A$, as long as the biconditional obeys certain laws, but if the biconditional is non-contraposable, we need to supplement the schema with "$\neg True(\langle A\rangle) \leftrightarrow \neg A$" to guarantee intersubstitutivity. Since Priest doesn't accept that supplementation, or intersubstitutivity, he can't give this easy answer to the charge that his "truth schema" is a pointless change of subject.

6.7 The Gödel "Paradox", Informal Provability, and Truth-Preservation

In the rest of the paper, I will largely be concerned with issues on which Priest thinks the dialetheic approach superior to the paracomplete. One concerns the Gödel incompleteness theorems.

Priest has been very insistent that Gödel's incompleteness theorems are genuinely paradoxical:

> In all honesty, Gödel's paradox is just as problematic for arithmetic as Russell's paradox is for set theory. (Priest 2010, p. 133)

And by this he seems to mean that the theorems give reason to believe that there are dialetheia *in the language of Peano Arithmetic*. In Priest (2006, p. 39) he says that there is a nonconstructive argument for dialetheia in arithmetic: the argument "does not produce dialetheias explicitly, but shows that there must be some." The claim is *not* that the Gödel sentence of PA (I'll assume conventions that make it unique, and designate it G_{PA}) has been shown to be dialetheic. Rather, I take him to reason as follows.

(i) He claims that we have a notion of informal proof that leads to contradiction, in licensing the conclusion that a sentence G^*_{PA} that asserts its own lack of informal provability is both informally provable and not informally provable. From this, he concludes (ii) that there are dialetheia involving the notion of informal provability. For the moment I will not contest (i) and (ii), though I'll come back to them. But from these, he goes on to argue that there are dialetheia *in the language of PA*. The argument (perhaps most explicit in Priest 2006, pp. 238–9) seems to be (iii) that this notion 'x is an informal proof of y' is effectively decidable; (iv) that therefore there is a predicate $IP(x, y)$ of Peano arithmetic that numeralwise expresses 'x is an informal proof of y' (though until we're clearer about informal proof we can't be sure which arithmetical predicate does this); and so (v) we can replace 'x is an informal proof of y' in G^*_{PA} by $IP(x, y)$, getting a dialetheic sentence G^{**}_{PA} in the language of PA. (It is because we don't know which arithmetic formula expresses informal proof that the argument for arithmetical dialetheia is nonconstructive.)

If this argument were correct, it would establish not merely that there are arithmetical dialetheia, but that there are arithmetical dialetheia of a very simple form: among the Σ_1^0 sentences. So even some of the theorems of extremely weak arithmetic theories like Robinson arithmetic would be dialetheia.[32]

This view (with or without the expansion in the previous footnote) strikes me as totally incredible. Of course one cannot argue against it without premises that someone who believed its conclusion is likely to regard as question-begging, but one can counter arguments offered in its favor. And even putting aside Priest's argument (steps (i) and (ii)) that there are dialetheia involving the notion of informal provability, the argument in steps (iii)–(v) that that would lead to dialetheia *within the language of arithmetic itself* just seems incorrect. We might equally argue directly from the Liar paradox to the conclusion that there are dialetheia within the language of PA, by considering the predicate $P(x)$ defined as 'x is $\langle Q \rangle$ and is true', where Q is a particular sentence asserting its own untruth and $\langle Q \rangle$ is a Gödel code of it in arithmetic:

> $P(x)$ has finite extension, so it's effectively decidable, so it's numeralwise expressible in arithmetic. Indeed, given the dialetheic resolution of the Liar paradox, it has both $\{\langle Q \rangle\}$ and $\emptyset$ as its extension, so it's numeralwise expressible *both* by the predicate '$x = \langle Q \rangle$' and by the predicate '$x \neq x$'. The first implies that if $\neg P(\langle Q \rangle)$ then $\langle Q \rangle \neq \langle Q \rangle$, and the second that if $P(\langle Q \rangle)$ then $\langle Q \rangle \neq \langle Q \rangle$; since both $P(\langle Q \rangle)$ and $\neg P(\langle Q \rangle)$, this establishes twice over that the claim that the Liar sentence is self-identical is dialetheic! (In fact, numeralwise expressibility within Robinson arithmetic would normally be taken to lead to the conclusion that $\langle Q \rangle \neq \langle Q \rangle$ is provable in that theory, so that that theory is inconsistent!)

I hope we all agree that even without the parenthetical addition, this argument is ludicrous, and I don't see how the argument to (v) given (ii) is any different. There's clearly something wrong with trying to convert dialetheia outside arithmetic to dialetheia within it via the notion of numeralwise expressibility, though the exact diagnosis of how the argument fails depends on a decision about how to extend the terms 'decidable' and 'numeralwise expressible' to a dialetheic context.

There is still the interesting question of steps (i) and (ii), the paradox involving G^*_{PA} rather than G^{**}_{PA}: that is, the paradox of informal proof itself, before Priest's argument that it can be arithmetized. I'm willing to accept that there is a pre-theoretic paradox here (maybe more than one), just as there are pre-theoretic Liar and Curry paradoxes the resolution of which requires giving up some natural assumptions about either truth or logic. If so, then we may need to weaken natural assumptions about informal proof as well as about truth and logic to avoid Post-inconsistency. I'm inclined to think that the weakenings needed to tame the truth paradoxes are enough: that the paradoxes of informal provability are basically just paradoxes of truth. Perhaps Priest would agree? Certainly, the argument for (i) and (ii) as Priest goes on to present it (2006, pp. 48–50) makes ineliminable use of the notion of truth, via the assumption that everything informally provable is true. If that is just a paradox of

---

[32] A further conclusion may not be far behind: that such extremely weak theories are deductively inconsistent. The idea would be to use the numeralwise expressibility claimed in (iv) to run an analog of the informal proof claimed in (i) within the weak arithmetic. I don't know whether Priest meant to endorse that further step, though pp. 238–9 of Priest (2006) are naturally read that way.

truth, then it's hard to see why it isn't as resolvable by paracomplete means as by dialetheic.

A full discussion of this would involve a discussion of how 'informally provable' is to be understood. A simple-minded approach, which I assume to be inadequate, would be to simply define 'informally provable' as 'both provable in T and true', where T is a formal system that captures our standards of informal proof. (If that were adequate, then the paradox of informal provability would certainly just be another paradox of truth that is easily handled by paracomplete means.) I'll sketch a less simple-minded approach to informal provability in a footnote (one that is in itself neutral between the paracomplete and the dialetheic).[33] But even without that, one can look at Priest's presentation of the paradox of informal provability (2006, pp. 48–50) and see whether that presentation supports the claim that the paradox is best resolved dialetheically.[34]

---

[33] One natural account of informal provability (which I think fits well with Priest's discussion, e.g. in making the usual Gödel sentence G_{PA} informally provable in PA even though not formally so) is that for a sentence to be informally provable from PA is for it to be formally provable in an appropriate expansion PA^T of PA which includes a theory of truth/satisfaction or of properties. Then not only the argument that informal provability leads to paradox, but the details of what is informally provable, will depend on the truth/satisfaction/property theory. In a theory with an unstratified truth predicate one might proceed roughly as follows:

- For each α, PA_α is PA with an added predicate 'True' and composition laws, and induction extended to include it, plus some formalization of "for all $\beta < \alpha$, every theorem of PA_β is true".

The proper formalization of this requires a system of ordinal notations, so the PA_α will be defined only for α less than some countable ordinal Γ (presumably a limit ordinal). [One natural candidate is the Feferman-Schütte ordinal Γ_0 (Feferman 1962), which will be the first one for which the ordering relation among notations for ordinals less than it can't be shown to be a well ordering by reasoning within $\cup\{PA_\beta : \beta < \alpha\}$. There may be arguments for an even earlier stopping point.] Then one can either

(i) take our informal theory to be $\cup\{PA_\alpha : \alpha < \Gamma\}$, or

(ii) take "our informal theory" to be vague, and its precisifications stratified, via an increasing sequence $\{\alpha_\xi : \xi < \Gamma\}$ of increasingly complex limit ordinals less than Γ.

(I'm inclined to think that (ii) is the better course.) In case (i), the informal theory we employ goes beyond the bounds of anything we can *recognize as* acceptable (though we may be able to recognize each of the fragments PA_{α_ξ} as acceptable); so have no reason to believe a Gödel sentence for the entire informal theory. In case (ii), the Gödel sentence involving the vague predicate 'informally provable' is itself vague, and its precisification involving the notion of informal provability$_{\xi_1}$ is only informally provable$_{\xi_2}$ when $\xi_2 > \xi_1$. In both cases, obvious arguments to paradox are blocked. And we'll be able to show that the usual Gödel sentence G_{PA} of PA is informally provable; as is the usual Gödel sentence of PA $+ G_{PA}$; and so on for a substantial transfinite sequence of iterated Gödel sentences.

[34] As remarked above, we must modify his presentation slightly to avoid the claim that the paradoxical argument is in PA. Rather, we must view the argument as done in a theory PA* that expands PA to include a truth or satisfaction predicate (or an ontology of properties and a property-instantiation predicate), and perhaps an informal provability predicate too though that might be definable using truth/satisfaction/instantiation. The new predicates are allowed in instances of the induction schema.

And the key to the argument for (i) and (ii) as Priest presents it is that it is part of our informal proof procedure to both recognize that all of its basic assumptions are true and to recognize that its rules are truth-preserving. From these, we can inductively establish the soundness claim, that everything informally provable is true. Using soundness, it is then easy to give an informal proof of the sentence G^*_{PA} that states its own informal unprovability. And from this informal proof, we can construct a proof of its own informal *provability*, i.e. of $\neg G^*_{PA}$. If this were all correct (and didn't depend on excluded middle), the only resolution would indeed be by a paraconsistent logic.

But is there really a good argument for the soundness of our entire informal theory? We needn't dwell here on the standard worry, of the gap between recognizing of each basic assumption that it is true and recognizing that all of them are true. For the fundamental problem with the argument for soundness lies in the assumption that the inferential rules are truth-preserving *even as applied to sentences that contain 'True' (and/or 'informally provable')*. Here we need to consider both (a) the rules of the basic logic and (b) the rules governing the notion of truth. (Also (c) any rules governing the notion of informal provability, but on the reductionist account suggested in note 33 there won't be any primitive rules regarding that, so I'll leave (c) aside.) In attempts to run the argument for soundness with a *classical Tarskian* truth predicate, it is the rules in (b) that are the issue: they can be shown to preserve truth as applied to sentences in the language of arithmetic or set theory without a truth predicate, but not throughout the whole language. In *naive truth theories* on the other hand, the rules in (b) are easily shown to preserve truth everywhere (when truth preservation is formulated in terms of a "genuine conditional" as below), and the rules in (a) are easily shown to preserve truth for sentences in the ground language, but the basic logical rules employed in the theory can't all be shown to preserve truth more generally. (The last part holds in *quasi-naive* theories as well.) Let's look at this non-classical case more closely.

In this non-classical context, there is a question about how the truth-preservation claim for a given rule of inference is to be formulated. Consider

(RP$_\supset$) $\forall \Gamma \forall y[\langle \Gamma, y \rangle$ is an instance of the rule and $(\forall x \in \Gamma)(True(x) \supset True(y))]$.

If that's what the claim that our rules "preserve truth" is taken to mean, then no paracomplete naive truth theorist can take the claim to be true. Indeed, when we reject excluded middle for some sentences of form $True(\langle A \rangle)$, we must reject that even the rule A/A is truth-preserving in the sense of (RP$_\supset$): for $True(\langle A \rangle) \supset True(\langle A \rangle)$ just amounts to $\neg True(\langle A \rangle) \vee True(\langle A \rangle)$.

Is the situation any better for the dialetheic theorist? Well, such a theorist can accept (RP$_\supset$). But the problem is that it won't be useful: because of the failure of Modus Ponens for $\supset$, (RP$_\supset$) won't allow us to infer from the truth of the premises of the application of a rule that its conclusion is true. For instance, suppose the rule under consideration is Modus Ponens for $\supset$, and consider the application $Q \supset \bot, Q \models \bot$ where Q abbreviates a Liar sentence and $\bot$ an absurdity. In the dialetheist's theory of truth, $True(\langle Q \supset \bot \rangle)$ and $True(\langle Q \rangle)$ both have value $\frac{1}{2}$ and $True(\langle \bot \rangle)$ has value 0, so $True(\langle Q \supset \bot \rangle) \wedge True(\langle Q \rangle) \supset True(\langle \bot \rangle)$ has value $\frac{1}{2}$, which is designated.

6 Paraconsistent or Paracomplete?

So according to the dialetheist, Modus Ponens for $\supset$ is "truth-preserving" by the standard (RP$_\supset$); and yet the dialetheist also holds that Modus Ponens for $\supset$ is a thoroughly unacceptable rule, in that accepting it would lead directly to Explosion.

The only reasonable ways to formulate truth-preservation of a rule, in either a paracomplete or a dialetheic theory, is to use one of the "genuine conditionals" of the theory in the formulation:

(RP$_{\ggg}$) $\forall \Gamma \forall y [[\langle \Gamma, y \rangle$ is an instance of the rule and $(\forall x \in \Gamma)(True(x))] \ggg True(y)]$.

And this conditional $\ggg$ needs to obey Modus Ponens, to avoid the sort of trivialization we've seen for (RP$_\supset$) in dialetheic theories. The good news is that both paracomplete theories and dialetheic theories have such conditionals with which to formulate truth-preservation. The bad news is that with any such conditional, (RP$_{\ggg}$) will fail for at least one rule that the theory takes to be valid. That's because as applied to Modus Ponens for $\ggg$, which we've assumed valid, (RP$_{\ggg}$) yields

(##) $[True(\langle A \ggg B \rangle) \wedge True(\langle A \rangle)] \ggg True(\langle B \rangle)$,

which together with Modus Ponens easily allows the proof of anything by a Curry argument, given anything close to naive assumptions about truth. (Given such assumptions, (##) is obviously equivalent to "Pseudo-Modus Ponens" $[(A \ggg B) \wedge A] \ggg B$, which results from the application of Conditional Introduction to Modus Ponens in just the way required for a Curry argument.)

From the foregoing it is evident that an argument for the truth of a Gödel sentence in a system that enriches PA or ZF with a naive truth predicate breaks down in a similar way in paracomplete and dialetheic logics: the apparent proof requires a truth-preservation assumption (RP$_{\ggg}$) *for a conditional $\ggg$ that obeys Modus Ponens*, and yet one can't have that (on minimally naive assumptions about truth) without generating Post-inconsistency (triviality) by a Curry argument. Priest's discussions sometimes give the impression that in a well-motivated paraconsistent dialetheic system one will be able to derive both the claim that its Gödel sentence is true and that it isn't, and that the paraconsistency will then come in to block the move from that contradiction to absurdities. But his case for that rests on the idea that we can combine Modus Ponens for $\ggg$ with (RP$_{\ggg}$), and as we've seen, that leads directly to absurdities by a Curry argument. Nontrivial dialetheic theories, like paracomplete ones, can't prove their own soundness.

In Priest (2010, p. 134) Priest suggests getting around this by adding the soundness claim as a separate axiom. And it's true that if one has a reasonable Post-consistent theory T (which of necessity can't prove its own soundness), one can add to it an axiom that says that everything provable in T is true. One can do this both when T is dialetheic and when it is paracomplete. But in neither case will the resultant theory T$^+$ prove *its own* soundness, it will only prove that of T. One can introduce a whole hierarchy of ever-stronger theories, going up to say the Feferman-Schütte ordinal; that's the PA$_\alpha$ construction sketched in note 33. But Priest's idea of an unstratified theory that proves its own soundness, without triviality, is as illusory in the dialetheic case as in the paracomplete. Of course, a good theory, whether paracomplete or dialetheic, won't declare itself *unsound*, but Priest's argument that the paradox must be resolved dialetheically requires that it declare itself sound, and that is something that even he cannot achieve without triviality.

6.8 Property Theory, Set Theory, and Model Theory

If one is attracted to a "naive" theory of truth and satisfaction, it is tempting to also go for a "naive" theory of properties: one where "properties are just shadows of predicates", so that every (parameterized) formula $A(x; o_1, \ldots o_k)$ of a language stands for a property $\lambda x A(x; o_1, \ldots o_k)$, and the property $\lambda x A(x; o_1, \ldots o_k)$ is instantiated by all and only those x such that $A(x; o_1, \ldots o_k)$. Can we obtain such a theory? Yes. Given a naive theory of satisfaction of either a paracomplete or dialetheic sort, it is a trivial task to convert it into a naive theory of properties, if we are willing to adopt very fine-grained identity conditions for properties. It is slightly less trivial to do this for coarser-grained identity conditions, but it can be done (as long as the conditions aren't taken to be *too* coarse): see Field (2018a).[35]

It is also somewhat tempting to look for a naive theory of classes: just like a naive theory of properties but with an axiom of extensionality. (If one could get such a theory, it needn't be to the exclusion of standard set theory: standard set theory is a natural mathematical construction in its own right.) An apparent motivation for a naive theory of classes is that there is a well-known awkwardness in standard set theory: many formulations and proofs are made more natural if one talks informally in terms of proper classes that the theory doesn't recognize. And if one modifies the theory by recognizing them, it looks like one is simply adding on a new layer of sets, just postponing the awkwardness to one that can only be overcome by "superclasses". But if we had naive classes (in a non-classical logic) *as well as* standard sets (in a classical fragment of the non-classical language, a fragment within which excluded middle and explosion don't need restriction), we could use them in place of classical proper classes: there would then be proper classes with proper classes as members, so there would be no need for a further level of superclasses. That certainly has appeal. (Though we could get something of the same benefit by the use of naive *properties* in place of classical proper classes, and a theory of naive properties is much easier to obtain.)

Can we get an adequate naive theory of classes? Ross Brady's early papers (e.g. Brady 1983) were notable for delivering simultaneously a naive theory of truth/satisfaction and what is in some sense a naive theory of classes. There were several different such theories, employing slightly different background logics, but each employed a special conditional. As I've already said, these Brady conditionals are clearly unsuited as the conditional for restricted quantification (as Brady himself evidently agrees, given his participation in the joint paper (Beall et al. 2006) which

---

[35] Of course, the notion of property is loose enough that there is no one right answer to the question of their identity conditions.

calls for a different conditional for this purpose); and they seem to me unsuitable as "ordinary" conditionals as well, for reasons given in Sect. 6.4. Nonetheless, it seems at first blush like a major advantage of his conditionals that they can be used for not just naive properties but naive classes, i.e. like naive properties but with extensionality.

In fact, though, the appearance of delivering extensionality is deceiving. It is true that Brady's construction delivers a rule that looks like an extensionality rule:

(Brady Extensionality Rule) $Class(a) \land Class(b) \land \forall u(u \in a \lll \ggg u \in b) \models \forall z(a \in z \lll \ggg b \in z)$.

Indeed, in some versions, it even delivers the conditional form

(Strong Brady Extensionality) $\models \forall x \forall y[[Class(x) \land Class(y) \land \forall u(u \in x \lll \ggg u \in y)] \ggg \forall z(x \in z \lll \ggg y \in z)]$.

But the failure of his $\ggg$ to be suitable for restricted quantification negatively affects the significance of this. In particular, the absence of the Weakening Rule $B \models A \ggg B$ means that we can't infer things from Brady extensionality that we'd expect to. For instance, we can't infer that there's at most one class with no members, or that there's at most one class in which everything is a member, or at most one in which the members are precisely the natural numbers. Indeed, we can easily construct counterexamples.[36]

Priest (2006, Chap. 18), Weber (2010) and others have considered other versions of naive theories of classes—in their cases, in a dialetheic framework—with the hope of overcoming this limitation by using conditionals that are more suited to restricted quantification. But the conditionals that they employ still diverge from those appropriate to restricted quantification, and they haven't managed to overcome the problem of "multiple empty sets". (The theories are also quite weak, when they don't turn out trivial.)

Harvey Lederman and I obtained some minor extensions of Brady's results (the best of which was in a logic that not only is paracomplete but also admits some dialetheias), but none which avoid the above problem. And Tore Fjetland Øgaard then proved a result that pretty rules out the prospects of doing much better: he showed that on any conditional meeting *very* minimal laws that seem required for extensionality to "mean what it ought to", the obvious formulation of a naive theory of classes is trivial *in either a paracomplete or paraconsistent logic*.[37] (See Field et al. 2017.)

---

[36]For instance, let U be $\{x : x = x\}$ and V be $\{x : \neg(x = x \ggg \neg(x = x))\}$. In the Brady theory, we can prove that everything is in U and everything is in V. (And no dialetheia are involved: we can reject the negation of these claims.) Nonetheless, we can define sets W for which we can prove that $U \in W$ and $V \notin W$ (and these claims aren't dialetheic either); for instance, where $W = \{y : \forall z(z = z \ggg z \in y)\}$. This doesn't violate "extensionality" as he defines it because U and V aren't "coextensive" as he defines it, despite both being universal sets.

[37]By "the obvious formulation of a naive theory" I mean one that includes

Abstraction Schema $\forall u_1 \ldots \forall u_n \forall z[z \in \{x : A(x, u_1 \ldots u_n)\} \lll \ggg A(z, u_1 \ldots u_n)]$

Priest has argued (e.g. the final section of his 2007) that obtaining a naive theory of classes is really important, in order to achieve a satisfactory model theory for one's overall theory of the world. If he's right about this, then the previous paragraphs suggest that we're *all* in trouble, the dialetheist as much as everyone else. But is he right?

Model theory is standardly done as part of classical set theory. And a large part of its point is in proofs of consistency or relative consistency: if you can prove in classical ZFC that a classical first-order theory T has a model if ZFC does, then (given the soundness theorem for classical logic) it follows in classical ZFC that T is deductively consistent if ZFC is. And this extends to theories in non-classical logics, provided that a soundness theorem for those logics can be established in ZFC: e.g. a soundness proof in ZFC for theories in a given axiomatic paracomplete logic of truth shows the deductive consistency of those theories in that logic. Similarly for paraconsistent logics, provided that 'consistency' is read as Post-consistency, i.e. nontriviality. Such proofs have clear value: they prove, to anyone who accepts classical set theory, that these paracomplete or paraconsistent logics of truth succeed at avoiding paradox.

Priest thinks that this far from exhausts the value of models: in particular, he thinks that we need an *intended* model of our overall theory, in an account of the meaning of our logical terms. This strikes me as far from obvious. Let us put aside the question of whether one ultimately needs anything more than an inferentialist account of our logical terms to explain their meaning. That is, let's grant, for the sake of argument, that an account of the meaning of logical terms requires notions like truth and satisfaction, perhaps combined with modal notions so that we can speak of the *conditions under which* a sentence would be true or a predicate satisfied. But to concede that is not to concede the need of a notion of models in the theory of meaning: for without talk of models, we already have a compositional theory of naive truth and satisfaction in a language with modality. (Adding modality to the languages that deal with naive truth for the Burgess conditionals is trivial.)[38] What I'm unclear about is what of importance one would get from an intended model

---

(for the same $\ggg$ as used in Extensionality), where we allow as instances of the schema formulas that themselves include the abstraction operator. It turns out that Øgaard's proof depends on this formulation, rather than one which instead merely has

Comprehension Schema $\quad \forall u_1 \ldots \forall u_n \exists y \forall z [z \in y \lll \ggg A(z, u_1 \ldots u_n)]$,

and which takes that schema to apply only to formulas in its own language. (The Comprehension Schema allows us to introduce abstracts for formulas not containing the abstraction operator, but by the italicized clause, one can't then assume that Comprehension applies to all formulas containing such abstracts.) Nothing in our paper *rules out* that there be a naive class theory with "genuine extensionality" that restricts itself to Comprehension rather than Abstraction. But nothing gives a whole lot of hope for one either, or suggests that the prospects for one are any better in a paraconsistent logic than in a paracomplete one. Moreover, I'm not sure what value Comprehension without Abstraction would have.

[38] It's important that the truth theory be naive: without that, the semantics would not be intuitively correct. And it's probably important that the account of truth and satisfaction be compositional, but as stressed earlier, one gets this automatically in the kinds of naive theory we've discussed.

that one wouldn't get from the compositional truth theory. The intended model has something that the compositional truth theory doesn't: an entity whose members are what the unrestricted quantifiers range over. But why is that important? Why not just say that these quantifiers range over everything, without there needing to be some entity of which everything is a member?

Moreover, even if one were to concede that we need some kind of notion of intended model, why think that the model be the sort of thing posited in a naive theory of classes as opposed to a naive theory of properties? Since any naive theory of truth and satisfaction automatically generates a naive theory of properties, there is no obvious difficulty in using such properties to get models, where there is a property attached to each predicate and the universal property serves as an entity describing the range of the unrestricted quantifiers. Indeed, I'd have thought that in an account of the meaning of a predicate it's more informative to ascribe a property to it than a set. And similarly, I'd have thought that in an account of universal unrestricted quantification it's more informative to invoke the universal property rather than the (alleged) universal class: after all, there might have been more things, or fewer, so the universal class doesn't get the modal features right.

In short, though I'm far from convinced that we need a notion of "intended model" for global theories whose quantifiers range over everything, there's no need to insist. If T is such a global theory, then an "intended model" for it given in a naive property theory would do as well or better than one given in a naive class theory, were a naive class theory possible.

And it's important to stress that whatever value a model theory for such a T given in a non-classical framework U might have, it would not serve all the purposes that a model theory for T given in classical set theory would serve. For in order to understand, and be convinced by, a proof in a non-classical theory U of the Post-consistency of T, you would already have to understand and accept the theory U. So if the task is to explain to an advocate of classical set theory what T comes to and to convince her of its Post-consistency, using a model theory that relies essentially on that same non-classical logic will not do the job. Again, it may or may not do other important jobs, but it won't do this one.

Of course, if an advocate of a non-classical theory offers an argument that her theory of truth does not induce Post-inconsistency (when applied to theories that are consistent in ω-logic), she wants her argument to be convincing *to herself* as well as to a classical logician. But this requires nothing extra, as long as her theory includes classical set theory, as my non-classical theory does. (That is, nothing extra is required as long as she (i) accepts the axioms of classical ZFC and (ii) accepts the rules of classical logic *when restricted to instances in the language of ZFC*, a language that does not include the truth predicate.) In that case, a proof of ω-conservativeness within classical set theory will also be a proof within the classical fragment of her non-classical theory. Such a proof, done in classical set theory, will convince the advocates of *any* theory (classical, paracomplete, or dialetheic) that accepts classical set theory as *an acceptable part of* mathematics.

It's important to stress that Priest himself quite properly uses classical constructions to show the coherence (ω-conservativeness) of naive truth in his own logics.

There is perhaps a special problem for him, raised by his suggestions (discussed in Sect. 6.7) that the set theory in which the constructions are done is inconsistent: in that case, there's a worry that he might be able to equally argue that naive truth is *not* coherent in his logic. Whether this is a serious worry for him, and whether it could be alleviated by a proof of ω-conservativeness entirely within his weaker logic without the full use of classical ZFC, are matters I will not discuss. They aren't issues for his dialetheism generally, but arise solely because of his view that the "Gödel paradox" is a paradox for our arithmetic and set-theoretic notions rather than just another paradox of truth.

6.9 Paradoxes of Denotation

Priest thinks that paracomplete approaches have special problems with paradoxes of denotation, but I think he is mistaken. 'Denotes' is easily defined in terms of 'true of': 't denotes x' means

> $\exists y(y=x$ and t is a closed singular term, and the identity sentence with t on the left and a variable on the right is true of y and nothing else).[39]

So any solution to the paradoxes of truth-of (or satisfaction) automatically yields a solution to the paradoxes of denotation.

Priest's claim, perhaps, is that there is an unsolved paradox even for truth-of, when the language contains a description operator. (Or a least number operator, but since that can be defined from a description operator, a treatment of the latter yields a treatment of the former.) But such an operator poses no special problem. Since descriptions needn't denote, the best treatment of a description operator is in the context of a free logic. In a reasonable such logic, we will have the following:

DD $\exists ! x A x \models A(\iota x A x)$.

Priest assumes the stronger

DD$_s$ $\exists x A x \models A(\iota x A x)$,

but this is highly dubious: "there are even natural numbers" is true, but not (I'd say) "the even natural number is even".

In a paradox like Berry's, the key step is getting from

(1) There are natural numbers not denoted by any term of length less than 10^6 characters in our current language (or rather, in a regimented version of our current language that excludes ambiguities and indexicals),

which is clearly true, to

(2) There is an N that is the smallest natural number not denoted by any term of length less than 10^6 characters in our current language.

---

[39] I'm taking 't denotes x' to be existence-entailing: no term denotes Santa Claus.

6 Paracomplete or Paracomplete?

Since if there is a smallest one it's unique, we can then use DD and a character count to get

(3) $\exists N[N$ is the smallest natural number not denoted by any term of length less than 10^6 characters in our current language, but N isn't denoted by 'the smallest natural number not denoted by any term of length less than 10^6 characters in our current language'].

And barring dialetheism, that would conflict with naive denotation.

But we block the paradox by blocking the move from (1) to (2). Even though the natural numbers are well ordered, the inference from $\exists n\, Fn$ to $\exists n[Fn \wedge (\forall m < n)\neg Fm]$ (where variables are restricted to the natural numbers) can't be expected to hold *without excluded middle for F*.[40] Indeed, that inference entails excluded middle for F. For let B be an arbitrary sentence, and Fn be

$$[n = 1] \vee [n = 0 \wedge B].$$

Then $F(1)$, so $\exists n\, Fn$. But what about $\exists n[Fn \wedge (\forall m < n)\neg Fm]$? Clearly, Fn entails $n = 1 \vee n = 0$, and $(\forall m < 1)\neg Fm$ entails $\neg F(0)$, so $\exists n[Fn \wedge (\forall m < n)\neg Fm]$ entails $\neg F0 \vee F0$, but that entails $\neg B \vee B$. Since B was arbitrary, we see that the move from (1) to (2) requires excluded middle. 'Denotes' is explained in terms of 'true of', so when F includes the word 'denotes' as applied to terms that include that predicate, excluded middle for it is suspect. This makes the inference from (1) to (2) suspect.

I'm not sure that Priest would contest that the step from (1) to (2) requires excluded middle: I don't see how he could, consistent with the semantics for $\exists$ that he himself accepts. Primarily he wants to bypass the inference from (1) to (2), by assuming that a proper logic of definite descriptions should include not just DD but DD_s. But as far as I can see, he gives no reason for that assumption; and as remarked above, insofar as condition DD_s strengthens DD it seems implausible.

Priest also argues that my own treatment of these paradoxes is undermined by the fact that I use classical logic in the model theory that I use to demonstrate the consistency of my logic. His point is that in each model M for my logic in the language L that includes 'denotes' and number theory, there will be a number n_M such that 'n_M is not 10^6-denoted by any term of L' gets designated value **1** in M but where for any smaller numeral m, 'm is not 10^6-denoted by any term of L' gets value less than **1**. So, he says, the description

(I) $\iota x(x$ is the smallest natural number not 10^6-denoted by any term of L)

ought to denote n_M. But this doesn't even make sense: there are many different models that fit the description of M; which one does he have in mind? Without specifying the particular model, he hasn't specified an n_M that the description ought to denote.

---

[40] All we can expect is the inference from $\exists n[Fn \wedge (\forall m < n)(Fm \vee \neg Fm)]$ to $\exists n[Fn \wedge (\forall m < n)\neg Fm]$. Of course, if F is a predicate for which excluded middle is assumed to hold generally, this yields the principle discussed in the text, for that F. The use of the least number principle within mathematics is not affected.

Besides not making sense, the claim is unmotivated: what we ought to expect is only that if we have a term *MOD* in the classical part of our language that denotes a specific model M_0, then

(II) $\iota x(x$ is the smallest natural number such that when its numeral is concatenated with the predicate 'not 10^6-denoted by any term of L', the result has value **1** in *MOD*)

should denote n_{M_0}.[41] And my account certainly gives *that*. But once we go to this model-theoretic context there is no Berry paradox: for there is no reason for supposing that the description (I) (which makes no mention of models) is co-denotational with the description (II) (which mentions a specific model M_0, and whose denotation will depend on the details of that model). Indeed, we shouldn't assume that the co-denotationality claim has a determinate answer, given that excluded middle for (I) can't be assumed.

Priest makes a suggestion I find interesting, about the Hilbert ϵ-operator. The *raison d'etre* for it in classical logic is the analog to Principle DD_s:

EPS$_{\epsilon,s}$ $\quad \exists x Ax \models A(\epsilon x Ax)$

And if we are attracted to this in a classical context, there is some *prima facie* plausibility in thinking that we could also have it in a non-classical context, without weakening it with an excluded middle assumption; that is, the full EPS$_{\epsilon,s}$, rather than merely

EPS$_\epsilon$ $\quad \exists x Ax, \forall x(Ax \vee \neg Ax) \models A(\epsilon x Ax)$.

But Priest points out that if we have such an operator in our language, and assume the full EPS$_{\epsilon,s}$, then we can run a Berry-like argument—not for 'the smallest number that isn't 10^6-definable in L' but for 'the Hilbert-chosen number that isn't 10^6-definable in L'. Point taken, but I don't think it's in the least clear that there's much cost to regarding an ϵ-operator (that obeys EPS$_{\epsilon,s}$) as illegitimate.

6.10 Epistemic Paradoxes and Paradoxes of Validity

There are epistemic paradoxes involving sentences like

- I don't (or shouldn't) accept this sentence
- I do (or should) reject this sentence
- I don't (or shouldn't) attach high credence to this sentence.

(See Caie 2012.) The paradoxes all turn on a variety of assumptions that can be doubted, e.g. they all assume perfect introspectibility of our epistemic states. I am nonetheless disinclined to completely dismiss them on the basis of the dubiousness

---

[41]This could be extended to terms *MOD* in the non-classical part of the language, though the formulation would require more care since we can't assume it determinate what such terms denote.

of these assumptions; I'm inclined to think, rather, that the sentences put pressure on the coherence of the epistemic notions that they employ, and should lead us to considerable refinement of these epistemic notions. Perhaps such a treatment can be given in a classical-logic framework; perhaps it requires non-classicality, either of a paracomplete or dialetheic sort or something else entirely. I don't at the moment see any reason to think that a dialetheic logic will prove better than a paracomplete one on this issue, but the fact is, I don't know. In any case, the issue is far too big to be treated in the present paper.

A notion that might or might not be regarded as epistemic, but in any case seems far more tractable than the epistemic notions above, is validity. There are paradoxes involving both 'valid' as a 1-place predicate of sentences and as a 2-place predicate of sentences (or a 1-place predicate of arguments):

I If Sentence I is valid then the Earth is flat.
II The argument from Sentence II to 'The Earth is flat' is valid.

The 1-place paradoxes, like I, *can* easily be handled by treating 'valid' along the lines of 'True': either as a paracomplete or a dialetheic predicate. Whether they *should* be handled in this way is debatable: there is some pressure, I think, to instead treat 'valid' as a classical predicate, in which case a different resolution of the paradoxes like I is required. (For instance, a treatment on which 'valid' behaves pretty much like 'provable' in one of the provability logics GL or GLS discussed in Boolos 1993.)

The 2-place validity paradoxes (like II) are more troublesome. One way to treat them is to take 'Valid(x,y)' to mean $\Box(True(x) \ggg True(y))$", where $\Box$ has a standard possible worlds semantics and $\ggg$ is one's favorite non-classical conditional. Since either a paracomplete or dialetheic semantics with a conditional can easily be extended to handle such a modal operator, one can then read a resolution of the 2-place validity paradoxes off from one's treatment of the truth paradoxes with conditionals. But again, it isn't obvious that this is the best way to resolve these paradoxes: I myself am inclined to favor one that treats the 2-place 'valid' as well as the 1-place as classical, and as behaving pretty much like 'provable from' in one of the provability logics GL or GLS. (Another alternative, for both 2-place and 1-place, is to treat 'valid' as stratified: this strikes me as far less objectionable than treating 'True' as stratified, for reasons I will not here discuss. One could also treat it as vague with stratified precisifications, along the lines of (ii) in note 33.)

Dialetheism seems to offer no special advantages with regard to the validity paradoxes. That's because, as Curry-like paradoxes, they purport to lead to absurd conclusions directly, rather than going via Explosion. In recent years there has been a move toward substructural resolutions of the validity paradoxes: in particular, toward resolving them by rejecting the rule of structural contraction. I'm highly skeptical of the need for any such resolution: see Field (2017). But there's no need to discuss that here, since substructuralism is neutral on the issue of paracomplete versus dialetheic.

6.11 Expressibility (1)

One of the main complaints that has been raised both against Priest's view and my own is that certain key notions that ought to be expressible according to these views come out inexpressible. I don't think either of us need to worry.

6.11.1 Dialetheism

In Priest's case what are undefinable are notions of "non-paradoxically true" and "non-paradoxically false" that *by stipulation* are to obey a certain adequacy condition framed in terms of acceptance and rejection. In the case of 'non-paradoxically true', the adequacy condition requires that it should be legitimate to accept "$\langle A \rangle$ is non-paradoxically true" exactly when it is legitimate to both accept "$\langle A \rangle$ is true" and reject "$\langle A \rangle$ is false". Given obvious truth rules (not requiring full naivety) together with the definition of falsity as truth of the negation, that's equivalent to saying that it should be legitimate to accept "$\langle A \rangle$ is non-paradoxically true" exactly when it is legitimate to both accept A and reject $\neg A$. And given Priest's acceptance of excluded middle, rejection of $\neg A$ requires acceptance of A, so there is a further simplification in the adequacy condition:

(C_T)   It should be legitimate to accept "$\langle A \rangle$ is non-paradoxically true" if and only if it is legitimate to reject $\neg A$.

For "non-paradoxically false" it's similar; indeed, equating a sentence being non-paradoxically false with its negation being non-paradoxically true (and assuming the redundancy of double negation), the adequacy condition (C_T) yields a corresponding one for "non-paradoxically false":

(C_F)   It should be legitimate to accept "$\langle A \rangle$ is non-paradoxically false" if and only if it is legitimate to reject A.

Prima facie, these (equivalent) "adequacy conditions" are natural: it would seem at first blush that a dialetheist must assert a sentence non-paradoxically true in order to preclude a hearer from thinking that while he believes the sentence, he also believes (or is agnostic about) its negation.

We can of course define '*solely* true' as 'true and not false', and '*solely* false' as 'false as not true'. But *sole* truth and *sole* falsehood don't satisfy the adequacy conditions that were stipulated for *non-paradoxical* truth and *non-paradoxical* falsehood. The reason is clear: it is the essence of an account like Priest's that the Liar sentence Q is both true and false, and also both not true and not false. So it is true and not false, hence solely true as that has been defined; it is also solely false.[42] But one can't

---

[42]On any *naive* dialetheic theory, sole truth is *equivalent* to truth, and sole falsity to falsity; so *any* dialetheia will also come out both solely true and solely false. With quasi-naivety this isn't quite so, but they are equivalent as regards the Liar sentence Q. (And even for the $Q*$ of Sect. 6.6, Priest's view entails that it is solely false as well as dialetheic.)

6 Paracomplete or Paracomplete?

possibly reject it or its negation, given that one accepts it. (Priest takes 'accept' and 'reject' to be non-dialetheic notions that are mutually exclusive.) So sole truth and sole falsehood won't satisfy the adequacy conditions stipulated for non-paradoxical truth and non-paradoxical falsehood.

And no other notion in Priest's theory will either, because (C_T) and (C_F) each would lead to serious problems within Priest's theory. Let's focus on (C_F). Given a notion of 'non-paradoxically false', we could construct a sentence W asserting its own non-paradoxical falsity, and then (C_F) would lead to embarrassing results.[43]

But the inexpressibility of (C_T) and (C_F) in Priest's theory is no cause for him to worry: his response ought to be (and I assume, is) that it isn't reasonable to demand of him that he come up with notions "non-paradoxically true" and "non-paradoxically false" that satisfy these "adequacy conditions", because the conditions are unreasonable.

Priest has pointed out that there is a way in his theory to go some way toward satisfying (C_F): define "non-paradoxical falsehood sub zero" by

$NPF_0(x): \quad True(x) \rightarrow \bot$,

where $\rightarrow$ is one of his modus-ponens-obeying conditionals. (The one used to restrict quantification is probably best, but it doesn't matter too much.) Like non-truth, this entails falsehood in Priest's logic (given excluded middle), but it's stronger than non-truth. And it satisfies the left to right of Condition (C_F), with the result that one cannot coherently assert any dialetheia to be NPF_0. (Perhaps we can even say of some dialetheia that they aren't NPF_0; that depends on the details of the conditional.) It also goes some way toward satisfying the right to left of (C_F), in that one can coherently assert 'The Earth is flat' to be NPF_0.

This however falls well short of the right to left of (C_F), because the dialetheist's treatment of typical Curry paradoxes is non-dialetheic. Consider the Curry sentence κ, equivalent to $True(\kappa) \rightarrow \bot$. On Priest's view, κ would have to count "non-paradoxically false" if there were a coherent such notion: we must reject its truth, since the assumption of its truth leads to $\bot$ via modus ponens. But here NPF_0 works *worse* than sole falsehood at capturing the intuitive idea of 'non-paradoxically false', for though we can assert that κ is solely false (false and not true), we can't assert that it is NPF_0: that amounts to asserting κ, and hence leads to contradiction. Whereas *sole* falsehood is too weak to satisfy Condition (C_F), NPF_0 is too strong.

The good news is that there is a way to handle this: we can say of κ that it is "non-paradoxically false sub 1":

$NPF_1(x): \quad [True(x) \rightarrow \bot] \vee [True(x) \rightarrow (True(x) \rightarrow \bot)]$.

---

[43] A preliminary argument that it is embarrassing: (C_F) would say of W that we should accept it if and only if we reject it. But the point of talk of rejection is to exclude acceptance: we can't accept and reject the same sentence (barring shift in meaning), as Priest agrees. So we must neither accept it nor reject it. But such agnosticism wouldn't bring much relief: if you have a view on which either accepting an additional claim or rejecting it leads immediately to trouble, it seems to me that the view is already in trouble. (An argument that doesn't rely on this objection to agnosticism will be given near the end of this subsection.)

Despite being weaker than NPF_0, NPF_1 also entails non-truth and hence falsehood in Priest's theory. The Liar sentence satisfies this by the first disjunct, and κ satisfies it by the second.

The bad news is that NPF_1 is also too strong as a replacement for the role of "non-paradoxically false". For consider the "iterated Curry" κ_1, equivalent to $True(\langle\kappa_1\rangle) \to (True(\langle\kappa_1\rangle) \to \bot)$. This can't be accepted on Priest's view, so it satisfies the intuitive conditions for being "non-paradoxically false", but he can't accept $NPF_1(\kappa_1)$. What we can accept is just $NPF_2(\kappa_1)$, where $NPF_2(x)$ has a third disjunct $True(x) \to (True(x) \to (True(x) \to \bot))$.

Long story short: there is a whole hierarchy of "non-paradoxical falsehood" predicates NPF_α, where the subscript is a notation for an ordinal in a well-behaved system of notations. Define $NPF_0(x)$ as above, and once NPF_α has been defined, define $NPF_{\alpha+1}(x)$ as $NPF_\alpha(x) \vee (True(x) \to NPF_\alpha(x))$. And when λ is (a notation for) a limit ordinal, we define $NPF_\lambda(x)$ to mean, roughly, "For some $\beta < \lambda$, $True(\langle NPF_\beta\rangle, x)$". Formulating this limit clause properly requires some care, but it is a routine exercise in systems of ordinal notation; and if 'True' behaves naively, the result is equivalent to the disjunction of all the $NPF_\beta(x)$ for β that precede λ. (And of course, we define NPT_β as NPF_β of the negation.) As β increases, NPT_β and NPF_β become weaker and weaker, and closer and closer to satisfying (C_T) and (C_F). Indeed for any sentence A except for some extremely pathological ones, we'll be able to find a sufficiently large α such that the assertibility of $NPF_\alpha(\langle A\rangle)$ coincides with the rejectability of A, and the assertibility of $NPT_\alpha(\langle A\rangle)$ coincides with the rejectability of $\neg A$. The main problem with attaining (C_T) and (C_F) is that no single α will do the trick for all sentences.[44]

One might perhaps hope for a single unramified NPF_∞ predicate that is in effect the disjunction of all the NPF_α, but it should be clear that there can't be one: if there were, we could use it to extend the hierarchy to a still weaker $NPF_{\infty+1}$,[45] thus defeating the claim that NPF_∞ included all of them. (Seeing in detail what blocks the formulation of such an NPF_∞ requires looking at the theory of ordinal notations, but basically it's the same as the reason why systems of ordinal notations can't be extended beyond the recursive ordinals.)

Indeed, there is no way that we can possibly get a predicate NPF that satisfies (C_F) in a dialetheic theory like Priest's, provided that it also satisfies

(*) $A, NPF(\langle A\rangle) \models \bot$

---

[44] Incidentally, it would really be somewhat more natural to define a sequence of weaker and weaker negation-like operators N^α, and to define $NPF_\alpha(x)$ to mean $N^\alpha True(x)$. The difference with what is in the text isn't substantial. It's worth stressing that either way, this is unlike stratified approaches to truth in that the N^α and NPF_α are all defined from the conditional; we do not need a hierarchy of *primitive* notions.

[45] Given a predicate NPF_∞, we define $NPF_{\infty+1}(x)$ in the manner above: $NPF_\infty(x) \vee (True(x) \to NPF_\infty(x))$. Letting the iterated Curry κ_∞ be equivalent to $NPF_\infty(\langle\kappa_\infty\rangle)$, we'd have to reject $NPF_\infty(\langle\kappa_\infty\rangle)$; but $True(\langle\kappa_\infty\rangle) \to NPF_\infty(\langle\kappa_\infty\rangle)$ by construction and so $NPF_{\infty+1}(\langle\kappa_\infty\rangle)$ is provable.

and other very minimal other conditions. (This supplements the argument given in note 43.) For consider a sentence W equivalent to $NPF(\langle W \rangle)$. Applying (*) to W, and using the equivalence together with structural contraction, we get

$$W \models \bot.$$

But since W leads to absurdity, we must reject it, and so by (C_F),

$$\models NPF(\langle W \rangle).$$

But by the equivalence again, that's just $\models W$, which with $W \models \bot$ means that we've been led by (C_F) to absurdity (triviality). There is no "non-paradoxical falsity" predicate that satisfies (C_F); if (*) is taken as a minimal condition on a "non-paradoxical falsity" predicate, then adding (C_F) is simply an incoherent demand to put on such a predicate. It's sort of like the incoherence of requiring both Modus Ponens and Conditional Introduction for a conditional in a naive theory.

But so what? The only obvious theoretical motivation for a demand that a dialetheist satisfy conditions (C_T) and (C_F) is the view that rejection is somehow a lesser notion than acceptance, and hence must ultimately be explained in terms of acceptance. And that is a view that almost any advocate of non-classical logic will reject.

Admittedly, the absence of general notions of non-paradoxical truth and non-paradoxical falsehood (as opposed to their legitimate stratified analogs) makes life awkward: Priest frequently uses such notions in informal statements of his position (for instance, in his explanations of how LP and other dialetheic logics work), and setting things right would often require a bit of circumlocution. But whoever thought that life is easy?

6.11.2 Paracompleteness

An analogous expressibility worry arises for paracomplete theories like mine. Instead of "non-paradoxical falsehood", the issue for me is what might be called "superdeterminate truth", where this is required to meet the following condition:

(C_{NSD}) It should be legitimate to accept "$\langle A \rangle$ is not superdeterminately true" exactly when it is legitimate to reject A.

(Formally this is exactly the same as Condition (C_F), I've just replaced 'non-paradoxically false' with 'not superdeterminately true'. The reason for altering the terminology is that in a paracomplete theory we reject different things than we do on a dialetheic view, which gives the condition a different feel in the two cases, and makes the difference in terminology natural.)

My response to the demand (C_{NSD}) is going to be like the response to (C_F) I've offered Priest: it's that we can go some way toward meeting (C_{NSD}), but that in full strength, the demand is unreasonable.

In analogy to the response to (C_F) that I've offered Priest, my approach is to partially meet (C_{NSD}) by specifying a sequence of predicates that successively approximate the illusory notion of superdeterminate truth. The idea is to first define a basic determinacy operator D. (In the book I took DA to be $A \wedge \neg(A \rightarrow \neg A)$; if we modify the $\rightarrow$ in the manner suggested at the end of Sect. 6.3.1 we'd need to define D somewhat differently. One might also consider using $\triangleright$ instead of $\rightarrow$ in the definition, though I prefer $\rightarrow$.) We can then iterate D a long way through the transfinite, using the truth predicate to handle the "infinite conjunctions" that we need at limit ordinals, much as we did in the dialetheic case; and we define 'Determinately$^\alpha$-True(x)' as $D^\alpha(True(x))$. (The slight disanalogy to how this was done for the NPF_α in the main text is entirely superficial: see note 44.) As α increases, 'Determinately$^\alpha$-True' comes closer and closer to meeting the demands of (C_{NSD}). For instance, we can say that the Liar sentence Q and the standard Curry sentence κ aren't Determinately^1-true, that κ_1 isn't Determinately^2-true, and in general that the iterated Curry κ_α isn't Determinately$^{\alpha+1}$-true. The attempt to define a superdeterminately operator as something like an infinite conjunction of all the Determinately$^\alpha$ fails, for reasons completely analogous to those we saw in the dialetheic case. And there is no other way to meet (C_{NSD}), given reasonable assumptions about how a superdeterminacy operator should behave. But this is no more worrisome in the paracomplete case than in the dialetheic (though as there, it means that we sometimes have to choose between awkwardness and sloppy formulation).

I'm willing to regard (C_{NSD}) as partially definitive of the idea of superdeterminacy, and to say that superdeterminacy is simply an illusory ideal. What about determinacy? I'm willing to simply define that (somewhat arbitrarily) as one of the D^α; say, as D^1. Note that if we do that, we can't say that for a sentence like κ_1, or a sentence that declares itself not to be determinately true, it's indeterminate whether it's true. All we can say is that *it's indeterminate whether it's indeterminate* whether it's true. And mostly I've been careful to say that, though in some informal remarks I've slipped.

Priest seems to badly misinterpret me here: he says that "the crucial difference" between the Field and Priest views is that

> Field has literally no way of expressing the notion of indeterminacy. .... [F]or him the notion is literally meaningless. The dialetheist has a way of expressing that something is false only — in the very words 'false and not true' (Priest 2010, p. 137).

In response, I repeat:

(a) I do explicitly define indeterminacy (and don't declare it meaningless). It's indeterminate *that A* iff $\neg DA$, i.e. iff $\neg A \vee (A \rightarrow \neg A)$, on the definition in the book under discussion; and it's indeterminate *whether A* if it's both indeterminate that A and indeterminate that $\neg A$. Using that, I can define such notions as it being indeterminate whether its indeterminate whether A. That isn't a species of indeterminacy, any more than it being indeterminate whether someone is bald is a species of his being bald.

(b) What I don't define is a notion of superdeterminacy that meets condition (C_{NSD}); that's because I think that (C_{NSD}) is an ill-motivated requirement that, in combination with other conditions natural to require of determinacy-type notions, is impossible to meet.

(c) Yes, the dialetheist can easily define "sole falsity", as falsehood without truth (or more simply, as non-truth). But that is hardly the issue. Those people who raised a worry for dialetheism about "sole falsity" were really worried about *non-paradoxical* falsity: they wanted a way to make an *assertion* that would express our *rejection* of, say, "The Earth is flat". (We can of course assert that we reject it, but what they wanted was a way to express the rejection in a non-autobiographical manner.) Calling the sentence not true is insufficient for this, in the dialetheist framework: witness Q. And there is a solution that works for "The Earth is flat" and similar sentences: for these sentences, we can assert $A \to \bot$ to reject A. But it doesn't work in every case, and we lack a completely uniform solution.

For a paracomplete theorist it's easy to express our rejection of 'The Earth is flat' by an assertion: we assert the negation. But there is the somewhat analogous problem of expressing our rejection of the Liar sentence by an assertion. And here too there is a local solution, but not a completely uniform one.

I don't see the slightest grounds for differentiating between the alleged need of a uniform "non-paradoxical falsity" predicate for dialetheist theories and the alleged need of a "superdeterminateness" predicate for paracomplete theories. I'm inclined to think that neither is a real problem, but the idea that one is a problem but not the other strikes me as a double standard.

6.11.3 *A Consequence for Validity*

We saw in Sect. 6.7 that (at least in a truth theory that accepts the minimal rules of (T-Elim) and (T-Introd), plus standard structural rules)[46] it is a consequence of the Curry paradox that good reasoning and necessary truth-preservation come apart. In such a truth theory, there will normally be at least one rule—typically including Modus Ponens in the case of non-classical theories and either (T-Elim) or (T-Introd) or both in the case of classical—that we take to be valid in the sense that we accept reasoning that accords with it, but we cannot accept that it preserves truth.[47] But surely, it might be thought, validity is the necessary preservation of *something* with a truthy nature. And what could it be?

A natural idea, for the paracomplete theorist who didn't know about the problem of making sense of a notion of superdeterminateness, would be that validity is the necessary preservation of superdeterminate truth. But of course, the paracomplete theorist can't say that: as we've seen, there's no sense to be made of superdeterminate truth.

A natural idea, for the dialetheist who didn't know about the problem of making sense of a notion of non-paradoxical falsity, would be that validity is the necessary preservation of *lack of non-paradoxical falsity*. That is, the idea would be that an

---

[46]The claim holds far more generally, but the discussion in Sect. 6.7 was restricted in this way, and I don't need any added generality here.

[47]We can avoid this situation only by having the converse situation, which is what a naive theorist gets if he or she evades the Curry Paradox by keeping Conditional Introduction. This route involves accepting the Curry sentence K and its equivalent $K \to \bot$, and regarding the instance of Modus Ponens taking us from these to $\bot$ as invalid in that reasoning in accordance with it is bad. At the same time, it involves regarding the inference as preserving truth, since given naivety, to say that it preserves truth is equivalent to K.

inference is valid if and only if necessarily, if the conclusion is non-paradoxically false then so is one of the premises. That's natural because from the dialetheist point of view, it's natural to say that the "Good Guys" are the sentences that aren't non-paradoxically false; and for an inference to be valid should be for it to necessarily preserve "Good Guy" status. But of course the dialetheist can't say that: as we've seen, there's no sense to be made of the notion of non-paradoxical falsehood. We can get very good approximations to that: for ordinals $\alpha < \Gamma_0$ (see note 33) we can define an inference to be α-*valid* if whenever the conclusion is NPF_α, so is one of the premises. If α is large enough this will work for most of the inferences we care to consider. (The paracomplete theorist could do the analogous thing, defining α-validity in terms of preservation of determinate$^\alpha$ truth.) But we can never get the notion of validity exactly right by these means.

A possible moral, for either paracomplete or dialetheist theories, is that there is no general notion of validity, only notions that are stratified (either in the way suggested above or in some other way). I could live with that, but I'm more inclined to a different moral: that an unstratified notion of validity needn't be jettisoned, but we must give up any account of it in terms of preservation of a truthy property.

Maybe this is something of a cost. (I'd think it more of a cost if classical theories didn't have similar problems!) But however one evaluates it, it seems clear that it is as much of a cost for the dialetheist as for the paracomplete theorist.

6.12 Expressibility (2): Model Theory and Revenge

6.12.1 Expressibility and Model Theory

One often hears it argued that the model theory standardly used for non-classical logics of truth (whether paracomplete or dialetheic) reveals expressibility limitations in those logics. (I'm speaking, at the moment, of model theories *given in classical logic* for the non-classical logics.) Such arguments take many forms. Here's a simple such argument:

> In the model theory we have an exclusive and exhaustive distinction between designated and undesignated values. Given that, we can easily define functions on the value space that take designated values to undesignated and undesignated to designated (e.g. which take all designated values to a particular undesignated one, and all undesignated to a particular designated one). Pick one such function f, and call a unary operator ~ of a language "genuine negation" if its semantics is given by f, i.e. if $|{\sim}A|$ is $f(|A|)$. But (given natural assumptions which I won't bother to enumerate) the laws of excluded middle and explosion *stated in terms of* ~ *instead of* ¬ are bound to hold. And that means that one can't have naive truth in a language with ~. So the fact that we do have naive truth in standard paracomplete and dialetheic logics rests on expressive limitations of those logics: they don't contain ~.

A quick response is that it would certainly be an expressive limitation if one couldn't assert the existence of such a function f, but obviously neither the paracomplete theorist nor the dialetheist is in *that* boat. It would also be an expressive limitation to concede that a linguistic operator like ~, properly evaluated by f, is a perfectly good linguistic operator, but nonetheless to refuse to employ a language in which it can be expressed. But I don't see how it is an "expressive limitation" to think that no good linguistic operator is properly evaluated by f; rather, it's a point of doctrine, one which both paracomplete theorists and dialetheists share.

There's a variety of related "expressive incompleteness" arguments that are based on set-theoretic model theories for non-classical logics. Set-theoretic model theory is the kind of model theory needed in consistency/ω-conservativeness proofs for non-classical logics, for it's what's required in order that these proofs convince a classical logician that the non-classical theories in question are coherent. (It's perfectly legitimate for an advocate of a non-classical theory to employ a model theory given in classical set theory, as long as that theory allows classical reasoning *in the 'True'-free fragment of the language,* which includes the language of standard set theory.)

In these classical proofs of the consistency of dialetheic or paracomplete theories, we introduce a space of semantic values and distinguish between designated and undesignated values. Since the model theory is classical, the division is sharp: in a given model, every sentence either has a designated value or doesn't, and the claim that it both does and doesn't must be rejected as absurd because it implies everything.

Obviously, the notion of having a designated value is a classical surrogate for the illusory notions that make no sense: the illusory notion of superdeterminateness, in the case of paracomplete theories; and the notion of not being non-paradoxically false, in the case of dialetheic. But unlike the illusory notions, the classical notions do make sense. And for either kind of logic, if the class of models is chosen appropriately then every valid inference will preserve designatedness in all appropriate models (and perhaps conversely, every inference that preserves designatedness in all appropriate models will be valid).

It might be thought that this gives us a different way to define superdeterminateness for paracomplete theories, or non-paradoxical falsity for dialetheic ones: the suggestion is that the paracomplete theorist identify superdeterminateness with *designatedness in the intended (paracomplete) model,* and that the dialetheist identify non-paradoxical falsity with **lack of** *designatedness in the intended (dialetheic) model*. But such a proposal would be thoroughly misguided:

> (1) We've already seen that talk of "the intended model" makes no sense even for the language of classical set theory (or the classical theory of sets and classes), if the model theory is developed in that same language. For we intend a theory of sets or classes to be talking about all sets or all classes; but (a) if model theory is done in set theory then models are sets, and standard set theories don't recognize a set of all sets; similarly (b) if model theory is done in a theory that admits proper classes then models are classes, and standard theories of classes don't recognize a class of all classes. It's doubtful that there is any model at all such that *even in the classical part of the language,* truth in the model coincides with genuine truth; if there are such models they are undefinable in set theory and highly unnatural.[48]

---

[48] No *set-theoretically definable* model can have this property, by Tarski's Theorem on the undefinability in a classical language of truth in that language. Hamkins (2003) shows that the question of

(2) Talk of "the intended model" is even worse in the context of models for an extended set theory (or theory of sets and classes) that includes a non-classical truth predicate. For models (in the model theories within classical set theory that are now under discussion) are classical entities; if reality is taken as non-classical, then obviously any model has to misrepresent reality. That is no problem once one recognizes the role that these models serve: as devices for proving consistency (or ω-conservativeness) even to a classical logician. But it shows that talk of "intended models" in this context is entirely misconceived.

The right way to think of things isn't that model theory allows us to define "truthy" notions like superdeterminateness or non-paradoxical falsity, the necessary preservation of which constitutes validity; rather, it's that we don't need the "truthy" notions to get an account of validity adequate to our needs in theorizing. (We don't need the modality either, we replace it by quantification over models.)

The "expressive incompleteness" arguments against dialetheic or paracomplete theories are often put by saying that these theories cannot express the notion of designated value for the logic they employ. But that's absurd: the notion of having a designated value in a given model M *is* expressible (if M itself is): it is defined (from M) in the classical set theory that is normally incorporated into the non-classical theory. What isn't expressible is a notion of "having designated value *in the intended model M*, but that isn't because of any inability to express designatedness, or even any inability to single out an important model ("the intended one"); rather, it's because the idea of "the intended model" simply makes no sense in this context, for reasons (1) and (2) just listed.[49]

Priest takes model-theoretic arguments more seriously than I do—not against dialetheism, but against paracomplete theories. I expect that in the dialetheic case, he'd respond to the above by saying that it's all well and good as far as it goes,[50] but

---

the existence of *undefinable* models with that property is independent of ZFC: you need a forcing construction to establish the consistency of the claim that such models exist.

[49]If there are "Hamkins models" of the sort mentioned in the previous footnote, then such models *are* inexpressible without expanding the classical part of the language. But this is so quite independent of the dialetheic or paracomplete truth theory. And in any case, such a model would approximate "intendedness" *only for 'True'-free sentences*. For sentences with 'True', designatedness is distinct from "truthy" notions in the theories in question here: notions like truth, determinate truth, and so forth are non-classical notions; whereas, as discussed in Sect. 6.11.3, designatedness is a different sort of notion used for getting the logic right.

[50]I'm not totally certain that he'd be so concessive, since in many places he seems to assume that model theory has a role incompatible with what I've suggested:

(i) As discussed above near the end of Sect. 6.9, Priest's argument that the paracomplete theorist has a problem with definite descriptions seems to turn on confusing classical model with reality. It is obviously true that in a given classical model, there will be a least number that satisfies the predicate 'is not 10^6-denoted in L'. But 'denote' (and hence '10^6-denoted') is a non-classical notion, and so the classical model is an imperfect guide to the non-classical reality. (In a *property-theoretic analog of* model theory, done in a paracomplete logic, there would be no argument for a least number that satisfies 'is not 10^6-denoted in L' in a given model.)

(ii) In his earlier discussions he tended to attribute to advocates of paracomplete theories the belief in "truth-value gaps", which led to complete confusion between those theories and theories in classical logic that genuinely posit truth-value gaps. (Priest was far from alone in this attribution, I regret to say.) Paracomplete theories, of course, *reject* the existence of gaps: they reject the claim

it is unduly focused on model theory *done in classical set theory*. I think he'd say that there is no problem for a dialetheist in developing a model theory *in a dialetheic set theory* in which

(a) there can be an intended model;

(b) in this intended model a given sentence can both have a designated value and not have a designated value;

(c) the claim that truth coincides with truth in the intended model is entailed by the theory (or at least, is true in the intended model).

Maybe this or something like it can be done. It's certainly a promissory note: it was, after all, by a classical set-theoretic construction of models that Priest has convinced us that his theory is nontrivial, and to my knowledge he hasn't even settled on the details of a non-classical "set theory" that we could use to make sense of dialetheic claims about models. (The quotation marks reflect the concern, expressed above, that no such "set theory" deserves the name, because it can have only a pale imitation of extensionality. But maybe we don't need any deep form of extensionality in the model theory—indeed, maybe we don't even need the pretext of extensionality, maybe a model theory using properties in place of sets would be adequate.)

Actually, I'm not entirely clear on what the demands on such a non-classical model theory are. It isn't clear why it couldn't just employ a 2-valued semantics whose only values are 'true' and 'not true' with only the former designated. Since the background logic is dialetheic, this doesn't lead to explosion: the Liar sentence can both have the designated value in the model and not have it, and that doesn't require that every sentence has this value. Also, it isn't clear why the evaluation clause for the conditional couldn't be trivial: a conditional is true in a model if and only if (if the antecedent is true in the model then so is the consequent), where of course the 'if...then' is the one used in the dialetheic logic. (You need to understand the logic to understand the model theory, but that seems to be so on any proposal to build the model theory on a non-dialetheic set theory or property theory.) Has all of Priest's work on the model theory of dialetheic logic been unnecessary?

I don't mean to be entirely dismissive: I think I dimly see the possibility of carrying out Priest's promissory note in a less cheap way than the one just envisioned. But to the extent that I see this in the dialetheic case, I do so in the paracomplete as well. The paracomplete analog is clear: to develop a model theory in a paracomplete property theory, in which (a) and (c) hold along with the rejection of the claim that

---

that there are sentences such that neither they nor their negation is true, since this is equivalent (in the logic common to all paracomplete and paraconsistent theories under discussion) to the claim that there are sentences such that both they and their negations are true. The way the confusion doubtless came about was in confusing truth with having designated value in some (unspecified) classical model. A parallel confusion would be one that attributed to the paraconsistent dialetheist the rejection of the claim that the Liar sentence Q isn't true, on the basis of their rejection of the claim that their value in models is undesignated. That would ignore the distinction between the paraconsistent dialetheist and the classical truth-value glut theorist who accepts that both Q and $\neg Q$ are true but doesn't accept $\neg Q$ (since he rejects the inference from $True(\langle \neg Q \rangle)$ to $\neg Q$). Paraconsistent dialetheist theories are totally different from classical glut theories; paracomplete theories are equally different from truth-value gap theories.

in the intended model posited in (a), a given sentence must either have a designated value or not have a designated value. Given that properties are just shadows of parameterized formulas, I suspect that it wouldn't be that hard to carry out such a program (though in this case too I'm unclear just what the constraints are, and on the value of carrying out the program). I'm not inclined to issue a promissory note on the basis of this dim perception, but I see no reason to think there's more of a problem in the paracomplete case than in the dialetheic.

6.12.2 Finale on Revenge

Priest (2007) puts the *prima facie* challenge raised by revenge arguments as follows:

> There is a uniform method for constructing the revenge paradox... [for a certain language L]. All semantic accounts [of the paradox] have a bunch of Good Guys (the true, the stably true, the ultimately true, or whatever). These are the ones we target when we assert. Then there's the Rest. The extended liar is a sentence, produced by a diagonalizing construction, which says of itself just that it's in the Rest. [p. 226].

He then goes on to argue that in dealing with this challenge there are only three options. One option (his Horn 3) is that the notion of being a Good Guy isn't expressible in any language. A second option (his Horn 2) is that the notion is expressible in some language, just not in the language L for which the notion of Good Guy has been constructed. The third option (his Horn 1) is that the notion of Good Guy is expressible in that language L, and thus [sic] we obtain a contradiction.

But the 'thus' in the third option (Horn 1) is unwarranted. For a sentence to be a member of the Rest is for it not to be a Good Guy. So the extended liar is a sentence that says "I am not a Good Guy". Priest's resolution of the extended liar here described is to accept both that it is a Good Guy and that it isn't—but not to accept things that the two together classically entail, such as that he's a fried egg. Obviously, this involves giving up a classical assumption involving negation, the rule of explosion. But there's a paracomplete alternative that Priest's discussion ignores: to *reject* that the sentence is a Good Guy and also *reject* that it isn't. This too involves giving up a classical assumption involving negation, the law of excluded middle. But Priest has set up the discussion so that *this* non-classical option is ruled out in advance.

It isn't that Priest doesn't consider paracomplete theories. He does, but he insists that they go under one of the inexpressibility options (Horns 2 and 3). In particular, he puts my position in Field (2008) under Horn 3. Evidently, he is here assuming that the Good Guys aren't the truths, but the superdeterminate truths. But rather than saying (what Priest's discussion seems to attribute to me) that there is an important but ineffable notion of being a Good Guy, the right thing to say is that *that* notion of Good Guy is just incoherent. (So yeah, it's not definable in our language, but that's a very good thing, just as it's a good thing that 'tonk' isn't definable.) There is, rather, a (fuzzy) transfinite sequence of better and better, with no clear limit. The approximating notions, of being a $Good_\alpha$ Guy, *are* all expressible in the language, i.e. fall under Horn 1 when that horn is freed of its unwarranted dialetheist presupposition.

His theory is analogous. Of course, he can say, as can I, that the Good Guys are just the truths. On that notion of Good Guy, obviously there is no inexpressibility issue for either of us since we both take 'True' as a primitive predicate. But if that's unacceptable for me it is for him too. A natural alternative is that the Good Guys are those sentences that aren't non-paradoxically false, on a conception of non-paradoxical falsity that in analogy to the conditions on superdefiniteness obeys (C_F). But that notion of Good Guy is likewise incoherent, so he must take it as (fortunately) inexpressible. Again, his theory provides a sequence of better and better approximations that *are* expressible.

I agree with Priest that for *coherent* notions of Good Guy, Horn 1 is the way to go, as long as one doesn't unwarrantedly build dialetheism into that horn.

6.13 Conclusion

Graham Priest has revolutionized the philosophy of logic by persistently, forcefully and to my mind persuasively making the case that there is strong reason to restrict classical logic to deal with the paradoxes of truth and related notions.

His preferred alternative is to allow the acceptance of contradictions in a paraconsistent logic that keeps the contradictions from spreading. This is not as radical a view as it is often portrayed. (I think that Priest's terminology, that he advocates "inconsistent" theories, has had a deleterious effect; many people hear that the view "advocates inconsistency" and dismiss it out of hand before bothering to learn about it, in a way they might not if told that it is a view on which the best way to resolve the paradoxes is in a logic that allows for the consistent acceptance of some contradictions.) Many of the objections that have been raised against the dialetheic view strike me as seriously off the mark, and there is much to be said in favor of the claim that it beats its classical rivals.

There are alternative non-classical resolutions of the paradoxes, of which I think the best are in a "paracomplete" logic whose conditional-free fragment is the dual of Priest's *LP*. (I'm taking *restricted* universal quantifiers to be outside the conditional-free fragment: even if they are taken as primitive rather than officially explained in terms of a conditional, they are themselves "conditional-like" in that a conditional can be defined from them.) This limited duality, I think, casts doubt on many arguments for one view over the other: arguments for paraconsistent dialetheic resolutions of the paradoxes over paracomplete resolutions, and arguments in the other direction. Indeed, before the introduction of conditionals (or universal restricted quantifiers), it isn't obvious that there is more than a verbal difference between the two views. I have long preferred the paracomplete approach over the dialetheic, but apart from the issue of conditionals, I'm hard pressed to see this as more than an aesthetic preference for one terminology over the other.

In line with this, much of this paper has been in an ecumenical spirit: not arguing for my preferred paracomplete approach over a dialetheic one, but simply exploring some threats that might be raised against either, and arguing (often against Priest) that they are no more of a threat to the paracomplete than to the dialetheic and not much of a threat against either.

But ecumenicalism has its limits. When it comes to conditionals and restricted quantifiers, I do see a serious difference between the prospects for the paracomplete and the dialetheic. A key issue is how restricted quantification (perhaps defined in terms of one sort of conditional '$\to$') interacts with ordinary conditionals (which I've symbolized as '$\triangleright$'). I have argued in early sections of the paper that on this issue, known paracomplete approaches do *much* better than known dialetheic ones, and indeed that there are limits in principle on how well a dialetheic one could do. I expect that there is considerable room for both paracomplete and dialetheic logicians to improve on the currently available theories of these matters, and that if the dialetheist were to abandon his or her focus on typical relevant logics, the gap between dialetheist and paracomplete could be considerably narrowed. But because of the in-principle limits, I think that this will remain an area in which the paracomplete approach does better.

References

Bacon, A. (2013). Curry's paradox and ω-inconsistency. *Studia Logica, 101*, 1–9.
Beall, J. C. (2009). *Spandrels of truth*. Oxford University Press.
Beall, J. C., Brady, R., Hazen, A., Priest, G., & Restall, G. (2006). Relevant restricted quantification. *Journal of Philosophical Logic, 54*, 587–598.
Boolos, G. (1993). *The logic of provability*. Cambridge University Press.
Brady, R. (1983). The simple consistency of a set theory based on the logic CSQ. *Notre Dame Journal of Formal Logic, 24*, 431–449.
Burgess, J. (1981). Quick completeness proofs for some logics of conditionals. *Notre Dame Journal of Formal Logic, 22*, 76–84.
Caie, M. (2012). Belief and indeterminacy. *Philosophical Review, 121*, 1–54.
Feferman, S. (1962). Transfinite recursive progressions of axiomatic theories. *Journal of Symbolic Logic, 27*, 259–316.
Field, H. (2008). *Saving truth from paradox*. Oxford University Press.
Field, H. (2014). Naive truth and restricted quantification. *Review of Symbolic Logic, 7*, 147–191.
Field, H. (2016). Indicative conditionals, restricted quantification and naive truth. *Review of Symbolic Logic, 9*, 181–208.
Field, H. (2017). Disarming a paradox of validity. *Notre Dame Journal of Formal Logic*.
Field, H. (2018a). Properties, propositions and conditionals. *Australasian Philosophical Review, 4*(1).
Field, H. (2018b). Reply to Zach Weber. *Australasian Philosophical Review, 4*(1).
Field, H., Lederman, H., & Øgaard, T. F. (2017). Prospects for a naive theory of classes. *Notre Dame Journal of Formal Logic*.
Hajek, P., Paris, J., & Sheperdson, J. (2000). The liar paradox and fuzzy logic. *Journal of Symbolic Logic, 65*, 339–346.
Hamkins, J. (2003). A simple maximality principle. *Journal of Symbolic Logic, 68*, 527–550.
Kripke, S. (1975). Outline of a theory of truth. *Journal of Philosophy, 72*, 690–716.
Lewis, D. (1974). *Counterfactuals*. Harvard University Press.
Lewis, D. (1981). Ordering semantics and premise semantics for counterfactuals. *Journal of Philosophical Logic, 10*, 217–234.
McGee, V. (1985). A counterexample to modus ponens. *Journal of Philosophy, 82*, 462–471.

Priest, G. (2002). Paraconsistent logic. In D. Gabbay & F. Guenther (Eds.), *Handbook of philosophical logic* (2nd ed., Vol. 6, pp. 287–303). Kluwer Academic Press.

Priest, G. (2005). Spiking the Field artillery. In J. C. Beall & B. Armour-Garb *Deflationism and paradox* (pp. 41–52). Oxford University Press.

Priest, G. (2006). *In Contradiction* (2nd ed.). Oxford University Press.

Priest, G. (2007). Revenge, Field and ZF. In J. C. Beall (Ed.), *Revenge of the liar* (pp. 225–233). Oxford University Press.

Priest, G. (2008). *Introduction to nonclassical logic* (2nd ed.). Cambridge University Press.

Priest, G. (2010). Hope fades for saving truth. *Philosophy, 85*, 109–140.

Restall, G. (1992). Arithmetic and truth in Łukasiewicz's infinitely valued logic. *Logique et Analyse, 139–40*, 303–312.

Stalnaker, R. (1968). A theory of conditionals. In N. Rescher (Ed.), *Studies in logical theory*. Blackwell Publishers.

Standefer, S. (2015). On artifacts and truth-preservation. *Australasian Journal of Logic, 12*(3), 135–158.

Weber, Z. (2010). Extensionality and restriction in naive set theory. *Studia Logica, 94*, 109–126.

Weber, Z. (2018). Property identity and relevant conditionals. *Australasian Philosophical Review, 4*(1).

Hartry Field is Silver Professor of Philosophy and University Professor at New York University. He specializes in metaphysics, philosophy of mathematics, philosophy of logic, and philosophy of science. He has had fellowships from the National Science Foundation, the National Endowment for the Humanities, and the Guggenheim Foundation. His books include *Science Without Numbers*, which won the Lakatos Prize, *Truth and the Absence of Fact*, and *Saving Truth from Paradox*.

Chapter 7
Priest's Anti-Exceptionalism, Candrakīrti and Paraconsistency

Koji Tanaka

Abstract Priest holds anti-exceptionalism about logic. That is, he holds that logic, as a theory, does not have any exceptional status in relation to the theories of empirical sciences. Crucial to Priest's anti-exceptionalism is the existence of 'data' that can force the revision of logical theory. He claims that classical logic is inadequate to the available data and, thus, needs to be revised. But what kind of data can overturn classical logic? Priest claims that the data is our intuitions about the validity of inferences. In order to make sense of this claim, I will appeal to the Madhyamaka Buddhist philosopher Candrakīrti. I will then pose a problem for Priest's anti-exceptionalism. Finally, I will suggest a way out of the problem for Priest. Whether or not he accepts my solution, I will let him decide.

Keywords Graham Priest · Anti-exceptionalism · Candrakīrti · Paraconsistency · Madhyamaka · Philosophy of logic

7.1 Anti-Exceptionalism

In the context of the philosophy of logic, *anti-exceptionalism* is the view that logic does not have any special status in relation to empirical sciences and that the methodology for theorising about logic is the same as that for the theorisation of empirical sciences.[1] According to this view, logic is not exceptional: it is revisable and not a priori, just like the theories of empirical sciences. This view is often associated with Quine.[2]

---

[1] See Hjortland (2017), Maddy (2002), Russell (2014, 2015), Williamson (2013).
[2] Quine's ultimate view on this issue might be complicated. Shapiro (2000), for instance, casts doubt on this interpretation of Quine.

K. Tanaka
School of Philosophy, Research School of Social Sciences, Australian National University, Canberra, Australia
e-mail: Koji.Tanaka@anu.edu.au

Priest also accepts anti-exceptionalism about logic (Priest 2006, 2014, 2016). Ironically, Priest's anti-exceptionalism is part of his argument against Quine's position on non-classical logic. Quine held that someone who denies classical logic and subscribes to non-classical logic is only changing the subject (Quine 1986). For him, someone who disagrees with the classical account of propositional connectives and quantifications means different things by 'not', 'if... then...' and so on. It is in response to this position of Quine that Priest develops his anti-exceptionalism in the service of defending paraconsistent logic (see also Priest 1979).

In this paper, I will examine Priest's anti-exceptionalist view of logic. He has taught me that the best respect you can pay to a philosopher is to take their work seriously by having a critical stance towards it. I will do exactly that. First, I will present what I take to be his anti-exceptionalist view. Second, I will show that the anti-exceptionalist view that Priest presents requires further specification. Third, I will fill a gap in Priest's presentation of his view by appealing to the Buddhist Madhyamaka philosopher Candrakīrti. I will then pose a problem for Priest's anti-exceptionalism. Fourth, I will suggest a way out of the problem for Priest. Whether or not he accepts my solution, I will let him decide.

7.2 Priest's Anti-Exceptionalism

Priest's anti-exceptionalism is part of his arguments for paraconsistent logic. A logic is said to be paraconsistent iff its consequence relation does not validate *ex contradictione quodlibet* (ECQ): $A, \neg A \models B$ for any A and B. An argument for paraconsistent logic is, thus, an argument against the classical principle ECQ.

Priest (1979) argues against ECQ by claiming that logic is a normative subject:

> [T]he notion of validity that comes out of the orthodox [classical] account is a strangely perverse one according to which ... any rule whose premises contain a contradiction [ECQ] is valid. By a process that does not fall far short of indoctrination most logicians have now had their sensibilities dulled to these glaring anomalies. However, this is possible only because logicians have also forgotten that logic is a normative subject: it is supposed to provide an account of correct reasoning. When seen in this light the full force of these absurdities can be appreciated. (Priest 1979: 297)

Let's agree that logic provides 'an account of correct reasoning'.[3] What exactly this amounts to is a controversial issue. For the sake of this paper, let's understand the claim that 'logic is normative' in the sense that it serves as the standard for evaluating one's reasoning as correct or incorrect.[4] It follows that reasoning is correct if it meets the standard set by logic. For instance, if logic declares the following inference as valid:

---

[3] This is, in fact, a controversial view. For an opposing view, see Harman (1986).

[4] According to MacFarlane (2000, 2002), this is Frege's view about the normativity of logic.

> A round square is round.
> A round square is not round.
> The moon is made of blue cheese.

reasoning to the conclusion on the basis of the premises counts as correct.

But how could the normative nature of logic be a reason for rejecting ECQ? How could it entail that the account of correct reasoning is paraconsistent? In order to answer these questions, Priest (1987) invites us to think of logic analogously to the theorisation of dynamics:

> [J]ust as with dynamics, so with logic, one needs to distinguish between reasoning or, better, the structure of norms that govern valid/good reasoning, which is the object of study, and our logical theory, which tries to give a theoretical account of this phenomenon. (p. 257)

Priest's analogy between dynamics and logic can be explained as follows. In the case of dynamics, we must make a distinction between the physical structure that governs the movement of physical objects and a theory about this structure. We theorise about the physical structure by first observing the movement of physical objects and then systematising the data we acquire through our observations. It is possible that an observation of the movement of an object does not match the prediction made by a theory. This is possible even in the case of a theory that has been largely accepted by a scientific community. If there is such a discrepancy, there are at least two options: reject the data (i.e. our observation) or modify (or dispose of) the theory. It is this second option that is important for Priest. When an anomaly is discovered, there must be an option to modify or reject the theory we currently have. How exactly this can be done is a matter of debate. Nevertheless, the important point is that the second option cannot be a priori ruled out in the case of dynamics or in any empirical science.

For Priest, this must be the case for logical theorisation too. When our observation does not match the system of logic, understood as a theory of the norms of correct/valid reasoning, we currently accept, we cannot always reject our observation as irrelevant or redescribe it so that it is no longer anomalous with one's logical theory (which is equivalent to a priori rejecting the data). For Priest (2016), we must consider such factors as adequacy to the data, simplicity, consistency, explanatory power, and avoidance of *ad hoc* elements. Putting aside the details of these factors, the important point for Priest is that the criteria for logical theorisation are basically the same as those of empirical sciences. In particular, just like the case of dynamics, rejecting or modifying one's logical theory in response to data cannot be ruled out a priori. There must be a possibility of modifying or disposing of one's logical theory based on new data.

By inviting us to understand the nature of logical theorisation analogously to the theorisation of empirical sciences, Priest places logic on the same footing as empirical science. In other words, logic has no exceptional status in relation to empirical science. This is Priest's anti-exceptionalism.

7.3 Logical 'Data'

Crucial to Priest's anti-exceptionalism as analysed above is the existence of 'data' that can force the revision of logical theory. What exactly is the kind of data that is capable of inciting logical revision? In the case of dynamics, it is easy to specify the kinds of data that might place a theory in doubt. If a moving object does not display the behaviour that is prescribed or predicted by the theory, we need to consider rejecting the data as an anomaly or modifying (or disposing of) the theory. But what can be the equivalent data in the case of logic? What could the data be which might lead to rejecting a logical theory? What kind of observation can undermine ECQ? An observation of someone simply reasoning badly cannot be a counterexample. As Priest claims, logic is normative: it is 'an account of correct reasoning' (ibid.). So we must be able to observe not only how we do in fact reason but how we *ought to* reason. What kind of observation could that be?

Priest (2016) answers this question in the following way. Consider the following vernacular inferences:

$$\frac{\text{John is in Rome.}}{\text{John is in Italy.}} \qquad \frac{\text{If John is in Rome he is in Italy.}}{\text{John is not in Rome.}}$$
$$\text{John is in Italy.} \qquad \qquad \text{John is not in Italy.}$$

Priest tells us that the inference on the left 'strike[s] us as correct' and the one on the right 'strike[s] us as invalid' (p. 355).

The inferences we just looked at are particular ones expressed vernacularly. What about inferential *forms* such as

$$\frac{A}{\text{If } A \text{ then } B}$$
$$B$$

where A and B are meta-variables for any formulas? Priest reminds us that we must be careful about the relationship between inferential forms and their particular instances. For instance

$$\frac{\text{If he were here he would be hopping mad.}}{\text{He would be hopping mad.}}$$

does not 'strike us as valid' (p. 356) even though it has the form of *modus ponens*. Or consider any Sorites argument. It involves multiple *modus ponens* even though it is paradoxical and the validity of particular Sorites argument may be in doubt.[5] There is a lot that can be said about the relationship between an inferential form and its instances. Despite this concern, Priest tells us that the form *modus ponens* as such 'strikes us as intuitively correct' (p. 356).

---

[5]In the case of Sorites argument, one may reject the premises instead of rejecting the inference in order to resolve the paradox. I am here raising a possible reaction one can have in response to Sorites arguments.

So, for Priest, the data that logical theorisation needs to accommodate are 'our intuitions'. As he writes:

> In the criterion of adequacy to the data, what counts as data? It is clear enough what provides the data in the case of empirical sciences: observation and experiment. What plays this role in logic? The answer, I take it, is our intuitions about the validity or otherwise of vernacular inferences [and inferential forms]. (p. 355)

For Priest, it is our observation of how people make intuitive judgements about inferences that provides data to be considered in logical theorisation. Thus, logic, as a theory, has to accord with our intuitions about the validity of various inferences: what should be accepted as valid/invalid is assessed against our intuitions.

However, an appeal to intuition is exactly what the defenders of the a priori nature of logic, such as Bealer (2000) and BonJour (1998), specify as the data that a theory of validity must be responsive to. They take our immediate intuitive judgements about validity as evidence for the validity/invalidity of inferences. Priest may not believe in the 'faculty' of rational intuitions to which they appeal. Nevertheless, for Priest as well as for Bealer and BonJour, the data that serves as evidence for establishing the validity/invalidity of inferences are our intuitions.[6] They both appeal to intuitions as evidence for logical theorisation.

How different is Priest's view about logical data from the view of the a priori defenders? Priest rejects the views of Bealer and BonJour (and others who defend the a priori nature of logic) (Priest 2016: §3). However, he does not explain how different his position is from that of the a priori defenders with respect to the data that logical theorisation must accommodate. Without any elaboration on the notion of intuitions that Priest appeals to, it is hard to understand the difference between his view and the views of those who subscribe to the a priori nature of logic. In what way, then, is Priest's view anti-exceptionalist? How different is his notion of intuitions from that used by the a priori defenders? In the next section, I will present one way to understand Priest's view and show that it is indeed anti-exceptionalism when it is construed in a certain way.

7.4 Candrakīrti's Take on Priest's Anti-Exceptionalism

I suggest that the best way to make sense of Priest's anti-exceptionalism is to appeal to the Buddhist philosopher Candrakīrti. Candrakīrti is one of the most, if not the most, influential philosophers in the Madhyamaka tradition after the 'founder' of the tradition Nāgārjuna. The Madhyamaka school is often considered to be the most influential school in the Buddhist philosophical traditions, especially in Tibet.[7] It

---

[6]Bealer and BonJour do not discuss our intuitions invalidating the currently accepted logic. That is, they are not anti-exceptionalists. Nevertheless, it is still the case that they as well as Priest appeal to intuitions as doing the groundwork for logical theorisation.

[7]While this is true in Tibet, it is actually not clear what the level of Madhyamaka's influence has been in China, Korea and Japan. Also, as contemporary scholars agree, Candrakīrti was a minor figure

is to this school that contemporary scholars are often drawn in bringing Buddhist material to contemporary philosophical discussions.[8]

What exactly Candrakīrti meant or even said is a matter of scholarly debate amongst Buddhist scholars. In this paper, I will largely follow the analysis of Tillemans (2011) who presents Candrakīrti as interpreting the notion of truth and knowledge in terms of what the people on the street accept. According to this analysis, Candrakīrti reduces truth and knowledge to mere opinions and beliefs. Truth is nothing more than what people on the street assent to and knowledge is nothing more than what they think. This is the case no matter whether they are opinions or beliefs about plumbing, empirical science or mathematics. Because of this, there is not much more to science than what can be expressed by ordinary notions like 'When wood, strings, and manual effort are present, sounds arise from musical instruments' (p. 155). I refer to this account of truth and knowledge as the *lokaprasiddha* (what is acknowledged by the world) account.

The *lokaprasiddha* account implies that theorising about what is true and what should count as knowledge requires investigating the opinions and beliefs of people on the street. What investigation must take place and how the results of the investigation should be understood are also matters of opinions and beliefs. But there is nothing special about how truth and knowledge should be expressed. They are expressible by ordinary notions.

This account is not as absurd as it might first appear, at least from a Mādhyamika point of view.[9] Crucial to Māhyamikas is the thought that anything that exists is empty of essential, intrinsic and independent property (*svabhāva*), i.e. the doctrine of emptiness. If an object does not possess this property, it may be said to exist but only because other things exist. For instance, a chariot might be said to exist only because of the existence of the wheels, the poles, the wheels, the body flagstaff, the yoke, the reins and the goad. A flame may be said to exist only because there is a certain amount of oxygen, fuel, etc. The main Madhyamaka thought is that everything whatsoever that can be said to exist is like that.

If a chariot can be said to exist only because of the existence of the wheels, the poles and so on, it may 'disappear' when one theorises about what a chariot is. For instance, a chariot may be 'reduced' to the wheels, the poles, etc. Opponents of Mādhyamikas, at least those of Nāgārjuna, employed reductive analyses to claim that a chariot and other macro-level objects do not really exist. Mādhyamikas follow such reductive analyses to their logical end and argue that no reductive analyses can bottom out because everything, not just macro-level objects, can be analysed away. If there is no end to a reductive analysis, however, employing such analyses to work out what is true cannot ultimately bear fruit. Assuming that reductive analysis is the

---

in India until his texts went to Tibet where he became an important thinker. I am here following the conventional wisdom of contemporary scholarship.

[8] For instance, the Cowherds (2011).

[9] I follow the contemporary convention of using 'Madhyamaka' as referring to the school of thought and 'Mādhyamika' as referring to the thinker who belongs to the school.

only mode of analysis in town,[10] there is not much point in engaging in analysis. Any attempt to work out why something is the way it is is unproductive as there is really nothing that can be analysed. In claiming to have truth or knowledge, an appeal to people's opinions and beliefs may be as good as anything else.

Now, Candrakīrti's *lokaprasiddha* account of truth and knowledge can be extended to the validity of inferences.[11] Following his *lokaprasiddha* account of truth and knowledge, we might think that whether or not an inference is valid is also just a matter of what the people on the street would accept. For instance, presented with an inference such as

> John is in Rome.
> If John is in Rome he is in Italy.
> John is in Italy.

we might say that it is valid. In so doing, we might be understood as simply expressing our opinions. We may not have any sustained reasons for why we believe it is valid. The only reason we can say to account for our response is simply that it strikes us as valid. Or, if there are additional reasons, they are merely additional opinions and beliefs.

According to this line of thought, in order to adjudicate which inference is valid, we just need to find out the opinions and beliefs of people on the street. This is *not* to say that we need to find out how people do, in fact, reason in order to work out what inferences are valid. That would be to confuse how people reason with how people ought to reason. What we need to find out is what people accept as 'an account of correct reasoning': what inferences should count as valid. On this account, there is no need to analyse logical concepts, the notion of validity or anything. All there is to logic is what can be expressed by things like 'This inference looks good to me' or 'That inference strikes me as valid'. If all there is to science is what can be expressed by ordinary notions like 'When wood, strings, and manual effort are present, sounds arise from musical instruments' (ibid.), all there is to logic can be expressed by ordinary notions like 'When you reason to the conclusion that John is in Italy from the premises that John is in Rome and that if John is in Rome he is in Italy, your reasoning strikes us as good'. According to the *lokaprasiddha* account of inference, there is nothing exceptional about validity in relation to empirical science.

This *lokaprasiddha* account fits nicely with Priest's view of logical data. In theorising about what inferences count as valid/invalid, Priest appeals to our intuitions. For him, what does the work of establishing an inference as valid or invalid is our intuition: what strikes us as valid or invalid. It is our pre-reflective judgements about

---

[10]Later Mādhyamikas seem to have developed different kinds of analyses based on different semantic accounts. See Tanaka (2014). But early Mādhyamikas seem to recognise only reductive analyses. In fact, without making this assumption, Nāgārjuna's argument for emptiness and Candrakīrti's *lokaprasiddha* account of truth and knowledge are implausible.

[11]In fact, Candrakīrti has an elaborate discussion of logical principles based on his *lokaprasiddha* account of truth. See Tillemans (2016). That discussion goes in a different direction from the current discussion about Priest's anti-exceptionalism. Here I deviate from Candrakīrti's thought about logical matters expressed in his writings.

inferences that provide us with data for logical theorisation. For him, these judgements are expressed as 'this inference strikes us as valid'. If we redescribe these pre-reflective judgements as opinions and beliefs that are expressible in ordinary notions, Priest's account boils down to a *lokaprasiddha* account.

Once Priest's account is redescribed as a *lokaprasiddha* account, we can show that his account is anti-exceptionalist and, thus, can be distinguished from a priori views as such Bealer's and BonJour's. If the data which we must account for in logical theorisation is people's opinions and beliefs, it is an empirical question what counts as valid inference. We need to survey what people think about various inferences. According to the *lokaprasiddha* account, how to conduct surveys and how to interpret the result of surveys are also empirical questions. So we have to investigate not only the opinions and beliefs of people about what inferences should count as valid but also those about the way to conduct surveys and to interpret the result of the surveys. However to be analysed, the result of such survey is unknown a priori. According to the *lokaprasiddha* account the data for logical theorisation is something that is not simply given to us; it is something that we must observe.

The data revealed by a survey may overturn an established logical theory if the survey contains a new statistic of opinions and beliefs about certain inferences. When a survey reveals that enough people reject an inference that is prescribed as valid according to a current logical theory or a theory that is hypothesised, we can either revisit and question the survey (i.e. reject the data) or reject the theory. People's opinions and beliefs may change in which case the currently accepted logical theory needs to be revised to take a new survey into account. So if the result of survey is what logical theorisation is responsive to, the *lokaprasiddha* account provides an anti-exceptionalist methodology that can make sense of the kind of data that can be used as evidence for logical theorisation.

Hence, once we redescribe Priest's view of logical data in terms of *lokaprasiddha*, we can show that his account is anti-exceptionalist. As is shown above, the data for logical theorisation under this account is available empirically in the way that the data for theorisation of dynamics is. Also, the *lokaprasiddha* account can make sense of revisability of logic. Priest's account as a *lokaprasiddha* can, thus, be shown to be an anti-exceptionalist view of logic.

7.5 Dismal Slough

The *lokaprasiddha* account of inferences was developed in analogy to (empirical) sciences. It is hard to imagine, however, that this account accommodates the opinions and beliefs people have about (empirical) sciences. It is very unlikely that the majority of people think of sciences as being expressible only in ordinary terms like: 'When wood, strings, and manual effort are present, sounds arise from musical instrument' (ibid.) at least in these modern days. It is more likely that people believe in the mechanisms underlying the sounds that arise from musical instrument and they think that the study of those mechanisms requires more than ordinary notions.

In fact, it is important for Priest that logical revision is triggered in a principled manner. The need for logical revision arises when certain criteria are met. And those criteria must be chosen carefully and they need to be examined thoroughly. It is most likely that Priest would not be happy to think of logical revision as simply a matter of the change of opinions and beliefs people happen to have at the time of survey. These considerations suggest that Priest requires something more than *lokaprasiddha*.

Tillemans' description of the *lokaprasiddha* account of truth and knowledge on which the *lokaprasiddha* account of inference is based may leave us wondering what is left of sciences and logic. The picture Tillemans (2011) paints of Candrakīrti's view is a 'dismal' one. As he puts it aptly, it entails a *dismal slough*.[12]

I think that there is another way to understand the *lokaprasiddha* account.[13] Candrakīrti reduces truth and knowledge to mere opinions and beliefs. He reaches such a view by following through the doctrine of emptiness to its logical end as he sees it. If there is nothing that ultimately exists, there is not really anything that we must accommodate in claiming to have truth and knowledge. But this is compatible with the idea that people accept there being a mechanism underlying the sounds of musical instruments. What is incompatible is the idea that there is an ultimate answer to what that mechanism is. For Candrakīrti, what that mechanism is is a question that can be answered only by appealing to the opinions and beliefs of people.

If we understand it this way, the *lokaprasiddha* account of truth and knowledge is not dismal. There is nothing in the account to redescribe what the sciences are or should be. It accepts what people think them to be. Similarly, the *lokaprasiddha* account of inference is not dismal. It does not radically redescribe what logic is or should be. All it does is opens up the possibility of revision in relation to the data that can be observed. It is this feature that makes it an account of anti-exceptionalism. And this feature of the account is not what is described as dismal.

7.6 Anti-Exceptionalism and Paraconsistency

As mentioned before, Priest has developed his anti-exceptionalism in the service of defending paraconsistent logic. How does anti-exceptionalism help him argue for paraconsistent logic? In order to answer this question, we must remember that an argument for paraconsistent logic is an argument against classical logic. In particular, a case for paraconsistent logic can be made by showing that ECQ ($A, \neg A \models B$ for any A and B) is invalid.

Priest's typical strategy to argue against ECQ is to provide counterexamples. For instance, Priest (2008) writes:

> Not only is this [ECQ] highly counterintuitive, there would seem to be definite counterexamples to it. There appear to be a number of situations or theories which are inconsistent, yet in which it is manifestly incorrect to infer that everything holds. (pp. 74–75)

---

[12] See also Tanaka (2015).
[13] In fact, Tillemans (2011) seems to acknowledge this. See also Tillemans (2016).

A counterexample to ECQ is a situation where not everything is true even when a contradiction obtains. One of the counterexamples is about visual illusions such as the waterfall effect where something stationary appears to move after seeing constant motion.

> [A] point in the visual field, say at the top, does not appear to move, for example, to revolve around to the bottom. Thus, things appear to move without changing place: the perceived situation is inconsistent. But not everything perceivable holds in this situation. For example, it is not the case that the situation is red all over. (pp. 75–76)

For Priest, a perceptual experience of waterfall effect is a counterexample to ECQ. The perceived situation is a contradictory one: things are perceived as moving without changing place. Yet not everything is perceived in such a situation. So a perception of waterfall effect is an instance where the premise of ECQ is affirmed but the conclusion is rejected. Thus, it provides the data in terms of which classical account of reasoning must be rejected. Because of the existence of such situations functioning as counterexamples, so Priest argues, ECQ and, thus, classical logic do not specify *correct* reasoning.

This strategy is, in fact, not available to Priest under the *lokaprasiddha* account. According to the *lokaprasiddha* account, there is no deadlock that we can all appeal to in theorising about what should count as valid/invalid. There may be no further reason that can be given for why an inference should count as valid. If so, the judgements about vernacular inferences may not depend on the judgements about inferential forms. For instance, the judgement that

> John is in Rome.
> If John is in Rome he is in Italy.
> ─────────────────────────
> John, is in Italy.

is valid may not depend on accepting *modus ponens* as valid. If this vernacular inference strikes one as valid, no further explanation, other than further opinions and beliefs, may be given for why it appears as valid. Under the *lokaprasiddha* account, this vernacular inference may be valid but not necessarily in virtue of its form. Similarly, the reason why *modus ponens* strikes us as valid may not be because of its instances. If *modus ponens* strikes us as valid, that validity does not necessarily spill over to its instances. According to the *lokaprasiddha* account, judgements about the validity of vernacular inferences may come apart from those about the validity of their forms.[14] One may reject the conclusion that the situation is red all over while accepting that things move without changing place; yet accept ECQ as a valid inferential form. Under the *lokaprasiddha* account, there is no such a thing as counterexample that can invalidate an inferential form. In order to argue against ECQ and, thus to argue for paraconsistency, Priest would have to show directly that ECQ, as an inferential form, is not accepted as valid according to the conventional wisdom of people.

---

[14] Many thanks go to an anonymous referee who brought out this feature of the *lokaprasiddha* account of validity.

So, under the *lokaprasiddha* account, in arguing for paraconsistent logic, one needs to show that people do not accept ECQ as valid. This is an empirical issue and requires a survey. All of the empirical studies in support of paraconsistency in the literature have been conducted with particular cases and with the assumption that a counterexample can show ECQ as invalid. In the above passage, Priest conducts an experiment on himself and describes a situation in support of paraconsistency. Ripley (2009) reports an experiment that shows that people accept contradictions in some situations. But, again, the experiment is conducted in relation to particular situations. The only 'experiment' I know of that supports paraconsistent logic is my own experience. I have taught large logic classes (sometimes 700+ students) to first-year university students at various places.[15] After presenting the classical definition of validity in terms of truth preservation, I ask them whether various inferential forms as well as vernacular inferences are valid or invalid. Except those students who have taken the definition to heart or those who have followed the lecture material carefully, most students consider ECQ to be invalid. It takes me some time and effort to explain to them that the (classical) definition of validity implies that ECQ is valid. My experience may not 'prove' anything on its own. However, it is possible (and I think that it is very likely) that people on the street do not accept ECQ as valid. Thus, under the *lokaprasiddha* account, anti-exceptionalism allows paraconsistent logic to be a live option.

7.7 Conclusion

Priest holds anti-exceptionalism about logic. That is, he holds that logic, as a theory, does not have any exceptional status in relation to the theories of empirical sciences. Priest's anti-exceptionalism is connected to his argument for paraconsistent logic. He claims that the classical principle ECQ is not adequate to the data. But what is the data that can lead to revision or overturning of logical theories? Priest claims that the data is our intuitions about the validity of inferences. In order to make sense of this claim, I appealed to the Madhyamaka Buddhist philosopher Candrakīrti. I applied Candrakīrti's *lokaprasiddha* account to Priest's view about the data that logical theories must respect. I have then shown the problematic nature of Priest's anti-exceptionalism as construed in terms of the *lokaprasiddha* account. I have, however, presented a way out for Priest. I have shown that Priest would have a coherent account of anti-exceptionalism that can be used to argue for paraconsistent logic if he accepted the *lokaprasiddha* account. Whether or not Priest is happy with my redescription of his view and my suggested solution, I will leave it to him to answer.

---

[15]My teaching experiences are limited to Australia and New Zealand. There might be some cultural differences but I am putting aside that issue from consideration for now.

References

Bealer, G. (2000). A theory of the a priori. *Pacific Philosophical Quarterly*, *81*, 1–30.
BonJour, L. (1998). *In defence of pure reason: A rationalist account of a priori justification*. Cambridge: Cambridge University Press.
Cowherds, T. (2011). *Moonshadows: Conventional truth in Buddhist philosophy*. Oxford: Oxford University Press.
Harman, G. (1986). *Change in view*. Cambridge, MA: MIT Press.
Hjortland, O. T. (2017). Anti-exceptionalism about logic. *Philosophical Studies*, *174*, 631–658.
MacFarlane, J. G. (2000). *What does it mean to say that logic is formal?*, Ph.D. dissertation, University of Pittsburgh.
MacFarlane, J. G. (2002). Frege, Kant, and the logic in logicism. *Philosophical Review*, *111*, 25–65.
Maddy, P. (2002). Naturalistic look at logic. *Proceedings and Addresses of the American Philosophical Association*, *76*, 61–90.
Priest, G. (1979). Two dogmas of quineanism. *Philosophical Quarterly*, *29*, 289–301.
Priest, G. (1987). *In contradiction*. Dordrecht: Martinus Nijhoff Publishers. (Expanded Edition published by Clarendon Press, Oxford, 2006.)
Priest, G. (2006). *Doubt truth to be a liar*. Oxford: Oxford University Press.
Priest, G. (2008). *An introduction to non-classical logic* (2nd ed.). Cambridge: Cambridge University Press.
Priest, G. (2014). Revising logic. In P. Rush (Ed.), *The metaphysics of logic* (pp. 211–223). Cambridge: Cambridge University Press.
Priest, G. (2016). Logical disputes and the a priori. *Logique et Analyse*, *59*, 347–366.
Quine, W. V. (1986). *Philosophy of logic* (2nd ed.). Cambridge, MA: Harvard University Press.
Ripley, D. (2009). Contradictions at the borders. In R. Nouwen, et al. (Eds.), *Vagueness in communication* (pp. 169–188). Dordrecht: Springer.
Russell, G. (2014). Metaphysical analyticity and the epistemology of logic. *Philosophical Studies*, *171*, 161–175.
Russell, G. (2015). The justification of the basic laws of logic. *Journal of Philosophical Logic*, *44*, 793–803.
Shapiro, S. (2000). The status of logic. In P. Boghossian & C. Peacocke (Eds.), *New essays on the a priori* (pp. 333–366). Oxford: Clarendon Press.
Tanaka, K. (2014). In search of the semantics of emptiness. In J.-L. Liu & D. Berger (Eds.), *Nothingness in Asian philosophy* (pp. 55–63). Routledge.
Tanaka, K. (2015). The dismal slough. In *Moonpaths: Ethics and emptiness* (pp. 43–53). The Cowherds, Oxford: Oxford University Press.
Tillemans, T. (2011). How far can a Mādhyamika Buddhist reform conventional truth? In *Moonshadows* (pp. 151–165). The Cowherds, Oxford: Oxford University Press.
Tillemans, T. (2016). *How do Mādhyamikas think?*. Boston: Wisdom Publications.
Williamson, T. (2013). *Modal logic as metaphysics*. Oxford: Oxford University Press.

Koji Tanaka is an Australian Research Council Future Fellow in the School of Philosophy, Research School of Social Sciences, at the Australian National University. His research focuses on logic, history and philosophy of logic, philosophy of language, Buddhist philosophy and classical Chinese philosophy and he has many publications in these areas.

Chapter 8
Looting Liars Masking Models

Diderik Batens

Abstract This paper does not raise objections but spells out problems that I consider at present unsolved within Priest's view on logic. In light of the state of scientific and other theories (Sect. 8.2) and in light of the character of natural languages (Sect. 8.3), Priest's central arguments do not seem convincing. Next, I offer some six independent obstacles for defining consistency, identifying models and describing the semantics and metatheory of **LP** (Sect. 8.4).

Keywords Paraconsistency · Dialetheism · Adaptive logics

8.1 Introduction

It must have been in the late 1970s that Graham first wrote to me. He was one of those planning the big black paraconsistency book (Priest et al. 1989) and collected some more papers. I sent mine around 1980. In 1982 we first met, both occasionally being in Pittsburgh, PA. Since then, we became friends. Now and then Graham passed by in Belgium, alone or with family. He was there in 1997 for the organization of the First World Congress of Paraconsistency, an event he had proposed to our research group on a previous visit. In 2000 I spent part of a sabbatical in his Brisbane department; Graham was a great help. Since then we regularly met, most often in conferences.

It will not come as a surprise that I was enthusiastic about the initiative that led to the present book. It is truly meet and just that Priest's work and Priest as a scholar are honoured with this book. Yet, there are two warnings that should prevent the reader from wrong expectations concerning my contribution. The first warning is that, while Priest and I have contributed to paraconsistent logic as a discipline, our views are rather different. One example concerns global versus local paraconsistency. Priest is convinced that there is a 'true logic' and that it is paraconsistent. My enthusiasm for paraconsistency is caused by the insight that scientists as well as philosophers of science need paraconsistent logics, actually several of them, in specific situations.

D. Batens (✉)
Centre for Logic and Philosophy of Science, Ghent University, Ghent, Belgium
e-mail: Diderik.Batens@ugent.be

© Springer Nature Switzerland AG 2019
C. Başkent and T. M. Ferguson (eds.), *Graham Priest on Dialetheism and Paraconsistency*, Outstanding Contributions to Logic 18,
https://doi.org/10.1007/978-3-030-25365-3_8

Disagreements may appear to come in handy for a contributor to this book. They offer an opportunity to highlight Priest's position and to discuss weaknesses and strengths. This is not without difficulties and my second warning pertains to them. Intellectual discussions are clearly essential to improving human knowledge and action. Yet, most public discussions, whether written or oral, are frustrating rituals. They are boring and inefficient sham fights, relying on a mistaken view of the value of research programmes and on a mistaken view of attaining knowledge. That, at the present moment, certain problems are unsolved and some objections unanswered need not mean much for the viability of an approach or for its future.[1] Unsolved problems may be solved in the future. The solutions may cause adjustments to the research programme and its theories, even adjustments in some rather central tenets. So even valid criticism that will never be retorted may be fully insignificant after all. Knowledge is attained by creative work, not by winning quarrels.

In which way can I sensibly contribute to the present volume? My plan is to describe some alternatives to aspects and parts of Priest's programme and to ask questions on the programme with the alternatives as background. If I succeed, the description should have some interest for the reader, independent of Priest's reaction, and should enable Priest to pick some topics which he might care to address for clarifying or pointing out a direction or presenting a solution.

While disagreeing, sometimes severely, with Priest in several central respects, I am deeply impressed by what he achieved over the years. It is hardly exaggerated to say that he developed dialetheism from a bunch of ideas plus a little bit of formal logic into a multifaceted, rich and connected set of philosophical and meta-mathematical theories, sided by a number of examples of paraconsistent mathematical theories. Next, while the disagreements will soon surface, as will the reasons for my doubts on the viability of his programme, I consider it of utmost importance that he continues to carry it out. If Priest succeeds, my view will need drastic readjustment and, even if it survives in changed form, I shall learn much about its weak spots and their circumvention. If Priest fails, it will be interesting and important to learn to which extent his views can be upheld and which problems can be given a solution in agreement with those views.

In what follows, I shall avoid digressions on the justification of my own views—this book is about Priest's—but I should still try to avoid misunderstandings. Occasionally, I shall have to refer to adaptive logics, but this is not the place to present background on those. Readers who do not believe me and are interested may follow the references. A particular difficulty is that Priest and I have criticized each other in previous papers. There were some misunderstandings, but commenting on those papers would only increase the confusion, besides being boring for the reader. So I have done my utmost to make the present paper self-contained, except that I expect the reader to be familiar with Priest's views or, where useful, to have a look at the

---

[1] Newton's achievement is not diminished by the fact that, from his days to the advent of relativity theory, several severe problems remained unsolved. Next, however significant the problems solved by early relativity theory, they were outnumbered by the problems solved by Newtonian mechanics. A sane view on scientific problem solving was elaborated by Laudan (1977).

passages from Priest's work that I refer to. And I hope I have learned from earlier misunderstandings and will avoid them better in the present paper.

The central topic of the present paper pertains to Priest's semantics for the logic **LP** and the connected understanding of semantics. The questions raised should not be read as criticism, but as a request for being instructed. As we all know, questions are not sensible and perhaps not even meaningful without some background. Everything one says, especially if it concerns philosophical background rather than, say a train timetable, has a variety of presuppositions. All of these cannot be made explicit, were it only because some are prejudices, but not making them explicit engenders misunderstanding. So, before getting to semantics, I shall comment on two background issues on which Priest and I have conflicting views. One concerns epistemological pluralism and its effect on logical pluralism, the other natural languages. I shall try to somewhat clarify both issues. More importantly, I shall attempt to show that the background views are sensible and viable. The aim is not to show that Priest's corresponding background views are mistaken, but merely that they are not obvious.

Let me end this section with a warning. I shall, as usual, write in a classical metalanguage. This means that an unqualified "not" stands for the negation of **CL** (Classical Logic) and that "false" stands for the classical negation of "true". In Sect. 8.4, however, on Priest's semantics, I shall use true and false as Priest does. To proceed differently would be too confusing. I shall remind the reader when we come to that section.

8.2 Logical Monism

This aspect of Priest's view is one of the hardest to make sense of for me. So I tried to avoid it. Yet, it is so central that I have to go into it in order to eliminate even the most elementary misunderstandings. I have stated my position and argued for it elsewhere and prefer to refer the reader there rather than repeating myself (Batens 2014, 2017). Still, I shall state here what is required for a correct understanding.

Taken in the broad sense of the term, a logic **L**, defined for a language $\mathcal{L}$, is a function that maps every set of sentences of $\mathcal{L}$ to a set of sentences of $\mathcal{L}$. So, where $\mathcal{W}$ is the set of sentences of $\mathcal{L}$ and $\wp(\mathcal{W})$ the power set of $\mathcal{W}$, $\mathbf{L}: \wp(\mathcal{W}) \to \wp(\mathcal{W})$. $\mathcal{L}$ is either a formal language or a fragment of a natural language. Aristotle's theory of the syllogism illustrates the latter.[2]

Once **L** is decently delineated, by a syntactic or semantic or procedural method, or by another good method that decently delineates **L**, a normative realm has been established. If **L** is, for example, **CL**, then it is correct to infer $A \supset B$ from $\neg(B \supset A)$ and it is not generally correct to infer A from the premises B and $A \supset B$. If **L** is

---

[2]Where a fragment of a natural language is involved, there is a striking structural circularity: the fragment comprises certain occurrences of sentences; the sentences need to have certain forms; in the occurrence, certain words need to have the meaning fixed by the theory. The importance of the theory will depend on properties of the fragment. The theory may be invoked as a stipulative definition.

intuitionistic logic, neither inference is generally correct. Other logics settle the meaning of ⊃ as detachable in both directions, and then the second inference is correct. So I list for future reference:

(1) Every logic defines its own normative realm.

Most logics in the aforementioned broad sense cannot even be delineated—there are uncountably many and only countably many can be delineated. Most of those that can be delineated are useless, for example, because some of their symbols are tonk-like.[3] Interesting logics fix the meaning of (at least some) interesting logical symbols and moreover have interesting meta-theoretic properties. Interesting logics are logics in a more narrow sense than the one considered, but this changes nothing to (1).

Presumably, Priest agrees with (1) in as far as it concerns formal systems or closure operations. Still, his view is that there is also logic in a more serious sense and for this he rejects (1). He coined the expression *logica ens* (Priest 2014c) to refer to that sense: the facts about consequence that are independent of our theories and of the way people actually reason; "what *is* actually valid: what really follows from what". The reference to this entity is (at least implicitly) present in nearly any of Priest's papers and books. I really tried to understand what *logica ens* might be, but it remains a mystery.

I surmise, but nothing more, that in order to reject (1) for *logica ens*, Priest locates *logica ens* not in a formal language, but in 'the vernacular'. I write that in single quotes because I find the claim baffling. In some publications "the vernacular" is used exclusively (Priest 2014c), in others, it only occurs occasionally and the standard term is "English" (Priest 2006b). I do not see that English might be constructed as a formal-system-like entity, an entity that resembles a formal language plus a logic fixing the meaning of the logical symbols, but that is more complex and more sophisticated. Neither does English seem to display much stability over time, or even space.[4] I rather see natural languages as malleable communication instruments, by which we understand each other, in small communities even on specialized matters like Priest's writings, mainly because we are able to clarify what we mean within the very language.

To some extent, Priest admits these complications (Priest 2014c, §4.2). He states (i) that the words involved in propositions about validity may change their meanings, whence the truth values of the involved sentences change rationally and (ii) that the meaning of logical constants in the vernacular may change rationally and hence that the propositions the vernacular is able to express may also change rationally. But in which way can such changes be combined with a *logica ens*?

---

[3]But caution is advisable. The logic **UCL** extends **CL** in that a **CL**-model is a **UCL**-model iff, in the model, every predicate of rank r has either the rth Carthesian product of the domain or the empty set as its extension (interpretation). Although no **UCL**-model 'fits' the real world, **UCL** is a sensible technical entity in some adaptive logics of inductive generalization (Batens and Haesaert 2001).

[4]Priest argues that *logica ens* may be modified (Priest 2014c), but it seems to me that the dynamics of natural languages reside at a 'deeper' level than the change Priest has in mind.

Suppose for a moment that there is a vernacular that agrees with Priest's view. Next, suppose that the set of sentences of this vernacular remained the same over the years, but that the implication of the vernacular has today a meaning different from the one it had twenty years ago. Let "old vernacular" and "new vernacular" refer to the vernacular before and after the meaning change. Let A be a sentence of this vernacular and Γ a set of such sentences. Suppose that A *really followed* twenty years ago from Γ in the old vernacular, whereas A does not *really follow* from Γ in the new vernacular. Does this not entail that A still *really follows* from Γ in the old vernacular? If so, has logical monism evaporated? Incidentally, even if the old and new implication can be combined into a single logic, this does not save logical monism.

It seems useful to briefly return to syllogistic.[5] To cut a long story short, every *categorial proposition* stands for an equivalence class of occurrences of sentences from the vernacular. Drunk logicians aside, no one ever says "All humans are mortal." Even those talking the vernacular would utter "Sooner or later a person dies" or "We'll all die" or "No one escapes death" or another decent English sentence. However, each such sentence S has also meanings that are not identical to the meanings of the categorial propositions in the equivalence class in which the occurrence of S belongs. This holds even for the categorial proposition itself. I'll spare you the examples of categorial propositions involving negations that cause trouble in most natural languages, idiolects and other vernaculars. Summarizing the central claim:

(2) Logics claimed to pertain to the vernacular actually pertain to a fragment of the vernacular that is selected by the normative realm of those logics.

A last but weighty topic considered in this section concerns the relation between logical monism and epistemic monism. The traditional Western idea, that knowledge is a unified and monolithic body, is hardly more than a very distant ideal. Sometimes it is proposed as an aim or programme—examples are the Vienna Circle's *Einheitswissenschaft* or some physicists' "theory of everything". Often, however, one reasons as if the ideal is already realized or nearby. Or one reasons as if it were the only sensible form of knowledge to pursue, or the only viable idea of 'finished knowledge'. And such unified knowledge body presumably presupposes a unique logic.

Actual knowledge is clearly very different from the so-called ideal. As I have argued for this elsewhere (Batens 2017), I heavily summarize. The present body of knowledge forms a patchwork of partial, incomplete and non-axiomatized theories. These theories stem from a large number of disciplines and subdisciplines. Most such theories are fully unrelated to other theories in the patchwork. Next, for nearly all mathematical theories and for most empirical theories, alternatives are available. Moreover, even a unique *concrete* framework for building a unified body of knowledge is absent—even for integrating economic theories, let alone for integrating them with cognitive psychology, let alone for integrating all that with quantum mechanics,

---

[5]The English Wikipedia page clarifies that the application of syllogistic to the vernacular requires a set of extremely artificial constraints. Comparing with Wikipedia pages in other languages is even more instructive. On the Plattdüütsch (Low Saxon) page, "mütt" (should) is used as a copula in one of the examples.

.... The situation is dramatically complicated *further* by the crucial role of methods and cognitive values.[6] There are theories about these as well as practices and both require a justification, which is about the theories and practices, not part and parcel of them. Next, the methodological theories often cause changes within descriptive theories and *vice versa*; this interaction may generate a complex dynamics involving many hard problems. Descriptive theories, methodological theories, and conceptual systems or languages, including logics, are modified and replaced. All such changes are unpredictable. What I mean is that it is plainly impossible to delineate beforehand a set to which the modified entities will belong.[7]

Suppose for a moment that our knowledge will evolve in the direction of a unified and monolithic body and that **L** will be the logic of this body. Is there any warrant that we, plodding and dabbling in our patchwork, should or even may do so—plodding and dabbling, I mean—in terms of **L**? I do not think so. All we can do is proceed in terms of the logics that underlie the diversity of theories and practices of our patchwork.

The supposition from the previous paragraph seems a bit unrealistic. One of the few things we seem to know is that the complexity of reality is thus that we shall get stuck forever with a patchwork and with a diversity of logics, even deductive logics. Presumably and hopefully, the patchwork will be much more integrated than it is today. This may be accomplished by specifying in a precise way the information flows between the different patches—these need neither be one-sided nor one-shot—and in which way the consequence sets of the enriched chunks are defined. Together with Bryson Brown, Priest has devised a rather general and embracing methodology (Brown and Priest 2004, 2015) which does precisely that and he has shown (Priest 2014b) that logical pluralists may apply the mechanism to their profit. Priest still remains faithful to logical monism. Yet, I wonder how he can be sure that humanity will not be stuck with a patchwork forever.

Until now, I was considering the position that the true logic might be determined by the hypothetical end state of our knowledge, provided this evolves into a unified and monolithic body. If our best knowledge at the end state is such a body, including methods and values and so on, this body should be considered our best hypothesis about the true structure of the world in all its aspects. The language in which the body will be expressed will be our best guess for a suitable language to codify our knowledge. Let us call this the empirical dimension of language.[8] So in this sense,

---

[6]So, yes, methods and values belong to our knowledge. A few positivist fossils aside, everyone agrees on that.

[7]A theory will presumably be required (i) to be a set of statements from a fully unspecified language and (ii) to be presented in a further unspecified way judged acceptable by the future competent community. Had Aristotle anything more specific to delineate the set from which twenty-first-century theories would be taken? And "theory" is an easy case as it need not involve meanings of 'referring' terms. Try redoing the exercise for the conceptual systems of social psychology or string theory.

[8]As in all language use that relates to the world, there is an interaction between conventional elements and properly empirical elements in the empirical dimension *of language*. This is not the place to spell that out.

there is an empirical aspect to the formal logic of the suitable language and to its conceptual system; the latter settles the informal logic of the language.[9]

Actually, Priest seems committed to a position that comes close to the one from the previous paragraph, except that he seems to think that the true logic can change over time (Priest 2006a, Chap. 10). I find this puzzling. Suppose that Priest's arguments for **LP** hold water. Let him throw in his preferred relevant implication. And let us not complain about all the logical terms (counterfactual implications and other conditionals, non-standard connectives, non-standard quantifiers and so on) that occur in natural languages—sloppily supposing that a conservative extension of **LP** takes care of all of them. But that does not make **LP** into the right logic on that position. Today we still have the patchwork. Where **CL** is taken to be the underlying logic, Priest might argue that the underlying logic is actually an inconsistency-adaptive logic—all of them define the same consequence set as **CL** if the theory is consistent (Batens 2015). But what about quantum mechanics? And what about many other patches, including methodological ones? What about constructive mathematical theories that clearly are patches of our present knowledge system? And so on. Take even set theory. Priest justly considers Zach Weber's Fregean set theory (Weber 2010a, b, 2013) as the best available paraconsistent set theory. This set theory, however, does not proceed in terms of Priest's favourite relevant implications. So in order to integrate this set theory into the knowledge body, Priest needs a mechanism that would just as well work for the patchwork, possibly one of the aforementioned 'chunk and permeate' mechanisms. Or else Priest needs to show that Weber's set theory survives the addition of *all* relevant implications that Priest allows within his language, viz. in the vernacular.[10] Moreover, Priest will still need to consider other set theories, like **ZF** and the other supposedly consistent ones, because these are not fragments of Zach Weber's. So, even if Priest is right in other respects, I do not see that he would, according to his own criteria, have identified 'the true logic' or even that he would have shown that such an entity exists.

There is an excellent reason to believe that we shall never get beyond the patchwork stage. We need to think *about* our knowledge in order to improve it or even to find out whether it is justified. Such thinking requires that we give up, at least for the sake of this argument, the knowledge elements about which we are thinking.[11] It is not essential for my argument, but let me add that this also applies to logic—the paragraph on the new and the old vernacular illustrates that. Anyway, we need the

---

[9]Informal logic is meant in the traditional sense here. Carnap (1947) tried to push informal logic into meaning postulates. This heavily restricts the possible meaning relations because it requires a theory rather than an underlying algebraic structure.

[10]The central equivalence sign in the abstraction axiom will still be Weber's, but every implication of the language may occur within the open formula in that axiom. This will impose restrictions on the interaction between the implications. Thus, where $\rightarrow$ is Weber's implication and $\Rightarrow$ *any* implication Priest allows in his language, the inference from $A \rightarrow (A \Rightarrow B)$ to $A \Rightarrow B$ should be invalid.

[11]This is a grave understatement. It is well known that creative scientific work requires a patchwork that is not only complex but also dynamic (Nickles 1980a, b, 1981; Meheus 2000).

patchwork and better hope that it remains available forever. Having a unified and monolithic theory locks one in a straitjacket nailed to a dungeon wall. It equates the perfect dogmatic closure.

8.3 Natural Languages

There are many convincing arguments for the view that we need paraconsistent logics. Yet, are Liar arguments really exciting? I shall not mention the common suspicions against them. In order to show, as Priest candidly does, by a Liar argument that explosive negations are tonk-like operators, one needs some presuppositions. One of Priest's presuppositions is clearly that natural languages are formal-system-like entities. The analogy seems fundamentally wrong to me. Is it not more appropriate to see natural languages as malleable communication instruments, as stated in Sect. 8.2? Again, I shall not discuss the topic, but rather stress that the view may be given a formal underpinning. In this way, I hope to show that there is a sensible alternative for Priest's view on natural language and that Priest's view is not obviously correct.

The idea of a malleable communication instrument is that the words of the language do not have a stable meaning, but may be given new meanings to express things that could not be expressed by the language as previously used. If a phrase is available that the 'speaker' knows to be ambiguous, it may be disambiguated. If no phrase (or word) is available, the speaker may give a special meaning to an existing one, possibly through a metaphor, or coin a new one. This may require an amount of communication about meanings rather than about the topic under discussion, but we all know how to do that. If the speaker and addressee can both interfere, the communication about meaning can often be drastically restricted. A reader of a text has to proceed more carefully, especially if the text is not written for this type of reader or is phrased clumsily.

It is quite obvious that communication about meanings takes place. That atomistic ideas, like those of Democritos, had remained part of the culture, made it easy for people like Dalton to propagate atomistic chemistry as an alternative for the existing qualitative chemistry. Galileo had a more difficult job to explain relative motion to his contemporaries. Ideas that did not belong to the logical space previously may originate and spread. In this way, *new possibilities* are generated.[12] While this malleability of non-logical terms is obvious from the history of the sciences, the situation is similar for logical terms. The word "and" may have a temporal meaning, in which case A-and-B is not a consequence of $\{A, B\}$. The word "if" is often meant as an equivalence, which is just one of the many reasons to be careful in interpreting Wason test results. And many more examples may be given.

---

[12]These are informal logical possibilities in the sense that they depend on meanings of non-logical terms. Nice relevant work on possibility was done by Nicholas Rescher (Rescher 2003, 2005; Almeder 2008).

Is it possible to communicate if words have no fixed meaning? We can because we do. This will be more convincing: it is possible to formally make sense of such languages and their use as I have shown in writing (Batens 2016) and shall now explain. That words do not have a fixed meaning does not prevent them from having a likely meaning, or a couple of them, possibly ordered. Let us proceed stepwise.

By way of a purely pragmatic decision, take **CL** as our preferred logic and hence as the norm. For the subsequent reasoning, start thinking about **CL**-models and allow them gradually to be transformed. If a model M allows for $v_M(\neg A) = 1$ while $v_M(A) = 1$, I shall say that M displays a *negation glut*.[13] A *negation gap* to the contrary is obtained when $v_M(A) = 0$, but nevertheless $v_M(\neg A) = 0$. A *conjunction glut* occurs in M if either $v_M(A) = 0$ or $v_M(B) = 0$ but nevertheless $v_M(A \wedge B) = 1$. A *conjunction gap* occurs in M if $v_M(A) = v_M(B) = 1$ but nevertheless $v_M(A \wedge B) = 0$. And so on for the other connectives, for the quantifiers and for identity. Summarizing, where ξ is the central logical symbol of **A**, if the model M is such that the conditions are sufficient for $v_M(\mathbf{A}) = 0$ in a **CL**-model M, but nevertheless $v_M(\mathbf{A}) = 1$, then there is a ξ-glut[14]; if the model M is such that the conditions are sufficient for $v_M(\mathbf{A}) = 1$ in a **CL**-model M, but nevertheless $v_M(\mathbf{A}) = 0$, then there is a ξ-gap.

It is possible to consider all logics that allow for gluts or gaps with respect to a logical symbol or for the combination of any selection of gluts with any selection of gaps. With respect to referring terms, a similar step may be taken, as was first suggested by Vanackere (1997). The idea is simple: replace within the language every non-logical term ξ by a numerically superscripted one ξ^i and next replace every $\Gamma \vdash A$ by $\Gamma^\dagger \vdash A^\ddagger$, obtained by replacing in every member of Γ and in A every non-logical term ξ by an indexed non-logical term ξ^i in such a way that no two identical non-logical terms ξ^i occur in the translation $\Gamma^\dagger \vdash A^\ddagger$—the accurate transformation is just a trifle more complicated.

So now we have logics that allow for gluts or for gaps or for ambiguities or for combinations of those—there is a naming system that identifies each combination. The logic that allows for all gaps and gluts and ambiguities is called **CLØI**. This is defined over the language in which the non-referring terms are superscripted. Zero logic, **CLØ**, is defined over the standard language: $\Gamma \vdash_{\mathbf{CLØ}} A$ iff $\Gamma^\dagger \vdash_{\mathbf{CLØI}} A^\ddagger$.

Note that (i) $p \wedge q \not\vdash_{\mathbf{CLØ}} p$ because there are models M, allowing for conjunction gluts, such that $v_M(p \wedge q) = 1$ and $v_M(p) = 0$, but also that (ii) $p \not\vdash_{\mathbf{CLØ}} p$ because $p^1 \not\vdash_{\mathbf{CLØI}} p^2$. So **CLØ** is justly called *zero logic*: nothing follows from anything, not even a premise from itself.

The fascinating bit happens when one goes adaptive, viz. when one moves to the logic **CLØ**$^\mathrm{m}$. Here abnormalities are minimized and in this case abnormalities

---

[13] Allowing for negation gluts in an indeterministic semantics is simple enough: remove the clause "if $v_M(A) = 1$, then $v_M(\neg A) = 0$" and keep "if $v_M(A) = 0$, then $v_M(\neg A) = 1$". In a deterministic semantics, the valuation function v_M is determined by the model, which here is $M = \langle D, v \rangle$ with D a set and v an assignment function. I refer to literature for details (Batens 2016).

[14] This may be phrased in a perfectly unambiguous way (Batens 2016). As the reader expected, **A** is a metametalinguistic variable. If you do not like meta talk: it is a variable for variables for formulas.

are the different gluts and gaps, and the ambiguities.[15] For any specific premise set, respectively set of non-logical axioms of a theory, $\mathbf{CL}\emptyset^m$ delivers a minimally abnormal interpretation of the premise set. And so do some other adaptive logics of the described family—those logics allow only for a selection of gluts or gaps or ambiguities.[16] Moreover, the application of $\mathbf{CL}\emptyset^m$ reveals which of the other logics of the family are suitable. Put differently, the application of $\mathbf{CL}\emptyset^m$ reveals the sensible minimally abnormal interpretations of premise sets. Some premise sets have minimally abnormal interpretations in logics that allow for one of a few selections of abnormalities; other premise sets have more minimally abnormal interpretations.[17]

Here are two of the many interesting facts about the logics in this family.

(3) If Γ has $\mathbf{CL}$-models, then $Cn_{\mathbf{L}^m}(\Gamma) = Cn_{\mathbf{CL}}(\Gamma)$ for every adaptive logic $\mathbf{L}^m$ of the family.
(4) If Γ has no $\mathbf{CL}$-models, whence $Cn_{\mathbf{CL}}(\Gamma)$ is trivial, then, some border cases aside,[18] there are adaptive logics $\mathbf{L}^m$ of the family such that $Cn_{\mathbf{L}^m}(\Gamma)$ is not trivial, but is a minimally abnormal interpretation of Γ.

The significance of these results is that one is able to handle languages in which the meaning of symbols is an empirical matter, viz. a matter determined by the text one is interpreting. So one may see this as a formal hermeneutics. Applying this to a natural language, logical words have a normal meaning, but may be used with a different meaning within a text. Non-logical words and phrases have normally the same meaning throughout a text, but some texts are ambiguous. When reading a text, we take it to be normal. If it turns out to be abnormal—the normal interpretation is then trivial—we shall try to find a maximally normal interpretation. It is very natural that there are several of these. All this nicely agrees with the behaviour of $\mathbf{CL}\emptyset^m$.

With respect to the application to natural languages, some people, especially those with sympathies for relevant implications, will complain that I chose $\mathbf{CL}$ as a 'the standard' for logical symbols. That choice was merely conventional. One does not need a new idea, only more work, to adjust the construction to another standard.

More work is required in other respects as well. For example non-logical linguistic entities are handled rather simplistically by $\mathbf{CL}\emptyset^m$, viz. as either ambiguous or unambiguous. A more realistic approach will associate several meanings with each such entity, and will give each of the meanings a definite weight. Such weights will obviously depend on the 'speaker', or rather on what we know about the speaker, on what

---

[15] Intuitively, there is an ambiguity in a model M when, in the superscripted language, non-logical symbols that merely differ in their superscript, like ξ^i and ξ^j, have a different meaning in M.

[16] The same holds for many adaptive logics outside the family. Infinitely many logics allow for certain gluts or gaps and there are several strategies to minimize abnormalities.

[17] Consider a premise set that has a minimally abnormal interpretation allowing for negation gluts and another one allowing for existential gaps. Any adaptive logic (from the family) that moreover allows for further abnormalities will obviously also define a minimally abnormal interpretation. However, those interpretations will be less interesting. Their consequence sets will contain many disjunctions of formulas where adaptive logics that allow for a smaller selection of abnormalities will have the disjuncts as consequences.

[18] One of the border cases is where Γ is itself the trivial set.

we know about whom the speaker considered to be the addressee, and all the further good old stuff from pragmatics and text pragmatics. This requires a sophistication of $\mathbf{CL\emptyset^m}$, but nothing that cannot be handled by an adaptive logic.

So considering a natural language as a malleable communication instrument is not just a muddy idea but may be explicated by formal means. Note that a lot of malleability is possible and that the approach allows for languages that are shared by innumerable linguistic communities, defined by region, specialization as well as by socioeconomic factors. Similarly, the approach makes it easier to handle such phenomena as dialect continua, for example, the one spreading from the Walloon region in the North to Portugal in the West and Southern Italy in the East, circumscribing five or more official 'languages'. The approach makes it easier to handle a number of further phenomena: that natural languages can be learned in the way that children learn them, that they evolve and so on. It also makes it easier to understand the sciences. For example, we can handle the patchwork formed by our knowledge as described in Sect. 8.2. Another example is that drastically new concepts may be spread so rapidly within the relevant scientific community.[19]

All this obviously does not prove that the formal-system-like approach to natural languages is mistaken. It does show, however, that there are other approaches and that some are capable of solving certain problems that were not yet solved on the formal-system-like approach.

Hopefully, the alternative approach also clarifies why I do not see the Liar as a major problem. Natural languages have several sorts of odd *occurrences* of sentences, that we keep outside the scope of our truth theory.[20] Adding another sort to house certain occurrences of the Liar sentence might be not too high a cost. The sort may be delineated by an adaptive criterion—concepts such as 'disjunctions of abnormalities' and several results may just be copied from the metatheory of adaptive logics. We still need paraconsistency of course, for so many reasons. Yet, we should be less afraid of explosive negations within a pluralistic setting.

8.4 Models

Let me remind the reader that, in this section, I shall follow Priest's way of using "true" and "false".

Since a long time, I felt a certain uneasiness with Priest's **LP**-models and tried to articulate it. There are two main elements to this. First, the crucial reference to natural language for the attack, in terms of the Liar, on explosive negation. A typical statement (Priest 2006b, p. 16) comes to this: If "notions necessary for the formulation of the [extended Liar] paradox [...] are not expressible in the language

---

[19] To explain in which way new concepts originate is a different matter. As $\mathbf{CL\emptyset^m}$ is described here it is not helpful for that purpose.

[20] It is a matter of occurrence indeed. Imagine "Mary loves Bill" written on the wall, and underneath it "This sentence is false", especially when written by another hand.

in question [of the Valuegappist]", then "the language for which the semantics has been given is not English, since these notions obviously are expressible in English." The second main element is a certain tension between on the one hand a classical or inconsistency-adaptive understanding of the semantics and on the other hand a paraconsistent understanding.

The original **LP**-semantics proceeded in terms of functions that assign to formulas values from $\{\{1\}, \{1, 0\}, \{0\}\}$, respectively denoting true, false and true, and false.[21] This is represented on the two first lines of the table.

Priest's semantics	$v(A) = \{1\}$	$v(A) = \{0, 1\}$	$v(A) = \{0\}$
Priest's terminology	A is true	A is true and false	A is false
classical terminology	$M \Vdash A$	$M \Vdash A$	$M \nVdash A$
	$M \nVdash \neg A$	$M \Vdash \neg A$	$M \Vdash \neg A$
bivalent semantics	$v(A) = 1$	$v(A) = 1$	$v(A) = 0$
	$v(\neg A) = 0$	$v(\neg A) = 1$	$v(\neg A) = 1$

The logic **LP** has also a bivalent semantics, assigning to formulas a value from $\{0, 1\}$. In the bivalent semantics, $v(\neg A)$ is obviously not a truth function of $v(A)$. Given the properties of **LP**, one has to characterize a deterministic model by specifying $v(A)$ as well as $v(\neg A)$ for every sentential letter A.[22] Semantic consequence is defined by Priest as follows: $\Gamma \vDash A =_{df}$ for all v, if $1 \in v(B)$ for all $B \in \Gamma$, then $1 \in v(A)$. The classical definition, proceeding in terms of the bivalent semantics, is very similar: $\Gamma \vDash^c A =_{df}$ for all v, if $v(B) = 1$ for all $B \in \Gamma$, then $v(A) = 1$.[23]

Both semantic systems are equivalent from a technical point of view.[24] To see that, it helps to realize that, in Priest's semantics, $1 \in v(\neg A)$ ($\neg A$ is true) iff $0 \in v(A)$ and $0 \in v(\neg A)$ ($\neg A$ is false) iff $1 \in v(A)$. The classical description of either semantics is transparent if one remembers that $M \Vdash A$ is read as "M verifies A" or "A is true in M" and $M \nVdash A$ is read as "M falsifies A" or "A is false in M". This "false" has a different meaning from Priest's, who considers the classical meaning wrong. There may be a similar difference with respect to "falsifying", but both descriptions can agree on "verifying". Note that the slash in $\nVdash$ stands for a classical negation, which is a tonk-like operator for Priest.

In the sequel of this section, I shall only talk about Priest's semantics and shall only use Priest's terminology, except where I explicitly state not to do so. I shall also, as in the above table, identify a model M with a valuation v.

---

[21] True and false here obviously stand for true-in-a-model and false-in-a-model.

[22] Priest's semantics has positive and negative extensions for predicates and these require no complications for the transition to the bivalent semantics. And the same trick would work for sentential letters.

[23] Actually, the implications in both definitions are different. For readers not familiar with the matter, just neglect that for the present paper.

[24] Actually, they are equivalent—for all Γ and A, $\Gamma \vDash A$ iff $\Gamma \vDash^c A$—provided Priest's semantics is consistent with respect to $\vDash$: there is no Γ and A such that $\Gamma \vDash A$ and $\Gamma \nvDash A$.

8 Looting Liars Masking Models

Relational semantics Where L is "L is false only",[25] L is true in the actual model, the model that corresponds to the state of reality, just in case its truth-value is $\{0\}$. By the T-schema and given what L states, $1 \in v(L) \leftrightarrow v(L) = \{0\}$. This readily gives one $1 \in \{0\}$ and $1 = 0$ and triviality. In 1993 Priest points out (Smiley and Priest 1993) that triviality is avoided when the conventional choice for functions is replaced by the choice for a relation. So, $x \in v(L)$ will be replaced by $\text{Rel}(L, x)$ where Rel is a binary relation between the set of formulas and $\{0, 1\}$. What was formerly noted as $1 \in v(L) \leftrightarrow v(L) = \{0\}$ and caused triviality, becomes

$$\text{Rel}(L, 1) \leftrightarrow (\text{Rel}(L, 0) \wedge \neg \text{Rel}(L, 1)). \tag{SL}$$

Call this (the modified or corrected formulation of) the Semantic Liar—it proceeds in terms of truth-in-a-model rather than in terms of truth. Priest requires that $\text{Rel}(L, 1) \vee \text{Rel}(L, 0)$ and that the equivalence in (SL) is a relevant contraposable equivalence, which here stems from the T-schema.

Priest points out (Priest 2006b, pp. 288–289) that, with this reformulation, a contradiction, viz.

$$\text{Rel}(L, 1) \wedge \neg \text{Rel}(L, 1) \wedge \text{Rel}(L, 0), \tag{8.1}$$

is still derivable, but that triviality is not.[26] I refer the reader to the original paper (Smiley and Priest 1993), but especially to Sect. 20.3 of the second edition of *In Contradiction* (Priest 2006b). There one obvious point is mentioned: that Disjunctive Syllogism is invalid with respect to the paraconsistent negation $\neg$. There also is a more informative point: that even when (dialetheic) sets like $\{x \mid \text{Rel}(A, x)\}$ are invoked, which by all means is a legitimate move, triviality does not follow. Originally, I was very enthusiastic about the reformulation. For one thing, it also answered some complaints I had had and had raised in writing. But then doubts returned.

Let us consider a graphic representation of the old **LP**-models with respect to a formula A—I'll keep the discussion at the propositional level as long as possible. Call this the *O-representation*:

A-type-1	A-type-2	A-type-3
$v(A) = \{1\}$	$v(A) = \{0, 1\}$	$v(A) = \{0\}$

One might naively think to translate this to the relational approach as follows:

A-type-1	A-type-2	A-type-3
$\text{Rel}(A, 1) \wedge \neg \text{Rel}(A, 0)$	$\text{Rel}(A, 1) \wedge \text{Rel}(A, 0)$	$\neg \text{Rel}(A, 1) \wedge \text{Rel}(A, 0)$

---

[25] That is: L is false and not true. Some will prefer "⌜L⌝ is false" in which ⌜L⌝ is a name for L. I shall continue to use formulas as names for themselves as this causes no confusion in the present paper.

[26] The contradiction is an obvious **LP**-consequence of $\text{Rel}(L, 1) \vee \text{Rel}(L, 0)$, (SL), and the **LP**-theorem $\text{Rel}(L, 1) \vee \neg \text{Rel}(L, 1)$.

However, this is clearly mistaken. A formula A may in the same model relate as well as not relate to a truth value. So there are eight A-types rather than three. Call the following the *R-representation*:

	1.1	1.2	2.1	2.2	2.3	2.4	3.1	3.2
Rel(A, 1)	×	×	×	×	×	×		
¬Rel(A, 1)		×			×	×	×	×
Rel(A, 0)			×	×	×	×	×	×
¬Rel(A, 0)	×	×		×		×		×

where $x \in \{0, 1\}$, at least one of Rel(A, x) and ¬Rel(A, x) holds in a model. Exactly one of them holds in models that are Rel-consistent with respect to A. Note that A-type-2.1 models are Rel-consistent but that both A and $\neg A$ are true in them.[27]

In the old semantics, A is true as well as false in some **LP**-models. However, the semantic machinery itself is apparently consistent: $v(A) = \{1\}$ and $0 \in v(A)$ jointly engender triviality; and few readers will have considered the case where $v(A) = \{1\}$ and $1 \notin v(A)$.[28] In order to avoid that an extended liar causes triviality, Priest moved to the relational semantics. This allowed him to take advantage of the paraconsistent negation within the semantic machinery and not only with respect to 'object-language' statements. For all we know, the move is effective. Moreover, it is natural from Priest's viewpoint, which considers the vernacular as a single entity, not as a hierarchy of languages. However, the move has also some unexpected effects.

Priest claims (Smiley and Priest 1993, p. 51): "The catastrophic results [viz. the triviality derived on the old approach] therefore appear as spin-offs of a conventional form of the representation used, not of the facts themselves". It certainly is correct that Rel-free formulas can be false, true or both in a model, just as before; and that the old 'representation' misrepresented 'the facts', as I now further clarify.

Trouble? What follows is put under a separate heading in order to introduce some conventions that pertain only to stuff within this division. The only entities considered here are propositional **LP**, the language of propositional **LP**, and the relational semantics of propositional **LP**. $\mathcal{S}$ will denote the set of sentential letters, $\mathcal{W}$ the set of propositional formulas built from $\mathcal{S}$.[29] Unless specified otherwise, $A, B \in \mathcal{W}$, $\Gamma \subseteq \mathcal{W}$ and $\Delta \subseteq \mathcal{W}$. An x as in Rel(A, x) is always a member of $\{0, 1\}$. The $\pm$ as in $\pm$Rel(A, x) will stand either for ¬ or for the empty string. Define $\text{Rel}_M^{\mathcal{S}}$ as the set of all expressions $\pm$Rel(A, x), in which $A \in \mathcal{S}$, that hold in M; define $\text{Rel}_M^{\mathcal{W}}$ as the corresponding set for $A \in \mathcal{W}$.

Where $A \in \mathcal{S}$, expressions $\pm$Rel(A, x) are semantically primitive. Provided that at least one of Rel(A, x) and ¬Rel(A, x) holds in M and that at least one of Rel($A, 0$)

---

[27] The numbering is of course conventional. In an A-type-$x.y$ model hold all statements true on the classical description of the A-type-x model (but now read paraconsistently).

[28] Yet paraconsistent set theories have some inconsistent sets, viz. sets of which certain entities are members as well as non-members.

[29] So I shall never write expressions like Rel(Rel($A, 1$), 1). Many results proven for $\mathcal{W}$ are provable in general, but this is a worry for later.

8 Looting Liars Masking Models

and Rel(A, 1) holds in M, nothing simpler forces ±Rel(A, x) to hold in M or prevents it from holding in M. The following fact results from this.

Fact 8.1 *For all M and for all $A \in \mathcal{S}$, there is a M' such that $\mathrm{Rel}_{M'}^{\mathcal{S}} = \mathrm{Rel}_{M}^{\mathcal{S}} \cup \{\pm\mathrm{Rel}(A, x)\}$.*

The usual induction leads from Fact 8.1 to Fact 8.2.

Fact 8.2 *For all M and for all $A \in \mathcal{W}$, there is a M' such that $\mathrm{Rel}_{M'}^{\mathcal{W}} \supseteq \mathrm{Rel}_{M}^{\mathcal{W}} \cup \{\pm\mathrm{Rel}(A, x)\}$.*

Definition 8.3 $\Gamma \vDash A$ iff A is true in all models in which all members of Γ are true.

In line with Priest, I take $\Gamma \nvDash A$ to obtain iff A is untrue in a model in which all members of Γ are true—A is *untrue* in M iff ¬Rel(A, 1) holds in M. Fact 8.2 entails the second halves of Facts 8.4 and 8.5.

Fact 8.4 *For some Γ and A, $\Gamma \vDash A$; for all Γ and A, $\Gamma \nvDash A$.*

Fact 8.5 *There are logical truths; nothing is a logical truth (for all A, $\nvDash A$).*

Let reflexive, monotonic and transitive be defined as usual for consequence relations.

Fact 8.6 *LP is reflexive and non-reflexive, monotonic and non-monotonic, transitive and non-transitive.*

Actually the matter is more extreme: $A \nvDash A$ for all A; for all Γ, Δ and A, if $\Gamma \vDash A$, then $\Gamma \cup \Delta \nvDash A$; for all Γ, Δ and A, if $\Gamma \vDash B$ for all $B \in \Delta$ and $\Delta \vDash A$, then $\Gamma \nvDash A$.

Definition 8.7 Γ is non-trivial iff there is a formula A such that $\Gamma \nvDash A$.

Fact 8.8 *Some Γ are trivial and all Γ are non-trivial.*

Definition 8.9 Γ is inconsistent iff, for some A, $\Gamma \vDash A$ and $\Gamma \vDash \neg A$.

Fact 8.10 *Some Γ are inconsistent and all Γ are consistent.*

Is any of these results bad? Maybe not. Maybe they clarify certain concepts from a dialetheic point of view. If dialetheism is right, we have to change certain insights and what precedes are just some more?

Of course, much of what precedes comes rather unexpected, definitely for some. Moreover, certain concepts seem to lose content. If, for all Γ and A, there is a countermodel for $\Gamma \vDash A$, then apparently the notion of a countermodel does not mean much. And what is the point of constructing a countermodel if the skies rain down countermodels? Does non-triviality mean much if every set of formulas is non-trivial? And most people speaking English definitely expect that the notion of consistency is consistent with the effect that no inconsistent set is also consistent.

Soundness also obtains odd properties. If $\Gamma \not\vDash A$, then soundness seems to require that $\Gamma \not\vdash A$ in some or other sense and I do not see any such sense for all Γ and A. Nothing in **LP**-proofs or in **LP**-tableaux seems to correspond to the universality of the absence of the semantic consequence relation. Is this unavoidable?

The matter is embarrassing in a personal way. Did I miss something? I have read most of Graham's writings and what precedes seems heavily contradicted there—I mean pragmatically contradicted. Even the proof of Fact 1 from *In Contradiction* is mistaken if what precedes is correct.[30] And what is the point of painstaking proofs of the non-triviality or consistency of a theory or fragment iff those concepts mean what they turn out to mean in Facts 8.8 and 8.10?

Another reason why the present results seem mistaken is that they overlook certain distinctions that anyone can see. While $\Gamma \not\vDash A$ holds for all Γ and A, there are Γ and A for which $\Gamma \vDash A$ holds and there are other Γ and A for which $\Gamma \vDash A$ doesn't hold. The problem is to express this distinction while remaining faithful to the dialetheic tenets. There is a similar problem in identifying or describing models—see below under the heading "Describing the semantics".

So is there a way around the present results? It seems relevant that, apparently, Priest wants to introduce only concepts that do not in any possible circumstances engender triviality themselves. Can this be realized without warranting non-triviality by definition? And how well can such concepts resemble the concepts available within the vernacular?

An infinite hierarchy? Complications may not end here. Call A a formula of thickness 0 if "Rel" does not occur in A and, where A is of thickness n, call $\text{Rel}(A, x)$ and $\neg \text{Rel}(A, x)$ formulas of thickness $n + 1$. As the *R-representation* shows, $\neg \text{Rel}(A, 1)$ is not only not a function of $\text{Rel}(A, 1)$. It is not a function of $\{\text{Rel}(A, 1), \text{Rel}(A, 0)\}$ either. Apparently, $\text{Rel}(A, x)$ entails $\text{Rel}(\text{Rel}(A, x), 1)$ and conversely.[31] Yet, by analogy to the above reasoning, $\neg \text{Rel}(\text{Rel}(A, 1), 1)$ is not a function of $\{\text{Rel}(\text{Rel}(A, 1), 1), \text{Rel}(\text{Rel}(A, 1), 0)\}$. This may be repeated *ad nauseam*. In other words, the *O-representation* of a **LP**-model concerns only formulas of thickness 0, and hence is terribly incomplete. Yet, the *R-representation* is hardly more complete. It fails as representation of the *full* truth status of A with respect to a **LP**-model.

Priest wants double negation and de Morgan properties to apply. So he presumably also wants to reduce the present infinity of ¬Rel-concatenations. A full reduction is obviously impossible. If, for example, one were to stipulate, for ↔ contraposable and detachable, $\neg \text{Rel}(A, x) \leftrightarrow \text{Rel}(A, 1 - x)$, then all A-type-2 models, which are characterized by $\text{Rel}(A, 1) \wedge \text{Rel}(A, 0)$, would by definition be

---

[30] The proof stems from the first edition and obviously has to be adapted to the relational semantics. However, precisely this is impossible in general. As, for every A, $\neg \text{Rel}(A, 1)$ holds in some model M, there cannot be a method to turn M into a **CL**-model in which A is false. Next, even metatheoretic proofs new to the second edition are in terms of the old set-theoretic semantics.

[31] Formulas in which Rel is iterated are required. Remember indeed that we are devising a theory about the semantics of English. The **LP**-semantics is the part that handles the traditional logical terms.

inconsistent with respect to Rel-expressions: $\text{Rel}(A, 1) \wedge \neg\text{Rel}(A, 1)$ as well as $\text{Rel}(A, 0) \wedge \neg\text{Rel}(A, 0)$ would hold.[32] Similarly, A-type-1 models and A-type-3 models would *by definition* be consistent with respect to Rel-expressions; contraposing: whether x is 0 or 1, $\text{Rel}(A, x) \wedge \neg\text{Rel}(A, x)$ would entail $\text{Rel}(A, 0) \wedge \text{Rel}(A, 1)$. But all that seems wrong.

If $\text{Rel}(A, 1) \wedge \neg\text{Rel}(A, 1)$, or $\text{Rel}(A, 0) \wedge \neg\text{Rel}(A, 0)$, there is clearly something unexpected the matter with the relation between A and its truth-value in the model. This situation is obviously possible, but it is different from, and definitely not a mere consequence of, A and $\neg A$ being both true in the model, which Priest codes as $\text{Rel}(A, 1) \wedge \text{Rel}(A, 0)$. The latter conjunction entails that something unexpected is the matter with A in the considered model, viz. that A behaves inconsistently. The preceding conjunctions entail that something unexpected is the matter with truth-in-a-model. A dialetheist should clearly allow for the possibility that our definitions of models result in inconsistent models. As noted before, that a contradiction is true in some models need not entail that truth-in-a-model is inconsistent in them. Similarly, considering a model of a consistent set of Rel-free formulas does not provide a warrant for the consistency of the set of Rel-formulas that hold in the model. Suppose, for example, that arithmetic—the set of formulas verified by the standard model—is inconsistent. In this case, there is an inconsistency in the method by which infinite predicative models are defined[33] and every infinite **CL**-model will presumably be trivial—even finite such models may be affected because the standard language schema has infinitely many schematic letters. So stipulating $\neg\text{Rel}(A, x) \leftrightarrow \text{Rel}(A, 1 - x)$, for $\leftrightarrow$ detachable, does not seem justifiable. The same obviously holds for stipulating $\neg\text{Rel}(A, x) \leftrightarrow \text{Rel}(\neg A, x)$. Priest's view on untruth (Priest 2006b, §4.9) agrees with this conclusion.

While I have argued against a certain way of reducing the infinity of $\neg$Rel-concatenations, a different way to reduce infinite concatenations to finite ones seems harmless. What I mean is requiring $\neg\text{Rel}(A, x) \leftrightarrow \text{Rel}(A, 1 - x)$, or even the slightly stronger $\neg\text{Rel}(A, x) \leftrightarrow \text{Rel}(\neg A, x)$, whenever A is itself a pure Rel-formula—a formula in which every non-logical symbol occurs within the first argument of Rel. The result of this change is that we shall still have the eight A-type models from the *R-representation*, but that there will be no further A-type models. Inconsistencies will spread upwards in nested formulas. Thus A-type-1.2 models will verify $\text{Rel}(\text{Rel}(A, 1), 1) \wedge \neg\text{Rel}(\text{Rel}(A, 1), 1)$ as well as all other entailed inconsistencies, such as $\text{Rel}(\neg\text{Rel}(A, 1), 1) \wedge \neg\text{Rel}(\neg\text{Rel}(A, 1), 1)$, but no further inconsistencies will be verified by any A-type-1.2 models. In as far as I can see, this approach still prevents an extended Liar from generating triviality.

Is the reduced approach to be preferred over the one that allows for infinite concatenations of consistency choices? This depends on what one wants. I prefer the infinite concatenation. The reason is that I am convinced that one can talk consistently about many an inconsistent theory T_1 and that, even if the 'theory', say T_2 in which one describes T_1 *needs* to be inconsistent, then, more often than not, there will be a

---

[32] One of them still holds if $A \leftrightarrow B$ is the 'material equivalence' $(\neg A \vee B) \wedge (A \vee \neg B)$.

[33] I am considering classical models. Yet, going paraconsistent does not remove the problem.

consistent T_3 that describes T_2. As I see it, whatever T_1, there is always a sufficiently high i such that T_i is consistent. Consider again A-type-1.2 models. Some of them verify an infinity of contradictions made up from A, $\text{Rel}(A, x_1)$, $\text{Rel}(\text{Rel}(A, x_1), x_2)$, $\text{Rel}(\neg\text{Rel}(A, x_1), x_2)$, ..., with $x_1, x_2, \ldots \in \{0, 1\}$. However, some model M_1 verifies all of the following formulas[34]

$$\begin{array}{ll} \text{Rel}(\neg\text{Rel}(A, 0), 1) & \neg\text{Rel}(\neg\text{Rel}(A, 0), 0) \\ \text{Rel}(\text{Rel}(A, 1), 1) & \neg\text{Rel}(\text{Rel}(A, 1), 0) \\ \text{Rel}(\text{Rel}(A, 0), 0) & \neg\text{Rel}(\text{Rel}(A, 0), 1) \\ \text{Rel}(\neg\text{Rel}(A, 1), 1) & \neg\text{Rel}(\neg\text{Rel}(A, 1), 0) \end{array}$$

and no contradiction outside $\text{Rel}(A, 1) \wedge \neg\text{Rel}(A, 1)$. If A has thickness 0 and nothing requires that there is an inconsistency in formulas of thickness 2 or more, then my preference would be to go for the consistent description of thickness 2.

The reader may note that the formulas of the different thicknesses belong to the same theory, viz. the theory describing a specific **LP**-model. That is correct. However, every formula that holds in the model and the thickness of which is i or less, is entailed by the set of formulas of thickness i that hold in a model. So, given the considered reduction, this set can be considered as a theory about the model. The general claim, in terms of $\{T_1, T_2, \ldots\}$ concerns also cases where these T_i are not themselves fragments of a single theory.

Priest will presumably prefer the aforementioned reduction to thickness 1 because of its similarity with the behaviour of negation within **LP**-formulas (of thickness 0). Presumably, he will disagree with my claims on the existence of a consistent T_i provided i is chosen large enough. Indeed, one can construct a Liar in any T_i. The day that I would be interested, I would do so and have my consistent description of it in T_{i+1}. Is that the true and final description? No. Will it do in all circumstances? No. Our knowledge is in a patchwork state anyway. And one can always accommodate sensible problems by moving up to the next theory. Or so I think.

Top-down spreading of inconsistency. A few paragraphs ago, I argued for the independence of two *loci* of inconsistency, inconsistency of formulas that contain typical terms from the semantic metalanguage and inconsistencies in other formulas. Priest agrees that there is no equivalence between both, but still holds that there is an implication.[35] In terms of Rel and a detachable implication, Priest accepts (8.2) but rejects (8.3).

$$\neg\text{Rel}(A, x) \rightarrow \text{Rel}(A, 1 - x) \quad (8.2)$$
$$\text{Rel}(A, x) \rightarrow \neg\text{Rel}(A, 1 - x) \quad (8.3)$$

Needless to say, I shall not defend (8.3) either. Inconsistent models verify for some A, $\text{Rel}(A, 1)$ as well as $\text{Rel}(A, 0)$. So affirming $\text{Rel}(A, x) \rightarrow \neg\text{Rel}(A, 1 - x)$ comes to spreading inconsistency without need. I think to have sufficiently clarified this before.

---

[34] Each formula to the left is entailed by the one to the right in Priest's understanding.

[35] The matter may be phrased in terms of truth, falsity and untruth.

But does the same not hold for (8.2) too? Obviously, if A is a linguistic entity of the right sort, in the present case a closed formula, we have

$$\text{Rel}(A, 1) \vee \text{Rel}(A, 0). \tag{8.4}$$

So, if the model verifies $\neg\text{Rel}(A, 1)$ *consistently*, it is bound to verify $\text{Rel}(A, 0)$. However, some models verify $\neg\text{Rel}(A, 1)$ inconsistently—they verify $\text{Rel}(A, 1) \wedge \neg\text{Rel}(A, 1)$. In the presence of (8.3), this entails $\text{Rel}(A, 1) \wedge \neg\text{Rel}(A, 1) \wedge \text{Rel}(A, 0)$ and hence comes to spreading inconsistency. So the reasons for rejecting (8.3) seem to apply just as well to (8.2); the first spreads inconsistency 'upwards', the second does it 'downwards'.[36]

In discussing the infinite hierarchy, I linked statements in which Rel occurs to our method to devise models, or to define models, whereas A was an arbitrary statement. Some will judge that, if our method to devise models goes wrong, then a general disaster should happen. This has a certain appeal. The method may very well cause triviality. So in that case (8.2) is quite all right. I doubt, however, that Priest will take this stand. The blazon of paraconsistency is precisely inconsistency without triviality—apologies for the infantile metaphor. So, the last argument for (8.2) is weak.

Describing the semantics Moving to the relational semantics generates a problem to identify models and groups of models within the **LP**-language. Take the *O-representation*. It seems to depict three possibilities with respect to a formula A: consistent truth, inconsistency, and consistent falsehood. Of course, one can say so in English. One can also say that A is true and not false in A-type-1 models, both true and false in A-type-2 models, and false and not true in A-type-3 models. What do these English statements mean? Given the kind of statements we are considering, these meanings have to be spelled out by logicians. That is precisely what devising the logic **LP** and describing its semantics is about. Remember the quote from the beginning of this section: "the language for which the semantics has been given is not English, since these notions obviously are expressible in English."

The distinction between the three A-types is rendered by $\text{Rel}(A, 1) \wedge \neg\text{Rel}(A, 0)$, $\text{Rel}(A, 1) \wedge \text{Rel}(A, 0)$ and $\neg\text{Rel}(A, 1) \wedge \text{Rel}(A, 0)$ respectively. The second one of these does its job perfectly—neglect for a moment formulas of thickness 2 and more. It identifies unambiguously models in which A is inconsistent. The two other 'descriptions' fail to do so. $\text{Rel}(A, 1) \wedge \neg\text{Rel}(A, 0)$ intends to describe A-type-1 models, but $\text{Rel}(A, 1) \wedge \neg\text{Rel}(A, 0)$ is also true in some A-type-2 models, viz. in A-type-2.2 models and A-type-2.4 models. Whichever convention one follows to handle formulas of higher thickness—the conventions were discussed before—moving there does not help to identify A-type-1 models. Adding formulas of thickness 1 that are verified by all A-type-1 models does not help either; these are also verified by A-type-2.2 models and A-type-2.4 models.

---

[36] Maybe Priest means (8.4) when he writes (8.2). In chapters stemming from the first edition of *In Contradiction* (Priest 2006b), $\leftrightarrow$ sometimes denotes a non-detachable implication and sometimes a detachable one; (8.3) spreads inconsistency on both readings.

Incidentally, pointing to the graphic representation of the model would not help either. We have seen that the representations are not unambiguous; different interpretations are possible. So one should be careful and require a clear description of which properties of models are represented by the representations. This includes a description of the models.

The complaint that Priest cannot identify the A-type-1 models and the A-type-3 models suggests that others can. One readily thinks of logicians that have no objections to explosive negations, like classical logicians or intuitionists. Priest has often pointed out (Priest 2006a, §6.3, 2006b, §20.4, 2014a, §10.4.2) that (what I briefly call) explosive logicians cannot *guarantee* consistency, or *force* consistency, or *rule out* inconsistency. Affirming the classical negation of A, the explosive logician would be committed to everything if she were to affirm A as well. And Priest continues to point out that the paraconsistent logician can do exactly the same thing, viz. by affirming $A \to \bot$, in which $\to$ is relevant, and hence detachable, and $\bot$ is the *falsum*, for which $\bot \to B$ is valid. Moreover, Priest invokes this possibility (Priest 2014a, §10.4.2) in a case similar to the one under discussion, which is to identify A-type-1 models and A-type-3 models. So let us try to complete the description of the A-types in those terms.

A first try might consist in modifying the models in such a way that $\neg A \to \bot$ holds in A-type-1 models and that $A \to \bot$ holds in A-type-3 models. But there is a problem. As we have three types of models, the following formula now holds: $\neg A \to \bot \lor (A \land \neg A) \lor A \to \bot$. So we can define a Liar formula Z such that $\text{Rel}(Z, 1) \leftrightarrow \text{Rel}(Z \to \bot, 1)$. Z-type-1 models and Z-type-2 models both verify $\text{Rel}(Z, 1)$. But then $\text{Rel}(Z \to \bot, 1)$ as well as $\text{Rel}(\bot, 1)$ hold in these models.[37] In Z-type-3 models holds $\text{Rel}(Z \to \bot, 1)$. So also $\text{Rel}(Z, 1)$ as well as $\text{Rel}(\bot, 1)$. So triviality results.

It is rather straightforward that the reasoning is mistaken. In order for A-type-1 models to verify arrow formulas, we need to update the semantics to a worlds semantics with a binary, or on Priest's present view ternary, accessibility relation, and we would have to adjust Rel accordingly. It is rather easy to see, however, that $\neg A \to \bot$ will not hold in the real world G of some A-type-1 models and that $A \to \bot$ will not hold in world G of some A-type-3 models. This is of course as it should be, because otherwise, we would have $A \vDash \neg A \lor (\neg A \to \bot)$ as well as $\neg A \vDash A \lor (A \to \bot)$. Now this is a touchy matter. Indeed, someone who combines the paraconsistent **LP**-negation $\neg$ with the classical negation $\sim$ definitely wants both $A \vDash \neg A \lor \sim \neg A$ as well as $\neg A \vDash A \lor \sim A$.[38] And someone who combines the paraconsistent **LP**-negation $\neg$ with a *detachable* classical implication $\supset$ will definitely want to have $A \vDash \neg A \lor (\neg A \supset \bot)$ as well as $\neg A \vDash A \lor (A \supset \bot)$.[39]

---

[37]The Liar formula itself does not entail triviality because Contraction (or Absorbtion) is not valid for the arrow. Triviality is engendered by (i) the decision that arrow-bottom formulas hold in the A-consistent models and (ii) the fact that the arrow is detachable.

[38]Where $\sim$ is intuitionistic negation, the expressions hold for finitistic A.

[39]In the presence of $\bot$, with its intended meaning that $\bot \vDash A$ holds, $A \supset \bot$ defines classical negation in case the implication is classical and defines intuitionistic negation in case the implication is intuitionistic. In the absence of $\bot$, it is possible to have a detachable classical or intuitionistic implication without having the disadvantages of an explosive negation. This is illustrated by the

Priest is willing to suppose counterfactually that classical negation makes sense (Priest 2006a, §6.3) and derives from this supposition that the classical logician—and forcibly also someone who takes classical negation to be sensible in certain contexts—can link affirming an inconsistency to a commitment to triviality, and actually cannot do more than this. This conclusion is clearly correct. If classical negation $\sim$ makes sense, introducing it in the *O-representation* is sufficient to identify the three types of models flawlessly. I mean, identify the models in terms of what they verify. Here is what the A-types of the *O-representation* verify if $\sim$ makes sense.[40]

A-type-1	A-type-2	A-type-3
A	A	$\sim A$
$\sim \neg A$	$\neg A$	$\neg A$

Obviously, every **LP**-model that is of two of the types is trivial: it assigns 0 as well as 1 to every formula. Priest claims that dialetheists can, just as much as explosive logicians, turn a commitment to inconsistency into a commitment to triviality. In view of what precedes, I do not see that this can be done in the present case.

I am *not* arguing that explosive logicians can do something that dialetheist cannot do. Explosive logicians separate theories from each other and separate languages from each other.[41] They have to pay a high price and the limitative theorems are part of that price. Dialetheists want to have one big theory, containing all knowledge in unified form and phrased, at least in principle and ultimately, in the vernacular. This, it seems to me, is the reason why they cannot identify the three A-type models. So $A \to \bot$ allows one, just as much as classical negation if it makes sense, to connect a formula to triviality,[42] but the dialetheist cannot use it in the present context because the unrestricted language allows for a Liar that engenders triviality.

Can one not apply the arrow-*falsum* method about the models rather than within the models? Is it possible to identify A-type-1 models by stating, for example, $\text{Rel}(A, 0) \to \bot$ instead of letting the models verify $A \to \bot$, which comes to stating $\text{Rel}(A \to \bot, 1)$. It seems to me that the question cannot be answered in the positive. Indeed, if $\text{Rel}(A, 0) \to \bot$ is true about A-type-1 models, then $(\text{Rel}(A, 0) \to \bot) \lor \text{Rel}(A, 0)$ is true about all three types of models. But then, it seems to me, this must be a truth about Rel, because the semantics is intended to be the semantics of

---

work of da Costa (da Costa 1963, 1974; da Costa and Alves 1977), Jaśkowski (1969), and others including myself. Priest and other dialetheists (and most relevantists) consider this a minor detail.

[40] Obviously $\sim\neg A \vDash A$ and $\sim A \vDash \neg A$. Moreover, the matter may be phrased in terms of classical implication and bottom. I leave it to the reader to figure out this as well as the intuitionistic case.

[41] It seems to me that the separation between object-language and metalanguage can be avoided, at least to some extent, provided one is willing to accept that theories are formulated in restricted fragments of a language.

[42] The difference, viz. that $A \lor \sim A$ is valid whereas $A \lor (A \to \bot)$ is not, is immaterial for this purpose.

the vernacular. The statement "either A is false or A is false implies everything" is not a true English statement.[43]

Consistency I now come to the most difficult point. In Chapters 16 and 18 of *In Contradiction*, Priest makes a number of claims that on the one hand resolve some of the questions I raised in the present section, but on the other hand seem to turn some of the other questions into insurmountable obstacles. In order to state the technicalities involved in Priest's point, I would have to explain the notion of collapsed models of Zermelo–Fraenkel set theory, **ZF**, and I do not have room to do so here. So I refer the reader to those chapters for details and for checking some of my claims. Moreover, let us suppose that there is a way to get around Facts 8.4–8.10, for otherwise the chapters are largely pointless anyway.

Let me start with a quote. "One may think of the metatheory of the logic, including the appropriate soundness and completeness proofs, as being carried out (as we know it can be) in **ZF**. According to the model-theoretic strategy, the results established in this way can perfectly well be taken to hold of the universe of sets, paraconsistently construed. The paraconsistent logician can, therefore, simply appropriate the results." (Priest 2006b, p. 259). So, according to Priest, the metatheory of **LP** can be carried out in **ZF**. Next, he claims that there is a paraconsistent set theory, call it **PZF**, that is obtained by 'collapsing' the **ZF**-models. Everything that holds in the original **ZF**-model, holds in the collapsed model, but more formulas, possibly some inconsistent ones, may hold in the collapsed model. The effect of the collapse is that the metatheory, which was phrased in **ZF**, is now phrased in **PZF**. It seems to me that there may be some loose ends, but let us suppose that Priest is right.

One still has to show that the theory has the same force. Phrased in **ZF**, the metatheory had material implication as its implication—write it as $\neg A \vee B$ for perspicuity. In **PZF**, however, Disjunctive Syllogism is invalid. Priest goes on to point out that the metatheory is consistent and that, reasoning in terms of **LPm**, which is basically an inconsistency-adaptive logic, "the disjunctive syllogism is perfectly acceptable provided the situation is consistent. Provided we do not have" $M \Vdash A$ and $M \nVdash A$, we can get from $M \Vdash A$ and $(M \nVdash A) \vee (M \Vdash B)$ to $M \Vdash B$.[44] It is worth spelling out the underlying reasoning and I shall spell it out in my words. (i) From $M \Vdash A$ and $(M \nVdash A) \vee (M \Vdash B)$ follows $((M \Vdash A) \wedge (M \nVdash A)) \vee (M \Vdash B)$. (ii) If interpreting the premises as consistently as possible allows one to consider the contradiction $(M \Vdash A) \wedge (M \nVdash A)$ as false, one may conclude to $M \Vdash B$. Let me rephrase this in semantic terms.[45] Consider the *models* that verify such expressions as $M \Vdash A$, the

---

[43] And fortunately so. If it were true, then, given what it is talking about, it should be valid. So it should be true in every model, which it is not.

[44] The expression $M \Vdash A$—Priest actually writes $\mathcal{I} \Vdash \alpha$—is read as "$A$ holds in M" by Priest. The same expression occurs as classical terminology in the very first table of the present section.

[45] The easiest way to understand the rest of the paragraph is to read it as a statement in classical terminology—so if a model both verifies and falsifies the same formula, then it is trivial.

models of the **LP**-semantics.[46] If no minimally inconsistent such *model*[47] verifies the contradiction $(M \Vdash A) \wedge (M \not\Vdash A)$, then all minimally inconsistent models verify $M \Vdash B$.

Step (ii) is the adaptive one. The motivation for it is that inconsistencies are normally false, so that we may consider them as false unless and until proven otherwise. In some cases, as the one Priest invokes here, the premises are consistent. So the consequence set coincides with the **CL**-consequence set: the minimally inconsistent **LP**-models of the premises are the consistent ones and these are the **CL**-models.[48]

The argument supposes that **ZF** is consistent, a claim that is obviously not provable in any absolute way. Of course, Priest may argue that he is not supposing more than the classicist is supposing, for example when the latter devises the metatheory of **CL**. That is a sensible defence for a dialetheist, but obviously not more than a defense. And, as argued before, stating that a model is consistent remains a problem, even if it does not prevent one from applying an inconsistency-adaptive logic. There are, however, other problems, of which I shall mention the most constructive one.

As Priest notes, that both $M \Vdash A$ and $M \Vdash \neg A$ hold does not make the semantics inconsistent, but $M \Vdash A$ and $M \not\Vdash A$ does. In which way does $M \Vdash A$ relate to other semantic expressions? It is typical that Priest reads (what I write as) $M \Vdash A$ as "A holds in M" and not, for example, the rather common "M verifies A". As Priest claims $\Vdash$ to behave consistently, he apparently wants $M \Vdash A$ whenever $\text{Rel}(A, 1)$. However, when does he want $M \not\Vdash A$? The easiest approach would go as follows, where Rel denotes truth-in-M.

$$\text{If } \text{Rel}(A, 1) \text{ then } M \Vdash A$$
$$\text{If } X \text{ then } M \not\Vdash A$$

But what should X be? Clearly $X=\text{Rel}(A, 0)$ is not all right, because then $\text{Rel}(A, 1) \wedge \text{Rel}(A, 0)$ would cause the semantics to be inconsistent, which is not what Priest wants. Setting $X=\text{Rel}(A, 0) \wedge \neg\text{Rel}(A, 1)$ would be all right *provided* Rel-expressions behaves consistently. But the Rel-version of the semantics was precisely introduced in order to allow inconsistent Rel-expressions and thus to escape the extended Liar provoked by the functional version of the semantics. So I see no X that would do. A wholly different possibility would be to define

$$M \Vdash A \text{ iff } \text{Rel}(A, 1) \tag{8.5}$$

---

[46] If the **LP**-semantics is indeed part of the semantics of English, then those *models* are simply the models from the **LP**-semantics. Yet, I shall keep italicizing "models" for perspicuity.

[47] A model of Γ is minimally inconsistent iff the set of inconsistencies that hold in it is not a proper superset of the set of inconsistencies that hold in another model of Γ.

[48] In more interesting cases the premises require certain but not all contradictions to be true. The **CL**-consequence set is then trivial, but the inconsistency-adaptive consequence set is, a few odd premise sets aside, non-trivial and moreover considerably richer than any consequence set defined by a paraconsistent Tarski logic. See the literature for details (Batens 2015).

in which "iff" is a non-contraposable equivalence—if it were contraposable, then $\text{Rel}(A, 1) \land \neg\text{Rel}(A, 1)$ would imply that $\Vdash$ behaves inconsistently. But how can one be sure that (8.5) forces $\Vdash$ to behave consistently? Suppose that $\text{Rel}(A, 1)$ and $\text{Rel}(A, 0)$. What prevents both $M \Vdash A$ and $M \nVdash A$ from being true?

Maybe Priest would argue that one should select the minimally inconsistent *models* of the **LP**-semantics—see above for the italics. In those *models* $\Vdash$ indeed behaves consistently. Nevertheless, the argument might be too swift.

As we have seen after introducing the semantic Liar (SL), the semantics of **LP** is not consistent. It verifies $\text{Rel}(L, 1) \land \neg\text{Rel}(L, 1)$. But then how can the semantics have been carried out in **ZF**? The more as the Disjunctive Syllogism was explicitly stated, in the original paper (Smiley and Priest 1993) as well as in the reworked version (Priest 2006b), to be invalid and that this was done to avoid triviality. The 'classical recapture' is indeed possible because the semantics can be carried out in **ZF**. But this is only possible because classicists are willing to operate within the patchwork. More specifically, and as stated before, they are willing to handle a restricted language in which no Liar can be formulated. An unrestricted language, however, seems to trivialize the semantics as phrased within **ZF**. If that is correct, a full recapture is excluded.

That the semantics is inconsistent does not prevent inconsistency-adaptive logics from efficiently doing their job. The Disjunctive Syllogism cannot be applied to some Rel-formulas because some of them are inconsistent. Yet Disjunctive Syllogism may be applied to other Rel-formulas (and perhaps to all $\vDash$-formulas), provided they are consistent in all minimally inconsistent *models*—basically, provided no extended Liar can be defined in terms of them.

8.5 In Conclusion

Allow me to repeat that my aim was not to argue, but to present some potentially interesting questions, some of which have hopefully not a standard or other ready answer. And, in case I have raised some interesting questions, that I leave it to Priest to select the ones he considers ripe for an answer.

The central difficulties outlined, especially in Sect. 8.4, seem to turn around a specific mixture of ingredients: a unified and monolithic knowledge system, a view on natural language and its logical terms as changeable but nevertheless formal-system-like, the choice of natural language, so conceived, as the language of the knowledge system, and the effects of Liars in such a scenery. Most dramatic seem the nasty effects of Liars on the semantics, notwithstanding the paraconsistent character of the environment. Actually, two different effects seem crucial.

The first effect is the inconvenience—it is not much more—that Liars apparently cause the semantics to be unavoidably inconsistent. The matter is related to the fact that several central meta-theoretic concepts—properties of formulas, sets of formulas, models, validity, the consequence relation, ...—turn out to be rather remote from the usual concepts as well as from the concepts implicit in Priest's writings. Given the complexity of the scenery, there may be changes that neutralize the inconvenience as well as the distance to the vernacular. And there is the alternative, to approach the inconsistent semantics by inconsistency-adaptive means. This approach seems free of foreseeable difficulties. Yet, the approach requires one to rework the semantics and the proof of its theorems. I see no way to 'translate' the version in terms of **ZF** and the set-theoretic **LP**-semantics.

The second effect concerns the possibility to identify models and groups (or types) of models and to express certain distinctions that are clearly there. One example is that, even if all sets of formulas are non-trivial in the sense of Fact 8.8, then nevertheless some Γ are trivial whereas others are not. The latter negation is different from the one in Fact 8.8. Means to remove this second effect seem in the same range as means to remove the first effect, but a single sweep might not be at hand.

Independent of the questions I raised and of their appropriateness, I would like to express my great appreciation for Graham's work. We have been raised in very different intellectual environments and our philosophical views are far apart, but contacts with Graham and with his writings have always been challenging and inspiring. It was a pleasure to find common tenets. It was a greater pleasure and a source of deeper insight to find points of disagreement, to consider them and to discuss them.

Acknowledgements I am grateful to the two referees for their helpful comments.

References

Almeder, R. (Ed.). (2008). *Rescher studies. A collection of essays on the philosophical work of Nicholas Rescher*. Frankfurt: Ontos Verlag.
Batens, D. (2014). Adaptive logics as a necessary tool for relativerationality. Including a section on logical pluralism. In E. Weber, D. Wouters, & J. Meheus (Eds.), *Logic, reasoning and rationality* (pp. 1–25). Dordrecht: Springer.
Batens, D. (2015). Tutorial on inconsistency-adaptive logics. In J.-Y. Béziau, M. Chakraborty, & S. Dutta (Eds.), *New directions in paraconsistent logic* (Vol. 152, pp. 3–38). *Springer Proceedings in Mathematics & Statistics*. Springer.
Batens, D. (2016). Spoiled for choice? *Journal of Logic and Computation, 26*(1), 65–95. E-published 2013. https://doi.org/10.1093/logcom/ext019.
Batens, D. (2017). Pluralism in scientific problem solving. Why inconsistency is no big deal. *Humana.Mente Journal of Philosophical Studies, 32*, 149–177.
Batens, D., & Haesaert, L. (2001). On classical adaptive logics of induction. *Logique et Analyse, 173–175*, 255–290 (appeared 2003).
Brown, B., & Priest, G. (2004). Chunk and permeate, a paraconsistent inference strategy. Part I: The infinitesimal calculus. *Journal of Philosophical Logic, 33*, 379–388.
Brown, B., & Priest, G. (2015). Chunk and permeate II: Bohr's hydrogen atom. *European Journal for Philosophy of Science, 5*(3), 297–314.
Carnap, R. (1947). *Meaning and necessity*. Chicago: University of Chicago Press.
da Costa, N. C. A. (1963). Calculs propositionnels pour les systèmes formels inconsistants. *Comptes rendus de l'Académie des sciences de Paris, 259*, 3790–3792.
da Costa, N. C. A. (1974). On the theory of inconsistent formal systems. *Notre Dame Journal of Formal Logic, 15*, 497–510.

da Costa, N. C. A., & Alves, E. H. (1977). A semantical analysis of the calculi C_n. *Notre Dame Journal of Formal Logic, 18*, 621–630.

Jaśkowski, S. (1969). Propositional calculus for contradictory deductive systems. *Studia Logica, 24*, 243–257.

Laudan, L. (1977). *Progress and its problems*. Berkeley: University of California Press.

Meheus, J. (2000). Analogical reasoning in creative problem solving processes: Logico-philosophical perspectives. In F. Hallyn (Ed.), *Metaphor and analogy in the sciences* (pp. 17–34). Dordrecht: Kluwer.

Nickles, T. (1981). What is a problem that we may solve it? *Synthese, 47*, 85–118.

Nickles, T. (1980a). Can scientific constraints be violated rationally? In *Scientific discovery, logic, and rationality* [18] (pp. 285–315).

Nickles, T. (Ed.). (1980b). *Scientific discovery, logic, and rationality*. Dordrecht: Reidel.

Priest, G. (2006a). *Doubt truth to be a liar*. Oxford: Clarendon Press.

Priest, G. (2006b). *In contradiction: A study of the transconsistent* (2nd ed.). Oxford: Oxford University Press (1st ed., 1987).

Priest, G. (2014a). Contradictory concepts. In E. Weber, D. Wouters, & J. Meheus (Eds.), *Logic, reasoning and rationality* (pp. 197–215). Dordrecht: Springer.

Priest, G. (2014b). Logical pluralism: Another application of chunk and permeate. *Erkenntnis, 29*, 331–338.

Priest, G. (2014c). Revising logic. In P. Rush (Ed.), *The metaphysics of logic* (pp. 211–223). Cambridge: Cambridge University Press.

Priest, G., Routley, R., & Norman, J. (Eds.). (1989). *Paraconsistent logic: Essays on the inconsistent*. München: Philosophia Verlag.

Rescher, N. (2003). *Imagining irreality*. Chicago and La Salle, Illinois: Open Court.

Rescher, N. (2005). *What if?*. New Brunswick, New Jersey: Transaction Publishers.

Smiley, T., & Priest, G. (1993). Can contradictions be true? *Proceedings of the Aristotelian Society, Supplementary Volumes, 67*, 17–33+35–54.

Vanackere, G. (1997). Ambiguity-adaptive logic. *Logique et Analyse, 159*, 261–280 (appeared 1999).

Weber, E., Wouters, D., & Meheus, J. (Eds.). (2014). *Logic, reasoning and rationality*. Dordrecht: Springer.

Weber, Z. (2010a). Extensionality and restriction in naive set theory. *Stadia Logica, 94*, 87–104.

Weber, Z. (2010b). Transfinite numbers in paraconsistent set theory. *Review of Symbolic Logic, 3*, 71–92.

Weber, Z. (2013). Notes on inconsistent set theory. In K. Tanaka, F. Berto, E. Mares, & F. Paoli (Eds.), *Paraconsistency: Logic and applications* (Vol. 26, pp. 315–328). *Logic, Epistemology, and the Unity of Science*. Dordrecht: Springer.

Diderik Batens is professor emeritus at the Centre for Logic and Philosophy of Science, Ghent University. He is a logician and epistemologist, and has a particular interest in adaptive logics and in paraconsistent logics. In the philosophy of science, he promotes a view that is fallabilist, dynamic and pluralist.

Chapter 9
Inferential Semantics, Paraconsistency, and Preservation of Evidence

Walter Carnielli and Abilio Rodrigues

Abstract Proof-theoretic semantics provides meanings to the connectives of intuitionistic logic without the need for a semantics in the standard sense of an attribution of semantic values to formulas. Meanings are given by the inference rules that, in this case, do not express preservation of truth but rather preservation of availability of a constructive proof. Elsewhere we presented two paraconsistent systems of natural deduction: the Basic Logic of Evidence (*BLE*) and the Logic of Evidence and Truth (LET_J). The rules of *BLE* have been conceived to preserve a notion weaker than truth, namely, *evidence*, understood as *reasons for believing in or accepting* a given proposition. LET_J, on the other hand, is a logic of formal inconsistency and undeterminedness that extends *BLE* by adding resources to recover classical logic for formulas taken as true, or false. We extend the idea of proof-theoretic semantics to these logics and argue that the meanings of the connectives in *BLE* are given by the fact that its rules are concerned with preservation of the availability of evidence. An analogous idea also applies to LET_J.

9.1 Introduction

As it is well-known, the so-called proof-theoretic semantics provides meanings to the connectives of intuitionistic logic without the need for a semantics in the standard sense, i.e., without the attribution of semantic values to formulas. The meanings are given by the deductive system itself, or more precisely, by the inference rules, that, in this case, do not express preservation of truth, but rather preservation of the availability of a constructive proof. The latter is a notion stronger than truth in

---

W. Carnielli (✉)
State University of Campinas, Campinas, Brazil
e-mail: walter.carnielli@cle.unicamp.br; walter.carnielli@gmail.com

A. Rodrigues
Federal University of Minas Gerais, Belo Horizonte, Brazil
e-mail: abilio@ufmg.br

© Springer Nature Switzerland AG 2019
C. Başkent and T. M. Ferguson (eds.), *Graham Priest on Dialetheism and Paraconsistency*, Outstanding Contributions to Logic 18, https://doi.org/10.1007/978-3-030-25365-3_9

the sense that a proposition may be true albeit no constructive proof of its truth is available.

In Carnielli and Rodrigues (2017), two paraconsistent systems of natural deduction were introduced: the Basic Logic of Evidence (*BLE*) and the Logic of Evidence and Truth (*LET$_J$*). Dually to intuitionistic logic, the natural deduction rules of *BLE* have been conceived to preserve a notion weaker than truth, namely, *evidence*, understood as *reasons for believing in and/or accepting* a given proposition. So, as in the case of intuitionistic logic, the meanings of the connectives in *BLE* are given by the fact that the rules are concerned with preservation of the availability of evidence. An analogous idea applies to *LET$_J$*, a logic of formal inconsistency and undeterminedness that extends *BLE* by adding rules that recover excluded middle and explosion for formulas in the scope of the connective ∘, that are assumed to behave classically.

Valuation semantics have been proposed for several paraconsistent logics. Given a language L, valuations are functions from the set of formulas of L to $\{0, 1\}$ according to certain conditions that somehow 'represent' the axioms and/or rules of inference. The semantics for da Costa's hierarchy C_n given in da Costa and Alves (1977), with some corrections in Loparic and Alves (1979), are 'non-truth-functional' valuation semantics. Later on, valuation semantics have been also proposed for da Costa's logic C_ω (Loparic 1986), and for several Logics of Formal Inconsistency (*LFI*s) (Carnielli et al. 2007; Carnielli and Coniglio 2016). In Loparic (2010), we find valuation semantics for intuitionistic logic, and in Carnielli and Rodrigues (2017) for *BLE* and *LET$_J$*. In these cases, the assignment of *1* and *0* to a proposition A means, respectively, that A holds and A does not hold. For example, in da Costa's logic C_1, the axiom

$$\neg\neg A \to A$$

has the following corresponding semantic clause:

$$v(\neg\neg A) = 1 \text{ implies } v(A) = 1,$$

that means nothing but:

if $\neg\neg A$ holds, then A holds.

Thus in this case there is nothing external to the formal system, no "intended meaning" acting as a motivation for the semantic clauses.

The main aim of this paper is to show that the basic idea of proof-theoretic semantics may be extended to the paraconsistent logics *BLE* and *LET$_J$*. Since we will expand the concept of proof-theoretic semantics to inferences not concerned with proofs, we will adopt here the more general expression *inferential semantics* to refer to both proof-theoretic semantics applied to intuitionistic logic and our proposal of extending it to paraconsistent logics. Second, we will argue that the valuation semantics proposed for *BLE* and *LET$_J$* in Carnielli and Rodrigues (2017), although useful from the technical point of view, are just mathematical mirrors of the proof environment, rather than semantics in the sense of assigning meanings to the formulas of the language.

The remainder of this paper is divided as follows. Section 9.2, a little provocative, points out some confusions that may result from acritically extending the standard distinction between syntax and semantics to nonclassical logics. It maybe, and we argue this is the case for *BLE* and *LET$_J$*, that meanings in a nonclassical logic are better explained not by providing a semantics to it, but by looking at what such a logic is intended to do. Section 9.3 recalls the ideas that motivated inferential (i.e., proof-theoretic) semantics for intuitionistic logic. In Sects. 9.4 and 9.5, we argue that an inferential semantics can be provided for *BLE*, and with some adjustments also for *LET$_J$*. Section 9.6 discusses the role of valuation semantics for intuitionistic logic and for the logics *BLE* and *LET$_J$*, arguing that, once the meanings have been given by the rules, the respective valuation semantics are better regarded as useful mathematical tools that yield decidability and other technical results.

9.2 Some Wrong Ideas in Nonclassical Logics

9.2.1 Syntax and Semantics

According to the standard view, syntax is concerned with the formal properties of linguistic expressions without regard to their meanings or interpretation. Syntax includes formulas, axioms, rules of inference and proofs—in sum: manipulation of symbols according to certain rules. Outside logic, the word 'semantics' has a wide sense, and has to do with meaning in general. In logic, however, semantics has to do with the interpretation of formal languages. Although there is no rigid limit between these two senses of the word 'semantics', they do not coincide, and some confusion may arise when they are used without distinction. So, from now on, when we want to emphasize that we are talking about semantics in the latter, strict sense, we will write 'Semantics', with a capital *S*. A Semantics may not provide a semantics.

The central concepts of an extensional Semantics are denotation and truth. By semantic value, we mean the 'entity' that is attributed to an expression of a formal language when this language is interpreted by a Semantics. For example, classical propositional logic attributes the semantic values *true* and *false* to propositional letters, which stand for propositions. First-order logic attributes objects from the domain to constants, sets to predicates, and *true* and *false* for propositions.

It is not easy to draw a dividing line between semantic and syntactic methods of proof. Indeed, sometimes this distinction makes no sense. But there should be no disagreement that carrying out an intuitionistic natural deduction proof is a syntactical activity, while checking classical truth-tables is a semantical activity.

Thus, semantics deals with meaning in general, while Semantics deals with the semantic values of the linguistic expressions of a formal system. Unless language is used to talk about language itself, meanings and semantic values are nonlinguistic entities. One may be inclined to say that these entities in some sense belong to the world. However, this may be a hasty conclusion.

There is a tension in the use of the word 'semantics' in logic that goes back to Tarski, who created model theory as a rigorous and systematic discipline whose subject matter is Semantics (but not semantics). According to Tarski,

> semantics [is] the totality of considerations concerning those concepts which ... express certain connexions between the expressions of a language and the objects and states of affairs referred to by these expressions. As typical examples ... *denotation, satisfaction,* and *definition.* (...) The concept of truth also ... is to be included here, at least in its classical interpretation according to which 'true' signifies the same as 'corresponding with reality'. (Tarski 1956b, p. 401)[1]

What Tarski is doing in model theory is Semantics, with a big capital S, but he informally explains what he is doing as if it were semantics. From the quotation above, it seems that the objects and states of affairs related to linguistic expressions, and upon which the truth of propositions depend, belong to reality. But a good amount of the work done in model theory has very little to do with meaning, nor with the 'world', or 'reality', in any minimally intuitive sense of these words. Model theory is concerned with structures that satisfy formulas of a given language. Satisfaction and truth are interdefinable with respect to atomic sentences: the object a satisfies the formula Px just in case Pa is true. But Tarski's motivation, as literally declared by himself, was to express the notion of truth as correspondence, that he calls the classical concept of truth (see quotation above and also Tarski 1956a, p. 153). Now, let us consider the nonstandard models of Arithmetic. They satisfy the axioms of first-order Arithmetic and are the subject matter of some existential statements. We say that the Peano axioms are *true* in the nonstandard models. But in what sense are we entitled to say that these models exist? In what sense we may say that these models are entities that belong to the world? Which world? In fact, one has to be a naive Platonist to believe in the existence of the entities that logicians make use to interpret linguistic expressions of formal systems. Nevertheless, sometimes people continue to talk about Semantics indistinctly, as if it had to do with relations between language and reality, as if it were concerned with meanings.

9.2.2 The Penchant for Completeness

The classical concept of logical consequence is primarily semantical. The pre-theoretical idea is that an argument is valid when it is not possible for the premises to be true and the conclusion false. Tarski's definition of logical consequence is an attempt to express this idea in precise terms: Γ logically implies A if and only if for every model M, if the sentences of Γ are true in M, then A is true in M. In classical logic, whether a given syntactical characterization of logical consequence coincides with the semantical is a relevant question. Consider, on the one hand, the first-order fragment of Frege's axiomatic system presented in the *Begriffsschrift* (1879) and,

---

[1] See also Tarski (1944, Sect. 5, 1956a, p. 252 and 1969, p. 63).

on the other, Tarski's semantic definition of logical consequence. It is indeed a big question as to whether or not they have the same extension.

Actually, the Semantics of classical logic to a great extent does explain the meaning of its expressions. The exception is the semantic value of sentences. It was Frege who established in his paper *On Sense and Reference* that in an extensional Semantics the reference (i.e., the semantic value) of a sentence is its truth-value. And the strangeness with which this thesis has been received by Frege (1980, pp. 150–1) and Dummett (1973, pp. 181ff) is a good example of the confusion between Semantics and semantics. The fact that the semantic value of the sentence 'Hesperus is Phosphorus' is the truth-value *true* does not explain its meaning, but is the correct choice for an extensional theory of logical consequence whose central notion is preservation of truth.

Indeed, the proof of completeness in classical logic is an important result, among other reasons, because the Semantics of classical logic, leaving aside the old controversy about truth-values as references of sentences, somehow explains the meanings of its expressions. The syntactical and the semantic presentation of classical logic, so to say, have 'different senses but the same reference'. But it is a mistake to think that, given a formal system F, a complete and sound Semantics for F always explains the meanings of the expressions of F and has an independent motivation. It may be that the Semantics for F is nothing but a way to prove technical results about F. What makes a formal system interesting or worth studying is not the fact that a Semantics has been found to it.

This penchant for completeness, however, is not the only mistaken idea in non-classical logics. Another is to take too seriously the distinction between syntax and semantics, and to think that in dealing with syntax the meanings of the symbols involved do not matter and that a Semantics is necessary to give meanings to the symbols. According to this view, syntax alone is not enough to make us capable of 'talking about reality'. Thus, once a deductive system is given, a central task is to find a Semantics and to prove that the system is sound and complete with respect to such Semantics. Of course, completeness is important because it is a technical tool, among other things, to provide counterexamples and thus to prove non-derivability in the system. But the idea that completeness shows that the formal system 'corresponds to the intended semantics' is a mistake—which intended semantics? It may be that this intended semantics just happens to be adequate but has no intuitive appeal independently of the formal system. Completeness actually shows that the formal system corresponds to the Semantics (not semantics).

9.3 Inferential Semantics

As it is well-known, the idea that meanings in logic can be provided by the syntactical activity is the starting point of the inferential, or proof-theoretic, semantics, an approach originating in natural deduction for intuitionistic logic.

> The term "proof-theoretic semantics" was introduced to stand for an approach to meaning based on what it is to have a proof of a sentence. (...) in contrast to a truth-conditional meaning theory, one should explain the meaning of a sentence in terms of *what it is to know* that the sentence is true, which in mathematics amounts to having a proof of the sentence. (Prawitz 2016, pp. 5–6, our emphasis.)

According to this view, meaning does not depend on truth and denotation anymore. The ontological bias in the relationship between meaning and truth gives way to the epistemic bias in the quotation above: the point is what is a sufficient condition to *know* that the sentence is true. This idea fits the rejection of excluded middle, a principle that introduces a realist ingredient in classical logic. A classical truth-table expresses the conditions for the truth of a proposition $A \vee \neg A$ independently of being known which one between A and $\neg A$ is true, and these truth-conditions constitute the meaning of $A \vee \neg A$. On the other hand, inferential semantics explains the meaning in terms of the inference rules. The 'link to reality', so to speak, is given by the deductive system: more precisely, by the introduction rules. So, from this perspective, to know that $A \vee \neg A$ is true is to have either a proof of A, or a proof of $\neg A$.

The origin of the idea that meanings are given by the natural deduction rules is in Gentzen (1935), where the system *NJ* for intuitionistic logic is presented. There we find the often-quoted passage below:

> The introductions represent, as it were, the 'definitions' of the symbols concerned, and the eliminations are no more, in the final analysis, than the consequences of these definitions. This fact may be expressed as follows: in eliminating a symbol, we may use the formula with whose terminal symbol we are dealing only 'in the sense afforded by the introduction [rule] of that symbol'. (Gentzen 1935, p. 80).

Let us return to disjunction. The classical semantic clause says that $A \vee B$ is true if and only if at least one among A and B is true. On the other hand, the introduction rules for disjunction say that a proof of A, or a proof of B, is a sufficient condition for having a proof of the disjunction $A \vee B$. Intuitionistically, a disjunction cannot be obtained otherwise. From this perspective, the meaning of the connective $\vee$ is given by how it is used in inferences where the point is not preservation of truth, but rather preservation of availability of a constructive proof—we will return to this point in Sect. 9.5.1 below.

9.3.1 Intuitionistic Natural Deduction

Natural deduction has been conceived by Gentzen as "a formalism that reflects as accurately as possible the actual logical reasoning involved in mathematical proofs" (Gentzen 1935, p. 74). Natural deduction rules, indeed, intend to be similar to inferences that occur in intuitive, informal reasoning. As we have just seen, Gentzen emphasizes the role of the introduction rules, but does not say how an elimination rule can be obtained from the respective introduction rule. This point is addressed by Prawitz in the so-called *inversion principle*. Consider an elimination rule $E*$ that has B as consequence. Deductions that satisfy the sufficient condition for deriving the major premise of $E*$, together with deductions of the minor premises, already "contain" a deduction of B (Prawitz 1965, p. 33). As an example, let us see how to obtain the elimination rule for $\vee$,

$$\cfrac{A \vee B \quad \overset{[A]}{\underset{C}{\vdots}} \quad \overset{[B]}{\underset{C}{\vdots}}}{C} \vee E$$

A derivation of A or a derivation of B is a sufficient condition for concluding the major premise $A \vee B$:

$$\cfrac{\overset{\vdots}{A}}{A \vee B} \quad \text{or} \quad \cfrac{\overset{\vdots}{B}}{A \vee B}$$

The minor premises of $\vee$-elimination are deductions of C from A and from B:

$$\overset{A}{\underset{C}{\vdots}} \quad \text{and} \quad \overset{B}{\underset{C}{\vdots}}$$

We have just to put them together to get a derivation of C:

$$\overset{\vdots}{\underset{\underset{C}{\vdots}}{A}} \quad \text{or} \quad \overset{\vdots}{\underset{\underset{C}{\vdots}}{B}}$$

Definition 9.1 (*Intuitionistic logic INT*) Consider the propositional language L_0 defined in the usual way over the set of connectives $\{\wedge, \vee, \rightarrow, \bot\}$. S_0 is the set of of formulas of L_0. $\neg A$ is defined as $A \rightarrow \bot$. Roman capitals stand for meta-variables for formulas of L_0. The following natural deduction rules define the system *INT* of intuitionistic propositional logic:

$$\cfrac{A}{A \vee B} \vee I \quad \cfrac{B}{A \vee B} \quad \cfrac{A \quad B}{A \wedge B} \wedge I$$

$$\cfrac{\overset{[A]}{\underset{B}{\vdots}}}{A \rightarrow B} \rightarrow I \quad \cfrac{A \vee B \quad \overset{[A]}{\underset{C}{\vdots}} \quad \overset{[B]}{\underset{C}{\vdots}}}{C} \vee E$$

$$\cfrac{A \wedge B}{A} \wedge E \quad \cfrac{A \wedge B}{B} \quad \cfrac{A \rightarrow B \quad A}{B} \rightarrow E$$

$$\frac{\bot}{A} \ EXP$$

Definition 9.2 (*Positive intuitionistic logic PIL*) Consider the propositional language L_1 defined in the usual way over the set of connectives $\{\wedge, \vee, \rightarrow\}$. Dropping the rule *EXP* from the system *INT* above yields *PIL*, positive intuitionistic propositional logic.

The introduction rules and the respective elimination rules obtained from the inversion principle intend to be in accordance with how intuitionistic mathematicians use the logical connectives. So, according to this view, there is a descriptive ingredient in intuitionistic logic: its subject matter is reasoning, and the 'part of reality', so to speak, that is being represented by natural deduction rules is mathematical thought.

9.3.2 A Brief Digression: Kripke Models Do Not "Explain the Meanings" in Intuitionistic Logic

At first sight, Kripke models for intuitionistic logic intend to play a role similar to the semantics of classical logic: a nonlinguistic device that provides meanings and explains the deductive system. Kripke's approach is based on the standard distinction between syntax and semantics and has been successful for alethic modal logics, but we cannot say the same with respect to intuitionistic logic. In Kripke (1965), regarding the intuitive interpretation of the semantics Kripke just presented, we read:

> We intend the nodes **H** to represent points in time (or "evidential situations"), at which we may have various pieces of information. If, at a particular point **H** in time, we have enough information to prove a proposition A, we say that $\phi(A, \mathbf{H}) = T$; if we lack such information, we say that $\phi(A, \mathbf{H}) = F$. If $\phi(A, \mathbf{H}) = T$ we can say that A has been verified at the point **H** in time. (...)
> To assert $\neg A$ intuitionistically in the situation **H**, we need to know at **H** not only that A has not been verified at **H**, but that it cannot possibly be verified at any later time. (Kripke 1965, pp. 98–99)

Excluded middle fails because to refute a proposition A at some node (i.e., moment of time, or stage) **H**, it is necessary that A receives the value F in all future nodes $\mathbf{H}'$. The point is not to express that a proposition is true or false *simpliciter*, as in classical logic, but rather to emphasize the fact that in some moments of time there may be no information (i.e., no proof) available for some proposition A. Our knowledge about some propositions may improve, and it may happen that a proposition A about which we know nothing at all today will be proved true tomorrow. Kripke models for intuitionistic logic intend to express the mind of an idealized mathematician who proves theorems as time passes.

There is no doubt that Kripke models are a powerful tool for establishing several properties of intuitionistic logic. But it has some insuperable difficulties. We will mention only one, and that is enough to reject Kripke models as a semantics that provides meanings to intuitionistic logic. Suppose that our mathematician is working on Heyting Arithmetic (Peano axioms plus first-order intuitionistic logic). Since each node (i.e., stage) in a Kripke model is classical, the respective domain cannot be empty, and our mathematician cannot start from nothing. So, the scenario of a mathematician with an empty mind, like a sheet of white paper, is not allowed—but it should be! The worse problem, however, is that in all stages the totality of natural numbers has to be already available in order to guarantee the truth of the axioms, a requirement that does not fit the basic ideas and motivations of intuitionism! Therefore, Kripke models for intuitionistic logic cannot play the role of a nonlinguistic device that explains the meaning of intuitionistic logic.

9.3.3 Reconciling Intuitionistic and Classical Logic

Recall the claim that intuitionistic logic is not talking about truth, but rather about constructive proofs. So, it seems that it is reasonable to emphasize that intuitionistic logic is incomplete with respect to the semantics that expresses the notion of preservation of truth. In this case, incompleteness is not a defect, since the point is precisely that there are truths not yet proved, and truths that cannot be proved by any constructive means.[2]

Brouwer (1907), motivated by considerations on the nature of mathematical knowledge, established the conceptual basis of intuitionistic logic. The principle of excluded middle is not valid in intuitionistic logic, and its rejection is due to the view that mathematical objects are not discovered but rather created by the human mind. From Brouwer's point of view, reconciling classical and intuitionistic approach in mathematics was not possible. But even in Heyting, a former student of Brouwer, we find a softer position, according to which intuitionistic logic as an investigation of mathematical objects as mental constructions does not imply that such objects do not exist independently of such constructions, nor that they cannot be investigated by other means. So, one may well be a realist about mathematical objects but still have an interest in intuitionistic logic. Classical and intuitionistic logic do not need to exclude the legitimacy of each other.[3]

Indeed, Prawitz presents an 'ecumenical system' that combines classical and intuitionistic connectives. The basic idea is that "the classical as well as the intuitionistic codification of deductive practice is fully justified on the basis of *different meanings* attached to the involved expressions" (Prawitz 2015, p. 28, our emphasis). In other words, and in line with the view defended here, classical and intuitionistic connectives are not talking about the same thing.

---

[2] An example is the Goldstein theorem whose proof is independent of Peano Arithmetic (formal number theory), and so cannot be proved by constructive means.
[3] We defend this view in Carnielli and Rodrigues (2016, Sect. 3).

When the classical and intuitionistic codifications attach different meanings to a constant we need to use different symbols, and I shall use a subscript c for the classical meaning and i for the intuitionistic. The classical and intuitionistic constants can then have a *peaceful coexistence* in a language that contains both. ((Prawitz 2015, p. 28), our emphasis.)[4]

Reconciling classical and intuitionistic logic is possible because they talk differently about different things, respectively, preservation of truth, and preservation of the availability of a constructive proof. This view depends on assigning an epistemic, rather than ontological, character to intuitionistic logic and fits the basic idea of inferential semantics.

9.4 Paraconsistency from an Epistemic Viewpoint

So far, two main points have been presented: (i) The standard (and classical) distinction between syntax and semantics, in the sense that the latter is responsible for providing meanings to the expressions of a formal system, does not apply to intuitionistic logic and may not apply to some nonclassical logics; (ii) Proof-theoretic semantics provides what is needed in order to "explain the meaning" of intuitionistic logic. Now we will argue that the same line of reasoning may be extended to paraconsistency.

The work on inferential semantics has been motivated by, and developed for, intuitionistic logic. The framework of intuitionistic logic is mathematics, where a constructive proof of A implies the truth of A. But some contexts of reasoning are concerned neither with truth (classical logic), nor with constructive proofs (intuitionistic logic). This is the case of the paraconsistent logics *BLE* and LET_J, designed to express *preservation of evidence*.

A paraconsistent logic is a logic such that the principle of explosion is not valid. Thus, a paraconsistent logic is able to formally represent contradictory but nontrivial contexts of reasoning. Contradictions that occur in such contexts do not need to mean that a given proposition A and its negation are true, nor that A is both true and false. We have already argued elsewhere that such contradictions should be understood from an epistemic point of view.[5] A non-dialetheist answer to the question

What does it mean to say that both A and $\neg A$ hold in a given context?

---

[4]The same idea of a "peaceful coexistence" between intuitionism and classicism is found in Dubucs (2008).
[5]For a more detailed account of paraconsistency as preservation of evidence and the duality between paraconsistent and paracomplete logics understood as concerned, respectively, with a notion weaker and stronger than truth, we refer the reader to Carnielli and Rodrigues (2016, Sect. 4) and (2017, Sects. 1 and 2).

has to be the following: A and $\neg A$ enjoy a property weaker than truth. The notion of evidence, understood as reasons for believing in/accepting a proposition, fits this idea. Conflicting evidence occurs when there are simultaneously reasons for accepting A and reasons for accepting $\neg A$, both nonconclusive.

9.4.1 The Logics BLE and LET$_J$

Now we turn to the logics *BLE*, the Basic Logic of Evidence, and *LET$_J$*, the Logic of Evidence and Truth.

Definition 9.3 (*The Basic Logic of Evidence BLE*) Consider the propositional language L_2 defined in the usual way over the set of connectives $\{\wedge, \vee, \rightarrow, \neg\}$. S_2 is the set of of formulas of L_2. The following natural deduction rules together with introduction and elimination rules for $\rightarrow, \wedge$ and $\vee$ (*PIL*) define the logic *BLE*:

$$\frac{\neg A}{\neg(A \wedge B)} \neg \wedge I \quad \frac{\neg B}{\neg(A \wedge B)} \qquad \frac{\neg(A \wedge B) \quad \begin{array}{c}[\neg A]\\ \vdots \\ C\end{array} \quad \begin{array}{c}[\neg B]\\ \vdots \\ C\end{array}}{C} \neg \wedge E$$

$$\frac{A \quad \neg B}{\neg(A \rightarrow B)} \neg \rightarrow I \qquad \frac{\neg(A \rightarrow B)}{A} \neg \rightarrow E \quad \frac{\neg(A \rightarrow B)}{\neg B}$$

$$\frac{\neg A \quad \neg B}{\neg(A \vee B)} \neg \vee I \qquad \frac{\neg(A \vee B)}{\neg A} \neg \vee E \quad \frac{\neg(A \vee B)}{\neg B}$$

$$\frac{A}{\neg\neg A} \text{ DNI} \qquad \frac{\neg\neg A}{A} \text{ DNE}$$

The logic *LET$_J$*, defined below, is a logic of formal inconsistency and undeterminedness that extends *BLE* by means of rules that recover the validity of explosion and excluded middle, thus restoring classical logic for formulas marked with ∘. As we have seen, *BLE* intends to express a notion weaker than truth, that we call evidence, but *BLE* is not able to express preservation of truth.[6] However, there are contexts in which we have to deal simultaneously with propositions already established as either true or false, and others for which only nonconclusive evidence is available. The logic *LET$_J$* is able to divide such contexts in smaller contexts according to the property ascribed to propositions, truth or nonconclusive evidence, and so it can deal simultaneously with evidence and truth.

---

[6]*BLE* ends up being equivalent to Nelson's logic *N4* (the propositional fragment of N^-, see Almukdad and Nelson (1984) and Odintsov (2008)), but has been conceived for a different purpose (see Carnielli and Rodrigues (2017), especially Sect. 5.3).

Definition 9.4 (*The logic of evidence and truth LET_J*) Consider the propositional language L_3 defined in the usual way over the set of connectives $\{\wedge, \vee, \rightarrow, \neg, \circ\}$. S_3 is the set of of formulas of L_3. The logic of formal inconsistency and undeterminedness LET_J is defined by adding to *BLE* the rules below:

$$\frac{\circ A \quad \overset{[A]}{\underset{B}{\vdots}} \quad \overset{[\neg A]}{\underset{B}{\vdots}}}{B} PEM^\circ \qquad \frac{\circ A \quad A \quad \neg A}{B} EXP^\circ$$

In LET_J, classical logic holds for formulas in the scope of the unary connective ∘, called a classicality operator, in the following sense: if $\circ A_1$, $\circ A_2$, ..., $\circ A_n$ hold, classical logic holds for all formulas that depend only on $A_1, A_2, \ldots, A_n$ and are formed with $\rightarrow$, $\wedge$, $\vee$ and $\neg$ (see Carnielli and Rodrigues (2017) Fact 18). So, the connective ∘ in LET_J can also be understood as a *context switch* that divides propositions into two different contexts, according to the underlying logic.

9.5 Back to Inferential Semantics

In this section, we return to the topic of inferential semantics. We start by recalling some points in the historical development of proof-theoretic/inferential semantics in order to show that it has been conceived within the context of proofs in constructive Mathematics, which is not concerned with truth but, rather, with constructive proofs. An analogous line of reasoning is extended to *BLE* and LET_J, conceived to be applied in contexts concerned with evidence, and both evidence and truth. We then show how to provide an inferential semantics to *BLE* and LET_J.

9.5.1 The Role of Context in Inferential Semantics

The beginning of the 1930s has been an enabling environment for the foundational ideas of inferential semantics. The explanation of the meanings of intuitionistic connectives given by the Brouwer–Heyting–Kolmogorov (*BHK*) interpretation can be traced back to the works of Kolmogorov (1932), and Heyting (1930, 1934).

In Heyting (1930, p. 307) says that an assertion of *p* means, in classical logic, that *p* is true, while a *Brouwerian assertion* of *p* means that it is known how to prove *p*, and he adds that 'to prove' means 'to prove by construction'. An independent but similar idea is found in Kolmogorov (1932) where Kolmogorov presented intuitionistic logic as a formal system concerned not with truth but rather with *solutions of problems*:

> In addition to theoretical logic, which systematizes a proof schemata for theoretical truths, one can systematize the schemata of the solution of problems (Kolmogorov 1932, p. 328).

9 Inferential Semantics, Paraconsistency, and Preservation of Evidence

And so he explains the propositional connectives as follows:

> If a and b are two problems, then $a \wedge b$ designates the problem "to solve both the problems a and b", while $a \vee b$ designates the problem "to solve at least one of the problems a and b". Furthermore, $a \to b$ is the problem "to solve b provided the solution of a is given ... $\neg a$ designates the problem "to obtain a contradiction provided that the solution of a is given" (Kolmogorov 1932, p. 329).

Heyting (1934) mentions Kolmogorov's explanation in terms of problems:

> Chaque variable représente un problème ; il n'explicite pas ce concept, qu'on peut interpréter comme la demande d'effectuer une construction mathématique qui satisfasse à certaines conditions (Heyting 1934, p. 17).[7]

The solution of a problem, thus, is a mathematical construction satisfying certain conditions, that is, an intuitionistic proof. Heyting's explanation of the meanings of intuitionistic connectives explicitly appears later in Heyting (1956):

> It will be necessary to fix, as firmly as possible, the meaning of the logical connectives; I do this by giving necessary and sufficient conditions under which a complex expression can be asserted (Heyting 1956, p. 97).

He continues and explains the $\wedge$, $\vee$, $\neg$ and $\to$ as follows, just changing Kolmogorov's *problems* by *assertibility conditions* [pp. 97–98]:

- $A \wedge B$ can be asserted if and only if both A and B can be asserted;
- $A \vee B$ can be asserted if and only if at least one of the propositions A and B can be asserted;
- $\neg A$ can be asserted if and only if we possess a construction which from the supposition that a construction A were carried out, leads to a contradiction;
- $A \to B$ can be asserted, if and only if we possess a construction C, which, joined to any construction proving A (supposing that the latter be effected), would automatically effect a construction proving B.

If we change assertibility conditions by proofs, we get the *BHK*-interpretation as presented by van Dalen and Troelstra (1988, p. 9).

It is clear, then, that the idea that the meanings of intuitionistic connectives are to be explained in terms of solutions of problems/assertibility conditions/proofs (that we express here as *availability of a constructive proof*) was drawn up from the idea that intuitionistic logic is not about truth but rather about constructive reasoning in Mathematics. The meanings come from how the connectives are used in the context of constructive Mathematics, where an assertion of A means that a constructive proof of A is available.

Dummett (1991) discusses the conditions under which introduction and elimination natural deduction rules are suited to provide meanings for the logical connec-

---

[7] We quote here from the French edition of Heyting's text originally publish in German in 1934.

tives. The background is Dummett's well-known concerns about a compositional theory of meaning. He says that in the context of a verificationist meaning-theory "[t]he hope is that this [an explanation of the meaning of a connective ( )] can be done by appealing to the introduction rule or rules for the connective ( ) in a natural deduction formalisation of logic" (Dummett 1991, p. 216). But of course some constraints have to be made to avoid *tonk*-like arguments—we return to this point soon. Dummett calls *harmony* the property ascribed to the rules of a given connective if its elimination rule can be "read off" from its introduction rule. Harmony, in this sense amounts to satisfying Prawitz' inversion principle. So, he proposes that if the introduction/elimination rules for a connective ∗ are harmonious, such rules may be taken as an explanation of the meaning of ∗ (Dummett 1991, pp. 246ff).

Actually, the issue is not so simple, and a lot of technical and conceptual work has already been done on this topic. Although it is not the aim of this paper to go deeply into this topic, that will be dealt with in detail elsewhere, we agree that harmony is convincing as a condition for rules providing meanings. However, we strongly disagree that there can be self-justifying logical rules "uncontaminated by any ideas foreign to proof theory" as Dummett (1991, p. 245) claims. Proof-theoretic considerations, like the ones carried out by Dummett, have to be contextualized: the inferences are always relative to some argumentative context that cannot be separated from the rules because the meanings do not, and cannot, depend solely on the rules. A previous condition for the rules and the respective formal system is the kind of argumentative context where such rules are used. Our point is that the property being preserved by propositions in inferences is an essential feature of an argumentative context, if the latter is going to be represented, or formalized, by means of a set of inferences allowed in it. Different properties give rise to different rules, and so to different formal systems. And further, the property being preserved in an argumentative context has a role in establishing the meanings of the linguistic expressions of the respective formal system.

That classical and intuitionistic logic are suitable to different contexts is clear from the quotations above. Our proposal here, again, is that this idea can be extended and successfully applied to the logic *BLE*, by considering that the property ascribed to propositions is evidence, and with some adjustments also to LET_J, that deals with contexts concerned with evidence and truth. Notice that this approach is totally immune to the *tonk* argument introduced by Prior, who intended to show that defining natural deduction rules for a logical operator is not enough to provide its meaning— but of course not! The connective *tonk* does not have any meaning in a context of preservation of truth, nor in a context of constructively solving a mathematical problem, nor in a context concerned with nonconclusive evidence. The rules of *tonk* preserve none of the properties ascribed to propositions in classical, intuitionistic and paraconsitent logics.

9.5.2 An Inferential Semantics for BLE and LET$_J$

First of all, let us call attention to the fact that all elimination rules of *BLE*, including the negative ones, can be obtained by means of the Prawitz inversion principle from the respective introduction rules. Take $\neg \rightarrow E$ as an example. The sufficient condition for introducing a negated implication $\neg(A \rightarrow B)$ are A and $\neg B$ together. So, from $\neg(A \rightarrow B)$ it may be conclude A as well as $\neg B$. The reader can easily check that the same line of reasoning can be applied to $\neg \wedge E$ and $\neg \vee E$. The inversion principle also applies to double negation: the sufficient condition for concluding $\neg\neg A$, given by the rule *DNI*, is A, and so the latter may be obtained from the former by the rule *DNE*. Therefore, we conclude that *BLE* satisfies the harmony condition.[8]

The inferential semantics for intuitionistic logic starts from a primitive notion of proof. The *BHK*-interpretation takes the notion of a constructive proof of an atom p as given, and establishes what it means to have a proof of molecular formulas in terms of its constituents. The explanation of meaning so obtained is compositional in the following sense: the meaning of a formula $A \wedge B$, for example, depends on the meanings of A, B and on the meaning of conjunction, that is given by the rule $\wedge I$.

The logic *BLE* is able to represent four scenarios with respect to a proposition A:

1. There is no evidence for the truth nor for the falsity of A: A does not hold and $\neg A$ does not hold.
2. There is only evidence that A is true: A holds and $\neg A$ does not hold.
3. There is only evidence that A is false: $\neg A$ holds and A does not hold.
4. There is evidence for the truth and for the falsity of A: A holds and $\neg A$ holds.

So, differently from the *BHK*-interpretation for intuitionistic logic in order to provide an inferential semantics for *BLE*, we need two primitive notions: positive and negative evidence. This is because evidence for truth (respectively, falsity) is different from absence of evidence for falsity (respectively, truth), and so positive and negative evidence are not complementary to each other.

In *BLE*, the falsity of A is represented by $\neg A$, and when $\neg A$ holds it means that there is negative evidence for A. This same evidence, however, is positive evidence for $\neg A$. So, we propose the following *Evidence Principle*:

κ is negative evidence for A iff κ is positive evidence for $\neg A$ (EP)

and an *Evidence Interpretation* for *BLE*:

[E1]   Positive evidence κ for $A \wedge B$ amounts to positive evidence κ_1 for A and positive evidence κ_2 for B;

[E2]   Positive evidence κ for $A \vee B$ amounts to positive evidence κ_1 for A or positive evidence κ_2 for B;

---

[8]Another remarkable feature of *BLE* is the symmetry between the introduction and elimination positive rules for $\wedge$ and $\vee$ on the one hand, and the introduction and elimination negative rules for the dual operators $\vee$ and $\wedge$ on the other (there is no dual of $\rightarrow$ in *BLE*).

[E3] Positive evidence κ for $A \rightarrow B$ is given when the supposition that there is positive evidence κ_1 for A leads to the conclusion that there is evidence κ_2 for B;

[E4] Negative evidence κ for $A \wedge B$ amounts to negative evidence κ_1 for A or negative evidence κ_2 for B;

[E5] Negative evidence κ for $A \vee B$ amounts to negative evidence κ_1 for A and negative evidence κ_2 for B;

[E6] Negative evidence κ for $A \rightarrow B$ amounts to positive evidence κ_1 for A and negative evidence κ_2 for B;

[E7] Negative evidence κ for $\neg A$ amounts to positive evidence κ for A.

Given the principle *EP*, 'negative evidence for A' may be changed to 'positive evidence for $\neg A$', thus making it clear that these clauses correspond closely to the rules $\neg \wedge I, \neg \vee I$, and $\neg \rightarrow I$, in clauses [E4], [E5], and [E6]. Clause [E7] deserves some remarks. It implies that positive evidence for A is also positive evidence for $\neg\neg A$. The idea is that any evidence that A is true is also evidence that it is false that A is false ($\neg\neg A$), and vice versa. Notice that this reading is in accordance with the fact that we consider the notions of truth and falsity subjected to classical logic (this is precisely the point of recovering classical logic by means of *LET$_J$*).[9]

A remarkable feature of the explanation of the meanings of the connectives given above is that the semantics so obtained is compositional. In *BLE*, the meanings of p and $\neg p$ are taken as primitive: when p (respectively, $\neg p$) holds, this means that there is positive (respectively, negative) evidence for p. The meaning of $A * B$, $* \in \{\wedge, \vee, \rightarrow\}$, is given by the meanings of A, B and the meaning of the respective connective, given by the introduction rule $*I$. The meaning of $\neg A$ is primitive if A is a propositional letter, and if $A = \neg B$, it depends on the meaning of B and on the rule *DNI*. Otherwise, if $A = B * C$, the meaning of $\neg A$ depends on B, on C, and on the respective rule $\neg * I$.

Now we turn to the logic *LET$_J$*. We start with the meaning of ∘. There are no introduction rules for ∘ in *LET$_J$* because a formula ∘A has to be introduced from *outside* the formal system. The idea is that the particular manner that establishes a given proposition A as true (or false) is not a problem of logic, but rather it is the user of the system who establishes A as true (respectively, false) and marks A with ∘. Accordingly, ∘A expresses the fact that the evidence for A (respectively, $\neg A$) was considered to be conclusive. So, we propose that the meaning of ∘A is given by the particular way by which the truth (respectively, the falsity) of A has been established (through means that are external to the formal system (see Carnielli and Rodrigues 2017, Sect. 4.2)).

---

[9]The logic *BLE*, besides being equivalent to Nelson's *N4*, is also equivalent to the propositional fragment of the refutability calculus proposed by López-Escobar (1972), where an explanation of refutation as a primitive notion in number theory that is similar to the clauses [E4]–[E7] above can be found (López-Escobar 1972, pp. 363ff). This is not a surprise, though, since the logics are equivalent, but his motivations are completely different from ours. Regarding the motivations and the "discovery" of *BLE* and *LET$_J$*, see Sect. 5.3 and footnote 21 of Carnielli and Rodrigues (2017).

Recall that *LET_J* divides propositions into two groups: the first group is a context governed by *BLE*, and the second is a context subject to classical logic *CPL*. The latter may be defined in a natural deduction system by adding to the positive intuitionistic logic (Definition 9.2) the rules *EXP* and *PEM* (see Sect. 9.4.1). It is well-known that the classical meanings of the logical connectives $\wedge$, $\vee$, $\rightarrow$ and $\neg$ cannot be given by the introduction rules because negation violates the harmony condition. Indeed, neither *EXP* nor *PEM* can be taken as an introduction rule for negation. There are other ways of defining classical negation, but none satisfies the harmony condition. However, it can be said that the classical meanings of the connectives are given by the inferences allowed in the wide sense that all classical inferences are truth-preserving. Since, as we argued, meanings do not depend solely on the inference rules but also on the property being preserved in the context under discussion, we propose to expand the idea of inferential semantics by saying that the meaning of a connective $*$ in classical logic is given globally by all the truth-preserving inferences in which $*$ occurs. It is obvious that, conceived in this way, the meanings of the connectives will be the same as the meanings given by the classical truth-tables, and so compositionality is preserved.

9.6 Valuation Semantics for Intuitionistic and Paraconsistent Logics

Having argued that the meaning of the logical connectives of the paraconsistent logics *BLE* and *LET_J*, analogously to intuitionistic logic, may be explained with the aid of inferential semantics, we turn now to the valuation semantics that has been proposed for several paraconsistent logics, including *BLE* and *LET_J* in Carnielli and Rodrigues (2017), and for intuitionistic logic in Loparic (1986). Valuation semantics, in these cases, are Semantics with a capital *S*, but not semantics, because they do not provide meanings. They are better seen as mathematical tools that represent the inference rules and/or the axioms in such a way that some technical results, including decision procedures, can be obtained.

9.6.1 Valuation Semantics for Intuitionistic Logic

The clauses of a valuation semantics for positive intuitionistic logic *PIL* (Definition 9.2) appear for the first time in Loparic (1986), where a semantics for da Costa's C_ω, an extension of *PIL*, is presented. We just need to add a clause for $\bot$ to the clauses of $\wedge$, $\vee$ and $\rightarrow$ to obtain a sound and complete valuation semantics for intuitionistic logic *INT* (Definition 9.1). Soundness and completeness, in this case, means only that the attribution of 0s and 1s according to the semantic clauses correctly represents the deductive system.

Definition 9.5 A *semi-valuation* s for *INT* is a function from the set S_0 of formulas to $\{0, 1\}$ such that:

(i) if $s(A) = 1$ and $s(B) = 0$, then $s(A \to B) = 0$;
(ii) if $s(B) = 1$, then $s(A \to B) = 1$;
(iii) $s(A \wedge B) = 1$ iff $s(A) = 1$ and $s(B) = 1$;
(iv) $s(A \vee B) = 1$ iff $s(A) = 1$ or $s(B) = 1$;
(v) $s(\bot) = 0$.

Definition 9.6 A *valuation* for *INT* is a semi-valuation for which the condition below holds:

(Val) For all formulas of the form $A_1 \to (A_2 \to \ldots \to (A_n \to B)\ldots)$ with B not of the form $C \to D$:
if $s(A_1 \to (A_2 \to \ldots \to (A_n \to B)\ldots)) = 0$, then there is a semi-valuation s' such that for every i, $1 \leq i \leq n$, $s(A_i) = 1$ and $s(B) = 0$.

The semantic clauses for $\wedge$, $\vee$ and $\to$ have been conceived in such a way that they represent the inferences in terms of the attribution of 0 and 1 to formulas, 0 means that the formula *does not hold* and *1* means that the formula *holds*. There is no motivation independent of the formal system. $\bot$ always receives 0, since it never holds. It is not difficult to see that clauses (iii) and (iv) correspond to the introduction rules for $\wedge$ and $\vee$. The clauses for $\to$ are three, (i), (ii) and *Val* that represent, respectively, *modus ponens*, the fact that a 'vacuous' assumption may be discharged, and the equivalence between $\Gamma, A \vdash B$ and $\Gamma \vdash A \to B$). This semantics is sound, complete, and provides a decision procedure for validity (and so theoremhood) in *INT*, illustrated by the quasi-matrices[10] below.

Example 9.7 $p \vee \neg p$ is invalid in *INT*.

p	0		1
$\bot$	0		0
$p \to \bot$	0	1	0
$p \vee (p \to \bot)$	0	1	1

Example 9.8 $p \to (q \to p)$ is valid in *INT*.

p	0				1		
q	0		1		0		1
$q \to p$	0	1	0		1		1
$p \to (q \to p)$	0	1	1	0	1	1	1
	s_1	s_2	s_3	s_4	s_5	s_6	s_7

---

[10] Quasi-matrices are nondeterministic matrices that represent the nonfunctional valuation semantics like the ones presented here. Quasi-matrices appear, for example, in da Costa and Alves (1977), Loparic (1986, 2010), Carnielli et al. (2007).

Notice that in Example 9.8 above, the semi-valuations s_1 and s_4 are not valuations, since they do not satisfy the clause *Val* of Definition 9.6.

9.6.2 Valuation Semantics for BLE and LET$_J$

Definition 9.9 (*Valuation semantics for BLE and LET$_J$*) (1) A valuation semantics for *BLE* is obtained from the semantics of *INT* (Definitions 9.5 and 9.6) by dropping clause *(v)* of Definition 9.5 and adding the clauses below:

(v) $s(A) = 1$ iff $s(\neg\neg A) = 1$,
(vi) $s(\neg(A \wedge B)) = 1$ iff $s(\neg A) = 1$ or $s(\neg B) = 1$,
(vii) $s(\neg(A \vee B)) = 1$ iff $s(\neg A) = 1$ and $s(\neg B) = 1$,
(viii) $s(\neg(A \to B)) = 1$ iff $s(A) = 1$ and $s(\neg B) = 1$.

(2) A valuation semantics for *LET$_J$* is obtained by adding the clause below to the semantics of *BLE*:

(ix) if $s(\circ A) = 1$, then $\big(s(A) = 1$ if and only if $s(\neg A) = 0\big)$.

It is easily seen that the semantic clauses for negation represent the deductive system when we consider the negative rules of *BLE* as they would be formulated axiomatically:

DNI and DNE: $A \leftrightarrow \neg\neg A$
$\neg \wedge$I and $\neg \wedge$E: $\neg(A \wedge B) \leftrightarrow (\neg A \vee \neg B)$
$\neg \vee$I and $\neg \vee$E: $\neg(A \vee B) \leftrightarrow (\neg A \wedge \neg B)$
$\neg \to$I and $\neg \to$E: $\neg(A \to B) \leftrightarrow (A \wedge \neg B)$

If we read *1* and *0*, respectively, as 'holds' and 'does not hold', the clause *(vii)* of Definition 9.9 says that $\neg(A \vee B)$ holds if and only if $\neg A \wedge \neg B$ holds. The clause *(ix)* says that if $\circ A$ holds, then classical conditions for negation also hold for *A*, and this is precisely the point of the rules *PEM$^\circ$* and *EXP$^\circ$*, to recover, respectively, excluded middle and explosion for formulas marked with $\circ$. The decision procedures for both *BLE* and *LET$_J$* are illustrated by the quasi-matrices below.

Example 9.10 $p \to (\neg p \to q)$ is invalid in *BLE*.

p	0						1					
$\neg p$	0		1				0				1	
q	0	1	0	1			0	1	0	1		
$\neg p \to q$	0	1	1	0	1	1	0	1	1	0	1	1
$p \to (\neg p \to q)$	0	1	1	1	0	1	1	0	1	1	0	1
	s_1	s_2	s_3	s_4	s_5	s_6	s_7	s_8	s_9	s_{10}	s_{11}	s_{12}

In the Example 9.10 above, the semi-valuation s_{11} turns out to be a valuation that acts as a counterexample.

Example 9.11 op $\to$ $(p \vee \neg p)$ is valid in LET_J.

p	0				1		
$\neg p$	0		1		0		1
op	0	0	1	0	1	1	0
$p \vee \neg p$	0	1	1	1	1	1	1
op $\to$ $(p \vee \neg p)$	0	1	1	1	1	1	1
	s_1	s_2	s_3	s_4	s_5	s_6	s_7

In the Example 9.11 above, the semi-valuation s_1 is not a valuation, since the clause *Val* of Definition 9.6 is not satisfied.

The valuation semantics above is sound, complete, and provide decision procedures for both *BLE* and *LET$_J$*. However, they are not compositional. The semantic value of $\neg A$ in *BLE* is not functionally determined by the value of A, neither the semantic value of $A \to B$ when both A and B receive 0. For this reason, even though they faithfully represent the deductive system, they cannot explain the meanings of the expressions in a satisfactory way. The same drawback applies to the valuation semantics for intuitionistic logic.

9.7 Final Remarks

The fact that truth is a semantical concept justifies the classical standard distinction between syntax and semantics. But things may be different in the realm of nonclassical logics. Indeed, two common mistakes in nonclassical logics are to maintain that the standard distinction between syntax and semantics applies without further ado and, consequently, to search for completeness based on the assumption that the deductive system needs to be 'explained' by a semantics—by naively copying classical logic.

Inferential semantics are motivated by the idea that meanings may be given by the rules of inference without the need for an external notion of meaning. This is obviously related to the well-known Wittgensteinian dictum that the meaning of a word is given by its use in language. Indeed, the widespread ideas that meaning depends on language use, that the context is an essential feature of language use, and therefore that meaning always takes place within a context, are compelling, but these ideas do not go too far without a nontrivial explanation of what a context of language use is. The prospect of giving a detailed and exhaustive analysis of different contexts in natural language is, at best, a very difficult task. But in logic, we may consider that different kinds of contexts of language use are given by different properties ascribed to propositions. Our proposal here places an additional ingredient in the original idea of inferential semantics: an account of the context in terms of the property of propositions that is preserved in inferences.

An essential feature of inferential semantics for intuitionistic logic is that its rules are conceived for contexts where the property of propositions being considered is availability of a constructive proof. We argued here that an inferential semantics for the paraconsistent logic *BLE* can be provided by considering availability of evidence as the property of propositions at stake in contexts governed by *BLE*. The logic LET_J is able to deal simultaneously with contexts concerned with evidence and with truth, and we suggested that the meanings of classical connectives in the contexts concerned with truth may be explained not by their introduction rules, but by all truth-preserving inferences made with them.

There is, however, a great deal of work to be done. The proof theory of *BLE* and LET_J is still to be worked out, and so it is still to be established to what extent the proof theory of these logics fits the conceptual aspects of them presented here and in Carnielli and Rodrigues (2017). Besides, several conceptual aspects of inferential semantics have to be analyzed and discussed with respect to *BLE* and LET_J. But we believe that the ideas proposed here will expand a field of study in logic and philosophy of logic that, until now, does not include paraconsistent logics.

9.8 Post Scriptum

The view on paraconsistency endorsed by us is radically opposed to dialetheism. As it is well-known, dialetheism is the view on paraconsistency, chiefly defended by Graham Priest, according to which there exist dialetheias, i.e., true contradictions, pairs of propositions A and $\neg A$ such that both are true. Assuming that the falsity of A is the truth of $\neg A$, dialetheism claims that there are propositions that are both true and false (see e.g., Priest and Berto 2013). Let us recall that a true contradiction would be made true by an object a and a property P such that a has and does not have the property P simultaneously and in the same respect.[11] Actually, dialetheism is a rather controversial position, and it is fair to say that it is seriously discussed in philosophical circles only. Indeed, people from Mathematics and empirical sciences are not very concerned as to whether or not there are true contradictions—they just take for granted that there are not.

It should be emphasized that the position defended in Carnielli and Rodrigues (2017) and further developed here implies the rejection of dialetheism because the informal interpretation proposed for LET_J does not allow true contradictions. This point is discussed in Carnielli and Rodrigues (2017, Sect. 4.2 and Fact 16), but the basic idea is as follows: in LET_J the truth and the falsity of a proposition A is expressed, respectively, by $A \wedge \circ A$ and $\neg A \wedge \circ A$. So, the simultaneous truth and

---

[11] These two conditions exclude contradictions due to properties instantiated in different moments of time and contradictions that depend on two different perspectives. Heraclitean oppositions, thus, are not dialetheias, and contradictions in history, like the ones asserted by Hegelians, do not qualify as dialetheias either.

falsity of A yields triviality, since from $\wedge E$ we get A, $\neg A$ and $\circ A$, and now from EXP° any B is proved.

The epistemic approach to paraconsistency is based on the view that contradictions do not belong to reality but rather to thought and language. This view is not new in paraconsistency, but as far as we know, a formal system together with an intended intuitive interpretation that expressly prohibits true contradictions had not been presented yet. LET_J and its informal reading has been conceived to be such a formal system. We hope that the present paper has contributed to make still more convincing our proposal of giving an intuitive and non-dialetheist reading of paraconsistency.

Acknowledgements The first author acknowledges support from *FAPESP* (Fundação de Amparo à Pesquisa do Estado de São Paulo, thematic project *LogCons*) and from a *CNPq* (Conselho Nacional de Desenvolvimento Científico e Tecnológico) research grant. The second author acknowledges support from *FAPEMIG* (Fundação de Amparo à Pesquisa do Estado de Minas Gerais, grants PEP 157-16 and 701-16). We would like to thank Eduardo Barrio, the audience of the Workshop CLE-BsAs Logic Group (Buenos Aires, April 2016), and two anonymous referees for some valuable comments on a previous version of this text.

References

Almukdad, A., & Nelson, D. (1984). Constructible falsity and inexact predicates. *The Journal of Symbolic Logic*, *49*(1), 231–233.
Brouwer, L. E. J. (1907). On the foundations of mathematics. In A. Heyting (Ed.), *Collected works* (Vol. I). North-Holland Publishing Company (1975)
Carnielli, W. A., & Coniglio, M. E. (2016). *Paraconsistent logic: Consistency contradiction and negation*. Springer.
Carnielli, W. A., & Rodrigues, A. (2016). Paraconsistency and duality: Between ontological and epistemological views. In *Logica yearbook 2015*. College Publications.
Carnielli, W. A., & Rodrigues, A. (2017). An epistemic approach to paraconsistency: A logic of evidence and truth. *Synthese*.
Carnielli, W. A., Coniglio, M. E., & Marcos, J. (2007). Logics of formal inconsistency. In Gabbay & Guenthner (Eds.), *Handbook of philosophical logic* (Vol. 14). Springer.
da Costa, N. C. A., & Alves, E. (1977). A semantical analysis of the calculi C_n. *Notre Dame Journal of Formal Logic*, *18*(4), 621–630.
Dubucs, J. (2008). Truth and experience of truth. In M. van Atten, et al. (Eds.), *One hundred years of intuitionism*. Birkhäuser Verlag.
Dummett, M. (1973). *Frege: Philosophy of language*. Harvard University Press.
Dummett, M. (1991). *The logical basis of metaphysics*. Harvard University Press.
Frege, G. (1980). Letter to Frege. In *Philosophical and mathematical correspondence* (H. Kaal, Trans.) (pp. 150–151). University of Chicago Press.
Gentzen, G. (1935). Investigations into logical deduction. In M. E. Szabo (Ed.), *The collected papers of Gerhard Gentzen*. North-Holland Publishing Company (1969).
Heyting, A. (1930). On intuitionistic logic. In M. Paolo (Ed.), *From Brouwer to Hilbert: The debate on the foundations of mathematics in the 1920s*. Oxford University Press (1998).
Heyting, A. (1934). *Les Fondements des Mathématiques, Intuitionnisme, Théorie de la Démonstration* (p. 1955). Paris: Gautier-Villars.
Heyting, A. (1956). *Intuitionism: An introduction*. North-Holland Publishing Company.
Kolmogorov, A. (1932). On the interpretation of intuitionistic logic. In M. Paolo (Ed.), *From Brouwer to Hilbert: The debate on the foundations of mathematics in the 1920s*. Oxford University Press.

Kripke, S. (1965). Semantical analysis of intuitionistic logic I. In J. Crossley & M. Dummett (Eds.), *Formal systems and recursive functions*. Amsterdam: North-Holland Publishing.
Loparic, A. (1986). A semantical study of some propositional calculi. *The Journal of Non-classical Logic, 3*(1), 73–95.
Loparic, A. (2010). Valuation semantics for intuitionistic propositional calculus and some of its subcalculi. *Principia, 14*(1), 125–133.
Loparic, A., & Alves, E. (1979).The semantics of the systems C_n of da Costa. In A. I. Arruda, N. C. A. da Costa, & A. M. Sette (Eds.), *Proceedings of the III Brazilian Conference on Mathematical Logic, Recife, 1979* (pp. 161–172). São Paulo: Brazilian Logic Society.
López-Escobar, E. G. K. (1972). Refutability and elementary number theory. *Indagationes Mathematicae, 34*, 362–374.
Odintsov, S. (2008). *Constructive negations and paraconsistency*. Springer.
Prawitz, D. (1965). *Natural deduction: A proof-theoretical study*. Dover Publications (2006).
Prawitz, D. (2015). Classical versus intuitionistic logic. In *Why is this a Proof? Festschrift for Luiz Carlos Pereira*. College Publications.
Prawitz, D. (2016). On the relation between Heyting's and Gentzen's approaches to meaning. In Piecha et al. (Ed.), *Advances in proof-theoretic semantics*. Springer.
Priest, G., & Berto, F. (2013). Dialetheism. *Stanford Encyclopedia of Philosophy*. http://plato.stanford.edu/archives/sum2013/entries/dialetheism/.
Tarski, A. (1944). The semantic conception of truth and the foundations of semantics. *Philosophy and Phenomenological Research, 4*, 341–376.
Tarski, A., & (1956a). The concept of truth in formalized languages,. (1936). In *Logic, Semantics. Metamathematics*: Clarendon Press.
Tarski, A., & (1956b). The establishment of scientific semantics,. (1935). In *Logic, Semantics. Metamathematics*: Clarendon Press.
Tarski, A. (1969). Truth and proof. *Scientific American, 63–70*, 75–77.
van Dalen, D., & Troelstra, A. S. (1988). *Constructivism in mathematics: An introduction*. Amsterdam: Elsevier.

Chapter 10
A Model-Theoretic Analysis of Fidel-Structures for mbC

Marcelo E. Coniglio and Aldo Figallo-Orellano

Abstract In this paper, the class of Fidel-structures for the paraconsistent logic **mbC** is studied from the point of view of Model Theory and Category Theory. The basic point is that Fidel-structures for **mbC** (or **mbC**-structures) can be seen as first-order structures over the signature of Boolean algebras expanded by two binary predicate symbols N (for negation) and O (for the consistency connective) satisfying certain Horn sentences. This perspective allows us to consider notions and results from Model Theory in order to analyze the class of **mbC**-structures. Thus, substructures, union of chains, direct products, direct limits, congruences and quotient structures can be analyzed under this perspective. In particular, a Birkhoff-like representation theorem for **mbC**-structures as subdirect products in terms of subdirectly irreducible **mbC**-structures is obtained by adapting a general result for first-order structures due to Caicedo. Moreover, a characterization of all the subdirectly irreducible **mbC**-structures is also given. An alternative decomposition theorem is obtained by using the notions of weak substructure and weak isomorphism considered by Fidel for C_n-structures.

Keywords Fidel structures · Paraconsistent logics · Logics of formal inconsistency · Model theory · Birkhoff decomposition theorem

---

M. E. Coniglio (✉)
Institute of Philosophy and the Humanities (IFCH), Centre for Logic,
Epistemology and the History of Science (CLE), University of Campinas
(UNICAMP), Campinas, Brazil
e-mail: coniglio@cle.unicamp.br

A. Figallo-Orellano
Centre for Logic, Epistemology and the History of Science (CLE),
University of Campinas (UNICAMP), Campinas, Brazil
e-mail: aldofigallo@gmail.com

Department of Mathematics, National University of the South (UNS),
Bahía Blanca, Argentina

© Springer Nature Switzerland AG 2019
C. Başkent and T. M. Ferguson (eds.), *Graham Priest on Dialetheism and Paraconsistency*, Outstanding Contributions to Logic 18,
https://doi.org/10.1007/978-3-030-25365-3_10

10.1 Paraconsistency and Non-deterministic Semantics

Paraconsistency is the study of logic systems having a negation $\neg$ which is not *explosive*, that is, there exist formulas α and β in the language of the logic such that β is not derivable from the contradictory set $\{\alpha, \neg\alpha\}$. In other words, the logic has contradictory but nontrivial theories.

There are several approaches to paraconsistency in the literature since the introduction in 1948 of Jaskowski's system of *Discussive logic* (see Jaśkowski 1998), such as Relevant logics, Adaptive logics, Many-valued logics, and many others.[1] The well-known 3-valued logic **LP** (*Logic of Paradox*) was introduced by Priest in (1979) with the aim of formalizing the philosophical perspective underlying G. Priest and R. Sylvan's *Dialetheism* (see, for instance, Priest and Berto 2017; Priest 2007). As it is well-known, the main thesis behind Dialetheism is that there are *true contradictions*, that is, that some sentences can be both true and false at the same time and in the same way. Since then, the logic **LP** was intensively studied and developed by several authors proposing, for instance, extensions to first-order languages and applications to Set Theory (see, among others, Restall 1992; Weber 2010; Omori 2015).

The publication in 1963 of N. da Costa's Habilitation thesis *Sistemas Formais Inconsistentes* (*Inconsistent Formal Systems*, in Portuguese, see Costa 1993) constitutes a landmark in the history of paraconsistency. In that thesis, da Costa introduces the hierarchy C_n (for $n \geq 1$) of C-systems. This approach to paraconsistency differs from others, as it is based on the idea of locally recovering the classical reasoning (in particular, the explosion law for negation) by means of a derived unary connective of *well-behavior*, $(\cdot)^\circ$. Being so, a contradiction is not explosive in general in such systems (namely, $\alpha, \neg\alpha \nvdash \beta$ for some α and β). But assuming additionally that α is well-behaved, then such a contradiction must be trivializing; namely, $\alpha, \neg\alpha, \alpha^\circ \vdash \beta$ for every β. The idea of C-systems was afterward generalized by Carnielli and Marcos in (2002) through the class of *Logics of Formal Inconsistency*, in short **LFI**s. In such logics, da Costa's well-behavior derived connective $(\cdot)^\circ$ is replaced by a (possibly primitive) *consistency* unary conective $\circ$. The basic idea of **LFI**s, as in da Costa's C-systems, is that $\alpha, \neg\alpha \nvdash \beta$ in general, but $\alpha, \neg\alpha, \circ\alpha \vdash \beta$ always. Of course the C-systems are particular cases of **LFI**s.

Giving a semantical interpretation for C-systems, and for **LFI**s in general, is not a simple task: most of the **LFI**s introduced in the literature (see Carnielli and Marcos 2002; Carnielli et al. 2007; Carnielli and Coniglio 2016) are not algebraizable by means of the standard techniques of algebraic logic (including Blok and Pigozzi's method, see Blok and Pigozzi 1989). Being so, the use of semantics of a non-deterministic character has shown to be a useful alternative way for dealing with **LFI**s. Several non-deterministic semantical tools were introduced in the literature in order to analyze such systems, allowing so decision procedures for them: non-truth-functional bivaluations (proposed by da Costa and Alves in 1976, 1977), possible-

---

[1]For a good introductory article on Paraconsistency see Priest et al. (2016) and the references therein.

translations semantics (proposed by Carnielli in 1990), and non-deterministic matrices (or Nmatrices), proposed by Avron and Lev in (2001). In particular, the Nmatrix semantics can be analyzed from a general (non-deterministic) algebraic perspective through the notion of *swap structures* (see Carnielli and Coniglio 2016, Chap. 6; Coniglio et al. 2017). However, as shown in Avron (2007), not every **LFI** can be characterized by a single finite Nmatrix. In particular, da Costa's logic C_1 cannot be characterized by a single finite Nmatrix.

Another interesting approach to non-deterministic semantics for nonclassical logics was proposed for Priest in (2014), through the notion of *plurivalent semantics*. Let $\mathcal{M} = \langle \mathcal{A}, D \rangle$ be a matrix semantics for a propositional signature Ξ (that is, $\mathcal{A}$ is an algebra over Ξ and D is a non-empty subset of the domain A of $\mathcal{A}$, called the set of *designated values*). Then, a plurivalent semantics over $\mathcal{M}$ is a pair $\mathcal{M}_\triangleright = \langle \mathcal{M}, \triangleright \rangle$ such that $\triangleright \subseteq \mathcal{V} \times A$ is a relation from the set $\mathcal{V}$ of propositional variables to A. It is assumed that, for every $p \in \mathcal{V}$, there is $a \in A$ such that $p \triangleright a$.[2] Given $\mathcal{M}_\triangleright$, a non-deterministic (or plurivalent) interpretation $[\cdot]_\mathcal{M}^\triangleright$ for the algebra of formulas over Ξ generated by $\mathcal{V}$ is defined recursively: $[p]_\mathcal{M}^\triangleright = \{a \in A : p \triangleright a\}$, if $p \in \mathcal{V}$; and $[c(\varphi_1, \ldots, \varphi_n)]_\mathcal{M}^\triangleright = \bigcup \{c^\mathcal{A}(a_1, \ldots, a_n) : a_i \in [\varphi_i]_\mathcal{M}^\triangleright \text{ for } 1 \leq i \leq n\}$, for every n-ary connective c in Ξ. Given a set of formulas $\Gamma \cup \{\varphi\}$, Γ *infers* φ in $\mathcal{M}_\triangleright$, denoted by $\Gamma \models_\mathcal{M}^\triangleright \varphi$, if $[\varphi]_\mathcal{M}^\triangleright \cap D \neq \emptyset$ whenever $[\beta]_\mathcal{M}^\triangleright \cap D \neq \emptyset$ for every $\beta \in \Gamma$. The plurivalent consequence relation $\models_p^\mathcal{M}$ generated from $\mathcal{M}$ is defined as expected: $\Gamma \models_p^\mathcal{M} \varphi$ iff $\Gamma \models_\mathcal{M}^\triangleright \varphi$ for every $\mathcal{M}_\triangleright$. Priest have shown in (2014) how to produce a family of plurivalent logics related to the first-degree entailment *FDE*. Additionally, in (2016) he applies the notion of plurivalent semantics in order to analyze Indian Buddhist logic from the pespective of formal logic.

Before all these efforts, M. Fidel has proved in 1969 (despite it was only published in 1977, see Fidel 1977), for the first time in the literature, the decidability of da Costa's calculi C_n by means of a novel algebraic-relational class of structures called C_n-structures. Afterward, this kind of structure was called *Fidel-structures* or **F-structures** (see Odintsov 2003). A C_n-structure is a triple $\langle \mathcal{A}, \{N_a\}_{a \in A}, \{N_a^{(n)}\}_{a \in A} \rangle$ such that $\mathcal{A}$ is a Boolean algebra with domain A and each N_a and $N_a^{(n)}$ is a non-empty subset of A. The intuitive meaning of $b \in N_a$ and $c \in N_a^{(n)}$ is that b and c are possible values for the paraconsistent negation $\neg a$ of a and for the well-behavior $a°$ of a in C_n, respectively. The use of relations instead of functions for interpreting the paraconsistent negation $\neg$ and the well-behavior connective $(\cdot)°$ of the C_n calculi is justified by the fact that these connectives are not truth-functional. That is, they cannot be characterized by means of truth functions, and so the use of relations seems to be a good choice, constituting the first non-deterministic semantics for da Costa's calculi C_n proposed in the literature. Given that every C_n can be characterized by **F**-structures over the 2-element Boolean algebra, as Fidel has shown in (1977), this result evidences the greater expressive power of **F**-structures with respect to N matrices. A discussion about this topic can be found in Carnielli and Coniglio (2016, Sects. 6.6 and 6.7).

---

[2]The general case in which this conditions is dropped is briefly analyzed by Priest in (2014).

Fidel-structures can be defined for a wide class of logics, not only paraconsistent ones. Concerning **LFI**s, Fidel-structures were defined in Carnielli and Coniglio (2014, 2016, Chap. 6) for several **LFI**s which are weaker than da Costa's C_1, starting from a basic but very interesting **LFI** called **mbC**.

This paper proposes the study of Fidel-structures for **mbC** from the point of view of Model Theory. Under this perspective, substructures, union of chains, direct products, direct limits, congruences, and quotient structures can be defined. In particular, a generalization to first-order structures of Birkhoff's representation theorem for algebras, due to Caicedo, can be obtained for **mbC**-structures (see Theorem 10.56). Moreover, a characterization of the subdirectly irreducible **mbC**-structures is given in Theorem 10.55. This representation theorem is compared to a similar one, obtained by Fidel for the calculi C_n.

10.2 The Logic mbC

The logic **mbC** is the most basic **LFI** analyzed by Carnielli et al. in (2007), which later on it was studied by several authors. This logic is based on propositional positive classical logic and, despite its apparent simplicity, it enjoys extremely interesting features. This section is devoted to briefly describe the logic **mbC** for the reader's convenience.

Consider the propositional signature $\Sigma = \{\wedge, \vee, \rightarrow, \neg, \circ\}$ for **LFI**s, and let $\mathcal{V} = \{p_i : i \in \mathbb{N}\}$ be a denumerable set of propositional variables. The algebra of propositional formulas over Σ generated by $\mathcal{V}$ will be denoted by $For(\Sigma)$.

Definition 10.1 (*Logic* **mbC**, Carnielli et al. 2007) The calculus **mbC** over the language $For(\Sigma)$ is defined by means of the following Hilbert calculus:

(A1) $\quad \alpha \rightarrow (\beta \rightarrow \alpha)$
(A2) $\quad (\alpha \rightarrow \beta) \rightarrow ((\alpha \rightarrow (\beta \rightarrow \gamma)) \rightarrow (\alpha \rightarrow \gamma))$
(A3) $\quad \alpha \rightarrow (\beta \rightarrow (\alpha \wedge \beta))$
(A4) $\quad (\alpha \wedge \beta) \rightarrow \alpha$
(A5) $\quad (\alpha \wedge \beta) \rightarrow \beta$
(A6) $\quad \alpha \rightarrow (\alpha \vee \beta)$
(A7) $\quad \beta \rightarrow (\alpha \vee \beta)$
(A8) $\quad (\alpha \rightarrow \gamma) \rightarrow ((\beta \rightarrow \gamma) \rightarrow ((\alpha \vee \beta) \rightarrow \gamma))$
(A9) $\quad \alpha \vee (\alpha \rightarrow \beta)$
(A10) $\quad \alpha \vee \neg \alpha$
(A11) $\quad \circ \alpha \rightarrow (\alpha \rightarrow (\neg \alpha \rightarrow \beta))$
(MP) $\quad$ *From α and $\alpha \rightarrow \beta$ infer β*

It is worth noting that **mbC** is obtained from a calculus for the positive cassical logic **CPL**$^+$ by adding the axiom schemas (A10) and (A11), concerning the paraconsistent negation $\neg$ and the consistency operator ∘. As observed in Carnielli and Marcos (2002), each calculus C_n is a particular case of **LFI** in which the consistency connective ∘ is defined in term of the others: for instance, $\circ\alpha \stackrel{\text{def}}{=} \alpha^\circ = \neg(\alpha \wedge \neg\alpha)$ in C_1. The logic C_1 can be seen as an axiomatic extension of **mbC** up to language (see Carnielli et al. 2007; Carnielli and Coniglio 2016).

Being weaker than C_1, the logic **mbC** cannot be characterized by any standard algebraic semantics, even in the wide sense of Blok and Pigozzi (1989). Moreover, it cannot be characterized by a single finite logical matrix. From this, it is clear that alternative semantics for **mbC** are necessary. In Carnielli and Coniglio (2014) and (2016), Fidel-structures for **mbC** and for several axiomatic extensions of it were presented. However, the formal study of the properties of the class of **F**-structures for such **LFI**s was never developed, in contrast with the study of the **F**-structures for C_n carry out by Fidel in (1977).

In the following sections, the class of **F**-structures for **mbC** will be studied as a basic but relevant example through an original approach based on elementary concepts of Model Theory and Category Theory.[3]

10.3 Fidel-Structures for mbC

In this section, the class of **F**-structures for **mbC** introduced in Carnielli and Coniglio (2014) and (2016) will be recast in the language of first-order structures. This constitutes a novel approach to Fidel-structures in general.

Consider classical first-order theories defined over first-order signatures based on the following logical symbols: the connectives $\wedge, \vee, \rightarrow$, and $\sim$ (for conjunction, disjunction, implication, and negation, respectively), the quantifiers $\forall$ and $\exists$, and the symbol $\approx$ for the equality predicate which is always interpreted as the identity relation. Assume also a denumerable set $V_{ind} = \{v_i : i \in \mathbb{N}\}$ of variables. The letters u, w, z will be used to refer to arbitrary variables. If Ξ is a first-order signature then Ξ-**str** denotes the category of Ξ-structures, that is, the category of first-order structures over the signature Ξ.

Definition 10.2 The *signature for* **F**-*structures for* **LFI**s is the first-order signature Θ composed by the following symbols:

(i) Two binary predicate symbols N and O, for the paraconsistent negation and the consistency operator, respectively;
(ii) Two binary function symbols $\sqcap$ and $\sqcup$, and an unary function symbol $-$, for Boolean meet, Boolean join, and Boolean complement, respectively;
(iii) Two constant symbols **0** and **1** for the bottom and the top element, respectively.

Observe that the subsignature Θ_{BA} of Θ obtained by dropping the predicate symbols N, O is the usual signature for Boolean algebras.

---

[3]For general notions on Model Theory and Category Theory the reader can consult Chang and Keisler (2012) and Mac Lane (1978), respectively.

Definition 10.3 An F-structure for **mbC** (in short, an **mbC**-structure) is a Θ-first-order structure

$$\mathcal{E} = \langle A, \sqcap^{\mathcal{E}}, \sqcup^{\mathcal{E}}, -^{\mathcal{E}}, \mathbf{0}^{\mathcal{E}}, \mathbf{1}^{\mathcal{E}}, N^{\mathcal{E}}, O^{\mathcal{E}} \rangle$$

such that:

(a) the Θ_{BA}-reduct $\mathcal{A} = \langle A, \sqcap^{\mathcal{E}}, \sqcup^{\mathcal{E}}, -^{\mathcal{E}}, \mathbf{0}^{\mathcal{E}}, \mathbf{1}^{\mathcal{E}} \rangle$ of $\mathcal{E}$ is a Boolean algebra; that is, $\mathcal{A}$ satisfies the usual equations axiomatizing Boolean algebras in the signature Θ_{BA}, see for instance (Chang and Keisler 2012, Example 1.4.3)[4];
(b) $\mathcal{E}$ satisfies the following Θ-sentences:

   (i) $\forall u \exists w N(u, w)$,
   (ii) $\forall u \exists w O(u, w)$,
   (iii) $\forall u \forall w (N(u, w) \to (u \sqcup w \approx \mathbf{1}))$,
   (iv) $\forall u \forall w (N(u, w) \to \exists z (O(u, z) \wedge ((u \sqcap w \sqcap z) \approx \mathbf{0})))$.

The class of **mbC**-structures will be denoted by **FmbC**.

Notation 10.4 *From now on, an **mbC**-structure $\mathcal{E}$ will be denoted by $\mathcal{E} = \langle \mathcal{A}, N^{\mathcal{E}}, O^{\mathcal{E}} \rangle$ such that $\mathcal{A} = \langle A, \sqcap^{\mathcal{E}}, \sqcup^{\mathcal{E}}, -^{\mathcal{E}}, \mathbf{0}^{\mathcal{E}}, \mathbf{1}^{\mathcal{E}} \rangle$ is a Boolean algebra. We will frequently write $\#^{\mathcal{A}}$ instead of $\#^{\mathcal{E}}$ for $\# \in \{\sqcap, \sqcup, -, \mathbf{0}, \mathbf{1}\}$. Given an **mbC**-structure $\mathcal{E}$ and $a, b \in A$, alternatively we can write $b \in N_a^{\mathcal{E}}$ instead of $N^{\mathcal{E}}(a, b)$. Similar notation will be adopted for the predicate symbol O, alternatively writing $b \in O_a^{\mathcal{E}}$ instead of $O^{\mathcal{E}}(a, b)$. This is in line with the traditional presentation of **F**-structures where the non-truth-functional unary connectives are interpreted by families of non-empty subsets of the domain of the structure (recall the definition of C_n-structure outlined in Sect. 10.1).*

The intuitive reading for $b \in N_a^{\mathcal{E}}$ and $c \in O_a^{\mathcal{E}}$ is that b is a *possible negation* $\neg a$ of a, and that c is a *possible consistency* $\circ a$ of a coherent with b. This is justified by Definition 10.6 below.

Remark 10.5 Observe that $N^{\mathcal{E}}(\mathbf{0}^{\mathcal{E}}, b)$ iff $b = \mathbf{1}^{\mathcal{E}}$, for every **F**-structure $\mathcal{E}$ for **mbC**. On the other hand, if $(\mathbf{1}^{\mathcal{E}}, \mathbf{1}^{\mathcal{E}}) \in N^{\mathcal{E}}$ then $(\mathbf{1}^{\mathcal{E}}, \mathbf{0}^{\mathcal{E}})$ must belong to $O^{\mathcal{E}}$. Indeed, it must exists $z \in A$ such that $(\mathbf{1}^{\mathcal{E}}, z) \in O^{\mathcal{E}}$ and $\mathbf{1}^{\mathcal{E}} \sqcap \mathbf{1}^{\mathcal{E}} \sqcap z = \mathbf{0}^{\mathcal{E}}$. Therefore $z = \mathbf{0}^{\mathcal{E}}$ and so $(\mathbf{1}^{\mathcal{E}}, \mathbf{0}^{\mathcal{E}}) \in O^{\mathcal{E}}$.

Recall now the semantics for **mbC** defined from **F**-structures (see Carnielli and Coniglio 2014, 2016). Given an **mbC**-structure $\mathcal{E}$, consider the Boolean implication defined as usual by $a \Rightarrow^{\mathcal{E}} b \stackrel{\text{def}}{=} -^{\mathcal{E}} a \sqcup^{\mathcal{E}} b$ for every $a, b \in A$. For $\# \in \{\wedge, \vee, \to\}$ let $\widehat{\#}^{\mathcal{E}}$ be the corresponding operation in $\mathcal{E}$, that is: $\widehat{\wedge}^{\mathcal{E}} = \sqcap^{\mathcal{E}}; \widehat{\vee}^{\mathcal{E}} = \sqcup^{\mathcal{E}};$ and $\widehat{\to}^{\mathcal{E}} = \Rightarrow^{\mathcal{E}}$. As stated above, $For(\Sigma)$ denotes the algebra of formulas for the logic **mbC**.

Definition 10.6 A *valuation* over an **mbC**-structure $\mathcal{E}$ is a map $v : For(\Sigma) \to A$ satisfying the following properties, for every formulas α and β:

---

[4]In difference with several authors, we admit the trivial one-element Boolean algebra, see Remark 10.39.

(1) $v(\alpha\#\beta) = v(\alpha)\widehat{\#}^{\mathcal{E}} v(\beta)$, for $\# \in \{\wedge, \vee, \to\}$;
(2) $v(\neg\alpha) \in N^{\mathcal{E}}_{v(\alpha)}$ (that is, $N^{\mathcal{E}}(v(\alpha), v(\neg\alpha))$);
(3) $v(\circ\alpha) \in O^{\mathcal{E}}_{v(\alpha)}$ (that is, $v(\circ\alpha)$ is such that $O^{\mathcal{E}}(v(\alpha), v(\circ\alpha))$) and $v(\alpha) \sqcap^{\mathcal{E}} v(\neg\alpha) \sqcap^{\mathcal{E}} v(\circ\alpha) = \mathbf{0}^{\mathcal{E}}$.

Observe that item (3) of the previous definition is well-defined by item (b)(iv) of Definition 10.3 and item (2) of the last definition.

The semantical consequence relation associated to **F**-structures for **mbC** is naturally defined.

Definition 10.7 Let $\Gamma \cup \{\alpha\} \subseteq For(\Sigma)$ be a finite set of formulas.
(i) Given a Fidel-structure $\mathcal{E}$ for **mbC**, we say that α is a semantical consequence of Γ (w.r.t. $\mathcal{E}$), denoted by $\Gamma \Vdash^{\mathbf{mbC}}_{\mathcal{E}} \alpha$, if, for every valuation v over $\mathcal{E}$: $v(\alpha) = 1$ whenever $v(\gamma) = 1$ for every $\gamma \in \Gamma$.
(ii) We say that α is a semantical consequence of Γ (w.r.t. Fidel-structures for **mbC**), denoted by $\Gamma \Vdash^{\mathbf{mbC}}_{\mathbf{F}} \alpha$, if $\Gamma \Vdash^{\mathbf{mbC}}_{\mathcal{E}} \alpha$ for every **F**-structure $\mathcal{E}$ for **mbC**.

Theorem 10.8 (Soundness and completeness of **mbC** w.r.t. **F**-structures, Carnielli and Coniglio 2014, 2016) *Let $\Gamma \cup \{\alpha\}$ be a finite set of formulas in $For(\Sigma)$. Then: $\Gamma \vdash_{\mathbf{mbC}} \alpha$ iff $\Gamma \Vdash^{\mathbf{mbC}}_{\mathbf{F}} \alpha$.*

Moreover, let $\mathbb{A}_2$ be the two-element Boolean algebra with domain $\{0, 1\}$, and let $\Vdash^{\mathbf{mbC}}_{\mathbf{F}_2}$ be the semantical consequence relation with respect to the class of **mbC**-structures defined over $\mathbb{A}_2$. By adapting the proof of Carnielli and Coniglio (2016, Theorem 6.2.16) it is easy to obtain the following result.

Theorem 10.9 (Soundness and completeness of **mbC** w.r.t. **F**-structures over $\mathbb{A}_2$) *Let $\Gamma \cup \{\alpha\}$ be a finite set of formulas in $For(\Sigma)$. Then: $\Gamma \vdash_{\mathbf{mbC}} \alpha$ iff $\Gamma \Vdash^{\mathbf{mbC}}_{\mathbf{F}_2} \alpha$.*

The latter result gives us a decision procedure for checking validity in **mbC**.

10.4 On the Axiomatization of mbC-Structures

In this section we briefly discuss the class **FmbC** of **F**-structures from the point of view of the syntactic form of its axioms (recall Definition 10.3). From this, some well-known results from Model Theory can be applied to **FmbC** in a direct way, as we shall see along this paper.

Recall from Chang and Keisler (2012, Page 407) that a *basic Horn formula* over a first-order signature Ξ is a formula of the form $\sigma_1 \vee \ldots \vee \sigma_n$ (for $n \geq 1$) such that at most one formula is atomic and the rest is the negation of an atomic formula over Ξ. In particular, formulas of the form $\sigma_1 \wedge \ldots \wedge \sigma_n \to \sigma_{n+1}$, where each σ_i is atomic, are (logically equivalent to) basic Horn formulas. A *Horn formula* over Ξ is any formula over Ξ built up from basic Horn formulas by using exclusively the connective $\wedge$ and the quantifiers $\forall$ and $\exists$. A *Horn sentence* is a Horn formula with no free variables.

Remark 10.10 (**FmbC as a Horn theory**) From Definition 10.3, it is easy to see that **FmbC** can be axiomatized by means of Horn sentences. Indeed, the axioms of Boolean algebras are sentences of the form $\forall x_1 \cdots \forall x_n \sigma$, where σ is an atomic formula. On the other hand, axioms (b)(i) and (b)(ii) are of the form $\forall u \exists w \sigma$, where σ is an atomic formula. Axiom (b)(iii) is of the form $\forall u \forall w (\sigma_1 \to \sigma_2)$, where σ_1 and σ_2 are atomic formula. Finally, axiom (b)(iv) is logically equivalent to a sentence of the form $\forall u \forall w \exists z ((\sigma_1 \to \sigma_2) \wedge (\sigma_1 \to \sigma_3))$ where each σ_i is atomic. This means that **FmbC** can be axiomatized by Horn sentences.

Now, recall from Chang and Keisler (2012, Pages 142–143) that an *universal-existencial sentence*, or a $\forall\exists$-*sentence*, or a Π_2^0-*sentence* over a first-order signature Ξ, is a sentence of the form $\forall x_1 \cdots \forall x_n \exists y_1 \cdots \exists y_k \sigma$, where σ is a formula over Ξ without quantifiers.

Remark 10.11 (**FmbC as a Π_2^0-theory**) It is immediate to see that the class of **mbC**-structures can be axiomatized by means of $\forall\exists$-sentences (that is, Π_2^0-sentences). Indeed, the axioms of Boolean algebras are of the form $\forall x_1 \cdots \forall x_n \sigma$, where σ is an atomic formula. On the other hand, axioms (b)(i) and (b)(ii) are of the form $\forall u \exists w \sigma$, where σ is an atomic formula. Axiom (b)(iii) is of the form $\forall u \forall w \sigma$, where σ is without quantifiers. Finally, axiom (b)(iv) is logically equivalent to a sentence of the form $\forall u \forall w \exists z \sigma$ where σ has not quantifiers. This means that **FmbC** can be axiomatized by Π_2^0-sentences over signature Θ.

Finally, it is worth noting that **FmbC** can also be axiomatized by sentences over signature Θ of the form $\forall x_1 \cdots \forall x_n (\sigma_1 \to \exists y_1 \cdots \exists y_k \sigma_2)$, where σ_1 and σ_2 are positive formulas (that is, built up from conjunctions and disjunctions only) without quantifiers.

The fact that **FmbC** can be axiomatized by sentences of a special form will be used along this paper, as it will be pointed out.

10.5 Homomorphisms and Substructures

From the definitions of the previous section, the category of **mbC**-structures can be defined in a natural way.

Definition 10.12 Let $\mathcal{E} = \langle \mathcal{A}, N^{\mathcal{E}}, O^{\mathcal{E}} \rangle$ and $\mathcal{E}' = \langle \mathcal{A}', N^{\mathcal{E}'}, O^{\mathcal{E}'} \rangle$ be two **mbC**-structures. An **mbC**-*homomorphism* h from $\mathcal{E}$ to $\mathcal{E}'$ is a homomorphism $h : \mathcal{E} \to \mathcal{E}'$ in the category of Θ-structures.

Remark 10.13 By definition, an **mbC**-homomorphism $h : \mathcal{E} \to \mathcal{E}'$ is a function $h : \mathcal{A} \to \mathcal{A}'$ satisfying the following conditions, for every $a, b \in \mathcal{A}$:

(i) $h : \mathcal{A} \to \mathcal{A}'$ is a homomorphism between Boolean algebras,
(ii) if $N^{\mathcal{E}}(a, b)$ then $N^{\mathcal{E}'}(h(a), h(b))$,
(iii) if $O^{\mathcal{E}}(a, b)$ then $O^{\mathcal{E}'}(h(a), h(b))$.

From the notions above, it is defined a category $\mathbb{F}\mathbf{mbC}$ of **mbC**-structures having the class **FmbC** of **F**-structures as objects and with **mbC**-homomorphisms as its morphisms. Clearly, it is a full subcategory of the category Θ-**str** of Θ-structures. For every **mbC**-structure $\mathcal{E} = \langle \mathcal{A}, N^{\mathcal{E}}, O^{\mathcal{E}} \rangle$, the identity homomorphism given by the identity mapping over A will be denoted by $id_{\mathcal{E}} : \mathcal{E} \to \mathcal{E}$. If $h : \mathcal{E} \to \mathcal{E}'$ and $h' : \mathcal{E}' \to \mathcal{E}''$ are two homomorphisms then the composite homomorphism from $\mathcal{E}$ to $\mathcal{E}''$ will be denoted by $h' \circ h$. As a consequence of the definitions, well-known basic notions and results from Model Theory can be applied to $\mathbb{F}\mathbf{mbC}$. The only detail to be taken into account in the constructions is that **mbC**-structures are first-order structures over Θ satisfying certain Θ-sentences, as it was discussed in Sect. 10.4.

Definition 10.14 Let $\mathcal{E} = \langle \mathcal{A}, N^{\mathcal{E}}, O^{\mathcal{E}} \rangle$ and $\mathcal{E}' = \langle \mathcal{A}', N^{\mathcal{E}'}, O^{\mathcal{E}'} \rangle$ be two **mbC**-structures. The structure $\mathcal{E}$ is said to be a *substructure* of $\mathcal{E}'$, denoted by $\mathcal{E} \subseteq \mathcal{E}'$, if the following conditions hold:

(i) $\mathcal{A}$ is a Boolean subalgebra of $\mathcal{A}'$ (which will be denoted as $\mathcal{A} \subseteq \mathcal{A}'$),[5]
(ii) $N^{\mathcal{E}} = N^{\mathcal{E}'} \cap (A')^2$ and $O^{\mathcal{E}} = O^{\mathcal{E}'} \cap (A')^2$.

Remark 10.15 The notion of substructure considered in Model Theory (and, in particular, in Definition 10.14) differs slightly from that used in some areas of Mathematics. For instance, in Graph Theory a graph $H = \langle V, E \rangle$ (where V and E denote, respectively, the set of vertices and edges) is said to be a *subgraph* of another graph $G = \langle V', E' \rangle$ provided that $V \subseteq V'$ and $E \subseteq E'$. Then, it can be possible to have a subgraph H of a graph G such that H is not a substructure of G seen as first-order structures (over the signature of graphs having, besides the equality predicate symbol $\approx$, a binary predicate symbol for the *edge* relation). Indeed, H is a substructure of G if and only if H is what is called in Graph Theory a *induced* subgraph of G.

In Category Theory and in Universal Algebra the subobjects (the subalgebras, respectively) are characterized by means of the the notion of monomorphism. Recall from Category Theory (see, for instance, Mac Lane 1978) that a *monomorphism* in a category $\mathbb{C}$ is a homomorphism $h : A \to B$ in $\mathbb{C}$ such that, for every pair of parallel homomorphisms $h', h'' : C \to A$ for which $h \circ h' = h \circ h''$ in $\mathbb{C}$, it is the case that $h' = h''$. On the other hand, recall the following notion from Model Theory (see, for instance, Chang and Keisler 2012).

Definition 10.16 Let $\mathfrak{A}$ and $\mathfrak{A}'$ be two first-order structures over a signature Ξ. A homomorphism $h : \mathfrak{A} \to \mathfrak{A}'$ in Ξ-**str** is an *embedding* if it is injective and, for every n-ary predicate symbol P and every $(a_1, \ldots, a_n) \in |\mathfrak{A}|^n$, $(a_1, \ldots, a_n) \in P^{\mathfrak{A}}$ if and only if $(h(a_1), \ldots, h(a_n)) \in P^{\mathfrak{A}'}$.[6]

---

[5]This means that $A \subseteq A'$, $0^{\mathcal{A}} = 0^{\mathcal{A}'}$, $1^{\mathcal{A}} = 1^{\mathcal{A}'}$ and, for every $a, b \in A$: $a\#^{\mathcal{A}}b = a\#^{\mathcal{A}'}b$ and $-^{\mathcal{A}}a = -^{\mathcal{A}'}a$ for $\# \in \{\sqcap, \sqcup\}$. Note that we write $s^{\mathcal{A}}$ instead of $s^{\mathcal{E}}$ when s correspond to a symbol of the subsignature Θ_{BA} of Θ.

[6]Since in Model Theory the equality predicate $\approx$ is considered as a predicate symbol which is always interpreted as the standard equality, the injectivity of an embedding is a consequence of the definition.

It is well-known that a homomorphism $h : \mathfrak{A} \to \mathfrak{A}'$ in Ξ-**str** is an embedding if and only if it is a monomorphism in the subcategory Ξ-**emb** of Ξ-**str** formed by Ξ-structures as objects and embeddings as morphisms. Clearly, if Ξ has only function symbols besides the identity predicate $\approx$ (this is the case of Universal Algebra) then Ξ-**emb** is the subcategory of Ξ-**str** in which every morphism is a monomorphism. Moreover, the induced substructures correspond to the subobjects in Ξ-**str**.

Remark 10.17 As a direct consequence of the definitions, a homomorphism $h : \mathcal{E} \to \mathcal{E}'$ in $\mathbb{F}$**mbC** is an embedding if and only if it is an injective homomorphism where conditions (ii) and (iii) of Remark 10.13 are replaced by

(ii)' $N^{\mathcal{E}}(a, b)$ if and only if $N^{\mathcal{E}'}(h(a), h(b))$;
(iii)' $O^{\mathcal{E}}(a, b)$ if and only if $O^{\mathcal{E}'}(h(a), h(b))$.

Clearly, $\mathcal{E} \subseteq \mathcal{E}'$ if and only if $\mathcal{A} \subseteq \mathcal{A}'$ and the inclusion map $i : \mathcal{A} \to \mathcal{A}'$ induces an embedding $i : \mathcal{E} \to \mathcal{E}'$ of Θ-structures.

Remark 10.18 From the definitions above, it is clear that in the category $\mathbb{F}$**mbC** the monomorphism (that is, the substructures in Θ–**str**) are not strong enough: in order to obtain a substructure in $\mathbb{F}$**mbC** the monomorphism must be, in addition, an embedding. Consider, for instance, the **mbC**-structures $\mathcal{E} = \langle \mathbb{A}_2, N^{\mathcal{E}}, O^{\mathcal{E}} \rangle$ and $\mathcal{E}' = \langle \mathbb{A}_2, N^{\mathcal{E}'}, O^{\mathcal{E}'} \rangle$ defined over the two-element Boolean algebra $\mathbb{A}_2$ such that $N^{\mathcal{E}} = \{(0, 1), (1, 0)\}$, $O^{\mathcal{E}} = \{(0, 1), (1, 1)\}$ and $N^{\mathcal{E}'} = O^{\mathcal{E}'} = \{(0, 1), (1, 0), (1, 1)\}$. Clearly, the identity $h : \{0, 1\} \to \{0, 1\}$ induces a monomorphism $h : \mathcal{E} \to \mathcal{E}'$ in $\mathbb{F}$**mbC** (since it is an injective homomorphism in $\mathbb{F}$**mbC**). But $\mathcal{E}$ *is not* a substructure of $\mathcal{E}$' since, for instance, $(1, 0) \in O^{\mathcal{E}'} \cap |\mathcal{E}|^2$ but $(1, 0) \notin O^{\mathcal{E}}$.

A weaker notion of substructure was considered in the literature of Model Theory under the name of *weak substructures*. Thus, $\mathfrak{A}$ is said to be a *weak substructure* of $\mathfrak{A}'$ in Ξ-**str** provided that $|\mathfrak{A}| \subseteq |\mathfrak{A}'|$ and $P^{\mathfrak{A}} \subseteq P^{\mathfrak{A}'}$, for every predicate symbol P. This is equivalent to say that the inclusion mapping $i : |\mathfrak{A}| \to |\mathfrak{A}'|$ induces a homomorphism $i : \mathfrak{A} \to \mathfrak{A}'$ in Ξ-**str**. For instance, in Graph Theory, a graph H is a subgraph of a graph G provided that H is a weak substructure of G, as observed in Remark 10.15. In $\mathbb{F}$**mbC** it can be considered a intermediate notion between weak substructures and substructures (as defined in Model Theory).

Definition 10.19 Let $\mathcal{E} = \langle \mathcal{A}, N^{\mathcal{E}}, O^{\mathcal{E}} \rangle$ and $\mathcal{E}' = \langle \mathcal{A}', N^{\mathcal{E}'}, O^{\mathcal{E}'} \rangle$ be two **mbC**-structures. We say that $\mathcal{E}$ is a *weak substructure* of $\mathcal{E}'$ in $\mathbb{F}$**mbC**, denoted by $\mathcal{E} \subseteq_W \mathcal{E}'$, if $\mathcal{A}$ is a Boolean subalgebra of $\mathcal{A}'$, $N^{\mathcal{E}} \subseteq N^{\mathcal{E}'}$, and $O^{\mathcal{E}} \subseteq O^{\mathcal{E}'}$. This is equivalent to say that the inclusion map $i : \mathcal{A} \to \mathcal{A}'$ induces an injective homomorphism $i : \mathcal{E} \to \mathcal{E}'$ in $\mathbb{F}$**mbC**.

Remark 10.20 Observe that if a Ξ-structure $\mathfrak{A}$ is a substructure of $\mathfrak{A}'$ then it is a weak substructure of $\mathfrak{A}'$. On the other hand, if $\mathcal{E}$ is a substructure of an **mbC**-structure $\mathcal{E}$' (hence, a weak substructure) in the category Θ-**str** then $\mathcal{E}$ is not necessarily an **mbC**-structure and so it is not necessarily a weak substructure of $\mathcal{E}'$ in $\mathbb{F}$**mbC**. For instance, consider the four-element Boolean algebra $\mathbb{A}_4$ with domain

$FOUR = \{0, a, b, 1\}$ (observe that $\mathbb{A}_4$ is isomorphic, as a Boolean algebra, to the powerset of $\{0, 1\}$) and let $\mathcal{E}' = \langle \mathbb{A}_4, N^{\mathcal{E}'}, O^{\mathcal{E}'} \rangle$ be the **mbC**-structure over $\mathbb{A}_4$ such that $N^{\mathcal{E}'} = \{(0, 1), (a, b), (b, a), (1, a)\}$ and $O^{\mathcal{E}'} = \{(0, a), (a, 0), (b, 0), (1, 0)\}$. Let $\mathcal{E} = \langle \mathbb{A}_2, N^{\mathcal{E}}, O^{\mathcal{E}} \rangle$ be the Θ-structure with domain $\{0, 1\}$ such that $N^{\mathcal{E}} = \{(0, 1)\} = N^{\mathcal{E}'} \cap (\{0, 1\})^2$ and $O^{\mathcal{E}} = \{(1, 0)\} = O^{\mathcal{E}'} \cap (\{0, 1\})^2$. Then $\mathcal{E}$ is a substructure of $\mathcal{E}'$ which is not an **mbC**-structure, hence it is not a weak substructure of $\mathcal{E}'$ in $\mathbb{F}$**mbC**.

Finally, it is worth noting that if $\mathcal{E}$ and $\mathcal{E}'$ are two **mbC**-structures such that $\mathcal{E}$ is a weak substructure of $\mathcal{E}'$ then every valuation over $\mathcal{E}$ (recall Definition 10.6) is a valuation over $\mathcal{E}'$.

It is interesting to notice that any **mbC**-structure $\mathcal{E}$ over $\mathbb{A}_2$ can be seen as a substructure (in $\mathbb{F}$**mbC**) of an **mbC**-structure $\mathcal{E}(\mathcal{A})$ over $\mathcal{A}$, for any Boolean algebra $\mathcal{A}$ with more than two elements.

Proposition 10.21 *Let $\mathcal{A}$ be a Boolean algebra with more that two elements. Given an **mbC**-structure $\mathcal{E} = \langle \mathbb{A}_2, N^{\mathcal{E}}, O^{\mathcal{E}} \rangle$ defined over $\mathbb{A}_2$ let $\mathcal{E}(\mathcal{A}) \stackrel{def}{=} \langle \mathcal{A}, N_{\mathcal{A}}^{\mathcal{E}}, O_{\mathcal{A}}^{\mathcal{E}} \rangle$ be the Θ-structure defined as follows:*

- $N_{\mathcal{A}}^{\mathcal{E}} = N^{\mathcal{E}} \cup \{(a, \sim a) : a \in |\mathcal{A}| \setminus \{0, 1\}\}$;
- $O_{\mathcal{A}}^{\mathcal{E}} = O^{\mathcal{E}} \cup \{(a, 1) : a \in |\mathcal{A}| \setminus \{0, 1\}\}$.

*Then, $\mathcal{E}(\mathcal{A})$ is an **mbC**-structure over $\mathcal{A}$ such that $\mathcal{E} \subseteq \mathcal{E}(\mathcal{A})$ in $\mathbb{F}$**mbC**.*

Proof It is straightforward from the definitions. □

As a consequence of the latter result, combined with Theorem 10.9, it can be seen that the logic of the **mbC**-structures over $\mathcal{A}$ is exactly **mbC**.

Theorem 10.22 (Soundness and completeness of **mbC** w.r.t. F-structures over a nontrivial Bolean algebra $\mathcal{A}$) *Let $\mathcal{A}$ be a nontrivial Boolean algebra (that is, $0^{\mathcal{A}} \neq 1^{\mathcal{A}}$). Let $\Vdash_{\mathbf{F}_{\mathcal{A}}}^{\mathbf{mbC}}$ be the semantical consequence relation with respect to the class $\mathbf{F}_{\mathcal{A}}$ of **mbC**-structures defined over $\mathcal{A}$, and let $\Gamma \cup \{\alpha\}$ be a finite set of formulas in $For(\Sigma)$. Then: $\Gamma \vdash_{\mathbf{mbC}} \alpha$ iff $\Gamma \Vdash_{\mathbf{F}_{\mathcal{A}}}^{\mathbf{mbC}} \alpha$.*

Proof If $\mathcal{A} = \mathbb{A}_2$ then the result follows by Theorem 10.9. Assume now that $\mathcal{A}$ has more than two elements. The 'only if' part is a consequence of Theorem 10.8. Now, suppose that $\Gamma \nvdash_{\mathbf{mbC}} \alpha$. By Theorem 10.9, there exists an **mbC**-structure $\mathcal{E}$ over $\mathbb{A}_2$, and a valuation v over $\mathcal{E}$ such that $v[\Gamma] \subseteq \{1\}$ but $v(\alpha) = 0$. By Proposition 10.21, there exists an **mbC**-structure $\mathcal{E}(\mathcal{A})$ over $\mathcal{A}$ such that $\mathcal{E} \subseteq \mathcal{E}(\mathcal{A})$. In particular, $\mathcal{E}$ is a weak substructure of $\mathcal{E}(\mathcal{A})$. By the last observation in Remark 10.20 it follows that v is also a valuation over $\mathcal{E}(\mathcal{A})$. This shows that $\Gamma \nVdash_{\mathbf{F}_{\mathcal{A}}}^{\mathbf{mbC}} \alpha$. □

It is worth noting that Fidel considered in (1977) the notion of **F**-substructures stated in Definition 10.19 in order to obtain a decomposition result for **F**-structures for da Costa's calculi C_n in terms of irreducible structures. The adaptation of this result to **mbC**-structures will be briefly analyzed in Sect. 10.8.4.

Recall that an *epimorphism* in a category $\mathbb{C}$ is a homomorphism $h : A \to B$ such that, for every pair of parallel homomorphisms $h', h'' : B \to C$ such that $h' \circ h = h'' \circ h$, it is the case that $h' = h''$, see for instance Mac Lane (1978). Then:

Proposition 10.23 *A homomorphism $h : \mathcal{E} \to \mathcal{E}'$ is an epimorphism in $\mathbb{F}\mathbf{mbC}$ if and only if h is onto as a mapping.*

Recall now that an *isomorphism* in a category $\mathbb{C}$ is a homomorphism $h : A \to B$ such that there exists a homomorphism $h' : B \to A$ where $h' \circ h = id_A$ and $h \circ h' = id_B$, see for instance Mac Lane (1978). Then:

Proposition 10.24 *A homomorphism $h : \mathcal{E} \to \mathcal{E}'$ is an isomorphism in $\mathbb{F}\mathbf{mbC}$ if and only if h is an embedding which is onto, that is, h is a bijective embedding.*

This means that, in the category $\mathbb{F}\mathbf{mbC}$ of **mbC**-structures, h is an isomorphism if and only if it is both a monomorphism and an epimorphism.

10.6 Union of Chains of mbC-Structures

As a consequence of the proposed approach to **F**-structures as being a class of Θ-structures axiomatized by a set of sentences of a certain form (recall Sect. 10.4), some basic results from Model Theory can be applied to its study. In this section, the union of chains of **mbC**-structures will be analyzed.

Definition 10.25 A *chain* of **F**-structures for **mbC** is a family $(\mathcal{E}_\lambda)_{\lambda < \mu}$ for an ordinal μ such that $\mathcal{E}_\xi \subseteq \mathcal{E}_\lambda$ whenever $\xi < \lambda < \mu$.

A chain can be displayed as $\mathcal{E}_0 \subseteq \mathcal{E}_1 \subseteq \ldots \subseteq \mathcal{E}_\lambda \subseteq \ldots$ for $\lambda < \mu$. From Model Theory, it is known that, for every chain $(\mathcal{E}_\lambda)_{\lambda < \mu}$ of **F**-structures for **mbC** (seen as first-order structures), there exists a Θ-structure $\mathcal{E}$ which is its union, such that $\mathcal{E}_\lambda \subseteq \mathcal{E}$ for every $\lambda < \mu$. The structure $\mathcal{E}$ is defined as follows:

(i) $A = \bigcup_{\lambda < \mu} A_\lambda$;
(ii) $N^{\mathcal{E}} = \bigcup_{\lambda < \mu} N^{\mathcal{E}_\lambda}$ and $O^{\mathcal{E}} = \bigcup_{\lambda < \mu} O^{\mathcal{E}_\lambda}$;
(iii) for $a, b \in A$: $a \sqcap^{\mathcal{E}} b = a \sqcap^{\mathcal{E}_\lambda} b$; $a \sqcup^{\mathcal{E}} b = a \sqcup^{\mathcal{E}_\lambda} b$; and $-^{\mathcal{E}} a = -^{\mathcal{E}_\lambda} a$, if $a, b \in A_\lambda$;
(iv) $\mathbf{0}^{\mathcal{E}} = \mathbf{0}^{\mathcal{E}_0}$ and $\mathbf{1}^{\mathcal{E}} = \mathbf{1}^{\mathcal{E}_0}$.

Observe that items (ii) and (iii) are well-defined given that, for $\xi < \lambda < \mu$, $\mathcal{E}_\xi \subseteq \mathcal{E}_\lambda$ and so $A_\xi \subseteq A_\lambda$. Clearly $\mathbf{0}^{\mathcal{E}} = \mathbf{0}^{\mathcal{E}_\lambda}$ and $\mathbf{1}^{\mathcal{E}} = \mathbf{1}^{\mathcal{E}_\lambda}$ for every $\lambda < \mu$.

Given that, as observed in Remark 10.11, the class $\mathbb{F}\mathbf{mbC}$ of **mbC**-structures can be axiomatized by means of $\forall\exists$-sentences (that is, Π_2^0-sentences), that class is closed under union of chains. This is a consequence of Chang and Keisler (2012, Theorem 3.2.3). In other words, the following result holds.

Proposition 10.26 *Let $(\mathcal{E}_\lambda)_{\lambda < \mu}$ be a chain of **mbC**-structures, and let $\mathcal{E}$ be its union. Then $\mathcal{E}$ is the least **mbC**-structure having every $\mathcal{E}_\lambda$ as a substructure.*

The last result can be interpreted in terms of the propositional logics generated by single **mbC**-structures (see Proposition 10.28 below). Indeed: recall from

Definition 10.7 the propositional consequence relation $\Vdash_{\mathcal{E}}^{mbC}$ generated by an **mbC**-structure $\mathcal{E}$.[7] Then, the following useful result can be stated.

Lemma 10.27 *Let $\mathcal{E}$ and $\mathcal{E}'$ be two **mbC**-structures such that $\mathcal{E} \subseteq \mathcal{E}'$. Then $\Vdash_{\mathcal{E}'}^{mbC} \subseteq \Vdash_{\mathcal{E}}^{mbC}$.*

Proof Suppose that $\mathcal{E} \subseteq \mathcal{E}'$ and let $\Gamma \cup \{\alpha\}$ be a finite set of formulas in $For(\Sigma)$ such that $\Gamma \Vdash_{\mathcal{E}'}^{mbC} \alpha$. Consider a valuation v over $\mathcal{E}$ such that $v[\Gamma] \subseteq \{\mathbf{1}^{\mathcal{E}}\}$. Then, the mapping $\bar{v} : For(\Sigma) \to A'$ such that $\bar{v}(\gamma) = v(\gamma)$ for every $\gamma \in For(\Sigma)$ is a valuation over $\mathcal{E}'$ such that $\bar{v}[\Gamma] \subseteq \{\mathbf{1}^{\mathcal{E}'}\}$. By hypothesis, $\bar{v}(\alpha) = \mathbf{1}^{\mathcal{E}'}$ whence $v(\alpha) = \mathbf{1}^{\mathcal{E}}$. This means that $\Gamma \Vdash_{\mathcal{E}}^{mbC} \alpha$. □

From this, Proposition 10.26 can be interpreted in terms of the propositional logics associated to **mbC**-structures.

Proposition 10.28 *Let $(\mathcal{E}_\lambda)_{\lambda < \mu}$ be a chain of **mbC**-structures, and let $\mathcal{E}$ be its union. Then $\Vdash_{\mathcal{E}}^{mbC} = \bigcap_{\lambda < \mu} \Vdash_{\mathcal{E}_\lambda}^{mbC}$.*

Proof It is clear from Lemma 10.27 that $\Vdash_{\mathcal{E}}^{mbC} \subseteq \Vdash_{\mathcal{E}_\lambda}^{mbC}$ for every $\lambda < \mu$, whence $\Vdash_{\mathcal{E}}^{mbC} \subseteq \bigcap_{\lambda < \mu} \Vdash_{\mathcal{E}_\mu}^{mbC}$. Now, let $\Gamma \cup \{\alpha\}$ be a finite set of formulas in $For(\Sigma)$ such that $\Gamma \not\Vdash_{\mathcal{E}}^{mbC} \alpha$. Then, there is a valuation v over $\mathcal{E}$ such that $v[\Gamma] \subseteq \{\mathbf{1}^{\mathcal{E}}\}$ and $v(\alpha) \neq \mathbf{1}^{\mathcal{E}}$. Since $\Gamma \cup \{\alpha\}$ is a finite set, then $v[\Gamma \cup \{\alpha\}] \subseteq A_n$ for some ordinal number $n < \mu$. Define a mapping $\bar{v} : For(\Sigma) \to A_n$ such that $\bar{v}(\gamma) = v(\gamma)$ whenever $\gamma \in \Gamma \cup \{\alpha\}$ and, for every formula $\gamma \notin \Gamma \cup \{\alpha\}$:

(0) if $\gamma \in \mathcal{V}$ then $\bar{v}(\gamma)$ is arbitrary;
(1) if $\gamma = \gamma_1 \# \gamma_2$ then $\bar{v}(\gamma) = \bar{v}(\alpha) \#^{\mathcal{E}_n} \bar{v}(\beta)$, for $\# \in \{\wedge, \vee, \to\}$;
(2) if $\gamma = \neg \gamma_1$ then $\bar{v}(\gamma)$ is such that $N^{\mathcal{E}_n}(\bar{v}(\gamma_1), \bar{v}(\gamma))$;
(3) if $\gamma = \circ \gamma_1$ then $\bar{v}(\gamma)$ is such that $O^{\mathcal{E}_n}(\bar{v}(\gamma_1), \bar{v}(\gamma))$ and $\bar{v}(\gamma_1) \sqcap^{\mathcal{E}_n} \bar{v}(\neg \gamma_1) \sqcap^{\mathcal{E}_n} \bar{v}(\gamma) = \mathbf{0}^{\mathcal{E}_n}$.[8]

It is clear by the very definition that $\bar{v}$ is a valuation over $\mathcal{E}_n$ such that $\bar{v}[\Gamma] \subseteq \{\mathbf{1}^{\mathcal{E}_n}\}$ and $\bar{v}(\alpha) \neq \mathbf{1}^{\mathcal{E}_n}$. This means that $\Gamma \not\Vdash_{\mathcal{E}_n}^{mbC} \alpha$. Therefore $\bigcap_{\lambda < \mu} \Vdash_{\mathcal{E}_\lambda}^{mbC} \subseteq \Vdash_{\mathcal{E}}^{mbC}$. □

10.7 Congruences and Quotient Structures

In this section, the well-known lattice isomorphism between the lattice of filters over a Boolean algebra $\mathcal{A}$ and the lattice of Boolean congruences over $\mathcal{A}$ will be extended to **mbC**-structures. Additionally, the quotient **mbC**-structures by **mbC**-congruences will be defined.

---

[7] The propositional consequence relation $\Vdash_{\mathbf{K}}^{mbC}$ generated by a class **K** of **mbC**-structures should not be confused with the first-order consequence relation $\models_{\mathbf{K}}$ defined over first-order Θ-sentences as follows: $\Upsilon \models_{\mathbf{K}} \sigma$ if, for every $\mathcal{E} \in \mathbf{K}$, it holds: $\mathcal{E} \models \sigma$ whenever $\mathcal{E} \models \varrho$ for every $\varrho \in \Upsilon$.

[8] In order to define $\bar{v}$ by induction on the complexity of γ, the complexity measure on $For(\Sigma)$ must be defined in a way such that $\circ \beta$ has a complexity degree strictly greater than that of $\neg \beta$, for every $\beta \in For(\Sigma)$.

Definition 10.29 Let θ be a relation on an **mbC**-structure $\mathcal{E} = \langle \mathcal{A}, N^\mathcal{E}, O^\mathcal{E} \rangle$. Then θ is said to be an **mbC**-congruence over $\mathcal{E}$ if the following conditions hold:

(i) θ is a Boolean congruence over $\mathcal{A}$;[9]
(ii) if $(x, x'), (y, y') \in \theta$ and $N^\mathcal{E}(x, y)$ then $N^\mathcal{E}(x', y')$;
(iii) if $(x, x'), (y, y') \in \theta$ and $O^\mathcal{E}(x, y)$ then $O^\mathcal{E}(x', y')$.

Given a Boolean algebra $\mathcal{A}$, let $Con_B(\mathcal{A})$ be the set of Boolean congruences defined on $\mathcal{A}$. It is well-known that the poset $(Con_B(\mathcal{A}), \subseteq)$ partially ordered by the inclusion relation is a distributive lattice.

Definition 10.30 Let $\mathcal{A}$ be a Boolean algebra, and let $F \subseteq A$. Then F is a *filter over* $\mathcal{A}$ if the following holds: (i) $\mathbf{1}^\mathcal{A} \in F$; (ii) if $x, y \in F$ then $x \sqcap y \in F$; and (iii) if $x \in F$ and $x \leq y$ then $y \in F$. We denote by $F(\mathcal{A})$ the set of filters over $\mathcal{A}$.

The following is a well-known result.

Theorem 10.31 *Given a Boolean algebra $\mathcal{A}$, there exists a lattice isomorphism between $F(\mathcal{A})$ and $Con_B(\mathcal{A})$ given by $F \mapsto R(F)$, where $R(F) = \{(x, y) \in A^2 : x \sqcap z = y \sqcap z \text{ for some } z \in F\}$. The inverse mapping is given by $\theta \mapsto [\mathbf{1}^\mathcal{A}]_\theta$, where $[a]_\theta$ denotes the θ-equivalence class of $a \in A$.*

Definition 10.32 Given an **mbC**-structure $\mathcal{E} = \langle \mathcal{A}, N^\mathcal{E}, O^\mathcal{E} \rangle$, a set $F \subseteq A$, is said to be an **mbC**-*filter* if the following conditions hold:

(i) F is a filter over the Boolean algebra $\mathcal{A}$;
(ii) $R(F)$ verifies conditions (ii) and (iii) of Definition 10.29, where $R(F)$ is defined as in Theorem 10.31.

Let us denote by $F_{\mathbf{mbC}}(\mathcal{E})$ and by $Con_{\mathbf{mbC}}(\mathcal{E})$ the set of **mbC**-filters and the set of **mbC**-congruences over a given **mbC**-structure $\mathcal{E}$, respectively. Thus, the following result holds.

Theorem 10.33 *Let $\mathcal{E} = \langle \mathcal{A}, N^\mathcal{E}, O^\mathcal{E} \rangle$ be an **mbC**-structure. Then, there exists a lattice isomorphism between $F_{\mathbf{mbC}}(\mathcal{E})$ and $Con_{\mathbf{mbC}}(\mathcal{E})$.*

Proof It is proved by an easy adaptation of proof of the corresponding result for Boolean algebras. □

Now, we are going to define quotient **mbC**-structures. Let $\mathcal{E}$ be an **mbC**-structure, and let θ be an **mbC**-congruence on it. Then A/θ is a Boolean algebra with the operations induced from $\mathcal{A}$; this Boolean algebra will be denoted by $\mathcal{A}/\theta$. Consider the following relations over A/θ induced from $\mathcal{E}$:

$$N^{\mathcal{E}/\theta} \stackrel{\text{def}}{=} \{([x]_\theta, [y]_\theta) \in A/\theta \times A/\theta : (x, y) \in N^\mathcal{E}\}$$

---

[9]That is, θ is an equivalence relation which is preserved by the operations of the Boolean algebra $\mathcal{A}$.

and
$$O^{\mathcal{E}/\theta} \stackrel{\text{def}}{=} \{([x]_\theta, [y]_\theta) \in \mathcal{A}/\theta \times \mathcal{A}/\theta \;:\; (x, y) \in O^{\mathcal{E}}\}.$$

From Definition 10.29, it follows that $(x, y) \in N^{\mathcal{E}}$ if and only if $([x]_\theta, [y]_\theta) \in N^{\mathcal{E}/\theta}$; the same holds for the predicate O. From this, it is easy to check that $\mathcal{E}/\theta = \langle \mathcal{A}/\theta, N^{\mathcal{E}/\theta}, O^{\mathcal{E}/\theta}\rangle$ is an **mbC**-structure. Now, consider the canonical projection $q : \mathcal{A} \to \mathcal{A}/\theta$ given by $q(x) = [x]_\theta$. It is clear that q is a homomorphism of Boolean algebras. Moreover, it is an **mbC**-homomorphism $q : \mathcal{E} \to \mathcal{E}/\theta$ which is onto, that is, an epimorphism in $\mathbb{F}$**mbC** (by Proposition 10.23).

It is well-known that given a Boolean homomorphism $h : \mathcal{A} \to \mathcal{A}'$, the relation $Ker(h) = \{(x, y) \in \mathcal{A} \times \mathcal{A} \;:\; h(x) = h(y)\}$ is a Boolean congruence. This allows us to prove the so-called *First Isomorphism theorem* for Boolean algebras. In order to generalize this result to **mbC**-structures, the homomorphisms must satisfy additional coherence properties.

Definition 10.34 Let $h : \mathcal{E} \to \mathcal{E}'$ be an **mbC**-homomorphism.[10] Then h is said to be *congruential* if it satisfies the following:

$$\text{If } N^{\mathcal{E}}(x, y) \text{ and } (x, x'), (y, y') \in Ker(h) \text{ then } N^{\mathcal{E}}(x', y'),$$

and

$$\text{If } O^{\mathcal{E}}(x, y) \text{ and } (x, x'), (y, y') \in Ker(h) \text{ then } O^{\mathcal{E}}(x', y').$$

Proposition 10.35 *Let $h : \mathcal{E} \to \mathcal{E}'$ be an **mbC**-homomorphism. Then, h is congruential if and only if the relation $Ker(h) = \{(x, y) \in \mathcal{A} \times \mathcal{A} \;:\; h(x) = h(y)\}$ is an **mbC**-congruence.*

Proof It follows from the very definitions. □

Theorem 10.36 (First Isomorphism theorem) *Let $h : \mathcal{E} \to \mathcal{E}'$ be an **mbC**-homomorphism which is congruential. Then, there is a unique **mbC**-monomorphism $\overline{h} : \mathcal{E}/Ker(h) \to \mathcal{E}'$ such that $\overline{h} \circ q = h$. In particular, if h is surjective then $\overline{h}$ is an isomorphism between $\mathcal{E}/Ker(h)$ and $\mathcal{E}'$.*

Proof As mentioned above, it is well-known that the Boolean congruence $\theta = Ker(h)$ is such that there is a unique Boolean monomorphism $\overline{h} : \mathcal{A}/\theta \to \mathcal{A}'$ with $\overline{h} \circ q = h$. Thus, in view of Proposition 10.35, it suffices to prove that $\overline{h}$ is an **mbC**-monomorphism. The details are left to the reader. □

---

[10]Observe that, in particular, $h : \mathcal{A} \to \mathcal{A}'$ is a Boolean homomorphism.

10.8 Birkhoff's Decomposition Theorems for mbC-Structures

In this section, Caicedo's generalization to first-order structures (see Caicedo 1981) of Birkhoff's decomposition theorem for algebras (see Birkhoff 1944) will be applied to the specific case of **mbC**-structures. In order to adapt Caicedo's result to the present framework, some important constructions over **mbC**-structures will be analyzed: direct and subdirect products, as well as the associated notion of subdirectly irreducible **mbC**-structures. Finally, in Sect. 10.8.4, a related decomposition result will be obtained by adapting the notions introduced by Fidel in (1977) in terms of weak substructures, recall Definition 10.19. Once again, the fact that **FmbC** can be axiomatized by means of sentences of a special form (recall Sect. 10.4) will we used, by adapting well-known results from Model Theory.

10.8.1 Direct Products of mbC-Structures

As observed in Remark 10.10 the class **FmbC** of **mbC**-structures can be axiomatized by means of Horn sentences. Then, as a consequence of Chang and Keisler (2012, Proposition 6.2.2), the class **FmbC** is closed under (arbitrary) direct products. That is:

Theorem 10.37 *The category* $\mathbb{F}\mathbf{mbC}$ *has arbitrary products.*[11]

For the reader's convenience the standard construction of products in $\mathbb{F}\mathbf{mbC}$ will be given in Definition 10.38 below.

Given a family $\{\mathcal{A}_i\}_{i \in I}$ of Boolean algebras, consider the standard construction $\mathcal{A} = \prod_{i \in I} \mathcal{A}_i$ of its product such that its support is $\prod_{i \in I} A_i = \{x \in (\bigcup_{i \in I} A_i)^I : x(i) \in A_i \text{ for every } i \in I\}$,[12] and the operations are defined pointwise; in particular, $\mathbf{0}^{\mathcal{A}}(i) = \mathbf{0}^{\mathcal{A}_i}$ and $\mathbf{1}^{\mathcal{A}}(i) = \mathbf{1}^{\mathcal{A}_i}$ for every $i \in I$. If $I = \emptyset$ then $\prod_{i \in I} \mathcal{A}_i$ is the trivial one-point Boolean algebra $\mathbb{A}_\perp$ (see Remark 10.39). The canonical i-projection $\pi_i : \prod_{i \in I} \mathcal{A}_i \to \mathcal{A}_i$ is given by $\pi_i(x) = x(i)$ for every $x \in \prod_{i \in I} A_i$ and every $i \in I$.

Definition 10.38 Let $\mathcal{E}_i = \langle \mathcal{A}_i, N^{\mathcal{E}_i}, O^{\mathcal{E}_i} \rangle$ (for $i \in I$) be an **mbC**-structure. The *direct product* of the family $\{\mathcal{E}_i\}_{i \in I}$ is the structure $\prod_{i \in I} \mathcal{E}_i = \langle \prod_{i \in I} \mathcal{A}_i, N^{\prod_{i \in I} \mathcal{E}_i}, O^{\prod_{i \in I} \mathcal{E}_i} \rangle$ defined as follows:

(i) $\prod_{i \in I} \mathcal{A}_i$ is the standard product of the family $\{\mathcal{A}_i\}_{i \in I}$ of Boolean algebras;
(ii) $(x, y) \in N^{\prod_{i \in I} \mathcal{E}_i}$ if and only if $(x(i), y(i)) \in N^{\mathcal{E}_i}$ for every $i \in I$;
(iii) $(x, y) \in O^{\prod_{i \in I} \mathcal{E}_i}$ if and only if $(x(i), y(i)) \in O^{\mathcal{E}_i}$ for every $i \in I$.

---

[11] In addition, also as a consequence of Chang and Keisler (2012, Proposition 6.2.2), it follows that the class **FmbC** is closed under reduced products; in particular, it is closed under ultraproducts.

[12] As usual, if I and Z are two sets, then Z^I denotes the set of mappings from I to Z.

It is an easy exercise to check that the construction described in Definition 10.38 is, in fact, the product of the given family of **mbC**-structures (that is, $\langle \prod_{i \in I} \mathcal{E}_i, \{\pi_i : i \in I\}\rangle$ satisfies the universal property of the product in $\mathbb{F}\mathbf{mbC}$).

Remark 10.39 Observe that the product of the empty family of **mbC**-structures is the terminal object $\mathbf{1}_\perp = \langle \mathbb{A}_\perp, N^{\mathbf{1}_\perp}, O^{\mathbf{1}_\perp}\rangle$ given by the one-element Boolean algebra $\mathbb{A}_\perp$ with domain $A_\perp = \{*\}$, and where $N^{\mathbf{1}_\perp} = O^{\mathbf{1}_\perp} = \{(*,*)\}$. Note that $\mathbf{0}^{\mathbb{A}_\perp} = \mathbf{1}^{\mathbb{A}_\perp} = *$.

10.8.2 Subdirect Products and Subdirectly Irreducible mbC-Structures

Recall from Remark 10.17 the characterization of embeddings in the category of **mbC**-structures. As discussed in Sect. 10.5, the substructures in $\mathbb{F}\mathbf{mbC}$ are defined by means of embeddings. It is worth noting that in Caicedo (1981) the embeddings are called *substructure monomorphisms*.

Definition 10.40 (*Subdirect product*, Caicedo 1981) For $i \in I$ let $\mathcal{E}_i = \langle \mathcal{A}_i, N^{\mathcal{E}_i}, O^{\mathcal{E}_i}\rangle$ be an **mbC**-structure. A *subdirect product* of the family $\{\mathcal{E}_i\}_{i \in I}$ is an embedding $h : \mathcal{E} \to \prod_{i \in I} \mathcal{E}_i$ (for some **mbC**-structure $\mathcal{E}$) such that $\pi_i \circ h$ is onto for every $i \in I$. It is also called a *subdirect decomposition* of $\mathcal{E}$.

Definition 10.41 (*Subdirectly irreducible structures*, Caicedo 1981) An **mbC**-structure $\mathcal{E}$ is said to be *subdirectly irreducible* (s.i.) in $\mathbb{F}\mathbf{mbC}$ if for every subdirect descomposition $h : \mathcal{E} \to \prod_{i \in I} \mathcal{E}_i$ in $\mathbb{F}\mathbf{mbC}$ with $I \neq \emptyset$, there is an $i \in I$ such that $\pi_i \circ h$ is an isomorphism in $\mathbb{F}\mathbf{mbC}$.

Theorem 10.42 (Caicedo 1981, Lemma 1) *Let* $\mathcal{E} = \langle \mathcal{A}, N^\mathcal{E}, O^\mathcal{E}\rangle$ *be an* **mbC**-*structure such that* $\mathcal{E} \neq \mathbf{1}_\perp$. *Then,* $\mathcal{E}$ *is subdirectly irreducible in* $\mathbb{F}\mathbf{mbC}$ *if and only if there exists a predicate* $P \in \{N, O, \approx\}$ *and* $(x, y) \in A^2$ *such that* $(x, y) \notin P^\mathcal{E}$, *and for every onto homomorphism* $h : \mathcal{E} \to \mathcal{E}'$ *in* $\mathbb{F}\mathbf{mbC}$ *which is not an isomorphism,* $(h(x), h(y)) \in P^{\mathcal{E}'}$.

Remark 10.43 It is interesting to characterize, in terms of Theorem 10.42, the **mbC**-structures which *are not* subdirectly irreducible. Thus, it follows from the previous result that an **mbC**-structure $\mathcal{E} = \langle \mathcal{A}, N^\mathcal{E}, O^\mathcal{E}\rangle \neq \mathbf{1}_\perp$ is not subdirectly irreducible in $\mathbb{F}\mathbf{mbC}$ if only if the following conditions hold:

1. For every $(x, y) \in A^2$, $(x, y) \notin N^\mathcal{E}$ implies that there exists an onto homomorphism $h : \mathcal{E} \to \mathcal{E}'$ in $\mathbb{F}\mathbf{mbC}$ which is not an isomorphism, such that $(h(x), h(y)) \notin N^{\mathcal{E}'}$ (hence $\mathcal{E}' \neq \mathbf{1}_\perp$);
2. For every $(x, y) \in A^2$, $(x, y) \notin O^\mathcal{E}$ implies that there exists an onto homomorphism $h : \mathcal{E} \to \mathcal{E}'$ in $\mathbb{F}\mathbf{mbC}$ which is not an isomorphism, such that $(h(x), h(y)) \notin O^{\mathcal{E}'}$ (hence $\mathcal{E}' \neq \mathbf{1}_\perp$);
3. For every $(x, y) \in A^2$, $x \neq y$ implies that there exists an onto homomorphism $h : \mathcal{E} \to \mathcal{E}'$ in $\mathbb{F}\mathbf{mbC}$ which is not an isomorphism, such that $h(x) \neq h(y)$ (hence $\mathcal{E}' \neq \mathbf{1}_\perp$).

In the rest of this section, the problem of characterizing the subdirectly irreducible **mbC**-structures will be considered.

Proposition 10.44 *The terminal **mbC**-structure* $\mathbf{1}_\perp = \langle \mathbb{A}_\perp, N^{\mathbf{1}_\perp}, O^{\mathbf{1}_\perp} \rangle$ *is subdirectly irreducible in* $\mathbb{F}\mathbf{mbC}$.

Proof It follows from the definition of $\mathbf{1}_\perp$ and from Definition 10.41: the only subdirect decomposition of $\mathbf{1}_\perp$ is $id_{\mathbf{1}_\perp}$, where the codomain of $id_{\mathbf{1}_\perp}$ is the product $\mathbf{1}_\perp$ of the empty family of **mbC**-structures. □

Now, let us consider the general case for **mbC**-structures defined over an arbitrary Boolean algebra $\mathcal{A} \neq \mathbb{A}_\perp$.

Given a Boolean algebra $\mathcal{A}$, let $\mathcal{E}_\mathcal{A}^{max} \stackrel{def}{=} \langle \mathcal{A}, N_\mathcal{A}^{max}, O_\mathcal{A}^{max} \rangle$ be the greatest **mbC**-structure defined over $\mathcal{A}$, where $N_\mathcal{A}^{max} = \{(x, y) \in A^2 : x \sqcup y = 1\}$ and $O_\mathcal{A}^{max} = A^2$. By simplicity, $\mathcal{E}_{\mathbb{A}_2}^{max}$ will be denoted by $\mathcal{E}_2^{max} = \langle \mathbb{A}_2, N_2^{max}, O_2^{max} \rangle$, recalling that $\mathbb{A}_2$ denotes the two-element Boolean algebra. Observe that $\mathcal{E}_{\mathbb{A}_\perp}^{max}$ is $\mathbf{1}_\perp$.

Lemma 10.45 *Let* $h : \mathcal{E}_2^{max} \to \mathcal{E}$ *be an **mbC**-homomorphism such that* $\mathcal{E} = \langle \mathbb{A}_2, N^\mathcal{E}, O^\mathcal{E} \rangle$ *is defined over* $\mathbb{A}_2$. *Then h is the identity map,* $\mathcal{E} = \mathcal{E}_2^{max}$ *and h is an isomorphism in* $\mathbb{F}\mathbf{mbC}$.

Proof Since h is, in particular, a Boolean homomorphism, $h(0) = 0$ and $h(1) = 1$, hence h is the identity map. Since h preserves the predicates and h is the identity map, $P_2^{max} = h[P_2^{max}] \subseteq P^\mathcal{E} \subseteq P_2^{max}$, and so $P^\mathcal{E} = P_2^{max}$ for $P \in \{N, O\}$. □

Proposition 10.46 $\mathcal{E}_2^{max}$ *is subdirectly irreducible in* $\mathbb{F}\mathbf{mbC}$.

Proof Consider in Theorem 10.42 the predicate symbol N and the point $(0, 0)$. By item (b)(iii) of Definition 10.3, $(0, 0) \notin N_2^{max}$. Let $h : \mathcal{E}_2^{max} \to \mathcal{E}$ be an onto homomorphism which is not an isomorphism in $\mathbb{F}\mathbf{mbC}$. In particular, $h : \mathbb{A}_2 \to \mathcal{A}$ is an onto homomorphism of Boolean algebras which is not an isomorphism in $\mathbb{F}\mathbf{mbC}$, hence $\mathcal{A}$ is the one-point Boolean algebra $\mathbb{A}_\perp$ or $\mathcal{A}$ is $\mathbb{A}_2$. By Lemma 10.45 $\mathcal{A}$ must be $\mathbb{A}_\perp$ (otherwise h will be an isomorphism in $\mathbb{F}\mathbf{mbC}$). From this $\mathcal{E}$ is the terminal **mbC**-structure $\mathbf{1}_\perp$ and so $(h(0), h(0)) \in N^\mathcal{E} = \{(*, *)\}$. By Theorem 10.42, it follows that $\mathcal{E}$ is subdirectly irreducible in $\mathbb{F}\mathbf{mbC}$. □

The following is a well-known result concerning Boolean algebras which will be useful for our purposes.

Proposition 10.47 *Let $\mathcal{A}$ be a Boolean algebra such that $\mathcal{A} \neq \mathbb{A}_\perp$.*
(1) Let $x, y \in A$ such that $x \neq y$. Then, there exists a homomorphism $h : \mathcal{A} \to \mathbb{A}_2$ of Boolean algebras such that $h(x) \neq h(y)$.
(2) Let $x \in A$ such that $x \neq 1$. Then, there exists a homomorphism $h : \mathcal{A} \to \mathbb{A}_2$ of Boolean algebras such that $h(x) = 0$.
(3) If $h : \mathcal{A} \to \mathcal{A}'$ is a non-injective homomorphism of Boolean algebras, there exists $z \in A$ such that $z \neq 0$ but $h(z) = 0$.
(4) If $h : \mathcal{A} \to \mathcal{A}'$ is a non-injective homomorphism of Boolean algebras, there exists $z \in A$ such that $z \neq 1$ but $h(z) = 1$.

From now on, the cardinal of a set X will be denoted by $card(X)$.

Proposition 10.48 *Let $\mathcal{E}_\mathcal{A}^{max}$ an **mbC**-structure over a Boolean algebra $\mathcal{A}$ such that $card(A) > 2$. Then, $\mathcal{E}_\mathcal{A}^{max}$ is not subdirectly irreducible in $\mathbb{F}$**mbC**.*

Proof Let $(x, y) \notin N_\mathcal{A}^{max}$. Then, $x \sqcup y \neq 1$ and so, by Proposition 10.47(2), there exists a homomorphism $h : \mathcal{A} \to \mathbb{A}_2$ of Boolean algebras such that $h(x \sqcup y) = 0$, whence $h(x) = h(y) = 0$. Clearly h induces an onto homomorphism $h : \mathcal{E}_\mathcal{A}^{max} \to \mathcal{E}_2^{max}$ in $\mathbb{F}$**mbC** which is not an isomorphism (since $card(A) > 2$), such that $(h(x), h(y)) = (0, 0) \notin N_2^{max}$.

Now, let $(x, y) \in A^2$ such that $x \neq y$. By Proposition 10.47(1), there exists a homomorphism $h : \mathcal{A} \to \mathbb{A}_2$ of Boolean algebras such that $h(x) \neq h(y)$. It is immediate to see that h induces an onto homomorphism $h : \mathcal{E}_\mathcal{A}^{max} \to \mathcal{E}_2^{max}$ in $\mathbb{F}$**mbC** which is not an isomorphism, such that $h(x) \neq h(y)$.

Finally, there is no $(x, y) \in A^2$ such that $(x, y) \notin O_\mathcal{A}^{max}$. Therefore, it follows that $\mathcal{E}_\mathcal{A}^{max}$ is not s.i. in $\mathbb{F}$**mbC**, by Remark 10.43. $\square$

Proposition 10.49 *(1) Let $(x, y) \in N_\mathcal{A}^{max} \setminus \{(0, 1)\}$. Then $\mathcal{E}_{\mathcal{A}, N(x,y)}^{max} \stackrel{def}{=} \langle \mathcal{A}, N_\mathcal{A}^{max} \setminus \{(x, y)\}, O_\mathcal{A}^{max} \rangle$ is an **mbC**-structure which is subdirectly irreducible in $\mathbb{F}$**mbC**.*

*(2) Let $(x, y) \in A^2 \setminus \{(1, 0)\}$. Then $\mathcal{E}_{\mathcal{A}, O(x,y)}^{max} \stackrel{def}{=} \langle \mathcal{A}, N_\mathcal{A}^{max}, O_\mathcal{A}^{max} \setminus \{(x, y)\}\rangle$ is an **mbC**-structure which is subdirectly irreducible in $\mathbb{F}$**mbC**.*

Proof (1) Let $\mathcal{E} \stackrel{def}{=} \mathcal{E}_{\mathcal{A}, N(x,y)}^{max}$. Clearly, $\mathcal{E}$ is an **mbC**-structure. In order to apply Theorem 10.42, consider the predicate symbol N and the point (x, y). By definition, $(x, y) \notin N^\mathcal{E}$. Now, suppose that we have an onto **mbC**-homomorphism $h : \mathcal{E} \to \mathcal{E}'$ which is not an isomorphism in $\mathbb{F}$**mbC**. Then, $h : \mathcal{A} \to \mathcal{A}'$ is an onto homomorphism of Boolean algebras. There are two cases.

Case 1: h is an isomorphism of Boolean algebras. Then, the only reason for h (an isomorphism of Boolean algebras) not being an isomorphism in $\mathbb{F}$**mbC** is that there exist $(c, d) \in A^2$ and a predicate symbol $P \in \{N, O, \approx\}$ such that $(c, d) \notin P^\mathcal{E}$ but $(h(c), h(d)) \in P^{\mathcal{E}'}$. Given that $O^\mathcal{E} = A^2$ it follows that $P \neq O$. On the other hand, since h is injective then $P \neq \approx$. Therefore, $P = N$ and $(c, d) = (x, y)$, by definition of $\mathcal{E}$. This implies that $(h(x), h(y)) \in N^{\mathcal{E}'}$ (and $\mathcal{E}' = \mathcal{E}_{\mathcal{A}'}^{max}$).

Case 2: h is not an isomorphism of Boolean algebras. Then, h is not injective. By Proposition 10.47(3), there exists $z \in A$ such that $z \neq 0$ but $h(z) = 0$. There are two subcases to analyze.

Case 2.1: Either $z \not\leq x$ or $z \not\leq y$.

Case 2.1.1: $z \not\leq x$. Let $x' = x \sqcup z$. Then, $x' \neq x$ (otherwise, if $x = x'$ then $z = z \sqcap (x \sqcup z) = z \sqcap x$ and so $z \leq x$, a contradiction). Clearly $h(x') = h(x)$. Moreover, $x' \sqcup y = (x \sqcup z) \sqcup y = 1$ (since $x \sqcup y = 1$). Hence $(x', y) \neq (x, y)$ such that $(x', y) \in N^\mathcal{E}$ and so $(h(x), h(y)) = (h(x'), h(y)) \in N^{\mathcal{E}'}$ since h is homomorphism in $\mathbb{F}$**mbC**.

Case 2.1.2: $z \not\leq y$. As in the proof of Case 2.1.1 it can be shown that $(h(x), h(y)) \in N^{\mathcal{E}'}$ (now by taking the pair $(x, y \sqcup z)$).

Case 2.2: Both $z \leq x$ and $z \leq y$. Let $x' = x \sqcap \sim z$. Then $x \neq x'$ (otherwise, if $x = x'$ then $z = z \sqcap x = z \sqcap (x \sqcap \sim z) = 0$, a contradiction). On the other hand $x' \sqcup y = (x \sqcap \sim z) \sqcup y = (x \sqcup y) \sqcap (\sim z \sqcup y) = 1 \sqcap (\sim z \sqcup y) = \sim z \sqcup y = \sim z \sqcup (y \sqcup z) = 1$. Hence $(x', y) \neq (x, y)$ such that $(x', y) \in N^{\mathcal{E}}$ and so $(h(x), h(y)) = (h(x'), h(y)) \in N^{\mathcal{E}'}$ since h is homomorphism in $\mathbb{F}\mathbf{mbC}$.

By Theorem 10.42, $\mathcal{E}$ is s.i. in $\mathbb{F}\mathbf{mbC}$.

(2) The proof is analogous to that of (1), but now considering the predicate symbol O and the point (x, y). $\square$

Remark 10.50 In Proposition 10.49 item (1), the point $(x, y) = (0, 1)$ was not considered because of the first observation in Remark 10.5 (namely, $(0, 1) \in N^{\mathcal{E}}$ for every $\mathcal{E}$). Similarly, in Proposition 10.49(2) it is required that $(x, y) \neq (1, 0)$ because of the second observation in Remark 10.5. Indeed, since $(1, 1) \in N_{\mathcal{A}}^{max}$ then $(1, 0)$ must belong to $O_{\mathcal{A}}^{max} \setminus \{(x, y)\}$ in order to get an $\mathbf{mbC}$-structure.

Proposition 10.51 *Let* $\mathcal{E}_{\mathcal{A}}^{max*} \stackrel{def}{=} \langle \mathcal{A}, N_{\mathcal{A}}^{max} \setminus \{(1, 1)\}, O_{\mathcal{A}}^{max} \setminus \{(1, 0)\}\rangle$. *Then,* $\mathcal{E}_{\mathcal{A}}^{max*}$ *is an* $\mathbf{mbC}$-*structure which is subdirectly irreducible in* $\mathbb{F}\mathbf{mbC}$.

Proof It is easy to see that $\mathcal{E}_{\mathcal{A}}^{max*}$ is indeed an $\mathbf{mbC}$-structure. We will use Theorem 10.42 in order to show that $\mathcal{E} = \mathcal{E}_{\mathcal{A}}^{max*}$ is subdirectly irreducible in $\mathbb{F}\mathbf{mbC}$. Thus, consider the predicate symbol O and the point $(1, 0) \in A^2$. Clearly, $(1, 0) \notin O^{\mathcal{E}}$. It will be shown that, for every onto $\mathbf{mbC}$-homomorphism $h : \mathcal{E} \to \mathcal{E}'$ which is not an isomorphism, $(1, 0) = (h(1), h(0)) \in O^{\mathcal{E}'}$. Thus, let $h : \mathcal{E} \to \mathcal{E}'$ be an onto $\mathbf{mbC}$-homomorphism which is not an isomorphism. There are two cases.

Case 1: h is an isomorphism of Boolean algebras. By adapting the argument used in the **Case 1** of the proof of Proposition 10.49, and taking into account that $\mathcal{E}$ is obtained from $\mathcal{E}_{\mathcal{A}}^{max}$ by removing just one point from $N_{\mathcal{A}}^{max}$ and just one point from $O_{\mathcal{A}}^{max}$, there are two subcases to analyze.

Case 1.1: $\mathcal{E}'$ is isomorphic to $\mathcal{E}_{\mathcal{A}}^{max}$ in $\mathbb{F}\mathbf{mbC}$ via h. That is, $\mathcal{E}' = \mathcal{E}_{\mathcal{A}'}^{max}$.

Case 1.2: $\mathcal{E}'$ is isomorphic to $\mathcal{E}_{\mathcal{A}, N(1,1)}^{max}$ in $\mathbb{F}\mathbf{mbC}$ via h (recall Proposition 10.49). That is, $\mathcal{E}' = \mathcal{E}_{\mathcal{A}', N(h(1), h(1))}^{max} = \mathcal{E}_{\mathcal{A}', N(1,1)}^{max}$.

In both cases, $(h(1), h(0)) \in O^{\mathcal{E}'}$. Observe that the case that $\mathcal{E}'$ is isomorphic to $\mathcal{E}_{\mathcal{A}, O(1,0)}^{max}$ is not allowed, given that $(1, 1) \in N^{\mathcal{E}'}$ and so $(1, 0)$ must be in $O^{\mathcal{E}'}$ (see Remark 10.5).

Case 2: h is not an isomorphism of Boolean algebras. Then, h is not injective. By Proposition 10.47(4), there exists $z \in A$ such that $z \neq 1$ but $h(z) = 1$. From this, $(z, 0) \in O^{\mathcal{E}}$ (since $(z, 0) \neq (1, 0)$) and so $(h(z), h(0)) = (1, 0) \in O^{\mathcal{E}'}$, since h is an homomorphism in $\mathbb{F}\mathbf{mbC}$. That is, $(h(1), h(0)) \in O^{\mathcal{E}'}$.

By Theorem 10.42, $\mathcal{E}$ is s.i. in $\mathbb{F}\mathbf{mbC}$. $\square$

Proposition 10.52 *Let* $\mathcal{E} = \langle \mathcal{A}, N^{\mathcal{E}}, O^{\mathcal{E}} \rangle$ *be an* $\mathbf{mbC}$-*structure over* $\mathcal{A}$ *such that* $\mathrm{card}(N_{\mathcal{A}}^{max} \setminus N^{\mathcal{E}}) \geq 1$ *and* $\mathrm{card}(O_{\mathcal{A}}^{max} \setminus O^{\mathcal{E}}) \geq 1$, *and* $\mathcal{E} \neq \mathcal{E}_{\mathcal{A}}^{max*}$ (*see Proposition 10.51*). *Then* $\mathcal{E}$ *is not subdirectly irreducible in* $\mathbb{F}\mathbf{mbC}$.

Proof Once again, Remark 10.43 will be used in order to prove that $\mathcal{E}$ is not subdirectly irreducible. Let $(x, y) \in N_{\mathcal{A}}^{max} \setminus N^{\mathcal{E}}$ and $(z, t) \in O_{\mathcal{A}}^{max} \setminus O^{\mathcal{E}}$. Let $(c, d) \notin N^{\mathcal{E}}$, and let $\mathcal{E}' = \langle \mathcal{A}, N^{\mathcal{E}}, O_{\mathcal{A}}^{max} \rangle$. It is easy to see that $\mathcal{E}'$ is an **mbC**-structure. Then, the identity map $h : A \to A$ induces an onto homomorphism $h : \mathcal{E} \to \mathcal{E}'$ which is not an isomorphism, since $(h(z), h(t)) = (z, t) \in O^{\mathcal{E}'}$ but $(z, t) \notin O^{\mathcal{E}}$. Observe that $(h(c), h(d)) = (c, d) \notin N^{\mathcal{E}'}$, since $N^{\mathcal{E}'} = N^{\mathcal{E}}$.

Now, let $(c, d) \notin O^{\mathcal{E}}$. We have two subcases to analyze.

Case 1: $N^{\mathcal{E}} = N_{\mathcal{A}}^{max} \setminus \{(1, 1)\}$. Then $O^{\mathcal{E}} \neq O_{\mathcal{A}}^{max} \setminus \{(1, 0)\}$ since, by hypothesis, $\mathcal{E} \neq \mathcal{E}_{\mathcal{A}}^{max*}$.

Case 1.1: $(c, d) = (1, 0)$. Then $(1, 0) \notin O^{\mathcal{E}}$ and so there exists $(z, t) \neq (1, 0)$ such that $(z, t) \notin O^{\mathcal{E}}$ (given that $\mathcal{E} \neq \mathcal{E}_{\mathcal{A}}^{max*}$). Thus, consider $\mathcal{E}' = \langle \mathcal{A}, N^{\mathcal{E}}, O^{\mathcal{E}} \cup \{(z, t)\} \rangle$. It is clear that $\mathcal{E}'$ is an **mbC**-structure such that the identity map $h : A \to A$ is an onto homomorphism $h : \mathcal{E} \to \mathcal{E}'$ which is not an isomorphism, since $(h(z), h(t)) = (z, t) \in O^{\mathcal{E}'}$ but $(z, t) \notin O^{\mathcal{E}}$. Observe that $(h(c), h(d)) = (c, d) = (1, 0) \notin O^{\mathcal{E}'}$.

Case 1.2: $(c, d) \neq (1, 0)$. Observe that $N_{\mathcal{A}}^{max} = N^{\mathcal{E}} \cup \{(1, 1)\}$. Let $\mathcal{E}' = \langle \mathcal{A}, N_{\mathcal{A}}^{max}, O^{\mathcal{E}} \cup \{(1, 0)\} \rangle$. It is easy to prove that $\mathcal{E}'$ is an **mbC**-structure. In addition, the identity map $h : A \to A$ is an onto homomorphism $h : \mathcal{E} \to \mathcal{E}'$ which is not an isomorphism, since $(h(1), h(1)) = (1, 1) \in N^{\mathcal{E}'}$ but $(1, 1) \notin N^{\mathcal{E}}$. Observe that $(h(c), h(d)) = (c, d) \notin O^{\mathcal{E}'}$ given that $(c, d) \neq (1, 0)$.

Case 2: $N^{\mathcal{E}} \neq N_{\mathcal{A}}^{max} \setminus \{(1, 1)\}$. Let $(x, y) \in N_{\mathcal{A}}^{max} \setminus N^{\mathcal{E}}$ such that $(x, y) \neq (1, 1)$ (such point must exists, by hypothesis). Then $x \sqcap y \neq 1$. Let $\mathcal{E}' = \langle \mathcal{A}, N^{\mathcal{E}'}, O^{\mathcal{E}'} \rangle$ such that $N^{\mathcal{E}'} \stackrel{def}{=} N^{\mathcal{E}} \cup \{(x, y)\}$ and $O^{\mathcal{E}'}$ is defined according to the following subcases.

Case 2.1: If $d = 0$ then $O^{\mathcal{E}'} \stackrel{def}{=} O^{\mathcal{E}} \cup \{(x, \sim(x \sqcap y))\}$. Observe that $(x, \sim(x \sqcap y)) \neq (c, d)$ since $x \sqcap y \neq 1$ (hence, $\sim(x \sqcap y) \neq 0 = d$).

Case 2.2: If $d \neq 0$ then $O^{\mathcal{E}'} \stackrel{def}{=} O^{\mathcal{E}} \cup \{(x, 0)\}$. Clearly $(x, 0) \neq (c, d)$.

Observe that $\mathcal{E}'$ is an **mbC**-structure: indeed, if (x, w) is the new point added to $O^{\mathcal{E}}$ in $O^{\mathcal{E}'}$ then $(x \sqcap y) \sqcap w = 0$. Thus, the property required in Definition 10.3(b)(iv) is satisfied for the new point (x, y) added to $N^{\mathcal{E}}$ in $N^{\mathcal{E}'}$. Moreover, the identity map $h : A \to A$ is an onto homomorphism $h : \mathcal{E} \to \mathcal{E}'$ which is not an isomorphism, since $(h(x), h(y)) = (x, y) \in N^{\mathcal{E}'}$ but $(x, y) \notin N^{\mathcal{E}}$. Note that $(h(c), h(d)) = (c, d) \notin O^{\mathcal{E}'}$ given that the point added to $O^{\mathcal{E}}$ in each case is different to (c, d).

Finally, if $(c, d) \in A^2$ such that $c \neq d$ then the identity map $h : A \to A$ is an onto homomorphism $h : \mathcal{E} \to \mathcal{E}_{\mathcal{A}}^{max}$ which is not an isomorphism, such that $h(c) \neq h(d)$.

By Remark 10.43, $\mathcal{E}$ is not subdirectly irreducible. □

Proposition 10.53 *Let $\mathcal{E} = \langle \mathcal{A}, N^{\mathcal{E}}, O^{\mathcal{E}} \rangle$ be an **mbC**-structure over $\mathcal{A}$ such that $N^{\mathcal{E}} = N_{\mathcal{A}}^{max}$ and $card(O_{\mathcal{A}}^{max} \setminus O^{\mathcal{E}}) \geq 2$. Then, $\mathcal{E}$ is not subdirectly irreducible in* $\mathbb{F}$**mbC**.

Proof By hypothesis, there exist $(x, y) \neq (z, t)$ such that $(x, y), (z, t) \in O_{\mathcal{A}}^{max} \setminus O^{\mathcal{E}}$. Let $(c, d) \notin N^{\mathcal{E}}$. Then, the identity map $h : A \to A$ is an onto homomorphism

$h : \mathcal{E} \to \mathcal{E}_{\mathcal{A}}^{max}$ which is not an isomorphism, since $(h(z), h(t)) = (z, t) \in O_{\mathcal{A}}^{max}$ but $(z, t) \notin O^{\mathcal{E}}$. Observe that $(h(c), h(d)) = (c, d) \notin N^{\mathcal{E}}$.

Now, let $(c, d) \notin O^{\mathcal{E}}$. By hypothesis, there exists $(c', d') \neq (c, d)$ such that $(c', d') \notin O^{\mathcal{E}}$. Let $\mathcal{E}' = \langle \mathcal{A}, N^{\mathcal{E}}, O^{\mathcal{E}} \cup \{(c', d')\} \rangle$. Clearly $\mathcal{E}'$ is an **mbC**-structure such that the identity map $h : A \to A$ is an onto homomorphism $h : \mathcal{E} \to \mathcal{E}'$ which is not an isomorphism, since $(h(c'), h(d')) = (c', d') \in O^{\mathcal{E}'}$ but $(c', d') \notin O^{\mathcal{E}}$. Observe that $(h(c), h(d)) = (c, d) \notin O^{\mathcal{E}'}$. The case for the identity predicate $\approx$ is treated as in the proof of Proposition 10.52. □

Proposition 10.54 *Let $\mathcal{E} = \langle \mathcal{A}, N^{\mathcal{E}}, O^{\mathcal{E}} \rangle$ be an **mbC**-structure over $\mathcal{A} \neq \mathbb{A}_2$ such that $O^{\mathcal{E}} = O_{\mathcal{A}}^{max}$ and $card\bigl(N_{\mathcal{A}}^{max} \setminus N^{\mathcal{E}}\bigr) \geq 2$. Then $\mathcal{E}$ is an **mbC**-structure which is not subdirectly irreducible in $\mathbb{F}$**mbC**.*

Proof The proof is analogous to that of Proposition 10.53. □

Observe that, if $(x, y), (z, t) \in N_2^{max}$ such that $(x, y) \neq (z, t)$ then $\mathcal{E} = \langle \mathbb{A}_2, N_2^{max} \setminus \{(x, y), (z, t)\}, O_2^{max} \rangle$ is *not* an **mbC**-structure. This is why is required that $\mathcal{A} \neq \mathbb{A}_2$ in the last proposition.

By combining the previous results, it is possible to determine, among all the **mbC**-structures, which of them are s.i. and which of them are not.

Theorem 10.55 *Let $\mathcal{E}$ be an **mbC**-structure defined over a Boolean algebra $\mathcal{A}$. Then $\mathcal{E}$ is subdirectly irreducible in $\mathbb{F}$**mbC** if and only if exactly one of the following conditions holds:*

(1) $\mathcal{E} = \mathbf{1}_\perp$; or
(2) $\mathcal{E} = \mathcal{E}_2^{max}$; or
(3) $\mathcal{E} = \mathcal{E}_{\mathcal{A}, N(x,y)}^{max} \stackrel{def}{=} \langle \mathcal{A}, N_{\mathcal{A}}^{max} \setminus \{(x, y)\}, O_{\mathcal{A}}^{max} \rangle$ for some $(x, y) \in N_{\mathcal{A}}^{max} \setminus \{(0, 1)\}$; or
(4) $\mathcal{E} = \mathcal{E}_{\mathcal{A}, O(x,y)}^{max} \stackrel{def}{=} \langle \mathcal{A}, N_{\mathcal{A}}^{max}, O_{\mathcal{A}}^{max} \setminus \{(x, y)\} \rangle$ for some $(x, y) \in A^2 \setminus \{(1, 0)\}$; or
(5) $\mathcal{E} = \mathcal{E}_{\mathcal{A}}^{max*} \stackrel{def}{=} \langle \mathcal{A}, N_{\mathcal{A}}^{max} \setminus \{(1, 1)\}, O_{\mathcal{A}}^{max} \setminus \{(1, 0)\} \rangle$.

Proof It is a direct consequence of Propositions 10.44, 10.46, 10.48, 10.49, 10.51, 10.52, 10.53 and 10.54. □

10.8.3 Subdirect Decomposition Theorem for mbC-Structures

Finally, we are ready to obtain a decomposition theorem for **mbC**-structures in terms of subdirectly irreducible structures. It is an instance of a general theorem obtained by Caicedo in (1981).

Indeed, as it was observed at the end of Sect. 10.4, the class $\mathbb{F}$**mbC** of **mbC**-structures can be axiomatized by sentences over signature Θ of the form $\forall x_1 \cdots \forall x_n(\sigma_1 \to \exists y_1 \cdots \exists y_k \sigma_2))$, where σ_1 and σ_2 are quantifier-free positive formulas over Θ. Thus, by combining Corollary 5 and Theorem 4 in Caicedo (1981), the following result is obtained.

Theorem 10.56 (Birkhoff-Caicedo's decomposition theorem for **mbC**-structures) *Any nontrivial **mbC**-structure $\mathcal{E} = \langle \mathcal{A}, N^{\mathcal{E}}, O^{\mathcal{E}} \rangle$ is a subdirect product of at most $\aleph_0 + card(A)$ nontrivial subdirectly irreducible structures.*

The relationship between Birkhoff-Caicedo's decomposition theorem for **mbC**-structures and the usual Birkhoff's decomposition theorem for varieties of algebras is not immediate. In the case of **mbC**-structures, it seems that there are more than necessary s.i. structures. Indeed, the single **mbC**-structure $\mathcal{E}_2^{max}$ is enough in order to characterize the logic **mbC**.

Theorem 10.57 (Soundness and completeness of **mbC** w.r.t. $\mathcal{E}_2^{max}$)
Let $\Gamma \cup \{\alpha\}$ be a finite set of formulas in $For(\Sigma)$. Then: $\Gamma \vdash_{\mathbf{mbC}} \alpha$ iff $\Gamma \Vdash_{\mathcal{E}_2^{max}}^{\mathbf{mbC}} \alpha$.

Proof The 'only if' part is a consequence of Theorem 10.8. Conversely, suppose that $\Gamma \nvdash_{\mathbf{mbC}} \alpha$. By adapting the proof of Carnielli and Coniglio (2016, Theorem 6.2.16), there exists an **mbC**-structure $\mathcal{E}$ over $\mathbb{A}_2$ and a valuation v over $\mathcal{E}$ such that $v[\Gamma] \subseteq \{1\}$ but $v(\alpha) = 0$. Given that $\mathcal{E}$ is a weak substructure of $\mathcal{E}_2^{max}$, it follows that v is also a valuation over $\mathcal{E}_2^{max}$, by the last observation in Remark 10.20. This shows that $\Gamma \nVdash_{\mathcal{E}_2^{max}}^{\mathbf{mbC}} \alpha$. □

By a similar argument, from Theorem 10.22 it can be proven the adequacy of **mbC** w.r.t. the **mbC**-structure $\mathcal{E}_{\mathcal{A}}^{max}$, for any Boolean algebra $\mathcal{A}$ with more than two elements.

Theorem 10.58 (Soundness and completeness of **mbC** w.r.t. $\mathcal{E}_{\mathcal{A}}^{max}$)
Let $\mathcal{A}$ be a Boolean algebra with more than two elements, and let $\Gamma \cup \{\alpha\}$ be a finite set of formulas in $For(\Sigma)$. Then: $\Gamma \vdash_{\mathbf{mbC}} \alpha$ iff $\Gamma \Vdash_{\mathcal{E}_{\mathcal{A}}^{max}}^{\mathbf{mbC}} \alpha$.

Proof The 'only if' part is a consequence of Theorem 10.8. Now, assume that $\Gamma \nvdash_{\mathbf{mbC}} \alpha$. By Theorem 10.22, there exists an **mbC**-structure $\mathcal{E}$ over $\mathcal{A}$ and a valuation v over $\mathcal{E}$ such that $v[\Gamma] \subseteq \{1\}$ but $v(\alpha) = 0$. Since $\mathcal{E}$ is a weak substructure of $\mathcal{E}_{\mathcal{A}}^{max}$, it follows that v is also a valuation over $\mathcal{E}_{\mathcal{A}}^{max}$, by the last observation in Remark 10.20. Therefore $\Gamma \nVdash_{\mathcal{E}_{\mathcal{A}}^{max}}^{\mathbf{mbC}} \alpha$. □

In the next subsection, an alternative decomposition theorem for **mbC**-structures will be presented, by adapting a decomposition theorem due to Fidel. As it will be argued below, this alternative decomposition theorem is closer to the traditional Birkhoff's decomposition theorem of Universal Algebra.

10.8.4 Fidel's Decomposition Theorem for mbC-Structures

In 1977, Fidel (see Fidel 1977) obtained for the first time the decidability of the hierarchy of paraconsistent calculi C_n of da Costa (see Costa 1993) in terms of certain **F**-structures called C_n-structures. In that paper, Fidel showed that every C_n-structure is weakly isomorphic to a weak substructure of a product of a special

C_n-structure defined over the two-element Boolean algebra $\mathbb{A}_2$ called **C**, see Fidel (1977, Theorem 8). By a weak isomorphism, we mean a homomorphism which is bijective as a mapping.[13]

By adopting the notion of weak substructure in $\mathbb{F}$**mbC** (recall Definition 10.19), it is possible to obtain another decomposition theorem for **mbC**-structures, alternative to the one given in Theorem 10.56. Thus, it will be shown in Theorem 10.60 that each **mbC**-structure is weakly isomorphic to a weak substructure in $\mathbb{F}$**mbC** of a product of **mbC**-structures over $\mathbb{A}_2$. The same result can be rephrased without using the notion of weak isomorphism (see Theorem 10.65 below).

Definition 10.59 Let $h : \mathcal{E} \to \mathcal{E}'$ be an **mbC**-homomorphism. Then h is said to be a *weak isomorphism* if h is a bijective mapping.

Theorem 10.60 (Weak subdirect decomposition theorem for **mbC**-structures) *Let $\mathcal{E}$ be an **mbC**-structure. Then, there exists a set I such that $\mathcal{E}$ is weakly isomorphic to a weak substructure of $\prod_{i \in I} \mathcal{E}_i$, where each $\mathcal{E}_i$ is defined over $\mathbb{A}_2$ for every $i \in I$.*

Proof Let $\mathcal{E} = \langle \mathcal{A}, N^{\mathcal{E}}, O^{\mathcal{E}} \rangle$ be an **mbC**-structure. By Birkhoff's representation theorem for Boolean algebras (1935), there exists a set I and a monomorphism of Boolean algebras $h : \mathcal{A} \to \prod_{i \in I} \mathcal{A}_i$ such that $\mathcal{A}_i = \mathbb{A}_2$ for every $i \in I$. If $I = \emptyset$ then $\mathcal{A} \simeq \mathbb{A}_\perp$ and so $\mathcal{E}$ is the terminal **mbC**-structure $\mathbf{1}_\perp$, hence the result holds with $I = \emptyset$. Now, suppose that $I \neq \emptyset$. For each $i \in I$ consider the structure $\mathcal{E}_i = \langle \mathbb{A}_2, N^{\mathcal{E}_i}, O^{\mathcal{E}_i} \rangle$ such that $N^{\mathcal{E}_i} = \{((\pi_i \circ h)(a), (\pi_i \circ h)(b)) : (a, b) \in N^{\mathcal{E}}\}$ and $O^{\mathcal{E}_i} = \{((\pi_i \circ h)(a), (\pi_i \circ h)(b)) : (a, b) \in O^{\mathcal{E}}\}$. Since $(0, 1) \in N^{\mathcal{E}}$ and $(1, b) \in N^{\mathcal{E}}$ for some $b \in A$ then $(0, 1) \in N^{\mathcal{E}_i}$ and $(1, (\pi_i \circ h)(b)) \in N^{\mathcal{E}_i}$. Analogously, it is proved that $(0, x) \in O^{\mathcal{E}_i}$ and $(1, y) \in O^{\mathcal{E}_i}$ for some $x, y \in TWO$. Suppose now that $(x, y) \in N^{\mathcal{E}_i}$. Then, there exists $(a, b) \in N^{\mathcal{E}}$ such that $x = (\pi_i \circ h)(a)$ and $y = (\pi_i \circ h)(b)$. Since $\mathcal{E}$ is an **mbC**-structure, there exists $c \in A$ such that $(a, c) \in O^{\mathcal{E}}$ and $a \sqcap^{\mathcal{E}} b \sqcap^{\mathcal{E}} c = 0$. From this, $(x, (\pi_i \circ h)(c)) \in O^{\mathcal{E}_i}$ such that $0 = (\pi_i \circ h)(0) = x \sqcap^{\mathbb{A}_2} y \sqcap^{\mathbb{A}_2} (\pi_i \circ h)(c))$. This shows that $\mathcal{E}_i = \langle \mathbb{A}_2, N^{\mathcal{E}_i}, O^{\mathcal{E}_i} \rangle$ is an **mbC**-structure, for every $i \in I$. Now, let $\mathcal{E}' = \prod_{i \in I} \mathcal{E}_i$ be the direct product in $\mathbb{F}$**mbC** of the family $\{\mathcal{E}_i\}_{i \in I}$, see Definition 10.38. Clearly, h is an **mbC**-homomorphism from $\mathcal{E}$ to $\mathcal{E}'$: if $(a, b) \in N^{\mathcal{E}}$ then $((\pi_i \circ h)(a), (\pi_i \circ h)(b)) \in N^{\mathcal{E}_i}$ for every $i \in I$. Hence, $(h(a), h(b)) \in N^{\mathcal{E}'}$ by Definition 10.38. Analogously, if $(a, b) \in O^{\mathcal{E}}$ then $(h(a), h(b)) \in O^{\mathcal{E}'}$. Consider now the structure $\mathcal{E}'' = \langle h(\mathcal{A}), N^{\mathcal{E}''}, O^{\mathcal{E}''} \rangle$ defined as follows: $N^{\mathcal{E}''} = \{(h(a), h(b)) : (a, b) \in N^{\mathcal{E}}\}$ and $O^{\mathcal{E}''} = \{(h(a), h(b)) : (a, b) \in O^{\mathcal{E}}\}$. It is easy to see that $\mathcal{E}''$ is an **mbC**-structure: in order to prove that property (b)(iv) of Definition 10.3 holds, suppose that $N^{\mathcal{E}''}(x, y)$. Then, there exists a unique $(a, b) \in A^2$ such that $x = h(a)$, $y = h(b)$ and $N^{\mathcal{E}}(a, b)$. From this, there exists $c \in A$ such that $O^{\mathcal{E}}(a, c)$ and $a \sqcap^{\mathcal{E}} b \sqcap^{\mathcal{E}} c = 0$. This means that $O^{\mathcal{E}''}(x, h(c))$ such that $x \sqcap^{h(\mathcal{A})} y \sqcap^{h(\mathcal{A})} h(c) = h(0) = 0$. The properties (b)(i)–(iii) are proved analogously. Since h is injective, it follows that $\mathcal{E}$ is weakly isomorphic to $\mathcal{E}''$ such that $\mathcal{E}'' \subseteq_W \prod_{i \in I} \mathcal{E}_i$, recalling Definition 10.19 of weak substructure in $\mathbb{F}$**mbC**. □

---

[13]Such homomorphisms are called *isomorphisms* in Fidel (1977, Definition 6).

In order to better understand the real significance of the latter decomoposition result, it is convenient to introduce some definitions in the general framework of Model Theory. First, observe that the notion of weak isomorphism of Definition 10.59 makes sense in any class of first-order structures.

Definition 10.61 Let $\mathfrak{A}$ and $\mathfrak{A}'$ be two first-order structures over a signature Ξ. A homomorphism $h : \mathfrak{A} \to \mathfrak{A}'$ in Ξ-**str** is said to be a *weak isomorphism* of Ξ-structures if it a bijective mapping.

The notion of direct image of a structure by a homomorphism can be weakened in a suitable way.

Definition 10.62 Let $\mathfrak{A}$ and $\mathfrak{A}'$ be two first-order structures over a signature Ξ, and let $h : \mathfrak{A} \to \mathfrak{A}'$ be a homomorphism in Ξ-**str**. The *weak image* of $\mathfrak{A}$ by h is the Ξ-structure $h(\mathfrak{A})_w$ with domain $h[|\mathfrak{A}|]$; for each function symbol f of arity n, f is interpreted in $h(\mathfrak{A})_w$ by restricting the domain and image of $f^{\mathfrak{A}'}$ to $h[|\mathfrak{A}|]$; the constants are interpreted as in $\mathfrak{A}'$[14]; and for every n-ary predicate symbol P and every $(b_1, \ldots, b_n) \in \big(h[|\mathfrak{A}|]\big)^n$, $(b_1, \ldots, b_n) \in P^{h(\mathfrak{A})_w}$ if and only if there exists $(a_1, \ldots, a_n) \in |\mathfrak{A}|^n$ such that $(h(a_1), \ldots, h(a_n)) = (b_1, \ldots, b_n)$ and $(a_1, \ldots, a_n) \in P^{\mathfrak{A}}$.

Recall the notion of weak substructure in Ξ-**str** considered right before Definition 10.19. The proof of the following results is straightforward.

Proposition 10.63 *Let* $h : \mathfrak{A} \to \mathfrak{A}'$ *be a homomorphism in* Ξ-**str**. *Then* $h(\mathfrak{A})_w$ *is a weak substructure of the direct image* $h(\mathfrak{A})$ *of* $\mathfrak{A}$ *by* h, *hence it is a weak substructure of* $\mathfrak{A}'$.

Proposition 10.64 *Let* $h : \mathfrak{A} \to \mathfrak{A}'$ *be a weak isomorphism in* Ξ-**str**. *Then* $\mathfrak{A}$ *is isomorphic (via h) to* $h(\mathfrak{A})_w$, *a weak substructure of* $\mathfrak{A}'$.

The last result relates weak isomorphisms with embeddings (recall Definition 10.16) in a clear way. Using the previous notions and results, Theorem 10.60 can be recast as follows (here, $\mathfrak{A} \simeq \mathfrak{A}'$ denotes that the Ξ-structures $\mathfrak{A}$ and $\mathfrak{A}'$ are isomorphic in Ξ-**str**).

Theorem 10.65 (Weak subdirect decomposition theorem for **mbC**-structures, version 2) *Let* $\mathcal{E}$ *be an* **mbC**-*structure. Then, there exists a set* I *and an* **mbC**-*structure* $\mathcal{E}'$ *such that* $\mathcal{E} \simeq \mathcal{E}' \subseteq_W \prod_{i \in I} \mathcal{E}_i$, *where* $\mathcal{E}_i \subseteq_W \mathcal{E}_2^{max}$ *for every* $i \in I$.

From Theorem 10.9 we know that the **mbC**-structures defined over the two-element Boolean algebra are enough to semantically characterize the logic **mbC**. Then, Theorem 10.65 reflects in a precise way the relationship between the semantical structures and the logic in a similar way to the traditional algebraic approach to logic. In that sense, Theorem 10.65 is more informative than Theorem 10.56.

---

[14]Observe that the interpretation of function symbols and constants is well-defined since f is a homomorphism of first-order structures, hence it is an algebraic homomorphism.

10.9 Concluding Remarks

In this paper, the class of **F**-structures for **mbC** was analyzed under the perspective of Model Theory. This approach is based on the observation that **F**-structures are nothing more than first-order structures satisfying specific Horn sentences of its underlying language, as it was seen in Sect. 10.4. Under this broad perspective, the present study could be adapted to the study of the class of **F**-structures for another **LFI**s as the ones proposed in Carnielli and Coniglio (2014) and (2016, Chap. 6). The fact that all these **LFI**s are axiomatic extensions of **mbC** imposes additional restrictions to the corresponding class of **F**-structures (which are still axiomatized by Horn sentences of the same kind), and so it would be expected that the class of irreducible structures should be reduced. Being so, Caicedo's version of Birkhoff's decomposition theorem (recall Theorem 10.56) would keep closer to Fidel's one (recall Theorems 10.60 and 10.65). As observed at the end of Sect. 10.8, it can be argued that Fidel's result reflects the meaning of Birkhoff's decomposition theorem for algebras in a more faithful way than the theorem obtained by using the notions from Model Theory, which reveals some limitations of the model-theoretic approach to **F**-structures. The fact that the notion of weak substructures plays a fundamental role in Fidel's decomposition theorem suggests that new concepts and tools should be developed for standard Model Theory in order to analyze the (meta)theory of propositional non-algebraizable logics under the perspective of **F**-structures.

Related to this, there is another interesting topic of future research, now concerning the development of a new approach to Model Theory for first-order **LFI**s (and non-algebraizable logics in general) by means of Fidel-structures. To fix ideas, consider the first-order version of **mbC**, namely the logic **QmbC** (see Carnielli et al. 2014). This logic can be semantically characterized by the so-called *paraconsistent Tarskian structures*, which are (standard) first-order structures together with a paraconsistent two-valued **mbC**-valuation extended naturally to first-order languages. Such valuations could be replaced by valuations over a given **F**-structure for **mbC** (in particular, an **mbC**-structure over the two-element Boolean algebra). This perspective, which generalizes the standard approach to Model Theory over ordered algebras, open interesting lines of research. Thus, the notions and results obtained here for **F**-structures for **mbC** could be adapted for the development of the new Model Theory based on **F**-structures for **QmbC**, as well as for other quantified non-algebraizable logics. In particular, interesting results on Model Theory for quantified **LFI**s such as the Keisler–Shelah Theorem for **QmbC** obtained by Ferguson in (2018) can be adapted to the proposed framework.

Finally, the connections between Priest's plurivalent semantics (recall Sect. 10.1) and other non-deterministic semantics, specifically Nmatrices and Fidel-structures, deserves future research. Moreover, a formal study of plurivalent semantics from the perspective of Model theory and Universal Algebra is an interesting task to be done.

Acknowledgements We are grateful to X. Caicedo and to the anonymous referees for their remarks, suggestions, and criticisms, which helped to improve the paper. The first author was financially supported by an individual research grant from CNPq, Brazil (308524/2014-4). The second author acknowledges support from a postdoctoral grant from FAPESP, Brazil (2016/21928-0).

References

Avron, A., & Lev, I. (2001). Canonical propositional Gentzen-type systems. In: R. Gore, A. Leitsch, & T. Nipkow (Eds.), *Proceedings of the 1st International Joint Conference on Automated Reasoning (IJCAR 2001)*, LNAI (Vol. 2083, pp. 529–544). Springer.

Avron, A. (2007). Non-deterministic semantics for logics with a consistency operator. *Journal of Approximate Reasoning, 45*, 271–287.

Birkhoff, G. (1935). On the structure of abstract algebras. *Proceedings of the Cambridge Philosophical Society, 31*, 433–454.

Birkhoff, G. (1944). Subdirect unions in universal algebra. *Bulletin of the American Mathematical Society, 50*(10), 764–768.

Blok, W. J., & Pigozzi, D. (1989). Algebraizable logics. In *Memoirs of the American mathematical society* (Vol. 77(396)). American Mathematical Society, Providence, USA.

Caicedo, X. (1981). The subdirect decomposition theorem for classes of structures closed under direct limits. *Journal of the Australian Mathematical Society, 30*, 171–179.

Carnielli, W. A. (1990). Many-valued logics and plausible reasoning. In G. Epstein (Ed.,) *Proceedings of the Twentieth International Symposium on Multiple-Valued Logic, Charlotte, NC, USA* (pp. 328–335). The IEEE Computer Society Press.

Carnielli, W. A., & Coniglio, M. E. (2014). Swap structures for **LFI**s. *CLE e-Prints, 14*(1).

Carnielli, W. A., & Coniglio, M. E. (2016). Paraconsistent logic: Consistency, contradiction and negation. In *Logic, epistemology, and the unity of science* series (Vol. 40) Springer.

Carnielli, W. A., Coniglio, M. E., & Marcos, J. (2007). Logics of formal inconsistency. In D. M. Gabbay & F. Guenthner (Eds.), *Handbook of philosophical logic* (2nd ed., Vol. 14, pp. 1–93). Springer.

Carnielli, W. A., Coniglio, M. E., Podiacki, R., & Rodrigues, T. (2014). On the way to a wider model theory: Completeness theorems for first-order logics of formal inconsistency. *The Review of Symbolic Logic, 7*(3), 548–578.

Carnielli, W. A., & Marcos, J. (2002). A taxonomy of C-systems. In W. A. Carnielli, M. E. Coniglio, & I. M. L. D'Ottaviano (Eds.), *Paraconsistency: The logical way to the inconsistent* (pp. 1–94). Proceedings of the 2nd World Congress on Paraconsistency (WCP 2000), Lecture Notes in Pure and Applied Mathematics (Vol. 228). New York: Marcel Dekker.

Chang, C. C., & Keisler, H. J. (2012). *Model theory* (3rd ed.). In *Dover books on mathematics series*. Dover Publications.

Coniglio, M. E., Figallo-Orellano, A., & Golzio, A. C. (2017). Non-deterministic algebraization of logics by swap structures. *Logic Journal of the IGPL*, to appear. Preprint available at arXiv:1708.08499v1 [math.LO].

da Costa, N. C. A. (1993). Sistemas Formais Inconsistentes (Inconsistent Formal Systems, in Portuguese). Habilitation thesis, Universidade Federal do Paraná, Curitiba, Brazil, 1963. Republished by Editora UFPR, Curitiba, Brazil.

da Costa, N. C. A., & Alves, E. H. (1976). Une sémantique pour le calcul C_1 (in French). *Comptes Rendus de l'Académie de Sciences de Paris (A-B), 283*, 729–731.

da Costa, N. C. A., & Alves, E. H. (1977). A semantical analysis of the calculi C_n. *Notre Dame Journal of Formal Logic, 18*(4), 621–630.

Ferguson, T. M. (2018). The Keisler-Shelah theorem for QmbC through semantical atomization. *Logic Journal of the IGPL*, to appear.

Fidel, M. M. (1977). The decidability of the calculi C_n. *Reports on Mathematical Logic, 8*, 31–40.
Jaśkowski, S. (1998, July). Rachunek zdań dla systemów dedukcyjnych sprzecznych (in Polish). *Studia Societatis Scientiarun Torunesis – Sectio A, I*(5), 57–77. Translated to English as "A propositional calculus for inconsistent deductive systems". *Logic and Logical Philosophy, 7*, 35–56, 1999. Proceedings of the Stanisław Jaśkowski's Memorial Symposium, held in Toruń, Poland, July 1998.
Mac Lane, S. (1978). Categories for the working mathematician. In *Graduate texts in mathematics* series (Vol. 5, 2nd. ed.). Springer.
Odintsov, S. (2003). Algebraic semantics for paraconsistent Nelson's logic. *Journal of Logic and Computation, 13*(4), 453–468.
Omori, H. (2015). Remarks on naive set theory based on LP. *The Review of Symbolic Logic, 8*(2), 279–295.
Priest, G. (1979). Logic of paradox. *Journal of Philosophical Logic, 8*, 219–241.
Priest, G. (2007). Paraconsistency and dialetheism. In D. M. Gabbay & J. H. Woods (Eds.), *Handbook of the history of logic* (The Many Valued and Nonmonotonic Turn in Logic) (Vol. 8, pp. 129–204). North Holland.
Priest, G. (2014). Plurivalent logics. *Australasian Journal of Logic, 11*, Article no. 1.
Priest, G. (2016). None of the above: The Catuṣkoṭi in Indian Buddhist Logic. In J.-Y. Beziau, M. Chakraborty, & S. Dutta (Eds.), *New Directions in paraconsistent logic* (pp. 517–527). Proceedings of the 5th World Congress on Paraconsistency (5th WCP), Springer Proceedings in Mathematics & Statistics series (Vol. 152). New Delhi: Springer.
Priest, G., & Berto, F. (2017). Dialetheism. In E. N. Zalta (Ed.), *The Stanford encyclopedia of philosophy* (Spring 2017 ed.). https://plato.stanford.edu/archives/spr2017/entries/dialetheism/.
Priest, G., Tanaka, K., & Weber, Z. (2016). Paraconsistent logic. In E. N. Zalta (Ed.), *The Stanford encyclopedia of philosophy* (Winter 2016 ed.). https://plato.stanford.edu/archives/win2016/entries/logic-paraconsistent/.
Restall, G. (1992). A note on naive set theory in LP. *Notre Dame Journal of Formal Logic, 33*(3), 422–432.
Weber, Z. (2010). Transfinite numbers in paraconsistent set theory. *The Review of Symbolic Logic, 3*(1), 71–92.

Marcelo E. Coniglio is Full Professor in Logic at the Philosophy Department of the Institute of Philosophy and the Humanities (IFCH) of the University of Campinas (UNICAMP) and Director of the Centre for Logic, Epistemology and the History of Science (CLE) of UNICAMP. He was President of the Brazilian Logic Society (SBL) from 2014 to 2017. He obtained his Ph.D. in Mathematics from the University of São Paulo (Brazil) in 1997. He has published more than 60 refereed articles and published and edited 6 books on different subjects of Nonclassical Logic, with emphasis on Paraconsistent logics and Combination of Logics.

Aldo Figallo-Orellano is Postdoctoral fellow at the Centre for Logic, Epistemology and the History of Science (CLE), University of Campinas (UNICAMP), Brazil and Assistant Professor at the Departament of Mathematics of the National University of South, Argentina. He is graduated in Mathematics at the National University of San Juan, Argentina in 2001. He got his M.Sc. and Ph.D. in Mathematics from the National University of South in 2007 and 2014, respectively. He works in algebraic logic and has published more than ten papers in this subject. In the past years, he was working on non-deterministic semantic for non-algebraizable logics.

Chapter 11
Unity, Identity, and Topology: How to Make Donuts and Cut Things in Half

A. J. Cotnoir

Abstract Priest's 2014 theory of unity and identity, based on a paraconsistent logic, has a wide range of applications. In this paper, I apply his theory to some puzzles concerning mereology and topology. These puzzles suggest that the classical mereotopology needs to be revised. I compare and contrast the Priest-inspired solution with another, based on classical logic, that requires the co-location of boundaries. I suggest that the co-location view should be preferred on abductive grounds.

Keywords Unity · Identity · Topology · Mereology · Composition · Change · Fission · Transitivity · Leibniz's Law · Symmetry

In a short unpublished note, Gödel once remarked:

> [A]t least intuitively, if you divide a geometrical line at a point, you would expect that the two halves of the line would be mirror images of each other. Yet, this is not the case if the geometrical line is isomorphic to the real numbers. (Putnam 1994, p. 3)

Because a division is *exhaustive* the "center" point must fall either in the left half or in the right. And because a division is *exclusive* this point cannot be in both halves, leaving one half *open* in that it does not contain its boundary, and the other side *closed* in that it contains its boundary.

How strange. Which side is the lucky one? Which side gets to have its own boundary as a part? Any attempt to answer would surely be arbitrary. That is, we feel an intuitive pull toward a certain kind of symmetry: if there is no principled difference between two objects, then there is no principled difference between their boundaries, either. When one object is open and another is closed, there should be some reason as to why. What is strange is not merely that it is *possible* to divide a line in such an asymmetric way, but rather that it is *impossible* to do so symmetrically.

In this essay, I want to explore two possible solutions to this puzzle. The first is a paraconsistent one based on the metaphysics of identity advanced by Priest (2014). The second is a consistent solution that requires massive amounts of co-location. I

A. J. Cotnoir (✉)
University of St Andrews, St Andrews, Scotland
e-mail: ac117@st-andrews.ac.uk

© Springer Nature Switzerland AG 2019
C. Başkent and T. M. Ferguson (eds.), *Graham Priest on Dialetheism and Paraconsistency*, Outstanding Contributions to Logic 18,
https://doi.org/10.1007/978-3-030-25365-3_11

apply the principles behind these solutions to a number of related topological puzzles, and evaluate their prospects.

11.1 How to Cut Things in Half

To begin, it is worth noting that this puzzle is not merely a puzzle about geometrical lines, but may be generalized two- and three- (and higher) dimensional material objects. Casati and Varzi (1999) present a related puzzle about continuous material objects in three dimensions: imagining dissecting a solid sphere made of perfectly homogenous matter. Which of the two half-spheres will be (topologically) closed, containing the boundary as a part? Symmetry considerations make it appear arbitrary which way the boundary will go, but it must go one way or the other—it cannot belong to both. (Of course it is possible for part of the boundary to break one way and another part to break the other, leaving the half-spheres partly open and partly closed. But in this case, the symmetry problem simply reappears at a smaller scale.)

Casati and Vazi (1999) go on to suggest a novel solution to the puzzle.

> Which of the two half-spheres will be closed? This is an embarrassing question. But it arises, we submit, only on the basis of an incorrect model of what happens topologically when a process of cutting takes place. Topologically, the cutting of an object is no bloodstained process — there is no question of which severed halves keeps the boundary, leaving the other open and bleeding (as it were). Rather, topologically, the explanation is simply that the outer surface of the sphere is progressively deformed until the sphere separates into two halves [...] Eventually the right and left portions split, and we have two, *each with its own complete boundary*. A long continuous process results in an abrupt topological change. (p. 87)

This strategy recommends thinking of cutting a sphere in half rather as a smooth deformation of the boundary. In Fig. 11.1, we imagine from step 1 to step 2 is a mere deformation of the boundary of the sphere inward toward its center. This process continues until step 3, where the two halves are connected by a single "hinge" point. Casati and Varzi suggest that the real magic occurs at step 4, the moment when the two halves actually split into two closed entities. They admit that topology cannot deliver a complete explanation of the transition between 3 and 4: "Of course, there is something deeply problematic about the magic moment of separation. [...] [The puzzle] can only disappear on a more complete assessment, which mereotopology simply cannot deliver. (p. 88)" So, even if this were a correct description of the process of cutting, we still would not have a full explanation; in one sense, the mystery still remains. It is important to note that this picture of cutting is consistent with *classical mereotopology*, the standard formal theory of parts, wholes, and their boundaries.[1]

But is this even a correct description of what happens? To my mind, this description leaves out an important aspect of the process. There is not one but *two* important

---

[1] By "*classical mereotopology*" I mean the system called *General Extensional Mereotopology with Closure* (GEMTC) by Casati and Varzi (1999, p. 59).

11 Unity, Identity, and Topology: How to Make Donuts ...

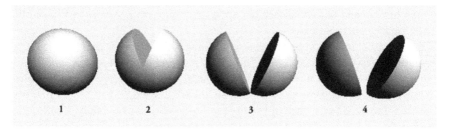

Fig. 11.1 Cutting a sphere

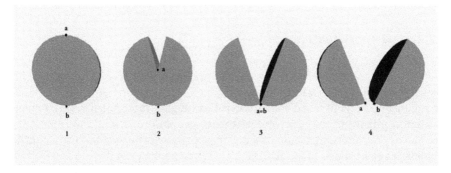

Fig. 11.2 Cutting a sphere (cross section)

topological changes here. The first change occurs from step 2 to step 3, when the two halves of the object go from being topologically *continuous* ("you can go from one half to the other half without ever leaving the interior") to being merely *contiguous* ("you can go from one half to the other without going through the exterior").[2] The second change occurs from step 3 to step 4, when the halves go from being contiguous to being separated. Both changes, though, are equally mysterious.

The mystery is easily seen if we track two point-sized parts of the object: a and b on the top and bottom of the sphere (Fig. 11.2). As we can see, at the change from continuity to contiguity (step 2 to step 3), the top surface and bottom surface of the sphere come into contact and hence a and b merge together to become one single point-sized part. (Alternatively, we might claim that one of a or b ceases to exist while the other survives. But of course there would be no principled metaphysical explanation as to which is which. It would also be strange for this nonexistent point-sized object to pop back into existence again at step 4.) The challenge is to explain how such a change can happen.

If a and b are a single point-sized part, to which half of the sphere does it belong? If it is part of the left half, then the right half is partly topologically open, and vice versa if it is part of the right half. Here, we see the reintroduction of the same asymmetry with cutting a geometrical line. Perhaps the symmetrical thing to say is that it is part

---

[2]These definitions are from Casati and Varzi (1999, p. 80).

of *both* halves, and that the dissection up to this point has not (yet) partitioned the sphere into two disjoint parts.

But then the next step becomes more puzzling. In the change from contiguity to separation (step 2 to step 3), a single point-sized part can either split left or split right, in which case we have the same asymmetry as before. The only symmetric explanation of this separation requires a single point-sized part to fission into two distinct point-sized parts, each of which partly comprised the boundary of one of the disjoint halves of the sphere. This fission also cries out for an explanation, and none seems forthcoming.

11.1.1 A Consistent Solution

I want to recommend an alternative model which gives more intuitive picture on which these mysterious transitions are avoided. A very natural thought is to think that, rather than identifying a with b at step 3, we ought to accept that a and b remain numerically distinct point-sized objects that become *co-located* with one another. So, the halves go from being continuous to being contiguous by virtue of parts of their boundaries coinciding—since there is no space separating them there are no interior parts of the sphere separating them, and thus one cannot traverse from one half to the other without leaving the interior. At step 4, of course, these two distinct point-sized parts cease to be co-located. So the halves change from being contiguous to separated by virtue of their boundaries failing to coincide; the space separating them is part of the exterior of the sum of the sphere's halves, and hence one cannot traverse from one half to the other without traversing the exterior. This alternative model offers a simple explanation of both topological changes, without having to explain how two point-sized objects could become one or how one point-sized object could become two. Such metaphysical fusion and fission never occurs. We only require the possibility of the co-location of distinct point-sized objects. This conception of co-located boundaries has a robust history; it was defended by Brentano (1988), revived and developed by Chisholm (1984, 1993), and formalized by Smith (1997).

This model is not available in many standard mereotopologies. For example, it cannot be accepted in systems that define parthood in terms of topological connection thus: x is part of y iff every z connected to x is connected to y.[3] Since a and b are co-located they are connected to exactly the same things, and hence must be parts of each other. But then by the antisymmetry of parthood, a and b are identical. The model also cannot be accepted by anyone who accepts classical mereology together with the following principle of location: x is part of y iff the location of x is a subregion of the location of y.[4] On this principle, the co-location of a and b

---

[3] These are the SMT systems of Casati and Varzi (1999).

[4] This is the mereological harmony principle called 1ρ by Uzquiano (2019, p. 204). Oppenheim and Putnam (1958) endorse the principle, as does Markosian (2014).

entails their mutual parthood, which again by antisymmetry yields their identity.[5] This model also cannot be accepted in classical point-set topology which entails that any self-connected object (like the sphere) cannot be partitioned into two disjoint closed objects.[6] But there are well-behaved mereotopologies that accommodate such co-location, chief among them is the formal theory in Smith (1997). I've argued that the view has explanatory virtues in solving this sort of puzzle, but these virtues should be weighed against the costs of theoretical revision.

11.1.2 A Paraconsistent Solution

Paraconsistent solutions to the problems of boundaries have begun to see the light of day.[7] Mereotopology based in paraconsistent logic was first developed in Weber and Cotnoir (2015); the theory allows (*contra* classical mereotopology) for connected objects to be divided exclusively and exhaustively into two closed parts.

More recently, Priest (2014) has developed a paraconsistent theory of identity and parthood that can be directly applied to the puzzle about how to cut a continuous object symmetrically in half. Priest's primary aim in developing the view is to provide a metaphysical explanation of *unity*, as when the parts of a given whole are united to form that whole. As the concept of unity has connections throughout the history of philosophy (both East and West), Priest (2014) applies his view to a wide range of issues: Frege's problem of the unity of propositions (Chaps. 1 and 9), the problem of universals, i.e., the One and the Many (Chaps. 3 and 7), Heidegger's nothingness (Chap. 4), the Self (Chap. 11), ontological dependence (Chap. 12), and others besides.

To see the general shape of his theory of unity, take an object x and its parts $y_1, \ldots, y_n$. Since x is genuinely counted as a single *whole* rather than a mere plurality of its parts, there must be something that unifies $y_1, \ldots, y_n$ to constitute x. A very natural answer here is that what unifies the parts to form a whole is a relation: the *composition* relation. Priest himself rejects this answer: as relations in general do not unify, we are owed some special explanation as to why this relation unifies and that seems not too far removed from the very thing we wanted to explain.

So, to provide an explanation, Priest postulates the existence of a special class of objects that constitute this unity, which I will *unity particles* or u-parts for short.[8] A unity particle u for an object x cannot merely be another (distinct) object alongside its parts $y_1, \ldots, y_n$, since that would simply set off a vicious Bradley-style regress.

---

[5]Of course the rejection of the antisymmetry of parthood is available. See, e.g., Cotnoir (2010).
[6]See, e.g., Hocking and Young (1961, p. 14).
[7]Early suggestions along these lines can be found in Priest (2006, Chap. 11).
[8]Priest (2014) following his earlier coinage (2002) calls them "gluons". The name is unfortunate: "gluon" is already the name for a fundamental particle of physics, the gauge boson for the strong force which is responsible for holding matter together by binding quarks into protons and neutrons. Of course, Priest isn't literally talking about gluons in the physicist sense since his gluons are responsible for the unity of all parts into any whole. Turner (2015) suggests the name is a mischievous wink at current physics. Anyway, I avoid all talk of gluons in what follows.

After all, what makes $u, y_1, \ldots, y_n$ a true unity and not a mere plurality? Another unity particle? No, this would simply defer the explanation for unity indefinitely. Unity particles must be indistinct from the objects they unify.

Priest suggests, then, that a given unity particle is identical to each part of the object it unifies. That is, if u is the u-part for x, then $u = y_1$ and $u = y_2$ and so on. Doesn't this mean that $y_1 = y_2$? Well, it would if identity were transitive, but Priest rejects transitivity of identity. To see why, we need a brief detour through paraconsistent logic.

In paraconsistent logics (like the variant Priest 2014, Sect. 2.10 uses second-order LP), interpretations of unary predicates include both an extension and an anti-extension, where every object in the domain must be in either the extension or the anti-extension but it could be in both. Then a sentence like "Fa" is true in an interpretation when the denotation of a is in the extension of F, false if the denotation of a is in the anti-extension of F, and both true and false whenever it is in both.

Now, Priest argues at length that identity should be defined in the usual Leibnizian way using the LP material conditional.[9]

Identity $\quad x = y$ iff $\forall X(Xx \equiv Xy)$

Note that this is a metalinguistic definition of the identity relation rather than an a biconditional definition stated in the object language; you should read "$a = b$" simply as shorthand for the second-order LP formula "$\forall X(Xa \equiv Xb)$".

The LP material biconditional is not transitive; $A \equiv B$ and $B \equiv C$ do not entail $A \equiv C$. Countermodel: A true, B is inconsistent, and C (just) false. Similarly, identity fails to be transitive as $\forall X(Xa \equiv Xb)$ and $\forall X(Xb \equiv Xc)$ do not entail $\forall X(Xa \equiv Xc)$. Countermodel: exactly one unary property P comprising an extension containing just (the denotations of) a and b, and an anti-extension containing just (the denotations of) b and c.

This theory of identity allows unity particles to unite the parts of a whole (by being identical to each of them), without forcing each part to be identical with any other part. Of course, if a u-part u is identical with some (consistent) part a, we will have that a and u share all their properties. And if u is identical to another (consistent) part b, they too will share all their properties. But it needn't follow that a and b share all their properties, as u itself may well be inconsistent.

Every composite object, then, has a unity particle. Are unity particles unique? Priest (2014, p. 20) argues as follows: suppose u and u' are both u-parts of x. Then u and u' are parts of x and so $u = u'$ and $u' = u$. One wonders, though, whether this is sufficient for uniqueness in a robust sense of each composite object having exactly one unity particle. Consider, for example, an object a composed of b and c, and call its associated u-part u_a. But suppose further that b is itself composed further of d and e, and call its associated u-part u_b. Then u_b is part of b and b is part of a. By the transitivity of parthood (accepted by Priest 2014, p. 89), u_b is part of a. But then $u_b = u_a$, so the unity particle for an object may be identical to the unity particle for a distinct object, casting doubt on whether identity is sufficient for "uniqueness".

---

[9]See Sects. 2.6–2.7 of Priest (2014).

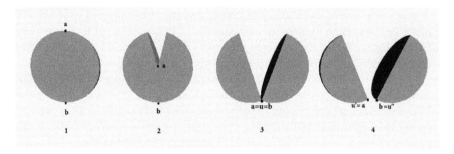

Fig. 11.3 Cutting a sphere 2 (cross section)

(The only option here appears to be to reject that u-parts are themselves parts of the objects they unite or reject the transitivity of parthood.)

Let's return now to the puzzle about cutting a sphere. According to Priest's theory of identity, the sphere by virtue of being a unitary object has an associated unity particle, call it u. And point-sized parts a and b of the sphere are each identical to u. When the sphere is transformed from a continuous whole (steps 1 and 2) to a contiguous whole (step 3) it is connected by a single "hinge" particle. But even at step 3, it is still a unified whole and so u still exists. A natural thing to say is that the hinge particle is u itself. In fact, given that every part of the sphere is identical to u, the hinge cannot fail to be u. And for that reason, at stages 1 and 2 we have it that $a = u = b$, so there's no barrier to affirming this at step 3. Now, of course, what does change is whether $a = b$. At steps 1 and 2, $a \neq b$ (compatible with $a = u = b$), after all they are discernible with respect to their spatial location.[10] At step 3, Priest can perfectly well affirm that $a = b$; after all, the only prior grounds we had for distinguishing them (their locations) now fail to do so.[11] Insofar as $\forall X (Xa \equiv Xb)$, then the transition from continuity to contiguity (step 2 to step 3) has an explanation (Fig. 11.3).

What about the transition from the sphere going from a contiguous object to two separated objects (step 3 to step 4). The answer must be that because there is no longer a single unified whole, the unity particle u goes out of existence. And so the single hinge particle connecting the two halves of the sphere ceases to connect them because it ceases to exist.

Hang on, we might think, what about a and b? Do they cease to exist? Apparently not, since both halves of the sphere remain completely bounded—there's no point-sized hole in the boundary of either. But how can u go out of existence when things that were identical to it stick around?

---

[10]Typically, only purely qualitative properties are thought to feature in Leibniz's Law. As locational properties aren't purely qualitative, they wouldn't typically be allowed to play a distinguishing role. Priest, however, rejects this restriction (2014, p. 23) allowing locational differences to play a distinguishing role. (See also Priest 2014, Sect. 2.7.)

[11]Priest (2014, p. 26) accepts that identities are temporary.

I'm not sure I have much to say to answer this question on behalf of Priest. But it is not as if this is an unintended consequence of Priest's view of unity—it *is* his view of unity. A u-part unifies some objects by virtue of being identical to them. They cease to be unified when the u-part ceases to be identical to them by virtue of the u-part ceasing to exist. How exactly one *expresses* this ceasing to exist is another question. For any composite object (any object having more than one part) its u-part will be non-self-identical.[12] Since the parts are distinct by being discernible via some property, the u-part must be discernible from itself, and hence non-self-identical. But the underlying LP logic yields that non-self-identical things do not exist, i.e., $u \neq u$ entails that $\neg \exists x (x = u)$.[13] So, in a sense, u-parts never exist, and as a result it is hard (impossible?) to express the situation when a u-part did exist, when $\exists x (x = u)$ goes from being both true and false to being (just) false.

Expressive limitations aside, the view does have something to say about the puzzle, and in particular offers a solution to the problem that does not require point-particles to be co-located or leave their fusion and fission completely unexplained. The solution comes at a fairly hefty revisionary cost: accepting inconsistent objects, nonexistent objects, non-transitive identity, widespread failures of the substitutivity of identicals, temporary and contingent identity, failures of modus ponens, and more. These costs are largely independent on one another. But Priest has provided further motivations for those revisions in other work.[14]

11.2 How to Make a Donut

I want to turn now to a related puzzle involving a different sort of topological change: turning a continuous sphere made of homogenous material into a donut. In keeping with the previous example, we can imagine smoothly deforming the surface of the sphere (step 1 to step 2), until the top surface and bottom surface meet at a point in the center, forming a "horn torus" (step 3). This point of connection then "breaks" as it were (much like the hinge point between the two halves of the sphere) leaving a hole in the middle, forming a "ring torus" (step 4) (Fig. 11.4).

How exactly to make sense of such a change in keeping with the solutions to the previous puzzles is not immediately obvious. The mystery becomes more puzzling when we track points a and b as before. The old mystery about how a and b could become identical (at step 3) remains. But the second transition (step 3 to step 4) is made even more puzzling. Here, a and b must split into distinct point parts. But in this case it is not just two boundary points (as before), but an (uncountable) infinite

---

[12] See Fact 1, Priest (2014, p. 27).
[13] See Fact 2, Priest (2014, p. 27).
[14] See, e.g., Priest (1995, 2002, 2005, 2006, 2010a, b, c).

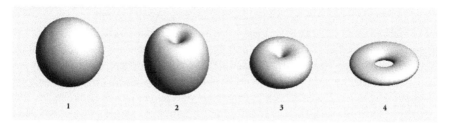

Fig. 11.4 Making a donut

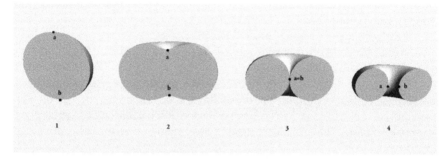

Fig. 11.5 Making a donut (cross section)

number of them that serve as a boundary around the circumference of the hole (Fig. 11.5). How exactly are we supposed to explain such a radical fission?[15]

11.2.1 A Paraconsistent Solution?

To what extent does the paraconsistent solution generalize? The explanation for the first key transition between steps 2 and 3 ports over nicely. Priest's theory of identity allowed us to explain the fusion between a and b since at step 3 there is no longer anything to distinguish them.

But the explanation for the second key transition between steps 3 and 4 does not transfer very well at all. First, let's try to explain why a and b "break" and a hole appears. In the previous example, the explanation for the "break" was that the sphere's unity particle went out of existence. But as the donut is still a singular object, it would seem that the u-part for it still exists. So the explanation for the topological change cannot be that the u-part goes out of existence.

---

[15] Of course there are other ways of making a donut. We could insist that the correct model involves simply punching out a portion of matter from the middle of the sphere, leaving a donut and its remainder (similar to the lumps of dough sometimes sold as "donut holes"); this method would not be all that different from the model of cutting the sphere in half. The puzzle isn't that there's no way of making a donut, but that there's seemingly no explanation for a very natural way of doing so.

On the other hand, it's possible that the unity particle for the sphere *does* go out of existence, if one takes seriously the thought that the sphere itself goes out of existence *qua* sphere. That is, the unitary object undergoes a substantial change from sphere (at step 1) to donut (at step 4), and so somewhere along the way the unity particle for the sphere is replaced by a numerically distinct unity particle for the donut. But when? Even at step 2 we arguably no longer have a (true) sphere, but something spheroid (maybe it is a deformed sphere?). At step 3, we already have something that can truly be called a torus, but maybe not a donut. One wonders then, what exactly are the identity conditions on unity particles? Priest (2014, Sect. 3.4) develops unity particles by analogy with (instances of) Aristotelian *forms*; we are told they come into existence when a unified object does and go out of existence when the object does. But what sorts of changes can they survive?

No matter what the answer here, we also need an explanation of how infinitely (uncountably) many points split from a/b in order to form a closed-loop boundary. This is particularly puzzling since it would appear the donut gains a lot of new point-sized parts *ex nihilo*. Of course, Priest is not committed to mereological essentialism—objects (and their u-parts) can survive the addition or subtraction of their material parts. The problem is rather a question of where the boundary points come from.

One way around the conclusion that the donut gains a bunch of new parts from nowhere is to insist that the new parts are not "new" but that they are all simply identical to a (and b for that matter), even at step 4. This would require postulating that the inner boundary of the hole is composed of a single multi-located particle. This single particle would be radically inconsistent, having all sorts of incompatible location properties. And if this is the account of fission that takes place in the puzzle of the donut, why isn't it also what happens between a and b in the original puzzle of the sphere? That is, are a and b identical post-split there too? If so, then contra appearances, the two halves of the sphere have parts in common. And so it would appear there's no symmetrical way dividing an object into two nonoverlapping parts.

Another option is to admit that the points are "new" (i.e., distinct), but insist that they aren't "from nowhere". They come from a/b. I suggest this not because I find it particularly explanatory, but only because it seems to be a natural consequence of Priest's (2014, Sect. 5.7) account of fission.

> Suppose we have an amoeba. Call it a. At some time t, it divides down the middle to form two new amoebas, b and c. After t, where is a? If it exists at all, it must be either b or c; it has not transmigrated elsewhere. But by symmetry, if it is b it is also c and vice versa. Hence either it has gone out of existence, or it is both b and c. Now it is difficult to suppose it has gone out of existence. It has not, after all, died. So it is both b and c. [...] After t, "a" denotes b and c, even though these are distinct. Hence, $a = b$ and $a = c$. (For the same reason, $a = a$ is false — as well as true — after t.) Notice that we cannot apply the substitutivity of identity to infer that $b = c$ because, as we saw [...] this principle of inference breaks down in the case of multiple denotation. (Priest 1995, p. 369–70)

Because Priest gives up the transitivity of identity (and similar instances of the substitutivity of identicals), he can endorse that each "new" boundary part is identical to a/b after the split. Indeed, the term "a" (or "b") comes to (plurally) denote the many boundary points of the donut hole. The transitivity of identity "fails to be truth preserving when the medial object has contradictory properties." Priest (2014, p. 67), and so those points may well be numerically distinct.[16]

But I am not convinced that the identity of a/b (pre-split) with the new boundary points can really suffice to provide an explanation of where they came from. Identity facts, on Priest's account, don't always explain the existence of new material parts. Consider, for example, a cup of coffee that mysteriously keeps refilling itself with new coffee after every sip that is taken. Such an object would be astounding—where does the new coffee come from? Imagine Priest's reply: the coffee has a u-part, u_c, and since each new portion of coffee is simply identical to u_c, it thus comes from u_c. I expect our astonishment would be undiminished.

11.2.2 A Consistent Solution?

The consistent solution outlined above can be generalized to the current puzzle. As before, the key idea involves coincident objects. The transition between step 2 and 3 is explained as before: a and b are two distinct point-sized parts that become co-located. The transition between step 3 and 4 would appear to require an uncountable infinitude of co-located boundary points stacked up at a/b. What reason could there be for that?

The Brentano–Chisholm theory of boundaries supplies us just such a reason.[17] Spatial locations have what Brentano called the *plerosis* or "fullness" of the point in space. How "full" a spatial location depends on how many boundary parts are located there. And how many are located there depends on how many such parts there need to be in order to ensure that everything has a boundary.[18] For example, the spatial point bisecting a line segment could serve as a boundary in two directions, and so there are two *plerotic* parts co-located there. A spatial point on the surface of a sphere (such as the location of a in step 1) is potentially a location for a boundary in infinitely many directions. So, there are infinitely many point-sized plerotic parts stacked up there. Similarly, the location of centerpoint of the horn torus (step 3) could serve as a boundary in infinitely (uncountably) many directions. And so, there are also uncountably many plerotic parts co-located at exactly that point. What happens

---

[16]In the case of splitting the sphere, we wouldn't want to apply this solution, which would commit us to say that term "u", which denotes the u-part for the sphere pre-split comes to denote a and b post-split. But if that were so, then u doesn't go out of existence even when the sphere does.

[17]The following presentation of these ideas owes much to Smith (1997).

[18]It should be noted that this reverses the ordinary direction of explanation. Typically we'd want to determine facts about boundaries from facts about what pointy objects there are around. On the Brentano–Chisholm view, we start with the condition that everything has a boundary and work back to the number of points. This aspect of the account is not strictly necessary for the co-location solution to go through. All that is needed is for there to be uncountably many co-located points—the underlying explanation for that fact needn't be as the Brentano–Chisholm view assumes. (Thanks to an anonymous referee for raising the issue.)

at the transition (step 3 to step 4)—when the donut "breaks" and a hole appears—is that all of these preexisting boundary parts spread out into their respective directions, continuing to serve as the boundary particles of their respective objects.

This view raises symmetry considerations anew: what explains which point-part goes where after the split? The Brentano–Chisholm view explains the underlying asymmetry by the ontological dependence of boundary parts on their wholes. Boundary parts cannot exist except as proper parts of the objects they serve as boundaries for.[19] This is not a brute asymmetry, since the boundary of an object is *defined in terms* of the object and essentially dependent on it. It's not clear that this kind of "directionality" is problematic. Or at the very least, it is not the *same* problem as the asymmetries with which we began.

As usual, there are no perfect solutions in philosophy, but there are many good ones. And not all good solutions are equally good. They must be adjudicated on the usual criteria for theory choice. One key criterion: to what extent does the theory require widespread revisions to theoretical commitments in distal parts of philosophy? On this score, at least, it would appear that the paraconsistent view as developed by Priest fares worse than co-location view. After all, the co-location view is consistent, requires no nonexistent objects, with a transitive, necessary, atemporal identity relation that satisfies the substitutivity of identicals, and a detachable conditional. Of course, there are other competing criteria: the ability to solve puzzles across a wide range of cases should weigh against theoretical revision. But given the number of other potential applications of these theories, a full evaluation cannot be carried out here.

We have seen that there is reason to agree with Dummett (2000, p. 505) when he writes

> The classical model is to be rejected because it fails to provide any explanation of why what appears to intuition to be impossible should be impossible. It allows as possibilities what reason rules out.

How exactly to fix the classical model is up for debate. It seems to me that if we want to cut continuous things exactly in half, we need to allow that point-sized objects can be co-located.

Acknowledgements The research and writing of this paper was supported by a 2017–2018 Leverhulme Research Fellowship from the *Leverhulme Trust*.

References

Brentano, F. (1988). *Philosophical investigations on space, time and the continuum*. London: Croom Helm.
Casati, R., & Varzi, A. C. (1999). *Parts and places: The structures of spatial representation*. Cambridge (MA): MIT Press.

---

[19] See Chisholm (1984).

Chisholm, R. M. (1984). Boundaries as dependent particulars. *Grazer philosophische Studien, 10*, 87–95.
Chisholm, R. M. (1993). Spatial continuity and the theory of part and whole: A brentano study. *Brentano Studien, 4*, 11–24.
Cotnoir, A. J. (2010). Anti-symmetry and non-extensional mereology. *The Philosophical Quarterly, 60*(239), 396–405.
Dummett, M. (2000). Is time a continuum of instants? *Philosophy, 75*(4), 497–515.
Hocking, J. G., & Young, G. S. (1961). *Topology*. Dover.
Markosian, N. (2014). A spatial approach to mereology. In S. Kleinscmidt (Ed.), *Mereology and location* (pp. 69–90). Oxford University Press.
Oppenheim, P., & Putnam, H. (1958). Unity of science as a working hypothesis. *Minnesota Studies in the Philosophy of Science, 2*, 3–36.
Priest, G. (1995). Multiple denotation, ambiguity, and th strange case of the missing amoeba. *Logique et Analyse, 38*, 361–373.
Priest, G. (2002). *Beyond the limits of thought* (2nd ed.). Oxford: Oxford University Press.
Priest, G. (2005). *Towards non-being*. Oxford: Oxford University Press.
Priest, G. (2006). *In contradiction. A study of the transconsistent* (2nd ed.). Oxford University Press.
Priest, G. (2010a). A case of mistaken identity. In J. Lear & A. Oliver (Eds.), *The force of argument*, chap. 11. Routledge.
Priest, G. (2010b). Inclosures, vagueness, and self-reference. *Notre Dame Journal of Formal Logic, 51*(1), 69–84.
Priest, G. (2010c). Non-transitive identity. In R. Dietz & S. Moruzzi (Eds.), *Cuts and clouds: Vagueness, its nature and its logic*, chap. 23. Oxford: Oxford University Press.
Priest, G. (2014). *One, being an investigation into the unity of reality and of its parts, including the singular object which is nothingness*. Oxford: Oxford University Press.
Putnam, H. (1994). Peirce's continuum. In K. Ketner (Ed.), *Peirce and contemporary thought: Philosophical inquiries*. Fordham University Press.
Smith, B. (1997). Boundaries: An essay in mereotopology. In L. Hahn (Ed.), *The philosophy of Roderick Chisholm* (pp. 534–561). Library of Living Philosophers. Open Court, LaSalle.
Turner, J. (2015). Review of graham priest 'one: Being an investigation into the unity of reality and of its parts, including the singular object which is nothingness'. *Notre Dame Philosophical Reviews*.
Uzquiano, G. (2011). Mereological harmony. In K. Bennett & D. Zimmerman (Eds.), *Oxford studies in metaphysics* (Vol. 6, pp. 199–224). Oxford University Press.
Weber, Z., & Cotnoir, A. J. (2015). Inconsistent boundaries. *Synthese, 192*(5), 1267–1294.

A. J. Cotnoir is Senior Lecturer in the Department of Logic and Metaphysics at the University of St Andrews. He works primarily in metaphysics and philosophical logic. He is the co-editor of a collection entitled *Composition as Identity* (Oxford University Press, 2014) and a co-author of *Mereology* forthcoming with Oxford University Press. He has published widely on mereology, identity, truth, and nonclassical logics in leading international journals.

Chapter 12
Contradictory Information: Better Than Nothing? The Paradox of the Two Firefighters

J. Michael Dunn and Nicholas M. Kiefer

Abstract Prominent philosophers (Bar-Hillel and Carnap, Popper, Floridi) have argued that contradictions contain either too much or too little information to be useful. We dispute this with what we call the "Paradox of the Two Firefighters." Suppose you are awakened in your hotel room by a fire alarm. You open the door. You see three possible ways out: left, right, straight ahead. You see two firefighters. One says there is exactly one safe route and it is to your left. The other says there is exactly one safe route and it is to your right. While the two firemen are giving you contradictory information, they are also both giving you the perhaps useful information that there is a safe way out and it is not straight ahead. We give two analyses. The first uses the "Opinion Tetrahedron," introduced by Dunn as a generalization of Audun Jøsang's "Opinion Triangle." The Opinion Tetrahedron in effect embeds the values of the "Belnap-Dunn 4-valued Logic" (Truth, Falsity, Neither, Both) into a context of subjective probability generalized to allow for degrees of belief, disbelief, and two kinds of uncertainty—that in which the reasoner has too little information (ignorance) and that in which the reasoner has too much information (conflict). Jøsang had only a single value for uncertainty. We also present an alternative solution, again based on subjective probability but of a more standard type. This solution builds upon "linear opinion pooling." Kiefer had already developed apparatus for assessing risk using

---

[1] Graham and Richard Routley (later Sylvan) invented the word "dialethism" (now spelled by others "dialetheism," closer to the Greek root for "dual truth"), expressing a stronger concept than "paraconsistency," a word introduced by Francisco Miró Quesada in 1976. See Priest et al. (1989, p. xx).
[2] Dunn (1976), p. 157: We are not claiming that there are sentences that are in fact both true and false. We are merely pointing out that there are situations where people suppose, assert, believe, etc. contradictory sentences to be true, and we therefore need a semantics that expresses the truth conditions of contradictions in terms of the truth values that the ingredient sentences would have to take in order for the contradictions to be true.

J. M. Dunn (✉)
Indiana University, Bloomington, USA
e-mail: dunn@indiana.edu

N. M. Kiefer
Cornell University, Ithaca, USA
e-mail: nicholas.kiefer@cornell.edu

expert opinion, and this influences the second solution. Finally, we discuss how these solutions might apply to "Big Data" and the World Wide Web.

Keywords Bayes · Conditional probability · Contradiction · Dialetheism · Information · Linear opinion pooling · Opinion Tetrahedron · Paradox · Paraconsistent logics · Relevance logic · Subjective probability

12.1 Introduction

Graham Priest has certainly been among the strongest advocates for paraconsistent logics (logics that tolerate contradictions without leading to triviality) and has been the strongest advocate, bar none, for dialetheism (the view that there are true contradictions). These are not just complicated, concealed, contradictions that someone, perhaps almost everyone, believes to be true, but rather contradictions that are actually true, and where we have good reason to believe they are true.[1]

We believe that there are good uses for paraconsistent logics, but we break ranks with Graham on dialetheism. We believe that there can be contradictions in our information—theories, belief systems, databases, The World Wide Web, whatever—but not in the "real world." Dunn (1976) built on this "epistemic" motivation in a four-valued approach to the first-degree entailments of the systems E and R of Anderson and Belnap.[2] Contradictions are to be treated as temporary problems that need to be lived with until a solution, or at least a work-around is found. But they are not to be embraced. Belnap (1977a, b) has also emphasized this in two papers with his talk of "told values" and inconsistent databases. And of course, as the referee has reminded us, "inconsistency may be difficult to detect (the general problem is NP-hard), and so it may well be that a single agent holds inconsistent information without realizing it is inconsistent."

The referee also pointed out that an agent might hold inconsistent information without *believing* it is true, saying this is apparent in the case of artificial agents such as a computer database. He also pointed out that according to some philosophers of science, e.g., Popper (1934, 1959), there is no need to believe that a scientific theory is true in order to adopt it as the best so far. Of course, this would seem to be a reason for a paraconsistent logic in the extreme case where the best theory so far is inconsistent.[3]

One motivation for paraconsistent logics is that they can provide safety from the logical paradoxes, say the Russell Paradox. But it turns out that the Curry Paradox

---

[3] Popper (1934, p. 50, 1959, p. 72) says: "The requirement of consistency plays a special rôle among the various requirements which a theoretical system, or an axiomatic system, must satisfy. ...
In order to show the fundamental importance of this requirement it is not enough to mention the obvious fact that a self-contradictory system must be rejected because it is 'false'. We frequently work with statements which, although actually false, nevertheless yield results which are adequate for certain purposes. ...(An example is Nernst's approximation for the equilibrium equation of gases.) But the importance of the requirement of consistency will be appreciated if one realizes that a self-contradictory system is uninformative. It is so because any conclusion we please can be derived from it."

forces one to adopt a logic so weak that it becomes basically unusable. See Meyer et al. (1979). We think that the avoidance of the paradoxes by adopting a set theory with limited comprehension, say Zermelo–Fraenkel set theory, is the best approach. In this paper, we will give an example of how a more commonplace contradiction might motivate the use of a paraconsistent logic, and how such a contradiction can be useful and have value without it being true. We shall also explore an alternative approach using the statistical method of *linear opinion pooling*. We shall then give a brief comparison of the two, showing how the first might be viewed as "the quick fix," whereas the second might be viewed as a more permanent solution.

Bar-Hillel and Carnap (1953, p. 229) wrote: "It might perhaps, at first, seem strange that a self-contradictory sentence, hence one which no ideal receiver would accept, is regarded as carrying with it the most inclusive information. …A self-contradictory sentence asserts too much; it is too informative to be true." Floridi (2011) wants a "strong semantic theory of information" that avoids what he labels as the Bar-Hillel–Carnap Paradox (BCP). But he says contradictions contain zero information, giving among other reasons that "inconsistent information is obviously of no use to a decision maker."[4] Various paraconsistent logicians have argued for theories of information wherein contradictions could contain differing amounts of information (one from the other). But even though JMD has thought about these things for years, the Paradox of the Two Firefighters (below) occurred to him only within the last several years. He would be embarrassed to admit this except for the fact that an example like this seems not to have been discussed in the literature surrounding contradictions.

We provocatively label this example as "The Paradox of the Two Firefighters" because of the various reactions we have received. It seems that almost everyone thinks they have a solution to it.[5] We shall discuss some of these solutions below, including of course our own, politely positioned at the end. Every paradox should of course have at least two solutions—that is what makes it a paradox.[6] And so, of course, we offer two solutions, both fundamentally based on subjective probability.

---

[4]Note that Popper (1934, 1959) quoted in the previous footnote can be viewed as agreeing with Floridi that contradictions contain no information.

[5]JMD has given talks on "The Two Firefighters" first at the Logica conference (2012), and then at the University of Munich Center for Mathematical Philosophy, the Modal Logic Workshop on Consistency and Structure at the Center for Formal Epistemology of Carnegie Mellon University, the School of Cognitive Science at Jadavpur University in Kolkata (India), the Logic Group at the University of Alberta, and the Info-Metrics Institute at The American University. It was at this last that JMD and NMK decided to write this joint paper. We want to thank Aparajita Karmakar, David Makinson, John Winnie, and an anonymous referee for their insightful comments. But we do not mean to say that we have taken everything they said into account.

[6]A short list of "solutions" to the Russell Paradox includes beside Zermelo–Fraenkel set theory, NBG, MK, ML, etc., and theory of types (Russell's own fix).

12.2 The Paradox of the Two Firefighters

Suppose you check into your hotel late at night, dead tired after a long flight, and are taken to your room on an upper floor by a bellhop. This is your first visit to this hotel. You go into and out of the elevator with him and feel lucky that the elevator is just a few doors away from your room. As you get off of the elevator, the bellhop tells you that your room is at the intersection of three hallways, and that it is important to read the fire safety instructions in your room since the halls are being repainted and the directional signs have been temporarily removed. The bellhop also explains that in case of fire the elevator will be out of service. When you get in your room you are so tired that you right away get ready for bed and fall fast asleep in minutes. Your last thought, before falling asleep, is that you will read the instructions the next morning. But in the middle of the night you are awakened by a fire alarm. Your look around and your room is on fire. You open the door and run outside. The door closes and you did not have the key with you. Only then do you remember that you never read the instructions. You see three possible ways out, just as the bellboy had mentioned. There is a hallway to your left, a hallway to your right, and also one straight ahead. Of course you know you cannot use the elevator, since as the bellhop explained the elevator will not operate in the event of a fire, and you can clearly see the flashing sign above it saying "Out of Order." Accepting all of this, what do you do in each of the following two scenarios?[7]

Scenario 1. You see two firefighters. One says there is exactly one safe route and it is to your left. The other says there is exactly one safe route and it is to your right. Contradictory information!

Scenario 2. You find no one to give directions. Incomplete information! Nothing!

Which scenario would you prefer? We think it is obvious that a rational agent would prefer to be in Scenario 1. While the two firefighters are giving you contradictory information, they do agree on one thing. There is a safe way out and it is not straight ahead. You may choose to accept the one point on which the two agree. In this case, three choices have been reduced to two, thereby increasing your odds for survival. Now you have to pick from those two and run, hoping to find the exit. On the other hand, the fact that the firefighters disagree indicates that at least one is wrong about which hallway you should take; perhaps one or both are wrong about the single exit as well. Still, you may choose between the R and L hallways, thinking they are more likely to offer an escape. We now turn to an analysis.

---

[7] We set things up so that everything starts off symmetrically (distributed) between the three choices. We have tried to throw a "veil of ignorance" over the situation so that you can bring no special knowledge to weigh the a priori odds of one choice over the other two. You can of course imagine that one of the firefighters is older and obviously more experienced than the other, and then we will add to the setup that they are twins, that they are wearing identical uniforms, with the same badges, can hand you the same certificates of training and of service, etc.

12.3 Preliminary Analysis

We shall now consider various reactions to the Paradox of the Two Firefighters that JMD has heard after his oral presentations, and in doing so defend that it gives an example of how contradictory information can sometimes be better than nothing (and relevant to decision-making). First a few abbreviations:

L: Left is a way out.
R: Right is a way out.
S: Straight is a way out.
O: There is only one way out.

What the first firefighter says can be symbolized by $(L\&O)$, and what the second firefighter says by $(R\&O)$. Note that $(L\&O)$ entails $\neg R$. So $(L\&O)$ is inconsistent with R, and hence $(R\&O)$.

One standard reaction has been to point out that $(L\&O)\&(R\&O)$ is not a "real contradiction," like $L\&\neg L$. It is only a partial contradiction, $(L\&O)$ and $(R\&O)$ are only contraries. We do not want to embarrass anyone by crediting them with this reaction, and in fairness their reaction was only on first hearing the Paradox of the Two Firefighters. But our reply is altogether obvious after a little reflection: contraries are even a stronger form of contradiction.

$(L\&O)\&(R\&O)$ entails $L\&\neg L$.

In traditional logic, contradictories are distinguished from contraries and subcontraries. Contraries cannot both be true, although they can both be false. Subcontraries cannot both be false, although they can both be true. Contradictories cannot both be true, and they also cannot both be false. L and $\neg L$ are contradictories. $(L\&O)$ and $(R\&O)$ are contraries, not subcontraries.

A second reaction is to point out that the firefighters made separate statements contradicting each other. There is no contradiction until they are conjoined. So we can avoid the contradiction by rejecting the adjunction rule A, B entail $A\&B$. This is precisely what Jaśkowski (1948) did with his Discursive Logic. This logic was one of the very first paraconsistent logics to be developed, and is sometimes called "Discussive Logic". It was proposed by Jaśkowski as a logic for discourse/discussion in which there are several participants making assertions that contradict each other.

Our reply is to say that it is important to consider the two statements together, and if that is not conjunction then we don't know what conjunction means. If we had reason to believe one more than the other, we might just disregard that other. But in the situation as described we have equal reason to believe each. Also, it is only by considering the two statements that we see that the two firefighters agree on O.

Another reaction is for someone to claim that we are not dealing with a conjunction of $L\&O$ and $R\&O$, but rather a disjunction.

On the face of it, this isn't so. If two people are testifying at a trial, and one says P and the other says Q, there is, in general, no reason to treat this as a disjunction. But there might be if they are contradicting each other. The most extreme case of this would be that one says P and the other says $\neg P$.

One firefighter says $L\&O$. The other says $R\&O$. But suppose the first firefighter then says "We can't both be right, but I am certain one of us is right." This makes perfect sense, retreating to what in correspondence David Makinson has described as using their disjunction as a "fall-back" reading or integration of the conflicting inputs.

So even if neither of the firefighters speaks up, you could reason more or less the same way. Both seem experts who have up-to-date info about the fire, and this is all the information you have readily (and safely) available. So perhaps at least one of them is right, i.e., $(L\&O) \vee (R\&O)$. $L\&O$ entails $\neg S$, $R\&O$ entails $\neg S$, therefore $\neg S$. In this case, you must choose between R and L. But both could be wrong, in that $\neg O$ holds and both L and R work. A happy event for you.

Makinson is known for the "AGM" (Alchourrón et al. 1985) notion of "belief revision." The key idea is that if one believes P, and then learns something Q that entails $\neg P$, that one retreats to a consistent set of beliefs that includes Q (and hence excludes P). Makinson's response to the Paradox of the Two Firefighters is in the same spirit.

It appears though that the Paradox of the Two Firefighters does not involve belief revision per se. That might be the case if you heard say first from Fireman 1 that $L\&O$ and believed it, but then heard from Firemen 2 that $R\&O$ and thereby somehow came to believe that instead. But that is not what happened. You do not know which to believe. Further, believe revision depends on the order in which the information arrives, and we want you to think that in our Scenario 1 the two firefighters arrive simultaneously and shout and point in different directions at the same time.

Makinson's idea seems right in spirit, but does not literally use belief revision. The referee has pointed out to us that the topic of merging inconsistent information has been investigated recently under the heading of "belief merging," which is related to but more general that "belief revision." See Konieczny and Pérez (2011), and D'Alfonso (2016).

Whether one is revising or merging beliefs, problems arise if you are in to what Hewitt (2009, 2015) calls a "pervasively inconsistent" background (think of the World Wide Web as an obvious example). Then unless one is going to fix everything at once, there is no consistent set of beliefs to drop back to. Real trouble would raise its head if we allowed the irrelevant implication of classical logic sometimes called "Explosion":

$A\&\neg A$ entails B.

Tanaka (2005) discusses several attempts at belief revision in an inconsistent environment (motivating his own approach based on relevance logic by work of Priest, and Restall and Slaney). We mention here as well Mares (2002).

Let's leave this topic for another day and get back to our two firefighters and consider Scenario $1'$—the same as Scenario 1 (two firefighters) but you can clearly see that you cannot safely go straight ahead. Say there is a conflagration in the hallway straight ahead and/or iron bars have dropped down. There are then only two options, left and right. It now might seem that there is no reason to have a preference between Scenarios $1'$ and 2.

But wait, the two firefighters still agree that there is a way out. So you can risk running left or right as opposed to staying in your room and preparing for your death.

The essence of the Paradox of the Two Firefighters can be duplicated over and over again. Here is an example that is very close to home. A couple is about to leave their house. The driver reaches into a pocket and, not finding the keys, says that they must have been left in a jacket pocket in the closet. The other person seems to remember seeing them on the piano. Again, this is all useful information for the search. At least the couple agree that the keys are in the house. But it is contradictory information. We should point out one important difference between this situation and that of the two firefighters. If you misplace your keys you can presumably take the time to investigate where they are. We know you are probably late to get somewhere, but there is nothing like the pressure of needing to decide which hallway to take to avoid being burned to death.

We next present our two solutions to our paradox.

12.4 A Paraconsistent Approach

By a paraconsistent approach we mean one that invokes some paraconsistent logic so as to avoid concluding anything whatsoever from a relatively minor and specific contradiction such as the information you obtain from the two firefighters. There are by now a number of paraconsistent logics. For an informed and recent discussion of these see Priest et al. (2016).

The system FDE of First-Degree Entailments, due to Anderson and Belnap, is the one that is focused on in Dunn (2009). Dunn (1976) interpreted a four-valued semantics (originally due to Timothy Smiley) in terms of sentences sometimes being assigned the values both true and false, or neither true nor false, as well as of course the usual values exactly true and exactly false. One way this could be done was to view an assignment as a relation, rather than a function, and another way was to view it as assigning a subset of the set of the usual two truth values t and f. Belnap (1977a, b) went on to treat these as the "truth values" $B = \{t, f\}$ for Both, $N = \emptyset$ for Neither, $T = \{t\}$ for True, and $F = \{f\}$ for False.

But these four values are in a way too few. Dunn (2009) enriched this framework by the extending work of Jøsang (1997), who created what he called an "opinion triangle," which in hindsight can be viewed as taking the values T, F, N (no B) and arranging them as the vertices of an equilateral triangle, dropping an axis from each vertex to the center of the opposite side. These were then each outfitted with a uniform scale of degrees of belief, disbelief, and uncertainty from 0 to 1. Values (b, d, u) were normalized with the condition that $b + d + u = 1$, and visualized as points within the triangle. Jøsang viewed his opinion triangle as generalizing the idea of subjective probability, but allowing for a way to measure not just the degree of belief b but also the degree of uncertainty u and an independent degree of disbelief d. (With standard subjective probability the degree of disbelief would just be $1 - b$.)

We do not want to complicate things with too many details here. The interested reader can consult Dunn (2009) or Jøsang (1997). The interesting part for now is that Jøsang accounted for degrees of uncertainty in just one sense, that of "ignorance" (lack of evidence). But uncertainty can also arise from "conflict" (contradictory evidence). That is what motived Dunn to expand Jøsang's opinion triangle to an "opinion tetrahedron." In the diagram below, we picture how the four-element De Morgan lattice DM4 can be expanded to an infinite-valued tetrahedron.

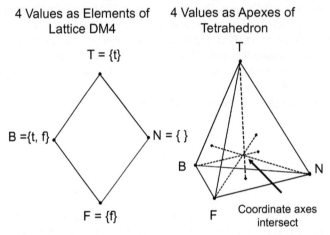

The quick idea then is that a sentence can be given a value (b, d, u, c), where b is degree of belief, d is degree of disbelief, u is degree of uncertainty (ignorance), and c is degree of conflict (contradiction). Thus before the firefighters arrive, you might evaluate each of R, S, and L as $(1, 0, 0, 0)$, optimistically assuming that there is no reason why any hallway would not lead to an exit stairway. Or you might more cautiously evaluate each as $(1/3, 0, 2/3, 0)$, assuming that at least one of the hallways must lead to an exit stairway. You put $2/3$ in the ignorance place and not in the disbelief place, because it is not that you disbelieve to any extent that the given hallway is not a way out—you just don't have any evidence that it is not.

After the firefighters show up and give their "pitches," the values presumably change for you. So if you are an optimist you disregard the conflict and focus on the fact that the two firefighters agree that straight is not an exit. So both R and L get the value $(1/2, 1/2, 0, 0)$, but S gets the value $(0, 1, 0, 0)$. On the other hand, if you are a pessimist you focus on the conflict and think that they must be incompetent and/or have flawed evidence, and you give both R and L say the value $(1/3, 0, 0, 2/3)$ and S something like the value $(\epsilon, 1 - \epsilon, 0, 0)$ (where ϵ is a small number that varies with your degree of pessimism). But in any event, the degree of belief in S shifts downward substantially after you listen to the two firefighters.

Another way to look at this is to consider the statement O that "There is only one way out."[8] Before the firefighters showed up you might have evaluated that as

---

[8] At this point, we might eliminate O as an atomic proposition and define it as $O = (R\&\neg S\&\neg L) \vee (\neg R\&S\&\neg L) \vee (\neg R\&\neg S\&L)$. Negation $\neg$, conjunction &, and disjunction $\vee$ are defined in Dunn

(3/8, 5/8, 0, 0), thinking that there are eight possible ways that the hallways might exit. Where 1 means "exit" and 0 means "no exit" and these can be represented as $(0, 0, 0), (0, 0, 1), \ldots, (1, 1, 1)$ (see the display in the next section). Only three of these represent possibilities where there is just one exit: $(1, 0, 0), (0, 1, 0)$, and $(0, 0, 1)$. So 3/8 represents your strength of belief and 5/8 represents your strength of disbelief. After the two firefighters show up and both say O without hesitation, you might if you take things at face value, evaluate it as $(1, 0, 0, 0)$. But remember we are talking about your subjective belief, so if you are cautious and concerned about their difference of opinion on L and R, you might want to reduce the belief value to less than 1, and put in some nonzero values for disbelief, uncertainty, and conflict. But presumably this will not differ much from the simpler $(1, 0, 0, 0)$.

12.5 A Probability Approach

The Bayesian decision-maker (you) will consider the messages from the firefighters as expressing their own beliefs about the possible escape paths, and will look for a way to combine this information with your own beliefs and come to a decision about the route. To set this up, let us cast the statements in terms of reports of probability distributions. Here, the distributions reported by the firefighters are rather trivial—the probabilities are zeros and ones—but the setting is useful. First, what is the space on which the probabilities are defined? There are three hallways, L, S, and R. Each can be an escape route or not, denoted by 1 or 0, respectively. Thus, there are 2^3 possibilities, $(L, S, R) = (0, 0, 0)$ to $(1, 1, 1)$; arrange these in lexicographic order and index the probabilities as $(p_1, \ldots, p_8)$:

(L, S, R)	Probability
(0, 0, 0)	p_1
(0, 0, 1)	p_2
(0, 1, 0)	p_3
(0, 1, 1)	p_4
(1, 0, 0)	p_5
(1, 0, 1)	p_6
(1, 1, 0)	p_7
(1, 1, 1)	p_8

(1976) (generalizing their definitions by Jøsang 1997) to allow for conflict as well as ignorance). The formula for calculating the value of a conjunction is somewhat complicated (and disjunction inherits this since it is defined from conjunction and negation in the usual way using De Morgan's Law). So it is probably wise for expository purposes not to use this definition of O here. We do point out that both conjunction and disjunction are associative and commutative so we can overlook grouping and order when forming them.

These are probability assessments reflecting the subjective views of the agents involved, you and the two firefighters. There is no question of these being "true" or "false." Of course, they can be contradictory.

The cases in which hallway L works are $(1, 0, 0)$, $(1, 0, 1)$, $(1, 1, 0)$, and $(1, 1, 1)$, so the probability that hallway L works is $p_L = p_5 + p_6 + p_7 + p_8$. For the decision at hand, the decision-maker is only interested in the probabilities p_L, p_S, p_R, three probabilities but not itself a probability distribution. Suppose you have no information at all about the relative likelihood of the hallways (e.g., an exit sign!), then it makes sense to assign probability 1/8 to each of the eight possible outcomes, resulting in aggregate probabilities of 1/2 associated with each hallway.[9]

In this setup, the first firefighter is reporting $P^1 = (p_L, p_S, p_R) = (1, 0, 0)$ and the second $P^2 = (0, 0, 1)$. You wish to combine this information with your own initial beliefs $P^0 = (1/2, 1/2, 1/2)$ to obtain posterior, updated beliefs P. A natural and attractive way to proceed is with a weighted average, the "linear opinion pool,"

$$P^* = w_0 P^0 + w_1 P^1 + w_2 P^2$$

where $w_0, w_1,$ and w_2 are the weights you assign to yourself (w_0) and the two firefighters (w_1, w_2). Each weight w_i is to be multiplied component-wise across the triple P^i, and then the results are added component-wise to obtain P^*.

The weights must add to 1, and it is plausible for you, being ignorant and fair minded, to assign the weights equally as 1/3 each.

We do a calculation to illustrate:

$$P^* = 1/3(1/2, 1/2, 1/2) + 1/3(1, 0, 0) + 1/3(0, 0, 1) \quad (12.1)$$
$$= (1/6, 1/6, 1/6) + (1/3, 0, 0) + (0, 0, 1/3) \quad (12.2)$$
$$= (3/6, 1/6, 3/6) = (1/2, 1/6, 1/2) \quad (12.3)$$

The weights, like the probabilities themselves, are subjective. They reflect your assessments of the relative reliability of the three information sources, and can depend on impressions (is one of them delirious?) but not on the probability assessments. More on this below. You may wish to give more weight to the firefighters' assessments, thinking that they are better informed than yourself, but it is unlikely that you will give your own assessment zero weight, if only to be certain that all three hallways are included in the posterior support.

The opinion pooling literature works with the entire probability distributions (here, over the 8 possible states) rather than the probabilities of events (combinations of states) as we have done here for simplicity. The classic reference on linear pooling is

---

[9]It might make sense here to agree that $p_1 = 0$, it is not the case that no escape is possible. In this case, you would assign probability 1/7 to each of the remaining possibilities and the resulting aggregate probabilities are 4/7. This is inessential to the argument.

Stone (1961); a review of the topic of combining probabilities is Genest and Zidek (1986).[10]

Proceeding with the specification that $P^0 = (1/2, 1/2, 1/2)$ and equal weights $w_i = 1/3$ we have

$$P^* = (1/2, 1/6, 1/2).$$

Thus, the contradictory information from the firefighters is informative in that the posterior probabilities on which your escape strategy will be determined are different from your prior probabilities, on the basis of which you would have decided in the absence of the firefighters.

Not only are the posterior probabilities different from the prior probabilities but also the ensuing decisions differ as well. In particular, with the information you will choose at random between hallways R and L. With only the prior, you would choose at random from all three hallways. So, from the point of view that information affects decisions, the contradictory opinions remain informative.

From a formal point of view, we can quantify uncertainty (crudely: we are summarizing a distribution with a single number) in terms of entropy. Entropy is a measure of uncertainty—lower entropy is less uncertainty. Here we must work with the full distribution over the eight possible states P_f and not simply the resulting summary probabilities affecting the decision. Thus, the prior distribution $P_f^0 = (1/8, \ldots, 1/8)$, an eight component vector, and the posterior distribution is

$$P_f^* = w_0(1/8, 1/8, 1/8, 1/8, 1/8, 1/8, 1, 8, 1/8) + w_1(0, 1, 0, 0, 0, 0, 0, 0) + w_2(0, 0, 0, 0, 1, 0, 0, 0)$$
$$= 1/3(1/8, 1/8, 1/8, 1/8, 1/8, 1/8, 1, 8, 1/8) + 1/3(0, 1, 0, 0, 0, 0, 0, 0) + 1/3(0, 0, 0, 0, 1, 0, 0, 0)$$
$$= (1/24, 9/24, 1/24, 1/24, 9/24, 1/24, 1/24, 1/24).$$

The entropy of a distribution P is $H(P) = -\sum_{i=1}^{K} p_i log_2 p_i$. We use log_2 as is customary for discrete distributions (but the inference clearly holds for any base). Note that $log_2(1/8) = -3$, and hence the entropy of the prior distribution $H(P_f^0) = 3.00$.

We next calculate the posterior distribution (noting that 1/24 occurs as a component six times in P_f^* and that 9/24 occurs two times):

---

[10] Alternatives to linear pooling include geometric and multiplicative approaches. These have the property that the support of the combined distribution is the intersection of the supports of the components. In this case, no escape. With linear pooling, the support is the union of the supports. NMK has found in risk management applications that differences in support for default rates for similar portfolios are common across experts. For the subjective approach to default prediction, see Kiefer (2007, 2011).

$$H(P_f^*) = -\{[6 \times (1/24 \times log_2(1/24))] + [2 \times (9/24 \times log_2(9/24))]\} \quad (12.4)$$
$$- \{(6/24 \times log_2(1/24))] + [(18/24 \times log_2(9/24))\} \quad (12.5)$$
$$= -\{[1/4 \times log_2(1/24)] + [3/4 \times log_2(9/24)]\} \quad (12.6)$$
$$= -\{(-4.58496\cdots/4) + [.75 \times (-1.41504\cdots)]\} \quad (12.7)$$
$$= -\{-1.145 + -1.058\} = 2.207519\ldots \quad (12.8)$$

Thus, from the entropy point of view, the contradictory firefighters provide a clear reduction in uncertainty, from 3 to approximately 2.21.

We are not given any information about the considerations underlying the firefighters' assessments of their probability distributions. Had they been in the building earlier, perhaps at different times when different hallways worked? Would that provide an argument that these apparently contradictory views are reasonable, or coherent in a logical sense? Alternatively, suppose that firefighter 1 was unaware even of the existence of the right-hand hallway, and firefighter 2 of the left? In that case, the two could be brought to consistency: each thinks only one of the two hallways of which he is aware works. These views would be consistent with a world in which the left and right hallways work but the middle did not. Do these explanations provide ground for a paraconsistency advocate to accept the situation?

We argue that the "contradictory" information is better than no information. Now, consider whether you would rather hear from one firefighter or two? Or three or four?

12.6 Comparison of the Paraconsistent and Probability Approaches

We start by saying that this discussion of the paraconsistent and probability approaches should be viewed as preliminary in nature. But we want to point out a few similarities and differences.

First, let us emphasize a similarity. Despite our labeling only one of these "the probability approach," both are based fundamentally on subjective probability. Subjective probabilities reflect a quantification of an individual's certainty/uncertainty about something. In this view, probability is not a property of a physical object or process, it is instead a continuous quantification of beliefs which does not simply divide them up in a two-valued way as "true" or "false."[11]

If two agents both believe something, but with different strengths, they do not really disagree. Person A says "I believe the keys are in a pocket of my jacket hanging in the closet, but I could be wrong." Person B says: "I believe the keys are

---

[11] Of course, the list of probabilities has to satisfy some properties if they are "coherent." Usually, we consider probabilities of zero or one, certainties, as likely to be unrealistic, or perhaps approximations for very small or large probabilities. In fact, these extremes arise all the time but are hidden. They arise in the (subjective) specification of the relevant event space, the list of what can happen. But in our application, these are the probabilities of interest, and the ordinary rules should apply.

on the piano, but I may be misremembering." This is not real disagreement. It is only if they were to start an argument over the precise probabilities that they might really disagree.

There is nothing contradictory about one individual having one set of probabilities and another individual a different set. Both of our firemen have coherent probabilities even though the information from the two is inconsistent, and we are not holding that the two contradictory views are both true.

Comparing the paraconsistent and probability approaches, we note both can be viewed as extend values beyond the two discrete values T and F, but they do so in different ways. The probability approach simply adds continuously many values in between: the unit interval [0, 1] and not just the pair set {0, 1}. The paraconsistent approach also adds continuously many values, but divides them up among four components (belief, disbelief, ignorance, conflict) and requires that they sum to 1.

Despite it being more elaborate, it could be that the paraconsistent approach is the best for a person in a hurry with no time to set up probability distributions, create weightings, and calculate entropy. That person can just implement "by the seat of their pants" intuitive (subjective) hunches about belief, disbelief, ignorance, and conflict. But conversely, the linear opinion pooling might be viewed as a more precise and traditional approach, though again resting at the bottom on subjective probabilities.

While we are talking about opinion pooling, the astute reader will have noticed that our example of the two firefighters has *two* firefighters, and that the inconsistency between L and R is divided between them. This is why linear opinion pooling is so appropriate. And it easy to see how the example might be extended in various ways to more than two firefighters, and how linear opinion pooling might be applied to such examples. Linear opinion pooling is based on the idea of combining various individual views, each of which is consistent but the combination might well be inconsistent.

Dietrich and List (2016) give practical situations in which possibly inconsistent information from multiple sources arises:

> A panel of climate experts may have to aggregate the panelists' conflicting opinions into a compromise view, in order to deliver a report to the government. A jury may have to arrive at a collective opinion on the facts of a case, despite disagreements between the jurors, so that the court can reach a verdict. In Bayesian statistics, we may wish to specify some all-things-considered prior probabilities by aggregating the subjective prior probabilities of different statisticians. In meta-statistics, we may wish to aggregate the probability estimates that different statistical studies have produced for the same events. An individual agent may wish to combine his or her own opinions with those of another, so as to resolve any peer disagreements.

But how can linear opinion pooling be applied if there is just one firefighter? How can just a single firefighter give contradictory information in a way that might be useful?

Dietrich and List give a nice example related to this too: "Finally, in a purely intra-personal case, an agent may seek to reconcile different 'selves' by aggregating their conflicting opinions on the safety of mountaineering, in order to decide whether to undertake a mountain hike and which equipment to buy."

They speak of two "selves" in quotes for what is popularly called "being of two minds." One might imagine that the single firefighter says something like this. "I have been up and down each of the three hallways. I am somewhat disoriented because of the fire, but I am sure that I remember that there is only one way out. Oh, and I also remember that it is the left hall. No, I also remember that it is the right hall. I am of two minds."

We can similarly modify the lost keys description so it is the driver and not the partner who seems to remember seeing the keys are on the piano, even though they have figured out that the keys must be in the jacket pocket.

12.7 Big Data and the World Wide Web

"Big Data"[12] is a hot topic. Large data sets are by their nature likely to be inconsistent. Indeed, this is the world of statistics. One observation will show that a certain medical treatment worked, the next observation will show it did not work. This is not regarded by statisticians as contradictory information (but is regarded as not enough information). Data sets might be carefully assembled and structured, or they might be assembled with no oversight and unstructured. The World Wide Web (WWW) is a particularly big and largely unstructured data set containing information from various sources, not all of which are intelligent, wise, or beneficent. But nonetheless most of us find it extremely useful in our daily and our scientific lives.

Suppose you want to find an answer to a certain yes/no question on the WWW. Which of the following scenarios do you prefer?

(A) You google and get no response—at least none that is relevant, including the various "ads."

(B) You google and get multiple conflicting responses.

We think we have all found that it is often better to find contradictory information on a search topic rather than finding no information at all. It might be nice to find that there is at least active interest in the topic, but one might be able to do something to assess the accuracy of the various responses by appraising the credentials of the informants (including both authority and motives), counting their relative number, assessing their arguments, trying to reproduce their experimental results, discovering their authoritative sources, etc. Any or all of these might apply to a particular case. In terms of the probability approach, these considerations affect the subjective specification of the weights used to pool the information.

It is worth emphasizing that the utility of contradictions is due not just to their content but also to their pragmatic context. Some of the tools might be taken as

---

[12] Lohr (2013) argues that the term "Big Data" should be credited to John Mashey, the Chief Scientist at Silicon Graphics in the 1990s, when he gave a number of unpublished talks on the subject in the late 1990s—see, e.g., http://static.usenix.org/event/usenix99/invited_talks/mashey.pdf.

"logical fallacies." For example, counting the number of sources can be interpreted as an Argument from Repetition. From the probability viewpoint, this is a matter of statistical dependence—dependent data are typically less informative than independent observations. The aim here is not to prove assertion P by the number of sources that say it, but rather to determine the subjective likelihood that P holds. An improvement would be to check for duplications (one source merely repeating another source) or other types of statistical dependence (where one source may not repeat another but is influenced), and to have "trust" weightings based on reliability (expertise, honesty, lack of bias, track record, etc.).

Audun Jøsang, and many others, have written extensively and intelligently on the subject of trust. Jøsang (1997) connects this to his "Opinion Triangle" to deal with subjective probability that allows for ignorance in belief. We suggest that the discussion of trust should be extended to include conflict, as well as ignorance, and perhaps then that the Opinion Tetrahedron could be used in place of the Opinion Triangle, or that linear opinion pooling could be used.

We doubt that when Graham Priest first began to think about paraconsistency in the late 1970s he had any idea about developing it as a tool for Big Data and the Internet. Much of the current emphasis on Big Data has to do with putting together larger and larger data sets by pooling existing data sets. While this is basically a good idea, it is of course a good bet to lead to inconsistencies. This emphasizes the importance of contradictory information, paraconsistency, and pooling.

References

Alchourrón, C. E., Gärdenfors, P., & Makinson, D. (1985). On the logic of theory change: Partial meet functions for contraction and revision. *The Journal of Symbolic Logic, 50*, 510–30.
Bar-Hillel, Y., & Carnap, R. (1953). Semantic information. *The British Journal for the Philosophy of Science, 4*, 147–157.
Belnap, N. D. (1977a). A useful four-valued logic. In G. Epstein & J. M. Dunn (Eds.), *Modern uses of multiple-valued logic* (pp. 8–37). Dordrecht: D. Reidel Publishing Co.
Belnap, N. D. (1977b). How a computer should think. In G. Ryle (Ed.), *Contemporary aspects of philosophy* (pp. 30–55). Stockfield: Oriel Press Ltd.
D'Alfonso, S. (2016). Belief merging with the aim of truthlikeness. *Synthese, 193*, 2013–2034.
Dietrich, F., & List, C. (2016). Probabilistic opinion pooling. In C. Hitchcock & A. Hájek (Eds.), *Oxford handbook of probability and philosophy*. Oxford University Press. Print version 2016, online version 2017.
Dunn, J. M. (1976). Intuitive semantics for first degree entailments and coupled trees. *Philosophical Studies, 29*(149), 168.
Dunn, J. M. (2009). Contradictory information: Too much of a good thing. *Journal of Philosophical Logic, 39*, 425–452.
Floridi, L. (2011). *The philosophy of information*. Oxford University Press.
Genest, C., & Zidek, J. V. (1986). Combining probability distributions: A critique and an annotated bibliography. *Statistical Science, 1*, 114–135.

Hewitt, C. (2009). Inconsistency robustness in foundations: Mathematics self proves its own consistency and other matters. https://arxiv.org/ftp/arxiv/papers/0907/0907.3330.pdf.
Hewitt, C. (2015). Formalizing common sense reasoning for scalable inconsistency-robust information coordination using direct logic reasoning and the actor model. In C. Hewitt & J. Woods assisted by Jane Spurr (Eds.), *Inconsistency robustness*. London: College Publications.
Jaśkowski, S. (1948). Rachunek zdań dla systemów dedukcyjnch sprzecznych. *Studia Societatis Scientiarum Torunensis*, I, Toruń 1948, 57–77. Reprinted in English as "A Propositional Calculus for Inconsistent Deductive Systems," in *Studia Logica*, 24, 1969, 143–157 and in: *Logic and Logical Philosophy*, 7, 1999, 35–56.
Jøsang, A. (1997). Artificial reasoning with subjective logic. In *Proceedings of the Second Australian Workshop on Commonsense Reasoning*, Perth.
Kiefer, N. M. (2007). The probability approach to default probabilities. *Risk*, 146–150.
Kiefer, N. M. (2011). Default estimation, correlated defaults and expert information. *Journal of Applied Econometrics*, 26, 173–192.
Konieczny, S., & Pérez, R. P. (2011). Logic based merging. *Journal of Philosophical Logic*, 40, 239–270.
Lorh, S. (2013). The origins of 'Big Data': An etymological detective story. *The New York Times*, Feb. 1, 2013. https://bits.blogs.nytimes.com/2013/02/01/the-origins-of-big-data-an-etymological-detective-story/?_r=0.
Mares, E. (2002). A paraconsistent theory of belief revision. *Erkenntnis*, 56, 229–246.
Meyer, R. K., Routley, R., & Dunn, J. M. (1979). Curry's paradox. *Analysis*, 39, 124–128.
Popper, K. R. (1934). *Logik Der Forschung: Zur Erkenntnistheorie Der Modernen Naturwissenschaft*. Vienna: J. Springer. English translation (1959) *The Logic of Scientific Discovery*. London: Hutchinson.
Priest, G. (1979). Logic of paradox. *Journal of Philosophical Logic*, 8, 219–241.
Priest, G., Routley, R., & Norman, J. (Eds.). (1989). *Paraconsistent logic: Essays on the inconsistent*. Munich: Philosophia Verlag GmbH.
Priest, G., Tanaka, K., & Weber, Z. (2016, Winter). Paraconsistent logic. In E. Zalta (Ed.) *The stanford encyclopedia of philosophy*. https://plato.stanford.edu/archives/win2016/entries/logic-paraconsistent/.
Stone, M. (1961). The opinion pool. *Annals of Mathematical Statistics*, 32, 1339–1342.
Tanaka, K. (2005). The AGM theory and inconsistent belief change. *Logique et Analyse*, 48, 113–150.

J. Michael Dunn has done his research which included information-based logics, algebraic logic, proof theory, nonstandard logics (especially relevance logic), quantum computation/logic, and relations between logic and computer science. He has an A.B. in Philosophy, Oberlin College (1963), and a Ph.D. in Philosophy (Logic), University of Pittsburgh (1966). His career started at Wayne State University, and he retired from Indiana University Bloomington in 2007 after 38 years, where he is Emeritus Oscar Ewing Professor of Philosophy, Emeritus Professor of Computer Science and Informatics, and was founding Dean of the School of Informatics. He is the author of 5 books and over 100 articles. He has been an editor of *The Journal of Symbolic Logic* and chief editor of *The Journal of Philosophical Logic*. He has received many awards and is a Fellow of the American Academy of Arts and Sciences.

Nicholas M. Kiefer is the Ta-Chung Liu Professor at Cornell University, where he is a member of the departments of economics and statistical science. He is widely known for his theoretical and applied contributions in modeling of duration data, the development of dynamic models under uncertainty learning and the information, and financial market microstructure. Kiefer's current research includes applications in financial economics, risk management, AML policies, and fair lending. This work develops methods for quantifying expert information and combining this with data information in decision-making settings. The unifying theme of his work is the combined use of statistics and economic theory. Kiefer is an internationally recognized expert, having published over 100 journal articles, books, and reviews. He is a Fellow of the Econometric Society and a recipient of the John Simon Guggenheim Memorial Fellowship. His point of view is reflected in the book with Christensen *Economic Modeling and Inference*.

Chapter 13
Variations on the Collapsing Lemma

Thomas Macaulay Ferguson

Abstract Graham Priest has frequently employed a construction in which a classical first-order model $\mathfrak{A}$ may be *collapsed* into a three-valued model $\mathfrak{A}^\sim$ suitable for interpretations in Priest's *logic of paradox* (LP). The source of this construction's utility is Priest's *Collapsing Lemma*, which guarantees that a formula true in the model $\mathfrak{A}$ will continue to be true in $\mathfrak{A}^\sim$ (although the formula may *also* be false in $\mathfrak{A}^\sim$). In light of the utility and elegance of the Collapsing Lemma, extending variations of the lemma to other deductive calculi becomes very attractive. The aim of this paper is to map out some of the frontiers of the Collapsing Lemma by describing the types of expansions or revisions to LP for which the Collapsing Lemma continues to hold and a number of cases in which the lemma cannot be salvaged. Among what is shown is that the lemma holds for a strictly more expressive form of LP including nullary truth and falsity constants, that any conditional connective that can be added to LP without inhibiting the lemma must be theoremhood-preserving, and that the Collapsing Lemma extends to the *paraconsistent weak Kleene* logic PWK as well.

Keywords Collapsing Lemma · *Logic of paradox* · Model theory

13.1 Introduction

Since its introduction in Priest (1991), Graham Priest has frequently appealed to his *Collapsing Lemma* as a method of constructing inconsistent, first-order models while retaining some control over their theories. The lemma guarantees that the formulae true in a classical model will continue to be designated in a certain type of quotient of that model when the formulae are evaluated by the lights of the calculus Priest calls the *logic of paradox* LP, a system first introduced by Asenjo in papers such as Asenjo (1966) and Asenjo and Tamburino (1975).[1]

---

[1]I appreciate the input of a referee reminding me to also credit Asenjo here.

T. M. Ferguson (✉)
City University of New York, New York City, USA
e-mail: tferguson@gradcenter.cuny.edu

© Springer Nature Switzerland AG 2019
C. Başkent and T. M. Ferguson (eds.), *Graham Priest on Dialetheism and Paraconsistency*, Outstanding Contributions to Logic 18, https://doi.org/10.1007/978-3-030-25365-3_13

The lemma first appears in Priest (1991), in which Priest introduces the logic LP$_m$—the *minimal logic of paradox*—and studies its properties. In that paper, Priest investigates whether or not various versions of LP$_m$ (propositional, first-order) enjoy the property of *Reassurance*, i.e., the property that if the closure of a set of sentences Γ under LP consequence is nontrivial then its closure under LP$_m$ is also nontrivial. In particular, the Collapsing Lemma is introduced as an intermediate step in proving that Reassurance holds for first-order LP$_m$ for languages with finite signatures. Independently of its use in proving Reassurance, there are a number of ways in which the Collapsing Lemma has proven interesting.

For one, it has been a versatile tool for the construction of inconsistent yet nontrivial models extending classical theories. Beginning with Priest (1994), Priest applied the lemma in order to construct inconsistent and decidable models of true arithmetic including a greatest natural number; this project was more formally investigated in the papers, Priest (1997) and (2000). In Priest (2012), the lemma was brought to bear on Zermelo–Frankel set theory to produce models of ZF satisfying the what Priest labels the "Axiom of Countability," i.e., the formal statement that all sets are countable.

The lemma is also notable for the generality of its underlying construction. The technique Priest uses to build inconsistent models is closely related to a similar method to construct K_3 (i.e., strong three-valued Kleene logic) models from classical models due to Dunn (1979). Dunn's *Theorem in Three-Valued Model Theory* is dual to Priest's result as nothing false in the classical model becomes true in the quotient, a feature studied closely in Ferguson (2012). Further cousins to the Collapsing Lemma—many of which tie into Priest's project of *plurivalent logics* of Priest (2014)—have been introduced in Ferguson (2017).

Given the utility and versatility of Priest's Collapsing Lemma, it is reasonable to wonder what enrichments to—or variations on—the three-valued semantics for LP admit the Collapsing Lemma. It is well-known that the interpretation of the conditional connective $\rightarrow$ in LP is not *detachable*, that is, the inference of *modus ponens* is not admissible in LP. For this limitation, enriching the three-valued interpretation of LP by other connectives has been a subject of interest. Despite the existence of analogous constructions, the Collapsing Lemma is far from trivial; it fails, for example, in the system RM$_3$, which augments LP's interpretations of negation, conjunction, and disjunction with a detachable implication connective. One of the goals in this study is to consider such variations on the Collapsing Lemma in more detail.

13.1.1 Formal Preliminaries

We will review some preliminary definitions from many-valued model theory (most of which is a variant in one form or another of the presentations of Ferguson 2012, 2014, 2017).

The basic syntactic unit that determines a language is the notion of a *signature*, i.e., a collection of extralogical symbols that are interpreted as constants, function symbols, or relations.

Definition 13.1 A *signature* σ is a collection $\langle \mathbf{C}, \mathbf{F}, \mathbf{R} \rangle$ whose members are sets of *constants*, *function symbols*, and *relation symbols*, respectively.

The sets of closed and open terms are defined as usual. For example, the collection of closed terms of σ—call it $\mathbf{CTm}_\sigma$—is the closure of the set $\mathbf{C}$ under applications of the relation symbols in $\mathbf{F}$ and the collection of its open terms—call it $\mathbf{Tm}_\sigma$—is the closure of $\mathbf{C}$ and a set of variables $\mathbf{Var}$ under $\mathbf{F}$.

Each signature σ determines a *first-order language* $\mathscr{L}_\sigma$ which includes a binary equality relation and the connectives and quantifiers in first-order presentations of LP. Let $\mathbf{At}_\sigma$ represent the set of *atomic formulae* of σ, i.e., the standard collection of formulae of the form $t_0 = t_1$ or $R(t_0, \ldots, t_{n-1})$ for each n-ary symbol $R \in \mathbf{R}$ and terms $t_0, \ldots, t_{n-1} \in \mathbf{Tm}_\sigma$. Then we apply the standard recursive definition of a first-order language with the usual connectives and quantifiers.

Definition 13.2 For a signature σ, the language $\mathscr{L}_\sigma$ is defined in Backus–Naur form, where $\varphi \in \mathbf{At}$, $t_0, t_1 \in \mathbf{Tm}_\sigma$, and $x \in \mathbf{Var}$:

$$\psi ::= \varphi \mid t_0 = t_1 \mid \neg\psi \mid \psi \wedge \psi \mid \psi \vee \psi \mid \psi \to \psi \mid \forall x \psi(x) \mid \exists x \psi(x)$$

N.b. that in the sequel we will discuss *expansions* of LP in which the collection of usual logical connectives is enriched with novel operators. Properly speaking, the introduction of, e.g., a unary connective $\diamond$ should require a new definition of a particular language. Because the amendments to these definitions would be in such cases trivial, we will trust that context will communicate to the reader any required alterations to the definition of a language.

With that having been said, with the syntactic notion of a language in hand, it remains to discuss how to *interpret* such a language. The primary ingredient in such interpretations is the notion of a first-order logical matrix, a set of interpretations of logical connectives and quantifiers that will induce some consequence relation. Although we are concerned primarily with LP, in order to consider expansions with new connectives, we must be somewhat general.

Definition 13.3 A *first-order logical matrix* is a tuple
$\langle \mathcal{V}, \mathcal{D}, f^{\circ_0}, \ldots, f^{\circ_{j-1}}, f^{Q_0}, \ldots, f^{Q_{k-1}} \rangle$ in which

- $\mathcal{V}$ is a non-empty set of truth values
- $\mathcal{D} \subseteq \mathcal{V}$ is a non-empty set of designated truth values
- each n-ary connective $\circ$ has an interpretation $f^{\circ_i} : \mathcal{V}^n \to \mathcal{V}$
- each type $\langle 1 \rangle$ quantifier Q has an interpretation $f^{Q_i} : \wp^+(\mathcal{V}) \to \mathcal{V}$.

where $\wp^+(X)$ is the set of non-empty subsets of X.

Now, we may define the particular three-valued matrix that is characteristic of first-order LP.

Definition 13.4 The matrix $\mathfrak{M}_{\mathsf{LP}}$ is the matrix

$$\langle \mathcal{V}, \mathcal{D}, f_{\mathsf{LP}}^{\rightarrow}, f_{\mathsf{LP}}^{\wedge}, f_{\mathsf{LP}}^{\vee}, f_{\mathsf{LP}}^{\rightarrow}, f_{\mathsf{LP}}^{\forall}, f_{\mathsf{LP}}^{\exists} \rangle$$

in which $\mathcal{V} = \{\mathfrak{t}, \mathfrak{b}, \mathfrak{f}\}$ and $\mathcal{D} = \{\mathfrak{t}, \mathfrak{b}\}$ and the truth functions are defined in the usual way from the primitive truth functions:

$f_{\mathsf{LP}}^{\rightarrow}$	
$\mathfrak{t}$	$\mathfrak{f}$
$\mathfrak{b}$	$\mathfrak{b}$
$\mathfrak{f}$	$\mathfrak{t}$

$f_{\mathsf{LP}}^{\wedge}$	$\mathfrak{t}$	$\mathfrak{b}$	$\mathfrak{f}$
$\mathfrak{t}$	$\mathfrak{t}$	$\mathfrak{b}$	$\mathfrak{f}$
$\mathfrak{b}$	$\mathfrak{b}$	$\mathfrak{b}$	$\mathfrak{f}$
$\mathfrak{f}$	$\mathfrak{f}$	$\mathfrak{f}$	$\mathfrak{f}$

$$f_{\mathsf{LP}}^{\vee}(X) = \begin{cases} \mathfrak{t} & \text{if } X = \{\mathfrak{t}\} \\ \mathfrak{b} & \text{if } \mathfrak{b} \in X \text{ and } \mathfrak{f} \notin X \\ \mathfrak{f} & \text{if } \mathfrak{f} \in X \end{cases}$$

Clearly (and crucially), the two-valued classical logic can be obtained as a restriction of the set of truth values in $\mathfrak{M}_{\mathsf{LP}}$ to the set of classical values, i.e., $\{\mathfrak{t}, \mathfrak{f}\}$.

Further ingredients are necessary before we can define consequence in classical logic or LP, so we introduce the notion of a *three-valued structure* and *valuations* upon such structures.

Definition 13.5 A *glutty three-valued structure* $\mathfrak{A}$ is a structure $\langle A, \mathbf{C}^{\mathfrak{A}}, \mathbf{F}^{\mathfrak{A}}, \mathbf{R}^{\mathfrak{A}} \rangle$ where:

- for each $c \in \mathbf{C}$ is an interpretation $c^{\mathfrak{A}} \in A$
- for each m-ary $f \in \mathbf{F}$ there is a function $f^{\mathfrak{A}} : A^n \to A$
- for each n-ary $R \in \mathbf{R}$ is a function $\mathbf{R}^{\mathfrak{A}} : A^n \to \mathcal{V}$
- there is an identity function $f^{=} : A^2 \to \mathcal{V}$ such that $f^{=}(x, x) \in \mathcal{D}$.

Classical models can be identified with the class of all glutty three-valued models that are *consistent* in the sense that $\mathfrak{b}$ is an element of neither the range of $f^{=}$ nor of any $R^{\mathfrak{A}}$.

One final definition must be given before defining classical and LP valuations. Traditionally, we call a structure a *Henkin model* when for any element $a \in A$ there is a constant $\underline{a} \in \mathbf{C}$ such that $\underline{a}^{\mathfrak{A}} = a$. In Priest (1979) or (1991), Priest makes the assumption that all structures are Henkin in this sense, thereby winning a significant decrease in the complexity of definitions. Although we lose some generality by this assumption, the general case is easy to recover, and in the sequel, we follow Priest in this assumption.

Definition 13.6 Let $\mathfrak{M}$ be a matrix sharing $\mathcal{V}$ and $\mathcal{D}$ with $\mathfrak{M}_{\mathsf{LP}}$. Then a glutty three-valued structure $\mathfrak{A}$ with signature σ induces an $\mathfrak{M}$-*interpretation* $v_{\mathfrak{A}}$ that assigns each closed formula of the language $\mathscr{L}_{\sigma}$ a truth value. Where any term t is assumed to be closed, we have the following definition:

- $v_{\mathfrak{A}}(t_0 = t_1) = f^{=}(t_0, t_1)$
- $v_{\mathfrak{A}}(R(\vec{t})) = R^{\mathfrak{A}}(\vec{t}^{\mathfrak{A}})$
- $v_{\mathfrak{A}}(\star \varphi) = f^{\star}(v_{\mathfrak{A}}(\varphi))$ for unary connectives $\star$
- $v_{\mathfrak{A}}(\varphi * \psi) = f^{*}(v_{\mathfrak{A}}(\varphi), v_{\mathfrak{A}}(\psi))$ for binary connectives $*$
- $v_{\mathfrak{A}}(Qx\varphi(x)) = f^{Q}(\{v_{\mathfrak{A}}(\varphi(\underline{a}))) \mid a \in A\})$ for quantifiers Q

We will say that $\mathfrak{A}, v_{\mathfrak{A}} \vDash \varphi$ when a closed formula φ is assigned a designated value by $v_{\mathfrak{A}}$.

Finally, we have enough material to define consequence for first-order LP.

Definition 13.7 For closed formulae of a language $\mathscr{L}_\sigma$, $\Gamma \vDash_{\mathsf{LP}} \varphi$ if for every glutty three-valued structure $\mathfrak{A}$, $\mathfrak{A}, v_{\mathfrak{A}} \vDash \varphi$ whenever $\mathfrak{A}, v_{\mathfrak{A}} \vDash \psi$ for each $\psi \in \Gamma$.

By considering classical structures and classical logic itself to be the restrictions of glutty models and LP to classical truth values, we have classical consequence as a special case of LP consequence restricted to consistent models.

Definition 13.8 For closed formulae of a language $\mathscr{L}_\sigma$, $\Gamma \vDash_{\mathsf{CL}} \varphi$ if for every consistent structure $\mathfrak{A}$, $\mathfrak{A}, v_{\mathfrak{A}} \vDash \varphi$ whenever $\mathfrak{A}, v_{\mathfrak{A}} \vDash \psi$ for each $\psi \in \Gamma$.

This provides the basic model theory necessary to discuss the Collapsing Lemma.

13.1.2 The Collapsing Lemma

The statement of Priest's Collapsing Lemma relies on a construction through which a particular quotient of the domain of a classical structure $\mathfrak{A}$ determines a corresponding glutty structure $\mathfrak{A}^\sim$. In virtue of the generality of this construction, the ensuing quotient is called a *Priest collapse* in Ferguson (2017).

On a two-valued structure $\mathfrak{A}$, a *congruence relation* is an equivalence relation $\sim$ on the domain A that respects the behavior of functions. In other words, for any $f^\mathfrak{A}$—an n-ary interpretation of the function symbol f—and n-tuples $\langle a_0, \ldots, a_{n-1} \rangle$ and $\langle a'_0, \ldots, a'_{n-1} \rangle$, if $a_i \sim a'_i$ for all $i < n$, then $f^\mathfrak{A}(a_0, \ldots, a_{n-1}) \sim f^\mathfrak{A}(a'_0, \ldots, a'_{n-1})$. Where $\sim$ is such a congruence relation on the domain of a structure, we may describe its Priest collapse modulo $\sim$.

In the sequel, if $[\![a]\!]$ is the equivalence class of an element a under $\sim$, for a tuple $\vec{a} = \langle a_0, \ldots, a_{n-1} \rangle$, we will use the the notation $[\![\vec{a}]\!]$ to denote the tuple $\langle [\![a_0]\!], \ldots, [\![a_{n-1}]\!] \rangle$. Furthermore, we extend the notion of membership so that $\vec{a}' \in [\![\vec{a}]\!]$ if for each a'_i in $\vec{a}'$, $a'_i \in [\![a_i]\!]$.

Definition 13.9 Let $\sim$ be a congruence relation on the domain of a structure $\mathfrak{A}$ and let $[\![a]\!]$ represent the equivalence class of a under $\sim$. Then $\mathfrak{A}^\sim$—the *Priest collapse* of $\mathfrak{A}$ with respect to $\sim$—is defined as follows:

- $A^\sim = \{[\![a]\!] \mid a \in A\}$
- for a constant c, $c^{\mathfrak{A}^\sim} = [\![c^\mathfrak{A}]\!]$
- for a function f, $f^{\mathfrak{A}^\sim}([\![a_0]\!], \ldots, [\![a_{n-1}]\!]) = [\![f^\mathfrak{A}(a_0, \ldots, a_{n-1})]\!]$.

For the interpretation of relation symbols and identity, we appeal to the following schema:

- $P^{\mathfrak{A}^\sim}(\llbracket \vec{a} \rrbracket) = \begin{cases} \mathfrak{t} & \text{if for all } \vec{a}' \in \llbracket \vec{a} \rrbracket, P^{\mathfrak{A}}(\vec{a}') = \mathfrak{t} \\ \mathfrak{f} & \text{if for all } \vec{a}' \in \llbracket \vec{a} \rrbracket, P^{\mathfrak{A}}(\vec{a}') = \mathfrak{f} \\ \mathfrak{b} & \text{otherwise} \end{cases}$

In other words, the collapse associates each element with its equivalence class under $\sim$ and takes an atom $P(\llbracket c \rrbracket)$ to be designated precisely when the initial classical model thought the formula $P(c')$ was true for *some* c' such that $c' \sim c$.

The Collapsing Lemma then describes the preservation of truth from the interpretation of a classical structure to the LP-interpretation of its Priest collapse.

Lemma 13.1 (Priest) *For any classical structure $\mathfrak{A}$ with interpretation $v_{\mathfrak{A}}$ and Priest collapse $\mathfrak{A}^\sim$ with LP interpretation $v_{\mathfrak{A}^\sim}$:*

$$\text{if } \mathfrak{A}, v_{\mathfrak{A}} \vDash \varphi \text{ then } \mathfrak{A}^\sim, v_{\mathfrak{A}^\sim} \vDash \varphi$$

A problem inherent to any variation of the Collapsing Lemma is how to determine when a three-valued generalization of a classical connective is an "authentic" version of that connective. Recalling Carnielli et al. (2000), we will call a function *hyperclassical* if it meets the following definition.

Definition 13.10 An n-ary truth function f on $\mathcal{V}_{\mathsf{LP}}$ is *hyperclassical* if $f[\{\mathfrak{t},\mathfrak{f}\}^n] \subseteq \{\mathfrak{t},\mathfrak{f}\}$, i.e., if f maps n-tuples of classical values to classical values. A logical matrix is hyperclassical if all its truth functions are hyperclassical.

Following this definition, we identify any hyperclassical truth function as an instance of the same type of the classical truth function that it extends, irrespective of any other properties of the three-valued truth function. While there are obviously compelling criteria for what makes, e.g., a three-valued operator an "authentic conditional," we follow Carnielli in treating hyperclassicality as a property strong enough to establish a family resemblance. Hence, that $f_{\mathsf{LP}}^{\rightarrow}$ restricted to the classical truth values coincides with classical implication will be construed as sufficient grounds to recognize LP's interpretation of "$\rightarrow$" as an adequate generalization of the classical conditional, despite the failure of *modus ponens* or *modus tollens* in LP.

13.2 Collapsing and Unary Connectives

Since the introduction of LP, a subject of frequent investigation has been the enrichment of its language by novel, unary connectives. This is a very natural thing to study; for example, we may want to supplement the negation of LP with an additional negation-type operator, or operators corresponding to other semantic properties. We will consider a system to be an *expansion* of LP if it enriches the language of LP with a new syntactic element—a connective or quantifier, for example—that has a corresponding three-valued interpretation consonant with the matrix semantics for

LP. Two such expansions might be redundant yet distinct from one another if, for example, two novel connectives are interdefinable between the two systems.

If we consider only hyperclassical connectives, we are able to ensure that talk of the Collapsing Lemma remains meaningful in such expansions of LP. Suppose, for example, that some enrichment makes use of a novel unary connective ⋄ not found in the standard formulation of the classical language. By the functional completeness of classical logic, we are able to introduce a defined version of ⋄ and then may ask—in this *expanded* language—whether the Collapsing Lemma holds. (As we had suggested earlier, in cases in which the logical connectives of LP are enriched by new ones, we implicitly make appropriate adjustments to the definitions of a language or a model, as the case requires.)

Before we get to the matter of collapsing, we are able to conveniently whittle down the list of candidate unary operations for which the Collapsing Lemma may hold. Many of these candidates are disqualified by considering only those unary truth functions that are hyperclassical with respect to *some* classical truth function.

Observation 13.1 *There are twelve unary truth functions on V_{LP} that are hyperclassical.*

Proof There are four unary classical truth functions and each determines three classically appropriate three-valued truth functions by selecting a value for $\mathfrak{b}$. □

We can make some initial observations concerning the types of unary connectives that may be added to LP while preserving the Collapsing Lemma.

Lemma 13.2 *Let ⋄ be a unary propositional connective extending the classical language. Then there exist exactly six interpretations of ⋄ on V_{LP} that may be added to $\mathfrak{M}_{\mathsf{LP}}$ for which the Collapsing Lemma holds in the extended language (including a redundant $f_{\mathsf{LP}}^{\vec{\diamond}}$):*

	$f^{\diamond_0}$	$f^{\diamond_1}$	$f^{\diamond_2}$	$f_{\mathsf{LP}}^{\vec{\diamond}}$	$f^{\diamond_3}$	$f^{\diamond_4}$
$\mathfrak{t}$	$\mathfrak{t}$	$\mathfrak{t}$	$\mathfrak{t}$	$\mathfrak{f}$	$\mathfrak{f}$	$\mathfrak{f}$
$\mathfrak{b}$	$\mathfrak{t}$	$\mathfrak{b}$	$\mathfrak{b}$	$\mathfrak{b}$	$\mathfrak{b}$	$\mathfrak{f}$
$\mathfrak{f}$	$\mathfrak{t}$	$\mathfrak{t}$	$\mathfrak{f}$	$\mathfrak{t}$	$\mathfrak{f}$	$\mathfrak{f}$

Proof First, we establish that the above six interpretations do in fact preserve the Collapsing Lemma. Most cases are simple. For interpretations $f^{\diamond_0}$ and $f^{\diamond_1}$, we note that they return a designated value for all arguments and hence $\diamond \varphi$ will trivially be designated in a collapsed model. Similarly, because $f^{\diamond_3}$ and $f^{\diamond_4}$ return $\mathfrak{f}$ for all classical arguments, they ensure that $\diamond \varphi$ will never be designated in a classical model to begin with, meaning that $\diamond \varphi$ cannot be a counterexample to the lemma in these cases. Of course, the lemma holds for $f_{\mathsf{LP}}^{\vec{\diamond}}$ already, leaving us to examine only $f^{\diamond_2}$.

In this case, we add a clause to the Collapsing Lemma's proof on complexity of formulae. Suppose that a classical model $\mathfrak{A}$ assigns $\diamond \varphi$ the value $\mathfrak{t}$. This holds only if $\mathfrak{A}$ assigns φ the value $\mathfrak{t}$. Assuming the lemma holds for φ, then $\mathfrak{A}^{\sim}$ assigns φ either $\mathfrak{t}$ or $\mathfrak{b}$. In each case, $f^{\diamond_2}$ will assign $\diamond \varphi$ a designated value, ensuring that the inclusion of this interpretation preserves the Collapsing Lemma.

Counterexamples to the six hyperclassical unary truth functions not appearing in the above table can be calculated without difficulty. Consider, for example, an interpretation for $\diamond$ adhering to the following truth table:

	$f^\diamond$
t	t
b	t
f	f

Let $\mathfrak{A}$ be a classical structure with elements a and b for which $a \sim b$ and a unary predicate P such that $P^{\mathfrak{A}}(a) = \mathfrak{t}$ and $P^{\mathfrak{A}}(b) = \mathfrak{f}$. Then, in the classical structure the value assigned to $\neg \diamond P(\underline{b})$ will be $\mathfrak{t}$, a designated value. However, because $P(\underline{b})$ is assigned the value of $\mathfrak{b}$ in the collapsed model $\mathfrak{A}^\sim$, $\diamond P(\underline{b})$ will be evaluated as $\mathfrak{t}$ and $\neg \diamond P(\underline{b})$ will be evaluated as $\mathfrak{f}$. Hence, the Collapsing Lemma will fail if such a connective is added to **LP**.

The remaining five cases follow from very similar arguments. □

Lemma 13.2 will provide a great deal of utility in the sequel, as we proceed to look at particular interpretations of unary connectives.

13.2.1 Expansions of **LP** with Consistency-Type Connectives

The investigations into a unary operator corresponding to a notion of *classicality* or *consistency* spearheaded by da Costa (1974) have led to the formulation of several expansions of **LP** (or three-valued logics definitionally equivalent to such expansions). Such a connective has served as a cornerstone of the broader project of *logics of formal inconsistency*—deductive systems that, while paraconsistent, can use a consistency connective that serves to isolate and mark situations in which a formula behaves classically. Variations of the interpretation of additional unary connectives are also important insofar as they determine similar families of deductive systems; a unary *undeterminedness* operator, for example, corresponds to the family of *logics of formal undeterminedness* introduced by Marcos (2005).

One such three-valued hyperclassical connective is the *inconsistency* connective • studied by Carnielli et al. (2000). Carnielli et al. introduce a logic of formal inconsistency **LFI1** in the context of databases, so that a formula $\bullet \varphi$ indicates that φ appears in a database along with its own negation. The interpretation of the inconsistency connective is as follows.

Definition 13.11 The truth function $f^\bullet$ evaluating the inconsistency connective • is defined:

	$f^\bullet$
t	f
b	t
f	f

13 Variations on the Collapsing Lemma

While the three-valued interpretation LFI1 gives to "→" differs from that of LP, the interpretations of negation, conjunction, and disjunction coincide. However, it is implicit in the results of Carnielli et al. (2000)—and explicitly stated by Omori (2015)—that the implication of LFI1 can be defined from the connectives of LP with the aid of the inconsistency connective. Hence, inference in LFI1 is interdefinable with an expansion of LP including the inconsistency connective. Consider languages that extend those of LP by •. Then we define LP• as follows.

Definition 13.12 LP• is the logic induced by the matrix $\mathfrak{M}_{\mathsf{LP}\bullet}$:

$$\langle \mathcal{V}, \mathcal{D}, f_{\mathsf{LP}}^{\neg}, f^{\bullet}, f_{\mathsf{LP}}^{\wedge}, f_{\mathsf{LP}}^{\vee}, f_{\mathsf{LP}}^{\rightarrow}, f_{\mathsf{LP}}^{\forall}, f_{\mathsf{LP}}^{\exists} \rangle$$

We have already provided ourselves with sufficient tools to prove the following.

Lemma 13.3 *The Collapsing Lemma fails for* LP•.

Proof The truth function $f^{\bullet}$ appearing in Definition 13.11 does not appear among the unary truth functions found in Lemma 13.2. Hence, the addition of this interpretation of • to LP requires that the Collapsing Lemma fails. □

Due to the interdefinability of LFI1 and LP•, we also immediately infer the following.

Corollary 13.1 *The Collapsing Lemma fails for* LFI1.

The consistency/classicality property described by da Costa is not the only semantical property for which a three-valued, unary connective has been proposed. A further unary connective in the paraconsistent tradition is the *possibility* connective $\triangle$ corresponding to the following interpretation:

Definition 13.13 The truth function $f^{\triangle}$ evaluating the possibility connective $\triangle$ is defined:

	$f^{\triangle}$
t	t
b	t
f	f

In 2014, Sano and Omori have investigated the inclusion of this connective in the case of the four-valued logic $\mathsf{E}_{\mathsf{fde}}$, defining the logic BD$\triangle$ as its enrichment. Moreover, this investigation introduces as a special case the logic LP$\triangle$, i.e., the enrichment of LP by $\triangle$. In this special case, Sano and Omori identify this connective both with the well-known "Baaz Delta" of Baaz (1996)—whence their choice of notation—as well as the possibility connective of Itala D'Ottaviano's system J_3 described in, e.g., D'Ottaviano (1985).[2]

Formally, Sano and Omori (2014) introduces the logic LP$\triangle$ semantically by the following.

---

[2]Note that D'Ottaviano uses "∇" for the "possibility connective" and reserves "$\triangle$" for another use, defining it as ¬∇¬.

Definition 13.14 Let LPΔ be the logic induced by the matrix $\mathfrak{M}_{\mathsf{LP}\Delta}$:

$$\langle \mathcal{V}, \mathcal{D}, f_{\mathsf{LP}}^{\neg}, f^{\Delta}, f_{\mathsf{LP}}^{\wedge}, f_{\mathsf{LP}}^{\vee}, f_{\mathsf{LP}}^{\rightarrow}, f_{\mathsf{LP}}^{\forall}, f_{\mathsf{LP}}^{\exists} \rangle$$

Then, we are able to show that the following lemma serves to adumbrate a further boundary to the extent of the Collapsing Lemma.

Corollary 13.2 *The Collapsing Lemma fails for* J_3 *and* LPΔ.

Proof This follows from Lemma 13.3 due to the interdefinability of J_3 and LFI1 observed in Carnielli et al. (2000). □

These enriched variations of LP share a common property in that they are logics of formal inconsistency, suggesting that there is a difficulty reconciling the Collapsing Lemma with a level of expressivity sufficient to adequately capture a robust notion of consistency. We are capable of making a more general point, in fact, concerning truth-functional expansions of LP.

First, we define the notion of a logic of formal inconsistency in a more rigorous fashion. As described in the handbook article (Carnielli et al. 2007), a deductive system with a consequence relation $\vDash$ is identified as a logic of formal inconsistency when it meets the following criteria.

Definition 13.15 A deductive system is an LFI with respect to a negation $\neg$ if:

a there are Γ, φ, ψ such that $\Gamma, \varphi, \neg\varphi \nvDash \psi$, and
b there is a set of formulae $\bigcirc(p)$ depending only on p such that:

- there are φ, ψ such that $\bigcirc(\varphi), \varphi \nvDash \psi$ and $\bigcirc(\varphi), \neg\varphi \nvDash \psi$
- for all Γ, φ, ψ, we have: $\Gamma, \bigcirc(\varphi), \varphi, \neg\varphi \vDash \psi$

Now, if LFIs are characterized as a class of paraconsistent logics that satisfy some threshold of expressiveness, we find that the project of expanding the expressivity of LP past this threshold runs into the limitations of the Collapsing Lemma.

Corollary 13.3 *The Collapsing Lemma fails for any expansion of* LP *that is a logic of formal inconsistency.*

Proof Suppose that there exists such a set of formulae $\bigcirc(p)$ that qualifies the expansion as an LFI. Since this set depends on only parameter p the conjunction $\bigwedge \bigcirc(p)$ corresponds to a hyperclassical truth function $f^{\bigcirc} : \mathcal{V} \to \mathcal{V}$. Consequently, if the Collapsing Lemma is to hold, $f^{\bigcirc}$ must be coextensional with one of the six truth functions described in Lemma 13.2. In the case in which $f^{\bigcirc}$ is identified with $f^{\diamond_4}$ (described in Lemma 13.2), $\bigcirc(\varphi)$ behaves as a falsity constant, whence for all formulae φ the inference $\bigcirc(\varphi), \varphi \vDash \psi$ holds, contrary to the definition of an LFI. In all other cases, any model in which sentences $R(s)$ and $R(t)$ are assigned values $\mathfrak{b}$ and $\mathfrak{f}$, respectively, witnesses that $R(s), \neg R(s), \bigcirc(R(s)) \nvDash R(t)$. The failure of this inference means that the system in question is not *gently explosive*, in violation of the definition of an LFI. □

13.2.2 Adding Novel Negations

In an effort to increase their expressiveness, the languages of LP and equivalent three-valued systems have frequently been enriched with a further *classical* or *exclusion* negation operator for which the principle of explosion holds. The most common strategy for developing such a negation is to conflate the values of t and b and allowing both to stand in for truth *simpliciter*. Two interpretations of three-valued paraconsistent negations that conform to this strategy have been introduced in the literature. Borrowing De and Omori's notation of (2015), these are the following.

Definition 13.16 The truth functions for the classical negations $\neg_1$ and $\neg_2$ are:

	$f^{\neg_1}$	$f^{\neg_2}$
t	f	f
b	f	f
f	t	b

The interpretation called "$\neg_1$" by De and Omori is frequently encountered, as it can be recognized as the strong negation "$\sim$" by Carnielli et al. (2000) or as the notation "$\neg_*$" by D'Ottaviano (1985).

When added to LP, these negations are "classical" in the sense that their inclusion allows LP to semantically express the fact that a formula is *false and only false*. For example, the strategy is successful to the extent that replacing the interpretation of negation in $\mathfrak{M}_{\mathsf{LP}}$ with $f^{\neg_1}$ induces the classical consequence relation.

As before, we are met by an explicit boundary restricting the scope of the Collapsing Lemma.

Corollary 13.4 *The Collapsing Lemma fails for any expansion of* LP *including a classical negation.*

Proof By consulting Lemma 13.2 anew, we recognize that the only hyperclassical negation for which the Collapsing Lemma holds is the one LP already assigns to $\neg$. □

So far, we have only surveyed *negative* facts concerning the Collapsing Lemma, that is, that expanding LP to the extent that the expansion can express certain classical properties of falsity necessarily causes the lemma to fail. One might thereby be led to think that the lemma is incompatible with *any* increase in the expressivity of LP, or, at least, increases that permit a semantically correct characterization of falsity. Surprisingly, this is not in fact the case, leading us to a couple of *positive* facts concerning the Collapsing Lemma and expansions of LP.

In each of the foregoing cases, we have expanded the expressive power of LP by a unary connective such as a negation, consistency connective, or possibility connective. In each case, the additional expressivity allows LP to speak about a notion of "only true" or "only false," and it is tempting to attribute the failure of the Collapsing Lemma—at least in part—to the ability of the enriched system to express such exclusionary concepts.

In light of that, it is interesting to note that according LP *can be* enriched with top and bottom particles—i.e., nullary *verum* and *falsum* constants—while preserving the Collapsing Lemma. Thus, there are expansions of LP for which the Collapsing Lemma holds whose expressivity properly extends that of LP insofar as these systems are rich enough to accurately express that a formula φ is, e.g., "only true," although the utility of such nullary constants is relatively limited.

For example, Ryosuke Igarashi introduces an expansion of LP with a falsity constant in (2015), where $\bot$ corresponds to a zero-ary truth function whose value is always $\mathfrak{f}$. Formally, we can define Igarishi's system LP$\bot$ as follows.

Definition 13.17 LP$\bot$ is the logic induced by the matrix $\mathfrak{M}_{\mathsf{LP}\bot}$

$$\langle \mathcal{V}, \mathcal{D}, \bot, f_{\mathsf{LP}}^{\neg}, f_{\mathsf{LP}}^{\wedge}, f_{\mathsf{LP}}^{\vee}, f_{\mathsf{LP}}^{\rightarrow}, f_{\mathsf{LP}}^{\forall}, f_{\mathsf{LP}}^{\exists} \rangle$$

where $v(\bot) = \mathfrak{f}$ for all interpretations v.

Note that we have clear strategies for defining such a connective $\bot$ that do not run afoul of the Collapsing Lemma. We note for example that the scheme $\diamond(\forall x(x = x))$ (where $\diamond$ is interpreted by the function $f^{\diamond_4}$ described in Lemma 13.2) will be true in precisely the same interpretations as $\bot$. Hence, by Lemma 13.2, we conclude that:

Lemma 13.4 *The Collapsing Lemma holds for* LP$\bot$.

Note also that $\diamond(\forall x(x = x))$ acts as a definable truth constant $\top$ when $\diamond$ is interpreted by $f^{\diamond_0}$. Hence, by Lemma 13.2, the Collapsing Lemma holds for expansions of LP in which $\top$ is added as well if we were to identify a logic LP$\top$. In other words, the properties of being capable of expressing truth or falsity itself are not in and of themselves sufficient to cause the Collapsing Lemma to fail.

13.3 Implication Connectives

Even as Priest introduces and defends LP, he concedes that the interpretation that LP affords to "$\rightarrow$" falls short of the common expectations of a conditional connective. Most importantly, the conditional connective is not *detachable*—i.e., the inference $\varphi, \varphi \rightarrow \psi \models \psi$ fails—witnessed by models in which φ is evaluated as $\mathfrak{b}$ but ψ is evaluated as $\mathfrak{f}$.

Without a detachable interpretation of "$\rightarrow$," even one who endorses the primary strategy of LP might concede that some alternative interpretation is called for. As Priest admits:

> I will just say that our semantics may well be considered adequate for "$\neg$", "$\vee$", "$\wedge$" but need somehow to be extended to deal with "$\rightarrow$". In virtue of all the work in this area, this seems quite plausible. (Priest 1979, p. 232)

Following Priest's suggestion, multiple augmented versions of LP have appeared in which a stronger interpretation is afforded to the conditional connective. Considering the utility of the Collapsing Lemma, it would be desirable were it to hold for such modified versions of LP as well. Such systems—like the three-valued RM$_3$—are frequently encountered. It is thus worthwhile to examine the hyperclassical three-valued implications that can be added to LP for which the Collapsing Lemma will continue to hold.

13.3.1 Collapsing and Detachment

Disappointingly, if one revises LP to include a three-valued, detachable conditional one will be forced to abandon the Collapsing Lemma in the process. To show this, we show two intermediate facts concerning the properties that will necessarily hold of a hyperclassical conditional connective if its addition to LP will enjoy certain properties. First, we can isolate a property that any hyperclassical conditional interpreted by a function $f^\rightarrow$ must have in order for the Collapsing Lemma to hold.

Lemma 13.5 *Let $f^\rightarrow$ be a three-valued interpretation of a hyperclassical implication and let $\mathfrak{M}'_{LP}$ be the matrix in which $f^\rightarrow_{LP}$ in $\mathfrak{M}_{LP}$ is replaced by $f^\rightarrow$. Then the Collapsing Lemma holds of $\mathfrak{M}'_{LP}$ only if $f^\rightarrow(\mathfrak{b}, \mathfrak{f}) \neq \mathfrak{f}$.*

Proof Consider a classical structure $\mathfrak{A}$ with unary predicates P and R and constants s and t. Suppose that $P^\mathfrak{A}$ maps $s^\mathfrak{A}$ to $\mathfrak{t}$ and $t^\mathfrak{A}$ to $\mathfrak{f}$ while $R^\mathfrak{A}$ maps both $s^\mathfrak{A}$ and $t^\mathfrak{A}$ to $\mathfrak{f}$. In this model, $v_\mathfrak{A}(P(t) \to R(t)) = \mathfrak{t}$ as its antecedent is false.

Now, consider a collapsed structure $\mathfrak{A}^\sim$ for which $s^\mathfrak{A} \sim t^\mathfrak{A}$. Then for the interpretation $v_{\mathfrak{A}^\sim}$, it must be the case that both $v_{\mathfrak{A}^\sim}(P(t)) = \mathfrak{b}$ and $v_{\mathfrak{A}^\sim}(R(t)) = \mathfrak{f}$. Presuming that the Collapsing Lemma holds requires that $v_{\mathfrak{A}^\sim}(P(t) \to R(t)) \in \{\mathfrak{t}, \mathfrak{b}\}$, so $f^\rightarrow(v_{\mathfrak{A}^\sim}(P(t)), v_{\mathfrak{A}^\sim}(R(t)))$ cannot be $\mathfrak{f}$. $\square$

Second, we can isolate a property that must hold of a hyperclassical conditional in a three-valued and paraconsistent setting if it is to be detachable.

Lemma 13.6 *Let $f^\rightarrow$ be a three-valued interpretation of a hyperclassical implication that is detachable. Then $f^\rightarrow(\mathfrak{b}, \mathfrak{f}) = \mathfrak{f}$.*

Proof Consider a glutty three-valued model $\mathfrak{A}$ in which $v_\mathfrak{A}(P(t)) = \mathfrak{b}$ and $v_\mathfrak{A}(R(t)) = \mathfrak{f}$. If $f^\rightarrow(v_\mathfrak{A}(P(t)), v_\mathfrak{A}(R(t)))$ were not $\mathfrak{f}$, then it would be a designated value. In such a model, then, $v_\mathfrak{A}(P(t))$ and $v_\mathfrak{A}(P(t) \to R(t))$ are both designated while $v_\mathfrak{A}(R(t))$ is not. $\square$

Clearly, these two properties are incompatible. Hence, we conclude the following.

Corollary 13.5 *The Collapsing Lemma fails for any modification of $\mathfrak{M}_{LP}$ in which $f^\rightarrow_{LP}$ is replaced by a truth-functional, detachable conditional.*

Going forward, we are able to precisely identify the revisions to the conditional connective of LP that can be made without upsetting the Collapsing Lemma.

Observation 13.2 *There are precisely four hyperclassical implications that can replace $f_{\mathsf{LP}}^{\rightarrow}$ in $\mathfrak{M}_{\mathsf{LP}}$ for which the Collapsing Lemma holds (including $f_{\mathsf{LP}}^{\rightarrow}$ itself). Where $a_0, a_1 \in \{\mathfrak{t}, \mathfrak{b}\}$, each function $f_{a_0 a_1}^{\rightarrow}$ must conform to the following matrices:*

	$\mathfrak{t}$	$\mathfrak{b}$	$\mathfrak{f}$
$\mathfrak{t}$	$\mathfrak{t}$	$\mathfrak{b}$	$\mathfrak{f}$
$\mathfrak{b}$	a_0	$\mathfrak{b}$	$\mathfrak{b}$
$\mathfrak{f}$	$\mathfrak{t}$	a_1	$\mathfrak{t}$

Proof That $f^{\rightarrow}$ is assumed to be hyperclassical means that $f^{\rightarrow}$ must agree with the classical interpretation of the conditional for classical arguments, whence:

	$\mathfrak{t}$	$\mathfrak{b}$	$\mathfrak{f}$
$\mathfrak{t}$	$\mathfrak{t}$	$-$	$\mathfrak{f}$
$\mathfrak{b}$	$-$	$-$	$-$
$\mathfrak{f}$	$\mathfrak{t}$	$-$	$\mathfrak{t}$

Because we include $f_{\mathsf{LP}}^{\rightarrow}$, preserving the Collapsing Lemma requires that whenever either $\neg \varphi$ or ψ are assigned designated values, $\varphi \rightarrow \psi$ must be assigned a designated value. Hence, any value that can be added in the above blank spaces must be either $\mathfrak{t}$ or $\mathfrak{b}$. For similar reasons, we mandate that whenever both φ and $\neg \psi$ take designated values, $\neg(\varphi \rightarrow \psi)$ must take a designated value.

Hence, $f_{a_0 a_1}^{\rightarrow}(\mathfrak{t}, \mathfrak{b})$, $f_{a_0 a_1}^{\rightarrow}(\mathfrak{b}, \mathfrak{b})$, and $f_{a_0 a_1}^{\rightarrow}(\mathfrak{b}, \mathfrak{f})$ must always be $\mathfrak{b}$. On the other hand, we only require that the values of $f_{a_0 a_1}^{\rightarrow}(\mathfrak{b}, \mathfrak{t})$ and $f_{a_0 a_1}^{\rightarrow}(\mathfrak{f}, \mathfrak{b})$ are designated, whence *any* designated value may be assigned to these arguments. □

Now we are able to study these systems in more detail. For perspicuity, let us define the following logical matrices.

Definition 13.18 Each $\mathfrak{M}_{\mathsf{LP}}^{a_0 a_1}$ for $a_0, a_1 \in \{\mathfrak{t}, \mathfrak{b}\}$ is defined as the matrix:

$$\langle \mathcal{V}, \mathcal{D}, f_{\mathsf{LP}}^{\neg}, f_{\mathsf{LP}}^{\wedge}, f_{\mathsf{LP}}^{\vee}, f_{a_0 a_1}^{\rightarrow}, f_{\mathsf{LP}}^{\forall}, f_{\mathsf{LP}}^{\exists} \rangle$$

Again, *n.b.* that $\mathfrak{M}_{\mathsf{LP}}^{\mathfrak{tt}}$ is just $\mathfrak{M}_{\mathsf{LP}}$.

It is worth observing that the four matrices, in fact, induce distinct consequence relations. This fact, however, is not trivial, for which reason we describe explicit inferences with respect to which the matrices $\mathfrak{M}_{\mathsf{LP}}^{a_0 a_1}$ disagree.

Lemma 13.7 *The inference $(R(t) \rightarrow P(t)), \neg(R(t) \rightarrow P(t)) \models R(t)$ is valid only with respect to $\mathfrak{M}_{\mathsf{LP}}$ and $\mathfrak{M}_{\mathsf{LP}}^{\mathfrak{bt}}$.*

Proof In $\mathfrak{M}$ and $\mathfrak{M}_{\mathsf{LP}}^{\mathfrak{bt}}$, all solutions to $f(x, y) = f_{\mathsf{LP}}^{\neg}(f(x, y)) = \mathfrak{b}$ are those in which x is designated, whence in the above inference, $R(t)$ must be designated. In cases $\mathfrak{M}_{\mathsf{LP}}^{\mathfrak{tb}}$ and $\mathfrak{M}_{\mathsf{LP}}^{\mathfrak{bb}}$, consider a model in which $v(R(t)) = \mathfrak{f}$ and $v(P(t)) = \mathfrak{b}$; in both cases, the value of the premises will be $\mathfrak{b}$ while the value of the conclusion will be $\mathfrak{f}$. □

Lemma 13.8 *The inference* $(R(t) \to P(t)), \neg(R(t) \to P(t)) \vDash \neg P(t)$ *is valid only with respect to* $\mathfrak{M}_{\mathsf{LP}}$ *and* $\mathfrak{M}_{\mathsf{LP}}^{tb}$.

Proof In $\mathfrak{M}_{\mathsf{LP}}$ and $\mathfrak{M}_{\mathsf{LP}}^{tb}$, any case in which $f(x, y) = f_{\mathsf{LP}}^{\to}(f(x, y)) = \mathfrak{b}$ is one in which y is either $\mathfrak{b}$ or $\mathfrak{f}$, that is, one in which $f_{\mathsf{LP}}^{\to}(y)$ is designated. Hence, whenever $R(t) \to P(t)$ and $\neg(R(t) \to P(t))$ are both designated, $P(t)$ must be assigned either $\mathfrak{b}$ or $\mathfrak{f}$, and $\neg R(t)$ will take a designated value. In cases $\mathfrak{M}_{\mathsf{LP}}^{bt}$ and $\mathfrak{M}_{\mathsf{LP}}^{bb}$, consider a model in which $R(t)$ takes a value of $\mathfrak{b}$ while $P(t)$ takes a value of $\mathfrak{t}$. Both $R(t) \to P(t)$ and $\neg(R(t) \to P(t))$ will be assigned a value of $\mathfrak{b}$ in such a model although $\neg P(t)$ will be assigned the value of $\mathfrak{f}$. □

Hence, the deductive systems induced by these matrices are all distinct. Given the distinctness of each system's *semantic* consequence relation, it is also worth mentioning that each of these systems is implicitly provided with a sound and complete natural deduction calculus by Barteld Kooi and Allard Tamminga's schematic treatment of expansions of LP in Kooi and Tamminga (2012).

Furthermore, as a referee has pointed out, the truth functions $f_{a_0 a_1}^{\to}$ for $a_0, a_1 \in \{\mathfrak{t}, \mathfrak{b}\}$ are all definable in LP itself. Because of this fact, the logics induced by the matrices $\mathfrak{M}_{\mathsf{LP}}^{bt}$, $\mathfrak{M}_{\mathsf{LP}}^{tb}$, and $\mathfrak{M}_{\mathsf{LP}}^{bb}$ can be thought of as *fragments* of LP.

13.3.2 Collapsing and Tomova's Properties

While LP's interpretation of "$\to$" is not detachable, *modus ponens* remains admissible in a weaker sense: Whenever formulae φ and $\varphi \to \psi$ are *theorems* of LP, then one may infer that ψ, too, is a theorem. In this weaker sense, the conditional connective of LP is detachable.

A curious fact is that while Corollary 13.5 entails that the Collapsing Lemma *fails* for any modification of LP in which *modus ponens* is admissible, the *only* modifications of implication in LP that *preserve* the Collapsing Lemma enjoy *modus ponens* in this weaker sense. That is, if we revise the interpretation of implication in LP while preserving the Collapsing Lemma, implication *must continue to be theoremhood-preserving*.

To show this, consider a pair of properties of three-valued implication connectives described by Natalya Tomova. In a series of papers (Tomova 2012, 2015a, b), Tomova investigates the properties of three-valued implication connectives that are hyperclassical in the present sense. Following an observation due to Rescher (1969), Tomova describes two species of *naturalness* represented here with some minor changes in terminology, e.g., Tomova describes the present notion of hyperclassicality as "C-extending." Furthermore, Tomova uses rational numbers as truth values, for which there exists a natural linear ordering. For present purposes, define $\leq$ so that $\mathfrak{f} \leq \mathfrak{b} \leq \mathfrak{t}$. Then:

Definition 13.19 A truth function $f^\rightarrow$ for a three-valued implication is *natural* if:

1. $f^\rightarrow$ a hyperclassical conditional
2. if $f^\rightarrow(x, y) \in \mathcal{D}$ and $x \in \mathcal{D}$ then $y \in \mathcal{D}$
3. whenever $x \leq y$, $f^\rightarrow(x, y) \in \mathcal{D}$.

Insofar as its conditional connective is not detachable, LP is not *natural* in the above sense. However, the weaker sense of detachability that LP *does* enjoy is reflected in Tomova's *weaker* property.

Definition 13.20 A truth function $f^\rightarrow$ for a three-valued implication is *weakly natural* if:

1. $f^\rightarrow$ a hyperclassical conditional
2. if $\varphi \rightarrow \psi$ and φ are tautologies, then ψ is a tautology
3. whenever $x \leq y$, $f^\rightarrow(x, y) \in \mathcal{D}$.

Clearly, no variant of LP can have a natural implication in Tomova's sense while enjoying the Collapsing Lemma. In contrast—and somewhat surprisingly—*every* implicational variant of LP for which the Collapsing Lemma holds is properly *weakly* natural.

Observation 13.3 *For any modification of LP by a novel, three-valued conditional connective, the Collapsing Lemma holds only if the conditional is properly weakly natural.*

Proof We have observed from Corollary 13.5 that no such implication can be natural *simpliciter*, leaving us to consider only properly weakly natural implications. In Tomova (2015b), Tomova shows that when $\{t, b\}$ are designated, truth functions corresponding to the following matrices are weakly natural:

	t	b	f		t	b	f
t	t	b	f	t	t	b	f
b	a_0	a_1	a_2	b	a_0	a_1	b
f	t	t	t	f	t	b	t

where each $a_i \in \{t, b\}$. Each of the four collapsible implications in the foregoing are instances of these matrices.

Now, we can extend this observation a bit by providing a characterization of the three-valued hyperclassical conditional connectives for which the Collapsing Lemma holds.

If one takes the property of being weakly natural to be a desirable feature and one has bilateralist leanings—i.e., one takes the falsification of a conditional to be as important as its verification—then one might want an extended principle that reflects how conditionals interact with logical falsehoods. □

Definition 13.21 A truth function $f^\rightarrow$ for a three-valued implication is *natural with respect to falsity* in a system with $f^\rightarrow_{\mathsf{LP}}$ if:

13 Variations on the Collapsing Lemma

1. $f^{\rightarrow}$ is a hyperclassical implication
2. if $x \in \mathcal{D}$ and $f_{\mathsf{LP}}^{\rightarrow}(y) \in \mathcal{D}$, then $f_{\mathsf{LP}}^{\rightarrow}(f^{\rightarrow}(x, y)) \in \mathcal{D}$
3. whenever $y \lesssim x$, $f_{\mathsf{LP}}^{\rightarrow}(f^{\rightarrow}(x, y)) \in \mathcal{D}$.

The feature of being natural with respect to falsity is essentially the statement that the conditional is natural not in the sense of getting *truth conditions* right but, rather, in the sense of getting the conditional's *falsity conditions* correct.

Observation 13.4 *For a system replacing the truth function $f_{\mathsf{LP}}^{\rightarrow}$ with a three-valued truth function $f^{\rightarrow}$, the Collapsing Lemma holds if and only if $f^{\rightarrow}$ and any hyperclassical implication definable therefrom are* properly *weakly natural (i.e., weakly but not strongly natural) and natural with respect to falsity.*

Proof Left to right can be confirmed by observing that the four implications described in Observation 13.2 are properly weakly natural and natural with respect to falsity.

For right to left, from Observation 13.3, we know that only properly weakly natural implications can play the role of $\rightarrow$ in a modification of $\mathfrak{M}_{\mathsf{LP}}$ while maintaining the Collapsing Lemma. In Tomova (2015b), Tomova characterizes the properly weakly natural implications by describing sixteen matrices. Of these matrices, eleven truth functions $f^{\rightarrow}$ are defined so that either $f^{\rightarrow}(\mathfrak{b}, \mathfrak{b}) = \mathfrak{t}$ or $f^{\rightarrow}(\mathfrak{b}, \mathfrak{f}) = \mathfrak{t}$. In the first case, such matrices violate clause 2 of Definition 13.21 while in the second case, because $\mathfrak{f} \lesssim \mathfrak{b}$ in Tomova's terminology, these functions violate clause 3.

Of the remaining five matrices, one can be seen to fail to satisfy the criteria. The matrix for this truth function is:

	t	b	f
t	t	f	f
b	b	b	b
f	t	t	t

It can be easily confirmed that this truth function $f^{\rightarrow}$ is indeed both properly weakly natural and natural with respect to falsity. However, from this function we can define a function $f^{\rightarrow}(f_{\mathsf{LP}}^{\rightarrow}(x), f_{\mathsf{LP}}^{\rightarrow}(y))$ with the following characteristic matrix:

	t	b	f
t	t	b	f
b	b	b	f
f	t	t	t

This defined implication is indeed hyperclassical but the matrix does not appear among the weakly natural matrices described in Tomova (2015b). Hence, its incorporation *induces* a non-weakly natural implication, contradicting the property.

What remain are the matrices characteristic of the functions $f_{\mathsf{LP}}^{\rightarrow}$, $f_{\mathfrak{bt}}^{\rightarrow}$, $f_{\mathfrak{tb}}^{\rightarrow}$, and $f_{\mathfrak{bb}}^{\rightarrow}$, which by Observation 13.2 do not interrupt the Collapsing Lemma. □

These facts seem quite curious—there are 243 hyperclassical implications that could be added to **LP** but only a small, naturally definable subset of these are harmonious with the lemma—and are worthy of being investigated more closely. For now, we leave the matter of implication behind and proceed to make a few observations concerning the Collapsing Lemma and the extensional connectives of conjunction and disjunction.

13.4 Conjunction and Disjunction

While there has been a great deal of debate concerning the philosophical merits of particular three-valued interpretations of negation and implication in paraconsistent logic, one encounters far less controversy regarding appropriate formalizations of conjunction and disjunction. This is not, however, to say that no alternatives to LP's interpretations of conjunction and disjunction have appeared in the literature. Indeed, three-valued interpretations of conjunction and disjunction distinct from those of LP have been introduced alongside a vigorous philosophical defense in, e.g., Sören Hallden's internal and external accounts of these operations in Halldén (1949).

As in Sect. 13.3, we can exhaustively describe the truth functions that can replace $f_{\mathsf{LP}}^\wedge$ and $f_{\mathsf{LP}}^\vee$ without interfering with the Collapsing Lemma.

Observation 13.5 *There are precisely four hyperclassical conjunctions and four hyperclassical disjunctions that can replace $f_{\mathsf{LP}}^\wedge$ and $f_{\mathsf{LP}}^\vee$, respectively, in $\mathfrak{M}_{\mathsf{LP}}$ for which the Collapsing Lemma holds. For $a_0, a_1 \in \{\mathfrak{b}, \mathfrak{f}\}$ and $b_0, b_1 \in \{\mathfrak{t}, \mathfrak{b}\}$, the functions $f_{a_0 a_1}^\wedge$ and $f_{b_0 b_1}^\vee$ conform to the following matrices:*

$f_{a_0 a_1}^\wedge$	$\mathfrak{t}$	$\mathfrak{b}$	$\mathfrak{f}$	$f_{b_0 b_1}^\vee$	$\mathfrak{t}$	$\mathfrak{b}$	$\mathfrak{f}$
$\mathfrak{t}$	$\mathfrak{t}$	$\mathfrak{b}$	$\mathfrak{f}$	$\mathfrak{t}$	$\mathfrak{t}$	b_0	$\mathfrak{t}$
$\mathfrak{b}$	$\mathfrak{b}$	$\mathfrak{b}$	a_0	$\mathfrak{b}$	b_1	$\mathfrak{b}$	$\mathfrak{b}$
$\mathfrak{f}$	$\mathfrak{f}$	a_1	$\mathfrak{f}$	$\mathfrak{f}$	$\mathfrak{t}$	$\mathfrak{b}$	$\mathfrak{f}$

Proof In the above matrices, all that is required of a_0 and a_1 is that $f_{\mathsf{LP}}^\neg(a_0)$ and $f_{\mathsf{LP}}^\neg(a_1)$ are designated. Hence, either $\mathfrak{b}$ or $\mathfrak{f}$ will suffice in either place. Likewise, all that is required of b_0 and b_1 is that both are designated, whence one may select for each any value from $\{\mathfrak{t}, \mathfrak{b}\}$. □

Before proceeding, let us consider some attractive properties we might expect of hyperclassical conjunctions and disjunctions and how they bear on plausibility of particular solutions to the above matrices.

Clearly, commutativity of conjunction and disjunction will fail in case $a_0 \neq a_1$ or $b_0 \neq b_1$, respectively. For example, when $a_0 = \mathfrak{b}$ and $a_1 = \mathfrak{f}$, a conjunction $P(t) \wedge R(t)$ might be assigned $\mathfrak{b}$—a designated value—while $R(t) \wedge P(t)$ is assigned a value of $\mathfrak{f}$. There exist cases in which the failure of commutativity makes sense, e.g., in the so-called "lazy" or "short circuit" evaluations of McCarthy logic described in (1963) or Lisp logic described in Fitting (1994). However, such failures are specific to the very particular context of the execution of computer programs, and in the more general case, we will expect commutativity to hold.

If we add to a demand for commutativity an additional demand that DeMorgan's Laws hold between conjunction and disjunction, we are driven to only two options: $\mathfrak{M}_{\mathsf{LP}}$ itself, and $\mathfrak{M}_{\mathsf{LP}}$ modified so that conjunction and disjunction are interpreted by functions $f^{\wedge}_{\mathfrak{bb}}$ and $f^{\vee}_{\mathfrak{bb}}$.

Consideration of these matrices reveals several things. For one, in these alternative interpretations—along with the function $f^{\rightarrow}_{\mathfrak{bb}}$ encountered in Definition 13.18—the third value $\mathfrak{b}$ acts as a type of "contaminant" insofar as whenever *any* argument is $\mathfrak{b}$, the value of these functions is $\mathfrak{b}$. This property—called the "predominance of the atheoretical element" by Åqvist in (1962)—is defended by progenitors of logics of nonsense such as Halldén (in 1949) or Bochvar (in 1938, translated as 1981). This property is common to a treatment of paradox going back to the *Principia Mathematica* in which paradoxical sentences are treated as ungrammatical—and, consequently, in which the appearance of an ungrammatical subformula entails that a complex formula is ungrammatical as well.

The "classical fragment" of Sören Halldén's *logic of nonsense* C includes three-valued generalizations of each of the classical connectives and quantifiers that are intended to authentically capture the connectives used in the *Principia Mathematica*. More recently, this classical fragment has reemerged in the papers Ciuni (2015) and Ciuni and Carrara (2016), in which the system is labeled "PWK"—*paraconsistent weak Kleene logic*—to emphasize that its truth functions are identical with Kleene's weak three-valued matrices of Kleene (1950).

Note that neither Halldén (1949) nor Ciuni and Carrara provide three-valued semantics for quantification. For this, we appear to Malinowski's (2002), in which we have the following "weak" accounts of the universal and particular quantifiers.

Definition 13.22 The weak universal and particular quantifiers $f^{\forall}_{\mathsf{PWK}}$ and $f^{\exists}_{\mathsf{PWK}}$ are type $\langle 1 \rangle$ quantifiers determined by the following:

$$f^{\forall}_{\mathsf{PWK}}(X) = \begin{cases} \mathfrak{t} & \text{if } X = \{\mathfrak{t}\} \\ \mathfrak{b} & \text{if } \mathfrak{b} \in X \\ \mathfrak{f} & \text{otherwise} \end{cases} \qquad f^{\exists}_{\mathsf{PWK}}(X) = \begin{cases} \mathfrak{t} & \text{if } \mathfrak{t} \in X \text{ and } \mathfrak{b} \notin X \\ \mathfrak{b} & \text{if } \mathfrak{b} \in X \\ \mathfrak{f} & \text{otherwise} \end{cases}$$

where $X \in \wp^{+}(\mathcal{V})$.

Note that the two quantifiers are interdefinable given the interpretation of negation in LP. For example, for any non-empty set of truth values X, $f^{\rightarrow}_{\mathsf{LP}}(f^{\exists}_{\mathsf{PWK}}(X)) = f^{\forall}_{\mathsf{PWK}}(\{f^{\rightarrow}_{\mathsf{LP}}(x) \mid x \in X\})$.

Definition 13.23 The matrix $\mathfrak{M}_{\mathsf{PWK}}$ is defined as follows:

$$\langle \mathcal{V}, \mathcal{D}, f^{\rightarrow}_{\mathsf{LP}}, f^{\wedge}_{\mathfrak{bb}}, f^{\vee}_{\mathfrak{bb}}, f^{\rightarrow}_{\mathfrak{bb}}, f^{\forall}_{\mathsf{PWK}}, f^{\exists}_{\mathsf{PWK}} \rangle$$

Given the above description of PWK, between the clauses for negation in Priest's own proof of the Collapsing Lemma from Priest (1991) and the foregoing Observations 13.2 and 13.5, we have nearly enough material to prove that the Collapsing Lemma extends to Halldén's logic PWK.

Corollary 13.6 (Collapsing Lemma for **PWK**) *For a classical model $\mathfrak{A}$ and a Priest collapse $\mathfrak{A}^\sim$, let $v_{\mathfrak{A}^\sim}$ be a **PWK** valuation on $\mathfrak{A}^\sim$. Then*

$$\text{If } \mathfrak{A}, v_{\mathfrak{A}} \vDash \varphi \text{ then } \mathfrak{A}^\sim, v_{\mathfrak{A}^\sim} \vDash \varphi.$$

Proof The basis step and case of negation follows immediately from Priest's own proof in (1991). Furthermore, the particular cases of disjunction, conjunction, and implication have been explicitly taken care of in Lemmas 13.2 and 13.5. Thus, all that remains is that we prove that the property extends through the quantifiers. In fact, given the earlier observation of the interdefinability of the quantifiers in **PWK** by **LP**'s negation, we need only prove one case and let the other follow from interdefinability.

We therefore cover the fact that the property holds for existentially quantified sentences. Now, suppose that $v_{\mathfrak{A}}(\exists x \varphi(x)) = \mathfrak{t}$. Then (recalling that $\mathfrak{A}$ is assumed to be a Henkin structure) $\mathfrak{t} \in \{v_{\mathfrak{A}}(\varphi(\underline{a})) \mid a \in A\}$. By induction hypothesis, then, this entails that $\{v_{\mathfrak{A}^\sim}(\varphi(\underline{a})) \mid a \in A\}$ is either $\{\mathfrak{t}\}$ or $\{\mathfrak{t}, \mathfrak{b}\}$. But by consulting Definition 13.22, we note that this condition entails that $v_{\mathfrak{A}^\sim}(\exists x \varphi(x)) \in \mathcal{D}$, as needed. □

So we have a further positive result that the Collapsing Lemma holds for **PWK**, a paraconsistent logic that disagrees with **LP** with respect to the interpretation of all operators except negation. Consequently, the suite of tools used by Priest is available to the study of this type of paraconsistent logic as well.

13.5 Conclusions

Among the foregoing results, I think that three items merit special attention and deserve further investigation.

For one, that the Collapsing Lemma extends to the systems **LP⊥** and **LP⊤** (or even an enrichment with both truth and falsity constants) shows that one can strictly increase the expressivity of **LP** without losing the lemma. That the increased expressivity allows for these systems to authentically refer to "only truth" and "only falsity" means that the failure of the Collapsing Lemma for, e.g., **LFI1** is *not* directly attributable to **LFI1**'s ability to express these types of notions. Of course, the *utility* of **LP⊥**'s newfound expressivity is limited—that a formula $\varphi \leftrightarrow \bot$ is assigned a designated value does not mean that φ is evaluated as false—so **LFI1** enjoys some property beyond **LP**, but pinpointing the cause of the failure of the Collapsing Lemma is not settled merely by appeal to increased expressivity.

Second, while it is not surprising that enriching **LP** with a detachable conditional causes the Collapsing Lemma to fail, I *do* find it very surprising that any conditional that can be added to **LP** without inhibiting the lemma is *weakly natural*. For all its perceived deficiencies, that the conditional in **LP** is theoremhood-preserving (if not truth-preserving) is still a very strong property. That such a strong property *necessarily* holds for any hyperclassical conditional that can be "safely" added to **LP** is certainly no fluke. There is, therefore, a reason to suspect that investigating this fact in more detail may uncover deeper subtleties with respect to the Collapsing Lemma.

Finally, I think that establishing that the Collapsing Lemma holds for **PWK** is worth studying in more detail. Halldén's motivations for introducing **PWK** and his interpretation of the third value seem miles away from Priest's motives in introducing **LP**. Halldén, for example, interprets the third value as a representation of the *meaninglessness* of a statement and emphatically denies in Halldén (1949) that understanding a value to be designated is in any way akin to understanding that value as truthlike. Although both allow for a formula $\varphi \wedge \neg \varphi$ to be assigned a designated value in a model, where Priest sees a formula that is both true and false, Halldén sees a formula that is *neither* of these. That **PWK**'s philosophical foundations are so orthogonal to Priest's development of **LP** suggests that great differences will be found in the application of the Collapsing Lemma. One such divide, for example, is that if one rehearses Priest's finite, collapsed models of arithmetic, every formula with complexity of Δ_1^0 or greater will take a designated value. Explaining such phenomena seems like an intriguing endeavor.

Acknowledgements I am grateful to the insightful and constructive remarks of three referees, whose input was of great help and is much appreciated.

References

Åqvist, L. (1962). Reflections on the logic of nonsense. *Theoria*, 28(2), 138–157.
Asenjo, F. G. (1966). A calculus of antinomies. *Notre Dame Journal of Formal Logic*, 7(1), 103–105.
Asenjo, F. G., & Tamburino, J. (1975). Logic of antinomies. *Notre Dame Journal of Formal Logic*, 16(1), 17–44.
Baaz, M. (1996). Infinite-valued Gödel logics with 0-1-projections and relativizations. In P. Hájek (Ed.), *Gödel '96: Logical foundations of mathematics, computer science and physics* (pp. 23–33). Berlin: Springer.
Bochvar, D. A. (1938). On a three-valued logical calculus and its application to the analysis of contradictions. *Matematicheskii Sbornik*, 4(2), 287–308.
Bochvar, D. A. (1981). On a three-valued logical calculus and its application to the analysis of the paradoxes of the classical extended functional calculus. *History and Philosophy of Logic*, 2(1–2), 87–112.
Carnielli, W., Coniglio, M. E., & Marcos, J. (2007). Logics of formal inconsistency. In D. Gabbay & F. Guenthner (Eds.), *Handbook of philosophical logic* (Vol. 14, pp. 15–107). The Netherlands: Springer.
Carnielli, W., Marcos, J., & de Amo, S. (2000). Formal inconsistency and evolutionary databases. *Logic and Logical Philosophy*, 8, 115–152.
Ciuni, R. (2015). Conjunction in paraconsistent weak Kleene logic. In P. Arazim & M. Dank (Eds.), *The logica yearbook 2014* (pp. 61–76). London: College Publications.
Ciuni, R., & Carrara, M. (2016). Characterizing logical consequence in paraconsistent weak Kleene logic. In L. Felline, F. Paoli, & E. Rossanese (Eds.), *New developments in logic and the philosophy of science*. London: College Publications.
da Costa, N. (1974). On the theory of inconsistent formal systems. *Notre Dame Journal of Formal Logic*, 15(4), 497–510.

De, M., & Omori, H. (2015). Classical negation and expansions of Belnap-Dunn logic. *Studia Logica, 103*(4), 825–851.

D'Ottaviano, I. (1985). The completeness and compactness of a three-valued first-order logic. *Revista colombiana de matematicas, 19*(1–2), 77–94.

Dunn, J. M. (1979). A theorem in 3-valued model theory with connections to number theory, type theory, and relevant logic. *Studia Logica, 38*(2), 149–169.

Ferguson, T. M. (2012). Notes on the model theory of DeMorgan logics. *Notre Dame Journal of Formal Logic, 53*(1), 113–132.

Ferguson, T. M. (2014). On non-deterministic quantification. *Logica Universalis, 8*(2), 165–191.

Ferguson, T. M. (2017). Dunn-Priest quotients of many-valued structures. *Notre Dame Journal of Formal Logic, 58*(2), 221–239.

Fitting, M. (1994). Kleene's three-valued logics and their children. *Fundamenta Informaticae, 20*(1–3), 113–131.

Halldén, S. (1949). *The logic of nonsense*. Lund: Lundequista Bokhandeln.

Igarishi, R. (2015). Logic of paradox and falsity constant. Paper delivered at PhilLogMath 2015.

Kleene, S. C. (1950). *Introduction to metamathematics*. Princeton, NJ: D. Van Nostrand.

Kooi, B., & Tamminga, A. (2012). Completeness via correspondence for extensions of the logic of paradox. *Review of Symbolic Logic, 5*(4), 720–730.

Malinowski, G. (2002). Many-valued logic. In D. Jacquette (Ed.), *A companion to philosophical logic* (pp. 545–561). Oxford: Blackwell Publishing.

Marcos, J. (2005). Nearly every modal logic is paranormal. *Logique et Analyse, 48*(189–192), 279–300.

McCarthy, J. (1963). A basis for a mathematical theory of computation. In P. Braffort & D. Hirschberg (Eds.), *Computer programming and formal systems* (pp. 33–70). Amsterdam: North-Holland Publishing Company.

Omori, H. (2015). Remarks on naive set theory based on LP. *Review of Symbolic Logic, 8*(2), 279–295.

Priest, G. (1979). The logic of paradox. *Journal of Philosophical Logic, 8*(1), 219–241.

Priest, G. (1991). Minimally inconsistent LP. *Studia Logica, 50*(2), 321–331.

Priest, G. (1994). Is arithmetic consistent? *Mind, 103*(411), 337–349.

Priest, G. (1997). Inconsistent models for arithmetic I, Finite models. *Journal of Philosophical Logic, 26*(2), 223–235.

Priest, G. (2000). Inconsistent models for arithmetic II, The general case. *Journal of Symbolic Logic, 65*(4), 1519–1529.

Priest, G. (2012). A note on the axiom of countability. *Al-Mukhatabat, 1*(1), 27–31.

Priest, G. (2014). Plurivalent logics. *Australasian Journal of Logic, 11*(1), 2–13.

Rescher, N. (1969). *Many-valued logic*. New York: McGraw Hill.

Sano, K., & Omori, Hitoshi. (2014). An expansion of first-order Belnap-Dunn logic. *Logic Journal of the IGPL, 22*(3), 458–481.

Tomova, N. (2012). A lattice of implicative extensions of regular Kleene's logics. *Reports on Mathematical Logic, 47*, 173–182.

Tomova, N. (2015a). Natural implication and modus ponens principle. *Logical Investigations, 21*(1), 138–143.

Tomova, N. (2015b). Erratum to: Natural implication and modus ponens principle. *Logical Investigations, 21*(2), 186–187.

Thomas Macaulay Ferguson works as an ontologist for the Cyc artificial intelligence project and is an affiliate research scholar at the Saul Kripke Center at the City University of New York. He is the author of *Meaning and Proscription in Formal Logic* (Springer 2017) and, with Graham Priest, is coauthor of the supplemental *Dictionary of Logic* (Oxford University Press 2016). His research includes work on philosophical logic, metaphysics, and the philosophy of mathematics.

Chapter 14
Dialetheic Conditional Modal Logic

Patrick Girard

Abstract Standard modal logic for alethic modalities analyses modalities as ranging over all possible worlds (the Leibnizian universe). This leaves very little room in the space of worlds to entertain impossible things. My proposal is to liberate the Leibnizian universe and reinforce the relative aspect of possibility; worlds are possible with respect to some worlds, and impossible for others. The central idea is to isolate relative possibility (Kripke) from conditionality (Lewis/Stalnaker). To accommodate counterpossibles, I provide a dialetheic conditional modal logic, a theory that is dialetheic at every level, in the logic as well as in the set theory behind it.

Keywords Dialethism · Modal logic · Conditional logic · Counterfactuals · Counterpossibles

The orthodox modal logic for metaphysics models the *Leibnizian* interpretation of necessity as *truth in all possible worlds*. Kripke (1959) proved S5 to be complete for the *Leibnizian* interpretation. On its own, S5 doesn't differentiate models that have a single equivalence class and those that have multiple classes. A simple bisimulation argument (see Blackburn et al. 2001) establishes this:

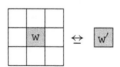

On the left-hand-side, w is related to all worlds that are in the shaded area; w doesn't *see* worlds in other equivalence classes. On the right-hand side is a perfect copy of the shaded area; worlds outside the equivalence class of w′ do not exist at all. A

---

[1] If you're not familiar with bisimulation, think of the picture as showing that any formula that can be invalidated at w can also be invalidated at w′ (see Priest 2008, Sect. 3.7.5); anything impossible at w is also impossible at w′, because w doesn't *see* any more worlds than w′.

P. Girard (✉)
University of Auckland, Auckland, New Zealand
e-mail: p.girard@auckland.ac.nz

© Springer Nature Switzerland AG 2019
C. Başkent and T. M. Ferguson (eds.), *Graham Priest on Dialetheism and Paraconsistency*, Outstanding Contributions to Logic 18,
https://doi.org/10.1007/978-3-030-25365-3_14

bisimulation between w and w' depicts the incapability of standard modal logic to distinguish the two models; everything that holds at w holds at w', and vice versa.[1] The received interpretation of metaphysical possibility is the one on the right: all possible worlds are equivalent, they are all accessible from one another.

The *Leibnizian* universe of possible worlds trivialises conditionals with impossible antecedents. Though snow is white, presumably it is not necessarily so. But what if it were necessarily black? According to Kripke (1980), water is necessarily H_2O, but could it be contingently H_2O, or necessarily H_3O? What if 13 had 2 and 6 as factors (cf. Baron et al. 2017)?

My proposal is to liberate the *Leibnizian* universe and reinforce the relative aspect of possibility; worlds are possible with respect to some worlds, and impossible for others. I will develop a *conditional modal logic* in which conditionals travel across equivalence classes of relative possibility. The central idea is to isolate *relative possibility* (objectified with a modal operator $\Diamond \varphi$) from *conditionality* (objectified with a conditional[2] operator $\varphi \boxdot\!\!\to \psi$). I like to think of the resulting models graphically using the representation of conditionals as Lewis (1973) systems of spheres:

The shaded area around w represents the class of worlds that are relatively possible with respect to w. Those are the ones that can be analysed with $\Diamond$. Conditionals $\boxdot\!\!\to$ allow jumps outside the class of relative possibility around w to access *impossible* worlds, such as v, for the analysis of *counterpossibles*. This can all be done in orthodox twentieth-century logic (aka 'classical logic'), and I will show in the first part of the paper how to formulate and axiomatise conditional modal logic.

Twentieth-century logic also trivialises conditionals with inconsistent antecedents. What if the Russell set really was a set? What if the Liar sentence was true? I will show in the second part of the paper how to liberate the treatment of inconsistent (im)possibility with a dialetheic conditional modal logic, first at the object level, and second through and through, all the way down (or up?) to set theory, with a dialetheic dialetheic conditional modal logic.[3] Dialetheic Dialetheic conditional modal logic offers pleasant and natural solutions to metaphysical (or logical?) problems.

---

[2]I follow Stalnaker (1968) and talk about conditionals instead of counterfactuals like Lewis (1973). I take 'conditionals' to be a more general notion, counterfactuals being the special case of conditionals with false antecedents.

[3]Repetition intended, to stress that dialetheic logic is used in the object and meta levels.

14.1 Conditional Modal Logic

I present *conditional modal logic* (CML) in a way that facilitates the transition to dialetheic conditional modal logic below. The language is a standard propositional language with a set of atoms PROP, propositional connectives ($\neg, \wedge, \vee, \supset, \equiv$), modalities ($\Diamond, \Box$) and conditional operators ($\Box\!\!\rightarrow, \Diamond\!\!\rightarrow$).

Following Chellas (1975), I first present a *minimal* version of the logic. *Miminal conditional modal logic* (MCML) has no constraint on modals, analogously to K in basic modal logic. Models M for MCML have a domain of worlds W, a propositional valuation V assigning sets of worlds to atoms, and two selection functions:

$$R : W \longrightarrow \wp(W)$$
$$S : W \times \wp(W) \longrightarrow \wp(W).$$

R and S are, respectively, used to interpret modal and conditional operators with *relative possibility* and *selection* (Stalnaker) or *similarity* (Lewis). So $v \in R(w)$ if v is a relative possibility to w, and $v \in S(w, X)$ if v is an X-world that *differs minimally* (Stalnaker) or is *most similar* (Lewis) to w. Following Nolan (1997), models allow for comparison of *impossible* worlds in terms of similarity, for instance[4]:

Neither u nor v are *possible*, because neither are a relative possibility for w (they are not in the shaded box that stands for w's equivalence relation). But u is more similar to w than v is, because it stands in a *smaller* sphere around w.

The semantic definition is given by the following recursive definition of a *truth-set* $[\![\varphi]\!]^M$ (the set of worlds at which φ is true):

$$\begin{aligned}
w \in [\![p]\!]^M &\equiv w \in V(p) \\
w \in [\![\neg\varphi]\!]^M &\equiv w \in W \setminus [\![\varphi]\!]^M \\
w \in [\![\varphi \wedge \psi]\!]^M &\equiv w \in [\![\varphi]\!]^M \cap [\![\psi]\!]^M \\
w \in [\![\varphi \vee \psi]\!]^M &\equiv w \in [\![\varphi]\!]^M \cup [\![\psi]\!]^M \\
w \in [\![\varphi \supset \psi]\!]^M &\equiv w \in [\![\varphi]\!]^M \supset w \in [\![\psi]\!]^M \\
w \in [\![\varphi \equiv \psi]\!]^M &\equiv w \in [\![\varphi]\!]^M \equiv w \in [\![\psi]\!]^M \\
w \in [\![\Diamond\varphi]\!]^M &\equiv R(w) \cap [\![\varphi]\!]^M \neq \emptyset \\
w \in [\![\Box\varphi]\!]^M &\equiv R(w) \subseteq [\![\varphi]\!]^M \\
w \in [\![\varphi \Diamond\!\!\rightarrow \psi]\!]^M &\equiv S(w, [\![\varphi]\!]^M) \cap [\![\psi]\!]^M \neq \emptyset \\
w \in [\![\varphi \Box\!\!\rightarrow \psi]\!]^M &\equiv S(w, [\![\varphi]\!]^M) \subseteq [\![\psi]\!]^M
\end{aligned}$$

---

[4] I continue to represent models with equivalence classes for R and Lewis' systems of spheres S, as I find them more convivial, with apologies if you don't agree.

I will omit the superscript M unless required. Frame and model validity are defined in the standard way. A formula φ is *valid in a model* if $w \in \llbracket \varphi \rrbracket$ for every $w \in W$, and it is *valid in a frame* if it is valid in every model based on the frame (i.e. for every propositional valuation). Finally, a formula is valid in a set of frames if it is valid in every frame which is in the set.

MCML does not have any assumption on the accessibility relations R and S. A sound and complete axiomatisation[5] is simply CK + K, the combination of Chellas' (1980) axioms and rules K for $\Box$ and CK for $\Box\!\!\rightarrow$:

Axioms:
Tautologies
$\Diamond \varphi \equiv \neg \Box \neg \varphi$
$\varphi \Diamond\!\!\rightarrow \psi \equiv \neg (\varphi \Box\!\!\rightarrow \neg \psi)$

Rules:

$$\dfrac{\varphi \quad \varphi \supset \psi}{\psi} \text{ MP} \qquad \dfrac{(\varphi_1 \wedge \cdots \wedge \varphi_n) \supset \psi}{(\Box \varphi_1 \wedge \cdots \wedge \Box \varphi_n) \supset \Box \psi} \text{ RK}$$

$$\dfrac{\varphi \equiv \psi}{\varphi \,\Box\!\!\rightarrow\, \xi \equiv \psi \,\Box\!\!\rightarrow\, \xi} \text{ RCEA} \qquad \dfrac{(\varphi_1 \wedge \cdots \wedge \varphi_n) \supset \psi}{(\xi \,\Box\!\!\rightarrow\, \varphi_1 \wedge \cdots \wedge \xi \,\Box\!\!\rightarrow\, \varphi_n) \supset \xi \,\Box\!\!\rightarrow\, \psi} \text{ RCK}$$

MCML provides the bare bones of a conditional modal logic, but is not suitable for metaphysical investigation without further assumptions. The first step, to find reasonable assumptions for the selection functions, is simple enough. I will endorse Lewis' conditional logic VC for the conditional part and S5 for the modal part. The second step is where more important decisions need to be taken, namely on the interaction between the selection functions. We need to decide how conditionals might be restrained by modalities, and vice versa. Let's take this more slowly.

Williamson (2010) gives an argument 'for equivalences between statements of metaphysical possibility and necessity on the one hand and statements involving counterfactual conditionals on the other'. Since he adopts the orthodox *Leibnizian* view that conditionals and counterfactuals operate over the set of all possible worlds, the conclusion is not surprising. Indeed, that S5 modalities can be defined in conditional logic was noted both by Lewis (1973, p. 22) and Stalnaker (1968, p. 105). If metaphysical possibility and conditionals range over all possible worlds, then the equivalence amounts to a translation exercise between two formal systems. Williamson, however, relies on two interaction constraints between modalities and conditionals, which motivates an interesting distinction for what is to come. Williamson formulates the constraints syntactically:

nec $\Box(\varphi \supset \psi) \supset (\varphi \,\Box\!\!\rightarrow\, \psi)$
pos $((\varphi \,\Box\!\!\rightarrow\, \psi) \wedge \Diamond \varphi) \supset \Diamond \psi$

---

[5] The proof of soundness and completeness is entirely standard, and I won't bother you with the details.

His interest in nec and pos[6] is to *sandwich* conditionals between two modal conditions so as to 'yield necessary and sufficient conditions for necessity and possibility in terms of the counterfactual conditional'. By taking $\psi = \bot$ in nec and pos, one can easily derive conditional definitions of necessity and possibility:

$$\Box \varphi \equiv (\neg \varphi \,\Box\!\!\rightarrow\, \bot),$$
$$\Diamond \varphi \equiv \neg (\varphi \,\Box\!\!\rightarrow\, \bot)$$

Hence, to the metaphysician committed to nec and pos, 'the impossible is that which counterfactually implies a contradiction; the possible is that which does not'. Williamson further concludes that 'metaphysically modal thinking is logically equivalent to a special case of counterfactual thinking. Whoever has what it takes to understand the counterfactual conditional and the elementary logical auxiliaries $\neg$ and $\bot$ has what it takes to understand possibility and necessity operators'. I disagree.[7]

What does it mean semantically to be committed to nec and pos? Neither are valid in MCML, as the reader can check.[8] But we can isolate the classes of frames that they define[9] those whose selection functions satisfy the following conditions[10]:

nec $\Box(\varphi \supset \psi) \supset (\varphi \,\Box\!\!\rightarrow\, \psi)$ ‖ NEC $S(w, X) \subseteq (R(w) \cap X)$
pos $((\varphi \,\Box\!\!\rightarrow\, \psi) \wedge \Diamond\varphi) \supset \Diamond\psi$ ‖ POS $(S(w, X) \subseteq Y \wedge R(w) \cap X \neq \emptyset) \supset (R(w) \cap Y \neq \emptyset)$

Crudely, NEC says that you can only conditionalise on accessible worlds. POS is harder to make sense of, but says roughly that consequents of conditionals are possible if their antecedents are.[11] We now have a choice as to which of NEC or POS to assume. POS but not NEC makes sense for impossibility. Whatever is a relative possibility ought to be usable in conditional reasoning. But conditional reasoning may surpass the realm of possibility. Having both together makes modalities redundant. Conditionals access (and only access) possible worlds. For an argument against NEC, I turn to the case of *countermathematicals*.

---

[6] Williamson formulates pos as the logically equivalent $(A \,\Box\!\!\rightarrow\, B) \supset (\Diamond A \supset \Diamond B)$, but this formulation will be better for us.

[7] For a thorough criticism of Williamson on counterpossibles, see Berto et al. (2017), who also propose a conditional modal logic similar to mine, but using a distinction between possible and impossible worlds instead of my more general notion of relative possibility. I don't like the twentieth-century distinction between possible and impossible worlds, or between *good* and *bad* worlds in general, as discussed in Girard and Weber (2015).

[8] For a counterexample to nec, take a model with two worlds w and v such that $v \notin R(w)$, $V(p) = \{v\}$, $v \in S(w, [\![p]\!])$ but $v \notin V(q)$. Then $\Box(p \rightarrow q)$ is vacuously true at w whereas $p \,\Box\!\!\rightarrow\, q$ is false. For a counterexample to pos, take a model with three worlds w, u and v such that $V(p) = \{u, v\}$, $V(q) = \{v\}$, $u \in R(w)$, and $v \in S(w, [\![p]\!])$.

[9] From Blackburn et al. (2001, Definition 3.2) A formula φ defines a class of frames K if for all frames F, $F \in K$ iff φ is valid in F.

[10] The proofs of definability are straightforward and left to the reader.

[11] A condition that implies but is not equivalent to POS is the converse of NEC: $(R(w) \cap X) \subseteq S(w, X)$. If we assumed this stronger condition in combination with NEC, we would reduce conditionals to strict implication, which would be a bad idea (cf., David 1973, Sect. 1.2).

A *countermathematical* is a conditional whose antecedent is a false mathematical claim. Countermathematicals offer the best motivation for rejecting NEC. According to Baron et al. (2017), countermathematicals are needed for a more general theory of explanation 'that covers cases of extra-mathematical explanation—explanations of non-mathematical or empirical facts by mathematical ones'. Their motivating example is taken from Baker (2005), and I follow suit, showing how a conditional modal logic can model the example in a natural way.

North American periodical cicadas have a life cycle of 13 or 17 years. The explanation for this is that 13 and 17 are prime numbers, and having a prime numbered life cycle is advantageous to the cicadas because it minimises encounters with predators. Now, consider the following two countermathematicals:

1. If in addition to 13 and 1, 13 had the additional factors 2 and 6, North American periodical cicadas would not have 13-year life cycles.
2. If in addition to 4 and 3, 12 had the additional factors 5 and 7, North American periodical cicadas would not have 13-year life cycles.

Given the parameters of the story, (1) is true, but (2) is false. To see this, consider the following model, taking c to stand for the proposition 'North American periodical cicadas have a 13-year life-cycle'.

$2 * 6 \neq 13$ $5 * 7 = 12$ $u : c$	
$2 * 6 \neq 13$ $5 * 7 \neq 12$ $w : c$	$2 * 6 = 13$ $5 * 7 \neq 12$ $v : \neg c$

World w is in the relative possibility equivalence class of the actual world (represented by the shaded area), one in which mathematics is the one we know and love. World v is a world in which we *twiddle* (see Baron et al. (2017) on twiddling) with mathematical facts just enough to get that $2 * 6 = 13$ while keeping everything else equal *as much as possible*.[12] Since in that world 13 is not a prime number, cicadas do not have a life cycle of 13 years, as per natural selection. In world u, however, we twiddle with mathematical facts just enough to get that $5 * 7 = 12$ while keeping everything else equal *as much as possible*, and in particular that 13 is a prime number, so cicadas preserve their life cycle of 13 years. Hence, $w \in [\![(2 * 6 = 13) \square\!\!\rightarrow \neg c]\!]$ and $w \notin [\![(5 * 7 = 12) \square\!\!\rightarrow \neg c]\!]$.

Now, as Baron et al. (2017) argue, keeping everything else equal *as much as possible* implies also keeping equal the consistency of mathematics. Assuming that

---

[12] 'So that is our basic procedure for handling the antecedent of a counterfactual implicated in a case of extra-mathematical explanation: hold as much fixed as you can within mathematics compatible with the twiddle, without inducing a contradiction', Baron et al. (2017), p.11.

mathematics is consistent in the actual world, both u and v are worlds in which mathematics is also consistent. This countermathematical contemplation is thus one that reaches outside the relative possibility of our world without invoking inconsistent worlds. Worlds u and v are consistent worlds, but they are impossible. They are impossible because of the necessity of mathematics. $2 * 6 \neq 13$ and $5 * 7 \neq 12$ are necessary mathematical truths. Were it the case that $2 * 6 = 13$, then it would be necessarily so, but this is impossible with respect to our world.

Finally, we can give the full conditional modal logic, its models and axiomatisation. Models M for conditional modal logic (CML) are the models of MCML whose selection functions satisfy POS, the S5 constraints for R and the VC constraints for S.[13]

As for the axiomatisation, on top of the MCML rules and axioms, the axioms for VC and S5 are well-known, and are easily shown to be sound and complete. I give the constraints on the selection functions and the corresponding axioms in parallel. The only additional axiom to VC + S5 is the interaction axiom pos that corresponds to the plausible constraint POS.[14]

T	$\Box \varphi \supset \varphi$		REF	$w \in R(w)$
4	$\Box \varphi \supset \Box\Box \varphi$		TRANS	$(v \in R(w) \land u \in R(v)) \supset u \in R(w)$
5	$\varphi \supset \Box \Diamond \varphi$		SYM	$v \in R(w) \supset w \in R(v)$
id	$\varphi \mathbin{\Box\!\!\rightarrow} \varphi$		ID	$S(w, X) \subseteq X$
cc	$(\varphi \mathbin{\Box\!\!\rightarrow} \psi \land \varphi \mathbin{\Box\!\!\rightarrow} \xi) \supset \varphi \mathbin{\Box\!\!\rightarrow} \psi \land \xi$		CC	$(S(w, X) \subseteq Y \land S(w, X) \subseteq Z) \supset (S(w, X) \subseteq Y \cap Z)$
ca	$(\varphi \mathbin{\Box\!\!\rightarrow} \xi \land \psi \mathbin{\Box\!\!\rightarrow} \xi) \supset \varphi \lor \psi \mathbin{\Box\!\!\rightarrow} \xi$		CA	$(S(w, X) \subseteq Z \land S(w, Y) \subseteq Z) \supset ((S(w, X) \cup S(w, Y)) \subseteq Z$
cso	$(\varphi \mathbin{\Box\!\!\rightarrow} \psi \land \psi \mathbin{\Box\!\!\rightarrow} \varphi) \supset (\varphi \mathbin{\Box\!\!\rightarrow} \xi \equiv \psi \mathbin{\Box\!\!\rightarrow} \xi)$		CSO	$(S(w, X) \subseteq Y \land S(w, Y) \subseteq X) \supset S(w, X) \subseteq Z \equiv S(w, Y) \subseteq Z$
cv	$(\varphi \mathbin{\Box\!\!\rightarrow} \xi \land \varphi \Diamond\!\!\rightarrow \psi) \supset \varphi \land \psi \mathbin{\Box\!\!\rightarrow} \xi$		CV	$(S(w, X) \subseteq Z \land S(w, X) \cap Y \neq \emptyset) \supset S(w, X \cap Y) \subseteq Z$
cs	$\varphi \land \psi \supset \varphi \mathbin{\Box\!\!\rightarrow} \psi$		CS	$w \in X \land w \in Y \supset S(w, X) \subseteq Y$
mp	$\varphi \mathbin{\Box\!\!\rightarrow} \psi \supset (\varphi \supset \psi)$		MP	$w \in X \supset w \in S(w, X)$
pos	$((\varphi \mathbin{\Box\!\!\rightarrow} \psi) \land \Diamond \varphi) \supset \Diamond \psi$		POS	$(S(w, X) \subseteq Y \land R(w) \cap X \neq \emptyset) \supset (R(w) \cap Y \neq \emptyset)$

This completes our twentieth-century analysis of conditional modal logic. In the rest of the paper, I will motivate and discuss dialetheic variants.

14.2 Dialetheic Conditional Modal Logic

The Liar sentence is necessarily true. Take an arbitrary world; then Liar is either true or false at that world. If it is true, we're done, and if it is false, then what it says is false, namely that it is not false, and so is true. Either way, Liar is true, and so necessarily true because we chose an arbitrary world. From Priest (2006, p. 4): 'a dialetheia is any true statement of the form: α and it is not the case that α'. Dialetheism is the view

---

[13] VC is Lewis's favourite counterfactual logic. I choose it here without a strong philosophical commitment. Assume the limit assumption for simplicity. The choice is entirely modular and you can pick your favourite conditional logic if you don't like VC; no problem.

[14] The axiom pos is canonical for POS in a standard Henkin completeness proof adapted from Chellas (1975). Check or proceed without concern.

that there are dialetheias. In 2002, Graham Priest provides an argument that there are true contradictions *beyond the limits of thoughts*. The argument establishes that contradictions are inevitable; when you try and dissolve (resolve?) a contradiction at one limit, you generate another one at a different limit. I take this to establish that there are necessary dialetheias, such as Liar.

Can a contradiction be contingently true? Rather, how can a formula φ be contingently contradictory? This is easier to answer, since we only need to find a formula that is both true and false in some, but not all, accessible worlds. Vague predicates provide good candidates for contingent contradictions (cf. Cobreros et al. 2015; Priest 2010; Ripley 2011). Take the predicate 'is bald', and consider a standard Sorites story about Graham being bald, from Graham having no hair to having a full head of hair, with intermediate Grahams having one more hair. Graham-1 is definitely bald, and for some large enough n, Graham-n is definitely not bald. There must thus be a last bald Graham and a first non-bald Graham, call this a *boundary-Graham*. But due to the vague nature of baldness, there are multiple boundary-Grahams (see Weber 2010). Now, line-up all Grahams from totally bald to totally not bald, each in its own possible world. Since some Grahams are boundary-Grahams, there are corresponding possible worlds with boundary-Grahams, and in those worlds, Graham is both bald and non-bald. Consequently, 'Graham is bald' is a contingent dialetheia. 'Graham is bald', unlike 'This sentence is true', is not a necessary dialetheia. So there are contingent and (sometimes both) necessary dialetheias.

So there are necessary and contingent dialetheias, and we need to account for those. My proposal is to adapt conditional modal logic to dialetheism. The main goal of dialetheic conditional modal logic is to accommodate modal talk of dialetheias, as well as impossibilia, in dialetheic conditional modal *models*. Models themselves have to be dialetheic.

The most common approach to providing semantics for deviant logics is to create models in orthodox set theory. But this set theory doesn't tolerate inconsistency, so epicycles are needed to provide semantics for deviant logics, especially those that tolerate inconsistency like dialetheic logic. Perhaps the most common epicycle is the introduction of multiple truth values (cf. Priest 2008). For dialetheic modal conditional logic, however, I will follow (Priest 2016) and split the propositional valuation into its *positive* and *negative* parts— like two overlapping solar systems.

The language of LP doesn't have a detachable arrow. Models M have a domain of worlds W, two selection functions $R : W \longrightarrow \wp(W)$ and $S : W \times \wp(W) \longrightarrow \wp(W)$, a *positive* propositional valuation V^+, and a *negative* propositional valuation V^-. R and S are respectively used to interpret modal and conditional operators. The semantic definition is given by the following recursive definition of a *truth-set* $[\![\varphi]\!]^{+M}$ (the set of worlds at which φ is true) and a *falsity-set* $[\![\varphi]\!]^{-M}$ (the set of worlds at which φ is false):

$$
\begin{aligned}
w \in [\![p]\!]^{+M} &\equiv w \in V^+(p) \\
w \in [\![\neg\varphi]\!]^{+M} &\equiv w \in V \in [\![\varphi]\!]^{-M} \\
w \in [\![\varphi \land \psi]\!]^{+M} &\equiv w \in [\![\varphi]\!]^{+M} \land w \in [\![\psi]\!]^{+M} \\
w \in [\![\varphi \lor \psi]\!]^{+M} &\equiv w \in [\![\varphi]\!]^{+M} \lor w \in [\![\psi]\!]^{+M} \\
w \in [\![\Diamond\varphi]\!]^{+M} &\equiv R(w) \cap [\![\varphi]\!]^{+M} \neq \emptyset \\
w \in [\![\Box\varphi]\!]^{+M} &\equiv R(w) \subseteq [\![\varphi]\!]^{+M} \\
w \in [\![\varphi \diamondsuit\!\!\rightarrow \psi]\!]^{+M} &\equiv S(w, [\![\varphi]\!]^{+M}) \cap [\![\psi]\!]^{+M} \neq \emptyset \\
w \in [\![\varphi \,\square\!\!\rightarrow \psi]\!]^{+M} &\equiv S(w, [\![\varphi]\!]^{+M}) \subseteq [\![\psi]\!]^{+M}
\end{aligned}
\quad \Big\| \quad
\begin{aligned}
w \in [\![p]\!]^{-M} &\equiv w \in V^-(p) \\
w \in [\![\neg\varphi]\!]^{-M} &\equiv w \in V \in [\![\varphi]\!]^{+M} \\
w \in [\![\varphi \land \psi]\!]^{-M} &\equiv w \in [\![\varphi]\!]^{-M} \lor w \in [\![\psi]\!]^{-M} \\
w \in [\![\varphi \lor \psi]\!]^{-M} &\equiv w \in [\![\varphi]\!]^{-M} \land w \in [\![\psi]\!]^{-M} \\
w \in [\![\Diamond\varphi]\!]^{-M} &\equiv R(w) \subseteq [\![\varphi]\!]^{-M} \\
w \in [\![\Box\varphi]\!]^{-M} &\equiv R(w) \cap [\![\varphi]\!]^{-M} \neq \emptyset \\
w \in [\![\varphi \diamondsuit\!\!\rightarrow \psi]\!]^{-M} &\equiv S(w, [\![\varphi]\!]^{+M}) \subseteq [\![\psi]\!]^{-M} \\
w \in [\![\varphi \,\square\!\!\rightarrow \psi]\!]^{-M} &\equiv S(w, [\![\varphi]\!]^{+M}) \cap [\![\psi]\!]^{-M} \neq \emptyset
\end{aligned}
$$

As before, I will use the simplified notation $[\![\cdot]\!]^+$ and $[\![\cdot]\!]^-$. To make the logic dialetheic, we simply allow V^+ and V^- to overlap, so that worlds may be in the truth and falsity sets of a formula. Frame and model validity are defined in the standard way. A formula φ is *valid in a model* if $w \in [\![\varphi]\!]$ for every $w \in W$, and it is *valid in a frame* if it is valid in every model based on the frame (i.e. for every propositional valuation). Finally, a formula is valid in a set of frames if it is valid in every frame in it.

So far so good, but there's a problem, noted by Martin (2014): the actual world is an impossible world. The argument is as follows. Since $\neg(\varphi \land \neg\varphi)$ is valid, $\Box\neg(\varphi \land \neg\varphi)$ is also valid. But since modalities dualise,[15] we get that $\neg\Diamond(\varphi \land \neg\varphi)$. 'Given that an impossible world is a world w where propositions that cannot possibly be true are true, any world w at which a contradiction is true is going to be an impossible world, according to the modal semantics given above' (Martin 2014, p. 66). But dialetheists are committed to true contradictions in the actual world, so they are committed to the actual world being impossible.[16] This is absurd, because the actual world, if anything, is possible. And this is bad, because one of the driving motivation for conditional modal logic, in this paper anyway, is to use $\Diamond$ as the measure of possibility. To fix the problem, Martin introduces an additional epicycle in models to block the duality of modalities. The epicycle is to enforce 'exclusivity of truth and falsity for modal formulae' (p. 73). The solution, namely to block duality of modalities, is the one I will also adopt below, but as a natural feature of the dialetheic models instead of an ad hoc patch. But first, let me provide one more motivation for adopting dialetheic meta-theory: advanced modalising.

Whereas basic modalising is about ordinary objects, like atoms, people, planets or galaxies (what if the Earth had three moons?), advanced modalising (Divers 1999, 2014) is about entities that are not ordinary objects, like sets and worlds (what if there were only one possible world?). Advanced modalising problems arise at the intersection of logic and metaphysics (Jago 2016). Several problems can be addressed with a dialetheic conditional modal logic, but one is salient, namely, the existence of the set of all worlds (Forrest and Armstrong 1984; Nolan 1996; Pruss 2001).

According to Forrest and Armstrong (1984), Lewis is committed to the following principles: 1) distinct worlds do not have any parts in common and 2) for any objects in any worlds, there exists a world that contains any number of duplicates of all of those objects (the principle of recombination). These two principles generate a

---

[15] $w \in [\![\Box\neg\varphi]\!]^+ \Rightarrow R(w) \subseteq [\![\neg\varphi]\!]^+ \Rightarrow R(w) \subseteq [\![\varphi]\!]^- \Rightarrow w \in [\![\Diamond\varphi]\!]^- \Rightarrow w \in [\![\neg\Diamond\varphi]\!]^+$.

[16] Since the problem is about the duality of modalities, it arises just the same in a formulation with three truth values.

cardinality problem for modal realism. The problem can easily be understood with the help of the inclosure schema of Priest (2002). Simply take Ω to be the set of all worlds, and a diagonaliser δ that takes a set of worlds and returns a world that combines all the atoms of the worlds in the set. Applied to the set of all worlds, the diagonaliser returns a world that has more atoms than it has, because of Cantor's theorem; contradiction. By the principle of uniform solution, the set of all worlds is a dialetheia.

The problem of the set of all worlds is only a problem for orthodox metaphysicians that assume explosive set theory and logic. The problem does not exist for dialetheic metaphysicians, so long as their theory is developed dialetheically; hence, the need for dialetheic dialetheic modal conditional logic, to which I now turn.

14.3 Dialetheic Dialetheic Modal Conditional Logic

By dialetheic dialetheic modal conditional logic, I mean a conditional modal logic as presented above, but entirely constructed dialetheically, including the set theory. So as to not be working in the void, here's the background propositional logic I am assuming. Any set-theoretical reasoning in the rest of the paper is to be regulated by it.[17]

Axioms:
$\varphi \to (\psi \to \varphi)$
$(\varphi \to (\psi \to \chi)) \to (\psi \to (\varphi \to \chi))$
$(\varphi \to \psi) \to ((\chi \to \varphi) \to (\chi \to \psi))$
$\varphi \vee \neg \varphi$
$\neg(\varphi \wedge \neg\varphi)$
$\varphi \leftrightarrow \neg\neg\varphi$
$\forall x \varphi \to \varphi$
$\varphi \to \exists x \varphi$

$\varphi \to \varphi \vee \psi$
$\psi \to \varphi \vee \psi$
$(\varphi \to \chi) \to ((\psi \to \chi) \to (\varphi \vee \psi \to \chi))$
$(\varphi \to (\psi \to \chi)) \to (\varphi \wedge \psi \to \chi)$
$\varphi \to (\psi \to \varphi \wedge \psi)$
$\forall x(\varphi \to \psi) \to (\exists x \varphi \to \psi)$ (x not free in ψ)
$\forall x(\varphi \to \psi) \to (\varphi \to \forall x \psi)$ (x not free in φ)

Rules: $\dfrac{\varphi \quad \varphi \to \psi}{\psi} \qquad \dfrac{\varphi}{\forall x \varphi}$

The set theory I assume is naive, and satisfies our good old set-theoretical friends:

Extensionality $\forall z(z \in x \leftrightarrow z \in y) \to x = y$
Comprehension $\exists y \forall x(x \in y \leftrightarrow \varphi)$

I will rely on the usual set-theoretical operations, but defined in the logically closed theory:

$X \subseteq Y := \forall z(z \in X \to z \in Y) \quad X \cap Y := \{z : z \in X \wedge z \in Y\}$
$X \setminus Y := \{z : z \in X \wedge \neg(z \in Y)\} \quad X \cup Y := \{z : z \in X \vee z \in Y\}$

---

[17] As I want to keep the focus on modalities and conditionals, I won't go into motivating the propositional logic here, and refer you to (Weber et al. 2016) instead.

14 Dialetheic Conditional Modal Logic

The language of dialetheic dialetheic modal conditional logic (DDCML) is a standard propositional language with atoms PROP $\cup \{\bot\}$,[18] propositional connectives ($\neg, \wedge, \vee, \rightarrow, \leftrightarrow$), modal ($\Diamond, \Box$) and conditional ($\Box\!\!\rightarrow, \Diamond\!\!\rightarrow$) operators. As above, I'll start with a minimal version of the logic, called MDDCML. Models M for MDDCML have a domain of worlds W, two selection functions $R : W \longrightarrow \wp(W)$ and $S : W \times \wp(W) \longrightarrow \wp(W)$, and a truth function $[\![\cdot]\!]^M$ that assigns subsets of W to every formula. R and S are respectively used to interpret modal and conditional operators. The truth function is constrained by the following conditions:

$$[\![p]\!]^M \subseteq W$$
$$w \in [\![\bot]\!]^M \leftrightarrow w \in [\![\varphi]\!]^M \text{ for every } \varphi$$
$$w \in [\![\neg\varphi]\!]^M \leftrightarrow w \in W \setminus [\![\varphi]\!]^M$$
$$w \in [\![\varphi \wedge \psi]\!]^M \leftrightarrow w \in [\![\varphi]\!]^M \cap [\![\psi]\!]^M$$
$$w \in [\![\varphi \vee \psi]\!]^M \leftrightarrow w \in [\![\varphi]\!]^M \cup [\![\psi]\!]^M$$
$$w \in [\![\varphi \rightarrow \psi]\!]^M \leftrightarrow w \in [\![\varphi]\!]^M \rightarrow w \in [\![\psi]\!]^M$$
$$w \in [\![\varphi \leftrightarrow \psi]\!]^M \leftrightarrow w \in [\![\varphi]\!]^M \leftrightarrow w \in [\![\psi]\!]^M$$
$$w \in [\![\Diamond\varphi]\!]^M \leftrightarrow \exists v \in R(w) \cap [\![\varphi]\!]^M$$
$$w \in [\![\Box\varphi]\!]^M \leftrightarrow R(w) \subseteq [\![\varphi]\!]^M$$
$$w \in [\![\varphi \Diamond\!\!\rightarrow \psi]\!]^M \leftrightarrow \exists v \in S(w, [\![\varphi]\!]^M) \cap [\![\psi]\!]^M$$
$$w \in [\![\varphi \Box\!\!\rightarrow \psi]\!]^M \leftrightarrow S(w, [\![\varphi]\!]^M) \subseteq [\![\psi]\!]^M$$

So MDDCML looks very much like MCML, but don't be deceived by how things look. You shouldn't read the semantics in the standard recursive way. Instead of having a propositional valuation on atoms that is lifted to arbitrary formulas recursively, the truth function is defined over all formulas in a circular way. Think of the truth function as being regimented by axiom schemas. The arrow $\rightarrow$ does not contrapose or contract, so the information can only be read in a positive way. Negation doesn't secure duality of modalities, as in MCML. This means that an axiomatisation of the normal modal fragment of the logic wouldn't have the usual dual axioms $\Diamond\varphi \leftrightarrow \neg\Box\neg\varphi$.

For a proper dialetheic conditional modal logic, we add all of the VC and S5 constraints of conditional modal logic from the previous section, as well as POS, but formulated with the present logic. So here's a dialetheic version of the constraints:

T	$\Box\varphi \rightarrow \varphi$		REF	$w \in R(w)$
4	$\Box\varphi \rightarrow \Box\Box\varphi$		TRANS	$v \in R(w) \wedge u \in R(v) \rightarrow u \in R(w)$
5	$\varphi \rightarrow \Box\Diamond\varphi$		SYM	$v \in R(w) \rightarrow w \in R(v)$
id	$\varphi \Box\!\!\rightarrow \varphi$		ID	$S(w, X) \subseteq X$
cc	$(\varphi \Box\!\!\rightarrow \psi \wedge \varphi \Box\!\!\rightarrow \xi) \rightarrow \varphi \Box\!\!\rightarrow \psi \wedge \xi$		CC	$(S(w, X) \subseteq Y \wedge S(w, X) \subseteq Z) \rightarrow (S(w, X) \subseteq Y \cap Z)$
ca	$(\varphi \Box\!\!\rightarrow \xi \wedge \psi \Box\!\!\rightarrow \xi) \rightarrow \varphi \vee \psi \Box\!\!\rightarrow \xi$		CA	$(S(w, X) \subseteq Z \wedge S(w, Y) \subseteq Z) \rightarrow ((S(w, X) \cup S(w, Y)) \subseteq Z$
cso	$(\varphi \Box\!\!\rightarrow \psi \wedge \psi \Box\!\!\rightarrow \varphi) \rightarrow (\varphi \Box\!\!\rightarrow \xi \leftrightarrow \psi \Box\!\!\rightarrow \xi)$		CSO	$(S(w, X) \subseteq Y \wedge S(w, Y) \subseteq X) \rightarrow S(w, X) \subseteq Z \leftrightarrow S(w, Y) \subseteq Z$
cv	$(\varphi \Box\!\!\rightarrow \xi \wedge \varphi \Diamond\!\!\rightarrow \psi) \rightarrow \varphi \wedge \psi \Box\!\!\rightarrow \xi$		CV	$(S(w, X) \subseteq Z \wedge S(w, X) \cap Y \neq \emptyset) \rightarrow S(w, X \cap Y) \subseteq Z$
cs	$\varphi \wedge \psi \rightarrow \varphi \Box\!\!\rightarrow \psi$		CS	$w \in X \wedge w \in Y \rightarrow S(w, X) \subseteq Y$
mp	$\varphi \Box\!\!\rightarrow \psi \rightarrow (\varphi \rightarrow \psi)$		MP	$w \in X \rightarrow w \in S(w, X)$
pos	$((\varphi \Box\!\!\rightarrow \psi) \wedge \Diamond\varphi) \rightarrow \Diamond\psi$		POS	$(S(w, X) \subseteq Y \wedge R(w) \cap X \neq \emptyset) \rightarrow (R(w) \cap Y \neq \emptyset)$

---

[18] I add the constant $\bot$ to be able to identify triviality. Unlike in twentieth-century logic, it cannot be defined by a contradictory formula.

As I've promised above, since modalities don't dualise, we do not have the problem raised by Martin. Whereas $\Box\neg(\varphi \wedge \neg\varphi)$ is valid, $\neg\Diamond(\varphi \wedge \neg\varphi)$ isn't; the actual world is possible even if it is inconsistent. We do not need epicycles to block the duality of modalities and the ensuing impossibility of the actual world. That the actual world is not impossible is a pleasant natural outcome of DDCML.

We definitely do not want to have NEC in DDCML, because of the trivial world, which is now a world just as much as any other world. If the trivial world exists, hopefully it is not a relative possibility to our world,[19] because then our world is also trivial (Humberstone 2011). Indeed, $\Diamond\bot \to \bot$ is valid in DCML.[20] But there are things that are strictly false: 'there are only two humans on Earth' is false. So the actual world is not in the equivalence class of worlds that contain the trivial world, on pain of absurdity.

It should now be easy to see that the conditional triviality principle $\Diamond\bot \,\square\!\!\to\, \bot$ is also valid in DCML. But notice that it is also valid in CML! The reason it is valid in CML, however, is that the trivial world does not exist, so conditionalising on the possibility of triviality is vacuous, and so trivial. This is not too surprising, since $\Diamond\bot \to \bot$ is also valid in orthodox S5. The advantage of DDCML over CML is to give proper credit (and place, in some other equivalence class of worlds) to the trivial world, while offering natural solutions to problems which have generated a plethora of epicycles. The DDCML metaphysical map of the universe of worlds is much cleaner.

14.4 Conclusion

A theory of impossibility can be modelled in orthodox twentieth-century logic, so long as conditionals are allowed to reach outside of the space of relative possibility of worlds, against the orthodox *Leibnizian* view. But the actual world is dialetheic, so conditional modal logic does not get at the heart of the metaphysics. For that, we need dialetheic dialetheic modal conditional logic, a logic accompanied by a suitable dialetheic set theory, one that is regulated by the same logic. This is a new joint venture for logicians and metaphysicians. As Kant (2004) famously said:

> Any doctrine of nature will contain only as much proper science as there is mathematics capable of application there.

Well, this is also true in philosophy:

> Any doctrine of philosophy will contain only as much proper metaphysics as there is logic capable of application there.

---

[19] With apologies to Kabay (2008).

[20] Assume that $w \in [\![\Diamond\bot]\!]$, then $R(w) \cap [\![\bot]\!] \neq \emptyset$, so there exists a world x such that $x \in R(w) \wedge x \in [\![\bot]\!]$. But $x \in [\![\bot]\!]$ implies that $x \in [\![\Box\varphi]\!]$ for every φ, so $R(x) \subseteq [\![\varphi]\!]$ for every φ. Furthermore, $x \in R(w)$ implies by 5 that $w \in R(x)$. So there is a world x such that $w \in R(x) \wedge R(x) \subseteq [\![\varphi]\!]$ for every φ, so $w \in [\![\varphi]\!]$ for every φ. Therefore $w \in [\![\bot]\!]$.

I propose dialetheic dialetheic modal conditional logic as a good starting point. Let's just call it dialetheic conditional modal logic, shall we?

> Perhaps there will be a time when logicians will look with incredulity at the naivety of their predecessors who thought that consistency was a *sine qua non* of reasoning, rationality, and similar notions.
>
> I do not claim that the equipment is perfect—far from it. But it at least allows us to make a start on exploring this new and unfamiliar terrain—the transconsistent.
>
> — Graham Priest, "In Contradiction"

Acknowledgements Graham, you've been a source of inspiration for several scholars that have known or read you, and you are a model of perseverance against adversity. Above all, you are a good friend, and I'm very pleased to be able to contribute this paper in your honour. I would like to thank Zach Weber, Fred Kroon, Joy Britten and Michael Hillas for valuable comments on a previous draft of this paper.

References

Baker, A. (2005). Are there genuine mathematical explanations of physical phenomena? *Mind*, *114*(454), 223–238.

Baron, S., Colyvan, M., & Ripley, D. (2017). How mathematics can make a difference. *Philosophers Imprint*, *17*(3). University of Michigan Press.

Berto, F., French, R., Priest, G., & Ripley, D. (2017, August). Williamson on counterpossibles. *Journal of Philosophical Logic*.

Blackburn, P., de Rijke, M., & Venema, Y. (2001). *Modal logic*. New York: Cambridge University Press.

Chellas, B. F. (1980). *Modal logic: An introduction*. Cambridge; New York: Cambridge University Press.

Chellas, B. F. (1975). Basic conditional logic. *Journal of Philosophical Logic*, *4*(2), 133–153.

Cobreros, P., Egré, P., Ripley, D., & van Rooij, R. (2015, August). Pragmatic interpretations of vague expressions: Strongest meaning and nonmonotonic consequence. *Journal of Philosophical Logic*, *44*(4), 375–393.

Divers, J. (1999). A genuine realist theory of advanced modalizing. *Mind*, *108*(430), 217–240.

Divers, J. (2014). The modal status of the Lewisian analysis of modality. *Mind*, *123*(491), 861–872.

Forrest, P., & Armstrong, M. D. (1984, June). An argument against David Lewis' theory of possible worlds. *Australasian Journal of Philosophy*, *62*(2), 164–168.

Girard, P., & Weber, Z. (2015) Bad worlds. *Thought: A Journal of Philosophy*.

Humberstone, L. (2011, January). Variation on a trivialist argument of Paul Kabay. *Journal of Logic, Language and Information*, *20*(1), 115–132.

Jago, M. (2016, July). Advanced modalizing problems. *Mind*, *125*(499), 627–642.

Kabay, P. D. (2008). *A defense of trivialism*.

Kant, I. (2004). *Prolegomena to any future metaphysics that will be able to come forward as science: With selections from the critique of pure reason*. Cambridge texts in the history of philosophy (rev. ed.). Cambridge, UK; New York: Cambridge University Press.

Kripke, S. A. (1959). A completeness theorem in modal logic. *The Journal of Symbolic Logic*, *24*(1), 1–14.

Kripke, S. A. (1980). *Naming and necessity*. Library of philosophy and logic (revised and enlarged ed.). Oxford: Blackwell.

Lewis, D. K. (1973). *Counterfactuals*. Library of philosophy and logic. Oxford: Blackwell.
Martin, B. (2014, September). Dialetheism and the impossibility of the world. *Australasian Journal of Philosophy*, 1–15.
Nolan, D. (1996). Recombination unbound. *Philosophical Studies*, *84*(2–3), 239–262.
Nolan, D. (1997). Impossible worlds: A modest approach. *Notre Dame Journal of Formal Logic*, *38*(4), 535–572.
Priest, G. (2002). *Beyond the limits of thought*. Oxford: Oxford University Press.
Priest, G. (2006). *In contradiction*. Oxford Scholarship Online, Oxford University Press.
Priest, G. (2008). *An introduction to non-classical logic: From if to is*. Cambridge introductions to philosophy (2nd ed.). Cambridge; New York: Cambridge University Press.
Priest, G. (2010, January). Inclosures, vagueness, and self-reference. *Notre Dame Journal of Formal Logic*, *51*(1), 69–84.
Priest, G. (2016). Thinking the impossible. *Philosophical Studies*, *173*(10), 2649–2662.
Pruss, A. R. (2001). The cardinality objection to David Lewis's modal realism. *Philosophical Studies*, *104*(2), 169–178.
Ripley, D. (2011). Contradictions at the borders. In R. Nouwen, R. van Rooij, U. Sauerland, & H.-C. Schmitz (Eds.), *Vagueness in communication* (pp. 169–188). Springer.
Stalnaker, R. C. (1968). A theory of conditionals. In W. L. Harper, R. Stalnaker, & G. Pearce (Eds.), *IFS*, number 15 in The University of Western Ontario Series in Philosophy of Science (pp. 41–55). The Netherlands: Springer. https://doi.org/10.1007/978-94-009-9117-0_2.
Weber, Z. (2010, October). A paraconsistent model of vagueness. *Mind*, *119*(476), 1025–1045.
Weber, Z., Badia, G., & Girard, P. (2016). What is an inconsistent truth table? *Australasian Journal of Philosophy*, *94*(3), 533–548.
Williamson, T. (2010, March). Modal logic within counterfactual logic. In B. Hale & A. Hoffmann (Eds.), *Modality: metaphysics, logic, and epistemology*. Oxford University Press.

Patrick Girard is originally from Québec, Canada, and is now a New Zealand citizen. He completed his Ph.D. in Philosophy at Stanford University in 2008, specialising in Logic, and is now a senior lecturer in Philosophy at the University of Auckland.

Chapter 15
Priest on Negation

Lloyd Humberstone

Abstract What conception of negation a dialetheist might have, in holding that a statement and its negation can both be true, has been the subject to considerable debate. Several of the issues in play in this area—such as the unique characterization of negation, and the interplay between contrariety and subcontrariety—are broached here by considering some positions taken on them by Graham Priest and assorted critics (in particular, Hartley Slater and Jean-Yves Béziau). Some of the more intricate points, as well as detailed discussions of commentators (including those just mentioned as well as Heinrich Wansing) on Priest are handled in explicitly labelled Digressions and in two Appendices.

Keywords Unique characterization of connectives · Negation · Logical relations · Contraries and subcontraries · Paraconsistency

15.1 Introduction and Background

Let us begin by recalling (i) that a paraconsistent logic is one according to which not just anything follows from an arbitrarily chosen contradiction, $A \wedge \neg A$, or—what we need not distinguish from this for present purposes (setting to one side what are called non-adjunctive logics[1])—follow from a formula A taken together with its negation $\neg A$, (ii) that one motivation for the study of such logics is for their use as the background logic in the development of inconsistent but non-trivial theories and

---

[1] See any survey of paraconsistent logic—for example, Priest and Routley (1989) (especially, Sects. 2.1, 3.1), Priest (2002) (Sect. 4.2), or Sect. 2 of Béziau (2002).

L. Humberstone (✉)
Department of Philosophy and Bioethics, Monash University,
Melbourne, VIC 3800, Australia
e-mail: lloyd.humberstone@monash.edu

© Springer Nature Switzerland AG 2019
C. Başkent and T. M. Ferguson (eds.), *Graham Priest on Dialetheism and Paraconsistency*, Outstanding Contributions to Logic 18, https://doi.org/10.1007/978-3-030-25365-3_15

(iii) that one position in conspicuous need of paraconsistent logic, so understood, is dialetheism, according to which the whole truth about the world is just such an inconsistent though non-trivial theory.[2]

In a characteristically forthright publication (Slater 1995), Hartley Slater argued that the motivation for paraconsistent logic from dialetheism, or from any variation on this position which finds semantic expression in the existence of (perhaps world- or index-relative) assignments of both the values True and False to formulas, undercuts itself, since if any prospect of the joint truth of A and $\neg A$ is on the cards, this means that $\neg$ does not deserve to be regarded as symbolizing negation. After all, negation should turn a statement into something incapable of being true when that statement is true as well as something incapable of being false when that statement is false. The first requirement is not satisfied by the envisaged semantic treatment of the putative paraconsistent logic even if the second is, as in Priest's favoured gluts-but-no-gaps position.[3] Or in traditional parlance, negation should form a contradictory rather than just a subcontrary of what is negated.

To this objection Priest has his own reply, which we will get to in Sect. 15.2.[4] The preamble to the reply as it appears in response not to Slater (1995) but to the similarly oriented Slater (2007a), does not seem to begin too well.[5] The following is from the latter reply, Priest (2007), p. 467:

> Slater and I agree that negation, whatever it is, is a contradictory-forming operator (cfo). It is the relation that obtains between pairs such as 'Socrates is mortal', 'Socrates is not mortal' and 'Some person is mortal', 'No person is mortal'. The crucial question, then, is what exactly, this relationship amounts to.
>
> Traditional logic—by which I mean logic in the Aristotelian tradition—characterizes the relation in a familiar way. A and B are contradictories if you must have one or the other, but you can't have both. That is, $\Box(A \vee B)$ and $\neg\Diamond(A \wedge B)$. Hence, A and $\neg A$ are contradictories if we have $\Box(A \vee \neg A)$ and $\neg\Diamond(A \wedge \neg A)$, that is, $\Box\neg(A \wedge \neg A)$. Consider any propositional logic, the modal extension of which satisfies Necessitation (if $\vdash A$ then $\vdash \Box A$), as it should. Then if it is such that both:
>
> (1) $A \vee \neg A$
> (2) $\neg(A \wedge \neg A)$
>
> are logical truths, $\neg$ is a cfo. Since *LP* satisfies these conditions, its negation symbol is a cfo. Note that this is not true of either intuitionistic logic or some paraconsistent logics, such as

---

[2]More explicitly, what is at issue here is non-trivialist dialetheism, since if trivialism—the view that all statements are true—were correct then there would be no equally pressing need to curtail the inference from contradictions to arbitrary statements. Priest's spelling of *dialetheism* with the 'e' before the '-ism' is employed here and in *dialetheist*; but I will follow Priest's etymological critics in writing *dialethic* rather than *dialetheic*, though more for reasons of euphony than historical fidelity.

[3]Favoured for current purposes, that is; with other considerations in mind Priest has given a sympathetic hearing to gluts and gaps and more—as in the exotic five-valued logic of Sect. 4.2 of Priest (2010), or Sects. 24.5–6 of Priest (2015), or, most fully in Sects. 4–5 of Priest (2014).

[4]Other replies to Slater (1995) include Béziau (2006), Brown (1999), Paoli (2003) and Restall (1997), from some of which we will be hearing in what follows.

[5]Slater replies to this response in Slater (2007b).

the da Costa C-systems.⁶ In the first of these, (1) is not a logical truth; in the second, (2) is not. This is why the charge that the negation operator in those logics is not really a cfo gets its bite.⁷

I read this as saying that for a one-place connective # to count as a cfo in a logic S it is necessary and sufficient that we have, for an arbitrary formula A of the language of that logic, a language presumed equipped with binary connectives $\wedge$ and $\vee$ that:

(1′) $\vdash_S A \vee \#A$
(2′) $\vdash_S \#(A \wedge \#A)$

and Priest concludes from the fact that with S as LP and # as $\neg$ these conditions are satisfied, that $\neg$ is a cfo in LP. No doubt the proposed definition of being a cfo—a confusing abbreviation used here for the last time in view of the fact that 'contradictory' and 'contrary' both begin with a 'c'—should also place some conditions on the behaviour of $\wedge$ and $\vee$ appearing in these conditions to secure that they live up to expectations as symbolizing conjunction and disjunction. But even with such an addition in place, this is an unpromising point from which to launch a reply to Slater, because if we take S instead as classical propositional logic with some functionally complete set of primitive connectives we could take # to be the connective associated with the constant-true 1-ary truth function (the *Verum* function, as it is sometimes called), and (1′) and (2′) would be satisfied. (Putting in terms of (1) and (2) themselves: we could interpret $\neg$ as expressing this truth function rather than the usual negation truth function.) Whatever slack (if any) there might be in the notion of a contradictory-forming operator, I take it that this Verum connective is not such an operator.⁸ So, with an overgenerous definition like this of a contradictory-forming operator in hand, showing that negation-according-to-LP counts as one, will not seem particularly impressive to anyone sympathetic to Slater's critique of the paraconsistent enterprise, or more specifically the dialethically motivated version of that enterprise.

This unintended interpretation of $\neg$ could be avoided if we took as unstated background assumptions certain principles as governing this connective and its relations to conjunction and disjunction: for example, if the presumption was that $\neg$

---

⁶This point about da Costa's logics was made by Priest and Routley in their (1989), as is quoted by Slater in (1995) with a view that Priest is vulnerable to the same criticism as he and Routley had levelled against da Costa. According to Béziau (2002, p. 473, n. 6), 'Later on Priest recognized that we should rather consider erroneous his original argument against da Costa's negation than to think that Slater's generalized argument is right.' This is not easily reconciled with the fact that the passage quoted from Priest with its remarks about da Costa above appeared in a publication dated 5 years later than that in which Béziau makes this remark.

⁷In this quotation, I have added parentheses around the '1' and '2', for convenient back reference. Priest has a similar discussion in other places, such as Chapter 4 ('Contradiction') of Priest (2006b), which had itself appeared (more or less) as Priest (1999); see also Priest (2002), p. 379.

⁸Béziau (2002), using 'LNC' as Priest does and 'EC' ('ex contradictione') for the principle that (for all A, B): $A, \neg A \vdash B$, writes (p. 477): 'In fact the question is still open to know if we can find an intuitive interpretation of an operator which obeys EC but not LNC or obeys LNC and not EC.' It is not clear what 'intuitive' means here, but the Verum interpretation satisfies LNC without satisfying EC, and a 'Falsum' interpretation satisfies EC without satisfying LNC.

is De Morgan negation, meaning that for all A, B, we have $\neg\neg A$ and A equivalent, as were $\neg(A \vee B)$ and $\neg A \wedge \neg B$, and also $\neg(A \wedge B)$ and $\neg A \vee \neg B$. The first of these would already disqualify the Verum interpretation of $\neg$. Taken altogether, though, they have the potentially awkward effect, emphasized, for example, in Brady (2004) (especially, Sect. 2), that (1′) and (2′) end up being equivalent to each other.[9] Since we do not want to bring conditional and biconditional connectives into the story,[10] we can think of the talk of equivalence here as involving a consequence relation $\vdash$, so that A and B are equivalent according to $\vdash$ ('$A \dashv\vdash B$') when $A \vdash B$ and $B \vdash A$.[11] ((1) and (2) are then understood as having a tacit '∅' on the left; as usual, we continue to keep this tacit.) Then the present point is that where $\vdash$ is any substitution-invariant consequence relation extending $\vdash_{\mathsf{FDE}}$, for any formula A, we have $A \vee \neg A \dashv\vdash \neg(A \wedge \neg A)$ (rather than the weaker claim that for any such $\vdash$, we have $\vdash A \vee \neg A$ for all A if and only if we have $\vdash \neg(A \wedge \neg A)$ for all A). What allowed the status of (1) and (2) to come apart in intuitionistic and da Costa style paraconsistent logics was the failure of these background assumptions, but with them in place, we seem to lose the distinction between the two conditions (1) and (2). We can still distinguish A and B's being subcontraries according to a consequence relation $\vdash_S$[12]:

(1″) $\vdash_S A \vee B$

from their being contraries according to $\vdash_S$:

(2″) $\vdash_S \neg(A \wedge B)$

and we can say that # is a subcontrary-forming operator or a contrary-forming operator according to $\vdash_S$, iff we have respectively (1‴) or (2‴)—for all formulas A in either case:

(1‴) $\vdash_S A \vee \#A$
(2‴) $\vdash_S \neg(A \wedge \#A)$.

The problem with (1) and (2) from Priest's discussion arises when we take # as $\neg$ itself is that $\neg$ is then playing a double role in (2), both as the connective used in characterizing contrariety—as in (2″) and (2‴)—and as the connective whose

---

[9] This point is also made near the top of p. 89 in Wansing's note (2006), whose subtitle coincides with my title.

[10] In Chap. 6 of Priest (2006a) there is a discussion of how one might supplement LP's resources with an entailment-expressing conditional, and the topic is broached several times in Priest (2006b)—for example, in Sect. 5.9.

[11] The notion of a consequence relation is assumed to be familiar; see, for example, p. 15 of Shoesmith and Smiley (1978).

[12] Some may feel that more needs to be said about subcontrariety and contrariety than is said by these formulations, so as to make the two relations mutually exclusive. This issue is discussed in Appendix 1 at the end of the paper; in fact, Priest himself takes this alternative line (the 'say more' line, as it is called in that Appendix). In the meantime, we assume they overlap, and that their intersection is the relation of being contradictories.

contrariety-forming credentials we are invoking that characterization to assess. ¬ is trying to play the first of these roles at its initial (main) occurrence in (2) and the second of them at its later (embedded) occurrence. If we don't use two different devices for these two jobs, we end up with the unsatisfactory (2′), which, as already observed, turns out to be satisfied whenever we have $\vdash_S \#A$ for all A—not what we had in mind for a contrary-forming operator.

We want to exclude such unintended interpretations of the negation connective, even if there is a sense (see Sect. 15.3) in which without a change of logical framework, other unintended interpretations are inevitable. And Priest makes a parenthetical remark at the start of Sect. 4.4 of Priest (2006b), immediately after remarking that we want to use ¬ to form the contradictory of a statement, to the effect that '(c)ontradictories, unlike contraries and subcontraries, are unique—at least to within logical equivalence'. While this, on one natural reading of it, is not the same as excluding all but a single semantic interpretation, the demand that negation should be uniquely characterized by the logical principles governing it, is another to which we shall return in Sects. 15.5 and 15.6. But first, some attention to Priest's reaction to Slater's objection is called for.

15.2 Contradictions Spread

In many places, and for our purposes p. 79 of Priest (2006b) is as good a source to cite as any other, Priest points out that there is no reason for a paraconsistent logic not to have as provable, and for the dialetheist not to accept as true, the Law of Non-Contradiction in the form given in (2) in the passage quoted from Priest in Sect. 15.1. This simply means that a particular inconsistent theory based on the logic in question, say including the theorems A and $\neg A$, will have among its theorems not only the contradiction $A \wedge \neg A$ but also the further contradiction resulting from conjoining this with the outright deliverances of the underlying logic $(A \wedge \neg A) \wedge \neg(A \wedge \neg A)$. Similarly in the case of the dialetheist accepting the one contradiction, there will be a commitment to a further contradiction, and then another—whose first conjunct is the previous contradiction and whose second is its negation, and so on. As Restall puts it, in his reply to Slater Restall (1997), 'contradictions in paraconsistent logics can *spread*'—though in the present formulation it is paraconsistently based theories rather than logics that this point applies to in the first place, since it arises when the paraconsistent logic is itself consistent.[13]

---

[13]It may be that Restall and Priest both accept that some of what the present author thinks of as non-logical theories, such naive semantics or naive set theory, are themselves to be thought of as part of the logic. For example, Priest's concern with getting a non-conservative extension (in fact, of nonconservativity in the most extreme form: triviality) by the joint imposition of two sets of principles when either set taken separately gives a conservative extension—for an uncontroversially 'purely logical' example of which, see p. 568 of Humberstone (2011a)—involves a rule governing a nullary connective together with an instance of the T-schema at Priest (2006b), p. 90, treats the two as being on a par in this respect. The principle governing the new connective, $*$ is that we

The spread Restall had in mind was not quite that just illustrated—from one non-modal contradiction to another (and another ...)—but, instead, a distinctively modal version, whose raw ingredients we saw in the passage quoted from Priest in Sect. 15.1. Slightly varying the dialectic, suppose our paraconsistent theorist says that (for example) the Russell set is an element of itself and it is not an element of itself, and gets the reply that since the theorist's own (modal) logic says that it is not possible for a statement and its negation to be true, it cannot be that—now abbreviating for convenience—$R \in R \land \neg(R \in R)$, since the logic requires that $\neg\Diamond(R \in R \land \neg(R \in R))$. But since whatever is the case is possible, there is also a commitment to $\Diamond(R \in R \land \neg(R \in R))$.

As both Priest (*in propria persona*) and Restall (on behalf of the dialetheist) envisage things, this is plausibly regarded as just another case of the inevitable spread. (The actual dialectical situation in Restall (1997) is that this case of spread shows that a weak modal dialetheism according to which some contradictions are possibly true collapses, in the current logical setting, into the less restrained standard dialetheist position that some contradictions *are* true.[14])

From the perspective of the hierarchy of languages, the above examples are cases of what might be termed horizontal spread. Priest is also happy to accept—though he would not welcome the label since he thinks of any such hierarchy as representing premature panic in the face of semantic paradox—cases of *vertical* spread: from the object language to the metalanguage. More particularly, Priest is happy for the logic of the metalanguage giving a semantic account of paraconsistent logic to itself be paraconsistent, saying that dialetheists, like intuitionists take themselves to be...

---

should have $A \leftrightarrow * \vdash B$, concerning which Priest says: "Triviality then follows from the instance of the T-schema $T\langle * \rangle \leftrightarrow *$." (As it happens the appeal to the T-schema is not in this case needed, since we can simply substitute $*$ itself for A in the condition $A \leftrightarrow * \vdash B$. But the *way* Priest cites the T-schema, perhaps not noticing the redundancy of the current appeal to it, suggests that it is being treated as a logical principle just like the unfortunate $*$ principle.) A warning about the word 'spread' in this connection: it is used quite differently in, for example, Sylvan and Urbas (1993) and (especially) Sect. 4 of Brady (2004), where the principle—rejection of which characterized paraconsistency—that contradictions have all statements as consequences, variously known elsewhere as *Explosion*, *Ex Falso Quodlibet*, *Ex Contradictione Quodlibet* (see Footnote 8), is referred to as the 'Spread Law'. On Restall's usage, the spread is from one contradiction to further contradictions, but not to triviality. Restall is sympathetic to the paraconsistent part of the story, which allows for this 'but not to triviality', though not to the dialetheist part, which would have us assenting to the contradictions in question.

[14]Closely related lines of thought can be found in more recent discussions, such as those of Asmus (2012) and Martin (2015). A related argument couched in possible worlds terms rather than in modal terms appeared in Lewis (1986), p. 7, note 3; of course Lewis had no sympathy for either the weaker or the *prima facie* stronger form of dialetheism, and was for this debate on the other side—doing the staring—of the famous incredulous stare: see Lewis (2004). (Priest describes a reaction of Stewart Shapiro's in these terms in Sect. 17.8 of Priest 2006a.) There is a corresponding issue about trivialism—of whether a weak modal form can be prevented from spreading to an outbreak of trivialism *simpliciter*, discussed, for example, in Humberstone (2011b).

...giving an account of the correct behaviour of certain logical particles. Is it to be supposed that their account of this behaviour is to be given in a way that they take to be incorrect? Clearly not. The same logic must be used in both 'object theory' and 'metatheory'.[15]

Such (as one says) homophonic semantic treatments come in two flavours: truth-theoretic and model-theoretic—a distinction Priest clearly explains in the previous chapter of Priest (2006b).[16] Homophonic incarnations of both styles of semantic theory have considerable interest and the model-theoretic case has raised philosophical attention in several areas.[17] But one place a dialethically motivated paraconsistent metalogic is not going to be of any assistance is in replying to someone like Slater who is deeply sceptical of the dialethic aspect of the enterprise.[18] We return to this presently, after looking at a passage aimed at undermining Slater's claim that the charge he (and Routley) had made against da Costa applies to his own logic (see Footnote 6 above).

Priest argues (2006b, pp. 79f. and 85f.) that the intuitionist and the paraconsistent logicians can each make use of a homophonic semantic theory (in the present case for simplicity in truth-theoretic form) with 'T' as a truth predicate and angle brackets as a way of forming a name for the statement (over which, in 2006, $\alpha, \beta, \ldots$ rather than the $A, B, \ldots$ above, are used as variables they enclose). In particular, concerning the homophonic clause for negation in the two cases, Priest says that the intuitionistic treatment still leaves negation only forming contraries and not contradictories while in the paraconsistent case if forms not just subcontraries but—*pace* Slater—contradictories. At p. 84f. he writes:

> In the case of someone who endorses the idea that there are truth value gaps, so that α and $\neg\alpha$ may both fail, one may object, as I did, that $\neg$ is not playing the role of a contradictory-

---

[15]Priest (2006b), p. 98.

[16]On p. 81: 'In this section, I have talked of truth. I have said nothing about truth-in-an-interpretation, as required, for example, for a model-theoretic account of validity. It is important to distinguish these two notions, for they are often confused. The first is a property (or at least a monadic predicate); the second is a (set-theoretic) relation. It is natural enough to suppose that truth is at least coextensive with truth-in-$\mathcal{I}$, where $\mathcal{I}$ is someone's privileged interpretation (set). And this may provide a constraint on the notion of truth-in-an-interpretation. But it, even together with an account of truth, is hardly sufficient to determine a theory of truth-in-an-interpretation. It does not even determine, for example, how to conceptualize an interpretation.'

[17]A smattering of discussions, listed chronologically: Dummett (1991) (especially, pp. 26–29, 33–37, 55–60 and 66), Humberstone (1996a), Brady and Rush (2009) (especially, Sect. 5), and Williamson (2017). Dummett's reply to Priest's rhetorical question, 'Is it to be supposed that their account of this behaviour is to be given in a way that they take to be incorrect?' is that the discussion should make use of only of principles accepted by both parties to such debates. This is admittedly going to make communication difficult in the present instance. Priest accepts the inference from 'It is impossible that A' to 'It is impossible that A and not-B', but not the further inference to 'Necessarily, if A then B'. One may then suggest (oversimplifying here for the sake of a structural point) that an argument be defined as *valid** in non-conditional terms, as one concerning which it is impossible that the premises are true and the conclusion not be, rather than as one concerning which it is necessary that if the premises are true then the conclusion is. Now an argument with contradictory premises is valid*, but from the validity* of an argument and the truth of its premises, the truth of the conclusion does not follow, by Priest's lights.

[18]Littmann and Simmons voice similar misgivings at p. 318 of Littmann and Simmons (2004).

forming operator. A genuine such operator is one given by the truth conditions:

$$T \langle \neg \alpha \rangle \text{ iff } \neg T \langle \alpha \rangle$$

> It is therefore natural to suppose that a dual objection can be made if one takes it that α and $\neg \alpha$ both be true. Dialetheic negation is merely a subcontrary-forming operator. The displayed clause defines the genuine contrary-forming operator.
>
> The situation is not the same, however. Given the notion of negation employed with gaps, the LEM and LNC fail. Given the conception of negation I have just described, they do not; so the negation is a contradictory-forming operator. It may just have surplus content as well.

This is rather hard to understand. Perhaps the sentence beginning 'Dialetheic negation' should run: 'Dialetheic negation, according to this supposition, is merely a subcontrary-forming operator', and the sentence after that seems to belong in the following paragraph, after the word 'however' (and perhaps to have 'contrary-forming' changed to 'contradictory-forming'). What then follows is presumably intended as a justification for this claim that the 'natural supposition' is mistaken—giving the reason that the situation is not the same (as that of intuitionistic negation), but it too is rather hard to follow, in part because the reference to the case of gaps had better subsume the case of intuitionistic logic if it is to bear on the disanalogy, and not just the three-valued Kleene logic (i.e. the logic determined by the matrix whose algebra coincides with Priest's *LP* matrix, but which designates only the 'top' value, rather than the top two), and in this case it is *not* true that 'both LEM and LNC fail', since we do have (2) from Sect. 15.1 provable in intuitionistic logic. (The common fate of these principles remarked on in that section was premised on $\neg$ being a De Morgan negation, which intuitionistic negation of course is not.) And the talk about surplus content at the end of the passage is not something I have been able to understand.[19]

Judging from the remarks quoted in the footnote just flagged, the talk of surplus content seems to be connected with paraconsistency more than dialetheism, and

---

[19] It is introduced on p. 83 of Priest (2006b) thus: 'As we observed in 4.4, the fact that $\neg \Diamond (\alpha \wedge \neg \alpha)$ holds does not rule out $\alpha \wedge \neg \alpha$ holding too. This does not mean that $\neg$ is not a contradictory-forming operator. It just means that there is more to negation than one might have thought. Let us call this more, for want of a better phrase, its *surplus content*. The classical view is to the effect that negation does not have surplus content: any such content would turn into the total content of everything since $\alpha \wedge \neg \alpha \vdash \beta$. But the classical view has been called into question by dialetheists'. I cannot understand this explanation of what *surplus content* is supposed to mean. Interestingly, as Sam Butchart pointed out to me, the phrase does not appear in the source text for Chap. 4, namely Priest (1999). Similarly, the phrase appears seven times in Chap. 5 of Priest (2006b), though not at all in the source material (Priest 1990). The first of these occurrences are on the opening page (p. 88) of the chapter: 'In the last chapter I argued for a certain account of negation. The account, whilst respecting traditional features of the notion of negation, such as its being a contradictory-forming operator, nonetheless allowed for contradictions to be true without triviality—that is, as I put it there, for negation to have surplus content'. Again, the metaphor seems to be the wrong way round: if the surplus content comprised the unwanted consequences of a negated statement when taken together with the statement negated, it would be *lacking* rather than *having* surplus content that was desirable in negation treated paraconsistently. (Yablo 2014, Sect. 6.1, employs the phrase surplus content when paraphrasing the argument of Popper and Miller 1983 against inductive probability, but this seems to be an unrelated use of the same terminology.)

despite its title, Slater (1995) does not defend the claim that from a statement and its negation, every statement follows. Instead, he asks whether a statement and its negation can both be true, and suggests that the answer must be *no* because part of what is required—the contrariety condition—for a connective to count as negation is that it forms from a statement something which *cannot* be true along with that statement. The 'spread' reply is that this is not being denied: they can't both be true. But then there is the disconcerting dialetheist addition: and also, they *can* be both be true (because, for example, they *are* both true). As I say, vertical spread like this is not going to persuade anyone sceptical of the enterprise, characteristically frustrated at the fact that there is *nothing* the acceptance of which commits the dialetheist to rejecting a given claim.[20] It would be better to explain to Slater how something of the form $A \wedge \neg A$ can be true other than by saying that it can be true because it can be the case that A is true and A is not true, but in such a way that $\neg$ does have something recognizable as the expected contrariety effect and not just the subcontrariety effect. It is simply bad form to contradict oneself in such mixed (dialetheist and non-dialetheist) company, at least if the aim is cross the unintelligibility gap.

This is just a special case, most recently recalled in Girard and Weber (2015), of Bob Meyer's famous description (from Meyer 1985) of heterophonic Routley–Meyer semantics (the metalogic being thoroughly classical) for relevant logic(s) as aiming to 'preach to the Gentiles in their own tongue'. Intensionality is foreign to standard mathematical discourse, so possible worlds semantics explains it in extensional terms, replacing the alien and potentially problematic 'necessarily' with familiar universal quantification over the elements of a Kripke model (perhaps with a suitable accessibility relation restricting the quantifier). Relevance is also something to come to terms with rather than to take for granted, so again we have a complicated variant in the Routley–Meyer model theory just mentioned with its ternary variation on the binary accessibility relation of the Kripke semantics, due to Routley and Meyer, with the Routley–Routley *-function (for negation) thrown in. Or we could just use the latter for *FDE* (no conditional connective present); the Urquhart semantics might be cited instead in this connection, with its greater intuitive appeal but somewhat looser fit with the logics it is meant to be a semantics for (though with some tweaking it is possible to correct the failure of fit in respect of disjunction: see Humberstone 1987). Alternatively for the case of negation in relevant logic, we could use the four-valued semantics. And for *LP* we have one of its three-valued reducts. Now devotees of *LP* can explain how a contradiction $A \wedge \neg A$ can be true (on an evaluation) without actually contradicting themselves in the process and leave the

---

[20]This issue arises in the context of dialogue with a dialetheist with whom you have a difference of opinion but who, disconcertingly, agrees with everything you say. It is given some attention in Ripley (2011) and, briefly, at the end of Horn and Wansing (2016), which provides a useful survey of several aspects of negation (though see also Wansing 2001 for some more technical material). Further references and discussion can be found in Priest (2006a), p. 291*ff.*

audience baffled.[21] The homophonic route has just this latter effect.[22] However, as we shall be able to appreciate after some semantic apparatus has been introduced in the following section, there really is no reply to (essentially) Slater's objection other than via this route, at least if we want to treat logics as consequence relations—or even (more refinedly) as sets of sequent-to-sequent rules in what is called in Sect. 15.5 the SET-FMLA framework. See especially Footnote 35. But we close the present section with an elaboration of Footnote 21, which, however, presumes a familiarity with the Dunn–Belnap semantics for *FDE*, described in Sect. 15.4.

Digression. A classic example of a well-motivated disinclination to think of functional completeness as something to be aimed for is provided by Stephen Blamey's restriction (in 2002) to monotonic functions as the operations in three-element matrices for partial logic compatible with the motivating idea that the third value represents what from a bivalent perspective is a truth value gap. Many combinatorially conceivable operations in the Dunn–Belnap semantics are similarly incompatible with the motivating idea of loosening up the Boolean semantics no more than is necessary to allow for gaps and gluts. See the discussion in p.22*f* of Humberstone (2014), for example, where a proposal of A. J. Dale is criticized as not supplying a suitable semantic interpretation for an intelligible connective in this setting. The idea is that for an *n*-ary connective #, here doubling as the name of the associated matrix operation $\#(X_1, \ldots, X_n)$, there should be a metalinguistic compound using conjunction and disjunction of the 2*n* atomic constituents $T \in X_i$, $F \in X_i$ ($i = 1, \ldots, n$) providing necessary and sufficient conditions for it to be the case that $T \in \#(X_1, \ldots, X_n)$, with the conditions obtained for $F \in \#(X_1, \ldots, X_n)$ obtained from this by interchanging conjunction and disjunction, and T and F, in this compound. We again get the kind of monotone property (with respect to $\subseteq$, which amounts to what is sometimes called the 'information ordering') from Blamey's discussion.[23] (The sets X_i here will take the form $h(A_i)$ in giving the Dunn–Belnap semantic account of $h(\#(A_1, \ldots, A_n))$.) Although Priest uses this feature as a technical tool in proving results about *LP* (to which it applies as for *FDE* except that we discard ∅—alias ***n***—from consideration), for example, in establishing 'Fact 1' on p. 50 of Priest (2006a), he does not seem to

---

[21] However, the use of matrix methodology should in no way be thought of as buying into the idea of functional completeness as a desideratum. This will frequently take us outside of what should be expressible given the particular motivations in play. See the Digression at the end of this section for illustration of this issue.

[22] It is not just over the issue of the vertical spread of contradictions that Priest seems to be following a motto: when the going gets tough, go homophonic. Note 30 of Chap. 5 of Priest (2006b), dealing with a suggestion of A. Everett by objecting to contraction for 'if…then …' in the metalanguage, is somewhat alarming. As Priest says there, he had treated the matter in somewhat similar terms in Priest (1996), where it is conditional proof that is called into question. But here we are discussing a Kripke style model theory with an accessibility relation which is supposed to be handling the non-classical conditional connective, so one does not expect to be debarred from reasoning *entirely classically* about what is true at which points in the models.

[23] As a referee reminds me, one can then take an interest in a restricted notion of functional completeness: definability of all the functions meeting whatever restrictions are considered desirable; cf. Wansing (1993) and references therein for deployment of a similar idea outside of the narrower confines of matrix semantics.

make as much of it as he might, perhaps because he is not particularly interested in distilling the essence of the Dunn–Belnap semantics. Taken as a constraint on the enterprise, however, it would immediately rule out the pathological connective mentioned in note 20 of Chap. 5 of Priest (2006b) (an example credited to Stephen Read), as well as a less obvious case from Denyer (1989), trying to make trouble for Priest's views by foisting on him a binary connective ('$') whose semantic interpretation violates the constraint: we have $\$(\{T\}, \{F\}) = \{F\}$ and $\$(\{T, F\}, \{F\}) = \{T\}$, or in the abbreviated notation $\$(t, f) = f$ and $\$(b, f) = t$, violating monotony since $t \subseteq b$ so we should have had $\$(t, f) \subseteq \$(b, f)$, but do not, since $f \not\subseteq t$. (Priest replies to Denyer in Priest 1989 but does not take up this aspect of the situation.) Denyer, having given a three-valued truth table for $ from which the above details were extracted, writes (Denyer 1989, p. 260):

> If dialetheic logic really is so impoverished as to have no such connective, dialetheic logicians will hardly wish to leave it in that state. (…) Dialetheists will not, I trust, object to introducing connectives by their truth tables; nor will they say that such a truth table as this is somehow improper.

This seems to presume that expressive adequacy requires functional completeness, as though there were no particular motivating ideas to constrain what might count as legitimate semantic behaviour for a connective. It would be like taking the Kripke semantics for intuitionistic logic and saying to an intuitionist using it to make the logic intelligible to one with no such sympathies, and saying: 'look, since we have all these points in the models and you have told us that the negation of a formula is true at a point when the formula itself fails to be true at all accessible points, you are committed to making sense of a new connective which attaches to a formula to making something which is true at a point just in case the formula to which it attaches fails to be true at that point itself'. The proposal would, of course, be politely declined since the compound formed with its aid would not be persistent ('hereditary') in the models, which would then undermine the unrestricted correctness of some of the rules to which the intuitionist is committed (including a Reductio ad Absurdum rule for instance, mentioned again in Footnote 57 below, along with a restricted version of the rule). **End of Digression.**

15.3 Carnapian Phenomena

A matter left unresolved in Sect. 15.1 was the awkward business of the entanglement with ¬ as the connective used in the general formulation of contrariety and ¬ as a connective concerning about whose contrary-forming status one might enquire. Consequence relations were nominally in play in that discussion but the proposals under discussion, variations on (1) and (2) offered in a passage quoted from Priest (2007)—though a similar passage occurs at in Chap. 4 of Priest (2006b) (p. 78), not quoted because it gives only the modal formulation—actually attended only to the consequences of the empty set. In Chap. 5 of Priest (2006b), we as it were leapfrog right over consequence relations in all their glory and into generalized (or 'multiple-conclusion') consequence relations, in which setting we can dispense not only with the structural role played by $\neg$ in the characterization of contrariety, but even with $\wedge$ and $\vee$ themselves. Here we use $\Vdash$ for such a relation, assuming (as in the case of consequence relations $\vdash$) that the defining conditions for being a generalized consequence relation are familiar.[24] It will be clear (to anyone not already familiar with generalized consequence relations) from the semantic remarks about valuations below according to such a relation $\Vdash$, A and B are subcontraries when they are related as on the left here, and contraries when they are related as on the right:

$$\varnothing \Vdash A, B \quad \text{and} \quad A, B \Vdash \varnothing.$$

From now on (as with consequence relations) we omit the '$\varnothing$' in such formulations.[25] The important point to note is that neither of these would be well formed as a condition on (formulas A, B and) a consequence relation $\vdash$, in the case of subcontrariety because there are too many formulas on the right, and in the case of contrariety because there are too few: but see the Digression below. But we have now, as the remarks on valuations below will make clear, avoided the 'main connective' use of $\neg$ in formulating contrariety, absorbing it into the structural machinery (by using $\varnothing$ on the right), so that applying the criterion of contrariety to A and $\neg A$, does not involve the double role Sect. 15.1 complained that negation was playing.[26]

One-place connectives σ and κ are accordingly said in Humberstone (2005a) to form subcontraries and contraries, respectively, according to $\Vdash$ if for all A in the language of $\Vdash$ we have:

$$\Vdash A, \sigma(A) \quad \text{and} \quad A, \kappa(A) \Vdash.$$

Although these conditions do not, as we have already quoted Priest as observing, uniquely characterize σ or κ, they allow us to draw conclusions about the relations

---

[24] For the defining conditions, see Shoesmith and Smiley (1978), p. 36, or Humberstone (2011a), p. 73.

[25] Other standard notational liberties are also taken, writing such things as '$\Gamma, A \Vdash B, C$' for '$\Gamma \cup \{A\} \Vdash \{B, C\}$'.

[26] Restall (2004) makes a similar use of empty right-hand sides with $A \wedge \neg A \Vdash \varnothing$ as a version of the Law of Non-Contradiction, though he writes $\vdash$ rather than $\Vdash$ (and $\sim$ rather than $\neg$: this applies for the passage about to be quoted), and does not use the terminology of consequence relations in the constrained way it is being used here (see Footnote 11), saying on p. 74, for example, that according to classical logic understood as classical propositional consequence we have LEM in the form '$A \vdash B \vee \neg B$', and that 'endorsing classical logic *very nearly* assures that we are committed to each instance of $B \vee \neg B$'. The insistence on the 'A' of the left and of the italicized phrase here will come as quite a surprise to many readers. Restall goes on to say that 'the situation with the law of non-contradiction is completely dual', giving this as $A \wedge \neg A \vdash B$. But what we have with consequence relations *sensu stricto* is precisely an asymmetrical arrangement not answering at all well to the demands of duality: there must be exactly one formula on the right of the $\vdash$ but there can be fewer or more on the left. This underlies the 'Carnapian phenomena' to be recalled presently; they were first isolated in Carnap (1943), especially, Sects. 15–18

between statements and arbitrary contraries or subcontraries of those statements. A simple example, following from these conditions via the 'Cut' condition on generalized consequence relations, include:

(3) $\kappa(A) \Vdash \sigma(A)$ (4) $A \Vdash \sigma(\kappa(A))$ (5) $\kappa(\sigma(A)) \Vdash A$.

Thus the idea of negation as a contradictory-forming operator is served by laying down the two conditions above for $\neg = \sigma = \kappa$. (Priest expounds this idea with slightly different—conditional rather than unconditional versions of the subcontrariety and contrariety conditions—in Chap. 5 of Priest 2006b.) Having used them to purify these conditions of any explicit reference to $\neg$ (or indeed $\wedge$, $\vee$) we turn to Carnap's interest in such conditions, after an intermezzo on corresponding purifications in the case of consequence relations themselves.[27]

Digression. Apropos of the point from the opening paragraph of this section, made that for consequence relations the contrariety and subcontrariety conditions above for formulas A, B, relative to a *generalized* consequence relation $\Vdash$ have either too many or too few formulas on the right, let us remark (familiarly enough) that the closest approximations for consequence relations that we can impose are, for contrariety, (i) that $A, B \vdash C$ for all formulas C and, for subcontrariety, (ii) that for all sets of formulas Γ and all formulas C, if $\Gamma, A \vdash C$ and $\Gamma, B \vdash C$ then $\Gamma \vdash C$. This suffices for the completeness half of the claim that $\vdash$ is determined by a class of valuations on each of which A, B are respectively never both true and never both false, but they do not suffice to make every valuation which is consistent with a $\vdash$ satisfying them behave in either of these two ways. (The semantic terminology used here is defined below.) Conspicuously the consequence relational formulation of contrariety, (i) above, amounts to *Ex Falso*/explosion to any statement as long as the negation of A is a contrary of A, and so *a fortiori* when negation is a contradictory-forming operator (according to $\vdash$), something further elaborated in semantic terms in Footnote 35 below, where the pertinent aspects of a formulation from Jean-Yves Béziau are extracted.

Béziau's discussion, however, also contains several strands that do not seem to bear as pertinently on the Slater–Priest debate. For example, on p. 19 of Béziau (2006) one reads:

> It is clear that inside classical logic, there are a lot of subcontrary-forming relations; however, the question is: are paraconsistent negations part of these subcontrary- forming relations? And the answer is: No. Because these negations are not definable in classical logic.

Similarly, in Béziau (1996), we have:

> It is important to note, against Slater, that the paraconsistent negations in the logics of da Costa and Priest cannot be characterized by the notion of subcontrary, at least by the classical notion of subcontrary, in particular because these paraconsistent negations cannot be defined in classical logic.

---

[27] Note that (3)–(5) all make sense with $\Vdash$ written as $\vdash$ and taken to be a consequence relation. For further examples and discussion, see Humberstone (2005a) and Sect. 8.11 of Humberstone (2011a).

The question of whether formulas are subcontraries or contraries—and thus of whether this or that connective is a subcontrary-forming operator or contrary-forming operator (or both, in which case it is a contradictory-forming operator)—has nothing to do with classical logic, however. We can ask this question of any connective relative to any consequence relation, using the definition just given (as we can in the case of generalized consequence relations, using the definitions given before the present Digression). In the first place, we can always ask whether two formulas—for example, a conditional and its converse—are subcontraries according to the intermediate logic LC, identified with its consequence relation $\vdash_{LC}$, getting in this case an affirmative answer. We can ask the question (getting the same answer) even—to illustrate the idea in an $\vee$-free setting—for the implicational fragment of this logic (discussed in Bull 1962). Accordingly, we can ask of a 1-ary connective in the language of a consequence relation whether applying it to a formula always yields a subcontrary of that formula, according to that consequence relation. Repeating a familiar example (see Sect. 15.5 below), with $\neg$ for dual intuitionistic negation and $\vdash_{HB}$ for the consequence relation of Rauszer's 'H–B' or Heyting–Brouwer logic[28] (combining standard and dual intuitionistic negation—associated by Rauszer with the names of Heyting and Brouwer, respectively—along with standard and dual intuitionistic implication, as well as conjunction and disjunction), we have $\vdash_{HB} A \vee \neg A$, and since this secures the desired result that $\Gamma, A \vdash_{HB} C$ and $\Gamma, \neg A \vdash_{HB} C$ imply for any Γ, C, that $\Gamma \vdash_{HB} C$, '$\neg$' is indeed a subcontrary-forming operator (according to $\vdash_{HB}$).[29]

Similarly, Béziau's claim (on the same page) that 'intuitionistic negation is not a contrary-forming relation', based on the fact that intuitionistic logic is what Béziau, rather confusingly, describes as 'strictly stronger' than classical logic, meaning that it is discriminatorily—rather than deductively—stronger than classical logic, though he also says that classical logic is definable in intuitionistic logic, meaning: faithfully embedded by a definitional translation. This would only be the case if one were thinking of intuitionistic logic as a set of provable formulas, whereas for the rest of his discussion Béziau is clearly thinking of logics as consequence operations/relations. (See Wójcicki 1988 Theorem 2.6.9, as contrasted with Theorem 2.6.8.).[30] There may

---

[28] See Rauszer (1980); while Rauszer writes the dual negation by horizontally inverting the standard negation symbol, here a vertically inverted form has been chosen.

[29] In fact, as is also well known, '$\neg$' forms the *strongest* subcontrary of what it applies to: $\vdash_{HB} A \vee B$ implies $\neg A \vdash_{HB} B$ for arbitrary A, B. Further information may be found in 8.11 and 8.22 of my (Humberstone 2011a). To the list of references supplied at p. 1250 of Humberstone (2011a) should be added Priest (2009), in which Priest discusses the relation between dual intuitionistic negation and da Costa's negation. Discussion of contrariety and contradictoriness connected with another optional extra from the negation menu for intuitionistic logic—strong negation ('constructible falsity')—is provided in Wansing (2001, 2006).

[30] I may be misinterpreting Béziau here, since editorial remarks at p. 57 of Béziau et al. (2007) clearly show familiarity with Wójcicki's observation (announced initially in 1970). (For certain fragments, one does have the definability of the corresponding classical fragment in the intuitionistic consequence relation, as C. A. Meredith famously showed in the case of the implicational fragment. But since the current topic is negation, it can hardly be any such positive fragment that Béziau has in mind.) A second point: the contrariety claim for intuitionistic negation Béziau describes as an

also be a further ingredient in Béziau's reluctance to acknowledge that Priest's negation forms subcontraries or that intuitionistic negation forms contraries and that arises from his taking 'subcontraries' to mean (as we would say): subcontraries which are not also contraries, and 'contraries' to mean: contraries which are not subcontraries—*mere* contraries as it is put in Appendix 1, *q.v.* for discussion of this issue. That may also be the reason Wansing (2001, 2006), has in mind, since on the opening page of the latter paper we read that 'the present writer (Wansing 2001) holds that intuitionistic negation fails to form contrary pairs but forms contradictory pairs'. On the usage followed in our discussion, any contradictories are contraries—though not 'mere' contraries—but intuitionistic negation fails the (similarly understood) subcontrariety condition for forming contradictories. However, this remark of Wansing's has here been quoted out of context, and his real reason for holding intuitionistic negation to be a contradictory-forming operator is presented in Appendix 2 at the end of the paper. **End of Digression**.

Following a usage of Dana Scott's, we call a valuation v (a function from the set of formulas of whatever language is under consideration to the two-element set $\{T, F\}$) *inconsistent* with a consequence relation $\vdash$ just in case for some set Γ of formulas, and some formula B, for which $\Gamma \vdash B$, we have $v(A) = T$ for all $A \in \Gamma$ and $v(B) = F$, and inconsistent with a generalized consequence relation $\Vdash$ just in case for some sets Γ, Δ of a formulas for which $\Gamma \vdash B$, we have $v(A) = T$ for all $A \in \Gamma$ and $v(B) = F$ for all $B \in \Delta$. Otherwise v is *consistent* with $\vdash$ or $\Vdash$, respectively. Next, we say that a consequence relation $\vdash$ (respectively, generalized consequence relation $\Vdash$) is *determined by* a set V of valuations when for all Γ, B: $\Gamma \vdash B$ iff for no $v \in V$, $v(A) = T$ for each $A \in \Gamma$ while $v(B) = F$ (respectively, all Γ, Δ: $\Gamma \Vdash \Delta$ iff for no $v \in V$, $v(A) = T$ for each $A \in \Gamma$ while $v(B) = F$ for all $B \in \Delta$). Finally, for this way of telling the story,[31] we need to introduce some vocabulary for properties of and operations on valuations as understood here. The properties are the properties of being #-Boolean for any n-place connective # which is traditionally associated with a particular truth function $f_\#$: these are the valuations on which for all formulas $A_1, \ldots, A_n$, we have $v(\#(A_1, \ldots, A_n)) = f_\#(v(A_1), \ldots, v(A_n))$. Thus, for example, the $\vee$-Boolean valuations are those verifying (i.e. assigning T to) a disjunction precisely when they verify some disjunct, the $\neg$-Boolean valuations are those valuations v for which $v(\neg A) \neq v(A)$ for all A, and so on. Finally, we have the binary operation we shall write here as $\cdot$ of conjunctive combination which is defined thus: for valuations u, v the valuation $u \cdot v$ is the unique valuation verifying exactly those formulas verified by both u and v. Then a preliminary (and readily checked) observation is that if u, v are valuations consistent with a consequence relation $\vdash$,

---

erroneous conclusion someone might reach by means of an argument similar to Slater's. Even if Béziau is wrong about the erroneousness of the conclusion, there is a good point here: why didn't Slater complain as loudly that intuitionistic negation doesn't deserve the name, on the grounds that it does not form contradictories because it fails to deliver subcontraries?

[31] As in Humberstone (2015), especially, Sect. 2, or 1.13–1.17 of Humberstone (2011a), in which suitable references to Carnap and others can be found.

then so is their conjunctive combination $u \cdot v$.[32] This is usefully strengthened by adding an infinitary form of conjunctive combination we shall write as $\prod$ where for a set V of valuations V, $\prod(V)$ is the valuation u such that for all formulas A, $u(A) = T$ iff for all $v \in V$, $v(A) = T$.[33]

Then Carnap's observation can be put by saying that if V is any set of valuations consistent with a consequence relation, so is the valuation $\prod(V)$, but that this is not so in the case of generalized consequence relations. Since, for example, the conjunctive combination of two $\vee$-Boolean valuations need not be $\vee$-Boolean, there are valuations consistent with the consequence relation—call it $\vdash_{CL}$—of classical propositional logic (with a stock of connectives taken to include $\vee$) which are not $\vee$-Boolean. (This can't happen with $\wedge$-Boolean valuations.) A more precise definition of $\vdash_{CL}$ could be supplied in proof-theoretic terms, but to deserve the name it had better amount to this: the consequence relation determined by the class of all #-Boolean connectives (for every # of the language for which this notion makes sense—i.e. which is associated with some bivalent truth function). These aberrant non-$\vee$-Boolean valuations still have to verify any disjunction at least one of the disjuncts of which they verify (or they would not be consistent with $\vdash_{CL}$, since $A \vdash_{CL} A \vee B$ and $B \vdash_{CL} A \vee B$). But they do not have to verify only the disjunctions which at least one disjunct verified: if we have $u(A \vee B) = T$, it suffices that for some Boolean valuations v_0, v_1, we have $u = v_0 \cdot v_1$ and $v_0(A) = T$ and $v_1(B) = T$. (Here a Boolean valuation is one which is #-Boolean for every connective # of the language.)

This last may seem a tall order: what if we take A as $\top$, true on all Boolean valuations, and B as $\bot$, true on none of them? Then if u is a Boolean valuation $u(A \vee B) = u(\top \vee \bot)$ and we won't be able to find Boolean valuations v_0, v_1 respectively verifying $\top, \bot$, because no such valuation is available for the $\bot$-case. But the condition given—the existence of Boolean valuations (whose conjunctive combination was the given valuation) verifying the respective disjuncts—was only claimed to be sufficient, not also to be necessary. If we want that, we have to focus on arbitrary conjunctive combinations of Boolean valuations (which of course include all the Boolean valuations themselves) to start with, and say that one of these verifies a disjunction just in case it can be written as a conjunctive combination of such valuations themselves verifying the respective disjuncts, and this we can easily do by noting that our Boolean u is itself $u \cdot v_T$, where $u(\top) = T$ and $v_T(\bot) = T$, v_T being the valuation assigning T to every formula. This is consistent with the consequence relation $\vdash_{CL}$—indeed with any consequence relation—because, vacuously, $\varnothing$ is a set of valuations consistent with any consequence relation, and $v_T = \prod(\varnothing)$. While the problem with disjunctions means that a consequence relation $\vdash$ telling us that $\vdash A \vee B$ does not succeed in forcing every valuation consistent with $\vdash$ to verify either A or B, even though that is the closest a consequence relation can come to saying

---

[32]In Humberstone (1996b) binary operations are considered which are—as it is put there—Galois dual to familiar sentence connectives in the way in which conjunctive combination is to conjunction.

[33]These are essentially familiar to philosophers as supervaluations over V, except that like all valuations here, they are bivalent and assign F to all formulas not verified by every $v \in V$ (rather than just those falsified by every $v \in V$).

outright that A and B are subcontraries, and assumes that the consequence relation is what is called in Humberstone (2011a) ∨-classical.[34] In particular even though our interest in $\vdash_{CL}$ may arise through its being determined by the set of Boolean valuations and $A \vee \neg A$ is (for any A) true on all such valuations, duly reflected in that fact that $\vdash_{CL} A \vee \neg A$, none of this is enough to force every valuation consistent with $\vdash$ to verify either a formula or its negation, so if this is what we wanted out of a consequence relation—be it $\vdash_{CL}$ or $\vdash_{LP}$ or whatever paraconsistent consequence relation you favour—we are not going to be able to get what we wanted, unless we pass to generalized consequence relations, and for essentially Carnapian reasons: this allows for the exclusion of unintended interpretations of the Boolean vocabulary. (Every valuation consistent with a generalized consequence relation ⊩ satisfying the condition inset above concerning σ must verify either A or else $\sigma(A)$, for any formula A. Similarly every valuation consistent with the generalized consequence relation determined by the class of Boolean valuations—⊩$_{CL}$, as we might call it—is Boolean, including ∨-Boolean because now we can say not only $A \Vdash A \vee B$ and $B \Vdash A \vee B$ but also $A \vee B \Vdash A, B$.)

These themes from Carnap have over the past 10 years found their way into the debate over paraconsistency in Marcos (2007) and Oller (2014, 2016), and in both cases, it is subcontrariety, as above, but contrariety that has been at the centre of attention, via the valuation v_T noted to be consistent with every consequence relation. (However, Béziau, who also makes this v_T observation in the middle of p. 23 of Béziau 2006, does mention, in addition, the corresponding point about subcontraries—that a formula and its negation are not subcontraries over the class of all valuations consistent with $\vdash_{CL}$.) This valuation v_T is of course non-Boolean—specifically, neither ⊥-Boolean nor ¬-Boolean—since it verifies every formula and its negation, whereas we wanted no formula and its negation to be true on any of our valuations. As in the previous case, it's not that the logic, thought of as a consequence relation, is incomplete as a result: we are talking about the classical consequence relation $\vdash_{CL}$, determined by the class of all Boolean valuations, which, if we thought of it as proof-theoretically characterized, amounts to the combination of soundness with completeness with respect to the class of Boolean valuations. It's that it has the expressive weakness Carnap diagnosed in every consequence relation: it does not force the valuations that respect it ('are consistent with it') to be Boolean, since the class of consistent valuations is inevitably closed under conjunctive combinations, which the Boolean valuations themselves are not. Oller (2014) throws up as a challenge modelled on Slater's argument—perhaps to be thought of as undermining that argument by parity of form and its (for Slater) unwanted conclusion—that negation in the mouth of the dialetheist is not really negation in the following form: even negation in the mouth of classical 'monaletheist' is not really negation, since A and $\neg A$ are not contraries on a $\vdash_{CL}$-consistent valuation verifying both of them.

---

[34] A roundabout *conditional* way—already mentioned (apropos of subcontrariety according to a consequence relation in general) under (ii) in the opening paragraph of the Digression in Sect. 15.3—which would amount to this for a $\vdash$ which is not ∨-classical would be to say that for any set Γ and formula C, if $\Gamma, A \vdash C$ and $\Gamma, B \vdash C$, we have $\Gamma \vdash C$.

(In fact, there is just the one such valuation, namely v_T, in view of the fact that $A, \neg A \vdash_{CL} B$.) Since, if we are doing logic by focussing on consequence relations these non-Boolean aspects of the consistent valuations are inescapable, the most that can be hoped for is that one's logic is determined by *some* class of valuations all of which are $\neg$-Boolean.[35] Accordingly, here only the current observation's bearing on the foregoing discussion will be addressed.

In Sect. 15.1 I criticized Priest for suggesting that (1) and (2) quoted in that section sufficed for the identification of $\neg$ as expressing negation, on the ground that we could just as well interpret $\neg$ as a connective for the 1-ary constant true ('Verum') connective, and went on to say that '[t]his unintended interpretation of $\neg$ could be avoided if we took as unstated background assumptions certain principles', amounting to the basic principles of De Morgan logic. In particular, of course, this interpretation would not be compatible with validating the principle that $\neg\neg A \vdash A$ (or $\neg\neg A \vdash_S A$, to use the notation in play there). As we have just been recalling in this section, it is not possible using a consequence relation to enforce the negation truth function as the interpretation of $\neg$ on every valuation consistent with it, but this is a matter of expressive incompleteness ('ineffability', as the title of Marcos 2007 suggests) whereas the problem of leaving open an alternative truth functional construal of $\neg$ satisfying (1) and (2) is a matter of semantic incompleteness: things which should be forthcoming, such as the Double Negation equivalences, simply fail to be.

Indeed—if I may be permitted an *ad hominem* paragraph here—at p. 81f of Priest (2006b), Priest explicitly justifies these and the other De Morgan equivalences (i.e. what are traditionally called De Morgan's Laws), and the Double Negation case is worth looking at; in this passage LEM and LNC are the necessitated version of (the

---

[35]Essentially this is what Béziau suggests in the Remark, just alluded to, in the middle of p. 23 of Béziau (2006). This Remark is appended to a Theorem, the key part of which we may extract here as follows: if $\vdash$ is a consequence relation on some language with 1-ary # as one of its connective and $\vdash$ is determined by at least one class V of valuations with the property that for no $v \in V$ and formula A, do we have $v(A) = v(\#A) = T$—never mind whether $\vdash$ is also determined by other classes of valuations (such as the class of all valuations consistent with $\vdash$) which do not share this feature with V—then for all formulas A, B, we have $A, \#A \vdash B$. This observation is more or less immediate from the definitions of the terms deployed in it: if $A, \#A \not\vdash B$ and $\vdash$ is determined by V, then there must be $v \in V$ with $v(A) = v(\#A) = T$ (and also $v(B) = F$), so V is not as advertised (A and $\#A$ are not 'contraries over V', being both verified on v). Short of paraconsistently homophonic maneuvering in the metalogic, this result is not negotiable, and as Béziau notes, this can be regarded as a way of presenting Slater's point. Here I have extracted the contrariety condition from Béziau's formulation, which incorporates also subcontrariety, and amounts to this: if $\vdash$ is determined by some class of valuations V such that $v(A) \neq v(\#A)$ for all formulas A and all $v \in V$ then # satisfies the condition already mentioned—that $A, \#A \vdash B$—as well as the 'subcontrariety' condition: whenever $\Gamma, A \vdash B$ and also $\Gamma, \#A \vdash B$, we have $\Gamma \vdash B$. In other words, reading # as $\neg$, $\vdash$ is $\neg$-classical, as it is put in Humberstone (2011a) (though Béziau uses a Tarski-style characterization of $\neg$-classicality which is more economical but less transparent); the condition that (for all A) $v(A) \neq v(\neg A)$ for $v \in V$ is the condition that V contains only $\neg$-Boolean valuations. Incidentally, these formulations with the explicit existential quantification—$\vdash$ determined by *some* V such that ...—are fine in the converse direction. (The 'absent converse' mentioned in Béziau's Remark is to the converse of a formulation to the effect that where a class of valuations V determines $\vdash$, if V contains only $\neg$-Boolean valuations then $\vdash$ is $\neg$-classical.)

formulas involved in) (1) and (2) from Sect. 15.1, and in rendering it I have changed α, β to A, B for continuity with the discussion above:

> So far, we have met two of the classical laws of negation, LEM and LNC, a third, the law of double negation (LDN) is simply derivable. The relation of being contradictories is symmetric. That is, if B is the contradictory of A, then A is the contradictory of B. In particular, A is the contradictory of $\neg A$. Hence $\neg\neg A$ just is A.

This last sentence really means that $\neg\neg A$ is equivalent to A rather than is A itself. The argument makes an essential play with the already remarked on uniqueness of contradictories by contrast with (mere) contraries and subcontraries—and again, this is to be understood as uniqueness to within equivalence. But (1) and (2)—or (LEM) and (LNC)—do not confer this crucial feature on $\neg$, which as we have already noted, can be understood as far as they are concerned both as negation and as the Verum connective. It follows that there is not, to within equivalence, a unique 'contradictory' for an arbitrary A, if this simply meant something satisfying the conditions imposed on $\neg$ by (1) and (2).[36] On the other hand, the informal argument just quoted from Priest is readily converted into a proof[37] using the characterization of negation as a contradictory-forming operator in the sense of satisfying the condition imposed above on σ and also that imposed on κ, relative to a generalized consequence relation $\Vdash$: this is what we get by combining (4) and (5) in the case in which $\sigma = \kappa = \neg$. But there is one obvious difficulty about them in a paraconsistent setting, and that is that they were laid down as conditions governing σ and κ according to a generalized consequence relation, part of the definition of which is that one can 'weaken' on the left—no problem—but also on the right—big problem. This means that whenever we have contraries A and $\kappa(A)$ according to $\Vdash$, we also have for any B: $A, \kappa(A) \Vdash B$. But this issue is most conveniently addressed in Sect. 15.5, after we attend to some leftover issues from the present discussion which are less intimately connected Carnapiana.

15.4 Semantic Addenda

The preceding discussion needs to address a thought that may have occurred to some of those seeing all the talk of bivalent valuations in Sect. 15.3 above. Isn't LP supposed to be a *three*-valued logic? The answer to this is that yes, of course, by definition $\vdash_{LP}$ is the consequence relation determined by a three-element matrix and certainly the matrix evaluations are homomorphisms from the language of LP

---

[36] Here I have been interpreting the passage quoted from Priest above as saying, incorrectly, LDN is derivable from LEM and LNC. A more generous—if somewhat vaguer—reading, suggested to me by Sam Butchart, is also possible: after mentioning LEM and LNC, what Priest says about LDN is that it is 'simply derivable': he does not explicitly say, as my interpretation presumes, that it is 'simply derivable *from these two principles*'.

[37] We do not even need to appeal to the commutativity of $\wedge$ and $\vee$ which underlies the remark about the contradictoriness relation being symmetric, since these connectives are out of the picture.

(thought of as an algebra[38]) to the algebra of the matrix. Such a matrix valuation h induces a bivalent valuation v_h by putting, for any formula A, $v_h(A) = T$ if $h(A)$ is one of the designated elements of the matrix, and otherwise $v_h(A) = F$. The (generalized) consequence relation determined by the matrix is then the (generalized) consequence relation determined in the sense explained earlier by the class of valuations v_h for h a matrix evaluation. The role of the many values of many-valued logic is not to compete with truth as the thing validity requires us to preserve, but to allow for distinctions for the sake of doing compositional semantics when there is a failure of designation functionality. (This is the perspective introduced in Dummett 1959a, the essential role of the bivalent valuations being later emphasized in their different ways by Roman Suszko and also Dana Scott. For further elaboration, see my (1998), or Sect. 2.11 in Humberstone (2011a) and for bibliographical references, especially the first new paragraph on p. 210 of the latter.)

In the present case the elements of the (algebra of the) *LP* matrix are often thought of, after the fashion of the Dunn–Belnap semantics for *FDE*, as non-empty sets of 'classical values', and denoted by something like $\{T\}$, $\{T, F\}$ and $\{F\}$ (for Dunn and Belnap we throw in $\varnothing$ as well), or $\{1\}$, $\{1, 0\}$ and $\{0\}$, or just *t*, read 'true only', *b* read 'both', and *f* 'false only', (with *n* for 'neither' thrown into the mix for the Dunn–Belnap semantics).[39] The algebra of the matrix is familiar as that for the three-element Kleene matrix and also Łukasiewicz's three-element matrix when attention is restricted to conjunction, disjunction and negation.[40] The designated elements are those which have T (or 1) as an element, i.e. the values *t* and *b*, to use the last mentioned nomenclature. Further, the assignment to a formula of any value containing T (respectively, F) as an element is registered by calling the formula true (respectively, false); to avoid confusion with the values of the functions v_h let us call the elements of the sets making up the values of h (*t* etc.) in this case 1 and 0 rather than T and F. Indeed this is the best way of motivating the particular algebra of the *FDE* matrix in Dunn (1976)—'a conjunction is true iff both conjuncts are true and false if at least one is false'—explaining matters more succinctly and naturally then just presenting the than four-valued table for ∧ that embody this motivating idea, and likewise in the case of ∨ and ¬. But it does have its dangers when it comes to thinking about contrariety and subcontrariety.

---

[38]More specifically: as the absolutely free algebra of its signature, with compounding via the primitive connectives as the fundamental operations.

[39]Priest uses a variant on the last notation in Priest (1979), with *t*, *p* (for *paradoxical*) and *f* for *t*, *b* and *f*. In Priest (2006a) he uses the $\{1\}$, $\{1, 0\}$, $\{0\}$ notation. In numerous other places, including material added in the section edition of Priest (2006a) (e.g. p. 288*f*.), as well as (2014), Priest prefers not to use sets of classical truth values at all but to avoid functions which would assign such sets of values in favour of relations relating formulas (potentially plurally) to the elements of those sets. This 'preference for the plural' extends well beyond the present manifestation of it: see Priest (1995).

[40]In fact, finding more emphasis in Kleene on the underlying algebra, in Humberstone (2011a) I described the *LP* matrix and the familiar ('strong') Kleene matrix both as Kleene matrices and called them **K**$_1$ and **K**$_{1,2}$ respectively, the subscripts reflecting which of the three elements denoted by 1, 2, and 3, were taken as designated.

The danger I have most particularly in mind was drawn to my attention by Sam Butchart, who pointed out that if being true and being false are no longer taken to be mutually exclusive[41] (and jointly exhaustive, one might add to bring the full four-valued scheme into play), then what might otherwise be taken to be mere reformulations now emerge as non-equivalent possibilities. Thus while Slater (1995) is happy to concede that Priest's negation does at least form subcontraries, this requires us to give priority to the 'at least one must be true' formulation of subcontrariety over the bivalently equivalent 'can't both be false' formulation, since in the present non-bivalent understanding of the terms, A and $\neg A$ *can* both be false, since with a matrix evaluation h such that $h(A) = \boldsymbol{b}$, we do have both $1 \in h(A)$ and $0 \in h(A)$.

Now, since one can give whatever labels one wants to the elements of a matrix semantics, regarding the matrix as of purely instrumental value (for compositional purposes) in determining a consequence relation, what matters is not what value is given the name *truth*, or what values are collected together under the umbrella predicate *true*, but simply which values are designated. Thus one is led to the following formulation in semantic terms of contrariety and subcontrariety for matrix semantics: formulas A, B are contraries, respectively, subcontraries relative to matrix $\mathfrak{M}$, if for any $\mathfrak{M}$-evaluation h, at most one, respectively, at least one, of $h(A)$, $h(B)$ is a designated value. In terms of bivalent valuations, this means that for A and B to be contraries what needs to be forbidden is $v_h(A) = v_h(B) = T$, and for them to be subcontraries, that $v_h(A) = v_h(B) = F$. Note that if $\Vdash$ is the generalized consequence relation determined by the matrix in question, this coincides with the more syntactic definition of what it is for A and B to be contraries or subcontraries according to $\Vdash$ given in the previous section. A suggestion along these lines, though differently expressed, can be found in Béziau (2006), which begins as a reply to Slater's criticism of dialetheism (or of paraconsistent negation, as Slater puts it) and then moves on to a separate criticism of Priest. Confusingly for our purposes, $\boldsymbol{t}$, $\boldsymbol{b}$ and $\boldsymbol{f}$ are referred to as 1, $1/2$ and 0. I will translate this notation back into ours for the purposes of the following quotation, from p. 21 of Béziau (2006), at the same time replacing 'v' by 'h', and lower case a, b by A, B, and Béziau's 'T' (for a set of formulas) by 'Γ'.[42] The set of *LP*-evaluations Béziau calls **P**, concerning which he begins by reminding us that for all $h \in \mathbf{P}$ and all formulas A, we have '$h(A) = \boldsymbol{f}$ iff $h(\neg A) = \boldsymbol{t}$ and $h(A) = \boldsymbol{b}$ iff $h(\neg A) = \boldsymbol{b}$', continuing:

> Now if we want to interpret the discussed traditional notions in this context (more generally in the context of a logic with more than two values), we must fix what 'truth' is and what 'falsity' is. And it is clear that if we interpret truth by $\boldsymbol{t}$ and falsity by $\boldsymbol{f}$, then $\neg$ is a contradictory-forming relation. And that is apparently why Priest thinks that his paraconsistent negation is really a negation. But his argumentation is vitiated, as Slater himself confusedly perceived.

---

[41] Recall that we are setting aside as dialectically unhelpful the vertically homophonic move considered in Sect. 15.2: oh but they *are* mutually exclusive—it just happens that despite being mutually exclusive truth and falsity can both apply to the same thing.

[42] The definition of *Cn* in the quotation should be understood as prefaced by "for all $h \in \mathbf{P}$". This formulation involves a consequence operation instead of a consequence relation. For the latter purposes '$A \in Cn(\Gamma)$' would be written as '$\Gamma \Vdash_{\mathsf{LP}} A$'.

The reason why Priest's argumentation is wrong is the following: he considers as designated elements (in the sense of matrix theory) not only t but also b, as we can see when he defines the notions of logical truth and semantic consequence. The last is defined by

$$A \in Cn(\Gamma) \text{ iff for every } h \in \mathbf{P}, h(B) = f \text{ for one } B \in \Gamma, \text{ or,}$$
$$h(A) = t \text{ or } h(A) = b$$

This definition allows (us) to have $A \notin Cn(B, \neg B)$, for any atomic formulas A and B, and therefore to say that *LP* is paraconsistent. Had t been taken as the only designated value, *LP* would not have been paraconsistent.

Priest's conjuring trick is the following: on the one hand, he takes truth to be only t in order to say that his negation is a *contradictory-forming relation*, and on the other hand he takes truth to be b and t to define *LP* as a paraconsistent logic. However, it is reasonable to demand to someone to keep his notion of truth constant, whatever it is. Therefore we have only the two following possibilities, which show that Priest cannot run away: in one case *LP* is paraconsistent and its negation is only a subcontrary-forming relation from the point of view of **P**, in the other case *LP*'s negation is a contradictory-forming relation but *LP* is not paraconsistent.

It is the first of these two options that corresponds to the v_h-based suggestion concerning contrariety: on it we can have $v_h(A) = v_h(\neg A) = T$, so there is trouble in thinking of $\neg$ as contrary-forming and therefore as contradictory-forming (since it is subcontrary-forming: certainly we cannot have $v_h(A) = v_h(\neg A) = F$). What about the first option, of regarding $\neg$ as contradictory forming on the grounds that it flips us between t and f? This did not seem quite right as a summary of any of the several strands in Priest's thought about negation. The idea that the negation of something true should be false and vice versa, is not the same as the idea that the negation of something that is *true only* should be *false only* and vice versa. It is, rather the idea that the negation of something which is true whether or not it is also false (so: having the value t or b) should be false whether or not it is also true (so: should have one of the values f, b). This condition is certainly satisfied by the *LP* matrix. However, I am not at all sure that even as reformulated in this way, this idea has much to do with Priest's claim that $\neg$ in *LP* is a contradictory-forming operator, since the claim is also satisfied by the four-valued *FDE* matrix, so this would mean that the issue of whether A and $\neg A$ are contradictories has nothing to do with the Law of Excluded Middle. This does not sit well with at least one part of the final inset quotation from Priest in Sect. 15.2 above: 'Given the notion of negation employed with gaps, the LEM and LNC fail. Given the conception of negation I have just described, they do not; so the negation is a contradictory-forming operator'.

Similarly, Berto (2006) extracts a formulation from Priest (1979) and (2006a) to the effect that $\neg A$ is true if and only if A is false and $\neg A$ is false if and only if A is true, and goes on to observe (p. 249) that this could be assented to by a proponent of gaps without gluts. Berto quotes also from Priest (1998), in which Priest says that on his approach, 'one does not make the assumption, usually packed into textbooks of logic without comment, that truth and falsity in an interpretation are exclusive and exhaustive'. Taken together with the previous link between negation and falsity this means that the truth of A and that of $\neg A$ are not being taken to be mutually exclusive (not contraries) and not jointly exhaustive (not subcontraries)—which is essentially

Slater's point: despite his protestations, Priest's ¬ does not form contradictories. When convenient, Priest can change to a homophonic tack and exploit the vertical spread described in Sect. 15.2: Yes, truth and falsity are exclusive (which is our main interest here, rather than exhaustivity) but such-and-such particular statement—here selecting some dialetheia—is, all the same (!), both true and false.[43]

15.5 Unique Characterization: Basic Issues

While the topic of uniqueness came up in Sect. 15.1 in a quotation from Priest (2006b) I would like for the present section to begin by winding the clock back to 1989, the publication year of the encyclopedic Priest et al. (1989), or more accurately, to my own review of the work (Humberstone 1992) which reminds us of some of the earlier discussion of unique characterization. Here is the relevant passage:

> Priest and Routley note that the idea is that negation forms subcontraries rather than contradictories, and make a comparison with intuitionistic logic, in which contrariety (rather than subcontrariety) replaces contradictoriness. They complain that statements do not have unique subcontraries (p. 166), but of course the same goes for contraries: yet the usual rules of intuitionistic logic do indeed uniquely characterize negation. The reason is that these rules secure that the negation of statement is the (deductively) weakest contrary of that statement; thus a correct dualization requires consideration—even if this never occurred to da Costa—of rules making the negation of a statement its strongest subcontrary. Uniqueness is achieved and the result is what is commonly called dual intuitionistic logic. For paraconsistent purposes, it will be necessary to outlaw the usual intuitionistic negation in favour of its dual, rather than working with both as does Cecylia Rauszer, the pioneer of this topic. (...) Actually, if anything, considerations of unique characterization tell against the editors' favoured approach; the usual logical principles of relevant (and weaker) systems are insufficiently strong to afford such characterization for negation, in the sense that if analogous principles are laid down for a second, duplicate, negation connective, and then pooled with the original principles, the result is not a system in which the (original) negation of a formula and the duplicate negation are freely interreplaceable *salva demonstrabilitate*.[44] (The reviewer is not committed to the view that such unique characterization—a notion originating with Belnap and frequently confused with implicit definability—is mandatory; the point is only that it is unwise to raise objections against others to which one's own account is vulnerable.)

The reference to the editors' favoured approach is to the general Anderson–Belnap inspired tradition of relevant logic, which in its →-free form can be taken to be *FDE*,

---

[43]Does explaining negation as forming the weakest statement incompatible with a given statement (Berto 2006, p. 253) help with any of this? (Note that this characterization coincides with that of intuitionistic negation as forming the weakest contrary of a statement; compare dual intuitionistic negation—see Footnote 29 and the text to which it is appended, as well as the passage quoted at the start of Sect. 15.5, below—as forming a statement's strongest subcontrary. In the classical setting, the strongest subcontrary coincides with the weakest contrary, so there is nothing to choose between the two formulations.)

[44]In fact, the original passage contained at this point the phrase 'salva provabilitate', but as an astute referee observes, there is no such phrase. I apologize profusely to readers of other publications of mine featuring this piece of confused pseudo-Latin.

in which the implicational connective's role can be played by ⊢, though for present purposes it certainly needs to subsume the stronger gap-free version *LP*. Let us consider the issue of the unique characterization in the first instance for *FDE*, before considering *LP*, and further, consider it from the perspective of the Australian Plan rather than the American Plan[45]—that is, using models with the Routleys' *-function rather than the Dunn–Belnap four values (t, f, b, n), as this makes the description easier. We write down enough to secure that the least consequence relation satisfying the conditions we have written down is ⊢$_{FDE}$ in the sense that $A_1, \ldots, A_n$ ⊢$_{FDE}$ B just in case the implication from the conjunction of the A_i to B is a correct Anderson–Belnap first-degree entailment.[46] This means suitable distributive lattice principles for ∧ and ∨ and then De Morgan (including double negation principles) to govern ¬. Then write all this down coupled with duplicates of everything involving ¬ but for a new 1-place connective ¬′. The original principles are said to characterize ¬ uniquely if the combined principles suffice for it to follow that for all A, we have ¬A ⊣⊢ ¬′A; more precisely, since we kept ∧ and ∨ constant in passing to the reduplicated negation system, we say that ¬ is uniquely characterized in terms of {∧, ∨}.[47] In fact let us baptize the negation-reduplicated version of ⊢$_{FDE}$ we have just described, ⊢$_{FDE^2}$. In view of the symmetrical treatment of ¬ and ¬′ in the combined system, it is clear that to show that this is the case, it would suffice to show either direction—for instance, to just show that ¬A ⊢$_{FDE^2}$ ¬′A. To see that in fact this cannot be shown, we reach for a reduplicated version of the models of Routley and Routley (1972), which can be taken to be of the form $\langle W, *, \star, V \rangle$ in which W is a non-empty set on which $*$ and $\star$ are a one-place functions (for which we use the notation '$x*$' rather than '$*(x)$' (and similarly with $\star$) satisfying the conditions that for all $x \in W$, $x** = x = x^{\star\star}$ and V assigns a subset of W to each of a countable sequence of sentence letters $p_1, p_2, \ldots$ (essentially, stipulating which elements of the model is to count as true at.) For arbitrary formulas of the language constructed from the sentence letters by means of connectives ∧, ∨, and ¬ and ¬′, we define the truth of a formula A at an element $x \in W$ in a model $\mathcal{M}$ (notated '$\mathcal{M} \models_x A$') by induction on the complexity of A, thus:

- $\mathcal{M} \models_x p_i$ iff $x \in V(p_i)$;
- $\mathcal{M} \models_x B \wedge C$ iff $\mathcal{M} \models_x B$ and $\mathcal{M} \models_x C$;
- $\mathcal{M} \models_x B \vee C$ iff $\mathcal{M} \models_x B$ or $\mathcal{M} \models_x C$;
- $\mathcal{M} \models_x \neg B$ iff $\mathcal{M} \not\models_{x*} B$;
- $\mathcal{M} \models_x \neg' B$ iff $\mathcal{M} \not\models_{x\star} B$.

It is clear that whenever $A_1, \ldots, A_n$ ⊢$_{FDE^2}$ B, and we have a point at which $A_1, \ldots, A_n$ are true in one of these models, then B is true at that point. But just take a two-element such model $\mathcal{M} = \langle W, *, \star, V \rangle$ with $W = \{a, b\}$, $a* = a$, $b* = b$,

---

[45] This terminology at least once enjoyed considerable currency—as in Meyer and Martin (1986).

[46] This is a standard way of turning the Anderson–Belnap binary *FDE*-relation into a consequence relation; see Pynko (1995), p. 445, where the corresponding consequence *operation* is defined.

[47] For more discussion of the concepts issues involved here, see Sect. 4.3 of Humberstone (2011a).

$a^\star = b$, $b^\star = a$ and $V(p_1) = \{b\}$. Suppressing the reference to this model by writing $\mathcal{M} \models_x A$ as '$x \models A$', and also writing p for p_1, we have (i) $a \models \neg p$, since $a^\star \not\models p$ (recalling that $a^* = a$ and $a \notin V(p)$), and (ii) $a \not\models \neg' p$ since $a^\star \models p$ (as $a^\star = b \in V(p)$). Together (i) and (ii) mean that $\neg p \not\vdash_{FDE^2} \neg' p$. Thus *FDE* does not uniquely characterize $\neg$ (in terms of the remaining connectives).

Such, at any rate, is the sort of reason I had in mind in the passage quoted from the review of Priest et al. (1989), for saying that negation is not uniquely characterized by the principles governing it in relevant logic *à la* Anderson and Belnap, perhaps a hasty conclusion in view of the absence of their favourite connective, the (relevant) implicational connective, and with certain aspects of the above argument left vague in the interests of brevity, since it is not actually *FDE* but *LP* that we need to be attending to in the present case. One might expect a minor tweak on the above treatment would deliver an analogous result for *LP*, but certainly, the most obvious move in that direction does not get us where we want. For a simple semantic description parallelling that given above, *before* to reduplicating the star function, one would need to Australianize the essentially 'American Plan' three-valued matrix semantics. So our starting point would be the Routley–Routley models (from Routley and Routley 1972) with their single *-functions (subject to the involution condition above: $x^{**} = x$), except that we now want $A \vee \neg A$ to come out true throughout the models. (I say 'throughout' as though large choices of W may be forced on us, though because of the involution condition, the generated models—to which we can restrict attention without logical loss—have at most two elements in them. This would not be so for the models $\langle W, *, \star, V \rangle$ recently in play, since alternating applications of the two star functions can take us arbitrarily far from our starting point.)

Since excluded middles for A are *LP*-consequences (indeed, already *FDE*-consequences) of excluded middles for A's proper subformulas, it would suffice to make a suitable stipulation on the V component of the models securing the basis cases $p_i \vee \neg p_i$. So, the 'tweak' that this suggests is that we say just enough to make sure that every point verifies least one of these two disjuncts. So if $x \notin V(p_i)$ we need to make sure that $x^* \notin V(p_i)$, since if we have $x \notin V(p_i)$ and $x^* \in V(p_i)$, that will mean that not only is p_i not true at x, but that $\neg p_i$ isn't either. (Excuse me, Graham, while I contrapose, sticking to the line that the *raison d'être* of the model theory should be to block an insistence on a non-classical metalogic.) Thus if $x^* \in V(p_i)$, we must have $x \in V(p_i)$. Imposing this condition on (all $x \in W$ in) our models would be a disaster, however, since substituting x^* for x (and cancelling the two applications of the star function) delivers its converse. Any points x and x^* will now verify exactly the same formulas, defeating the purpose for distinguishing them, and we will not only get $A \vee \neg A$ true everywhere, as desired, but $A \wedge \neg A$ true nowhere, blocking the countermodels refuting the claim, for instance, that $p \wedge \neg p \vdash_{LP} q$. (Here we write p_1, p_2 as p, q.)

Let us recall that $\neg$ is not a congruential connective for $\vdash_{LP}$, in the sense that in general formulas A and B which are *LP*-consequences of each other need not have negations standing which are *LP*-consequence of each other. Diagnosis (well known): $A \vdash_{LP} B$ does not imply $\neg B \vdash_{LP} \neg A$; example: all excluded middles $A \vee \neg A$ are thus equivalent but their negations (being equivalent to $A \wedge \neg A$) are in general not.[48] This shows that we cannot provide a semantics for $\vdash_{LP}$ in terms of models truth preservation at each element of which is to match $\vdash_{LP}$, since the excluded middles would be true at the same points but their negations would not be. We could take a leaf out of Kripke's book—or rather his paper, Kripke (1965)[49]—and deal with the non-congruentiality of $\Box$ in some modal logics (familiar too from the Routley–Meyer semantics for relevant logics) by declaring some points in the models to be normal, and subject to special conditions not demanded of other points, and taking the consequence relation of interest to be that of truth preservation at all the normal points. The 'collapsing' argument at the end of the previous paragraph would not go through if we made our models be $\langle W, *, N, V \rangle$ with N a non-empty subset of W subject to the condition (6):

(6)   For all $x \in N$, if $x^* \in V(p_i)$ then $x \in V(p_i)$.

Imposing this condition (as well as the involution condition) on the models, our soundness observation would take the form: whenever $A_1, \ldots, A_n \vdash_{LP} B$, for any model $\mathcal{M} = \langle W, *, N, V \rangle$, and any $x \in N$, if $\mathcal{M} \models_x A_i$ for $i = 1, \ldots, n$, then $\mathcal{M} \models_x B$. I would imagine that there is a corresponding completeness result, though since the semantics is now much more complicated than the matrix semantics it would perhaps not be of much interest, though we should check that we have managed to avoiding validating the *Ex Falso* transitions of the previous attempt.

To that end, it suffices to make sure that $p \wedge \neg p$ can be true at a 'normal' point (since we are then free to arrange for q not to be). Thus for $a \in N$ in one of the above models we want $a \models p$ and $a \models \neg p$, the latter meaning that $a^* \not\models p$. So in setting up the model we need to have $a \in V(p)$ but $a^* \notin V(p)$. But this does violate our special condition (6). Instantiating the universal quantification in (6) to a we are not facing a counterexample to the condition, since that would involve having $a^* \in V(p_i)$ with $a \notin V(p_i)$, and we have things the other way round. On the other hand, recalling the collapse argument above, instantiating (6) instead to a^* and writing the resulting a^{**}

---

[48] Béziau (2002), p. 477: 'In Priest's logic *LP* and in da Costa and D'Ottaviano's logic *J3*, the formulas $p \vee \neg p$ and $q \vee \neg q$ are logically equivalent but not their negations and here again no philosophical justification for this failure has been presented'. This seems a bit harsh in the case of Priest, since the equivalence in the former case is a by-product of the decision to smooth gaps out of the picture to concentrate in gluts, and the latter brings with it the distinction between those contradictions which embody the gluts and those which are simply false. (In the quotation just given, I have changed the notation to match that in use here.)

[49] In fact, it is the use of non-normal worlds in models for properly quasi-normal modal logics in Chap. III of Segerberg (1971) that is closer in spirit, and it is the distinguished worlds rather than normal worlds there that are pertinent. (These enter into the definition of validity as the worlds unfalsifiability at which is significant, rather than entering into the inductive definition of truth at a world.) The use of 'non-normal' in the text echoes Priest (1992), though unless I misunderstand it, I am disagreeing with the parenthetical remark in the following passage from p. 301 of that paper: 'These constructions ensure that normal worlds are complete and consistent, respectively. (Note how these issues can be handled separately on this construction, in a way that they cannot be, using the * operator.)'

as a, so that we get: if $a \in V(p)$ then $a* \in V(p)$, does give a conditional to which the present case would be a counterexample—but the initial restriction on (6) now requires that $a* \in N$ and while we have had to assume $a \in N$, we can—and indeed we had better—keep $a*$ out of N.

While this secures the validity (=truth at normal points in all models) of the excluded middles $A \vee \neg A$ with A a propositional variable, we need this for arbitrary A, and so need to check by induction on the complexity of A that for any such A, the analogue of (6) holds for an arbitrary model $\mathcal{M} = \langle W, *, N, V \rangle$ satisfying (6) itself:

(6)$^+$ For all $x \in N$, if $\mathcal{M} \models_{x*} A$, then $\mathcal{M} \models_x A$.

The envisaged induction—working through the cases of $A = B \wedge C$, $A = B \vee C$ and $A = \neg B$ poses no difficulties, but there is trouble ahead.

The trouble comes when we try to duplicate this treatment of LP to show a failure of unique characterization for $\neg$ as in the FDE case, only now with models $\langle W, *, \star, N, V \rangle$, and satisfying both condition (6) and a similar condition in the case of $\star$. We want all excluded middles $A \vee \neg A$ and $A \vee \neg' A$ to end up valid, including the case in which A is constructed using $\neg$, $\neg'$ or both. So we will need (6)$^+$ as well as the analogous result for $\neg'$, but this time the induction—taking the case of (6)$^+$ itself by way of example—falters at the new inductive case ($B = \neg' A$), as the interested reader is invited to verify.

However, there is a simple matrix solution to hand.[50] We can expand the LP matrix with a second 1-ary operation for a candidate $\neg'$:

$$\neg' : t \mapsto f \quad b \mapsto f \quad f \mapsto t.$$

The consequence relation, $\vdash$, determined by this expanded matrix now satisfies the De Morgan and double negation equivalences as well as LEM, for the new negation $\neg'$ as well as for the original $\neg$, but $\neg p \nvdash \neg' p$ (consider $h(p) = b$).

What does this show about whether LP uniquely characterizes negation (to within equivalence, in terms of $\wedge$ and $\vee$)? To get more precise about this we need to introduce the idea of a *sequent*. For the moment, since we are concentrating on consequence relations, we can think of a sequent as a candidate element of such a relation, i.e. as a pair consisting of a set of formulas and an individual formula, except that rather than writing $\langle \Gamma, A \rangle$ for such a pair, we use the more suggestive notation $\Gamma \succ A$. (Other notations such as '$\Gamma \Rightarrow A$' and '$\Gamma : A$' are also common; '$\succ$' is from Blamey 2002.) For a more complete match with common terminology, one might add the restriction that Γ is finite, and for a more explicit way of registering the choice as to what logical framework we have opted to work with, call the whole thing (as in Humberstone 2011a) a SET-FMLA sequent, with a view to making room for SET-SET sequents, for example, of the form $\Gamma \succ \Delta$ standing to generalized consequence relations as SET-FMLA sequents do to consequence relations. (We return to the matter of alternative logical frameworks below.)

---

[50]For this solution, I am indebted to Sam Butchart.

It is really a collection of sequent-to-sequent rules (including the zero-premiss case), here presumed always to be substitution-invariant rules, which uniquely characterizes a connective, an m-ary connective #, say, when reduplicating those rules as above gives a set of rules sufficing for the proof of the sequent

$$\#(A_1, \ldots, A_n) \succ \#'(A_1, \ldots, A_n)$$

and the converse. More precisely—a complication already alluded to—we should add the rider 'in terms of such-and-such' to mention other connectives appearing unchanged in the #-reduplicated, and we should note that what has just been defined is unique characterization *to within equivalence*, a potentiality stronger demand being unique characterization to within synonymy, meaning that the left- and right-hand formulas of the sequent displayed above should be fully interchangeable in the combined system. This potentiality can be realized in non-congruential settings such as that of *LP* itself, but we are not attending to it separately here since we are aiming to show that even the weaker condition of unique characterization to within equivalence is not satisfied as far as negation is concerned. (When no explicit qualification is made, what will be meant by unique characterization below is: unique characterization to within equivalence.) Now, one special case arises when the rules governing the connective in which we are interested themselves are all zero-premiss sequent-to-sequent rules, and if we are starting with a consequence relation $\vdash$ we can take as such rules all substitution instances (in whatever language we are interpreting the rule as applied to) of sequents belonging to, together with the structural rules corresponding to the defining conditions for a consequence relation (in particular the non-zero-premiss rules of weakening and cut). In this case we can say that the initial consequence relation $\vdash$ uniquely characterizes the connective in question when the smallest substitution-invariant consequence relation extending $\vdash$ and $\vdash'$ (the latter being like $\vdash$ except having #' in place of #) contains the sequent above for all $A_1, \ldots, A_n$ of the combined language. The three-element matrix above shows that the consequence relation $\vdash_{LP}$ fails to characterize negation uniquely (to within equivalence, in terms of the connectives $\wedge$, $\vee$). (Note that the combined consequence relation here is substitution-invariant, since this is so for any matrix-determined consequence relation.)

There are several reasons why one might think that the fact that the consequence relation $\vdash_{LP}$ does not uniquely characterize negation is of little interest. After all, if we restrict attention to the restriction of this consequence relation to the language in which the sole connective is $\vee$, which coincides with the similarly restricted consequence relation obtained from $\vdash_{CL}$, this consequence relation does not uniquely charactacterize $\vee$.[51] This would not make one say that *CL* (classical logic) does not uniquely characterize $\vee$, because one of the rules needed for the unique characterization is the rule (or indeed just the $\Gamma = \Delta = \varnothing$ special case of this rule):

---

[51] See Humberstone (2011a) p. 599*f*.

$$\frac{\Gamma, A \succ C \quad \Delta, B \succ C}{\Gamma, \Delta, A \vee B \succ C}$$

which does not get to be inherited by the combined ($\vee$, $\vee'$) consequence relation because—the Carnapian theme returns—it is not embodied in any zero-premiss sequent-to-sequent rule. This just shows that there is more to unique characterization by sequent-to-sequent rules than is captured by the special case we get when we talk about a consequence relation uniquely characterizing a connective. The above rule and the zero-premiss rules (sequent schemata) $A \succ A \vee B$ and $B \succ A \vee B$ obviously characterize $\vee$ uniquely—and we don't have to say 'in terms of the remaining connectives', either—so we have not shown that negation is not uniquely characterized by the logic *LP* as long as we don't insist on flattening out our conception of what a logic is to think it can be captured by a mere consequence relation. We can, incidentally, use the above sequent-to-sequent rule governing $\vee$—the ($\vee$ Left) sequent calculus rule—or its natural deduction equivalent (again presented in sequent-to-sequent format):

$$\frac{\Gamma, A \succ C \quad \Delta, B \succ C \quad \Theta \succ A \vee B}{\Gamma, \Delta, \Theta \succ C}$$

in setting *LP* up as a collection of rules, reduplication of which is taken as pertinent to unique characterization, and still use the earlier matrix argument concerning negation. This is because, in the matrix given, the join ($\vee$) of two undesignated elements is undesignated, which suffices for the rule to preserve validity in the matrix.[52] (To put it another way: the set of designated elements is a prime filter.)

The previous paragraph concerned the question of unique characterization for $\vee$ in *LP tout court* rather than in terms of the remaining connectives, the latter having the answer that even the consequence relation $\vdash_{LP}$ uniquely characterizes $\vee$ in terms of $\wedge$ and $\neg$ because $\vee$ is (explicitly) definable in terms of $\wedge$ and $\neg$. This indicates a way in which, if it is only the relative ('in terms of') notion of unique characterization that one is interested in for a logic treating some—for definiteness, let us say—binary connective #, one can add another binary connective, §, say, to start with and boost the logic to secure that $A \mathsection B$ and $A \# B$ are equivalent for all formulas A and B. If one then asks whether # is uniquely characterized in terms of the remaining connectives the answer is a trivial 'yes' because the reduplication of # as #′ leaving § intact will produce a combined logic in which each of $A \# B$ and $A \#' B$ is equivalent to $A \mathsection B$, thereby making them equivalent to each other and thus securing uniqueness on the cheap. (This is just a particularly dramatic case of definability, since in the original logic $A \# B$ is definable as $A \mathsection B$.) However, dialectically our current situation with *LP* has been to urge that even using the relative notion of unique characterization, $\neg$ is not uniquely characterized by *LP* conceived as a logic in the framework SET-FMLA.

---

[52]More informatively: the above ($\vee$ Left) and $\vee$-Elimination preserve the property of holding on any given matrix evaluation, where a sequent $\Gamma \succ A$ *holds on h* iff it is not the case that $h(C)$ is designated for each $C \in \Gamma$ while $h(A)$ is undesignated.

(The latter is not a matter of identifying the logic with a consequence relation but with a set of sequent-to-sequent rules—or perhaps a family of such interderivable sets, and one of which will serve as a presentation of the logic—for which the sequents concerned are SET-FMLA sequents.)

A question may be raised, incidentally, about the unique characterization of $\wedge$, which, by contrast with $\vee$ (because we are in SET-FMLA) might be expected to be uniquely characterized by its zero-premiss rules (with the aid of the basic SET-FMLA structural rules). The question might arise because of the way the consequence relation $\vdash_{FDE}$, and so derivatively its extension $\vdash_{LP}$ was defined: $A_1, \ldots, A_n \vdash_{FDE} B$ iff $A_1 \wedge \ldots \wedge A_n$ stood in the binary relation of first-degree entailment to the formula B. This might create the impression that reduplicating with respect to $\wedge$ actually forces us to reduplicate the occurrences of $\wedge$ that are thereby concealed in the commas on the left. But no: familiar expectations are met, because having once specified the consequence relation of interest it is this consequence relation which is subject to reduplication: nothing is concealed in the commas at all, they just separate the elements of a set. The fact that we said which set was involved by making a reference to $\wedge$ does not mean that we are still mentioning $\wedge$ every time we talk about that set of formulas.

The framework SET-FMLA is not looking favourable as a setting for the unique characterization of negation in *LP*, an interesting companion observation to Béziau's observation that this framework is also not a suitable habitat for a contrary-forming but 'non-explosive' negation operator. (In Footnote 35, the point was made in terms of consequence relations rather than logics in SET-FMLA.) But issues of uniqueness are highly sensitive to variation—much more so than is suggested in the passage quoted from my (Humberstone 1992) at the start of this section—from one logical framework to another.

To illustrate this sensitivity consider the difference between SET-FMLA and SET-FMLA$_0$, where the latter permits at most one formula on the right of the sequent separator, a framework introduced (essentially) by Gentzen to give sequent calculus rules for intuitionistic rather than classical logic (the latter being treated essentially in SET-SET).[53] For present purposes we compare SET-FMLA with SET-FMLA$_0$ as frameworks in which to develop not intuitionistic but (Johansson's) *minimal* logic, and, recalling that $\neg$ in the latter logic is supposed to have the logical properties it has if taken as defined by $\neg A = A \rightarrow \mathbf{f}$, grafting this definition onto the $\{\wedge, \vee, \rightarrow\}$-fragment ('positive logic') of intuitionistic logic in SET-FMLA, but saying nothing special about the new nullary connective (sentential constant) $\mathbf{f}$. Remaining in SET-FMLA, one simply uses the definition just given of $\neg$, without—to repeat—providing any rules specifically governing $\mathbf{f}$ (in a natural deduction or a sequent calculus setting). If we ask about whether $\mathbf{f}$ is thereby uniquely characterized, the answer is of course negative, since having subjected $\mathbf{f}$ to no rules at all, adding a duplicate $\mathbf{f}'$ and subjecting it to the same rules does not suffice for a proof of $\mathbf{f} \succ \mathbf{f}'$. (Otherwise the proof could be mirrored to yield a proof of $p \succ q$.) But we can shift from SET-FMLA to SET-FMLA$_0$,

---

[53]This example is given to illustrate the same point in the Digression on p. 1192*f.* of Humberstone (2011a).

and, in the spirit of intuitionistic linear logic (more specifically, the sequent calculus for **ILL**), provide **f** with the sequent calculus rules:

$$(\textbf{f Left}) \quad \textbf{f} \succ \qquad\qquad (\textbf{f Right}) \quad \frac{\Gamma \succ}{\Gamma \succ \textbf{f}}$$

Now, we have said something special about **f**—not only with the Left rule but also in the Right rule, since on this approach one does not have the structural rule of Right Weakening. And although there is no difference between the SET-FMLA$_0$ sequents in which the succedent is not empty (and so also are also SET-FMLA sequents) thus provable from those in the straight SET-FMLA treatment, we have now managed to characterize **f** uniquely: for an **f**′ added and governed by analogous rules we have **f** ≻ **f**′ provable (and conversely, of course). The straightforward verification is left to the reader.[54]

We have the same situation if ¬ is taken as primitive and use these rules, again in SET-FMLA$_0$ without a right weaking rule[55]:

$$(\neg\, \text{Left}) \quad \frac{\Gamma \succ A}{\Gamma, \neg A \succ} \qquad\qquad (\neg\, \text{Right}) \quad \frac{\Gamma, A \succ}{\Gamma \succ \neg A}$$

These rules evidently characterize ¬ uniquely since the left rule gives us (from initial sequent $A \succ A$) $A, \neg A \succ$, and then applying a the right rule for a duplicate ¬′ delivers $\neg A \succ \neg' A$ (and conversely, interchanging the ¬ and ¬′ steps).

Another treatment sometimes offered for minimal logic in sequent calculus moves from SET-FMLA$_0$ to SET-FMLA$_0$, which being with it the non-uniqueness already mentioned, as well as a need to attend to the choice of primitives. For example, on p. 16 of Buss (1998), we read that 'minimal logic is formalized like *PJ*, except with the restriction that every succedent contain exactly one formula', where *PJ* is Buss's variant of Gentzen's intuitionistic sequent calculus *LJ* and is couched in a language taking ¬ rather than **f** (or ⊥) as primitive: but now the needed form of a premiss sequent for the (¬ Right) rule has been outlawed and there is no way to prove such sequents as $\succ \neg(p \wedge \neg p)$.

The uniqueness phenomenon is even more immediately evident in an approach which still uses sequent-to-sequent rules but which departs from sequent calculus

---

[54]One of the referees of the present paper expressed surprise at this formulation of minimal logic, wondering what it was really sound and complete, suggesting that a word is in order on the semantics in play. Using suitably adapted Kripke semantics for Johansson's logic (as in Segerberg 1968 or Sect. 2.3 of Odintsov 2008) the idea is to define a sequent $\Gamma \succ \Delta$ of SET-FMLA$_0$ to be *valid* just in case in every model, at every point x in the model, with Q for its set of queer or non-normal elements (at precisely which elements **f** is true, in any modal), if all formulas in Γ are true at x in the model then either $\Delta = \{D\}$ for some formula and D is true at x, or $\Delta = \emptyset$ and $x \in Q$. The valid sequents are then easily seen to coincide with those provable in the present proof system.

[55]This is again familiar from the literature. See, for example, the top of p. 95 of Curry (1957). A discussion of minimal logic and some close relations dear to Curry's heart emphasizing their paraconsistent potential can be found in Odintsov (2008).

proper in allowing complexity-increasing operational rules, and in particular two-way rules, such as taking, instead of (**f** Left) above, the vertically inverted version of (**f** Right)—an approach which is equivalent given the usual structural rules (which need not include right weakening for present purposes). Similarly in the case of ¬, using (¬ Right) along with its inverted form. In this case, to show unique characterization, we just use one direction of the rule to remove a connective and the other direction (in the reduplicated setting) to restore the duplicate. Such two-way rules (or rule pairs) have had a long history in logic going back to at least 1948, famous advocates of their use being Popper, Kneale, and Scott, and more recently Sambin.[56] They may lack some of the charms of sequent calculus rules proper (such as the sub-formula property for operational rules) but they do usefully foreground the unique characterization issue, as is noted, for example, with the rules ($\Box\uparrow$) and ($\Box\downarrow$) at p. 776 of Blamey and Humberstone (1991). (Or again, one could cite Belnap's 'display logic' for a similarly motivated treatment. See the exposition in Wansing 1998.) Indeed the discussion of that paper shows how without a proper specification of the logical framework involved and the set of sequent-to-sequent rules on offer for that framework, the question of whether, e.g. modal logic as such—or even a specific modal logic, such as **S5**—uniquely characterizes $\Box$ makes no sense. In the case of Blamey and Humberstone (1991), unique characterization is made possible for $\Box$ by a particularly blatant departure from more conventional frameworks: the creation of an extra 'modal premiss' position to house formulas which amounts to adding their necessitations to the usual premiss position on the left. In the case of minimal logic, it is the somewhat subtler matter of exploiting empty succedents arising in SET-FMLA$_0$. But the basic point is the same: if someone asks whether negation is uniquely characterized in minimal logic, one must inquire as to what logical framework the questioner has in mind, even once it has been made clear that it is absolute uniqueness rather than unique characterization in terms of other (preferably, definitionally independent) connectives. So much for correcting the crude tone taken in the passage is quoted from the review (Humberstone 1992) at the start of this section. It would be interesting to see whether negation can be equipped in paraconsistent logics with rules in SET-FMLA$_0$ (or SET-SET) which can similarly exploit (for instance) the empty succedent by suitably constraining Right Weakening, but we leave the matter here. (Yes, minimal logic is itself technically already a paraconsistent logic, though not one answering to the motivating idea, since it is, as we might say, 'negatively explosive'—see Priest 2002, p. 288*f.* or Odintsov 2008, p. 3.)

---

[56] See my (Humberstone 2011a), p. 151, for detailed bibliographical references and further discussion. Those references can be supplemented by mention of at least one new recruit: Restall (in preparation).

15.6 Unique Characterization: Some Refinements

The matrix argument (with $\neg'(b) = f$ but otherwise as for Priest's $\neg$) from the previous section for the non-uniqueness of LP's negation in the sense subsequently clarified as: $\vdash_{LP}$ does not uniquely characterize $\neg$, presents us with a $\neg'$ which is evidently not a paraconsistently suitable negation since, where $\vdash$ is the consequence relation determined by the expanded matrix, $p, \neg'p \vdash q$. This does not undermine the non-uniqueness proof but it raises a question. Suppose one thought it desirable to have unique characterization where possible for logical notions deemed fundamental, that negation was one such notion, but that paraconsistency was mandatory for anything's deserving to be considered as a negation connective. What if the only way to strengthen one's favoured logical treatment of negation so as to secure unique characterization conflicted with the paraconsistency desideratum? Then it would seem reasonable to consider a constrained version of unique characterization—uniqueness modulo paraconsistency, one might (albeit sloppily) put it: to fail to satisfy the revised conception of uniqueness, we should want to see a consequence relation in which both $\neg$ and $\neg'$ were paraconsistently suitable ('non-explosive') and yet yielded non-equivalent compounds from the same components. We shall show that $\neg$ is not uniquely characterized by $\vdash_{LP}$ even when this (apparently) weaker understanding of unique characterization is in force. But before doing so, it will be useful to air a philosophical disagreement between Priest and myself as to the significance of unique characterization, to which end we need to see Priest giving his half of the story as it bears on negation; this passage is from Chap. 12 ('Logical Pluralism') of Priest (2006b) (p. 198f.).

> Classical and intuitionist connectives have different meanings. (Indeed, some intuitionists, such as Dummett, even claim that some classical connectives have no meaning at all). Thus, take negation as an example. Classical and intuitionist negations have different truth conditions. But the difference in truth conditions entails difference in meaning. Hence, the two connectives have different meanings. Now, either vernacular negation is ambiguous or it is not. If it is not, then, since the different theories attribute it different meanings, they cannot both be right. We have a simple theoretical pluralism. The other possibility is that vernacular negation is ambiguous. Thus, it may be argued that vernacular negation sometimes (p.199) means classical negation and sometimes means intuitionist negation. But if this were right, we would have two legitimate meanings of negation, and the correct way to treat this formally would be to have two corresponding negation signs in the formal language, the translation manual telling us how to disambiguate when translating into formalese. In exactly this way, it is often argued that the English conditional is ambiguous, between the subjunctive and indicative. This does not cause us to change logics, we simply have a formal language with two conditional symbols, say $\supset$ and $>$, and use both. This is pluralism in a sense, but the sense is just one of ambiguity.
>
> In fact, I see no cogent reason to suppose that negation is ambiguous in this sense. If the account of negation of Chapter 4 is correct, negation is unique. Indeed, negation cannot have different meanings when reasoning about different kinds of things. This is because we can reason about different kinds simultaneously. We can, for example, reason from the claim that it is not the case that a and b have some particular property in common, where a and b are of different kinds. Moreover, there are other reasons as to why vernacular negation cannot be ambiguous between intuitionist and classical negation. If it were, as I have argued, we could have two formal negations. But it is well known that in the presence of classical

negation, many other important intuitionist distinctions collapse. For example, the intuitionist conditional collapses into the classical conditional.

At the end of this passage, there is a footnote reference acknowledging Harris (1982), as in my own discussion of this issue in Sect. 4.3 of Humberstone (2011a), though I suspect that these collapses were rather more widely known about than written at the time (Harris 1982) appeared.[57] So it is interesting to see Priest drawing such a different conclusion from them.[58] The line I took was, in effect, that the collapse considerations of the second paragraph of the above quotation undermine the claim at the start of the first paragraph: if classical and intuitionistic negation connectives have different meanings then it ought to be possible to combine them in the language of a single logic exploring their interrelations. But since the rules governing intuitionistic negation[59] are already strong enough to characterize it uniquely (as forming weakest contraries), in the combined logic with classical negation satisfying these rules *and more*, the 'more' spills over onto intuitionistic negation and we lose the contrast. (The classical rules are, as it is put in Humberstone (2011a), where several further examples of this phenomenon are cited, *stronger than needed* for unique characterization.) This description of the matter presumes some sympathy with the view that, perhaps subject to further conditions, the primitive rules governing a connective determine or reflect its meaning—not something one could expect to establish or refute once and for all. But it does add some interest to the question, raised at the start of this section, of illustrating the non-uniqueness of negation as it behaves according to $\vdash_{LP}$ with $\neg$ and its duplicate both being paraconsistently non-explosive.

Not having had much success in finding simple matrix argument for that conclusion with a small number (≤ 9, say), let me offer instead an 'off the shelf' solution, basing a matrix on a familiar algebra, even though this algebra will be twice as large as hoped. This is not at all to claim that no smaller matrix can be found with the desired effect—just that I was not able to find one. But let us back up and recall what

---

[57] I was certainly aware of the negation case in 1974 and on noticing it mentioned it to as many of my fellow graduate students at Oxford as would stand still long enough. The point is implicit in the restriction on the rule called RAA$_\neg$ on p. 174 of Humberstone (1979), which prevents the intuitionistic style negation in play there from collapsing into the simultaneously present classical style negation. (This was alluded to at the end of the Digression with which Sect. 15.2 concluded. A more sophisticated venture with a similar motivation can be found in Lucio 2000.) The material in Sect. 4.3 of Humberstone (2011a), incidentally, was originally presented at a splendid AAL conference in Perth in 1983, organized by Priest.

[58] There are several other aspects of the passage just quoted which might merit further attention than can be given to them here. For instance, the claim that 'Classical and intuitionist negations have different truth conditions' makes one wonder what semantic account is in play supplying truth condition to the classical negation of a statement and the intuitionistic negation of that same statement; it's not as though there is even, in the intuitionistic case a single option as to how to do the (heterophonic, model-theoretic) semantics: do we use the Kripke semantics, the Beth semantics, or something else again? What are 'the' truth conditions for intuitionistic disjunctions, for instance, ($\vee$ receiving a different treatment in Beth and Kripke)?

[59] Here we understand the rules as applying in a substitution-invariant way over the combined language, not subject to such restrictions as were alluded to in Footnote 57.

the desired effect is. Instead of writing the duplicating negation as '$\neg'$', which would be rather cumbersome in some of the diagrams that will be presented, I will write it as '$\sim$'. Thus what we will call $\vdash_{LP^2}$ is the smallest substitution-invariant consequence relation containing every sequent in $\vdash_{LP}$ as well as all such sequents which have $\neg$ replaced (throughout) by $\sim$. And what we want to show is that $\neg p \nvdash_{LP^2} \sim p$, by exhibiting a matrix the consequence relation $\vdash^+$, say, determined by which extends (not necessarily properly) $\vdash_{LP^2}$, retaining paraconsistency. (That is, we want neither $p, \neg p \vdash^+ q$ nor $p, \sim p \vdash^+ q$.) Since we want to show that $\neg p \nvdash_{LP^2} \sim p$ in this way, via the fact that $\neg p \nvdash^+ \sim p$ (and $\vdash_{LP^2} \subseteq \vdash^+$), $\neg p$ will need to be able to be assigned a designated value while $\sim p$ gets an undesignated value in our matrix.

To keep things simple, we hope to have the set of designated elements comprise a filter—indeed a prime filter—in the algebra of the matrix, which we will take to be a double De Morgan lattice, which is to say an algebra with fundamental operations $\wedge, \vee$ of rank 2 and $\neg, \sim$ of rank 1, the $\{\wedge, \vee\}$-reduct being a distributive lattice and each of the $\{\wedge, \vee, \neg\}$- and $\{\wedge, \vee, \sim\}$-reducts being a De Morgan algebra.[60] Because of LEM for $\sim$ (i.e. the fact that since $\vdash_{LP^2} A \vee \sim A$ we must have $\vdash^+ A \vee \sim A$, for all formulas A, and we will be working with a prime filter as our set of designated elements), this will mean the corresponding value of p is itself designated, so one idea is to follow the simple idea behind the *LP* matrix (with its elements *t*, *b*, *f*) and give p and $\neg p$ the very same designated value—call it a, which $\sim$ will have to map to an undesignated element, call it b. By the involution property, we know that $\sim b$ is, in turn, a, but we do not know what $\neg b$ is: calling it c, one simple option is to mirror the structure so far and just as a was a fixed point for $\neg$, have c be a fixed point for $\sim$. This third element c has to be a designated element of the matrix, both because of LEM for $\neg$, since b was not designated, and because of LEM for $\sim$, since any failure of a point left fixed by one of our negation operations to be designated would give a counterexample to LEM for that negation.

Thus, writing a line between x and y to indicate that the negation written above the line interchanges a and y, the story so far is that for our three elements a, b, c, we have

$$a \overset{\neg}{\text{—}} a \overset{\sim}{\text{—}} b \overset{\neg}{\text{—}} c \overset{\sim}{\text{—}} c$$

Accordingly, if we want to know what $\neg \sim \neg a$ is, we see that a is paired with itself by $\neg$, so $\neg \sim \neg a = \neg \sim a$, and then since a is $\sim$-paired with b, $\neg \sim a = \neg b$ and finally since b is $\neg$-paired with c, we conclude that $\neg \sim \neg a = \neg \sim a = \neg b = c$. Similarly, one might calculate $\neg \sim \neg \sim c = \neg \sim \neg c = \neg \sim b = \neg a = a$. But now we want to do such calculations in the presence of $\wedge$ and $\vee$ too, so we need to pack this information into a De Morgan lattice, the simplest way to do which is to consider the distributive

---

[60]This last means that $\neg$ and $\sim$ both have to satisfy the following identities here stated in a neutral notation, with $-\!\!: -(x \wedge -y) \approx -x \vee y, -(x \vee y) \approx -x \wedge -y$, and $-\!-x \approx x$. Cf. the 'double Ockham algebras' of Fang (2008). Ockham algebras are distributive lattices expanded by a 1-ary operation obeying the De Morgan identities, construed narrowly so to exclude the involution condition (i.e. the identity $-\!-x \approx x$). Hence the present terminology of double De Morgan algebras—or more accurately double De Morgan lattices. (See Footnote 61.)

Fig. 15.1 The distributive lattice freely generated by a, b, c

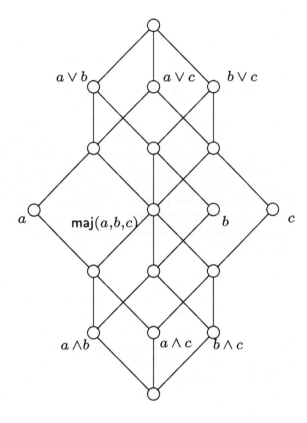

lattice freely generated by our a, b, c, since the De Morgan (and involution) identities will then tell us what the results of applying $\neg$ and $\sim$ to arbitrary elements have to be for us to have a De Morgan lattice on our hands. Let us take a preliminary glance at the lattice in question before expanding it to accommodate the two negations, as I will call the complementation like operations which interpret our negation connectives $\neg$ and $\sim$. (We use the same symbols for the connectives and the operations of this algebra, as also is the case with $\wedge$ and $\vee$.) The lattice is depicted in Fig. 15.1.

In Fig. 15.1, a few salient points have been labelled: the atoms and the co-atoms, as well as the free generators a, b, c (which in the expansion to be undertaken will be respectively $a, \sim a, \neg a$, as per the above discussion), as well as their companion maj(a, b, c), where maj(x, y, z) is the usual ('at least two-thirds') *majority* polynomial $(x \wedge y) \vee (x \wedge z) \vee (y \wedge z)$. As any lattice theory text will mention at this point: we could equally well write this in dual form, with the $\wedge$ and $\vee$ interchanged. Figure 15.2 is enlarged so that we can identify all the elements, providing at least one term for each of them, as well as to specify the behaviour of the two negations on these four middle-ranking elements. Broken lines indicate $\neg$-paired elements and wavy lines indicated $\sim$-paired elements. Even when these negations are defined for all elements, maj(a, b, c) will turn out to be the only one which is a fixed point for

15 Priest on Negation

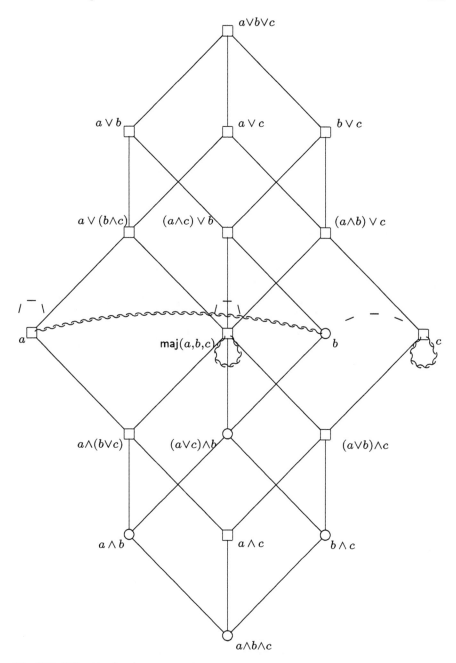

Fig. 15.2 Filling out the picture somewhat

both $\neg$ and $\sim$, represented in Fig. 15.2 by the differently styled loops (broken and wavy).

Something else has been incorporated into Fig. 15.2, in making some of the nodes square instead of circular. These are to be the designated elements of our matrix. Recall that these need to include $\neg a$ (which coincides with a) and c (alias $\neg\sim a$) but to exclude b ($=\sim a$), since the intention is to invalidate the transition from $\neg p$ to $\sim p$ by a matrix evaluation h with $h(p) = a$. For this reason, we have taken the filter generated by a, c (equivalently: the principal filter generated by $a \wedge b$) to comprise our set of designated elements. Recall that the 'h' for such evaluations is used because they are homomorphisms, which is why we use as the algebra an expansion of a free distributive lattice and not a free bounded distributive lattice, which has an extra 1 and 0, nullary fundamental operations as top and bottom elements, respectively above and below the $a \vee b \vee c$ and $a \wedge b \wedge c$ of Fig. 15.2.[61] The language of *LP* lacks corresponding nullary connectives (sentential constants) to play these all-implied and all-implying roles, and there is nothing to be gained by having the language—considered as an algebra—and the algebra of the matrix, differing in signature (similarity type) in that way.

The visual effect of completing Fig. 15.2 by fully diagramming the action of the two negations turned out to be one of buzzing confusion, and so we must resort to separate diagrams, one displaying the behaviour of $\neg$ in Fig. 15.3 and that of $\sim$ in Fig. 15.4. It is to be emphasized despite the two diagrams, these are not presented here as depicting two De Morgan lattices (though of course they could be), but as aspects of the presentation of one double De Morgan lattice.

Here, we just use the De Morgan identities (involution included) to work our way inward to the generators. So to calculate, for instance, what $\neg((a \vee b) \wedge c)$ should be, we reason:

$$\neg((a \vee b) \wedge c) = \neg(a \vee b) \vee \neg c$$
$$= (\neg a \wedge \neg b) \vee \neg c$$
$$= (a \wedge c) \vee b.$$

To find a suitable term letters are sometimes reordered to give one in 'alphabetical order' where possible (rewriting $b \vee a$ as $a \vee b$, for instance). Similar moves in the case of $\sim$ give rise to Fig. 15.4.

Thus, when the 'double De Morgan lattice' depicted in Figs. 15.3 and 15.4 (collectively) is taken together with the information about which elements are designated, what has been described is a finite matrix, though we have not written out the 18-line

---

[61] In the literature the appellation 'De Morgan algebra' is typically now reserved for the bounded version of a De Morgan lattice. The older terminology of quasi-Boolean algebras seems to have fallen out of use. ('Bounded' here means having the bounds in the signature, thereby restricting the options for homomorphisms and subalgebras; of course, all finite De Morgan lattices are *de facto* bounded in the sense of having top and bottom elements.) The numbers of quasi-varieties of De Morgan lattices and of De Morgan algebras are very different, as is reported in the middle of p. 321 of Rivieccio (2012).

15 Priest on Negation

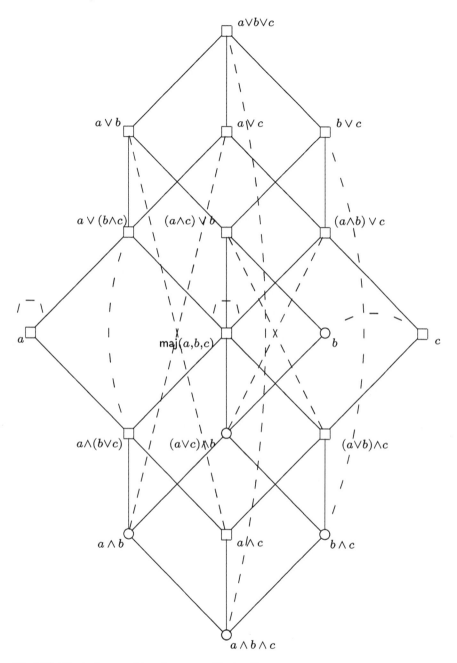

Fig. 15.3 Displaying ¬-pairings by means of broken lines

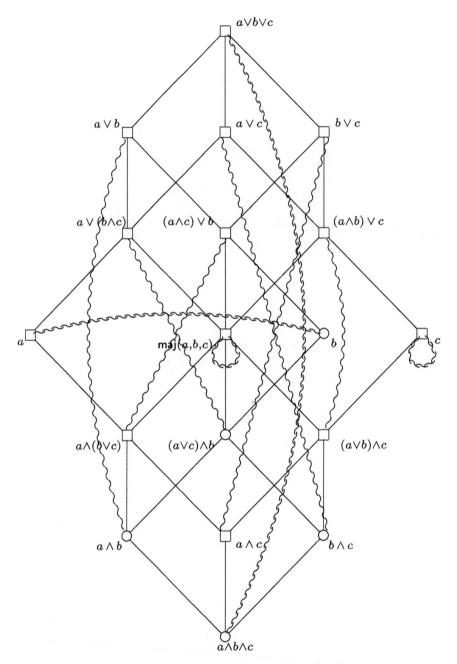

Fig. 15.4 Displaying ∼-pairings by means of wavy lines

tables for the negation operators or the 18-by-18 row-and-column tables for conjunction and disjunction. As foreshadowed above, the consequence relation determined by this matrix we denote by $\vdash^+$, and from the start we organized matters so as to have available a matrix evaluation h with $h(p) = a$, making $h(\neg p)$ designated and $h(\sim p)$ undesignated, so that $\neg p \not\vdash^+ \sim p$, and therefore $\neg p \not\vdash_{LP^2} \sim p$, since evidently $\vdash_{LP^2} \subseteq \vdash^+$. This last is the conclusion already obtained in Sect. 15.5 but this time we have cohabiting non-equivalent $\neg$ and $\neg'$ (written as $\sim$) each of which is paraconsistent. (This is evident from the fact that the corresponding matrix operations in Figs. 15.3 and 15.4 have fixed points among the designated elements.)

There is also another differentiating feature, aside from its being thoroughly paraconsistency-friendly, of this illustration of non-uniqueness for *LP*'s $\neg$ by comparison with that mentioned in the preceding section. As suggested in Footnote 48, there may be justifiable as well as unjustifiable failures of congruentiality, and it was there suggested that *LP*'s failures, arising over LEM, fall on the justified side of this division—by contrast with Da Costa's logics in which no two formulas are ever freely interreplaceable. The combined consequence relation $\vdash^+$ of our recent investigations shares this feature with $\vdash_{LP}$ itself, failures of congruentiality arising only as a result of (the two incarnations of) LEM. This is because the $\{\wedge, \vee\}$ and De Morgan equivalences (Double Negation included) are valid in the 18-element $\vdash^+$ matrix as a result of identities: the sequents involved are *algebraically* valid or, as it is put in 2.15 of Humberstone (2011a), 'indiscriminately' valid, meaning that they are valid in every matrix based on the same algebra. (By contrast, sequents $\succ A \vee -A$, where '$-$' is $\neg$ or $\sim$, owe their validity to the particular choice of designated elements.[62]) But in the case of the three-element matrix for $\neg$ and $\neg'$ in Sect. 15.5, with the equivalence of $\neg'\neg'p$ with p is valid only in virtue of the pattern of designation, since these formulas may receive distinct albeit equi-designated values. In particular, using (as ever) the same notation for matrix operations as for the connectives, $\neg'\neg'\boldsymbol{b} = \boldsymbol{t}$. (Recall that $\neg'(\boldsymbol{t}) = \neg'(\boldsymbol{b}) = \boldsymbol{f}$, $\neg'(\boldsymbol{f}) = \boldsymbol{t}$.) This opens the door to failures of congruentiality, in the present case witnessed by the fact that while $p \dashv\vdash \neg'\neg'p$, where $\vdash$ is the consequence relation determined by the matrix in question, we do not have $\neg p \dashv\vdash \neg\neg'\neg'p$, in particular because the sequent $\neg p \succ \neg\neg'\neg'p$ is invalid in the matrix, taking us, when $h(p) = \boldsymbol{b}$, from the designated value $\boldsymbol{b}$ on the left to the undesignated $\boldsymbol{f}$ on the right.

To close, we note a respect in which both the three-element matrix from Sect. 15.5 and the 18-element matrix of the present section are alike, both showing that *LP* does not uniquely characterize negation in a stronger sense than that given by saying that the associated consequence relation $\vdash_{LP}$ does not characterize negation uniquely. Recall from Sect. 15.5's discussion that, basically because a consequence relation is a collection of SET-FMLA sequents rather than SET-SET sequents, the facts that $\vdash_2 \supseteq \vdash_1$ and that $\vdash_1$ has the $\vee$-elimination or ($\vee$ Left) property—any consequence of each disjunct is a consequence of the disjunction—do not suffice for $\vdash_2$ to inherit

---

[62] It may be tempting to connect this with the idea that of *LEM* 'as a (possibly correct) substantive metaphysical principle not appropriately built into one's account of the logic of the connectives concerned', as it was put in the opening paragraph of Humberstone and Lock (1986).

this property. So if we are thinking of *LP* as given by a proof system with sequent-to-sequent rules securing it but then focussing on just the induced consequence relations, special steps must be taken to secure the property.[63] If the rule encoding this property—the sequent calculus rule ($\vee$ Left) or the natural deduction rule ($\vee$E)—is taken as one of those in the set of rules with which *LP* is identified, special steps need to be taken to ensure that the consequence of the 'combined' ($\neg + \neg'$) logic has the property. In the case of the consequence relations determined by both the three-element matrix and the 18-element matrix, this property is possessed, since the set of designated elements is in each case not just a filter but a prime filter of the underlying lattice.

Our investigation into whether negation is uniquely characterized in *LP* has considered several variations on what this might be taken to amount to, but none of the options has yielded an affirmative answer to this question. While such questions have an interest of their own, they also connect with another theme of Priest's, since we have not found an answer sitting well with his twofold conviction that (i) contradictories are unique and (ii) the $\neg$ of *LP* is a contradictory-forming operator.

Acknowledgements I am greatly indebted to Sam Butchart for helpful observations and suggestions which have improved this paper, to two referees for the detection and correction of numerous errors, and to Heinrich Wansing for additional assistance.

Appendix 1: Subcontraries and Mere Subcontraries

For the general point to be discussed here, the title could equally well have been (or included) 'Contraries and Mere Contraries', but that chosen reflects more specifically some aspects of the discussion of Slater (1995). Let us recall that there are two understandings of the terminology of contraries and subcontraries, distinguished as the 'say less' and 'say more' interpretations in Humberstone (2003), where numerous textual references and some further discussion may be found.[64] On the 'say less' approach, the most influential advocate for which was perhaps E. J. Lemmon,[65] semantically formulated, contraries/subcontraries are defined to be incapable of joint truth/falsity, or modulating to a consequence relation relativity, to be contraries/subcontraries relative to $\vdash$ when $\vdash$ is determined by some class of (bivalent) valuations no element of

---

[63] See p. 328 of Rivieccio (2012), where Font (1997) is observed not to have taken the necessary steps. The same point underlies Rivieccio's observation on the preceding page, that where $\vdash'$ is the smallest consequence relation $\vdash$ extending $\vdash_{\mathsf{FDE}}$ such that $A \wedge \neg A \vdash B$ for all A, B, we do not have $\neg p \wedge (p \vee q) \vdash' q$. The observation hangs on the absence of the qualification '$\vee$-classical' in characterizing $\vdash'$. (Rivieccio calls $\vdash_{\mathsf{FDE}}$ '$\vdash_{\mathcal{B}}$', in honour of Belnap.)

[64] And where, in fact, the 'say less' interpretation is explicitly labelled as such, though the phrase 'say more' for the alternative does not explicitly appear.

[65] See note 12 of my (Humberstone 2003) for evidence of this influence, and for an elaboration of Lemmon's treatment and its early reception, Humberstone (2013b), especially, Sects. 4 and 4.1. These papers provide further bibliographical details.

which verifies/falsifies both. This is the usage that has been followed in the present paper.

The 'say more' interpretation, understandably, says more, adding a requirement that the class of valuations in question should also provide at least one verifying both of the subcontraries (respectively, falsifying both of them) for them to deserve that description. Both are perfectly legitimate things to mean by the terminology, but the smoothest discussion of the issues with which we have been concerned takes the 'say less' approach. Contrariety and subcontrariety are then what are called *coercive* logical relations in Humberstone (2013b), by contrast with, for example, consistency, understood as the relation holding between *A* and *B* precisely when they are not contraries, is what is there called a *permissive* logical relation. The coercive relations are defined by a requirement—and here we put it syntactically—that something should be provable in the logic to which they are relativized, whereas the permissive relations such as consistency and independence are those obtaining in virtue of something's *not* being provable in the logic in question.[66] (In terms of quantification over valuations in a determining class, the coercive relations involve universal quantification and the permissive relations involve existential quantifications.) Contrariety on and subcontrariety on the 'say more' approach are *mixed* coercive and permissive logical relations, like the relation of unilateral implication (holding between *A* and *B* according to $\vdash$ when $A \vdash B$ and $B \nvdash A$), and having selected the 'say less' option, we can still make a concession to the alternative option by calling contraries which are not also subcontraries, *mere contraries*, and subcontraries which are not also contraries *mere subcontraries*.

Note that the definitions are just given of mere contrariety and mere subcontrariety could equivalently have been put in the following way: mere contraries are contraries which are not contradictories, and mere subcontraries are subcontraries which are not contradictories.[67] For this reason, it does not matter whether Slater's objection is put by saying, on the one hand, that *LP*'s negation turns a formula into a subcontrary but not (in the general case) into one which is a contradictory, or, on the other hand, that the *LP*'s negation turns a formula into a mere subcontrary, so nothing hangs in this debate on whether Slater had the 'say less' or the 'say more' understanding

---

[66]The claim about something being provable is something of an oversimplification. If we have an ∨-classical consequence relation $\vdash$ then the subcontrariety of A, B does find expression in the provability of their disjunction, but the 'purified' form of this condition given in (ii) of the Digression in Section is a *conditional* rather than an unconditional $\vdash$-condition, and even in the unconditional case—as we see from considering contrariety, the provability is a matter of provability of sequents rather than just formulas. If we were thinking of the purified version of subcontrariety for SET-SET sequents we do of course have the unconditional demand that $\succ A, B$ should be provable—formulated as the condition '$\Vdash A, B$' on generalized consequence relations at the start of Sect. 15.3.

[67]Here I am, in the usual 'say less' manner, taking *contradictory* to mean *subcontrary and contrary*, and in calling the formulations just cited equivalent am assuming the meta-logical analogue of the equivalence of $p \wedge \neg q$ with $p \wedge \neg(p \wedge q)$—an equivalence which does not hold for *LP* or *FDE*, leaving someone with Priest's inclinations (both logical and methodological) with the option, if they wanted to exploit it, of distinguishing two separate notions of mere contrariety and also of mere subcontrariety.

of the word *subcontrary* in mind. Béziau seems to think otherwise, and himself favours the 'say more' reading, as is also clear from Béziau (2016). This may be part of why he thinks that Slater's discussion is confused, since he attributes this interpretation to Slater also (see the Digression below) and it is perhaps also connected with his claim—quoted in the Digression in Sect. 15.3—that intuitionistic negation is not a contrary-forming operator. To illustrate in the latter case, for some choices of A, A and $\neg A$ will not be *mere* contraries, in the sense that they are contraries without being subcontraries: for example, take A as $p \wedge \neg p$. We resume the present discussion after an optional explanation of Béziau's grounds for attributing the 'say more' interpretation to Slater. (Appendix 2 addresses Heinrich Wansing's similar claim that negation in intuitionistic logic does not form what we are calling mere contraries.)

Digression. On the 'say more' interpretation, subcontraries and contradictories are mutually exclusive relations. Slater (1995) has the example of coming to call red things *blue* not amounting to changing their colour but only our terminology for attributing colours to them. This is meant to be analogous to coming to calling subcontraries *contradictories* not actually making them contradictories, so in view of the mutual exclusivity of *red* and *blue* as standardly understood, Béziau concludes that Slater assumes as standard an understanding of *subcontrary* and *contradictory* one which makes them likewise mutually exclusive, as the 'say more' interpretation does. But this may be a purely incidental feature of the example. Here is Béziau making a case for his reading of Slater, from p. 19 of Béziau (2006), beginning with a characterization of the 'say less' option:

> In this case, confusing subcontraries with contradictories would not be the same as switching red with blue, or cats with dogs, but rather would amount of confusing dogs with canines. Let us call *global confusion* this kind of error by contrast to the first one that we can call *switching confusion*. As Slater claims through his red and blue example that paraconsistent logicians are making a switching confusion rather than a global one, it seems implicit that he doesn't consider that all contradictories are subcontraries, neither do we here.

As I say, the fact that blue and red were chosen may just be for dramatic effect. In any case, although Slater did envisage a switch, calling red things blue and blue things red, the point of the example would have survived if red things came to be called blue but the blue things continued to be called blue. And this would have been more plausible since no party to the present debate could be described as calling contradictories *subcontraries* in the way that Slater accuses Priest of calling subcontraries *contradictories*. (The opening sentence of Béziau 1996 seems to offer such a description but this seems to be a matter of not being completely familiar with the 'calling Xs Ys' construction in English.) Despite its having (if I am right) no bearing on the merits of Slater's objection to dialetheism, it would have been interesting to know whether in writing Slater (1995) he had in mind the 'say less' or the 'say more' interpretation.[68] **End of Digression**.

---

[68] While finishing off the present contribution, I became increasingly curious about this. Unfortunately, by the time I tried to contact Slater for an answer to the question (May 2016), I heard from

The use of the word 'mere' above is a special *ad hoc* technicalization of the ordinary use. In the everyday sense the dual intuitionistic negation of a formula is not a 'mere' subcontrary, if that means there is nothing more informative to be said on the matter, since (as recalled more than once already) it specifically forms the *strongest* subcontrary of the formula in question (and similarly in the case of intuitionistic negation proper and mere contrariety, interchanging 'contrary' and 'subcontrary' and 'weakest' and 'strongest'). And in the case of intuitionistic logic itself, we note that there is always something stronger to be said than merely that two formulas are subcontraries, since by the Disjunction Property, one will be provable outright. Really this would be better described as the Subcontrariety Property, since the issue arises in ∨-free fragments of the logic (as was noted for *LC* in the Digression in Sect. 15.3): whenever formulas A, B are subcontraries in the sense there defined according to the fragment-specific consequence relation in question, one or other of them is a consequence of ∅. Indeed one can supply such an example for a fragment of classical logic too, taking the {↔}-fragment of $\vdash_{CL}$: whenever A and B, constructed using only ↔ from sentence letters, are subcontraries, at least one of A or B is a classical tautology.[69] Unlike the intuitionistic case, the Subcontrariety Property for the equivalential fragment of classical logic does not extend to the obvious ternary (or more) generalization of the relation of subcontrariety: for example, p, q and $p \leftrightarrow q$ are 'ternary subcontraries' (no Boolean valuation falsifying all of them), no one of which is tautologous. (However, the result does extend to allowing dependence on assumptions: If $\Gamma \vdash_{CL} A \vee B$ and the only connective appearing in $\Gamma \cup \{A, B\}$, is ↔, then either $\Gamma \vdash_{CL} A$ or $\Gamma \vdash_{CL} B$. A proof of this, which requires some apparatus going beyond that deployed for the $\Gamma = \emptyset$ case treated in Footnote 69, can be found in Humberstone 2019; see Proposition 1.4(ii) there.)

An attempt at precisifying something like the 'mereness' idea for coercive binary logical relations is set out in pp. 200–204 of my (Humberstone 2013b), which involves passage from such a relation, R, to the relation $\mu(R)$ of standing *minimally* in the relation R, with $\mu(R)$ defined to hold between A and B when (1) $R(A, B)$ and

---

a relative fielding his correspondence the sad news that it was too late: Hartley had suddenly fallen victim to an incapacitating and inoperable brain tumour.

[69]To see this, suppose that neither A nor B is tautologous and write A and B in the equivalent form of formulas $A' = p_1 \leftrightarrow p_2 \leftrightarrow \cdots \leftrightarrow p_m$ (parentheses omitted because of associativity) and $B' = q_1 \leftrightarrow q_2 \leftrightarrow \cdots \leftrightarrow q_n$, in which each sentence letter (the p_i and q_i) that occurs occurs exactly once. (We delete any letters occurring in pairs until only one occurrence remains, knowing that one occurrence must remain or the formula would have been tautologous: shades of Leśniewski here.) Noting that any such formula is false on any ↔-Boolean valuation in which exactly one of its sentence letters is false (indeed any odd number of those sentence letters), consider two cases: (i) for some i, j, $p_i = q_j$; and (ii) no p_i coincides with any q_j. In case (i), pick one shared sentence letter (a p_i which coincides with some q_j, that is) and consider a Boolean valuation assigning F to that sentence letter and T to all the others: each of A', B', and hence of A, B, is false on this valuation, contradicting the assumption of subcontrariety. For case (ii), pick any one of the variables p_i occurring in A' and one q_j occurring in B' and again assign the selected variables the value F while assigning T to all the other variables; as before, we get a Boolean valuation falsifying both A' and B'.

(2) there is no (coercive) binary logical relation $S \subsetneq R$ such that $S(A, B)$.[70] Thus if R is the relation of subcontrariety, its minimization $\mu(R)$, unlike R itself, never holds between contraries A and B, since if the relation of being contradictories held between them, that would then be a proper subrelation S of R for which $S(A, B)$, violating (2). The discussion of logical relations in Humberstone (2013b) has many features oriented specifically to the setting of classical logic but we can apply this idea to the Disjunction Property, to revert to this slightly misleading label, along with the name $R^\vee$ for the relation of subcontrariety, now to be understood as 'according to the consequence relation of intuitionistic logic'. As observed in the previous paragraph, in this setting subcontrariety is a somewhat degenerate affair, in that there is always something much more informative that can be said, since one or other of the relata is outright (intuitionistically) provable. In terms of the present apparatus this amounts to: $\mu(R^\vee) = \emptyset$. The reason is as follows. Writing $R^{①}$ (respectively, $R^{②}$) for the relation holding between any two formulas the first (respectively, the second) of which is *IL*-provable, the Disjunction Property means that $R^\vee = R^{①} \cup R^{②}$, so whenever $R^\vee(A, B)$ there is a proper subrelation of $R^\vee$, whether it be $R^{①}$ or $R^{②}$ (or indeed their intersection) which relates A and B; thus we can never have $\mu(R^\vee)(A, B)$. This example shows that being 'minimal subcontraries' does not coincide with being 'mere subcontraries' as defined above, since, for example, p and $q \to q$ are mere subcontraries (subcontraries which are not contraries—equivalently, contraries which are not contradictories) but not minimal subcontraries (since they stand in the relation $R^{②}$). Note that the present example can be understood with either classical or intuitionistic logic as the background setting—the Disjunction Property is not required. (For the intuitionistic case the appropriateness of our official definition of subcontrariety will not be taken for granted in Appendix 2 below.)

Finally, let us note that if one applies the 'mere' modifier in connection with contrary or subcontrary-forming operators, an ambiguity arises, as we illustrate in the contrariety case. A 1-ary operator # may be called a mere contrary-forming operator in order to convey that for every formula A, #A and A are mere contraries (contraries which are not contradictories). On the other hand one may by this same phrase intend to convey that # is a contrary-forming but not a contradictory-forming operator: in other words while for every formula A and #A are contraries, it is not the case that for every formula A, A and #A are contradictories. (To get back to the first—stronger—sense, replace this last *every* by *any*.) It is in this second sense that intuitionistic negation is a mere contrary-forming operator: see the example just before the Digression above. In Appendix 2 we look at the claim that (*pace* Priest among others), intuitionistic negation forms contradictories. These clarificatory remarks also apply if instead of taking a contrary-forming (or subcontrary-forming or contradictory) operator to be—as we have been presuming—a 1-place primitive or derived *connective*, but any function from formulas to formulas. In any extension of the normal modal logic **KD** (sometimes called **D**), for instance,Otherwise the intuitionist $\Box p$ and $\Box \neg p$ we can

---

[70]On the reasonable assumption that the intersection of any two logical relations is a logical relation, (2) can equivalently be formulated as (2'): for any (coercive) binary logical relation S such that $S(A, B)$, we have $R \subseteq S$.

ask whether these are contraries, and answer affirmatively. (The 'say more' advocate will have to hedge here and say, for example, that they are contraries in **KD** but not in its extension **KD!**,[71] where they are contradictories.) First, we can consider a (partial) function f which assigns to each formula $\Box A$ the formula $\Box \neg A$. This, in the extended sense of the phrase, would then count as a contradictory-forming (better: contradictory-yielding) operator, but it is what Humberstone (2011a) calls a non-connectival operation on formulas, in that there is no **KD**-definable connective # for which #$\Box A$ is **KD**-provably equivalent to $\Box \neg A$. Indeed with A as, e.g. p, $\Box \neg A$ is a paradigm case of contrariety from the modal square of opposition in *some* sense in which another mere contrary for all that has been said in this paper, namely $\neg(\Box p \vee \Box \neg p)$, is not: see Sect. 3 of my (Humberstone 2005b), for this and references to the work of Robert Blanché, who saw in this last formula together with $\Box p$ and $\Box \neg p$ a triangle of formulas all mutually contrary in the only sense (as Blanché felt) there is. What is involved here (further elaborated in Humberstone 2005b) may be connected with what Horn (1989) calls polar contrariety; the issue (as with the corresponding issue for subcontraries) has been kept out of sight in the present paper as not being of special significance in connection with Priest's discussion of negation and the reactions of commentators to that discussion. However, Horn does have a useful discussion, with historical references, to the main subject—'say less' versus 'say more'—on p. 36*f.* of Horn (1989).

Appendix 2: Wansing on Priest on Negation

At the end of the Digression in Sect. 15.3 it was mentioned that Wansing does not share Priest's view that intuitionistic negation forms mere contraries, preferring to describe it as genuinely contradictory-forming. Here we examine his reasons. For conformity with the rest of the notation used above, 'α' is replaced by 'A' and to avoid a conflict with the '(1), (2)' of Sect. 15.1, that numbering has been replaced by '(i), (ii)' in the following (somewhat ellipsed) quotation from p. 92*f.* of Wansing (2006), though the unparenthesized '1, 2' remain as in Wansing (2006). The references to Priest are to (1999), which this passage begins by summarizing—essentially the same as the material from Priest (2007), p. 467 (and Chap. 4 of Priest 2006b):

> The set $\{A, \neg A\}$ thus forms a contradictory pair if the LNC and the LEM are provable:
>
> (*i*) $\vdash \neg(A \wedge \neg A)$ and $\vdash A \vee \neg A$.
>
> A reformulated statement of the contradictoriness of $\{A, \neg A\}$ corresponding to the reading Priest attributes to traditional logic and common sense is:
>
> 1. A and $\neg A$ cannot both be true.
> 2. One of A and $\neg A$ must be true.

---

[71] Here we use the Chellas '!' from Chellas (1980), as in Humberstone (2011a) or (2016). Indeed we can say, speaking in the 'say less' manner, that among normal modal logics those in which $\Box p$ and $\Box \neg p$ are contraries are *precisely* the normal extensions of **KD**.

On another reading, $\{A, \neg A\}$ is a contradictory pair if

1. Necessarily, if A is true, then $\neg A$ is false.
2. A and $\neg A$ cannot both be false.

If the Deduction Theorem is assumed (to avoid using implication), this reading gives

(ii) $A \vdash \neg\neg A$ and $\vdash \neg(\neg A \wedge \neg\neg A)$.

If the contradictoriness of $\{A, \neg A\}$ is expressed by (i), intuitionistic negation fails to be contradictory-forming, and if the contradictoriness of $\{A, \neg A\}$ is expressed by (ii), intuitionistic negation does give rise to contradictions. As Priest (1999, p. 107) observes, in order to consider the truth conditions of negated statements, we 'need a definition of falsity. Let us define "A is false" to mean that $\neg A$ is true. (...) And the present definition is one that all parties can agree upon, classical, intuitionist and paraconsistent'. Since Priest regards intuitionistic negation to be not a contradictory-forming connective but a contrary-forming one, his explanation cannot mean that A is false iff its contradictory is true. Otherwise, the intuitionist as Priest conceives of him or her should not agree upon the definition of falsity. Thus, the notion of falsity in a logic Λ is expressed by the negation operation used in Λ (assuming that Λ comprises only one 'official' negation and taking into account that this may not be a contradictory-forming connective and hence not a negation in Priest's sense). The question thus is: When is $\{A, \neg A\}$ a contradictory pair?

Now, none of this—Priest's discussion or Wansing's development of it—is really methodologically optimal, as Section 1 of the present paper has tried to make clear. The question that should be asked is not what makes $\{A, \neg A\}$ a contradictory pair, but what makes $\{A, B\}$ in general to a contradictory pair, or more simply put, what it is for A and B to be contradictories. One then applies the general ruling to the special case in which B is $\neg A$. The most plausible general proposal to see as implicit in the above passage arises by putting 'B' in place of all of its occurrences of '$\neg A$'. A further moral of the discussion of the present paper is that by taking the 'say more' approach Priest and Wansing make it hard to describe the separate ingredients of their differing proposals permitted on the 'say less' approach with the simple vocabulary of contrariety and subcontrariety. (Wansing goes on to consider contrariety specifically in Sect. 3 of Wansing (2006) but this concerns the mixed coercive–permissive relation of mere contrariety rather than just the 'incompatibility' conception of contrariety of the 'say less' approach.) The first and second conjuncts of (i) are Priest's chosen ways of formulating the contrariety and subcontrariety conditions on A and B for the case of $B = \neg A$, whereas the first and second conjuncts of (ii) are Wansing's preferred formulations of the same two conditions. Moving to the general level, then, we may say that Wansing's preference for the contrariety and subcontrariety conditions for A, B according to a consequence relation $\vdash$ is given by:

Wansing contraries iff $A \vdash \neg B$; *Wansing subcontraries* iff $\vdash \neg(\neg A \wedge \neg B)$;

Priest contraries iff $\vdash \neg(A \wedge B)$; *Priest subcontraries* iff $\vdash A \vee B$.

Some criticisms made in Sect. 15.1 of Priest's discussion applies to Wansing's also: to bring to bear a criterion of contrariety or subcontrariety (or the two of them together, for contradictoriness) on the question of whether something and its negation stand in the chosen relation, the clean approach would formulate the various criteria without deploying that very connective in the formulation, and in fact we went on to prefer formulations purified not only of $\neg$ but also of $\wedge$ and $\vee$. (For example, for $\vdash$ with $\vee$ in its language the Priest subcontrariety condition will not deliver the intended results unless $A \vee B$'s consequences are the common consequences of A and B, since $\vdash$ may fail to be $\vee$-classical.) But Wansing here is not defending Priest's approach, so much as looking for 'Priest-style' criteria that might be preferable, and, let us recall, specifically noting that his preferred criteria rule that according to the consequence relation $\vdash_{IL}$ of intuitionistic logic, since for any formula A, A and $\neg A$ are Wansing contradictories in the sense of being both Wansing contraries and Wansing subcontraries, even though they are not Priest contradictories because they are not in general Priest subcontraries.[72]) (As we noted in Appendix 1, intuitionistic subcontrariety à la Priest—purged of any 'say more' trappings—is something of a degenerate relation because of the Disjunction Property.) Note that, by contrast with the case of subcontrariety, Wansing contrariety and Priest contrariety coincide for $\vdash = \vdash_{IL}$ and we use the terminology of Wansing contrariety for these equivalent formulations below (so that we can use 'Wansing contradictories' for pairs that are both Wansing contraries and Wansing subcontraries).[73]

Wansing does not say a lot in favour of his subcontrariety condition, which would of course coincide with the disjunctive version against the background of $\vdash_{CL}$ rather than $\vdash_{IL}$, though certainly it looks like a more direct transcription of—'2' under (i) in the quoted passage—'A and $\neg A$ cannot both be false'—once $\neg$ is connected, in the manner Priest suggests, with falsity. Whatever one wants to call it, there is a perfectly good logical relation here, holding—to purify the characterization—between A and B (according to $\vdash$) when the weakest contraries (according to $\vdash$) of A and B are themselves contraries. (Here we rely on the purified—and rather paraconsistency-unfriendly) characterization of contrariety given at the start of the Digression in Sect. 15.3.[74]) There are many families of discernible logical relations of interest each collapsing to a single familiar relation in the setting of $\vdash_{CL}$, though leading separate lives in the more discriminating habitat of $\vdash_{IL}$, such as the relation of what in Humberstone (2011a), e.g. Remark 4.22.10 on p. 555, is called pseudo-subcontrariety, holding between A and B in that order when $A \to B \vdash B$. (Formulation purifying away the $\to$ to get the intended general effect: whenever for all Γ if $\Gamma, A \vdash B$ then $\Gamma \vdash B$.) Intuitionistically (i.e. taking $\vdash$ as $\vdash_{IL}$) this relation is not symmetric, so the 'in that order' is essential. On the other hand any talk of subcontraries with or without the 'pseudo' prefix, is inclined to suggest symmetry because of the wording 'A and

---

[72] We recall also from Footnote 29 that Wansing discusses numerous proposed negation connectives—such as that associated with constructible falsity—in Wansing (2006) and their status as contrary and contradictory-forming operators, though here we consider only standard ('Heyting') intuitionistic negation. Later chapters of Odintsov (2008) also treat these Nelson style systems in some detail.

[73] From the perspective of the methodogical desideratum ('purity') we are putting on hold for this Appendix, Wansing's formulations in (a) and (b) are preferable, in that fewer object language connectives are involved.

[74] Spelling this out in more detail, we may say that A, B are Wansing subcontraries according to a consequence relation $\vdash$ when the following condition is satisfied: for any A', B' if A', $A \vdash C$ and B', $B \vdash C$ for all C, then A', $B' \vdash C$ for all C. Specializing to $\vdash = \vdash_{IL}$, this relation holds between formulas A and B iff $\vdash_{IL} \neg(\neg A \wedge \neg B)$.

B are subcontraries', so in Humberstone (2001) when A and B stand in this relation we say A *anticipates* B (according to $\vdash$); this paper looks into several aspects of this relation (see also Humberstone 2011a, p. 625f. for an update). Alternatively, one could symmetrize pseudo-subcontrariety by taking the intersection of the relation with its converse, which amounts to requiring the provability of

$$((A \to B) \to B) \wedge ((B \to A) \to A),$$

well known as the *definiens* in a definition of $A \vee B$ in the intermediate logic LC, mentioned already in the Digression in Sect. 15.3, though for this point one really must give the fundamental reference, Dummett (1959b). According to Proposition 1.1 in Humberstone (2002) two formulas A and B are implicit implicational converses in intuitionistic logic, in the sense that there are formulas $C \to D$ and $D \to C$ respectively equivalent to A and B, if and only if the conjunction inset above is intuitionistically provable. (Alternatively put: iff A and B are mutual pseudo-subcontraries according to $\vdash_{IL}$; for $\vdash_{CL}$ we can just say 'are subcontraries'—i.e. replace the above formula with $A \vee B$, which indeed we can already say at the level of the intermediate $\vdash_{LC}$, in view of the above definability point.) But let us return to Wansing's discussion of contradictoriness, to see him dismissing three other candidate characterizations (Wansing 2006, p. 83).

> Among the various classically equivalent Priest-style formulations of Aristotle's characterization of contradictoriness, there are three pairs of conditions that may be set aside:
>
> (a) $A \vdash \neg\neg A$, $\neg A \vdash \neg A$
>   (Necessarily, if A is true, then $\neg A$ is false. Necessarily, if A is false, then $\neg A$ is true.)
> (b) $\neg A \vdash \neg A$, $\neg\neg A \vdash A$
>   (Necessarily, if $\neg A$ is true, then A is false. Necessarily, if $\neg A$ is false, then A is true.)
> (c) $\vdash \neg(A \wedge \neg A)$, $\vdash \neg(\neg A \wedge \neg\neg A)$
>   (A and $\neg A$ cannot both be true. A and $\neg A$ cannot both be false.)
>
> In (a) and (b) the condition $\neg A \vdash \neg A$ holds because of a general property of derivability. In (c), one condition is an instantiation of the other. We may ask whether or not the remaining classically valid and classically equivalent pairs of conditions, exhibiting in addition to $\neg$ at most $\wedge$ and $\vee$, hold in various non-classical propositional logics.

The reference to 'the remaining classically valid and classically equivalent pairs of conditions' is perhaps to those collected in a table on the following page of Wansing (2006), though it could be interpreted quite generally as referring to arbitrary further pairs of conditions. As before, abstracting from the above formulations to the underlying binary relations involved in A and B's being contradictory according to the various proposals, we get *general* conditions on A and B which are all classically equivalent, which, for the special case on which the above passage concentrates in which B is $\neg A$, conditions are not just equivalent but classically valid (satisfied for arbitrary A, according to the consequence relation of classical logic, that is). And for a clearer focus on the issues, it pays to consider separately the contrariety and subcontrariety components in each case, and do so in the generalized setting—which is to say as conditions on arbitrary A and B, rather than just taking B as $\neg A$.

Table 15.1 Classically equivalent contrariety and subcontrariety conditions

Combination	Contrariety condition	Subcontrariety condition
(a)	$A \vdash \neg B$	$\neg A \vdash B$
(b)	$B \vdash \neg A$	$\neg B \vdash A$
(c)	$\vdash \neg(A \wedge B)$	$\vdash \neg(\neg A \wedge \neg B)$

This removes any inclination to think of the second condition for (c) as a redundant special case (substitution instance) of the first, since we now see that Table 1's contrariety and subcontrariety conditions as independent (still reading $\vdash$ as $\vdash_{IL}$), and in particular we can have the contrariety condition satisfied for a particular A, B while the subcontrariety condition is not. For example, take $A = p$ and $B = \neg p \wedge q$. Thus, in the terminology introduced above, these formulas are (intuitionistically) Wansing contraries but not Wansing subcontraries. (Conversely, keeping A as here and changing $\wedge$ to $\vee$ in B gives us a pair of Wansing subcontraries—though not Priest subcontraries, for the reason given in Appendix 1—which are not Wansing contraries.)

Table 15.1 separates out the general components in play in the passage from Wansing (2006) last quoted. Let us consider the various conditions with special reference to the choice of $\vdash$ as $\vdash_{IL}$, so as to connect with Wansing's proposal that $\neg$ is a contradictory-forming operator in intuitionistic logic. The first point that jumps out is that the three contrariety conditions are all equivalent, in the sense that for any A, B, any one of these conditions is satisfied (with $\vdash$ as $\vdash_{IL}$) if and only if any other is satisfied. We noted this already before in connection with the contrariety conditions for (a) and (c), temporarily baptized Wansing contrariety and Priest contrariety, so Wansing's own favoured proposal, combining Wansing contrariety with Wansing subcontrariety (the subcontrariety condition for (c) in Table 15.1) is in fact equivalent, in the intuitionistic case, to proposal (c), which Wansing writes 'may be set aside'. for the reason that 'in (c), one condition is an instantiation of the other'. Wansing was there, of course, looking at the special case in which B was $\neg A$, but we can see in Table 15.1 that the same holds in the general case: the subcontrariety condition is the special case or 'subschema'—though of course in this case not generally provable—of the contrariety condition, in that every instance of the former is an instance of the latter (though not conversely). But this is not an objection to the characterization of contrariety and subcontrariety in this way, since we would simply be saying that formulas are subcontraries (according to $\vdash_{IL}$) iff their negations are contraries—an entirely familiar state of affairs: think of the Square of Opposition. In the present case we can equally say, as we can there, that on proposal (c) formulas are contraries iff their negations are subcontraries, since the outer negations allow for double negation replacements—or indeed *any* classically sanctioned (propositional) replacements which would otherwise raise an intuitionist's eyebrow—in their scopes (by Glivenko).

That same consideration reminds us that while Wansing's conception of contrariety coincides with the Priest's conception, his subcontrariety condition differs from Priest's disjunctive conception—a point in its favour in view of the degenerate nature of the latter, even though it is Priest's rather than Wansing's notion that amounts in the present setting to subcontrariety according to the consequence relation $\vdash_{IL}$, as this was given above in 'connective-free' terms (last recalled in Footnote 34). Wansing's version can be thought of as replacing the reference to $A \vee B$ in Priest's with its double negation—intuitionistically equivalent to $\neg(\neg A \wedge \neg B)$. (Of course, the positions are here described in 'say less' terminology rather than that Priest and Wansing themselves officially adopt.) There is no need to exclude it on the grounds that the subcontrariety half follows from the contrariety half of the condition, since, as we have seen, this is not so in general for the two forms as defining as binary relations, neither of the binary contrariety and subcontrariety relations given in (c) of Table 15.1 is included in the other. And this is just as well, since it coincides with what Wansing wants to say in the intuitionistic case.

The subcontrariety conditions recorded for (a) and (b) are not, by contrast with the corresponding contrariety conditions, equivalent to each other (taking $\vdash$ as $\vdash_{IL}$), and are evidently close relatives of the pseudo-subcontrariety conditions mentioned above.[75] There is no reason why they should not be given consideration as subcontrariety-like conditions, preferably (again) in a form purified at least of the presence of $\neg$. This multiplicity of options arises for the same reason as we noted in Sect. 15.4 that when the true and the false no longer partition the field, 'what might otherwise be taken to be mere reformulations now emerge as non-equivalent possibilities'. In the case of subcontrariety in IL the situation is even more complicated, since we don't just have to distinguish 'at least one of A, B is true' (the $A \vee B$ version of subcontrariey) from 'not both A, B are false' (the Wansing version, under (c) in Table 15.1), but we have also these further options (using $(A \to B) \to B$, $\neg A \to B$, etc., once we are considering, as with Wansing subcontrariety, departures from the abstract characterization mentioned in Footnote 34 and earlier). These are, of course, repercussions of the generally greater discriminatory power of deductively weaker logics—the subject of Humberstone (2005c)—though Humberstone (2011a) docs suggest some guidance in looking for the most direct intuitionistic analogues of classical connectives in the shape of their topoboolean incarnations in the Kripke semantics (making the intuitionistic compound true at a point in a model if the corresponding classical compound is true at all accessible points, that is). See, for example, p. 786$f\!f\!$. of Humberstone (2011a), where this is applied in the case of exclusive disjunction ($\underline{\vee}$, say), a particularly significant connective for the present discussion since the Lemmon-style logical relation (along the lines of $R^{\vee}$ in Appendix 1) $R^{\underline{\vee}}$, induced by this connective in the classical setting is the relation of being contradictories.[76]

---

[75]$((A \to B) \to B)$ has $\neg A \to B$ as an intuitionistic consequence, though in general not conversely, and adding the converse in its general form would classicize $\vdash_{IL}$.

[76]By contrast, there seems no such uniquely salient non-classical analogue of exclusive disjunction in FDE, where the two pre-eminent candidates have their own merits: the strong exclusive disjunc-

We conclude with some general remarks on the subject of Wansing contradictories (in intuitionistic logic). Let us denote the set of Wansing contradictories of a formula A by $WCtd(A)$, recalling here the precise definition:

$$WCtd(A) = \{B : \vdash_{IL} \neg(A \wedge B) \wedge \neg(\neg A \wedge \neg B)\}.$$

As one expects, the relation here is symmetric: for any A, B, $A \in WCtd(B)$ if and only if $B \in WCtd(A)$. Another expectation—cf. the title of Cresswell (2008)—might be that contradictories are unique, i.e. that for any A, all formulas in $WCtd(A)$ are (IL-)equivalent, but it is easy to see that this expectation is not fulfilled for Wansing contradictories, since, for example, $WCtd(\neg p)$ contains both p and $\neg\neg p$. A certain interest accordingly attaches to the question of how much variation one might expect to find within $WCtd(A)$ for a given A.

For example, one can ask whether there is in general for any given formula, strongest or weakest Wansing contradictories in intuitionistic logic, and for assistance in pursuing that question, make the observation that the set of Wansing contradictories of a formula $WCtd(A)$, coincides with the set of formulas which are *classically* equivalent to $\neg A$. If A is, for example, $\neg p \vee q$ $WCtd(A)$ contains $\neg(\neg p \vee q)$, $p \wedge \neg q$, and (infinitely) many other non-IL-equivalent formulas. Relatedly, one could explore the inverse image under $\neg$ of a formula C in intuitionistic logic, where this is understood along the same lines as the corresponding notion in (classically based) modal logic, for which (Humberstone 2013a) introduces the notation, for a monomodal logic S, '$\Box^{-1}[C]^S$' to denote the set $\{B: \vdash_S \Box B \leftrightarrow C\}$, called the inverse image under $\Box$ of the formula C, according to S. For present purposes, one would be interested in the set $\neg^{-1}[C]^{IL}$, comprising those formulas whose negations are intuitionistically equivalent to C. This will be empty if C has no equivalent with $\neg$ as main connective, but in the simple case of $C = \neg p$ contains not only p and $\neg\neg p$, already mentioned, but also such things as $p \wedge (q \vee \neg q)$, not equivalent to either of them. More generally a little calculation reveals that for any formula B:

$$\neg^{-1}[\neg B]^{IL} = \{A : A \dashv\vdash_{CL} B\}.$$

We see that there is no IL-strongest formula in general in such a set, since given any formula B not containing the sentence letters $q_1, \ldots q_n, \ldots$ the sequence:

$$\begin{array}{c} B \\ (q_1 \vee \neg q_1) \wedge B \\ (q_2 \vee \neg q_2) \wedge ((q_1 \vee \neg q_1) \wedge B) \\ (q_3 \vee \neg q_3) \wedge ((q_2 \vee \neg q_2) \wedge ((q_1 \vee \neg q_1) \wedge B)) \\ \vdots \end{array}$$

---

tion of p with q, $(p \wedge \neg q) \vee (\neg p \wedge q)$, on the one hand, and the weak exclusive disjunction on the other $(p \vee q) \wedge (\neg p \vee \neg q)$. See p. 29 of Humberstone (2014). The two are equivalent in LP, as are the stronger and weaker FDE equivalence connectives contrasted there (originally in both cases, in fact, by A. J. Dale).

contains ever stronger formulas, as long as B is IL-consistent. There is a similar sequence of ever weaker formulas to establish the dual claim, obtainable from the above sequence by replacing all occurrences of $\wedge$ with $\rightarrow$ (and replacing the proviso on B with: 'as long as B is not IL-provable'). To tie this up with our discussion of Wansing contradictories, and clarify the recent use of the word 'relatedly', note that for any formula C,

$$WCtd(C) = \neg^{-1}[\neg\neg C]^{IL}$$

and thus

$$WCtd(C) = \{A : A \dashv\vdash_{CL} \neg C\}.$$

Such a characterization tells immediately, for example, that the set of Wansing contradictories of any given formula is closed under $\wedge$ and $\vee$. But with this we conclude our reflections on (some of the strands in) Wansing's discussion of Priest on negation.

References

Asmus, C. (2012). Paraconsistency on the rocks of dialetheism. *Logique et Analyse, 55*, 3–21.
Berto, F. (2006). Characterizing negation to face dialetheism. *Logique et Analyse, 49*, 241–261.
Béziau, J.-Y. (1996). Review of B. H. Slater (1995), *Mathematical Reviews*, MR1338752 (96e:03035) 1996. (Revised version on MathSciNet, 2007.)
Béziau, J.-Y. (2002). Are paraconsistent negations negations? In W. A. Carnielli, M. E. Coniglio, & I. M. L. D'Ottaviano (Eds.), *Paraconsistency: The logical way to the inconsistent* (pp. 465–486). New York: Marcel Dekker.
Béziau, J.-Y. (2006). Paraconsistent logic! (A Reply to Slater). *Sorites, 17*, 17–25.
Béziau, J.-Y. (2016). Disentangling contradiction from contrariety via incompatibility. *Logica Universalis, 10*, 157–170.
Béziau, J.-Y., He, H., Costa-Leite, A., Zhong, Y., & Ma, Y. (Eds.). (2007). *Handbook of the second world congress and school on universal logic UNILOG'07 (August 16–22 2007)*. China: Xi'an.
Blamey, S. (2002). Partial logic. In D. Gabbay & F. Guenthner (Eds.), *Handbook of philosophical logic, Vol. III: Alternatives to classical logic* (pp. 1–70). Dordrecht: Reidel. Reprinted in Gabbay & Guenthner (Eds.), *Handbook of philosophical logic* (2nd ed., Vol. 5, pp. 261–354).
Blamey, S., & Humberstone, L. (1991). A perspective on modal sequent logic. *Publications of the Research Institute for Mathematical Sciences, Kyoto University, 27*, 763–782.
Brady, R. (2004). On the formalization of the law of non-contradiction (pp. 41–48) in Priest (2004).
Brady, R., & Rush, P. (2009). Four basic logical issues. *Review of Symbolic Logic, 2*, 488–508.
Brown, B. (1999). Yes, Virginia, there really are paraconsistent logics. *Journal of Philosophical Logic, 28*, 489–500.
Bull, R. A. (1962). The implicational fragment of Dummett's LC. *Journal of Symbolic Logic, 27*, 189–194.
Buss, S. R. (1998). An introduction to proof theory. In Buss (Ed.). *Handbook of proof theory* (pp. 1–70). Amsterdam: Elsevier.
Carnap, R. (1943). *Formalization of logic*, first publ. 1943, reprinted as Volume II within *Introduction to semantics and formalization of logic*, 1961. Cambridge, MA: Harvard University Press.
Chellas, B. (1980). *Modal logic: An introduction*. Cambridge: Cambridge University Press. Reprinted (with corrections) 1988 and subsequently.
Cresswell, M. J. (2008). Does every proposition have a unique contradictory? *Analysis, 68*, 112–114.

Curry, H. B. (1957). *A theory of formal deducibility.* Notre Dame, Indiana: Notre Dame University Press.
Denyer, N. (1989). Dialetheism and trivialization. *Mind, 98,* 259–263.
Dummett, M. (1959a). Truth. *Proceedings of the Aristotelian Society, 59,* 141–162.
Dummett, M. (1959b). A propositional calculus with denumerable matrix. *Journal of Symbolic Logic, 24,* 97–106.
Dummett, M. (1991). *The logical basis of metaphysics.* Cambridge, MA: Harvard University Press.
Dunn, J. M. (1976). Intuitive semantics for first-degree entailments and "Coupled Trees". *Philosophical Studies, 29,* 149–168.
Fang, J. (2008). Commuting double Ockham algebras. *Science in China Series A: Mathematics, 51,* 185–194.
Font, J. M. (1997). Belnap's four-valued logic and De Morgan lattices. *Logic Journal of the IGPL, 5,* 413–440.
Girard, P., & Weber, Z. (2015). Bad worlds. *Thought, 4,* 93–101.
Harris, J. H. (1982). What's so logical about the "Logical" axioms? *Studia Logica, 41,* 159–171.
Horn, L. (1989). *A natural history of negation.* Chicago: University of Chicago Press.
Horn, L., & Wansing, H. (2016). Negation. In E. N. Zalta (Ed.), *Stanford encyclopedia of philosophy* (Spring 2016 ed.). http://plato.stanford.edu/archives/spr2016/entries/negation/.
Humberstone, L. (1979). Interval semantics for tense logic: Some remarks. *Journal of Philosophical Logic, 8,* 171–196.
Humberstone, L. (1987). Operational semantics for positive R. *Notre Dame Journal of Formal Logic, 29,* 61–80.
Humberstone, L. (1992). Review of Priest, Routley and Norman (Priest 1989) (and of J. M. Dunn and A. Gupta (eds.), Truth or Consequences (Kluwer: 1990)). *Australasian Journal of Philosophy, 70,* 362–366.
Humberstone, L. (1996a). Homophony, validity, modality. In B. J. Copeland (Ed.), *Logic and reality: Essays on the legacy of A. N. Prior* (pp. 215–236). Oxford: Clarendon Press.
Humberstone, L. (1996b). Classes of valuations closed under operations Galois-Dual to boolean sentence connectives. *Publications of the Research Institute for Mathematical Sciences, Kyoto University, 32,* 9–84.
Humberstone, L. (1998). Many-valued logics, philosophical issues. In E. Craig (Ed.), *Routledge encyclopedia of philosophy* (Vol. 6, pp. 84–91). London: Routledge.
Humberstone, L. (2001). The pleasures of anticipation: Enriching intuitionistic logic. *Journal of Philosophical Logic, 30,* 395–438.
Humberstone, L. (2002). Implicational converses. *Logique et Analyse, 45,* 61–79.
Humberstone, L. (2003). Note on contraries and subcontraries. *Noûs, 37,* 690–705.
Humberstone, L. (2005a). Contrariety and subcontrariety: The anatomy of negation (with Special Reference to an Example of J.-Y. Béziau). *Theoria, 71,* 241–262.
Humberstone, L. (2005b). Modality. In F. C. Jackson & M. Smith (Eds.), *The Oxford handbook of contemporary philosophy,* Chapter 20 (pp. 534–614). Oxford and New York: Oxford University Press.
Humberstone, L. (2005c). Logical discrimination. In J.-Y. Béziau (Ed.), *Logica universalis: Towards a general theory of logic* (pp. 207–228). Birkhäuser: Basel.
Humberstone, L. (2011a). *The connectives.* Cambridge MA: MIT Press.
Humberstone, L. (2011b). Variation on a trivialist argument of Paul Kabay. *Journal of Logic, Language and Information, 20,* 115–132.
Humberstone, L. (2013a). Inverse images of box formulas in modal logic. *Studia Logica, 101,* 1031–1060.
Humberstone, L. (2013b). Logical relations. *Philosophical Perspectives, 27,* 176–230.
Humberstone, L. (2014). Power matrices and Dunn-Belnap semantics: Reflections on a remark of Graham Priest. *Australasian Journal of Logic, 11,* 14–45.

Humberstone, L. (2015). Sentence connectives in formal logic. In E. N. Zalta (Ed.), *Stanford encyclopedia of philosophy* (Fall 2015 ed.). http://plato.stanford.edu/archives/fall2015/entries/connectives-logic/.
Humberstone, L. (2016). *Philosophical applications of modal logic*. London: College Publications.
Humberstone, L. (2019). Supervenience, dependence, disjunction. *Logic and Logical Philosophy, 28*, 3–135.
Humberstone, L., & Lock, A. (1986). Semicomplemented lattices and the finite model property. *Zeitschr. für math. Logik und Grundlagen der Math., 32*, 431–437.
Kripke, S. A. (1965). Semantical analysis of modal logic II. Non-normal modal propositional calculi. In J. W. Addison, L. Henkin, & A. Tarski (Eds.), *The theory of models* (pp. 206–220). Amsterdam: North-Holland.
Lewis, D. (1986). *On the plurality of worlds*. Oxford: Blackwell.
Lewis, D. (2004) Letters to Beall and Priest (pp. 176–177) in Priest (2004).
Littmann, G., & Simmons, K. (2004). A critique of dialetheism (pp. 314–335) in Priest (2004).
Lucio, P. (2000). Structured sequent calculi for combining intuitionistic and classical first-order logic. In H. Kirchner & C. Ringeissen (Eds.), *Frontiers of Combining Systems: Third International Workshop, FroCoS 2000, LNAI #1794* (pp. 88–104). Berlin: Springer.
Marcos, J. (2007). Ineffable inconsistencies. In J.-Y. Béziau, W. Carnielli, & D. Gabbay (Eds.), *Handbook of paraconsistency* (pp. 341–352). London: College Publications.
Martin, B. (2015). Dialetheism and the impossibility of the world. *Australasian Journal of Philosophy, 93*, 61–75.
Meyer, R. K. (1985). *Proving semantic completeness "Relevantly" for R*. Logic Research Paper #23 of the Logic Group. Department of Philosophy, RSSS, Australian National University.
Meyer, R. K., & Martin, E. P. (1986). Logic on the Australian plan. *Journal of Philosophical Logic, 15*, 305–332.
Odintsov, S. P. (2008). *Constructive negations and paraconsistency*. Dordrecht: Springer.
Oller, C. A. (2014). Is classical negation a contradictory-forming operator? *Notae Philosophicae Scientiae Formalis, 3*, 1–7.
Oller, C. A. (2016). Contradictoriness, paraconsistent negation and non-intended models of classical logic. In H. Andreas & P. Verdée (Eds.), *Logical studies of paraconsistent reasoning in science and mathematics* (pp. 103–109). Cham (Switzerland): Springer.
Paoli, F. (2003). Quine and Slater on paraconsistency and deviance. *Journal of Philosophical Logic, 32*, 531–548.
Popper, K., & Miller, D. (1983). A proof of the impossibility of inductive probability. (Letters to) *Nature, 302*, 687–688.
Priest, G. (1979). The logic of paradox. *Journal of Philosophical Logic, 8*, 219–241.
Priest, G. (1989). Denyer's $ not backed by sterling arguments. *Mind, 98*, 265–268.
Priest, G. (1990). Boolean negation and all that. *Journal of Philosophical Logic, 19*, 201–215.
Priest, G. (1992). What is a non-normal world? *Logique et Analyse, 35*, 291–302.
Priest, G. (1995). Multiple denotation, ambiguity, and the strange case of the missing amoeba. *Logique et Analyse, 38*, 361–373.
Priest, G. (1996). Everett's trilogy. *Mind, 105*, 631–47.
Priest, G. (1998). What is so bad about contradictions? *Journal of Philosophy, 95*, 410–426. (Appearing with minor changes as 'What's So Bad About Contradictions?' as pp. 24–38 in Priest et al. (2004).
Priest, G. (1999). What not? A defence of dialetheic theory of negation. In D. Gabbay & H. Wansing (Eds.), *What is negation?* (pp. 101–120). Dordrecht: Kluwer.
Priest, G. (2002). Paraconsistent logic. In D. Gabbay & F. Guenthner (Eds.), *Handbook of philosophical logic* (2nd ed., Vol. 6, pp. 287–393). Dordrecht: Kluwer.
Priest, G. (2006a). *In contradiction: A study of the transconsistent* (2nd ed.). Oxford: Clarendon Press. (First edn. Martinus Nijhoff, The Hague 1987).
Priest, G. (2006b). *Doubt truth to be a liar*. Oxford: Oxford University Press.

Priest, G. (2007). Reply to Slater. In J.-Y. Béziau, W. Carnielli, & D. Gabbay (Eds.), *Handbook of paraconsistency* (pp. 467–474). London: College Publications.
Priest, G. (2009). Dualising intuitionistic negation. *Principia, 13*, 165–184.
Priest, G. (2010). The logic of the catuskoti. *Comparative Philosophy, 1*, 24–54.
Priest, G. (2014). Plurivalent logics. *Australasian Journal of Logic, 11*, 2–13.
Priest, G. (2015). None of the above: The catuṣkoṭi in Indian Buddhist logic. In J.-Y. Béziau, M. Chakraborty, & S. Dutta (Eds.), *New directions in paraconsistent logic* (pp. 517–527). Springer.
Priest, G., Beall, J. C., & Armour-Garb, B. (Eds.). (2004). *The law of non-contradiction: New philosophical essays*. Oxford: Oxford University Press.
Priest, G., & Routley, R. (1989). Systems of paraconsistent logic (pp. 151–186) in Priest (1989).
Priest, G., Routley, R., & Norman, J. (Eds.). (1989). *Paraconsistent logic: Essays on the inconsistent*. Munich: Philosophia.
Pynko, A. J. (1995). Characterizing Belnap's logic via De Morgan's laws. *Mathematical Logic Quarterly, 41*, 442–454.
Rauszer, C. (1980). An algebraic and Kripke-style approach to a certain extension of intuitionistic logic. *Dissertationes mathematicae, 167*
Restall, G. (1997). Paraconsistent logics!. *Bulletin of the Section of Logic, 26*, 156–173.
Restall, G. (2004). Laws of non-contradiction, laws of the excluded middle, and logics (pp. 73–84) in Priest (2004).
Restall, G. Generality and existence I: Quantification and free logic. in preparation.
Rivieccio, U. (2012). An infinity of super-Belnap logics. *Journal of Applied Non-Classical Logics, 22*, 319–335.
Ripley, D. (2011). Negation, denial, and rejection. *Philosophy Compass, 6*, 622–629.
Routley, R., & Routley, V. (1972). Semantics of first degree entailment. *Noûs, 6*, 335–359.
Segerberg, K. (1968). Propositional logics related to Heyting's and Johansson's. *Theoria, 34*, 26–61.
Segerberg, K. (1971). *An essay in classical modal logic*. Uppsala: Filosofiska Studier.
Shoesmith, D. J., & Smiley, T. J. (1978). *Multiple-conclusion logic*. Cambridge: Cambridge University Press.
Slater, B. H. (1995). Paraconsistent logics? *Journal of Philosophical Logic, 24*, 451–454.
Slater, B. H. (2007a). Dialetheias are mental confusions. In J.-Y. Béziau, W. Carnielli, & D. Gabbay (Eds.), *Handbook of paraconsistency* (pp. 459–466). London: College Publications.
Slater, B. H. (2007b). Response to Priest. In J.-Y. Béziau, W. Carnielli, & D. Gabbay (Eds.), *Handbook of paraconsistency* (pp. 475–476). London: College Publications.
Sylvan, R., & Urbas, I. (1993). Paraconsistent classical logic. *Logique et Analyse, 36*, 3–24.
Wansing, H. (1993). Functional completeness for subsystems of intuitionistic propositional logic. *Journal of Philosophical Logic, 22*, 303–321.
Wansing, H. (1998). *Displaying modal logic*. Dordrecht: Kluwer.
Wansing, H. (2001). Negation. In L. Goble (Ed.), *The Blackwell guide to philosophical logic* (pp. 415–436). Basil Blackwell.
Wansing, H. (2006). Contradiction and contrariety: Priest on negation. In J. Malinowski & A. Pietruszczak (Eds.), *Essays in logic and ontology* (pp. 81–93). Amsterdam: Rodopi.
Williamson, T. (2017). Dummett on the relation between logics and metalogics. In M. Frauchiger (Ed.), *Truth, meaning, justification, and reality: Themes from Dummett* (pp. 153–176). Berlin: De Gruyter.
Wójcicki, R. (1988). *Theory of logical calculi*. Dordrecht: Kluwer.
Yablo, S. (2014). *Aboutness*. Princeton: Princeton University Press.

Lloyd Humberstone is Reader in Philosophy at Monash University, where he has been since 1975. He has had visiting positions at the University of Auckland, Princeton University and St. Andrews. His most recent book, *The Connectives*, is a deep study of the sentential connectives.

Chapter 16
From Iff to Is: Some New Thoughts on Identity in Relevant Logics

Edwin Mares

Abstract In this paper, I set out a semantics for identity in relevant logic that is based on an analogy between the biconditional and identity. This analogy supports the semantics that Priest has set out for identity in basic relevant logic and it motivates a version of the Routley–Meyer semantics in which identities can be viewed as constraints on the ternary relation that is used to treat implication.

Keywords Relevant logic · Identity · Non-classical logic · Formal semantics

---

This chapter is dedicated to Graham Priest. I first met Graham in the late 1980s, when I was a postdoc at the Australian National University. I owe Graham a substantial personal debt. He has always been supportive of my work and my career, and a friend. But I would like to say here that every philosophical logician owes Graham a sizeable debt. When I got into philosophical logic in the 1980s, the situation was not good. Logicians were leaving philosophy and going to computer science. One reason was that there were more jobs for computer scientists, but that wasn't the only reason. There were more exciting things to do in computer science. Graham, more than anyone else, changed that. With the publication of *In Contradiction* in 1987, a clear, sustained, and coherent case for paraconsistent logic came into being. That book also gave us a new paradigm of how to weave logic, epistemology and metaphysics together to the benefit of all three disciplines. David Lewis had published *Plurality of Worlds* a year earlier, but his logical views were much more conservative. *Plurality* argued that classical first-order logic is all we really need to talk about modality and other supposedly intentional phenomena. For those of us working in non-classical logics, Graham's work was an exemplar of how to motivate and apply our logical systems. The clarity of his work and the strength of his arguments (not just in that book) have made dialetheism a going concern in philosophy, and they have also brought the problem of alternative logics into the centre of contemporary philosophical debate. Now philosophical logic is a very exciting field and it is no longer isolated from the rest of philosophy. It is now really fun to be a philosophical logician. And that is due in no small part to Graham.

E. Mares (✉)
Victoria University of Wellington, Wellington, New Zealand
e-mail: Edwin.Mares@vuw.ac.nz

© Springer Nature Switzerland AG 2019
C. Başkent and T. M. Ferguson (eds.), *Graham Priest on Dialetheism and Paraconsistency*, Outstanding Contributions to Logic 18,
https://doi.org/10.1007/978-3-030-25365-3_16

16.1 Introduction

Identity has the following logical properties:
- identity is an equivalence relation: it is *transitive*, *reflexive*, and *symmetrical*;
- identity warrants substitutions: it provides us with inference licences of a particular sort: '$t = u$' justifies the inference from a sentence A to A', where A' results from the replacement of one or more occurrences of t with u.

It is not immediately clear, however, how these properties should be represented in the various relevant logics and whether the same principles of identity should be added to all relevant logics or whether very different principles are appropriate for different systems.

In this paper, I follow Priest (2001) in distinguishing between the basic relevant logic, N_*, and mainstream relevant logics, such as B, DJ, and R. I suggest that one set of principles is appropriate for basic relevant logic and another for all the mainstream systems. I motivate the different principles by their fit with the different semantic frameworks that these logics have. Basic relevant logic has a model theory due to Priest (1992), in which implication works very differently at normal worlds from the way it behaves at non-normal worlds. I suggest, following Priest, that identity statements should also behave in different ways in normal and non-normal worlds, in parallel to implication. Mainstream relevant logics have a semantics due to Richard Routley and Robert Meyer according to which implication has a uniform satisfaction condition across situations. I suggest that identity be treated in a fairly uniform way as well in mainstream relevant logic. I motivate this parallel between implication and identity for both basic and mainstream relevant logics by looking at the meaning of identity understood in terms of its inferential role.[1]

The plan of the paper is as follows. I begin by discussing and rejecting two definitions of identity from the literature. The first is the traditional definition which dates back to at least Russell according to which two things are identical if and only if they share all their properties. The second is due to Philip Kremer and uses J. M. Dunn's theory of relevant predication. After these are set aside, I look at a suggestion that the principles a logic contains concerning identity should parallel its principles concerning the biconditional. The parallel between which biconditionals can be proven in a system and which principles of identity are acceptable seems too restrictive, but it has some degree of plausibility. I suggest that a more reasonable approach is to look at the semantical treatment of implication and the biconditional for each relevant logic and give identity a parallel treatment. I call this the implication-identity analogy and use it to determine principles for the identity for basic relevant logic and other principles that are uniform across the various mainstream relevant logics. I give the formal semantics and axioms for the various mainstream systems (basic relevant logic is treated entirely in terms of its formal semantics). I then end with a brief discussion of the sort of metaphysics of identity that is compatible with the semantics.

---

[1] I discuss the differences between worlds and situations in Sect. 16.5.

The language that I use I call 'predicate logic'. This name usually denotes the language of first-order logic, but here I use it for a language with the usual propositional connectives, predicate and relation constants, and primitive singular terms. It does not have quantifiers. It seems to me that this should be properly called predicate logic, rather than a system the most interesting thing about which are the quantifiers.[2] I leave off the quantifiers because they introduce difficulties that have nothing to do with identity. Throughout, I use the notation $A(u/t)$ to mean any result of replacing zero or more occurrences of t with u in A.

16.2 The Classical Definition

One straightforward way of connecting identity to implication is to define identity using implication. Suppose that we have a second-order relevant logic and that we adopt the classical Russellian definition of identity:

$$t = u =_{df} \forall P(Pt \leftrightarrow Pu)$$

This definition has many virtues. First, strong forms of substitution and transitivity follow directly from it. Second, it can be used to show that identity is a logical constant in a strong sense. For suppose that there are two relation symbols, $=$ and $\approx$, that satisfy this definition. Then we can very easily prove that $t = u \leftrightarrow t \approx u$, and that $=$ and $\approx$ can be replaced for one another in any formula, yielding an equivalent formula.

In the presence of the standard principles of quantification, however, the classical definition leads to the following apparent paradox of material implication:

$$(*)\ t = u \rightarrow (A \rightarrow A),$$

for any formula A, regardless of whether A contains t or u. This thesis is explicitly rejected in Dunn (1987) and Mares (1992).

$(*)$ appears to be an intolerably irrelevant formula because $t = u$ and $A \rightarrow A$ need not share any non-logical content. That is, they may not share any individual terms nor any predicate constants. But, as Kremer (1999) has pointed out, $(*)$ is really is a paradox of material implication. For $t = u$, on this definition, is really a quantificational formula. I think Kremer is correct. If we accept

$$\forall P(Pt \rightarrow Pu) \rightarrow (A \rightarrow A)$$

then we cannot object to $(*)$ on the ground that it violates the principles that motivate relevant logic.

---

[2]I have heard predicate logic called 'one-half order logic', but I don't think that name really connotes anything.

We can use the classical definition. But we have to realise that, if we do, from a semantic point of view we relegate identities to a relatively minor role. In strong mainstream relevant logics, like R and E, we can derive the following:

$$\frac{\vdash A}{\vdash (A \to A) \to A}$$

Let's call this rule RN, since in the context of E, it is a form of necessitation. Thus, from (∗) and the transitivity of implication, we obtain:

$$\frac{\vdash A}{\vdash t = u \to A}$$

In all situations in which *any* identity holds, *all* the theorems of the logic also obtain. It is, however, a virtue of the semantics for relevant logics that some situations do not contain all the theorems. If we want to model agents' beliefs, theories that are not about logic, and other such phenomena, these 'non-logical' situations are very useful. They would be more useful if they could also include some identities, for many agents have beliefs about what is identical to what without believing in all the theorems of logic.

A similar sort of problem arises from adopting the classical definition even in weaker mainstream relevant logics that do not admit RN. As I show in Sect. 16.6, in any situation a in which $A \to A$ holds, if $A \to B$ is valid on the model, then $a \models A \to B$. So, the classical definition tells us that identities are only true at situations in which all the implicational theorems of the logic obtain. This seems overly restrictive. Hence I suggest abandoning the classical definition, at least for mainstream relevant logics.

16.2.1 Relevant Indiscernibility

Kremer (1999) uses Michael Dunn's theory of relevant predication to construct a theory of identity. In Dunn (1987, 1990), Anderson et al. (1992), Dunn develop's a theory that differentiates between an ordinary case of predication and a relevant predication. On this theory, an open formula $A(x)$ is relevantly predicated of an individual u if and only if u satisfies $A(x)$ by virtue of its being u. Relevant predication is represented by a variable binder ρ such that $\rho x(A(x))u$ means that A is relevantly predicated of u. Formally, Dunn's contextual definition of ρ is as follows:

$$\rho x(A(x))u =_{df} \forall x(x = u \to A(x))$$

16 From Iff to Is: Some New Thoughts on Identity in Relevant Logics

In addition, Dunn says that a predicate F expresses a property that always determines a case of relevant predication if and only if, for all values of x,

$$Fx \to \rho y(Fy)x$$

The notion of a relevant predicate, and especially the notion of a relevant property is used by Dunn, among other things, to provide a formal distinction between real cases of predication and so-called Cambridge predications. A case of Cambridge predication is

Donald is such that Socrates is wise.

Donald himself may have nothing to do with Socrates or wisdom. The truth-maker for this sentence is just the fact that Socrates is wise, which has nothing to do with wisdom. Donald satisfies the predicate 'being such that Socrates is wise' but it does not follow from his being Donald that Socrates is wise in the sense of relevant implication.

Kremer reverses Dunn's theory. On Dunn's theory, the notion of relevant predication is defined using identity. Kremer instead starts with a set of formulas that determine real properties of objects: $A(x), B(x), C(x), \ldots$. Kremer then says that individuals u and v are identical if and only if they are indiscernible relative to this set, that is, the perhaps infinite conjunction $(A(u) \leftrightarrow A(v)) \wedge (B(u) \leftrightarrow B(v)) \wedge (C(u) \leftrightarrow C(v)) \wedge \ldots$ holds.

Kremer's theory has intuitive appeal, but it does not work well when a weaker relevant logic is the background logic. Dunn introduced the notion of relevant predication with R as his background logic. In R we can infer from a thing's having a particular predicate relevantly to its having it in the standard way. By definition,

$$\rho x(Px)u \to \forall x(x = u \to Px) \tag{16.1}$$

Thus, instantiating,

$$\rho x(Px)u \to (u = u \to Pu). \tag{16.2}$$

But, in R, it is a theorem that, where B is provable, the rule MidT is derivable:

$$\frac{\vdash A \to (B \to C)}{\vdash A \to C}$$

So, from (16.2), the fact that $u = u$ is an axiom (and hence provable), and the transitivity of implication we obtain

$$\rho x(Px)u \to Pu. \tag{16.3}$$

In weaker relevant logics, such as B and DJ, however, MidT is not derivable.[3] And there would seem to be no deduction from a thing's having a property relevantly and just having that property.

Of course we could alter the definition of $\rho x(Px)u$ to be $Pu \wedge \forall x(x = u \to Px)$. Then we could derive Pu from $\rho x(Px)u$, but the expression $Pu \wedge \forall x(x = u \to Px)$ does not mean 'u has the property of being P by virtue of its being u'. Rather it says that Px is implied by $x = u$ and, coincidentally, Pu. In weaker relevant systems, the theoretical cohesion of relevant predication breaks down.

Therefore, it seems that the definition of a relevant property is not appropriate in the framework of weaker relevant logics. As such defining identity in terms of it may work well for stronger relevant logics, but is rather unmotivated for weaker systems. I think I have a theory of identity that works for both weak and strong systems. I put the topic of relevant predication and its relationship to identity aside and pursue a rather different course.

16.3 Iff Axioms and Identity

Another natural suggestion, made to me a long time ago by Bob Meyer, is that our principles for an identity for a logic should mirror the theorems of that logic that govern equivalence. For example, if the logic has as a theorem,

$$(p \leftrightarrow q) \to ((q \leftrightarrow r) \to (p \leftrightarrow r))$$

then it should contain the principle,

$$t = u \to (u = v \to t = v).$$

The rationale behind this parallel is that the biconditional is the logic's way of identifying propositions. It makes sense for it to use the same principles to identify things of other categories like individuals.

This idea requires for support the further idea that $A \leftrightarrow B$ means that A and B are the same proposition. The motivation behind certain relevant logics, such as the logic E of relevant entailment and Ross Brady's logic MC of meaning containment, support this idea. On Brady's view, for example, $A \to B$ means that the content of A contains the content of B. Thus, $A \leftrightarrow B$ means that the content of A and B contain one another. It is a small step to say that $A \leftrightarrow B$ means that A and B have the same content.

---

[3]Both of these logics have the depth relevance property. The depth of a subformula in a formula is the number of times it is within the scope of nested implications. If MidT were derivable, then we would be able to derive from the axiom $((p \to p) \to (p \to p)) \to ((p \to p) \to (p \to p))$ and the axiom $p \to p$ to $((p \to p) \to (p \to p)) \to (p \to p)$, which would violate depth relevance.

16 From Iff to Is: Some New Thoughts on Identity in Relevant Logics

With regard to the logic E, the case is similar. Wilhelm Ackermann, Alan Anderson, and Nuel Belnap, who created E did so in order to have a logic in which $A \to B$ means that B is derivable from A. So, $A \leftrightarrow B$, for them, means that A and B are derivable from one another. By the soundness and completeness theorems for E over the Routley–Meyer semantics, A and B are derivable from one another in E if and only if in every model for E, A and B are true at all the same situations. If we take propositions to be sets of situations, then, $A \leftrightarrow B$ expresses the fact that A and B are the same proposition.

For logics like R, on the other hand, that are supposed to capture a contingent notion of implication, there is no reason to think of the biconditional as representing the identity of propositions. On my interpretation of R, $A \to B$ means that, in a context, we have licence to infer from A to B. Surely the information that A and B are the same proposition would give us such licence, but it is not a necessary condition of our having an inference license that we believe that they are the same. Just because we have licence to infer from A to B and back again does not mean that A and B are the same.

So let us set aside logics in which the implication is supposed to be contingent, or is supposed not to represent some form of entailment, and proceed for the moment to discuss only logics that are supposed to codify an entailment.

Let's start with the transitivity of identity. We can distinguish a conjunctive form of transitivity, i.e.

$$\text{(CT)} \ (t = u \wedge u = v) \to t = v,$$

from a nested form, that is,

$$\text{(NT)} \ t = u \to (u = v \to t = v).$$

According to the parallel between the biconditional and identity, logics that contain the nested form of the transitivity of the biconditional, such as the logics E and R, should also contain NT. Logics that contain the conjunctive form of the transitivity of the biconditional, such as Ross Brady's logic MC, should contain CT.

In addition, there are rule forms of CT and NT. The rule form of CT, RCT, is

$$\frac{\vdash t = u \quad \vdash u = v}{\vdash t = v}$$

and the rule form of NT, RNT, is

$$\frac{\vdash t = u}{\vdash u = v \to t = v}$$

In the logic B, which is usually taken to be the weakest relevant logic, the derivability of RNT entails the derivability of RCT. In B the following rule can be derived:

$$\frac{\vdash A \leftrightarrow B}{\vdash (B \leftrightarrow C) \to (A \leftrightarrow C)}$$

So, by the parallel between the biconditional and identity, B$^=$ should contain RNT.

The issue concerning the substitution rule is more complicated. The version of substitution that I used in Mares (1992) was

$$(A \wedge t = u) \to A(u/t)$$

where A is implication free. Consider the parallel thesis concerning the biconditional:

$$(A \wedge (p \leftrightarrow q)) \to A(q/p)$$

where A is implication free. Let A just be p. Then we have

$$(p \wedge (p \leftrightarrow q)) \to q.$$

This is the thesis version of the detachment rule for the biconditional. It is difficult to motivate the acceptance of this thesis without also accepting *pseudo modus ponens* (pmp):

$$(p \wedge (p \to q)) \to q.$$

But pmp, as we know very well, added to B creates a logic that cannot be used to formalise naïve set theory or the naïve theory of truth. It allows the derivation of Curry's paradox from the naïve comprehension principle and from the unrestricted T-schema.

Moreover, this version of substitution cannot be justified on the parallel between identity and the biconditional. Weaker relevant logics (DJ and weaker systems) do not contain as theorems all instances of the schema

$$(A \wedge (p \leftrightarrow q)) \to A(q/p).$$

They do not, as I strongly hinted above, contain the theorem in which A is p. DJ contains the instances in which A is $r \leftrightarrow p$ or $r \to p$. But it doesn't have as a theorem $((\neg r \leftrightarrow \neg p) \wedge (p \leftrightarrow q)) \to (\neg r \leftrightarrow \neg q)$, since in DJ, $r \leftrightarrow q$ and $\neg r \leftrightarrow \neg q$ are not provably equivalent.

But all the weaker relevant logics, on the parallel between identity and the biconditional, do justify the following rule form of substitution (RSub):

$$\frac{\vdash t = u}{\vdash A \leftrightarrow A(u/t)}$$

where there are no restrictions on A.[4]

---

[4] RSub has some similarities to the one that Priest supports in (2008) but that is a semantic rule and is much stronger than RSub. I discuss this semantic rule in Sect. 16.4.

16 From Iff to Is: Some New Thoughts on Identity in Relevant Logics

One problem with RSub is that adding it to the axiom systems for any of the logics that I am discussing, as formulated in our current formal language, makes no difference to the set of provable formulas. The only provable identities in any of these logics are the simple self-identities of the form $t = t$, so the conclusion of sound instances of the rule, is always of the form $A \to A$ and this is already a theorem of the propositional logic.

If the language is extended by function constants to produce complex terms, then RSub might become more substantive. For example, if we have addition and number constants in our language as well as axioms sufficient to produce the usual quantifier-free theorems of arithmetic, we could make provable statements such as

$$2 + 3 = 7 \to 5 = 7.$$

But the precise treatment of functions in relevant logic is still a controversial matter, and I have no space (and no ideas) to pursue it here. Instead, in the next section, I leave behind the parallel between theorems governing biconditionals and identity and look at a semantic connection between (bi-)implication and identity.

There does seem to be an analogy between identity and the biconditional. Both are symmetrical and both *tell us about* licences to infer. All logically complex formulas give us licences to infer of a sort. A conjunction $A \wedge B$ tells us that we can infer A and B. But there is something different about an implication, $A \to B$. It does not just give us the licence to infer B from A, it tells us that there is this licence. Similarly, an identity statement $u = v$ does not just give us the licence to infer $A(u)$ from $A(t)$, it tells us that we can substitute u for t and the converse in any true statements. I call this the 'bi-implication-identity analogy', or merely the 'implication-identity analogy'. In what follows, I construct a theory based on the semantics for basic and then mainstream relevant logics that formalises this analogy.

16.4 Semantics of Identity I: Basic Relevant Logic

The first semantic framework I introduce is due to Priest. It is his semantics for N_* (Priest 1992, 2008). On this semantics, a frame is a quadruple $(W, N, *, f_\to)$ such that W is a set, N is a non-empty subset of W, $*$ is a unary operator on W, and $f_\to$ is a function from ordered pairs of subsets of W to subsets of W. N is the set of normal worlds and $*$ is the Routley star operator that is used to model negation. The one condition on these frames is that for $a \in W$, $a^{**} = a$.

The points of the frames of basic relevant logic are best understood as *worlds*. A world is a complete universe. The difference between normal worlds and non-normal worlds in this semantics is in their treatment of implication (and the biconditional). Let's think of a proposition as a set of worlds. At a normal world a, an implication holds between X and Y if and only if $X \subseteq Y$. If a is, on the other hand, a non-normal world, then X implies Y at a if and only if $a \in f_\to(X, Y)$. At normal worlds, the true implications are those that accurately represent entailment relations on the whole frame. At non-normal worlds, implications are made true randomly.

Priest gives a metaphysical reading of the meaning of implication in models for basic relevant logic. In these models, the true implications of normal worlds represent the true laws of logic of the model. Non-normal worlds, however, act as worlds of 'logic fictions'. In science fiction stories, the laws of physics may be different than in the actual world. Similarly, in a logic fiction story, the laws of logic may be different. Just as science fiction stories represent worlds that are nominally inaccessible from our own world, logic fictions represent worlds that are logically inaccessible from our own world.

On my interpretation of relevant logic, the logical connectives are understood in terms of their inferential behaviour, even in terms of their semantic characterisation. Giving Priest's reading of his models a slight inferential makeover, we obtain this: A true implication is a licence to infer from one proposition to another. On the semantics for basic relevant logic, these licences at normal worlds are accurate in the sense that they preserve truth in all worlds. At non-normal worlds, they may not be accurate. They may not preserve truth in all worlds and, in fact, they may not preserve truth in the worlds in which they are true. They might lead individuals in non-normal worlds astray.[5]

The method for understanding the role of implication in these models is first to determine its inferential role in normal worlds and then to view its role in non-normal worlds as the same, but in some non-normal worlds, it is ineffective in accomplishing this role. Let us apply the same technique to understanding identity. At normal worlds, identity statements should be accurate. That is, for all normal worlds, let us set $a \models t = u$ to be true if and only if t denotes the same individual as u. If we do so, then at normal worlds we can understand the inferential role of identity in terms of substitution. For if $t = u$ and A are both true at a normal world, then, given the natural extension of Priest's truth conditions for propositional logic to predicate (given below), then $A(u/t)$ is also true in that world. Thus, we can see identity statements as licences to infer from true sentences to substitution instances. At non-normal worlds, we understand identity in the same way, but it might not be accurate in some cases. In the formal semantics, at non-normal worlds random identities are made true.

Let us make this more formal. A frame for basic relevant predicate logic with identity is a sextuple $(W, N, *, I, f_\rightarrow, f_=)$ such that $W, N, *$ and $f_\rightarrow$ are defined as they are for propositional frames, I is a non-empty set (of individuals), and $f_=$ is a function from W to $\wp(I^2)$.

A basic relevant predicate logic model is a septuple $(W, N, *, I, f_\rightarrow, f_=, V)$ such that $(W, N, *, I, f_\rightarrow, f_=)$ is a basic relevant predicate logic frame and V is an assignment of individuals to terms and pairs of worlds and sets of n-tuples of individuals to predicate constants. As usual, I define $[\![A]\!]$ as the set of all worlds a such that $a \models A$.

---

[5]Having non-normal worlds of this sort is useful in representing various sorts of misinformation.

- $a \models P(t_1 \ldots t_n)$ if and only if $(V(t_1), \ldots, V(t_n)) \in V(P)(a)$;
- $a \models t = u$ if and only if $(V(t), V(u)) \in f_=(a)$;
- $a \models A \wedge B$ if and only if $a \models A$ and $a \models B$;
- $a \models A \vee B$ if and only if $a \models A$ or $a \models B$;
- $a \models \neg A$ if and only if $a^* \not\models A$;
- If $a \in N, a \models A \rightarrow B$ if and only if $[\![A]\!] \subseteq [\![B]\!]$ and if $a \notin N$, then $a \models A \rightarrow B$ if and only if $a \in f_\rightarrow([\![A]\!], [\![B]\!])$;
- If $a \in N, a \models t = u$ if and only if $V(t) = V(u)$ and if $a \notin N$, then $a \models t = u$ if and only if $(V(t), V(u)) \in f_=(a)$.

This is just Priest's semantics for basic relevant logic with identity in Priest (2008, §23.6).

By an easy induction on the structure of formulae, one can show that truths at normal worlds are closed under substitution:

Theorem 16.1 *If $a \in N$, $a \models t = u$, and $a \models A$, then $a \models A(u/t)$.*

Since at all normal worlds $A \rightarrow A$ is true, a semantic form of RSub is valid:

Corollary 16.2 *If $a \in N$ and $a \models t = u$, then $a \models A \rightarrow A(u/t)$.*

Together with the fact that identity is symmetrical at normal worlds, we can derive from Corollary 16.2, for any normal world a, if $a \models t = u$ then $a \models u = v \rightarrow t = v$.

Speaking of symmetry, the bi-implication-identity analogy motivates the view that identity is symmetrical at both normal and non-normal worlds. A bi-implication, $A \leftrightarrow B$ is defined as $(A \rightarrow B) \wedge (B \rightarrow A)$, and this entails at both normal and non-normal worlds $B \leftrightarrow A$. To make identities symmetrical we merely need to stipulate that $f_=(a)$ is symmetrical for all worlds a.

16.5 Semantics of Identity II: Mainstream Relevant Logics

Mainstream relevant logics have significantly different semantics from basic relevant logics. The points in models for mainstream relevant logics should be understood as world states, set-ups, or situations. I call them 'situations'. A situation is different from a world in that many situations can obtain in a single world. A world is a complete universe. A situation captures perhaps partial information about a world or set of worlds. For example, the information that is available to me in my lounge as I write this sentence constitutes a situation.

As opposed to the semantics for basic relevant logic, in the Routley–Meyer semantics for mainstream relevant logics, the satisfaction condition for implication is the same throughout the frame. At every situation a, the following holds:

$a \models A \rightarrow B$ if and only if for all situations x and y $(Raxy \wedge x \models A) \Rightarrow y \models B$

In this section, I set out a theory in which the satisfaction condition for identities mirrors in important respects this condition for implication. In order to motivate this theory, I first look at a philosophical interpretation of the ternary relation R used in the condition from 'On the Ternary Relation and Conditionality' (Beall et al. 2012).

One interpretation of the ternary relation in Beall et al. (2012) has to do with the notion of functional application. As I have said, I take the points in Routley–Meyer models to be situations and situations contain information. Some information can be applied to other information. The information that a light switch is connected in a particular wire and that these lead back to an operating electrical plant when applied to the information that the switch is pressed yields the information that a particular light turns on. When all the information in a situation is applied to all the information in another (or perhaps the same) situation the result is a collection of pieces of information. We say that $Rabc$ if and only if c contains all the information that results from the application of the information in a to the information in b. We say by way of abbreviation that a when applied to b yields situations like c. Thus we read $a \models A \to B$ as saying that if the information in a is applied to a situation that contains A it will result in situations that contain B.

My suggestion is to treat identity using the same notion of application. Suppose that one is in a situation A in which she has information such that when applied to a situation that contains the information that A, results in the information that B. Suppose also that in a there is the information that $t = u$. Then a contains the information that when applied to a situation that contains any instance of $A(u/t)$ results by telling us that some situation obtains that contains all instances of $B(u/t)$. On my view, identity is an *inference modifier*. The identities that are contained in a situation modify and extend the inferences that are licensed in that situation. To use Gilbert Ryle's metaphor of an inference ticket, identity provides the ticket holder with an upgrade. She is warranted to do more inferences: those that are substitution instances of the inferences licensed by the original ticket.

This notion of identity as an inference modifier can be represented axiomatically by the following version of the substitution scheme:

$$(\text{Sub} \to)\ (u = v \land (A \to B)) \to (A(u/t) \to B(u/t))$$

The implication, $A \to B$, is the original inference licence and $A(u/t) \to B(u/t)$ is the modified inference. Sub$\to$ has as an instance the following version of the transitivity principle:

$$(t = u \land (t = v \to t = v)) \to (u = v \to t = v)$$

This transitivity principle might look rather strange, but as a remark of Sect. 16.6 below shows, it says that in any normal situation in which $t = u$, $u = v \to t = v$. Thus, when restricted to normal situations, it is a very standard transitivity principle. The identity schema, $t = t$, and symmetry schema, $t = u \to u = t$, have to be added independently. I give the full axiom systems for the various mainstream relevant logics in Sect. 16.7.

It might seem that the substitution schema that I have suggested is too weak. The standard substitution axiom for relevant logics (Dunn 1987; Mares 1992; Priest 2008) is

$$\text{(FullSub)} \quad (t = u \wedge A) \to A(u/t).$$

My own Sub→ is an instance of FullSub. The problem with FullSub is that, in weak relevant logics, it gives identity powers in situations that are beyond those of implication and the biconditional. In logics, the semantics of which do not have a fully reflexive ternary relation (i.e. in which $Raaa$ does not always hold), the implicational information in a situation a does not always tell us about the closure properties of a itself. It happens in many situations in many models for weak relevant logics that $A \to B$ and A both obtain in a situation and B fails to obtain. It is only in this way that the principle of pseudo modus ponens, $((A \to B) \wedge A) \to B$, can be rejected. The rejection of pseudo modus ponens is essential for those like Priest who wish to use a weak relevant logic to construct a naïve theory of truth or a naïve set theory. Having pseudo modus ponens, together with some standard logical principles, allows the derivation of Curry's paradox. In order to maintain the parallel between the inferential properties of implication (and the biconditional) and those of identity, I, therefore, reject FullSub.[6]

16.6 Formal Semantics for Mainstream Relevant Logics with Identity

A propositional Routley–Meyer frame is a quadruple $(S, N, R, *)$ such that S is a non-empty set (of situations), N (the set of normal situations) is a non-empty subset of S, R is a ternary relation on S, and $*$ is a unary operator on S such that the following definition and semantic conditions hold:

$$a \leq b \ =_{df} \ \exists x (x \in N \wedge Rxab)$$

Semantic Conditions:

SC1 $\leq$ is a partial order;
SC2 If $a \leq b$ and $Rbcd$, then $Racd$;
SC3 If $a \leq b$ then $b^* \leq a^*$;
SC4 $a^{**} = a$.

---

[6]This argument is only meant to show why we need to reject FullSub if we reject pseudo modus ponens. It is not meant to show that pseudo modus ponens is not intuitive from the perspective of an informational interpretation of the semantics. The latter issue is not easily settled. But given the view that the R relation is supposed here to represent application, it is at least plausible that on some understanding of application, not every situation is closed under the application of itself to itself. A complete world is so closed, but it is not clear that every informational part of the world need be closed under self-application.

To extend frames to treat predicate calculus, I add a domain of individuals D. This, as usual, is a non-empty set. The semantics is a constant domain. Like Priest (2005, 2008), I do not assume that all objects *exist* in all situations. Rather, the domain of a situation contains both things that exist and things that do not exist.

To deal with identity, I have to add further machinery to the definition of frames. I take the elements of identity frames from my old paper, Mares (1992), although I change their properties somewhat and I use different (and, I hope, more reader friendly) notation here. As in Sect. 16.4, I use the function $f_=$ to pick out the sets of ordered pairs that are identified by situations. In addition, I use the function Ind to pick out for each situation the set of pairs of individuals that are made *indiscernible* at that situation. It is important in this semantics that $f_=$ and Ind be distinguished from one another. If the semantics were to make valid FullSub, then every a such that $(i, j) \in f_=(a)$ would also be such that $(i, j) \in Ind(a)$, but this is not the case here. Rather, the relationship between $f_=$ and Ind is somewhat more subtle.

Before I explain this relationship, however, I need to give the requisite properties to Ind. To this end, every Routley–Meyer identity frame satisfies the following postulates:

Ind1 If $(i, j) \in Ind(a)$ then $(i, j) \in Ind(a^*)$;
Ind2 If $Rabc$ and $(i, j) \in Ind(a)$, then there is a $b' \geq b$ such that $(i, j) \in Ind(b')$ and $Rab'c$;
Ind3 If $Rabc$ and $(i, j) \in Ind(a)$, then there is a $c' \leq c$ such that $(i, j) \in Ind(c')$ and $Rabc'$;
Ind4 If $(i, j) \in Ind(a)$ then $(j, i) \in Ind(a)$; if $(i, j) \in Ind(a)$ and $(j, k) \in Ind(a)$, then $(i, k) \in Ind(a)$.

The postulate Ind1 ensures that, if $a \models \neg A(t)$ and $(V(t), V(u)) \in Ind(a)$, then $a \models \neg A(u)$ as well. Ind2 and Ind3 treat implications. Ind2 forces $a \models A(t) \to B$ and $(V(t), V(u)) \in Ind(a)$ to imply that $a \models A(u) \to B$ and Ind3 does the same thing with regard to the consequents of implications.

I could drop Ind2 and Ind3. This would have the effect of making valid a restricted form of Sub→: $(t = u \wedge (A \to B)) \to (A(u/t) \to B(u/t))$, where A and B are implication free. This would bring the current theory closer to that of Mares (1992).

Here are the postulates for $f_=$:

$f_=1$ If $a \leq b$, then $f_=(a) \subseteq f_=(b)$;
$f_=2$ If $Rabc$ and $(i, j) \in f_=(a)$, then there is a $b' \geq b$ such that $(i, j) \in Ind(b')$ and $Rab'c$;
$f_=3$ If $Rabc$ and $(i, j) \in f_=(a)$, then there is a $c' \leq c$ such that $(i, j) \in Ind(c')$ and $Rabc'$;
$f_=4$ If $(i, j) \in f_=(a)$, then $(j, i) \in f_=(a)$;
$f_=5$ If $(i, j) \in f_=(a)$ and $(j, k) \in Ind(a)$, then $(i, k) \in f_=(a)$;
$f_=6$ If $a \in N$ then $(i, i) \in f_=(a)$.

$f_=2$ and $f_=3$ are crucial for the validity of Sub→. If FullSub is to be made valid instead, these two postulates can be replaced with the postulate that if $(i, j) \in f_=(a)$, then $(i, j) \in Ind(a)$.

16 From Iff to Is: Some New Thoughts on Identity in Relevant Logics

If we want to satisfy conjunctive transitivity (CT), we add:

$f_=7$ If $(i, j) \in f_=(a)$ and $(j, k) \in f_=(a)$, then $(i, k) \in f_=(a)$.

And to satisfy nested transitivity (NT) we add instead:

$f_=7'$ If $(i, j) \in f_=(a)$ and $Rabc$, if $(j, k) \in Ind(a)$, then $(i, k) \in f_=(a)$.

If we add neither of these postulates, then the form of transitivity that is valid is

$$(t = u \wedge (t = v \rightarrow t = v)) \rightarrow (u = v \rightarrow t = v).$$

All normal situations satisfy $t = v \rightarrow t = v$. So, if a is a normal situation, if $a \models t = u$, then $a \models u = v \rightarrow t = v$. Thus, a form of nested transitivity is valid on all Routley–Meyer identity frames. Standardly, in treatments of identity in relevant logic, however, CT holds throughout frames (Dunn 1987; Mares 1992; Priest 2008). I can justify the acceptance of CT using the idea of identities as inference modifiers in the following way. An identity $u = v$ tells us that we have licence to make certain inferences. If one also has the information that $t = u$, then we have an inference modifier on hand as well as an inference licence (after all, an inference modifier on this picture is a certain sort of inference licence). Modifying the licence $u = v$ using $t = u$ allows us to obtain $t = v$. This, as Theorem 16.4 below shows, is sufficient to show that CT should be valid. Thus, it seems reasonable to add $f_=7$ to the list of postulates defining a Routley–Meyer identity frame.

A Routley–Meyer identity model is a octuple $(S, N, R, *, I, f_=, Ind, V)$ such that $(S, N, R, *, I, f_=, Ind)$ is a Routley–Meyer identity frame and V is a function from the names of $\mathcal{L}$ to I and from n-place predicates of $\mathcal{L}$ and situations into the power set of I^n, such that if $a \leq b$ then $V(P, a) \subseteq V(P, b)$ and if $(i, j) \in Ind(a)$ then $< ..., i, ... > \in V(P)(a)$ if and only if $< ..., j, ... > \in V(P)(a)$. V determines a satisfaction relation, $\models$, between situations and formulae such that the following recursive clauses hold:

- $a \models P(t_1...t_n)$ if and only if $(V(t_1), ..., V(t_n)) \in V(P)(a)$;
- $a \models t = u$ if and only if $(V(t), V(u)) \in f_=(a)$;
- $a \models A \wedge B$ if and only if $a \models A$ and $a \models B$;
- $a \models A \vee B$ if and only if $a \models A$ or $a \models B$:
- $a \models \neg A$ if and only if $a^* \not\models A$;
- $a \models A \rightarrow B$ if and only if $\forall x \forall y((Raxy \wedge x \models A) \Rightarrow y \models B)$.

Where $\mathfrak{M}$ is a Routley–Meyer identity model, a formula A is valid on $\mathfrak{M}$, written '$\mathfrak{M} \models A$', if and only if $N \subseteq [\![A]\!]$, where N is the set of normal situations of $\mathfrak{M}$. A formula is valid over a class of models if it is valid on each member of that class.

The usual hereditariness lemma can be proven by a simple induction on the length of formulae:

Lemma 16.3 (Hereditariness) *if $a \leq b$ and $a \models A$, then $b \models A$.*

From Lemma 16.3 and the fact that (by SC1) in all frames, for any situation b there is some normal situation a such that $Rabb$, the following semantic entailment theorem can be derived:

Theorem 16.4 (Semantic Entailment) *In a Routley–Meyer identity model $\mathfrak{M}$, $\mathfrak{M} \models A \to B$ if and only if, for all $a \in S$, if $a \models A$ then $a \models B$.*

Lemma 16.5 shows that if for a situation a, individuals (i, j) are in $Ind(a)$, then i and j really are indiscernible at a.

Lemma 16.5 *If $(V(t), V(u)) \in Ind(a)$ and $a \models A$ then $a \models A(u/t)$.*

Proof Suppose that $(V(t), V(u)) \in Ind(a)$. By induction on the length of A:

Case 1. A is $P(v_1...v_n)$. This follows directly by the conditions on V.
Case 2. A is $t = v$ or $v = t$. Follows by $f_{=}5$ and $f_{=}4$.
Case 3. A is $B \wedge C$. Follows by the inductive hypothesis and the satisfaction condition for conjunction.
Case 4. A is $\neg B$. Follows by Ind1.
Case 5. A is $B \to C$. Suppose that $a \models B \to C$ and suppose that $Rabc$ and $b \models B(u/t)$. Then, by Ind2, there is a b' such that $Rab'c, b \leq b'$ and $(V(t), V(u)) \in Ind(b')$. By Ind4, $(V(u), V(t)) \in Ind(b')$. By the inductive hypothesis $b' \models B$. By Ind3, there is a $c' \leq c$ such that $Rab'c'$. By the satisfaction condition for implication, $c' \models C$. By the inductive hypothesis, $c \models C(u/t)$. But $c' \leq c$ and so, by Lemma 16.3, $c \models C(u/t)$. Generalising, by the satisfaction condition for implication, $a \models B(u/t) \to C(u/t)$. □

The following theorem shows that the substitution principle holds in all Routley–Meyer identity models.

Theorem 16.6 *If $a \models t = u$ and $a \models A \to B$, then $a \models A(u/t) \to B(u/t)$.*

Proof Suppose that $a \models t = u$ and $a \models A \to B$. Suppose also that $Rabc$ and $b \models A(u/t)$. Then by $f_{=}2$, there is a $b' \geq b$ such that $(V(t), V(u)) \in Ind(b')$ and $Rab'c$. By Lemma 16.3, $b' \models A(u/t)$, and by Lemma 16.5, $b' \models A$. By $f_{=}3$, there is a $c' \leq c$ such that $Rab'c'$ and $(V(t), V(u)) \in Ind(c')$. By the satisfaction condition for implication, $c' \models B$ and so, by Lemma 16.5, $c' \models B(u/t)$. By Lemma 16.3, $c \models B(u/t)$. Thus, by the satisfaction condition for implication, $a \models A(u/t) \to B(u/t)$. □

16.7 Logics

The logic that is sound and complete over the whole class of Routley–Meyer identity models is the logic $B^{=}$. Here is an axiomatisation of it:

Axioms

1. $A \to A$
2. $(A \wedge B) \to A$; $(A \wedge B) \to B$
3. $((A \to B) \wedge (A \to C)) \to (A \to (B \wedge C))$
4. $((A \to C) \wedge (B \to C)) \to ((A \vee B) \to C)$
5. $(A \wedge (B \vee C)) \to ((A \wedge B) \vee (A \wedge C))$
6. $\neg\neg A \to A$
7. $t = t$
8. $t = u \to u = t$
9. $(t = u \wedge (A \to B)) \to (A(u/t) \to B(u/t))$

Rules

Adj $\vdash A, \vdash B \Rightarrow \vdash A \wedge B$
MP $\vdash A \to B, \vdash A \Rightarrow \vdash B$
Suff $\vdash A \to B \Rightarrow \vdash (B \to C) \to (A \to C)$
Pref $\vdash B \to C \Rightarrow \vdash (A \to B) \to (A \to C)$
Cont $\vdash A \to \neg B \Rightarrow \vdash B \to \neg A$

If $f_=7$ or $f_=7'$ are added to the definition of a Routley–Meyer identity frame, then we need to add CT or NT respectively to the axiomatisation.

Soundness is quite easy to prove using Theorem 16.4 and the semantic postulates and conditions. Completeness can be proven using the techniques of Fine (1988) and Mares (1992).

Here are some axiom schemata that can be added to B to create some of the better known mainstream relevant logics. The conjunctive syllogism axiom (CS) can be added to B to produce DJ:

$$(\text{CS}) \;((A \to B) \wedge (B \to C)) \to (A \to C)$$

To make CS valid, the condition $Rabc \Rightarrow \exists x(Rabx \wedge Raxc)$ is added to the definition of a Routley–Meyer frame. To produce TW, which is the logic of ticket entailment without the principle of contraction, we add to B the transitivity schemes B and B' and the principle of contraction:

$$(\mathsf{B}) \; (B \to C) \to ((A \to B) \to (A \to C))$$

$$(\mathsf{B'}) \; (A \to B) \to ((B \to C) \to (A \to C))$$

$$(\text{Cont}) \; (A \to \neg B) \to (B \to \neg A)$$

The semantic postulates for these transitivity axioms are $\exists x(Rabx \wedge Rxcd) \Rightarrow \exists x(Racx \wedge Rbxd)$ and $\exists x(Rabx \wedge Rxcd) \Rightarrow \exists x(Raxd \wedge Rbdx)$. The semantic postulate for Cont is $Rabc \Rightarrow Rac^*b^*$. To obtain the logic T, add to TW the principle of contraction (W):

$$(W)\ (A \to (A \to B)) \to (A \to B)$$

The semantic condition for contraction is $Rabc \Rightarrow \exists x(Rabx \wedge Rxbc)$. To produce RW add to the logic TW the principle of permutation (C):

$$(C)\ (A \to (B \to C)) \to (B \to (A \to C))$$

The semantic condition for C is the so-called Pasch Postulate: $\exists x(Rabx \wedge Rxcd) \Rightarrow \exists x(Racx \wedge Rxbd)$. To obtain the logic R of relevant implication, add W to RW.

It would be interesting to see which if any postulates governing identity are changed by the addition of these axioms and semantic postulates. One change is that in R we can derive the *demodaliser* theorem:

$$(\text{DeModal})\ A \leftrightarrow ((A \to A) \to A)$$

An instance of Sub$\to$ is

$$(t = u \wedge ((A \to A) \to A)) \to ((A(u/t) \to A(u/t)) \to A(u/t)).$$

Thus, from DeModal and Sub$\to$, by substitution of equivalents, it follows that

$$(t = u \wedge A) \to A(u/t).$$

That is to say, FullSub is valid in R. That FullSub is valid in R makes sense from a semantic point of view. In R models, $Raaa$ always holds. Thus, each situation is closed under its implicational information. By the implication-identity analogy, every situation should also be closed under its identity information.[7]

By the implication-identity analogy, it would seem appropriate for logics that include CS also include the axiom NT. It does not seem, however, that the addition of the semantic clause $Rabc \Rightarrow \exists x(Rabx \wedge Raxc)$ automatically makes valid NT. As far as I can tell, we need to add $f_=7$ separately. Similarly, logics that contain B and B′ should, on the implication-identity analogy, include NT, but it does not seem that adding the semantic clauses for the transitivity axioms allows us to prove the soundness of NT.

16.8 The Metaphysics of Identity

In the metaphysics for modal logic, one key debate is between those who support a theory of necessary identity and those who advance a theory of contingent identity. In (2008, Chaps. 16 and 17), Priest presents these semantics, including a theory of

---

[7]In Mares (1992) I gave an argument due to Belnap that we should reject FullSub for R. I no longer find this argument very persuasive.

avatars to treat contingent identity.[8] With regard to relevant logics, I think it is useful to distinguish between three ways of treating identity:

- Contingent Identity: Identities vary from point to point and even vary from normal point to normal point.
- Weakly Necessary Identity: Identities are uniform among normal points but vary among non-normal points.
- Absolutely Necessary Identity: Identities are uniform among normal points and only the identities that hold in normal points can hold in non-normal points.

Contingent identity is the same as the theory of the same name applied to modal logic as is absolutely necessary identity. Priest and I support weakly necessary identity in the semantics for basic relevant logic (Priest 2008, §23.6 and Sect. 16.4 above).

Absolutely necessary identity comes in two forms. First we can hold that exactly the same identities hold in all points in the frame. This view is clearly incompatible with the standard aims of relevant logicians.[9] For it makes valid the schema $A \rightarrow t = t$. Second, we can hold the view that Priest labels 'the subset constraint" (Priest 2008, §26.4.4). According to the subset constraint, an identity holds at a non-normal world only if it holds at normal worlds. This is a more interesting thesis. Since an actual identity need not be satisfied at every point, we do not obtain $A \rightarrow t = t$.

But there are also problems with the subset constraint. Consider a model in which 'Hesperus is Mercury' does not hold in any normal situation. Then, in no non-normal point is Hesperus identical to Mercury. So we have valid on this model

> Hesperus is Mercury implies that the moon is made of green cheese.

In the intended model for a relevant logic, for each formula A, there should be at least one situation in which A is true (and conversely, at least one at which A fails to be true, so that not everything entails it). In keeping with this, we should abandon the subset principle, and with it an absolutely necessary identity.

Weakly necessary identity is a more attractive theory, and I think it does fit well with basic relevant logic. I also think that it is defensible to constrain the models for mainstream relevant logic to comply with weakly necessary identity. We merely add the condition that if $a \in N$, $f_=(a) = \{(i, i) : i \in I\}$. Minor fiddling with the definition of the canonical model yields a completeness proof in the style of Mares (1992).

I prefer, however, not to add this condition and to have models in which identity is contingent. Suppose that we have a model that obeys the added condition to make identity weakly necessary and suppose that on this model 'Hesperus is Venus' is satisfied by all normal situations. Then the following is valid on the frame:

---

[8] In (2005, §4.4) and (2008, §17.3.13), Priest also develops a view according to which the identity of a thing is a contingent property of a thing similar to the colour or shape of that thing and uses this to support contingent identity.

[9] Although Robert Goldblatt does add it to Fine's semantics for quantified relevant logic for technical reasons (Goldblatt 2011, Chap. 6).

> Hesperus is Mercury implies that Mercury is Venus.

This sentence, I think, is true. On my semantics, any normal situation in which Hesperus is Venus, satisfies this sentence. But I do not think the counterfactual version of it is true:

> If Hesperus were Mercury, then Mercury would be Venus.

In stating the counterfactual one is asking the audience to imagine what the world would be like if Hesperus were Mercury. They might imagine that the bright star someone pointed at in the evening turned out to be the closest planet to the sun and that there is another planet between the Earth and the sun that is Venus. On my semantics, if $t = u$ obtains in every normal situation, then if $t = v$ holds in any situation, so does $u = v$. In order to invalidate the counterfactual, it would seem that we need a situation in which we have Hesperus being identical to Mercury without its being identical to Venus. If we allow normal situations to satisfy different identities, then we can have situations in which Hesperus is Mercury and in which Hesperus fails to be identical to Venus.

Of the three positions with regard to the metaphysics of identity, I prefer contingent identity for mainstream relevant logics. In order to accept contingent identity, I need to have a way of understanding the domain allows them to be identical to one another at some normal situations or worlds and not at others. I could, for example, accept Priest's notion of an avatar. On this view, the members of the domain as things that are parts of normal objects. For example, 'Hesperus' and 'Venus' name things that in the actual world (and certain actual situations) are parts of the same thing. But at other worlds they are not. Or we could treat the members of the domain as 'worms' that have as parts normal individuals. The part of Venus that is in the actual world is the same as the part of Hesperus that is in the actual world. But in non-actual worlds (and non-actual situations) they have different parts. Or I could adopt Priest's theory of identity properties (Priest 2005, §4.4). In the actual world a thing has both the property of being identical to Hesperus and the property of being identical to Venus, but in other worlds something might have one of these properties without having the other. At any rate, there is a variety of theories on the market that make sense of contingent identity and that can be used to interpret the present semantics. I do not want to choose one metaphysics here, but rather just to indicate that there are ways of incorporating the present semantics into a coherent metaphysics of worlds, situations, and individuals.

16.9 Conclusion

In this paper, I have constructed a theory of identity for a wide range of relevant logics. The motivation for this theory is an analogy between implication and identity. Both implications and identity statements tell us about licences to make inferences. Two formal semantics are produced for identity. The formal semantics for basic relevant logics divides frames between normal worlds in which true identities tell us the real truth about identity: $t = u$ at a normal world if and only if t and u both refer to the same thing at that world. At non-normal worlds, identities obtain or fail randomly. In this way, we have a very clear analogy between the way implication and identity work in these models. And this theory is just the one that Priest constructs in (2008). The Routley–Meyer semantics for mainstream relevant logics treats implication using a ternary accessibility relation between situations. Identity in Routley–Meyer models is treated as an inference modifier. The identities that hold in a situation a constrain the situations to which a is accessible. If the implicational information in a tell us that from A one has the licence to infer that B, then if $t = u$ holds in a, it contains the information that from $A(u/t)$ one has the licence to infer that $B(u/t)$.

References

Anderson, A., Belanp, N. D., & Dunn, J. M. (1992). *Entailment: Logic of relevance and necessity* (Vol. II). Princeton: Princeton University Press.
Beall, Jc., Brady, R., Michael Dunn, J., Hazen, A. P., Mares, E., Meyer, R. K., et al. (2012). On the ternary relation and conditionality. *Journal of Philosophical Logic, 41*, 565–612.
Dunn, J. M. (1987). Relevant predication I: The formal theory. *Journal of Philosophical Logic, 16*, 347–381.
Dunn, J. M. (1990). Relevant predication II: Intrinsic properties and internal relations. *Philosophical Studies, 60*.
Fine, K. (1988). Semantics for quantified relevance logic. *Journal of Philosophical Logic, 14*, 27–59.
Goldblatt, R. (2011). *Quantifiers, propositions, and identity*. Cambridge: Cambridge University Press.
Kremer, P. (1999). Relevant identity. *Journal of Philosophical Logic, 28*, 199–222.
Mares, E. (1992). Semantics for relevance logic with identity. *Studia Logica, 51*, 1–20.
Priest, G. (1992). What is a non-normal world? *Logique et Analyse, 139*, 291–302.
Priest, G. (2001). *An introduction to non-classical logic* (1st ed.). Cambridge: Cambridge University Press.
Priest, G. (2005). *Towards non-being: The logic and metaphysics of intentionality*. Oxford: Oxford University Press.
Priest, G. (2008). *An introduction to non-classical logic: From if to is* (2nd ed.). Cambridge: Cambridge University Press.

Edwin Mares is a professor of philosophy at Victoria University of Wellington. He works mostly on non-classical logic and has written more than sixty articles about logics and their interpretations. He is also the author of *Relevant Logic: A Philosophical Interpretation* (Cambridge 2004), *A Priori* (Routledge 2011), and, with Stuart Brock, *Realism and Antirealism* (Routledge 2007).

Chapter 17
The Difficulties in Using Weak Relevant Logics for Naive Set Theory

Erik Istre and Maarten McKubre-Jordens

Abstract We discuss logical difficulties with the naive set theory based on the weak relevant logic DKQ. These are induced by the restrictive nature of the relevant conditional and its interaction with set theory. The paper concludes with some possible ways to mitigate these difficulties.

17.1 Introduction

The quest to realize a nontrivial naive set theory has seen many iterations. Achieving this goal requires us to give up some logical inferences that are regularly used in classical set theory. Contraction usually needs to go, disjunctive syllogism is suspect, and we might dispense with weakening. Each inference lost makes us rethink the strategies we use when proving theorems in mathematics. The hope is that the price we pay in logical inferences may allow us to circumvent classical limitations: infinite hierarchies of theories, the ad hoc division between sets and classes, and obscure formalized notions of truth (Priest 2006).

One such iteration of this project has been the application of weak relevant logics to naive set theory. We know that this family of logics produce nontrivial naive set theories thanks to a proof from Brady (2006). In addition, sustained effort to develop these set theories has been carried out by Weber (2010, 2012). At first glance, these developments suggest that a weak relevant logic naive set theory might be feasible.

But there are limitations to this theory. There are serious difficulties that arise in doing mathematics beyond these initial developments. This paper outlines some of these with examples to show where the pitfalls lie. It then concludes with some explorations that seem necessary to push the project forward.

17.2 DKQ

The particular logic of this type we'll be focusing on in this paper is DKQ. The reason for this is that this logic supports the work done by Weber on which we focus.[1] We provide the Hilbert-style axiomatic system presentation for this logic. The language is composed of lower case letters as variables, upper case letters as predicate letters, and the logical symbols $\to, \vee, \wedge, \exists, \forall, \neg$.

Axioms:

A1. $A \to A$
A2. $(A \wedge B) \to A$
A3. $(A \wedge B) \to B$
A4. $A \to (A \vee B)$
A5. $B \to (A \vee B)$
A6. $((A \to B) \wedge (A \to C)) \to (A \to (B \wedge C))$ (conjunctive syllogism)
A7. $((A \to C) \wedge (B \to C)) \to ((A \vee B) \to C)$ (proof by cases)
A8. $(A \wedge (B \vee C)) \to ((A \wedge B) \vee (A \wedge C))$ (distribution)
A9. $\neg\neg A \to A$ (double negation elimination)
A10. $(A \to \neg B) \to (B \to \neg A)$ (contraposition)
A11. $((A \to B) \wedge (B \to C)) \to (A \to C)$ (hypothetical syllogism)
A12. $A \vee \neg A$ (law of excluded middle)
A13. $\forall x A \to A[y/x]$, y free for x in A
A14. $\forall x(A \to B) \to (A \to \forall x B)$, x not free in A
A15. $\forall x(A \vee B) \to (A \vee \forall x B)$, x not free in A
A16. $A[y/x] \to \exists x A$, where y is free for x in A
A17. $\forall x(A \to B) \to (\exists x A \to B)$, where x is not free in B
A18. $(A \wedge \exists x B) \to \exists x(A \wedge B)$, where x is not free in A

Rules:

R1. $A, A \to B \vdash B$ (modus ponens)
R2. $A, B \vdash A \wedge B$
R3. $A \to B, C \to D \vdash (B \to C) \to (A \to D)$
R4. $A \vdash \forall x A$ (universal generalization)
R5. $x = y \vdash A(x) \to A(y)$ (substitution)

---

[1] Brady (2006) argues for a weak relevant logic without the law of excluded middle (LEM). The loss of LEM in a system already deprived of so many logical inferences makes it very difficult to prove much of anything. Many theorems in Weber's work explicitly rely on it.

17 The Difficulties in Using Weak Relevant Logics for Naive Set Theory

Meta-rules:

MR1. If $A \vdash B$, then $C \vee A \vdash C \vee B$.
MR2. If $A \vdash B$, then $\exists x A \vdash \exists x B$.

The meta-rules have the additional restriction that they cannot be used on $A \vdash B$ if universal generalization has been used on a free variable of A (Brady 2006; Weber 2012). The biconditional, $A \leftrightarrow B$ is shorthand for $(A \rightarrow B) \wedge (B \rightarrow A)$. The turnstile as in $\Gamma \vdash A$ has the usual meaning that there is a proof of A from the assumptions Γ. The additional axiom schemas for naive set theory are as follows.

Axiom 17.1 (*Unrestricted Comprehension*) $\exists y \forall x (x \in y \leftrightarrow A(x))$

Axiom 17.2 (*Axiom of Extensionality*) $\forall x \forall y \forall z ((z \in x \leftrightarrow z \in y) \leftrightarrow x = y)$

In our presentation of Hilbert-style proofs, proofs are a linear sequence of instances of the axiom schemas as well as the application of valid rules to those instances. In these proofs, each line has an attached annotation, often abbreviated, which details which axioms schemas and lines of the proof are used to reach each line in the proof. We are also allowed to use assumptions by annotating a line as an assumption. Each line of a proof is of the following form.

$$n.\ A\ |\ \text{Annotation}$$

In this paper, our primary concern is evaluating DKQ as a logic for naive set theory. We take any and all logical rules that we can to maximize its logical strength while maintaining non-triviality. Thus what the theory does have as logical inferences are not of concern. Rather, we are interested in what logical inferences the theory does not endorse. To provide short examples of what DKQ can do, we provide proofs of the De Morgan Laws and of axiom schemas for proof by contradiction.

Proposition 17.3 *The following De Morgan laws are valid:*

- $\vdash \neg(A \wedge B) \rightarrow \neg A \vee \neg B$
- $\vdash \neg(A \vee B) \rightarrow \neg A \wedge \neg B$

Proof We prove the contrapositive of the first law.

1. $\neg A \rightarrow (\neg A \vee \neg B)$ | $\vee$ Intro
2. $(\neg A \rightarrow (\neg A \vee \neg B)) \rightarrow (\neg(\neg A \vee \neg B) \rightarrow \neg\neg A)$ | Contraposition
3. $\neg(\neg A \vee \neg B) \rightarrow \neg\neg A$ | MP with 1, 2
4. $\neg\neg A \rightarrow A$ | Double Neg
5. $\neg(\neg A \vee \neg B) \rightarrow A$ | Hypo Syl, MP
6. $\neg(\neg A \vee \neg B) \rightarrow B$ | Similar to 5
7. $(\neg(\neg A \vee \neg B) \rightarrow A) \wedge (\neg(\neg A \vee \neg B) \rightarrow B) \rightarrow (\neg(\neg A \vee \neg B) \rightarrow (A \wedge B))$ | Conjunctive Syl
8. $\neg(\neg A \vee \neg B) \rightarrow (A \wedge B)$ | MP with 5, 6, 7

The second law is more straightforward.

1. $\neg(A \vee B) \rightarrow \neg A$ — $\vee$ Intro, Contrapos
2. $\neg(A \vee B) \rightarrow \neg B$ — $\vee$ Intro, Contrapos
3. $((\neg(A \vee B) \rightarrow \neg A) \wedge (\neg(A \vee B) \rightarrow \neg B)) \rightarrow (\neg(A \vee B) \rightarrow (\neg A \wedge \neg B))$ — Conjunctive Syl
4. $\neg(A \vee B) \rightarrow (\neg A \wedge \neg B)$ — MP with 1, 2, 3

□

Proposition 17.4 $A \rightarrow B, A \rightarrow \neg B \vdash \neg A$

Proof

1. $A \rightarrow B$ — Assumption
2. $A \rightarrow \neg B$ — Assumption
3. $(A \rightarrow B) \wedge (A \rightarrow \neg B) \rightarrow (A \rightarrow (B \wedge \neg B))$ — Conj Syl
4. $A \rightarrow (B \wedge \neg B)$ — MP w/ 1, 2, 3
5. $(A \rightarrow (B \wedge \neg B)) \wedge ((B \wedge \neg B) \rightarrow \neg(\neg B \vee B)) \rightarrow (A \rightarrow \neg(\neg B \vee B))$ — Conj Syl
6. $A \rightarrow \neg(\neg B \vee B)$ — MP w/ 4, 5
7. $A \rightarrow \neg(\neg B \vee B) \rightarrow (\neg B \vee B) \rightarrow (\neg A)$ — Contraposition
8. $\neg B \vee B$ — LEM
9. $\neg A$ — MP w/ 6, 7, 8

□

Proposition 17.5 $A \rightarrow \neg A \vdash \neg A$.

Proof

1. $A \rightarrow \neg A$ — Assumption
2. $\neg A \rightarrow \neg A$ — Axiom 1
3. $((\neg A \rightarrow \neg A) \wedge (A \rightarrow \neg A)) \rightarrow ((A \vee \neg A) \rightarrow \neg A)$ — Proof by Cases
4. $(A \vee \neg A) \rightarrow \neg A$ — MP with 1,2 ,3
5. $A \vee \neg A$ — LEM
6. $\neg A$ — MP with 4, 5

□

Our present focus is on those axiom schemas and inferences that DKQ *does not* have which make mathematical reasoning more difficult. Some notable axiom schemas that DKQ omits or cannot prove are contraction, disjunctive syllogism, weakening, and pseudo modus ponens:

Definition 17.6 The following are named axiom schemas.

- Contraction is $(A \rightarrow (A \rightarrow B)) \rightarrow (A \rightarrow B)$.
- Disjunctive syllogism is $(\neg A \vee B) \rightarrow (A \rightarrow B)$.
- Weakening is $B \rightarrow (A \rightarrow B)$.
- Pseudo modus ponens is $(A \wedge (A \rightarrow B)) \rightarrow B$.

DKQ also fails to validate the kind of deduction theorem that we are used to. That is, we do not get that $A \vdash B$ implies $\vdash A \to B$. Deriving a formula from an assumption no longer suffices to show that the assumption *relevantly implies* the formula. This is due to the stringent requirements placed on $\to$ and not tightening up the turnstile. The turnstile is still allowed to operate as we are used to—it is reflexive, monotonic and transitive.

Normally the turnstile only expresses a relation between formulas of the axiomatic system and is not a connective of the system. However, DKQ blurs this distinction with meta-rules which allows inferences based on the existence of established turnstile relationships. It is these meta-rules we first discuss.

17.3 The Mystery of Meta-rules

Meta-rules are included in DKQ as a means of alleviating the logical price paid for non-triviality. They express principles that we ordinarily think of as being derivable axiom schemas: MR1 is a sort of disjunctive weakening and MR2 is a sort of existential generalization. We show that MR1 would be derivable in classical logic.

Proposition 17.7 *Let the logic in consideration be classical. MR1 follows from the axiom schema for proof by cases and the deduction theorem.*[2]

Proof Assume that $A \vdash B$. If from A we can construct a proof of B, then we can construct a proof of $C \vee B$ by a few additional steps involving disjunction introduction, hypothetical syllogism, and modus ponens. Thus we have a proof that

$$A \vdash C \vee B.$$

Then the deduction theorem constructs a proof that

$$\vdash A \to C \vee B.$$

We can also construct a proof that

$$\vdash C \to C \vee B$$

by the axiom schema for disjunction introduction. Then conjunction introduction, and the proof by cases schema give that

$$\vdash C \vee A \to C \vee B.$$

---

[2]MR2 can also be derived in a similar way, using the deduction theorem, A16, and A17.

An assumption of $C \vee A$ and modus ponens gives us the result:

$$C \vee A \vdash C \vee B.\qquad \square$$

Note that we use proof by cases to establish MR1. As it turns out, working within the restricted framework of DKQ we require MR1 to establish proof by cases for DKQ "over the turnstile". This is useful when we have established the turnstile relation for formulas, but are unable to establish a relevant implication. This happens often in doing naive set theory, and it would be hard to do without MR1 (or MR2) for this reason.

Proposition 17.8 (Proof by Cases) *If $A \vdash C$ and $B \vdash C$, then $A \vee B \vdash C$.*

Proof This is a result of Meta-rule 1. From $A \vdash C$, we can derive that $A \vee B \vdash B \vee C$. From $B \vdash C$ we can derive that $B \vee C \vdash C \vee C$. Transitivity gives us that $A \vee B \vdash C \vee C$. The equivalence of $C \vee C$ and C finishes the proof. $\qquad \square$

The content of the meta-rules is mysterious. As we see in our proof of MR1, MR1 is making a claim about what proofs can be constructed. It asserts a turnstile relationship. In DKQ, we cannot construct the proof that validates this turnstile relationship. This is a mysterious property for a formal system to have. It asserts the existence of a proof without actually providing one. We turn to formal systems for the clarity they can provide about proofs, but the use of a meta-rule necessarily obscures transparency in proof.

Meta-rules are naturally unsatisfying for the prover working in a Hilbert-style system. Classically, they are typically either provable or eliminable (think Cut, for example); their use is tolerated since proof existence can be verified by meta-theorems. This does not seem to be the case for MR1 or MR2.

In particular, for a mathematical theory, this is undesirable. No theorem can be believably proved on the assertion that there is a proof out there but it cannot be constructed.[3] It does little to elucidate naive set theory when we make use of these meta-rules. That said, doing without the meta-rules does not seem like an option. DKQ needs all the strength it can muster while remaining nontrivial. We move on to more significant problems.

---

[3]Note the difference between a non-constructive proof and a non-constructed proof. The former is still a visible argument, the latter is not. Even an infinitary proof procedure offers the promise of some kind of visible proof.

17.4 The Turnstile Versus the Relevant Conditional

The lack of the deduction theorem and the strict provability of the relevant conditional are a difficult combination. It is worth remembering the intuitive appeal of the usual deduction theorem and what it implies about classical logic. We believe that the notion of "deduction" that we use in mathematical reasoning has three main properties: reflexivity, monotonicity, and transitivity. It should be no surprise that losing the connection between this notion of deducibility and our logic's conditional would cause some problems.

In DKQ, the failure of the deduction theorem is due to the turnstile being capable of different inferences than the conditional, e.g., the turnstile allows weakening, while the conditional has strong interoperability with the defined paraconsistent negation.[4] Note that we also have that modus ponens is defined as occurring over the turnstile. This avoids validating the axiom schema for pseudo modus ponens. Pseudo modus ponens leads to a proof of Curry's paradox (Beall and Murzi 2013).

It is possible to change the turnstile relation so that it mirrors the relevant conditional. This would push some kind of a theorem which had the proper form of the "deduction theorem". But this misses the point, it is not the mere form of a deduction theorem that we seek. We seek one using our usual notion of deducibility; we do not wish to change the notion of deducibility the turnstile currently represents to something other than "assume A, then deduce B". It really does seem to represent our usual means of deduction in mathematical theories, and it turns out to be nice to have around.[5]

There are many things that only ever seem provable over a turnstile, like the injectivity of a function or whether a set is irreflexive. These are defined in DKQ naive set theory as follows.

Definition 17.9 A function f is injective if $f(x) = f(y) \vdash x = y$.

Definition 17.10 A set a is irreflexive if $x \in a \vdash x \notin x$.

In the former case, injectivity needs to be defined over a turnstile since our substitution rule is also expressed with the turnstile. Any use of substitution necessarily breaks any relevant connection we are attempting to establish. In the latter case, strictly ordered requires proving irreflexivity, as in "a set a is irreflexive if for any $x \in a$, $x \notin x$". It is almost never the case that $x \in a$ has any relevant connection to $x \notin x$ except in the most trivial of cases, i.e., the intension of a includes a conjunct of $x \notin x$.

The further problem implied by having to define these two properties with the turnstile is that we can no longer define natural sets that collect the objects with that

---

[4]There's also the problem that the usual proofs of a deduction theorem for a logic rely on the conditional being capable of contraction.

[5]This is a strong reason why the presentation of DKQ is easier in a Hilbert-style axiomatic system: a natural deduction formalization of DKQ would contain a $\rightarrow_I$ rule that did not intuitively correspond to our usual understanding of conditional proof. Such a natural deduction system can be found in Brady (2006, Chap. 3).

property. One of the strengths of naive set theory is that it can express any collection we can think of. If we are forced to use the turnstile to express a property then we cannot adequately represent that relationship in the object language, and thus cannot construct a set term. Finally, in practice in the theory, the turnstile just turns out to be more useful. We can prove a lot more with it when doing DKQ naive set theory than we can with the relevant conditional. The conditional is then shoved to the side and only aids us in maintaining non-triviality. This problem is only compounded because it is the relevant conditional that has a strong connection with negation and if we cannot prove theorems with it, then we do not get to use these negation properties.

Even if we establish some theorems with the relevant conditional, it is not clear that it does enough with negation to adequately reason about set theory. Recall the axiom of extensionality,

$$\forall z(z \in x \leftrightarrow z \in y) \leftrightarrow x = y.$$

And consider its contraposed form

$$\neg \forall z(z \in x \leftrightarrow z \in y) \leftrightarrow x \neq y.$$

To get that two sets are not equal requires proving

$$\exists z(\neg(z \in x \to z \in y) \lor \neg(z \in y \to z \in x)).$$

To be able to prove this requires a counterexample inference:

$$(A \land \neg B) \to \neg(A \to B).$$

DKQ as defined does not have this, but we can add

$$A \land \neg B \vdash \neg(A \to B)$$

and still be covered by the non-triviality proof in Brady (1989).

The larger loss is that $\neg(A \to B)$ is no longer informative in the theory. We cannot have the implication $\neg(A \to B) \to (A \land \neg B)$ in the logic as contraposition would, in the context of DKQ and related logics, imply the conditional was material, i.e.,

$$\neg A \lor B \to (A \to B).$$

This[6] is a problem when we start a proof on the assumption that two sets are not equal. Returning to the contraposition of the axiom extensionality, this assumption

---

[6]It is not clear what the status of $\neg(A \to B) \vdash (A \land \neg B)$ is. A naive notion of relevance implies that $A \to B$ can fail to hold because A was not relevant to B rather than it being because A was true and B was false. This rule is not included in DKQ.

Note that in some nonclassical logics (e.g., Nelson's constructive logics), $A \land \neg B$ is equivalent to $\neg(A \to B)$, without forcing the conditional to behave materially. In the context of DKQ, the

of two nonequal sets amounts to a disjunction of two negated conditionals. This would ordinarily imply that the sets have a different element, but we cannot draw that conclusion with DKQ. This means that any proof which attempts to start on a negated conditional is doomed from the start.

This shows that the interrelationship of the turnstile and the relevant conditional in DKQ naive set theory is an odd one. The turnstile is more useful: it can prove more things. The relevant conditional prevents triviality and would allow us to make use of our paraconsistent negation, but relevant conditionals are notoriously difficult to establish. Yet, the relevant conditional's weaknesses for mathematical reasoning extend beyond this.

17.5 Weakening

The first problem we may observe that arises from the loss of weakening is the difficulties with establishing set equality. We would not normally think weakening had much to do with equality. But it has a unique role in set theory that we do not see in other theories where equality is a primitive. For example, consider an axiomatic theory given for Peano Arithmetic. The axioms for this theory include

$$Sx = Sy \to x = y$$

or

$$\neg \exists x (Sx = 0).$$

Equality proofs in PA typically do not refer to more elementary relations because equality in PA is as simple as it gets. Any use of weakening in PA really does weaken our position in being able to prove another equality, as it subjects us to another assumption (Meyer Unpublished).[7]

Since equality can be proven in set theory through the axiom of extensionality, which involves the more primitive connectives of $\in$ and $\to$, the provability of equality does rely on the properties of the conditional. Without weakening in set theory, we lose the ability to derive equality when we simply have that two sets "have the same stuff". With DKQ, we also need that the sets have "relevantly connected reasons for having the same stuff".

---

negation is De Morgan and we have Double Negation Elimination, so materiality of the conditional does follow.

[7] We also have evidence in the form of relevant Peano Arithmetic that PA can get away without weakening. This theory has been explored and seems capable of proving most of what we need in doing mathematics. This suggests that the loss of weakening did not inflict too grievous a wound. That said, it was found to not recover all of PA and was thus largely abandoned. See Friedman and Meyer (1992) for more on relevant arithmetic. It is worth noting that the work on this is notoriously difficult to find as a lot of it is unpublished.

It is worth considering a test-case. Consider sets $A = \{0, 1, 2\}$ and $B = \{0, 1, 2\}$. We would like to be able to identify these sets merely by their coincidence of elements. The way this coincidence can be formalized requires weakening. For given the particular $0 \in A$, we use weakening to conclude that $0 \in B \to 0 \in A$. When allowed to weaken, we do not have to rely on the particular reasons that A or B contain 0.[8] But without weakening, we do.[9]

This means that for any proofs of equality in this system, except in some trivial cases, we have to use the defining properties of the sets. In the introduced informal case, we can prove equality if we have defined A and B in the same way. But the lack of weakening admits a way to construct infinite classes of sets for which we cannot demonstrate equality but for which we also cannot demonstrate elements for which they differ—this can even be done for *every* class (infinite or not), adding much noise to the theory. So requiring formulaic identity quickly becomes too cumbersome to usefully categorize set-theoretic objects. Further, logically enforced formulaic identity appears to undermine our natural reason for a set theory: a nontrivial identification of intensions with their extensions.

Definition 17.11 (Brady 2006) Let the formula A contain only the connectives $\neg$, $\wedge$, $\vee$, and $\to$. Then *depth* is inductively defined as follows.

1. The subformula A of the formula A is of depth 0 in A.
2. If $\neg B$, $B \wedge C$, or $B \vee C$ is a subformula occurrence of A of depth d in A then B, and C in the cases of $\wedge$ and $\vee$, are of depth d in A.
3. If $B \to C$ is a subformula occurrence of A of depth d in A, then both of these occurrences of B and C are of depth $d + 1$ in A.

Definition 17.12 A logic is *depth relevant* if for all formulas A and B, if

$$\vdash A \to B$$

then A and B share a sentential variable in a subformula at the same depth.

Theorem 17.13 *DK is depth relevant (Brady 2006, p. 164).*

Corollary 17.14 *Let A and B be atomic formula, i.e., not containing any logical connectives, which need not be distinct. Then DKQ cannot prove $A \to (B \to B)$ or $\neg(B \to B) \to \neg A$.*

Proof Neither of the offered forms share a sentential variable. Since A and B are atomic, we do not need to be concerned that a quantifier inference allows this as well. Thus they cannot be derived in DKQ. □

---

[8] Note that when we name sets by their extension, we do not know their intension. This greatly reduces what the relevant conditional can do.

[9] This also suggests we intuitively think of the conditional that manages set equality as material and truth-functional. For a particular $z \in B$ that A shares, we have $z \in A$ and use disjunction introduction to get $z \notin B \vee z \in A$. On the other hand, for any $z \notin B$ we can still get $z \notin B \vee z \in A$. These two observations together gives $\forall z(z \notin B \vee z \in A)$.

Now, consider A and B defined by $A := \{x | x = \emptyset\}$ and $B := \{x | x = \emptyset \wedge (C \to C)\}$ where C is any atomic formula. We can find infinitely many such C's by taking atomic membership formula for sets defined by valid formula. For example take an atomic formula D and consider the valid forms $(D \to D)$ and $(D \to D) \to (D \to D)$ and so on. Using these formulas, take a set like $\{x | (D \to D)\}$, and then use the atomic membership formula $z \in \{x | (D \to D)\}$ for C. To derive that $A = B$ would require deriving that $x \in A \to x \in B$. Deriving this would require a direct *relevant* proof of $x = \emptyset$ to $C \to C$.[10] While Corollary 17.14 cannot completely rule these out as being provable in DKQ naive set theory, it does help to demonstrate that proving this is not likely to be possible in every instance.[11]

We also cannot possibly find something that is a member of A that is not a member of B. For if we find any $x \in A$, that means $x = \emptyset$ must be provable. And $C \to C$ is always provable, and so the conjunction $x = \emptyset \wedge (C \to C)$ is also provable. That is, we can prove

$$x \in A \vdash x \in B.$$

But without a deduction theorem we cannot guarantee $x \in A \to x \in B$ is also provable. Thus we cannot find a proof that these sets are equal, nor can we prove they are not equal.

One may object that the demands we are making are classically minded, without regard for the nonclassical logic within which we work. However, it is important to recall exactly what naive set theory is meant to accomplish. A large part of the attractiveness of naive set theory is that any intension can be identified by its extension and that we can reason about those intensions as extensions. The loss of weakening completely negates this. Without weakening, we entirely lose the ability to reason about the equivalence of intensions except when they are really truly equivalent at the formulaic level. But that is not the point of naive set theory!

The loss of weakening is a problem outside of proving equality as well. Mathematical proofs usually take advantage of assumptions which operate as "contextual information", like in the statement "if x and y are ordinals, $x \in On$ and $y \in On$, then $x \subseteq y$ or $y \subseteq x$". That x and y are ordinals is meant to be used to derive the result. There are a few different paths we could take for such a proof, but the following exhibits the problem.

Assume we attempt to show that $x \subseteq y$ directly after assuming some instances of LEM. This requires proving that $z \in x \to z \in y$. Since x was arbitrary, we do not know the set property that defined x and so we cannot move forward from $z \in x$.[12] We need to use that x is an ordinal to imply some formula $A(z)$ that was enough to conclude $z \in y$. That is, from establishing

---

[10] And the contraposition fares no better, since that would require a direct proof from $\neg(C \to C)$ to $x \neq \emptyset$.
[11] It also seems difficult in DKQ to produce proofs of propositions of the form $A \to (B \to C)$.
[12] That is, by assuming $z \in x$, we literally have no access to the defining property of the set x. The relevant conditional does not cooperate well in these cases.

$$x \in On \vdash A(z),$$

we would want to establish that z is enough to get $A(z) \to z \in y$. It is weakening which would allow us to say that

$$x \in On \vdash z \in x \to A(z)$$

and thus that
$$x \in On \vdash z \in x \to z \in y.$$

Without weakening, we can make no such inference. The use of the assumed information $x \in On$ breaks the connection needed for the relevant implication.[13]

Even if one is convinced that weakening plays an important role in set theory, there does not seem to be much we can do about the situation. The next section discusses different conditionals that may mitigate the downsides of the relevant conditional. It is plausible to wonder if we can use one of those to rewrite the axiom of extensionality. This apparently would weaken the strong connection needed between sets to make them equal. But it turns out that this strict form of the axiom of extensionality is also keeping triviality at bay.

Definition 17.15 Fix an arbitrary term t. Define $\bot := \forall y (t \in y)$.

Proposition 17.16 *Let $\Leftrightarrow$ be some conditional that satisfies weakening. Then DKQ naive set theory with the extensionality axiom as $\forall z (z \in x \Leftrightarrow z \in y) \Leftrightarrow x = y$ is trivial.*

Proof Consider the sets $U := \{x | x = x\}$ and $V := \{x | \exists y (x \in y)\}$. These sets are universal, i.e., for any $x, x \in U$. With weakening we can prove these sets equal. If $x \in V$ then weakening gives $x \in V \to x = x$. With the axiom of comprehension giving $x = x \to x \in U$, we can conclude that $x \in V \to x \in U$. For the other direction, we use that V is universal. Since $x \in V$ for any x, by weakening we have $x \in U \to x \in V$. Thus $U = V$.

This leads to triviality through Russell's set. We can prove that $R \neq R$. Thus $R \notin U$ and since $U = V$, $R \notin V$. That $R \notin V$ implies that $\forall y (R \notin y)$. In particular, $R \notin \{x | x \notin \{z | \bot\}\}$ which establishes $\neg (R \notin \{z | \bot\})$ and thus $R \in \{x | \bot\}$ by double negation elimination. The axiom of comprehension concludes that $\bot$. □

17.5.1 Working Around These Problems

To help mitigate some of these problems, we can add an additional conditional and leverage the usual $\bot$ based negation more heavily, that is we define a negation as $\sim A := A \to \bot$. Any properties that this negation satisfies then come from the proof

---

[13] This appears to be another example where weakening plays a unique role in set theory.

that $\bot \vdash A$ for an arbitrary A, and the properties of the conditional it is used with. Since this negation can be defined internally, the non-triviality result still applies.

A particular use case for this stronger negation is that it gives more explicit meaning to the negation of equality. Consider $\neg(x = y)$ and $\sim (x = y)$. As discussed before, the former is entirely uninformative at a logical level. It also cannot even tell us whether x and y really are two different sets with two different intensions. For $\neg(R = R)$ does happen in the theory, and clearly we do not want to think of R as countably distinct from R lest the singleton $\{R\}$ be infinite.[14] The other negation asserts that x and y must be distinct on pain of triviality. This can be more useful to know.

In the case of the conditional, we can attempt to add a conditional which behaves more normally, i.e., has weakening and allows a more usual version of conditional proof.[15] The idea of a separate weakening conditional in relevant logics is not new. One place it appears in the literature is in the development of relevant restricted quantification.[16] In this work, it takes the form of the "enthymematic" conditional. This conditional works by first adding a constant t, understood as the "conjunction of all theorems" to the language. This constant t satisfies

$$A \dashv\vdash t \to A.$$

The enthymematic conditional is defined as $A \mapsto B := (A \wedge t) \to B$. We can then prove the enthymematic conditional weakens.

Proposition 17.17 $B \vdash A \mapsto B$

Proof By the properties of t,

$$B \vdash t \to B.$$

Further, by transitivity of the conditional, from $(A \wedge t) \to t$ and $t \to B$ we get that $(A \wedge t) \to B$. This is what we want by definition of the enthymematic conditional, i.e.,

$$B \vdash A \mapsto B. \qquad \square$$

---

[14] Counting defined in the way we normally think of as in set theory, by mappings. The fact that $R \neq R$ should not suffice to generate injective mappings from arbitrarily sized sets which demonstrate that the singleton set $\{R\}$ has more than 1 set inside it. For example, there should not exist an injective mapping from a set containing two obviously distinct things like $\emptyset$ and $\{\emptyset\}$ to the singleton set $\{R\}$. The weak negation applied to equality fails to show that two sets are actually "distinct" in a meaningful way with regard to counting.

[15] Another conditional *will* take us beyond what is covered in Brady's non-triviality proof.

[16] Relevant restricted quantification would likely be useful for relevant naive set theory, however the proper framework for it isn't yet clear and adding it in would take us further from Brady's non-triviality result. However, we can achieve some of the properties that would be wanted from relevant restricted quantification with a new arrow. More information on restricted relevant quantification can be found in Beall et al. (2006).

The nice thing about the enthymematic conditional is that it resembles the relevant conditional as much as possible. This probably means that the naive set theory with it included is still nontrivial but this has not been proven.[17] The main difference between the enthymematic conditional and the relevant condition is that we lose access to the strong negation properties of the relevant conditional. For example, consider the contraposition of $A \mapsto B$. This becomes $\neg B \to \neg A \vee \neg t$. We have no inferences with $\neg t$ and so this is deductively inert.

The enthymematic conditional does suffice to prove some theorems outside the scope of the relevant conditional. However, these theorems are not necessarily that much of an improvement. All we have added is weakening and this is not enough to get a deduction theorem with it. Recall standard proofs of the deduction theorem use contraction or the equivalent inference of self-distribution. The enthymematic conditional is contraction free. Thus we still have the issue that if proving a property requires using assumptions, i.e., working over the turnstile, then we are unable to represent that property in the object language. Thus, the enthymematic conditional is still unsatisfactory for our purposes.

To get a conditional for DKQ naive set theory which has both weakening and a deduction theorem requires moving away from the relevant conditional.[18] I refer to work from Weber which defines such a "naive" conditional (Weber 2019; Caret and Weber 2015). This conditional is denoted $\Rightarrow$. We give it all of the axioms associated with the relevant arrow, except for those involving negation and conjunctive syllogism.[19] It also gets its own version of modus ponens, $A, A \Rightarrow B \vdash B$. We further assume that it weakens, $B \vdash A \Rightarrow B$.

The more complicated issue with including this as a primitive connective is that we have to encode a deduction theorem. Since the arrow is also meant to be contraction free, this means the turnstile relationship as we have defined it will not suffice since we consider collections of assumptions to be sets. This follows from the principle that multiple assumptions or "uses" of A can all be identified as a single instance in the relationship exhibited by $A \vdash B$.

Definition 17.18 The *use* of a formula A means that it occurs as a premise in the application of a rule.

A deduction theorem for the naive conditional cannot hold with *this* particular turnstile. Thus, we consider the turnstile relation in which the assumed collections of

---

[17]This is not to say that including it does not expose us to *any* danger of finding ourselves with a trivial naive set theory. Standard relevant semantics suggest that the addition of the t constant with its usual introduction of rule of $\vdash t$ may make it possible to prove $((A \to B) \wedge A \wedge t) \to B$. Then one could use a pseudo modus ponens type proof of Curry's paradox with a slight adjustment to the definition of the Curry set: $\{x | (x \in x \to p) \wedge t\}$. Thanks to Edwin Mares for showing me this revenge Curry.

[18]The hardest part in adding such a conditional in the present context is that it must be contraction free. See Beall and Murzi (2013) for further exposition on how completely we need to avoid contraction.

[19]We cannot have conjunctive syllogism without also validating the proof of Curry's that uses pseudo modus ponens.

formulas are multisets, which are capable of tracking multiple instances of a single object.

Definition 17.19 A *multiset* is a set which contains multiple distinct copies of the same element, e.g., where the following brackets represent multisets, $\{A, A\} \neq \{A\}$.[20]

Then this turnstile relationship, which is denoted $\vdash_M$ to separate it from the original, can sensibly track multiple uses or assumptions of the same formula in a proof. Thus we may write

$$A, A \vdash_M B$$

to mean that A was used or assumed twice. Note that this turnstile relationship still satisfies the properties of reflexivity, transitivity, and monotonicity and that if $A \vdash_M B$ holds then $A \vdash B$ also holds by definition. It is for this turnstile we assume a deduction theorem for the naive conditional, which can be understood as adding another meta-rule.

Axiom 17.20 (*Meta-Rule 3*) Let Γ be a multiset of formulas and A and B formulas. Let $\Rightarrow$ be the naive conditional. We assume the following meta-rule for this conditional: if $\Gamma, A, A \vdash_M B$ then $\Gamma, A \vdash_M A \Rightarrow B$.[21]

Though the naive conditional does not have much interaction with the paraconsistent negation, it admits some behaviors when interacting with the defined negation $A \Rightarrow \bot$ that make it look intuitionistic.

Proposition 17.21 *The following hold for $\Rightarrow$ and $\vdash_M$:*

- $A \Rightarrow B \vdash_M (B \to \bot) \Rightarrow (A \Rightarrow \bot)$
- $A \vdash_M (A \Rightarrow \bot) \Rightarrow \bot$
- $A, A \vdash_M (A \Rightarrow B) \Rightarrow ((A \Rightarrow (B \Rightarrow \bot)) \Rightarrow (A \Rightarrow \bot))$

Proof The first proposition is the result of hypothetical syllogism: take the instance

$$((A \Rightarrow B) \wedge (B \Rightarrow \bot)) \Rightarrow (A \Rightarrow \bot).$$

Then modus ponens and the deduction theorem gets the desired result.

The second proposition is the result of modus ponens. That is, we have

$$A, A \Rightarrow \bot \vdash \bot.$$

The deduction theorem completes the result.

---

[20] These can be formalized in the usual set theory as by tagging each element of a set as distinct. For example the multiset $\{A, A\}$ would be $\{\langle 0, A\rangle, \langle 1, A\rangle\}$ as a normal set.

[21] As with the other meta-rules, this meta-rule has an inherently non-constructive nature.

The third follows from repeated modus ponens, transitivity and the deduction theorem. We have
$$A, A \Rightarrow B \vdash B$$
and
$$A, A \Rightarrow (B \Rightarrow \bot) \vdash B \Rightarrow \bot.$$
Further since
$$\bot \vdash A \Rightarrow \bot,$$
we have that
$$B, B \Rightarrow \bot \vdash A \Rightarrow \bot.$$
Thus
$$A, A, A \Rightarrow B, A \Rightarrow (B \Rightarrow \bot) \vdash A \Rightarrow \bot.$$
The deduction theorem completes the result. □

The benefits of a naive conditional are twofold: first, it can absorb contextual information provided in a proof into the formal language; and second, anything defined with it is representable in the formal language. This makes it more familiar to work with. In the context of DKQ naive set theory without Weber's use of the enthymematic conditional, Istre (2017) provides some examples of a naive conditional in action.

17.6 Conclusion

The project of naive set theory is likely such that it will require trying a lot of different things. Figuring out how to work around a century of entrenched classical theory is not easy and the difficulties outlined above may be resolvable. Perhaps we might find some more logical strength or develop new insights into how to reason in this context. But it does seem unlikely given that even the recovery of weakening behavior with conditional proof can lead to disaster, as in Corollary 17.14. Still, the hope remains that we may find a way around some of the problems we see in classical set theory and still have a robust theory.

In the end, the point is that when developing a naive set theory, we should not get lost about what our real goal is: the development of an insightful theory about the connection between arbitrary intension and extension. As it stands now, DKQ naive set theory ventures too far from the spirit of the naive set theorist, and the working mathematician and imposes too many technical complications. But if these problems can resolved, it may offer a vast and interesting naive universe.

References

Beall, J. C., & Murzi, J. (2013). Two flavors of curry's paradox. *The Journal of Philosophy, 110*(3), 143–165.

Beall, J. C., Brady, R. T., Hazen, A. P., Priest, G., & Restall, G. (2006). Relevant restricted quantification. *Journal of Philosophical Logic*.

Brady, R. (1989). The non-triviality of dialectical set theory. In *Paraconsistent logic: Essays on the inconsistent* (pp. 437–470).

Brady, R. (2006). *Universal logic*. Stanford, California: CSLI Publications.

Caret, C., & Weber, Z. (2015). A note on contraction-free logic for validity. *Topoi, 34*(63).

Friedman, H., & Meyer, R. K. (1992). Whither relevant arithmetic? *The Journal of Symbolic Logic, 57*(3), 824–831.

Istre, E. (2017). *Normalized Naive Set Theory*. PhD thesis, University of Canterbury.

Meyer, R. K. Unpublished work.

Priest, G. (2006). *In contradiction: A study of the transconsistent* (Expanded ed.). Oxford: Clarendon.

Weber, Z. (2010). Transfinite numbers in paraconsistent set theory. *The Review of Symbolic Logic, 3*(1), 71–92.

Weber, Z. (2012). Transfinite cardinals in paraconsistent set theory. *The Review of Symbolic Logic, 5*(2), 269–93.

Weber, Z. (2014). Naive validity. *Philosophical Quarterly, 64*(254), 99–114.

Erik Istre is a nonclassical mathematician from Louisiana, United States. He has a proclivity for breaking things and trying to put them back together. With the rest of his time, he plays tabletop and board games and spends time with his dog Kaya.

Maarten McKubre-Jordens is a nonclassical mathematician from Christchurch, New Zealand. Beyond nonclassical mathematics, his interests include teaching, homebrewing, home improvement, and motorcycles. He lives in a cosy house on a quiet street, with an infinitely loving and patient wife, two kids, and a cat called Gus (short for Augustina).

Chapter 18
ST, LP and Tolerant Metainferences

Bogdan Dicher and Francesco Paoli

Abstract The strict-tolerant (ST) approach to paradox promises to erect theories of naïve truth and tolerant vagueness on the firm bedrock of classical logic. We assess the extent to which this claim is founded. Building on some results by Girard (Diss Math 136, 1976) we show that the usual proof-theoretic formulation of propositional ST in terms of the classical sequent calculus without primitive Cut is incomplete with respect to ST-valid metainferences, and exhibit a complete calculus for the same class of metainferences. We also argue that the latter calculus, far from coinciding with classical logic, is a close kin of Priest's LP.

18.1 Introduction

Graham Priest is well known for being an advocate of non-classical logics as alternatives to, not merely extensions of, classical logic. In particular, he defends paraconsistent logics from the backdrop of a commitment to dialetheism. He is also fond of classical mathematics and, to the extent that it is instrumental in delivering its goods, of classical reasoning. One of the important themes in the development of Priest's dialetheist programme is that of 'classical recapture', i.e. the task of recovering the outcomes of classical reasoning within a paraconsistent setting. Indeed, having shown what his 'logic of paradox' (LP) can do in this respect, Priest concludes that

> dialetheic logic gives us the full power of classical logic except where classical logic is demonstrably useless, and then more. (Priest 2006, p. 119)

Recently, the consensus that classical logic is 'useless' in inconsistent situations has been challenged. A novel approach, called *strict-tolerant* (ST for short) approach, put forward by Pablo Cobreros, Paul Egré, David Ripley and Robert van Rooij in

---

B. Dicher
Centre for Philosophy of the University of Lisbon, Lisbon, Portugal

F. Paoli (✉)
Department of Pedagogy, Psychology, Philosophy, University of Cagliari, Cagliari, Italy
e-mail: paoli@unica.it

a series of papers (Cobreros et al. 2012, 2013; Ripley 2013a, b), proposes to show that classical logic *can* handle the paradoxes. The idea is to take a standard sequent calculus for plain old classical first-order logic without primitive Cut, and extend it with rules for a fully disquotational truth predicate. The Cut rule is of course admissible (by Gentzen's *Hauptsatz*) in the *logical* fragment of the calculus, yet it fails to be so once the truth rules are added—and it is exactly these failures of Cut that save the day when we attempt to reproduce herein a derivation of the Liar (or of any other paradox). Model-theoretically, this perspective relies on a three-valued semantics, based on a certain non-preservational account of consequence defined in terms of the strong Kleene matrices—essentially, Girard's *three-valued interpretation* of the classical sequent calculus without cut (Girard 1976). There is some indication here that the approach lives in two worlds, or perhaps hovers somewhere in between them. The model-theoretic image suggests that there is some connection between ST and Priest's LP. On the proof-theoretic side of the fence, things appear to be reassuringly classical—at least as far as the logical basis is concerned. (We will get to the details of LP and ST in the next section.)

In this paper, we assess the justness of two claims made by the ST-theorists, or at least implicitly suggested in their papers. On the one hand, they show that, at the level of ST-valid inferences, there is an exact match between their classical proof-theoretic picture and their non-classical model-theoretic one, whence they contend that choosing between the two is something like a matter of logical taste. Along the lines of what Girard has done in the context of his work on the three-valued interpretation of the sequent calculus (Girard 1976, 1987), we will challenge this view by pointing out that, already at the propositional level, the classical proof system they pick is badly *incomplete* when it comes to encompassing all ST-valid *meta*inferences; we will also show what needs to be done in order to complete it. On the other hand, according to Cobreros et al., the viability of their approach shows that there is no need to give up classical logic unless you have to reason about such delicate matters as naïve truth or predicate vagueness; failures of Cut and transitivity can be confined to these domains, outside which we can happily resort to the well-trodden tools of classical logic. We will provide two arguments against this thesis (one of which is borrowed, in essence, from Barrio et al. 2015 and Pynko 2010). Their common upshot is that ST has a lot more in common with LP, and with dialetheism in general, than is apparent from the seemingly innocent classical clothes in which it has been proof-theoretically dressed.

Before explaining how we set about to do this, we pause for a brief disclaimer regarding some of our expositive choices. From here on, we will talk a lot about LP and ST. The reader is entitled to wonder what exactly do we mean by these labels. We would be tempted to reply that these acronyms denote 'logics', were it not for the fact that this would probably trigger a further question, namely, what do we mean by a logic. We can say this much from the outset: in the long-standing controversy between partisans of logics-as-collections-of-validities and advocates of logics-as-consequence-relations, we take sides with the latter. Having pointed out this, we observe that coming clean at this point with a precise definition of LP or ST as 'logics' would force us to settle, once and for all, a number of issues (for example: are we considering them as propositional or as first-order systems? Or else as theories with specific descriptive vocabulary? Are we privileging a single-conclusion or a multiple-conclusion account of consequence?) that we prefer to resolve contextually, at least until we are ready to reconsider the whole thing in more general terms in Sect. 18.5. Up to that point, we ask the reader to bear with our using these labels as vague names. In any case, the definitions of validity and consequence that matter for the formal development of our discourse will be given at the appropriate places.

Section 18.2 provides a brief overview of LP and ST and of the way in which they handle paradoxical sentences. As it turns out, some classically valid metainferences are lost in ST. This prompts us to investigate in more detail the notion of metainference, a task which we discharge in Sect. 18.3. We distinguish between a global and a local conception of metainferential validity, and argue that it is the latter that better represents metainferential validity in ST. Section 18.4 contains the main result of this paper. After showing that propositional ST is incomplete with respect to the class of all locally ST-valid metainferences, we present a calculus for which this completeness result holds. We then investigate the extent to which this calculus is just another incarnation of LP. We conclude that ST fails to deliver on the promise of being classical logic. To our minds, it is either LP or at least a paraconsistent logic closely related to it. Two arguments buttressing this claim are fleshed out in Sect. 18.5.

18.2 Preliminaries

18.2.1 LP and Dialetheism

All too often, the expressions 'logic' and 'logical theory' are conflated, to the effect that the latter acquires a meaning properly ascribed to the former. Priest's work is everywhere dotted with urges to take this distinction seriously. For, as Priest never ceased to remind us, our theory about a domain's logic *may* be wrong. (See the quotation below and also Priest 2005, Chap. 10, 2014.) That is, *logic*—the rules of proof employed when reasoning—and, respectively, *logical theory*—our theory about those rules of proof—may come apart. An early statement of this position is in Priest (1979, p. 225), where we read:

> The formal logician is essentially an applied mathematician. It is his job to construct mathematical systems which model (in the physicist's sense, not the logician's) some natural phenomenon. The phenomenon the logician is particularly interested in, is normal (naïve) reasoning carried out in a natural language.

Priest is also (in)famous for believing that, in some circumstances, correct vernacular reasoning forces us to accept contradictions (a view he sometimes characterises as *dialetheism*). Thus, an appropriate account of how we reason should reflect this feature of the vernacular. The simplest and probably most convincing example of a situation in which one must accept a contradiction is that of paradoxes such as the

Liar ('This sentence is not true'). We can get a better idea of what is at stake here by considering vernacular reasoning in the realm of mathematics. The same sort of deductive moves that are involved when we recognise the Liar to be both true and false is applied when we grasp the truth of the Gödel sentence. Only in this case, the model, i.e., our logical theory, handles the upshot of this reasoning rather differently. The theory that Gödel's theorem shows to be incomplete is not semantically closed, and this forces us to take a detour through the metalanguage. Were we to have let everything happen at object-language level, we would have had to make our theory semantically closed—just like natural languages are. That would also render the theory inconsistent. Hence the detour is the price one must pay if the theory is to be consistent. Nevertheless, having paid this price, we get an edulcorated theoretical representation of reasoning in mathematics. According to Priest, this is a wrong theory of the logic of *naïve proofs*:

> [W]e see that our naïve notion of proof appears to outstrip the axiomatic notion of proof precisely because it can deal with semantic notions. (Priest 1979, p. 223)

While we get a formal theory of *some* kind of proofs, we do not get a formal theory of proofs such as those which we actually build when reasoning about the Liar or the Gödel sentence. All this happens because we fear that we cannot deal with contradictions. Indeed, most logics have a very unsophisticated way of (mis)handling contradictions. This includes—or at least included until the advent of ST, see below, Sect. 18.2.2—classical logic. Contradictions allow the inferring of whatever. For, so goes the orthodox dogma, *ex contradictione sequitur quodlibet*. (See Priest et al. 2013; Sylvan 2000 for a brief history of how *ex contradictione* became a logical law.) However, this destructive inferential power of contradictions renders them all but useless within our logical theories. Nevertheless, the vernacular practice does struggle with contradictions; this is a lesson we learn from paradoxes such as the Liar. Thus, our theoretical account of the vernacular practice must accurately reflect this fact. Fear of contradictions, Priest argues, is unmotivated, for there are ways to get better theoretical accounts of naïve proofs without trivialising the consequence relation. Logics that can deal with the explosive character of a contradiction are a blessing in no disguise for our attempts at making sense of vernacular reasoning. (See however Steinberger 2016 for a different perspective.)

The *logic of paradox* (LP) introduced by Priest in his (1979), marks a significant moment in the struggle to rehabilitate non-trivial inconsistent reasoning. LP is introduced by tweaking the familiar strong Kleene logic in accordance with the belief that there are true contradictions. The middle value $\frac{1}{2}$ is read as a truth-value glut: a sentence that has value $\frac{1}{2}$ is both true and false. Valuations can assign to true sentences either the value 1 (true alone) or the value $\frac{1}{2}$ (both true and false); both these values are designated.

At the propositional level, all this can be made formally precise by means of the following definitions, some of which will be needed later in the paper.

Definition 18.1 If $\mathcal{L}$ is a propositional language, **FOR** $(\mathcal{L})$ will denote the absolutely free algebra of $\mathcal{L}$, with universe $FOR(\mathcal{L})$. A *sequent* on $\mathcal{L}$ is an ordered pair (Γ, Δ)

of finite, possibly empty subsets of $FOR(\mathcal{L})$, written $\Gamma \Rightarrow \Delta$ for ease of notation. $SEQ(\mathcal{L})$ is the set of all sequents on $\mathcal{L}$, and $SEQ^*(\mathcal{L})$ the set of all sequents $\Gamma \Rightarrow \Delta$ on $\mathcal{L}$ such that $\Gamma \cup \Delta \neq \emptyset$.

Hereafter, $\mathcal{L}_0$ will denote the propositional language with connectives $\neg$ (unary) and $\wedge$ (binary). The connectives $\vee, \rightarrow$ can be defined in $\mathcal{L}_0$ as usual: $\alpha \vee \beta := \neg(\neg\alpha \wedge \neg\beta)$ and $\alpha \rightarrow \beta := \neg(\alpha \wedge \neg\beta)$.

Consider the set $\{0, \frac{1}{2}, 1\} \subseteq \mathbb{Q}$ as a linearly ordered set under the usual order of rationals. This set can be viewed as the universe of an algebra **K** of language $\mathcal{L}_0$, where $\neg x = 1 - x$ and $x \wedge y = \min(x, y)$. **K** is usually called the 3-*element Kleene algebra*.

Definition 18.2 A *valuation* is a homomorphism $v : \textbf{FOR}(\mathcal{L}_0) \rightarrow \textbf{K}$. A valuation is *Boolean* if its range is $\{0, 1\}$.

Definition 18.3 If $X \subseteq FOR(\mathcal{L}_0)$ and $\alpha \in FOR(\mathcal{L}_0)$, α is a *consequence* of X in LP, which we write as $X \vDash_{LP} \alpha$, iff, for any valuation v, if $v(\beta) \in \{1, \frac{1}{2}\}$ for all $\beta \in X$, then $v(\alpha) \in \{1, \frac{1}{2}\}$.

With this interpretation of the logical connectives, it is no longer the case that a contradiction entails everything. A counterexample is provided by a valuation v which assigns $\frac{1}{2}$ to a premise α. Its negation $\neg\alpha$ will then be evaluated as $\frac{1}{2}$. So both premisses of *ex contradictione quodlibet* get designated values. But any formula β to which v doesn't assign 1 or $\frac{1}{2}$ will not follow from α and $\neg\alpha$. In theories over LP, contradictions do not explode.

Nevertheless, there is a trade-off here. As one gains the power to limit the deleterious effects of contradictions, one loses the power to do other things. For instance, in LP, one cannot infer β from α and $\alpha \rightarrow \beta$: any valuation which assigns 0 to the consequent of the conditional premiss but assigns $\frac{1}{2}$ to its antecedent is a counterexample. The material conditional, being no longer detachable, cannot plausibly be taken to be a conditional at all. So LP doesn't have a conditional. This is a symptom of a more worrisome phenomenon: while LP has the same theorems as classical logic, it has a weaker consequence relation. Thus, some classical entailments cannot be recovered in LP. This is a problem because, as Priest puts it:

> Intuitionism is a revisionist philosophy. It sees a good part of the reasoning of classical mathematics, particularly that concerning infinite totalities, as quite fallacious. It has therefore wished to debunk it. The programme of paraconsistent logic has never been revisionist in the same sense. By and large, it has accepted that the reasoning of classical mathematics is correct. What it has wished to do is to reject the excrescence of *ex contradictione quodlibet*, which does not appear to be an integral part of classical reasoning, but merely leads to trouble when reasoning ventures into the transconsistent. (Priest 2006, p. 221)

One has to agree that there's no better way of saving classical mathematics than by salvaging classical reasoning, while, nonetheless, avoiding the troubles generated by explosion. The *strict-tolerant approach* promises to do just that.

18.2.2 ST and the Nontransitive Approach to Paradox

The approach comes in two distinct philosophical flavours, depending on whether it is framed within a referentialist account of meaning, truth and consequence, or within an inferentialist perspective. Without much loss of generality, we unfold the following discussion in the latter vein. ST-theorists claim that assertion and denial can be either *strict* or *tolerant*, and that a finite set of formulas Δ validly follows from a finite set of formulas Γ if, whenever there is warrant to strictly assert all the formulas in Γ, there is warrant to tolerantly assert at least one formula in Δ. (A parallel formulation in terms of denial is also possible. See, besides the references in the previous section, Zardini 2013; Wintein 2013; Fjellstad 2016.) Strict and tolerant assertion can be given a model-theoretical treatment in terms of the same three-valued semantics that describes LP: it suffices to identify formulas that are strictly assertable with formulas that receive the value 1 under a certain valuation, and formulas that are tolerantly assertable with formulas that receive a value greater than $\frac{1}{2}$. Unlike consequence in LP, however, the resulting (multiple-conclusion) notion of validity will not be based on the idea of preservation of designated values. In propositional logic, all this can be rendered more precise along the following lines.

Definition 18.4 A valuation v *ST-satisfies* a sequent $\Gamma \Rightarrow \Delta$ on $\mathcal{L}_0$ (in symbols, $v \vDash_{ST} \Gamma \Rightarrow \Delta$) iff either there is $\gamma \in \Gamma$ s.t. $v(\gamma) \in \{0, \frac{1}{2}\}$ or there is $\delta \in \Delta$ s.t. $v(\delta) \in \{1, \frac{1}{2}\}$. A sequent $\Gamma \Rightarrow \Delta$ on $\mathcal{L}_0$ is *ST-valid* (in symbols, $\vDash_{ST} \Gamma \Rightarrow \Delta$) if $v \vDash_{ST} \Gamma \Rightarrow \Delta$ for all valuations v.

As Kazushige Terui correctly pointed out to us, Definition 18.4 is nothing but Jean-Yves Girard's *three-valued interpretation* of the sequent calculus LK for classical propositional logics (See below for a precise formulation of this calculus.) Girard (1987, Chap. 3) proves (although this is independently re-proved in Ripley 2013a) that the ST-valid sequents coincide exactly with the provable sequents of LK, and that this correspondence extends to the first-order case. Thus, Cobreros et al. claim that at the level of propositional or first-order *logic*, ST-validity and classical validity are one and the same thing. However, they also observe that one can develop non-trivial theories of naïve truth and vague predicates as first-order theories over the classical sequent calculus *without Cut*.

In order to understand why this last qualification is crucial, it pays to see how troubles arise in classical logic if a transparent truth predicate is added to it. Such a predicate is governed by the following sequent calculus rules:

$$(T\text{-L}) \; \frac{\alpha, \Gamma \Rightarrow \Delta}{T\langle\alpha\rangle, \Gamma \Rightarrow \Delta} \qquad (T\text{-R}) \; \frac{\Gamma \Rightarrow \Delta, \alpha}{\Gamma \Rightarrow \Delta, T\langle\alpha\rangle}$$

Now suppose that we want to reason classically with a paradoxical sentence such as the Liar (which we will represent symbolically as λ and assume to be *the same as* $\neg T\langle\lambda\rangle$). Then we can build the following sequent calculus derivation:

18 ST, LP and Tolerant Metainferences

$$\frac{\dfrac{\dfrac{\dfrac{T\langle\lambda\rangle \Rightarrow T\langle\lambda\rangle}{\neg T\langle\lambda\rangle \Rightarrow \neg T\langle\lambda\rangle}}{\dfrac{\lambda \Rightarrow \lambda}{\dfrac{T\langle\lambda\rangle \Rightarrow \lambda}{\Rightarrow \neg T\langle\lambda\rangle, \lambda}}}}{\Rightarrow \lambda} \quad \dfrac{\dfrac{\dfrac{\dfrac{T\langle\lambda\rangle \Rightarrow T\langle\lambda\rangle}{\neg T\langle\lambda\rangle \Rightarrow \neg T\langle\lambda\rangle}}{\dfrac{\lambda \Rightarrow \lambda}{\dfrac{\lambda \Rightarrow T\langle\lambda\rangle}{\neg T\langle\lambda\rangle, \lambda \Rightarrow}}}}{\lambda \Rightarrow}}{\Rightarrow}$$

By weakening in the conclusion sequent of this derivation, everything follows from everything. So the classical consequence relation is trivialised.

Up to its final step, this derivation deals only with the rules for negation and truth. The two subderivations are innocuous given the distinction between strict and tolerant assertion. Intuitively, they simply exhibit the peculiarity of paradoxical sentences. These are neither to be strictly asserted, nor to be strictly denied. They are, however, tolerantly assertable as well as tolerantly deniable.

In its final, and crucial, step, the derivation above puts together, via an application of Cut, the derivations of $\Rightarrow \lambda$ and $\lambda \Rightarrow$ respectively. Were Cut not to be available, there would be no way to derive the empty sequent. Certainly, Cut is admissible in the classical sequent calculus, but the cut elimination property for a sequent system is not necessarily inherited by all linguistic or axiomatic *extensions* of the calculus. So, what the above derivation shows is that the classical sequent calculus, formulated without Cut as a primitive rule, no longer has Cut as an admissible rule once it is endowed with the rules for T. This calculus, therefore, does *not* allow for the trivialising derivation of the empty sequent. Failures of Cut allow the ST-theorist to have the cake and eat it too. They can keep to an undiminished classical logical basis and endow it with an unrestricted Tarskian T-schema—or with a tolerance principle for predicate vagueness—without succumbing to the Liar or to the Sorites.

Notice also that ST retains *ex contradictione quodlibet*. A simple application of the left negation rule followed by an application of Weakening yields:

$$\frac{\dfrac{\alpha \Rightarrow \alpha}{\alpha, \neg\alpha \Rightarrow}}{\alpha, \neg\alpha \Rightarrow \beta}$$

for any α, β. So the 'excrescence' accused by Priest is still present on the body of 'naive proof'. (Here, we take the 'naivety' of proof to consist in the unrestricted reliance on the transparent truth predicate T.) Nevertheless, this excrescence is no longer a malign tumour. Or so it would seem.

As with every other attempt to accommodate reasoning with the paradoxes, there is a price for success. The rule of Cut is, by orthodox lights, an internal codification of the transitivity of the consequence relation. If Cut is no longer admissible in the calculus, then in ST we have failures of transitivity. There are two separate issues here. On the one hand, one might entertain worries of various kinds about dispensing with transitivity, despite the efforts made by the defenders of ST to assuage such worries. On the other hand, one might accuse Cobreros et al. of not keeping fully to their promise to stay classical. In this regard, however, they are ready to bite the bullet and admit that their approach is non-classical at the level of *metainferences*. Beside transitivity, many other metainferences that hold for classical logic fail for ST. We believe that this concept deserves further discussion, and it is exactly this problem that we tackle next.

18.3 Metainferences

18.3.1 What Is a Metainference?

Some questions immediately spring to our minds: first, (Q1) what is, exactly, a metainference? Next, (Q2) when can a metainference be considered *valid*, by ST lights? And finally, (Q3) are the calculi provided for ST closed under all such metainferences? Let us try to address these questions—we remark once more—by confining ourselves to the domain of *propositional* logic.

Let us start with Q1. In a number of places, Ripley (2013a, b) suggests that metainferences are properties that a consequence relation may or may not be closed under. In Cobreros et al. (2013, p. 849), the authors explicitly say that metainferences are "principles under which a consequence relation might (or might not) be closed". This is a passable account of metainferences, albeit one that underplays the intricacies behind Definition 18.4. As we have seen, the ST-valid sequents are the valid (or derivable) sequents of the classical sequent calculus. Notice, however, that in a sequent calculus all of the action takes place at the level of sequent-to-sequent rules, whereby from one or more sequents (intuitively understood as 'inferences') we derive more sequents (i.e., more 'inferences'). Which is to say, the action takes place at the level of metainferences. It is therefore only natural to account for metainferences as syntactic objects of the system under consideration. Moreover, as we shall see, doing so will have important benefits when it comes to making sense of the 'classicality' of the ST approach. In particular, having metainferences as *bona fide* syntactic objects, will go a long way towards providing a general enough vantage point from which to assess the consequence relation codified by the calculi under discussion (see below, Sect. 18.4.1).

So how should we understand metainferences *qua* syntactic objects? One option, embraced by Barrio et al. (2015, pp. 556–557), is to take a metainference to be a certain set of uniform substitutions of finite sets of schematic sequents. In effect, this merely precisifies the suggestion made in the preceding informal discussion of sequent-to-sequent rules. On this view, a metainference is a *rule* of a particular kind. Yet we want our conceptual apparatus to go a little bit further and to be able to make some more distinctions. While this 'substitutional' account of metainferences is clearly appropriate as a definition of meta*rule* of inference, it is, however, not the best choice when it comes to characterising metainferences proper. In the present context, it is metainferences, not metarules, that are crucial. When working in specific theories (such as theories of vagueness or truth) we may be interested in whether a certain inference or metainference is valid independently of it instantiating a valid rule or metarule. For instance, we may want to be able to single out (the rule-form of the) Tarski-equivalence for a particular sentence α, say the Liar, independently of it being an *instance* of a rule.

The most straightforward option is to take a metainference to be a nonempty set of sequents, one of which is labelled as its conclusion:

Definition 18.5 Let $\mathcal{L}$ be a propositional language. A *metainference* in $\mathcal{L}$ is an ordered pair (X, S), where $X \subseteq SEQ(\mathcal{L})$ and $S \in SEQ(\mathcal{L})$.

(This is not unheard of in the literature; see, for instance, French 2016, p. 120; Peregrin 2008, p. 271.)

18.3.2 Which Metainferences Are Tolerantly Valid?

We come next to our second question, regarding the validity of the metainferences. At first blush, and in keeping with the intuitive picture discussed above, one might be tempted to take metainferences to be valid whenever they preserve ST-validity. This appears to be the solution endorsed by ST-theorists. For instance, they remark (Cobreros et al. 2013, p. 852) that in the passage from S4 to S5 (where the second logic counts as a strengthening of the first), S5 loses some metainferences that are valid in S4. Thus the metainference that gives us $\Rightarrow \bot$ from $\Rightarrow \Diamond p \rightarrow \Box \Diamond p$ is S4-valid as the premise sequent itself is not valid. By contrast, this metainference is not S5-valid, because the premise sequent of the inference is S5-valid, whereas its conclusion takes value 0 under every valuation of the appropriate sort. Rohan French correctly suggested in correspondence that the fact that this metainference is considered to be S4-valid by the ST-theorists means that they cannot have in mind Barrio et al.'s notion of a metainference here, as this metainference is not valid in that sense. In particular, the substitution instance of the premise sequent where $s(p) = \bot$ is valid in S4 (actually, it is a theorem in every normal modal logic, $\Diamond \bot \rightarrow \Box \Diamond \bot$ being K-equivalent to $\Box \top \vee \Box \Diamond \bot$), while S4 is, of course, consistent, making the conclusion of the rule S4-invalid. In any case, call this the *global* conception of metainferential validity.

Once Definition 18.5 has been granted, this conception of metainferential validity is readily seen to be untenable. Recall that a metainference in $\mathcal{L}$, for us, is just is an ordered pair (X, S), where $X \subseteq SEQ(\mathcal{L})$ and $S \in SEQ(\mathcal{L})$. We do not identify metainferences with sets of such pairs closed under uniform substitution; nor do we assume that sequents are constituted by schematic formulas. If applied to ST, therefore, the validity preservation requirement would vacuously vindicate all sorts of undesirables: every one-premiss metainference $(\{S\}, S')$ such that S is an ST-invalid sequent and S' is arbitrary; atomic instances of fallacies like affirming the consequent; and, what is far worse from an ST viewpoint, atomic instances (with premisses $p \Rightarrow q$, $q \Rightarrow r$ and conclusion $p \Rightarrow r$) of transitivity.

Luckily enough, this is not the only conception of the validity of metainferences available. Indeed, we claim that, for such purposes as that of assessing the classicality of ST, we are better served by a *local* notion of metainferential validity. When assessing metainferential validity locally, one does not consider every valuation. Instead, one restricts one's attention to those valuations that validate the premiss-sequents of a metainference. We thus require a valid metainference to exhibit tolerant preservation over those valuations. The definitions below, which refer to concepts that have been introduced in Definition 18.4, render these two notions of metainferential validity explicit:

Definition 18.6 A metainference (X, S) in $\mathcal{L}_0$ is *globally ST-valid* (in symbols, $X \vDash_{ST}^g S$) if, whenever $\vDash_{ST} S'$ for all $S' \in X$, we have that $\vDash_{ST} S$.

Definition 18.7 A metainference (X, S) in $\mathcal{L}_0$ is *locally ST-valid* (in symbols, $X \vDash_{ST}^l S$) if for every valuation v, if $v \vDash_{ST} S'$ for all $S' \in X$, then $v \vDash_{ST} S$.

Clearly, $X \vDash_{ST}^l S$ implies $X \vDash_{ST}^g S$ but not conversely, the metainferences $(\{\Rightarrow p\}, \Rightarrow q)$ or even $(\{p \Rightarrow q, q \Rightarrow r\}, p \Rightarrow r)$ being counterexamples.

Our third question can now be rephrased as follows. We know that ST, even at the propositional level, lacks some *classically valid* metainferences, like instances of Cut. But is ST sound and complete with respect to all *locally ST-valid* propositional metainferences? In the next part of our paper we answer this question in the negative, and—building on (Girard 1987, Chap. 3)—we single out a calculus for which this completeness theorem holds.

18.4 A System for Tolerantly Valid Metainferences

18.4.1 The Blok-Jónsson Theory of Consequence

The generalisation of Tarski's account of logical consequence, due to Blok and Jónsson (2006), is an appropriate backdrop for our discussion. Their basic idea is that one and the same logic can be presented as a consequence relation over different sets of formal objects (say, formulas, sequents or equations), and that whenever we intend to single out a logic we should abstract away from the specifics of these presentations. According to the standard definition nowadays in use, if we are given a propositional language $\mathcal{L}$ with its attendant set of formulas $FOR(\mathcal{L})$, a (single-conclusion) *consequence relation* over $FOR(\mathcal{L})$ is a relation $\vdash \subseteq \wp(FOR(\mathcal{L})) \times FOR(\mathcal{L})$ obeying the principles of Reflexivity, Monotonicity, and Cut—see Definition 18.3 above for an example. Blok and Jónsson, however, observe that the same three principles govern many other contexts where we are reasoning from given premisses to a certain conclusion, irrespective of whether we are manipulating formulas or other (usually more complicated) objects in the course of our deductions. Consequently, they suggest to replace the set $FOR(\mathcal{L})$ in the Tarskian definition by an *arbitrary set A*:

18 ST, LP and Tolerant Metainferences

Definition 18.8 An *abstract consequence relation (acr)* over the set A is a relation $\vdash \subseteq \wp(A) \times A$ obeying the following conditions for all $a \in A$ and for all $X, Y \subseteq A$:

1. $X \vdash a$ if $a \in X$ *(Reflexivity)*;
2. If $X \vdash a$ and $X \subseteq Y$, then $Y \vdash a$ *(Monotonicity)*;
3. If $Y \vdash a$ and $X \vdash b$ for every $b \in Y$, then $X \vdash a$ *(Cut)*.

Henceforth, we write $X \vdash Y$ to mean that $X \vdash b$ for every $b \in Y$.

Just like derivability relations in Hilbert-style calculi are syntactic paragons of Tarskian consequence relations, *abstract proof systems* (see e.g. Metcalfe et al. 2009, Chap. 3)—axiom systems where elements of an arbitrary set act as stand-ins for formulas in their various capacities as axioms, members of derivations, etc.—serve as a primary source of acr's.

Definition 18.9 Let A be a set. An *inference* for A is an ordered pair (X, a), where $X \subseteq A$ is a finite or empty set (of premisses) and $a \in A$ is the conclusion. Inferences with no premisses are called *axioms*. An *(inference) rule* r for A is a set of inferences for A, which are called *instances* of r.

Usually, axioms and rules are given via schemata that use metavariables to stand for arbitrary members of A or sets based on A.

Definition 18.10 An *abstract proof system (aps)* C is an ordered pair (A, R) consisting of a set A and a set R of rules for A.

Observe that these definitions mesh very well with Definition 18.5 in the foregoing subsection. Any propositional Hilbert-style calculus over the language $\mathcal{L}$ can be viewed as an example of an aps (A, R), with $A = FOR(\mathcal{L})$. Likewise, any propositional sequent calculus (or sequent-style natural deduction calculus) over the language $\mathcal{L}$ can be regarded as an instance of an aps (A, R), with $A = SEQ(\mathcal{L})$. Moreover, finite metainferences, as defined in Definition 18.5, are special cases of inferences as defined in Definition 18.9. They are inferences for $SEQ(\mathcal{L})$, i.e. inferences (in the technical sense) where the objects we manipulate are themselves 'inferences' (in the intuitive sense).

We now have to link up the concepts of aps and acr. In other words, we have to define a suitable relation of derivability in aps's that can function as a primary example of abstract consequence relation.

Definition 18.11 Let $C = (A, R)$ be an abstract proof system. A C-*derivation* d of $a \in A$ from $X \subseteq A$ is a finite tree labelled by members of A such that:

1. a labels the root;
2. For each node labelled a_0, either $a_0 \in X$ or its child nodes are labelled $a_1, \ldots, a_n$ and $(\{a_1, \ldots, a_n\}, a_0)$ is an instance of a rule of C.

If there is a C-derivation d of a from X, then a is said to be C-*derivable* from X, written $d; X \vdash_C a$ or simply $X \vdash_C a$. a is a *theorem* of C iff $\emptyset \vdash_C a$.

It is not hard to prove the following result, which establishes the desired connection:

Theorem 18.12 *Let* $C = (A, R)$ *be an aps. Then* $\vdash_C$ *is an acr over* A.

As any sequent calculus, the calculus LK for classical propositional logic is an instance of an aps; therefore its derivability relation $\vdash_{LK}$ *on sequents* is an instance of an acr.

Definition 18.13 LK is the aps $(SEQ(\mathcal{L}_0), R)$, where R is the following set of rules (with the customary notational conventions):

$$(ID) \ \frac{}{\alpha \Rightarrow \alpha} \qquad (CUT) \ \frac{\Gamma \Rightarrow \Delta, \alpha \quad \alpha, \Pi \Rightarrow \Sigma}{\Gamma, \Pi \Rightarrow \Delta, \Sigma}$$

$$(WL) \ \frac{\Gamma \Rightarrow \Delta}{\alpha, \Gamma \Rightarrow \Delta} \qquad (WR) \ \frac{\Gamma \Rightarrow \Delta}{\Gamma \Rightarrow \Delta, \alpha}$$

$$(\neg L) \ \frac{\Gamma \Rightarrow \Delta, \alpha}{\neg \alpha, \Gamma \Rightarrow \Delta} \qquad (\neg R) \ \frac{\alpha, \Gamma \Rightarrow \Delta}{\Gamma \Rightarrow \Delta, \neg \alpha}$$

$$(\wedge L) \ \frac{\alpha, \beta, \Gamma \Rightarrow \Delta}{\alpha \wedge \beta, \Gamma \Rightarrow \Delta} \qquad (\wedge R) \ \frac{\Gamma \Rightarrow \Delta, \alpha \quad \Gamma \Rightarrow \Delta, \beta}{\Gamma \Rightarrow \Delta, \alpha \wedge \beta}$$

We will also use the following rules for the derived connective $\vee$, obtained by unpacking its definition and applying backwards the rules for $\wedge$ and $\neg$ to the conclusion-sequents below:

$$(\vee L) \ \frac{\alpha, \Gamma \Rightarrow \Delta \quad \beta, \Gamma \Rightarrow \Delta}{\alpha \vee \beta, \Gamma \Rightarrow \Delta} \qquad (\vee R) \ \frac{\Gamma \Rightarrow \Delta, \alpha, \beta}{\Gamma \Rightarrow \Delta, \alpha \vee \beta}$$

Observe that, by Definition 18.11, $\{\Gamma_i \Rightarrow \Delta_i\}_{i \in I} \vdash_{LK} \Gamma \Rightarrow \Delta$ iff $\Gamma \Rightarrow \Delta$ is provable in the calculus obtained from LK by taking the $\Gamma_i \Rightarrow \Delta_i$'s as additional axioms. Also, observe that the aps LK^-, whose set of rules coincides with R except for lacking the rule of Cut, is a *different* aps and gives rise to a *different* acr. For example, $p \Rightarrow q, q \Rightarrow r \vdash_{LK} p \Rightarrow r$, while $p \Rightarrow q, q \Rightarrow r \nvdash_{LK^-} p \Rightarrow r$. In view of Gentzen's Hauptsatz, both acr's have the same *theorems*: $\vdash_{LK} \Gamma \Rightarrow \Delta$ iff $\vdash_{LK^-} \Gamma \Rightarrow \Delta$.

Consider now a different aps:

Definition 18.14 LK^-_{INV} is the aps $(SEQ(\mathcal{L}_0), R')$, where R' is obtained from R by removing Cut and adding the following set of rules:

$$(\neg L^\partial) \ \frac{\Gamma \Rightarrow \Delta, \neg \alpha}{\alpha, \Gamma \Rightarrow \Delta} \qquad (\neg R^\partial) \ \frac{\neg \alpha, \Gamma \Rightarrow \Delta}{\Gamma \Rightarrow \Delta, \alpha}$$

$$(\wedge L^\partial) \ \frac{\alpha \wedge \beta, \Gamma \Rightarrow \Delta}{\alpha, \beta, \Gamma \Rightarrow \Delta} \qquad (\wedge R^\partial) \ \frac{\Gamma \Rightarrow \Delta, \alpha \wedge \beta}{\Gamma \Rightarrow \Delta, \alpha} \qquad \frac{\Gamma \Rightarrow \Delta, \alpha \wedge \beta}{\Gamma \Rightarrow \Delta, \beta}$$

Again, we have ∂-rules for the derived connective $\vee$:

$$(\vee L^\partial) \; \dfrac{\alpha \vee \beta, \Gamma \Rightarrow \Delta}{\alpha, \Gamma \Rightarrow \Delta} \qquad \dfrac{\alpha \vee \beta, \Gamma \Rightarrow \Delta}{\beta, \Gamma \Rightarrow \Delta} \quad (\vee R^\partial) \; \dfrac{\Gamma \Rightarrow \Delta, \alpha \vee \beta}{\Gamma \Rightarrow \Delta, \alpha, \beta}$$

LK^-_{INV} differs from both LK and LK^-. In fact, clearly $\vdash_{\mathrm{LK}^-} \subseteq \vdash_{\mathrm{LK}^-_{INV}} \subseteq \vdash_{\mathrm{LK}}$. However, the preceding inclusions are strict. On the one hand $p \Rightarrow q, q \Rightarrow r \vdash_{\mathrm{LK}} p \Rightarrow r$, while $p \Rightarrow q, q \Rightarrow r \nvdash_{\mathrm{LK}^-_{INV}} p \Rightarrow r$; on the other, $\Rightarrow p \wedge q \vdash_{\mathrm{LK}^-_{INV}} \Rightarrow p$, while $\Rightarrow p \wedge q \nvdash_{\mathrm{LK}^-} \Rightarrow p$. (Observe that we are presupposing the soundness part of Theorem 18.24 below in our non-derivability claim for LK^-_{INV}, while the one for LK^- can be verified by inspection of the rules.) Nevertheless, all these acr's have the same theorems.

Remark that, like every honest-to-goodness acr on sequents, both $\vdash_{\mathrm{LK}^-}$ and $\vdash_{\mathrm{LK}^-_{INV}}$ (as well as, of course, $\vdash_{\mathrm{LK}}$) have unrestricted Cut *on sequents*: If $\{\Gamma_i \Rightarrow \Delta_i\}_{i \in I} \vdash_{\mathrm{LK}^-} \Gamma \Rightarrow \Delta$ and $\{\Pi_j \Rightarrow \Sigma_j\}_{j \in J} \vdash_{\mathrm{LK}^-} \Gamma_i \Rightarrow \Delta_i$ for every $i \in I$, then $\{\Pi_j \Rightarrow \Sigma_j\}_{j \in J} \vdash_{\mathrm{LK}^-} \Gamma \Rightarrow \Delta$. The same holds true for $\vdash_{\mathrm{LK}^-_{INV}}$. However, Cut on formulas:

$$\dfrac{\Gamma \Rightarrow \Delta, \alpha \qquad \alpha, \Pi \Rightarrow \Sigma}{\Gamma, \Pi \Rightarrow \Delta, \Sigma}$$

is an *admissible*, though generally not a *derivable*, rule for both acr's.

18.4.2 A Belnap-Style Extension Lemma

Let us now wrap up what we have seen so far. We have singled out a definition of metainference. Moreover, we argued that a local account of metainferential validity is best suited for ST. We have raised the problem of finding a propositional calculus that is sound and complete with respect to metainferences of this kind. It is not too hard to see, though, that Cobreros et al.'s favourite pick—namely, LK^-—cannot be the right choice. The metainference whose unique premiss is $\Rightarrow p \wedge q$ and whose conclusion is $\Rightarrow p$ is out of its reach. In LK^- there is no way to decompose the conjunctive succedent in its premiss. On the other hand, LK^-_{INV} does not seem to be such a long shot. Indeed, a proof that LK^-_{INV} has the required property is at least *implicit* in Girard (1987, Theorems 3.1.9, 3.2.5). However, since the connections to, and implications for, the present framework are somewhat hard to extract from Girard's original proof, we believe that it is not pointless to include here a detailed proof of this result—actually, the one we had devised before learning from Terui that Girard's result was there at all.

In order to get into gear, we need some additional notions, which we collect in the subsequent definitions.

Definition 18.15 A *theory* is a set $T \subseteq SEQ(\mathcal{L}_0)$ that contains all the provable sequents of LK_{INV}^- and is closed w.r.t. to all the rules of LK_{INV}^-. A theory T is:

- *S-consistent*, if $S \notin T$;
- *prime*, if whenever $\Gamma \Rightarrow \Delta \in T \cap SEQ^*(\mathcal{L}_0)$, then either there is $\gamma \in \Gamma$ s.t. $\gamma \Rightarrow \in T$, or there is $\delta \in \Delta$ s.t. $\Rightarrow \delta \in T$;
- *complete*, if for all formulas α, either $\Rightarrow \alpha \in T$ or $\alpha \Rightarrow \in T$;
- *consistent*, if for all formulas α, not both $\Rightarrow \alpha \in T$ and $\alpha \Rightarrow \in T$.

Observe that any prime theory T is complete, because $\alpha \Rightarrow \alpha \in T$ for any $\alpha \in FOR(\mathcal{L}_0)$. Likewise, T is a theory iff it is deductively closed, i.e. $T \vdash_{LK_{INV}^-} S$ implies $S \in T$.

Definition 18.16 If $\Gamma \Rightarrow \Delta \in SEQ^*(\mathcal{L}_0)$, its *formula translation* $ft(\Gamma \Rightarrow \Delta)$ is defined as follows:

$$ft(\Gamma \Rightarrow \Delta) = \begin{cases} \neg\gamma_1 \vee \ldots \vee \neg\gamma_n \vee \delta_1 \ldots \vee \delta_m, \text{ if } \Gamma = \{\gamma_1, \ldots, \gamma_n\}, \Delta = \{\delta_1, \ldots, \delta_m\}; \\ \neg\gamma_1 \vee \ldots \vee \neg\gamma_n, \text{ if } \Gamma = \{\gamma_1, \ldots, \gamma_n\}, \Delta = \emptyset; \\ \delta_1 \vee \ldots \vee \delta_m, \text{ if } \Gamma = \emptyset, \Delta = \{\delta_1, \ldots, \delta_m\}. \end{cases}$$

The next technical lemma will be needed in the sequel.

Lemma 18.17 *(i) For all sequents $S_1, \ldots, S_n$ in $SEQ^*(\mathcal{L}_0)$, $\{S_1, \ldots, S_n\} \dashv\vdash_{LK_{INV}^-} \Rightarrow ft(S_1) \wedge \ldots \wedge ft(S_n)$;*
(ii) For all $X, X' \subseteq SEQ(\mathcal{L}_0)$ and for all $S, S' \in SEQ(\mathcal{L}_0)$, if $X \vdash_{LK_{INV}^-} S$ and $X' \cup \{S\} \vdash_{LK_{INV}^-} S'$, then $X \cup X' \vdash_{LK_{INV}^-} S'$;
(iii) For all $\alpha, \beta, \gamma \in FOR(\mathcal{L}_0)$, $\Rightarrow \alpha \wedge (\beta \vee \gamma) \vdash_{LK_{INV}^-} \Rightarrow \beta \vee (\alpha \wedge \gamma)$;
(iv) For all $\Gamma \Rightarrow \Delta, \Pi \Rightarrow \Lambda, \Theta \Rightarrow \Sigma \in SEQ(\mathcal{L}_0)$, $\Gamma \Rightarrow \Delta \vdash_{LK_{INV}^-} \Pi \Rightarrow \Lambda$ implies $\Gamma, \Theta \Rightarrow \Delta, \Sigma \vdash_{LK_{INV}^-} \Pi, \Theta \Rightarrow \Lambda, \Sigma$;
(v) For all $\alpha \in FOR(\mathcal{L}_0)$ and for all $\Gamma \Rightarrow \Delta$ in $SEQ^*(\mathcal{L}_0)$, $\Pi \Rightarrow \Lambda \in SEQ(\mathcal{L}_0)$, $\Rightarrow \alpha \vdash_{LK_{INV}^-} \Pi \Rightarrow \Lambda$ implies $\Rightarrow \alpha \vee ft(\Gamma \Rightarrow \Delta) \vdash_{LK_{INV}^-} \Gamma, \Pi \Rightarrow \Delta, \Lambda$.

Proof We confine ourselves to proving (i) and (v), leaving the remaining items as an exercise.

(i) We just spell out the case in which for all $i \leq n$, $S_i = \gamma_1^i, \ldots, \gamma_m^i \Rightarrow \delta_1^i, \ldots, \delta_r^i$. Add to LK_{INV}^- each S_i as a new axiom. Fix an arbitrary $i \leq n$. Applying several times $(\neg R)$ to S_i we obtain $\Rightarrow \neg\gamma_1^i, \ldots, \neg\gamma_m^i, \delta_1^i, \ldots, \delta_r^i$. From this, by various applications of $(\vee R)$ we get $\Rightarrow ft(S_i)$. Since i was arbitrary, we have obtained all of $\Rightarrow ft(S_1), \ldots, \Rightarrow ft(S_n)$, whence applications of $(\wedge R)$ deliver $\Rightarrow ft(S_1) \wedge \ldots \wedge ft(S_n)$. Conversely, add $\Rightarrow ft(S_1) \wedge \cdots \wedge ft(S_n)$ to LK_{INV}^- as a new axiom. By $(\wedge R^\partial)$ we derive $\Rightarrow ft(S_i)$ for all $i \leq n$. Several applications of $(\vee R^\partial)$ and $(\neg L^\partial)$ deliver each S_i for all $i \leq n$.

(v) Suppose $\Rightarrow \alpha \vdash_{LK_{INV}^-} \Pi \Rightarrow \Lambda$ and add $\Rightarrow \alpha \vee ft(\Gamma \Rightarrow \Delta)$ to LK_{INV}^- as a new axiom. By $(\vee R^\partial)$ we derive $\Rightarrow \alpha, ft(\Gamma \Rightarrow \Delta)$, whence by repeated applications of $(\vee R^\partial)$ and $(\neg L^\partial)$ we obtain $\Gamma \Rightarrow \alpha, \Delta$. Since $\Rightarrow \alpha \vdash_{LK_{INV}^-} \Pi \Rightarrow \Lambda$, by (iv) we conclude that $\Gamma \Rightarrow \alpha, \Delta \vdash_{LK_{INV}^-} \Gamma, \Pi \Rightarrow \Delta, \Lambda$. Applying (ii) to this yields our conclusion. ∎

We are now going to parlay an adaptation of Belnap's Prime Extension Lemma (Restall 2000, p. 92 ff.), a Lindenbaum-style construction used for completeness theorems in relevant logics, into a tool for guaranteeing that any S-consistent theory can be extended to a prime (and still S-consistent) one.

Definition 18.18 Let $X, Y \subseteq SEQ(\mathcal{L}_0)$. The ordered pair (X, Y) is a *d-pair* if there are no finite subsets $X' \subseteq X \cap SEQ^*(\mathcal{L}_0)$ and $\{\Gamma_i \Rightarrow \Delta_i\}_{i \leq n} \subseteq Y \cap SEQ^*(\mathcal{L}_0)$ s.t. $X' \vdash_{LK^-_{INV}} \Gamma_1, \ldots, \Gamma_n \Rightarrow \Delta_1, \ldots, \Delta_n$. A d-pair is *full* if $X \cup Y = SEQ(\mathcal{L}_0)$.

Observe that in any d-pair (X, Y), X and Y are disjoint; moreover, in any d-pair (X, Y) with $Y \neq \emptyset$, the empty sequent $\Rightarrow$ cannot belong to X (and therefore, if the d-pair is full, it must belong to Y). In fact, suppose otherwise; then, if $S \in Y$, it would follow that $\{\Rightarrow\} \vdash_{LK^-_{INV}} S$ by weakening. This contradicts the fact that (X, Y) is a d-pair.

Lemma 18.19 *If T is a theory and $S \notin T$, then $(T, \{S\})$ is a d-pair.*

Proof If $S \notin T$, then there is no finite subset $X \subseteq T \cap SEQ^*(\mathcal{L}_0)$ s.t. $X \vdash_{LK^-_{INV}} S$, from which the conclusion follows. ∎

Lemma 18.20 *If (X, Y) is a full d-pair, then X is a prime theory.*

Proof We first show that X is a theory. This is obvious if Y is empty. If Y is nonempty, then $X \subseteq SEQ^*(\mathcal{L}_0)$ by our previous remark. Suppose that $X \vdash_{LK^-_{INV}} S$ and that $S \notin X$. Then $X' \vdash_{LK^-_{INV}} S$, for some finite subset X' of X. Also, $S \in Y$, because (X, Y) is full. This contradicts the fact that (X, Y) is a d-pair.

To establish primeness, suppose that $\Gamma \Rightarrow \Delta \in X \cap SEQ^*(\mathcal{L}_0)$, that for no $\gamma \in \Gamma$ we have that $\gamma \Rightarrow \in X$, and that for no $\delta \in \Delta$ is it the case that $\Rightarrow \delta \in X$. Then $\gamma \Rightarrow \in Y$ for all $\gamma \in \Gamma$ and $\Rightarrow \delta \in Y$ for all $\delta \in \Delta$. However, since $\{\Gamma \Rightarrow \Delta\} \vdash_{LK^-_{INV}} \Gamma \Rightarrow \Delta$, this contradicts the fact that (X, Y) is a d-pair. ∎

The crucial property of d-pairs is their *extensibility*: the result of appending an arbitrary nonempty sequent to either their first or their second component is still a d-pair. This much is vouchsafed by the next proposition.

Proposition 18.21 *Let (X, Y) be a d-pair, and let $S \in SEQ^*(\mathcal{L}_0)$. Then at least one of $(X \cup \{S\}, Y)$, $(X, Y \cup \{S\})$ is a d-pair.*

Proof Suppose for contradiction that neither $(X \cup \{S\}, Y)$ nor $(X, Y \cup \{S\})$ is a d-pair, for $S = \Gamma \Rightarrow \Delta$. Then there exist finite families $\{\Pi_i \Rightarrow \Lambda_i\}_{i \leq n} \subseteq X \cap SEQ^*(\mathcal{L}_0)$ and $\{\Gamma_j \Rightarrow \Delta_j\}_{j \leq m} \subseteq Y \cap SEQ^*(\mathcal{L}_0)$ such that

$$\{\Pi_i \Rightarrow \Lambda_i\}_{i \leq n} \vdash_{LK^-_{INV}} \Gamma, \Gamma_1, \ldots, \Gamma_m \Rightarrow \Delta, \Delta_1, \ldots, \Delta_m.$$

Moreover, there exist finite families $\{\Pi'_k \Rightarrow \Lambda'_k\}_{k \leq p} \subseteq X \cap SEQ^*(\mathcal{L}_0)$ and $\{\Gamma'_l \Rightarrow \Delta'_l\}_{l \leq r} \subseteq Y \cap SEQ^*(\mathcal{L}_0)$ such that

$$\{\Pi'_k \Rightarrow \Lambda'_k\}_{k \leq p}, \Gamma \Rightarrow \Delta \vdash_{LK^-_{INV}} \Gamma'_1, \ldots, \Gamma'_r \Rightarrow \Delta'_1, \ldots, \Delta'_r.$$

By the former assumption and the properties of sequent theories,

$$\{\Pi_i \Rightarrow \Lambda_i\}_{i \leq n}, \{\Pi'_k \Rightarrow \Lambda'_k\}_{k \leq p} \vdash_{LK^-_{INV}} \Gamma, \Gamma_1, \ldots, \Gamma_m \Rightarrow \Delta, \Delta_1, \ldots, \Delta_m,$$
$$\{\Pi_i \Rightarrow \Lambda_i\}_{i \leq n}, \{\Pi'_k \Rightarrow \Lambda'_k\}_{k \leq p} \vdash_{LK^-_{INV}} \Pi'_k \Rightarrow \Lambda'_k \text{ for all } k \leq p,$$

whence by Lemma 18.17.(i)–(ii),

$$\{\Pi_i \Rightarrow \Lambda_i\}_{i \leq n}, \{\Pi'_k \Rightarrow \Lambda'_k\}_{k \leq p} \vdash_{LK^-_{INV}} \Rightarrow \mathit{ft}(\Gamma, \Gamma_1, \ldots, \Gamma_m \Rightarrow \Delta, \Delta_1, \ldots, \Delta_m) \wedge$$
$$\mathit{ft}(\Pi'_1 \Rightarrow \Lambda'_1) \wedge \ldots \wedge \mathit{ft}(\Pi'_p \Rightarrow \Lambda'_p).$$

Let $\alpha = \mathit{ft}(\Pi'_1 \Rightarrow \Lambda'_1) \wedge \ldots \wedge \mathit{ft}(\Pi'_p \Rightarrow \Lambda'_p)$. By Lemma 18.17.(ii)–(iii),

$$\{\Pi_i \Rightarrow \Lambda_i\}_{i \leq n}, \{\Pi'_k \Rightarrow \Lambda'_k\}_{k \leq p} \vdash_{LK^-_{INV}} \Rightarrow \mathit{ft}(\Gamma_1, \ldots, \Gamma_m \Rightarrow \Delta_1, \ldots, \Delta_m) \vee (\alpha \wedge \mathit{ft}(\Gamma \Rightarrow \Delta)).$$

However, applying once more Lemma 18.17.(i),

$$\Rightarrow \alpha \wedge \mathit{ft}(\Gamma \Rightarrow \Delta) \dashv\vdash_{LK^-_{INV}} \{\Pi'_k \Rightarrow \Lambda'_k\}_{k \leq p}, \Gamma \Rightarrow \Delta,$$

whereby, by Lemma 18.17.(ii),

$$\Rightarrow \alpha \wedge \mathit{ft}(\Gamma \Rightarrow \Delta) \vdash_{LK^-_{INV}} \Gamma'_1, \ldots, \Gamma'_r \Rightarrow \Delta'_1, \ldots, \Delta'_r.$$

In virtue of Lemma 18.17.(v),

$$\Rightarrow \mathit{ft}(\Gamma_1, \ldots, \Gamma_m \Rightarrow \Delta_1, \ldots, \Delta_m) \vee (\alpha \wedge \mathit{ft}(\Gamma \Rightarrow \Delta)) \vdash_{LK^-_{INV}} \Gamma_1, \ldots, \Gamma_m, \Gamma'_1, \ldots, \Gamma'_r \Rightarrow \Delta_1, \ldots, \Delta_m, \Delta'_1, \ldots, \Delta'_r.$$

From this, a final application of Lemma 18.17.(ii) yields

$$\{\Pi_i \Rightarrow \Lambda_i\}_{i \leq n}, \{\Pi'_k \Rightarrow \Lambda'_k\}_{k \leq p} \vdash_{LK^-_{INV}} \Gamma_1, \ldots, \Gamma_m, \Gamma'_1, \ldots, \Gamma'_r \Rightarrow \Delta_1, \ldots, \Delta_m, \Delta'_1, \ldots, \Delta'_r,$$

contradicting the fact that (X, Y) is a d-pair. ∎

We are now ready to embark on a Lindenbaum-type completion that extends any d-pair with nonempty second component to a full d-pair.

Theorem 18.22 *Any d-pair (X, Y) with $Y \neq \emptyset$ can be extended to a full d-pair.*

Proof Take any enumeration $S_1, S_2, \ldots$ of $SEQ(\mathcal{L}_0)$. Define a sequence of d-pairs as follows:

$$(X_0, Y_0) = (X, Y);$$
$$(X_{n+1}, Y_{n+1}) = \begin{cases} (X_n \cup \{S_n\}, Y_n), & \text{if this is a d-pair;} \\ (X_n, Y_n \cup \{S_n\}), & \text{otherwise.} \end{cases}$$

This sequence is well-defined. In fact, for all $i \in N$, if S_i is the empty sequent, then $(X_i, Y_i \cup \{S_i\})$ will be a d-pair whenever (X_i, Y_i) is one; otherwise, by Proposition 18.21, either $(X_i \cup \{S_i\}, Y_i)$ or $(X_i, Y_i \cup \{S_i\})$ will be a d-pair. The pair $(\bigcup X_k, \bigcup Y_k)$ is a d-pair, for suppose otherwise. Then there exist finite families $\{\Pi_i \Rightarrow \Lambda_i\}_{i \leq n} \subseteq \bigcup X_k \cap SEQ^*(\mathcal{L}_0)$ and $\{\Gamma_j \Rightarrow \Delta_j\}_{j \leq m} \subseteq \bigcup Y_k \cap SEQ^*(\mathcal{L}_0)$ such that

$$\{\Pi_i \Rightarrow \Lambda_i\}_{i \leq n} \vdash_{LK_{INV}^-} \Gamma_1, \ldots, \Gamma_m \Rightarrow \Delta_1, \ldots, \Delta_m.$$

Let p be the smallest index s.t. $\{\Pi_i \Rightarrow \Lambda_i\}_{i \leq n} \subseteq X_p$ and $\{\Gamma_j \Rightarrow \Delta_j\}_{j \leq m} \subseteq Y_p$. Then (X_p, Y_p) would not be a d-pair, a contradiction. Also, $(\bigcup X_k, \bigcup Y_k)$ is a full d-pair by construction, which establishes our claim. ∎

Corollary 18.23 *Let $S \in SEQ(\mathcal{L}_0)$. If T is an S-consistent theory, then there exists a prime and S-consistent theory T' such that $T \subseteq T'$.*

Proof By Lemma 18.19, $(T, \{S\})$ is a d-pair which can, in virtue of Theorem 18.22, be extended to a full d-pair (X, Y). By Lemma 18.20, X is the required prime and S-consistent theory T'. ∎

18.4.3 The Completeness Result

We now have all we need for the following completeness theorem: a finite metainference (X, S) in $\mathcal{L}_0$ is locally ST-valid if and only if S is LK_{INV}^--derivable from X.

Theorem 18.24 *Let $\{\Pi_i \Rightarrow \Lambda_i\}_{i \leq n} \cup \{\Gamma \Rightarrow \Delta\} \subseteq SEQ(\mathcal{L}_0)$. Then*

$$\{\Pi_i \Rightarrow \Lambda_i\}_{i \leq n} \vdash_{LK_{INV}^-} \Gamma \Rightarrow \Delta \text{ iff } \{\Pi_i \Rightarrow \Lambda_i\}_{i \leq n} \vDash_{ST}^l \Gamma \Rightarrow \Delta.$$

Proof From left to right, our theorem is proved by induction on the length of the LK_{INV}^--derivation of $\Gamma \Rightarrow \Delta$ from $\{\Pi_i \Rightarrow \Lambda_i\}_{i \leq n}$ (safely left to the reader).

From right to left, assume contrapositively that $\{\Pi_i \Rightarrow \Lambda_i\}_{i \leq n} \nvdash_{LK_{INV}^-} \Gamma \Rightarrow \Delta$. This means that $\Gamma \Rightarrow \Delta \notin T$, where T is the theory generated by $\{\Pi_i \Rightarrow \Lambda_i\}_{i \leq n}$. By Corollary 18.23, T can be extended to a prime theory T' such that $\Gamma \Rightarrow \Delta \notin T'$. We now define a mapping $v : FOR(\mathcal{L}_0) \to \{0, \frac{1}{2}, 1\}$ as follows: for any $\alpha \in FOR(\mathcal{L}_0)$,

$$v(\alpha) = \begin{cases} 1, & \text{if } \Rightarrow \alpha \in T', \alpha \Rightarrow \notin T'; \\ 0, & \text{if } \Rightarrow \alpha \notin T', \alpha \Rightarrow \in T'; \\ \frac{1}{2}, & \text{if } \Rightarrow \alpha \in T', \alpha \Rightarrow \in T'. \end{cases}$$

Since, as we already observed, any prime theory is complete, the mapping v is well-defined. Next, one checks that it is a valuation:

1. $v(\neg \alpha) = \neg v(\alpha)$;
2. $v(\alpha \wedge \beta) = v(\alpha) \wedge v(\beta)$.

(1) If $v(\neg \alpha) = 1$, then $T' \vdash_{LK_{INV}^-} \Rightarrow \neg \alpha$, whence by $(\neg L^\partial)$, $T' \vdash_{LK_{INV}^-} \alpha \Rightarrow$. If it were the case that $T' \vdash_{LK_{INV}^-} \Rightarrow \alpha$, by $(\neg L)$ we would have $T' \vdash_{LK_{INV}^-} \neg \alpha \Rightarrow$, which contradicts the fact that $\neg \alpha \Rightarrow \notin T'$. So $\alpha \Rightarrow \in T'$, $\Rightarrow \alpha \notin T'$, and this means $v(\alpha) = 0$, whence $\neg v(\alpha) = 1$. Similarly, if $v(\neg \alpha) = 0$, then $\neg v(\alpha) = 0$. Finally, if $v(\neg \alpha) = \frac{1}{2}$, then $T' \vdash_{LK_{INV}^-} \Rightarrow \neg \alpha, \neg \alpha \Rightarrow$, whereby we get $T' \vdash_{LK_{INV}^-} \Rightarrow \alpha, \alpha \Rightarrow$ by applying $(\neg R^\partial)$, $(\neg L^\partial)$. Thus, $v(\alpha) = \frac{1}{2}$ and $\neg v(\alpha) = \frac{1}{2}$.

(2) Suppose $v(\alpha \wedge \beta) = 1$. So $T' \vdash_{LK_{INV}^-} \Rightarrow \alpha \wedge \beta, T' \nvdash_{LK_{INV}^-} \alpha \wedge \beta \Rightarrow$. Applying $(\wedge R^\partial)$ twice, we obtain $T' \vdash_{LK_{INV}^-} \Rightarrow \alpha, \Rightarrow \beta$. If it were the case that $T' \vdash_{LK_{INV}^-} \alpha \Rightarrow$, then by (WL) $T' \vdash_{LK_{INV}^-} \alpha, \beta \Rightarrow$ and by $(\wedge L)$, $T' \vdash_{LK_{INV}^-} \alpha \wedge \beta \Rightarrow$, a contradiction. In sum, $\Rightarrow \alpha \in T'$, $\alpha \Rightarrow \notin T'$, which means $v(\alpha) = 1$. Similarly, $v(\beta) = 1$ and thus $v(\alpha) \wedge v(\beta) = 1$. Suppose now $v(\alpha \wedge \beta) = 0$. So $T' \vdash_{LK_{INV}^-} \alpha \wedge \beta \Rightarrow$, $T' \nvdash_{LK_{INV}^-} \Rightarrow \alpha \wedge \beta$. By $(\wedge L^\partial)$, $T' \vdash_{LK_{INV}^-} \alpha, \beta \Rightarrow$. If we had $T' \vdash_{LK_{INV}^-} \Rightarrow \alpha, \Rightarrow \beta$, then by $(\wedge R)$ $T' \vdash_{LK_{INV}^-} \Rightarrow \alpha \wedge \beta$, a contradiction. Thus w.l.g. $T' \nvdash_{LK_{INV}^-} \Rightarrow \alpha$. Since T' is a prime theory, $T' \vdash_{LK_{INV}^-} \alpha \Rightarrow$ and thus $v(\alpha) = 0 = v(\alpha) \wedge v(\beta)$. Finally, assume $v(\alpha \wedge \beta) = \frac{1}{2}$. So $T' \vdash_{LK_{INV}^-} \alpha \wedge \beta \Rightarrow$, $\Rightarrow \alpha \wedge \beta$. By $(\wedge L^\partial)$, $(\wedge R^\partial)$,

$$T' \vdash_{LK_{INV}^-} \alpha, \beta \Rightarrow, \Rightarrow \alpha, \Rightarrow \beta.$$

Thus $v(\alpha), v(\beta) \in \{1, \frac{1}{2}\}$. Since T' is prime, $T' \vdash_{LK_{INV}^-} \alpha, \beta \Rightarrow$ implies that either $T' \vdash_{LK_{INV}^-} \alpha \Rightarrow$ or $T' \vdash_{LK_{INV}^-} \beta \Rightarrow$. So w.l.g. $v(\alpha) = \frac{1}{2} = v(\alpha) \wedge v(\beta)$.

To establish the theorem, it remains to prove that $v \vDash_{ST} \Pi_i \Rightarrow \Lambda_i$ for all $i \leq n$, and that $v \nvDash_{ST} \Gamma \Rightarrow \Delta$. Suppose, *ex absurdo*, that for some $j \leq n$, $v \nvDash_{ST} \Pi_j \Rightarrow \Lambda_j$. Then $v(\pi) = 1$ for all $\pi \in \Pi_j$, and $v(\sigma) = 0$ for all $\sigma \in \Lambda_j$. In particular, $T' \nvdash_{LK_{INV}^-} \pi \Rightarrow$ for all $\pi \in \Pi_j$, and $T' \nvdash_{LK_{INV}^-} \Rightarrow \sigma$ for all $\sigma \in \Lambda_j$. However, this contradicts the fact that T' is a prime theory that contains $\Pi_j \Rightarrow \Lambda_j$. Finally, we argue by contradiction again and assume that $v \vDash_{ST} \Gamma \Rightarrow \Delta$. Thus, either there is $\gamma \in \Gamma$ such that $v(\gamma) \in \{0, \frac{1}{2}\}$ or there is $\delta \in \Delta$ such that $v(\delta) \in \{1, \frac{1}{2}\}$. This means that either there is $\gamma \in \Gamma$ such that $T' \vdash_{LK_{INV}^-} \gamma \Rightarrow$ or there is $\delta \in \Delta$ such that $T' \vdash_{LK_{INV}^-} \Rightarrow \delta$. In both cases, by weakening, we obtain $T' \vdash_{LK_{INV}^-} \Gamma \Rightarrow \Delta$, a contradiction. Thus, we have shown that $\{\Pi_i \Rightarrow \Lambda_i\}_{i \leq n} \nvDash_{ST}^l \Gamma \Rightarrow \Delta$, as was to be proved. ∎

18.5 ST and Dialetheism

It is now the time to come back to the questions we have raised in our introduction: what is a propositional logic? Can ST, at the propositional level, be rightfully identified with classical logic? Or is it closer to LP? In this final section, we consider in turn two possible characterisations of propositional logic and argue that, on the former, ST is nothing but LP in disguise, while on the latter, it is, at the very least, *different* from classical logic—and there's a sharp tang of dialetheism about it, at that.

18.5.1 ST Is LP

If a logic, in essence, has to be a consequence relation, then Sect. 18.4.1 would seem to provide us with a *prima facie* promising precisification of this idea. Moreover, this could bypass some difficulties raised by an uncritical adoption of the Tarskian definition. From this perspective, a logic *is* simply an acr. This is all well and good, were it not for the fact that Definition 18.8 makes too many unnecessary distinctions. Let us be reminded that one and the same logic, classical logic, can be introduced in many different ways. For instance, it can be introduced semantically via the equational consequence relation of the 2-element Boolean algebra. Or it can be introduced proof-theoretically via either a Hilbert-style calculus or the sequent calculus LK. However, the resulting acr's are, formally speaking, *different*—if for no other reason, at least because they are over different sets (of equations, formulas, sequents, respectively).

A way out of this problem is suggested by Blok and Jónsson, via their definition of *equivalence* for acr's. (As a matter of fact, they reserve the term 'equivalence' for a stronger concept, whose definition presupposes their (very abstract) account of substitution-invariant acr's. They use the term *similarity* for the relation defined below. Since this aspect of their work is irrelevant for our present purposes, we disregard this distinction.) Blok and Jónsson intend to specify formal conditions under which we are guaranteed that two acr's, possibly defined over different sets, are essentially identical.

Definition 18.25 Let $\vdash_1, \vdash_2$ be acr's over the sets A_1, A_2, respectively. We say that $\vdash_1$ and $\vdash_2$ are *equivalent* iff there exist a mapping $\tau : A_1 \to \wp(A_2)$ and a mapping $\rho : A_2 \to \wp(A_1)$ such that the following conditions hold for every $X \cup \{a\} \subseteq A_1$ and for every $b \in A_2$:

S1 $X \vdash_1 a$ iff $\tau(X) \vdash_2 \tau(a)$;
S2 $b \dashv\vdash_2 \tau(\rho(b))$.

(In S1, $\tau(X)$ is the union of all sets $\tau(b)$, for $b \in X$.) The maps τ and ρ are called *transformers*. In plain words, S1 expresses the fact that $\vdash_1$ is faithfully translatable via the mapping τ into $\vdash_2$, while S2 (given S1) amounts to the fact that the two mappings ρ and τ are mutually inverse. Readers familiar with abstract algebraic logic will have noticed that this definition resembles the definition of *algebraisability* (Blok and Pigozzi 1989; Font 2016). In fact, the algebraisability relation can be viewed as a particular case of equivalence (with substitution-invariant transformers) between a consequence relation on formulas and the equational consequence relation of some class of algebras.

Equivalence is a reflexive, symmetric and transitive relation. The idea is now to identify propositional logics with equivalence classes of acr's modulo this relation (perhaps with some stipulations needed to ensure that the class with respect to which we mod out is actually a set). At least in the case of classical logic, this delivers the expected results. In fact, let us define:

Definition 18.26 If $X \subseteq FOR(\mathcal{L}_0)$ and $\alpha \in FOR(\mathcal{L}_0)$, we say that α is a *consequence* of X in CL, and write $X \vDash_{CL} \alpha$, iff, for any Boolean valuation v, if $v(\beta) = 1$ for all $\beta \in X$, then $v(\alpha) = 1$.

It is clear from Definition 18.26 that $\vDash_{CL}$ is an acr over $FOR(\mathcal{L}_0)$ (and that so is the similarly defined $\vDash_{LP}$). It is not hard to show that:

Theorem 18.27 *The acr $\vdash_{LK}$ is equivalent to the acr $\vDash_{CL}$ via the mutually inverse transformers (for simplicity given here only for sequents whose antecedent and succedent are both nonempty):*

$$\tau(\alpha) = \{\Rightarrow \alpha\};$$
$$\rho(\gamma_1, \ldots, \gamma_n \Rightarrow \delta_1, \ldots, \delta_m) = ft\{\gamma_1, \ldots, \gamma_n \Rightarrow \delta_1, \ldots, \delta_m\}.$$

Since all the other above-mentioned acr's for classical propositional logic (and many more) fall in the same equivalence class, it would seem as though it does no harm to identify classical logic with this class. This squares nicely with Blok and Jónsson's insight that the formal objects we deal with matter much less than the *reasoning principles* applied when manipulating them.

If so, which equivalence class can be appropriately identified with ST? Theorem 18.24 points to a precise answer: since the acr $\vdash_{LK^-_{INV}}$ captures exactly the locally ST-valid metainferences, it is the equivalence class of this relation that lays the best claims to that role. And here, the ST-theorist faces a major threat to the alleged classicality of her system. Various collapse theorems of ST onto LP have been offered in the literature, most notably by Barrio and collaborators and by Pynko (Barrio et al. 2015; Pynko 2010). In particular, Pynko's result can be given an especially perspicuous form if we recast it in the Blok-Jónsson framework:

Theorem 18.28 *The acr $\vdash_{LK^-_{INV}}$ is equivalent to the acr $\vDash_{LP}$ via the same transformers as in Theorem 18.27.*

In sum, we have constructed the following argument:

1. A propositional logic is an equivalence class of a given acr modulo the relation introduced in Definition 18.25. More precisely, the logic defined by a given acr $\vdash$ is the class of all acr's that are equivalent to $\vdash$ according to Definition 18.25.
2. By Theorem 18.24, $\vdash_{LK^-_{INV}}$ is sound and complete with respect to locally ST-valid metainferences: in the case of finitely many premises, the relations $\vdash_{LK^-_{INV}}$ and $\vDash^l_{ST}$ coincide.

3. Therefore ST, which it is plausible to consider as the logic defined by $\vDash_{ST}^l$, can be identified with the equivalence class of $\vdash_{LK_{INV}^-}$.
4. By Theorem 18.27, $\vdash_{LK_{INV}^-}$ is equivalent to $\vDash_{LP}$.
5. LP is the logic defined by $\vDash_{LP}$.
6. Therefore, ST is the same logic as LP.

There is not much leeway for a response to the above argument on the part of the ST-theorist. Sure, she could go as far as to deny 5, perhaps by claiming, e.g., that $\vDash_{LP}$ does not do full justice to LP, which is better served by a multiple-conclusion relation. Alternatively, she could challenge 3 by casting doubts on the local notion of metainferential validity. Still, the safest option would seem to be that of denying 1. In fact, Definition 18.25 leaves much to be desired from the ST viewpoint, at least if the latter is paired with a *bilateralist* account of meaning and the validity of inferences, as in, e.g., Ripley (2013a). Consider the transformer τ in Theorem 18.27. Such a mapping is supposed to be saying which sequents translate formulas of $\mathcal{L}_0$. However, it strikes the eye that it does so by completely ignoring the left-hand side of such sequents. The partisan of ST could argue that τ is a 'bad' transformer in that it is unbalanced on the side of assertions. A 'good' transformer should respect the fundamental symmetry between assertions and denials, arising from the fact that neither is logically prior to the other. It could be contended that equivalence *à la* Blok and Jónsson is a necessary but not sufficient condition for identifying acr's, and that further constraints, designed to single out acceptable transformers, are needed. If acr's draw unnecessary distinctions, perhaps their equivalence classes have the opposite drawback of bundling up logics that must be kept apart.

18.5.2 ST Is Not CL

Let us concede that statement 1 above is dubious, at least for the sake of argument. Now the ball is in the bilateralist's court. She has to come up with a positive suggestion for identifying logics. She will probably agree that identifying logics with classes of theorems is too uncouth a criterion: presumably, she will be reluctant to conflate LP and classical logic. Thus the appropriate answer, whatever it might be, will have to be given within the logics-as-consequence-relations perspective.

Being an inferentialist, however, she will not be insensitive to the idea that consequence-relations are generated by means of inferences—some of which are taken to be determinative of the meaning of the logical operations of which they hold. (This is not the place to dwell on the details of this; for a monographic characterisation of this position see Francez 2015; for a bilateralist version of it see Restall 2005.) In turn, a specification of a consequence relation determines two classes of inferential roles that can be codified in it. There are *good* inferential roles. For instance, the inferential role specified by the rules for conjunction is *good* relative to the classical consequence relation. The same relation does not classify as *good* the inferential role determined by the rules:

$$\frac{\Gamma \Rightarrow \Delta, \alpha}{\Gamma \Rightarrow \Delta, \alpha \# \beta} \qquad \frac{\beta, \Gamma \Rightarrow \Delta}{\alpha \# \beta, \Gamma \Rightarrow \Delta}$$

These are the rules for Prior's *tonk* and, as it is well known, on a classical consequence relation that has Cut as an internal rule, *tonk* is a trivialising operator. If Cut is no longer available, then *tonk*, just like the T-predicate, no longer leads to triviality (Ripley 2015). This is not about criticising *tonk*; this is about establishing the following fact. Depending on how the consequence relation is specified, some inferential roles, as determined by rules, may or may not be *good*, where goodness is measured by the requirement that the relation is not to become trivial. We may follow Avron (1991) and try to characterise logics by both their consequence relation and the class of connectives that are good according to that consequence relation.

Let us follow this line of thought for a moment. Ripley claims that "the difference between CL and ST is in the T rules, not in the vocabulary involved" (Ripley 2013a, fn. 7, p. 146). This, certainly, is *a* difference; is it the only *logical* difference between the consequence relation of ST and that of CL? Recall that classical propositional logic can be equivalently presented in a language $\mathcal{L}_1$ that expands $\mathcal{L}_0$ by the two propositional constants $\top$ and $\bot$. Similarly, valuations can be defined as homomorphisms from **FOR** $(\mathcal{L}_1)$ to **K**, viewed as an algebra of this language; so, in particular $v(\top) = 1$ and $v(\bot) = 0$ for every valuation v. The constants $\top$ and $\bot$ are term definable: for every *Boolean* valuation v, $v(\top) = 1 = v(p_0 \vee \neg p_0)$ and $v(\bot) = 0 = v(p_0 \wedge \neg p_0)$, where p_0 is a fixed variable of $\mathcal{L}_1$. But this is no longer the case if v is an *arbitrary* valuation. Nevertheless, $\top$ and $\bot$ can be *added* to our language, enriching either LK$^-$ or LK$^-_{INV}$ with the new axioms $\Rightarrow \top, \bot \Rightarrow$ and obtaining a conservative extension thereof. Clearly, according to this consequence relation $\top$ and $\bot$ count as *good*. Moreover, in this calculus Cut is still an admissible rule. Is this all that a sequent system for ST should contain?

We will argue below that this isn't true. Such a sequent calculus should also contain a zero-ary connective λ, governed by the axioms $\Rightarrow \lambda$ and $\lambda \Rightarrow$. The reader can easily see how this inferential behaviour fits into the grander perspective. For λ, as defined by the two axioms above, already occurs, albeit in disguise, in the proof of Theorem 18.24. The interpretation which we use in order to prove that LK$^-_{INV}$ is complete with respect to the class of all locally valid ST-metainferences assigns the value $\frac{1}{2}$ only to those formulae α such that both $\Rightarrow \alpha$ and $\alpha \Rightarrow$ are in T'.

Ripley also claims that classical logic is apt to deal with 'any sentence whatsoever' (Ripley 2013a, §1.1, p. 1). This may well be true, yet the question we would like to ask is rather different. We want to know whether classical logic can deal with every type of assertion and denial to which the ST-theorists appeal. It is not at all transparent that this is the case. Is there any reason to suppose that, out of the box, classical logic can deal with *tolerant* assertion and denial? In one sense, this is unproblematic. We may think, as Ripley seems to do, of tolerant assertion and denial as *silent partners* of their strict counterparts. Thus, strictly asserting (denying) something, is, *eo ipso*, tolerantly asserting (denying) it. The converse, of course, is not the case, as shown by the paradoxes. However, classical logic by itself cannot track the difference between strict and tolerant assertion and denial. Left to its own devices, it would happily cut all over the place, which, under appropriate circumstances, has disastrous effects. We need to make sure that this 'classical' logic has the resources to discriminate between strict and tolerant assertion and denial. This should be a matter of logicality: after all, by ST lights, logic is about how assertions and denials, strict or tolerant, are constrained. Thus there should be an ST-internal syntactical mark of tolerance and of its ability to stand alone (i.e. not to be a mere accessory of a strict speech act). Just consider that CL has such marks for strict assertion and denial. These are the zero-ary connectives $\top$ and $\bot$. Their 'point' is best understood by considering their connection with the empty positions on either side of the sequent sign. A sequent with an empty antecedent says that everything clashes with denying the succedent. So $\top$ is the exemplary case of *strict assertability*. An empty succedent means that everything clashes with asserting the antecedent. So $\bot$ is the paragon of strict deniability. The connectives $\top$ and $\bot$ simply internalise within the calculus the 'meaning' of these special positions.

However, ST does not trade in *strict* assertion and denial alone. It is also concerned with *tolerant* assertion and *tolerant* denial. Yet as described so far, ST is syntactically incapable of distinguishing between strict and tolerant assertion and denial. This is not a debilitating limitation; it is, however, one that misrepresents the target of the formal system which is used to present ST. For, to repeat, that target is the interplay between tolerant assertion and denial just as much as it is the interplay between their strict versions. This is where the constant λ comes in, as an internal (syntactical) expression of tolerant assertability and deniability. But λ cannot inhabit a specification of the classical consequence relation that is closed under Cut. And we shouldn't expect it to, for classical logic proper has no tolerance for tolerant assertions and denials. Since the consequence relation of ST classifies λ as *good*, and that of CL classifies it as *bad*, we have further reasons to believe that they codify different logics.

18.6 Concluding Remarks

In this paper we have argued that ST is not classical logic. We believe that it is, in fact, Priest's LP. That this confusion has arisen is not at all surprising. The true identity of ST is revealed only from the very general perspective on how to identify logics *qua* consequence-relations developed by Blok and Jónsson. However, we believe that this is the correct vantage point. The way logics are presented can fog the matters— the reader has only to review our previous discussion of metainferences and their validity in order to convince herself of the cogency of this point. Of course, there is only so much philosophical subtlety that can be accommodated within such an abstract framework. We allow that, in principle, one may fault the perspective we have assumed on this ground. (Although we ourselves have failed to let ourselves be convinced by the reasons one may bring in defence of this move.) Even if that is the case, our safe retreat is ensured. The contention that ST is not classical logic still holds water. Compared to classical logic, ST has an extra connective: the liar.

Acknowledgements The authors gratefully acknowledge the support of Regione Autonoma Sardegna, within the Project CRP-78705 (L.R. 7/2007), "Metaphor and argumentation". Versions of this paper have been presented at the Navarra Workshop on Logical Consequence (Pamplona, May 2016), the Prague Workshop on Nonclassical Logics (December 2016) and at the Workshop on Consequence and Paradox: Between Truth and Proof (Tübingen, March 2017). We thank Pablo Cobreros, Petr Cintula and the other participants to these events for their precious remarks. We thank Eduardo Barrio, Nissim Francez, Rohan French, Dave Ripley and two anonymous reviewers for their insightful comments. Finally, we are extremely grateful to Kazushige Terui for his invaluable pointers to the literature on Girard's three-valued interpretation of the sequent calculus.

References

Avron, A. (1991). Simple consequence relations. *Information and Computation, 92*, 105–139.
Barrio, E., Rosenblatt, L., & Tajer, D. (2015). The logics of strict-tolerant logic. *Journal of Philosophical Logic, 44*(5), 551–571.
Blok, W. J., & Jónsson, B. (2006). Equivalence of consequence operations. *Studia Logica, 83*(1–3), 91–110.
Blok, W. J., & Pigozzi, D. (1989). *Algebraizable logics.* Memoirs of the AMS, number 396. Providence, RI: American Mathematical Society.
Cobreros, P., Egré, P., Ripley, D., & van Rooij, R. (2012). Tolerant, classical, strict. *Journal of Philosophical Logic, 41*, 347–85.
Cobreros, P., Egré, P., Ripley, D., & van Rooij, R. (2013). Reaching transparent truth. *Mind, 122*(488), 841–866.
Fjellstad, A. (2016). Naive modus ponens and failure of transitivity. *Journal of Philosophical Logic, 45*, 65–72.
Font, J. M. (2016). *Abstract algebraic logic: An introductory textbook.* London: College Publications.
Francez, N. (2015). *Proof-theoretic semantics.* London: College Publications.
French, R. (2016). Structural reflexivity and the paradoxes of self-reference. *Ergo, 3*(5), 113–131.
Girard, J.-Y. (1976). Three-valued logic and cut-elimination: The actual meaning of Takeuti's conjecture. *Dissertationes Mathematicae, 136.*
Girard, J.-Y. (1987). *Proof theory and logical complexity.* Napoli: Bibliopolis.
Metcalfe, G., Olivetti, N., & Gabbay, D. (2009). *Proof theory for fuzzy logics.* Berlin: Springer.
Peregrin, J. (2008). What is the logic of inference? *Studia Logica, 88*, 263–294.
Priest, G. (1979). The logic of paradox. *Journal of Philosophical Logic, 8*, 219–241.
Priest, G. (2005). *Doubt truth to be a liar.* Oxford: Oxford University Press.
Priest, G. (2006). *In contradiction* (1st ed., 1987, 2nd expanded ed., 2006). Oxford: Clarendon.
Priest, G. (2014). Revising logic. In P. Rush (Ed.), *The metaphysics of logic* (pp. 211–223). Cambridge: Cambridge University Press.
Priest, G., Tanaka, K., & Weber, Z. (2013). Paraconsistent logic. In E. N. Zalta (Ed.), *Stanford Encyclopedia of Philosophy*, Spring 2015 Edition. http://plato.stanford.edu/archives/spr2015/entries/logic-paraconsistent/.
Pynko, A. P. (2010). Gentzen's cut-free calculus versus the logic of paradox. *Bulletin of the Section of Logic, 39*(1–2), 35–42.
Restall, G. (2000). *An introduction to substructural logics.* London: Routledge.
Restall, G. (2005). Multiple conclusions. In P. Hajek, L. Valdes-Villanueva, & D. Westerstahl (Eds.), *Logic, Methodology and Philosophy of Science: Proceedings of the Twelfth International Congress.* London: Kings' College Publications (pp. 189–205).
Ripley, D. (2013a). Paradoxes and failures of cut. *Australasian Journal of Philosophy, 91*(1), 139–164.

Ripley, D. (2013b). Revising up: Strengthening classical logic in the face of paradox. *Philosophers' Imprint, 13*(5), 1–13.
Ripley, D. (2015). Anything goes. *Topoi, 34,* 25–36.
Steinberger, F. (2016). Explosion and the normativity of logic. *Mind, 125*(498), 385–419.
Sylvan, R. (2000). A preliminary Western history of sociative logics. In D. Hyde & G. Priest (Eds.), *Sociative logics and their applications.* Aldershot: Ashgate.
Wintein, S. (2013). On the strict-tolerant conception of truth. *Australasian Journal of Philosophy, 92*(1), 1–20.
Zardini, E. (2013). Naive modus ponens. *Journal of Philosophical Logic, 42,* 575–593.

Bogdan Dicher obtained his doctorate in philosophy in 2015 from the University of Melbourne, Australia. He held a postdoctoral appointment at the University of Cagliari, Italy, and since September 2017 he is an FCT postdoctoral fellow at the Centre for Philosophy of the University of Lisbon, Portugal. He works mainly on philosophy of logic, particularly on logical pluralism and inferentialism. Bogdan has published in *Mind, The Review of Symbolic Logic* and *Theoria.*

Francesco Paoli has been teaching logic and philosophy of science at the University of Cagliari since 2001. He wrote the book *Substructural Logics: A Primer* (Kluwer, Dordrecht, 2002) and several research papers on mathematical and philosophical logic, on universal algebra, and on the foundations of quantum mechanics. The philosophical problems that intrigue him the most are logical pluralism, vagueness, and the meaning of logical constants.

Chapter 19
Annotated Natural Deduction for Adaptive Reasoning

Patrick Allo and Giuseppe Primiero

Abstract We present a multi-conclusion natural deduction calculus characterizing the dynamic reasoning typical of Adaptive Logics. The resulting system AdaptiveND is sound and complete with respect to the propositional fragment of adaptive logics based on **CLuN**. This appears to be the first tree-format presentation of the standard linear dynamic proof system typical of Adaptive Logics. It offers the advantage of full transparency in the formulation of locally derivable rules, a connection between restricted inference-rules and their adaptive counterpart, and the formulation of abnormalities as a subtype of well-formed formulas. These features of the proposed calculus allow us to clarify the relation between defeasible and multiple-conclusion approaches to classical recapture.

Keywords Paraconsistent logic · Adaptive logic · Natural deduction · Classical recapture

19.1 Introduction

In this paper, we outline a multiple-conclusion natural deduction calculus in which the dynamics of standard (Fitch-style) dynamic proofs of Adaptive Logics (Batens 2007) can be reconstructed. Adaptive logics are a family of logics that can be used to formalize a wide range of defeasible reasoning-forms. Their consequence relations rely on the standard idea of interpreting premises as normally as possible through the selection of models of its premises, but it is only at the level of its proof-theory that its distinctive approach comes to the fore. Adaptive logics, namely, reconstruct

P. Allo
Centre for Logic and Philosophy of Science, Vrije Universiteit Brussel,
Oxford Internet Institute, University of Oxford, Oxford, UK
e-mail: patrick.allo@vub.be

G. Primiero (✉)
Department of Philosophy, University of Milan,
Milan, Italy
e-mail: Giuseppe.Primiero@unimi.it; gprimiero@gmail.com

© Springer Nature Switzerland AG 2019
C. Başkent and T. M. Ferguson (eds.), *Graham Priest on Dialetheism and Paraconsistency*, Outstanding Contributions to Logic 18, https://doi.org/10.1007/978-3-030-25365-3_19

defeasible reasoning patterns as dynamic proofs; proofs in which steps performed earlier may later be retracted when the assumptions they were based on no longer hold. In particular, an inconsistency-adaptive logic captures paraconsistent reasoning avoiding triviality in the face of inconsistency, while trying to make up for its deductive weakness by provisionally applying classical inference-rules when there is no explicit indication that inconsistencies are involved in that inference.

The dynamics of retracting earlier lines in a proof can be captured in a rather natural way in linear proof-formats, including standard axiomatic and Fitch-style natural deduction proofs, but is much less straightforward in a tree-like proof-format. Consider, for instance, the following retraction in an application of *Ex Contradictione Quodlibet*:

$$
\begin{array}{llll}
(1) & p & \text{Prem} & \emptyset \\
(2) & p \vee q & \text{Addition} & \emptyset \\
(3) & \neg p & \text{Prem} & \emptyset \\
(4) & q & \text{DS} & \{p\} \quad \boxtimes^5 \\
(5) & p \wedge \neg p & \text{Adjunction} & \emptyset
\end{array}
$$

Here, at line (4) disjunctive syllogism (DS) is applied on the condition that p behaves normally, i.e. that the contradiction $p \wedge \neg p$ hasn't been derived. When this contradiction is effectively derived at line (5), line (4) is marked (here and in the following by $\boxtimes$) and is from then on no longer assumed to be part of the proof. This type of reasoning illustrates the idea of provisional applications of classical inference-rules to paraconsistent logics that reject the disjunctive syllogism, but in which the restricted form $\phi \vee \psi, \neg \phi / \psi \vee (\phi \wedge \neg \phi)$ is retained.

Contrast this, now, with the following attempt to reconstruct a similar reasoning process in a Gentzen–Prawitz-style proof-tree:

$$
\text{DS}^* \ \frac{\dfrac{\Gamma \vdash p}{\Gamma \vdash p \vee q} \vee \text{I} \quad \Gamma \vdash \neg p \quad \Gamma \not\vdash p \wedge \neg p}{\Gamma \vdash q} \quad \frac{\Gamma \vdash p \quad \Gamma \vdash \neg p}{\Gamma \vdash p \wedge \neg p} \wedge \text{I}
$$
$$?$$

When in this proof an explicit contradiction is derived in the right-hand branch, the assumption of its invalidity (stated explicitly in the left-hand branch) no longer holds. In this format, however, the order used to construct the proof cannot be read off the proof itself (an issue that could easily be fixed). But also, more importantly, it isn't even clear what it might mean to retract the line where q is derived, since the result of removing that line from the proof is in itself no longer a well-formed proof.

The proof-format we propose solves this problem by making two changes: first, we add indices to judgements to keep track of stages in the construction of a proof; and second, we exploit the fact that judgements that are 'marked' at a certain stage do not have to be removed, because there is simply no need to prevent their implicit re-use since every assumption or premise should explicitly be written down in the place it is used. Instead, it is the derivation of the same judgement at a later stage that is (or maybe) blocked, because the original assumption that led to its initial derivation probably no longer holds. We therefore provide, for the first time, an appropriate Natural Deduction translation of adaptive reasoning, whose proofs have so far always been presented in a linear format. We do so by formulating a general approach to Adaptive Reasoning, which accommodates both strategies standardly used to retract previously made judgements. While the reformulation of the Reliability Strategy is a much easier task, Minimal Abnormality is a more daunting one given the complexity of its procedural translation in a proof-tree. Nonetheless, we show that it can in principle be done within our logic, although more efficient procedures might be devised.

Because this system uses multiple-conclusion judgements, it also explicitly captures the connection between unconditional derivations of certain disjunctions in the paraconsistent logic and the conditional deductions of one of their disjuncts in the adaptive logic. Moreover, the choice of modelling inconsistency-adaptive reasoning brings us closer to the original motivations for the development of adaptive logic (Batens 1989), but also allows us to engage with current philosophical debates of relevance to Graham Priest's work and in particular how one should best approach the question of *classical recapture* in paraconsistent logics. The latter problem can be summarized as follows. When one adopts a logic that is strictly weaker than classical logic, the question of how one should account for epistemically useful classical inference forms that are invalidated by one's preferred logic almost immediately arises. In the case of paraconsistent logic, this question is often deemed urgent, as the practical and epistemic usefulness of the inference forms that are lost, like the disjunctive syllogism, is almost undisputed. Inconsistency-adaptive logics present one possible answer to this challenge under the form of defeasible inference forms that allow one to use classical inference-steps on the condition that certain assumptions are not violated. It is also a response that Graham has endorsed (Priest 1991, 2006, Chap. 16). His specific proposal on how this should be implemented has, in recent years, become the focus of a renewed interest in the problem of how dialetheists should account for classical recapture. We contend that the combination of a multiple-conclusion calculus with the reconstruction of the defeasible dynamics of adaptive proofs can further clarify this debate.

The paper is structured as follows. We introduce in Sect. 19.2 a basic natural deduction system called **minimalND**, which acts as the Lower Limit Logic of our adaptive system. In Sect. 19.3, we extend the system to account for adaptive reasoning through the definition of an appropriate abnormal form of expressions and appropriate adaptive rules; the new system is called **AdaptiveND**. In Sect. 19.4, we define marking strategies to identify derivation steps that can no longer be assumed to hold in the tree. In Sect. 19.5 we define basic meta-theoretical properties. We return to comment on the challenge of classical recapture in Sect. 19.6.

19.2 minimalND

We start by defining the type universe for the $\{\neg, \rightarrow, \wedge, \vee\}$ fragment of intuitionistic propositional logic corresponding to minimal logic. We call this logic **minimalND**

$$\frac{}{A \in \mathsf{Prop}} \text{\scshape Atom} \qquad \frac{}{\bot \in \mathsf{Prop}} \bot \qquad \frac{\phi \in \mathsf{Prop}}{\neg \phi \in \mathsf{Prop}} \neg$$

$$\frac{\phi_1 \in \mathsf{Prop} \quad \phi_2 \in \mathsf{Prop}}{\phi_1 \rightarrow \phi_2 \in \mathsf{Prop}} \rightarrow \qquad \frac{\phi_1 \in \mathsf{Prop} \quad \phi_2 \in \mathsf{Prop}}{\phi_1 \wedge \phi_2 \in \mathsf{Prop}} \wedge$$

$$\frac{\phi_1 \in \mathsf{Prop} \quad \phi_2 \in \mathsf{Prop}}{\phi_1 \vee \phi_2 \in \mathsf{Prop}} \vee$$

Fig. 19.1 Formula formation rules

and use it as the equivalent of a Lower Limit Logic—the paraconsistent logic that governs the unconditional steps in a proof. Contrary to what is standard in an intuitionistic setting, we do not allow the deduction of $\bot$ from an explicit contradiction. Whereas $\bot$ can be eliminated via *Ex Falso Quodlibet*, there is no introduction-rule for $\bot$, and this is what makes our base-logic paraconsistent. It is only when the assumption of consistency is introduced that the connection between negation inconsistency and absolute inconsistency can provisionally be recreated.

We start by defining the syntax of our language:

Definition 19.1 (*minimalND*). Our starting language for minimalND is defined by the following grammar:

$$\begin{aligned}
&\mathsf{Type} := \mathsf{Prop} \\
&\mathsf{Prop} := A \mid \bot \mid \neg \phi \mid \phi_1 \rightarrow \phi_2 \mid \phi_1 \wedge \phi_2 \mid \phi_1 \vee \phi_2 \\
&\Gamma := \{\phi_1, \ldots, \phi_n\} \\
&\Delta := \{\phi_1, \ldots, \phi_n\}
\end{aligned}$$

The type universe of reference is the set of propostions Prop, construed by atomic formulas closed under negation, implication, conjunction, disjunction and allowing $\bot$ to express absolute contradictions. Formula formation rules are given in Fig. 19.1.

Definition 19.2 (Judgements). A multiple-conclusion minimalND judgement is of the form $\Gamma; \cdot \vdash_s \Delta$, where Γ is the usual set of assumptions, Δ is a set of formulas of the language and s is a positive integer.

The set Γ on the left-hand side of the derivability sign is to be read conjunctively. Similarly for the semicolon symbol, which is introduced here but is only used in Sect. 19.3 to separate standard assumptions in Γ from conditions (in the adaptive sense). At this stage, the symbol $\cdot$ following the semicolon is used to express an empty set of adaptive conditions. The set Δ and the comma (if it occurs) on the right-hand side of the derivability sign are both to be read disjunctively. This characterizes our calculus as multiple conclusion. Standard context formation rules are, as usual in a proof-theoretic setting, inductively given for both left- and right-hand

$$\frac{}{\cdot \vdash_s \text{wf}} \text{N}\textsc{il} \qquad \frac{\Gamma; \cdot \vdash_s \text{wf} \quad \phi \in \text{Prop}}{\Gamma, \phi; \cdot \vdash_{s+1} \text{wf}} \Gamma\text{-}\textsc{Formation}$$

$$\frac{\Gamma; \cdot \vdash_s \text{wf} \quad \phi \in \Gamma}{\Gamma; \cdot \vdash_{s+1} \phi} \textsc{Prem}$$

Fig. 19.2 Context formation rules

side set of formulas, as shown in Fig. 19.2, see e.g. Troelstra and Schwichtenberg (2000, pp. 5–6) and Pfenning (2004, Sect. 2.3) for a formulation closer to ours. Nil establishes the base-case of a valid empty context, we use wf as an abbreviation for 'well-formed'; Γ-Formation allows extension of contexts by propositions; Prem establishes derivability of formulas contained in context and this rule, in particular, defines the equivalent of the adaptive Premise rule.

The derivability sign is enhanced with a signature s that corresponds to a counter of the ordered derivation steps executed to obtain the corresponding ND-formula in a tree. This annotation only comes to use in the next extension of the calculus in Sect. 19.3.

The semantics of connectives is given in the standard proof-theoretic way by Introduction and Elimination Rules in Fig. 19.3. Introduction of → corresponds to conditional proof, while its elimination formalizes Modus Ponens. Rules for ∧ are standard; notice that ∨-Elimination makes the disjunctive reading of the comma on the right-hand side of the turnstile explicit. ⊥ can be eliminated by *Ex Falso*, but cannot be introduced. Dually, our paraconsistent negation ¬ can be introduced, but not eliminated.

Finally, we introduce in Fig. 19.4 a set of rules to enforce structural properties. WL is a Weakening on the left-hand side of the judgement: it allows the monotonic extension of assumptions preserving already derivable formulas. Notice that this rule requires an empty set of formulas ; · following Γ. As will become clear in the next section, this means that weakening is only valid when the set of *adaptive conditions* is empty, that is, when no provisional assumptions are made that depend on the premises. We do not need to formulate a WR rule for weakening of the set Δ of derivable formulas, as this can be obtained by a detour of ∨-Introduction and Elimination. CL for Contraction on the left allows elimination of repeated assumptions and EL for Exchange on the left is valid just by set construction, as there is no order. CR and ER do a similar job on the right-hand side of the judgement. Finally, Cut (also known as Substitution in some Natural Deduction Calculi) guarantees that derivations can be pasted together, and in general it requires that there are no clashes of free variables in Γ, Γ'.

The resulting system is equivalent to the propositional fragment of **CLuN**, the logic obtained by adding Excluded Middle to the positive fragment of classical logic.

$$\frac{\Gamma, \phi_1; \cdot \vdash_s \Delta, \phi_2}{\Gamma; \cdot \vdash_{s+1} \Delta, \phi_1 \to \phi_2} \to I \qquad \frac{\Gamma; \cdot \vdash_s \Delta, \phi_1 \to \phi_2 \quad \Gamma'; \cdot \vdash_{s'} \Delta', \phi_1}{\Gamma; \Gamma' \vdash_{\max(s,s')+1} \Delta, \Delta', \phi_2} \to E$$

$$\frac{\Gamma; \cdot \vdash_s \Delta, \phi_1 \quad \Gamma'; \cdot \vdash_{s'} \Delta', \phi_2}{\Gamma, \Gamma'; \cdot \vdash_{\max(s,s')+1} \Delta, \Delta', \phi_1 \wedge \phi_2} \wedge I \qquad \frac{\Gamma; \cdot \vdash_s \Delta, \phi_1 \wedge \phi_2}{\Gamma; \cdot \vdash_{s+1} \Delta, \phi_{i \in \{1,2\}}} \wedge E$$

$$\frac{\Gamma; \cdot \vdash_s \Delta, \phi_1}{\Gamma; \cdot \vdash_{s+1} \Delta, \phi_1 \vee \phi_2} \vee I \qquad \frac{\Gamma; \cdot \vdash_s \Delta, \phi_2}{\Gamma; \cdot \vdash_{s+1} \Delta, \phi_1 \vee \phi_2} \vee I \qquad \frac{\Gamma; \cdot \vdash_s \Delta, \phi_1 \vee \phi_2}{\Gamma; \cdot \vdash_{s+1} \Delta, \phi_1, \phi_2} \vee E$$

$$\frac{\Gamma; \cdot \vdash_s \Delta, \bot}{\Gamma; \cdot \vdash_s \Delta, \phi} \bot E \qquad \frac{\Gamma; \phi \vdash_s \Delta, \psi}{\Gamma; \cdot \vdash_{s+1} \Delta, \psi, \neg \phi} \neg I$$

Fig. 19.3 Rules for I/E of connectives

$$\frac{\Gamma; \cdot \vdash_s \Delta, \phi_1}{\Gamma, \phi_2; \cdot \vdash_{s+1} \Delta, \phi_1} \text{WL} \qquad \frac{\Gamma, \phi_1, \phi_1; \cdot \vdash_s \Delta, \phi_2}{\Gamma, \phi_1; \cdot \vdash_{s+1} \Delta, \phi_2} \text{CL} \qquad \frac{\Gamma, \phi_1, \phi_2; \cdot \vdash_s \Delta, \phi_3}{\Gamma, \phi_2, \phi_1; \cdot \vdash_{s+1} \Delta, \phi_3} \text{EL}$$

$$\frac{\Gamma; \cdot \vdash_s \Delta, \phi_1 \quad \Gamma', \phi_1; \cdot \vdash_{s'} \Delta', \phi_2}{\Gamma; \Gamma'; \cdot \vdash_{\max(s,s')+1} \Delta, \Delta', \phi_2} \text{CUT}$$

$$\frac{\Gamma; \cdot \vdash_s \Delta, \phi, \phi}{\Gamma; \cdot \vdash_{s+1} \Delta, \phi} \text{CR} \qquad \frac{\Gamma; \cdot \vdash_s \Delta, \phi_1, \phi_2}{\Gamma; \cdot \vdash_{s+1} \Delta, \phi_2, \phi_1} \text{ER}$$

Fig. 19.4 Structural rules

This is a very weak paraconsistent (but not paracomplete) logic that does not validate any of the usual De Morgan rules (Batens 1980), and which has been used as the Lower Limit Logic of one of the first adaptive logics.

Theorem 19.1 *minimalND is sound and complete w.r.t. to the propositional fragment of **CLuN**.*

Proof Soundness can be shown as usual, with the key step verifying that ($\neg$I) is sound in view of the completeness-clause for negation

$$\text{If } v(\phi) = \text{False, then } v(\neg \phi) = \text{True} \qquad \text{C}\neg$$

Completeness follows from the provability of all **CLuN**-axioms. Below, we only give the proofs for Excluded Middle and Peirce's Law.

$$
\begin{array}{c}
\text{PREM} \dfrac{}{p;\cdot\vdash_1 p} \\
\text{CR}\dfrac{\dfrac{\dfrac{\dfrac{\emptyset;\cdot\vdash_2 p,\neg p}{\emptyset;\cdot\vdash_3 p\vee\neg p,\neg p}\vee\text{I}}{\emptyset;\cdot\vdash_4 p\vee\neg p, p\vee\neg p}\vee\text{I}}{\emptyset;\cdot\vdash_5 p\vee\neg p}}{}
\end{array}
\neg\text{I}
$$

$$
\rightarrow\text{I}\dfrac{\text{WR}\dfrac{\dfrac{p;\cdot\vdash_1 p}{p;\cdot\vdash_2 p,q}}{;\cdot\vdash_3 p, p\to q}\quad \text{CR}\dfrac{\rightarrow\text{E}\dfrac{(p\to q)\to p;\cdot\vdash_4 (p\to q)\to p}{(p\to q)\to p;\cdot\vdash_5 p, p}\text{PREM}}{(p\to q)\to p;\cdot\vdash_6 p}}{;\cdot\vdash_7 ((p\to q)\to p)\to p}\rightarrow\text{I}
$$

□

19.3 AdaptiveND

We now extend minimalND to characterize a new logic called AdaptiveND to allow for inconsistency-adaptive reasoning. To this aim one needs

1. the explicit formulation of an Ω set of propositions;
2. the formulation of judgements including an *adaptive condition*;
3. the formulation of a rule that allows the derivation of new formulas independent from such an adaptive condition;
4. the formulation of a rule that allows the derivation of new formulas that depend on such an adaptive condition.

We offer accordingly new definitions for the syntax of this logic and the related form of judgements.

Definition 19.3 (*AdaptiveND*). The language of AdaptiveND is as follows:

$$
\begin{aligned}
&\text{Type} := \text{Prop} \\
&\text{Prop} := A \mid \bot \mid \neg\phi \mid \phi_1\to\phi_2 \mid \phi_1\wedge\phi_2 \mid \phi_1\vee\phi_2 \\
&\Gamma := \{\phi_1,\ldots,\phi_n\} \\
&\Delta := \{\phi_1,\ldots,\phi_n\} \\
&\Omega := \{\phi\wedge\neg\phi \mid \phi\in Prop\}
\end{aligned}
$$

Definition 19.4 (*Judgements*). An AdaptiveND-judgement is of the form $\Gamma;\Theta^-\vdash_s \Delta$, where:

1. the left-hand side of $\vdash_s$ has Γ as in minimalND;

Fig. 19.5 Ω formation rules

$$\frac{\phi \in \mathsf{Prop}}{(\phi \wedge \neg\phi) \in \Omega} \ \Omega\text{-FORMATION}$$

$$\frac{\Gamma; \cdot \vdash_s \mathsf{wf} \quad \phi \in \Omega}{\Gamma; \phi^- \vdash_{s+1} \mathsf{wf}} \ \text{Adaptive Condition-formation}$$

$$\frac{\Gamma; \Theta^- \vdash_s \mathsf{wf} \quad \phi \in \Omega}{\Gamma; (\Theta \cup \{\phi\})^- \vdash_{s+1} \mathsf{wf}} \ \text{Adaptive Condition-extension}$$

2. the semicolon sign on the left-hand side of $\vdash_s$ is conjunctive;
3. Θ refers to a finite subset of Ω, i.e. a set of formulas of a specific inconsistent logical form; we write ϕ instead of $\{\phi\}$ when Θ is the singleton $\{\phi\}$; below we introduce an appropriate Ω-formation rule;
4. the last place of the left-hand side context is always reserved for negated formulas of type Ω; we shall use ϕ^- to refer to the negation of ϕ, and Θ^- for $\{\phi^- \mid \phi \in \Theta\}$;
5. the right-hand side is in disjunctive form.

When the second place on the left-hand side of $\vdash$ is empty, we shall write $\Gamma; \cdot \vdash$, thus reducing to the form of a **minimalND**-judgement. Moreover, in **AdaptiveND**, the annotation on the proof stage **s** is optionally followed by one of the following two marks:

⊠ to mark that at the current stage some previously derived formula is retracted; the meaning of this annotation is given for the Reliability Strategy by a corresponding rule ⊠R, presented in Sect. 19.4.1; for the Minimal Abnormality Strategy, its meaning is given by two rules ⊠M1 and ⊠M2, presented in Sect. 19.4.3;

✓ to mark that at the current stage some previously derived formula is now finally derived, i.e. will no longer be marked by ⊠; the use of this annotation is formally given below in Definition 19.14.

We now introduce the rules for **AdaptiveND**. In Fig. 19.5, we describe the formation and use of formulas $\phi \in \Omega$. By Ω-**Formation**, the explicit contradiction $\phi \wedge \neg\phi$, with ϕ any proposition, is a formula of the Ω type. In the Adaptive tradition, a formula of type Ω is called an *abnormality* or *abnormal formula*. By **Adaptive Condition Formation**, given a valid context Γ and a formula ϕ of the Ω type, a context Γ followed by the Adaptive Condition that expresses the defeasible assumption that ϕ *is false*, is a well-formed context. This corresponds to the use of syntactic restrictions that are applied to the use of conditions as additional elements of a proof line in the standard linear format of adaptive logics. By **Adaptive Condition Extension**, a newly constructed formula of type Ω can be added to an existing nonempty Adaptive Condition.

In Fig. 19.6, the calculus is extended by introducing the conditional rule **RC**, which states that if a disjunction ψ, ϕ is derivable from Γ, with ϕ an abnormal formula, then ψ can also be derived alone under Γ and the Adaptive Condition that ϕ be false. Because the application of **RC** can be delayed by keeping formulae of type Ω on

Fig. 19.6 Conditional rule

$$\frac{\Gamma;\Theta^- \vdash_s \psi, \phi \quad \phi \in \Omega}{\Gamma;(\Theta \cup \{\phi\})^- \vdash_{s+1} \psi} \text{ RC}$$

Fig. 19.7 Unconditional rules

$$\frac{\Gamma;\Theta^- \vdash_s \phi_1 \quad \phi_1;\cdot \vdash_{s'} \phi_2}{\Gamma;\Theta^- \vdash_{max(s,s')+1} \phi_2} \text{ RU}$$

$$\frac{\Gamma;\Theta^- \vdash_s \phi_1 \quad \Gamma';\Theta'^- \vdash_{s'} \phi_2 \quad \phi_1,\phi_2;\cdot \vdash_{s''} \phi_3}{\Gamma,\Gamma';(\Theta \cup \Theta')^- \vdash_{max(s,s',s'')+1} \phi_3} \text{ RU2}$$

the right-hand side of the turnstile, the role of the unconditional rules of the standard calculus is subsumed under the **Cut** rule. The single and multi-premise versions of the unconditional rules displayed in Fig. 19.7 can thus be treated as derived rules as shown by the procedures for rewriting a succession of RC and RU applications as a succession of Cut and RC applications in Fig. 19.8.

The marking of a judgement in an **AdaptiveND** proof signals that the marked judgement cannot be used as a premise for any other rule. In that sense, the marking of judgements expresses a dead end in a proof-tree. To preserve information about retracted judgements within a proof, and to facilitate the practice of cutting and pasting proofs together without having to renumber the judgements, we extend the standard formalization of proofs as trees, and define adaptive proofs as sequences of proof-trees.

Definition 19.5 (*Proof-Tree*). A well-formed **AdaptiveND** tree is a finite proof-tree obtained by deriving **AdaptiveND** judgements from other **AdaptiveND** judgements where

1. the top leaves of the tree are instances of the **Prem** rule and
2. each next step is obtained by applying one of the **minimalND** proof-rules or one of the **AdaptiveND** proof-rules.

Definition 19.6 (*Adaptive Proof*). An **AdaptiveND** proof is a sequence $\langle T_i \rangle_{i \in I}$ of **AdaptiveND** trees with each $i \in I$ equal to the highest numbered judgement in T_i.

As a notational shorthand, we will sometimes include marked judgements as unused premises to incorporate dead ends in a single proof-tree, and include final judgements of an earlier proof-tree as top leaves of a new proof-tree.

The Adaptive strategies developed in the next Section have the aim of establishing which abnormal formulas can no longer be safely considered as conditions in the application of the Conditional Rule RC, in turn requiring the retraction of the previously derived formulas. To this aim, three elements need to be introduced:

1. minimal disjunctions of formulas of type Ω, denoted by $\bigvee(\Delta)^{min}$, with $\Delta \subset \Omega$;
2. the union of all abnormalities that occur as a disjunct of some $\bigvee(\Delta)^{min}$ derived up to a certain stage from Γ, denoted by $\text{UnRel}(\Gamma)$;

$$\dfrac{\dfrac{\Gamma; \cdot \vdash_1 \phi_1, \phi \quad \phi \in \Omega}{\Gamma; \phi^- \vdash_2 \phi_1} \text{RC} \quad \phi_1; \cdot \vdash_3 \phi_2}{\Gamma; \phi^- \vdash_4 \phi_2} \text{RU}$$

$$\Downarrow$$

$$\dfrac{\dfrac{\Gamma; \cdot \vdash_1 \phi_1, \phi \quad \phi_1; \cdot \vdash_2 \phi_2}{\Gamma; \cdot \vdash_3 \phi_2, \phi} \text{Cut} \quad \phi \in \Omega}{\Gamma; \phi^- \vdash_4 \phi_2} \text{RC}$$

$$\dfrac{\dfrac{\Gamma; \cdot \vdash_1 \phi_1, \phi \quad \phi \in \Omega}{\Gamma; \phi^- \vdash_2 \phi_1} \text{RC} \quad \dfrac{\Gamma'; \cdot \vdash_3 \phi_2, \phi' \quad \phi' \in \Omega}{\Gamma'; \phi'^- \vdash_4 \phi_2} \text{RC} \quad \phi_1, \phi_2 \vdash_5 \phi_3}{\Gamma, \Gamma'; \{\phi, \phi'\} \vdash_6 \phi_3} \text{RU2}$$

$$\Downarrow$$

$$\dfrac{\Gamma; \cdot \vdash_4 \phi_1, \phi \quad \dfrac{\dfrac{\Gamma; \cdot \vdash_1 \phi_2, \phi' \quad \phi_1, \phi_2; \cdot \vdash_2 \phi_3}{\Gamma', \phi_1; \cdot \vdash_3 \phi_3, \phi'} \text{Cut}}{\dfrac{\Gamma, \Gamma'; \cdot \vdash_5 \phi_3, \phi, \phi' \quad \phi \in \Omega}{\dfrac{\Gamma, \Gamma'; \phi^- \vdash_6 \phi_3, \phi' \quad \phi' \in \Omega}{\Gamma, \Gamma'; \{\phi, \phi'\}^- \vdash_7 \phi_3} \text{RC}} \text{RC}} \text{Cut}}$$

Fig. 19.8 Redundancy of unconditional rules

3. the set of choice-sets of a set of sets $\Delta_1, \ldots, \Delta_n$, corresponding to all $\bigvee(\Delta_i)^{min}$ derived up to a certain stage from Γ, denoted by $\Phi(\Gamma)$.

The rules in Fig. 19.9 establish these three constructions.

Rule MinDab says that a disjunctive formula of the Ω-type derived at some stage s of a derivation can be considered minimal at stage s' if at no previous stage $t < s'$ a shorter one was derived under the same context Γ. Note that the superscript *min* works as an annotation indicating that the Δ in the first premise of the MinDab rule satisfies the side-condition of the same rule, but it does not indicate that Δ^{min} and Δ are two different sets.

Rule UnRel says that given minimal disjunctions of abnormalities, each derived according to MinDab at stages s-n up to s-1, the union of all such formulas can be derived at stage s, defined as follows:

Definition 19.7 (*Set of unreliable formulas*). $\mathsf{UnRel}(\Gamma)$ derived at stage s from Γ denotes the conjunction of the union of some $\Delta_1, \ldots, \Delta_n$ derived from Γ up to stage $\mathsf{s\text{-}1}$ according to MinDab.

Keep in mind that $\mathsf{UnRel}(\Gamma)$ denotes a conjunction and not a set of formulae. As such, when it occurs on the right-hand side of the turnstile it should not be read as a disjunction of abnormalities (which is the intended reading for a set Δ of abnormalities in that position). The side-condition for the Rule UnRel ensures that each Δ_i^{min} used as a premise is still minimal at stage $\mathsf{s\text{-}1}$. Contrary to how unreliable formulae at a stage are defined in standard adaptive proofs, this rule does not require that all minimal disjunctions of abnormalities derived at a given stage need to be used as its premises, but the side-condition of this rule does refer to all disjunctions of abnormalities that occur at earlier stages of the proof.

Choice-sets provide selections of abnormalities that might turn out to be true at a stage s defined as follows:

Definition 19.8 (*Choice-set of* $\{\Delta_1, \ldots, \Delta_n\}$). $\Phi(\Gamma)$ derived at stage s from Γ denotes the set of choice-sets of some $\{\Delta_1, \ldots, \Delta_n\}$ derived from Γ up to stage $\mathsf{s\text{-}1}$ according to MinDab. We denote each such choice-set with $\mathsf{choice}_i(\{\Delta_1, \ldots, \Delta_n\})$.

The choice-set of an empty set of minimal disjunction of abnormalities is empty. This notion of choice-set is procedurally implemented by the rule Choice in Fig. 19.9, where again each premise must be intended as the conclusion of a MinDab rule and the side-condition ensures that each Δ_i^{min} used as a premise is still minimal at stage $\mathsf{s\text{-}1}$. Here too, it is not required to use all minimal Δ's that are already derived as premises. The Rule $\mathsf{MinChoice}$ allows the selection of a *minimal* choice-set out of $\Phi(\Gamma)$, denoted by $\mathsf{choice}_i(\{\Delta_1, \ldots, \Delta_n\})^{min}$.

$$\frac{\Gamma; \cdot \vdash_s \Delta \quad \Delta \subset \Omega \quad \text{with no } \Delta' \subseteq \Delta \text{ s.t. } \Gamma; \cdot \vdash_{t<s'} \Delta'}{\Gamma; \cdot \vdash_{s'} \Delta^{min}} \text{ MinDab}$$

$$\frac{\Gamma; \cdot \vdash_{s\text{-}n} \Delta_1^{min} \quad \ldots \quad \Gamma; \cdot \vdash_{s\text{-}1} \Delta_n^{min} \quad \text{with no } \Delta_i^{min} \subset \Delta_j^{min}}{\Gamma; \cdot \vdash_s \mathsf{UnRel}(\Gamma)} \text{ UnRel}$$

$$\frac{\Gamma; \cdot \vdash_{s\text{-}n} \Delta_1^{min} \quad \ldots \quad \Gamma; \cdot \vdash_{s\text{-}1} \Delta_n^{min} \quad \text{with no } \Delta_i^{min} \subset \Delta_j^{min}}{\Gamma; \cdot \vdash_s \Phi(\Gamma)} \text{ Choice}$$

$$\frac{\Gamma; \cdot \vdash_{s\text{-}1} \Phi(\Gamma) \quad \text{with no } \mathsf{choice}_j(\{\Delta_1, \ldots, \Delta_n\}) \in \Phi(\Gamma), \text{ s.t. } \mathsf{choice}_j \subset \mathsf{choice}_i}{\Gamma; \cdot \vdash_s \mathsf{choice}_i(\{\Delta_1, \ldots, \Delta_n\})^{min}} \text{ MinChoice}$$

Fig. 19.9 Deriving minimal disjunctions of abnormalities, unreliable formulae, and minimal choice-sets of abnormalities

The derivation of a minimal disjunction of abnormalities and the formation of their union and choice-set is a process that occurs along with the development of the proof-tree, and the marking procedure depends on the derivation of these types of judgements. Unlike for the standard definitions of unreliable formulae and minimal choice-sets, the conclusions of the rules UNREL and CHOICE do not necessarily coincide with the minimal choice-sets and sets of unreliable formulae as they are used in the standard proof-format. The following propositions guarantee that this does not lead to further complications: if an abnormality occurs in $\mathsf{UnRel}(\Gamma)$, but is not unreliable according to Γ, then $\mathsf{UnRel}(\Gamma)$ was derived from a Δ_i that isn't minimal; similarly, if an abnormality occurs in some $\mathsf{choice}_i(\{\Delta_1, \ldots, \Delta_n\})^{min}$, but isn't verified by a minimally abnormal model, then it was derived from a Δ_i that isn't minimal. The key to these results is that the side-conditions that apply to the premises of UNREL and CHOICE refer to all disjunctions of abnormalities that occur unconditionally in the proof, and not only to those used as premises.

Proposition 19.1 *If $\Gamma; \cdot \vdash_s \mathsf{UnRel}(\Gamma)$ occurs in a proof and Λ is the set of all Δ^{min} derived up to stage s that are minimal at s, then each conjunct of $\mathsf{UnRel}(\Gamma)$ is in $\bigcup(\Lambda)$.*

Proof Immediate from the fact that if $\Gamma; \cdot \vdash_{s-n+(i-1)} \Delta_i^{min}$ is a premise used to derive $\Gamma; \cdot \vdash_s \mathsf{UnRel}(\Gamma)$ then (i) $\Delta_i^{min} \in \Lambda$, and (ii) Δ_i^{min} is minimal at stage $s - 1$. □

Proposition 19.2 *If $\Gamma; \cdot \vdash_s \Phi(\Gamma)$ occurs in a proof and Λ is the set of all Δ^{min} derived up to stage s that are minimal at s, then each $\phi \in \Phi(\Gamma)$ is a subset of some choice-set from Λ.*

Proof If $\Gamma; \cdot \vdash_{s-n+(i-1)} \Delta_i^{min}$ is a premise used to derive $\Gamma; \cdot \vdash_s \Phi(\Gamma)$, adding a premise $\Gamma; \cdot \vdash_{s-n+(j-1)} \Delta_j^{min}$ for some Δ_j that was already derived at some stage $s - (n + m)$ would not lead to the violation of the side-condition for the application of CHOICE. Let $\Gamma; \cdot \vdash_s \Phi'(\Gamma)$ be the conclusion obtained by adding this premise, and note that each $\phi' \in \Phi'(\Gamma)$ can be obtained by extending a $\phi \in \Phi(\Gamma)$ with some member of Δ_j. □

Proposition 19.3 *If $\Gamma; \cdot \vdash_s \mathsf{choice}_i(\{\Delta_1, \ldots, \Delta_n\})^{min}$ is derived from $\Gamma; \cdot \vdash_{s-1} \Phi(\Gamma)$ and Λ is the set of all Δ^{min} derived up to stage s that are still minimal at s, then $\mathsf{choice}_i(\{\Delta_1, \ldots, \Delta_n\})^{min}$ is a subset of some minimal choice-set from Λ, and for each minimal choice-set $\mathsf{choice}_j(\{\Delta_1, \ldots, \Delta_n, \ldots, \Delta_{n+m}\})^{min}$ from Λ there is a $\mathsf{choice}_k(\{\Delta_1, \ldots, \Delta_n\})^{min} \in \Phi(\Gamma)$ such that $\mathsf{choice}_k(\{\Delta_1, \ldots, \Delta_n\})^{min} \subseteq \mathsf{choice}_j(\{\Delta_1, \ldots, \Delta_n, \ldots, \Delta_{n+m}\})^{min}$.*

Proof Assume $\Gamma; \cdot \vdash_s \mathsf{choice}_i(\{\Delta_1, \ldots, \Delta_n\})^{min}$ is obtained by an application of MINCHOICE to a judgement with $\Phi(\Gamma)$ as a consequent. Let Δ_{n+1} be minimal at stage s, and assume that $\Phi'(\Gamma)$ is derivable from judgements with $\Delta_1, \ldots \Delta_{n+1}$ as consequent. By Proposition 19.2, we know that $\mathsf{choice}_i(\{\Delta_1, \ldots, \Delta_n\})^{min}$ is a subset of one or more members of $\Phi'(\Gamma)$. Let $\psi_1, \ldots, \psi_m$ be an enumeration of the members of Δ_{n+1}. (i) If some $\psi_i \in \Delta_{n+1}$ is a member of $\mathsf{choice}_i(\{\Delta_1, \ldots, \Delta_n\})^{min}$, then the latter is also minimal in $\Phi'(\Gamma)$. (ii) If no $\psi_i \in \Delta_{n+1}$ is a member of

$\text{choice}_i(\{\Delta_1, \ldots, \Delta_n\})^{min}$, then, because Δ_{n+1} is not included in any Δ_i, any $\text{choice}_i(\{\Delta_1, \ldots, \Delta_n\})^{min} \cup \{\psi_i\}$ must be minimal in $\Phi'(\Gamma)$. □

19.3.1 A Simple Example

We present here a simple derivation in AdaptiveND, where $\Gamma = \{(\neg p \vee q), p, (p \rightarrow q), (p \rightarrow \neg p)\}$:

$$\cfrac{\cfrac{\cfrac{\overline{\Gamma; \cdot \vdash_1 (\neg p \vee q)}^{\text{PREM}}}{\Gamma; \cdot \vdash_2 \neg p, q}^{\vee E} \quad \overline{\Gamma; \cdot \vdash_3 p}^{\text{PREM}}}{\Gamma; \cdot \vdash_4 (p \wedge \neg p), q}^{\wedge I} \quad (p \wedge \neg p) \in \Omega}{\Gamma; (p \wedge \neg p)^- \vdash_5 q}^{\text{RC}}$$

In the above derivation, all judgements up to stage 4 are obtained by minimalND rules. Stage 5 derives a formula on condition of the abnormality $(p \wedge \neg p)$ being false. This corresponds to changing a multiple conclusion judgement at stage 4 into a single conclusion one at stage 5 by turning one of the conclusions into an adaptive condition. This move is justified by the syntactical form of the abnormality, stated as the side-condition $(p \wedge \neg p) \in \Omega$ for the application of the RC rule.

19.4 Adaptive Proofs and Rules for Marking

In standard Adaptive Logics, one introduces strategies to tell which applications of the RC rule should be retracted in view of the Minimal Disjunction of Abnormalities that have been derived. Adaptive Logics come with marking mechanisms that allow such retractions, according to different possible strategies. The two 'standard' strategies and their rationale are (Batens 2001):

- *Reliability*: once $\text{UnRel}(\Gamma)$ is derived at some stage s, *every* formula ψ derived at some prior stage s' on the assumption that some $\phi \in \text{UnRel}(\Gamma)$ is false, needs to be retracted;
- *Minimal Abnormality*: once $\Phi(\Gamma)$ is derived at some stage s, a formula ψ is marked if either (i) it is derived at some prior stage s' on an assumption Θ which intersects with every minimal choice-set $\text{choice}_i(\{\Delta_i, \ldots, \Delta_n\}) \in \Phi(\Gamma)$; or (ii) for some minimal choice-set $\text{choice}_i(\{\Delta_i, \ldots, \Delta_n\}) \in \Phi(\Gamma)$, there is no derivation of ψ on another condition Θ', such that the intersection of Θ' with choice_i is empty.

Fig. 19.10 Marking for reliability

$$\frac{\Gamma;\Theta^- \vdash_s \psi \quad \Gamma;\cdot \vdash_{s'} \mathsf{UnRel}(\Gamma) \quad \Theta \cap \mathsf{UnRel}(\Gamma) \neq \emptyset}{\Gamma;\Theta^- \vdash_{\max(s,s')+1\boxtimes R} \psi} \boxtimes R$$

19.4.1 Marking-Rule for Reliability

Reliability is the adaptive strategy that takes the most cautious interpretation of abnormalities: any formula that in view of the premises might behave abnormally, because it occurs in a minimal disjunction of abnormalities, is deemed unreliable and should not be assumed to behave normally. This means in practice that a formula ψ derived on the assumption that ϕ behaves normally will be 'marked' as soon as the unreliability of ϕ is established. The result of this marking is that ψ should no longer be treated as a formula that was derived.

In Fig. 19.10, we define a new rule $\boxtimes R$ that depends on the derivation of a set of unreliable formulas.

19.4.2 Extending the Example

Let us now extend the example from Sect. 19.3.1 with a new branch to illustrate the derivation step obtained by the Marking-Rule $\boxtimes R$. Let $\mathbb{D}$ be the derivation from our initial example that ended with the derivation at stage 5 of q in context Γ and with $(p \wedge \neg p)^-$ as a condition. We extend it now as follows:

$$\cfrac{\mathbb{D} \qquad \cfrac{\cfrac{\Gamma;\cdot \vdash_6 p \quad \Gamma;\cdot \vdash_7 p \to \neg p}{\Gamma;\cdot \vdash_8 \neg p} \to E \quad \Gamma;\cdot \vdash_9 p}{\Gamma;\cdot \vdash_{10} p \wedge \neg p} \wedge I}{\Gamma;(p \wedge \neg p)^- \vdash_5 q \qquad \Gamma;\cdot \vdash_{11 \boxtimes R} q} \boxtimes R$$

In this derivation, a new abnormality is derived at stage 10, namely, the same that is assumed to be false at stage 5. Note that we avoid the superfluous inference step from the formula $(p \wedge \neg p)$ to the corresponding singleton set of unreliable formulas; moreover, it is essential that this abnormality be derived under an empty condition, i.e. under context $\Gamma;\cdot$, as explained above for the required strict condition on WL. A difference between standard (i.e. linear) adaptive proofs and the proposed tree-style Natural Deduction derivation proof-format becomes evident here. In the former, a marking-rule implies the need to proceed backwards on the derivation, to mark all lines that were derived on an assumption that is shown to be violated and thus can no longer be considered derived. In the latter, on the other hand, there is no need to remove formulas because the result obtained at stage 5 cannot be reused in an extension of this proof. Instead a new derivation step is performed (stage 11), where the conclusion q is marked. Moreover, if we were ever to get again

$$\frac{\Gamma; \Theta^- \vdash_s \psi \qquad \Gamma; \cdot \vdash_{s'} \Phi(\Gamma) \qquad (\dagger)}{\Gamma; \Theta^- \vdash_{\max(s,s')+1} \boxtimes_{MA} \psi} \boxtimes M$$

with

$$\Theta \cap \text{choice}_i(\{\Delta_1, \ldots, \Delta_n\})^{min} \neq \emptyset \qquad (\dagger)$$

for each $\text{choice}_i(\{\Delta_1, \ldots, \Delta_n\})^{min} \in \Phi(\Gamma)$.

$$\frac{\Gamma; \Theta_1^- \vdash_s \psi \quad \ldots \quad \Gamma; \Theta_n^- \vdash_{s+n} \psi \quad \Gamma; \cdot \vdash_{s'} \text{choice}_i(\{\Delta_1, \ldots, \Delta_n\})^{min} \quad (\dagger) \quad (\ddagger)}{\Gamma; \Theta^- \vdash_{\max(s+n, s')+1} \boxtimes_{MA} \psi} \boxtimes M2$$

with

$$\Theta_1 \cap \text{choice}_i(\{\Delta_1, \ldots, \Delta_n\})^{min} \neq \emptyset,$$
$$\vdots \qquad (\dagger)$$
$$\Theta_n \cap \text{choice}_i(\{\Delta_1, \ldots, \Delta_n\})^{min} \neq \emptyset,$$

and

there is no $\Gamma; \Theta' \vdash_{t<s} \psi$ with $\Theta' \notin \{\Theta_1, \ldots, \Theta_n\}$. $\qquad (\ddagger)$

Fig. 19.11 Marking for minimal abnormality

$\Gamma; (p \wedge \neg p)^- \vdash_i q$, it would be obtained as the result of some new derivation $\mathbb{D}'$ and be the conclusion at some stage $i > 11$, where an additional step would again be required to mark it at a later stage.

19.4.3 Marking-Rules for Minimal Abnormality

Minimal abnormality is the marking strategy that reflects the following condition: a formula ψ derived on a condition Θ^- is retracted if, either (i) every minimal choice-set includes some condition in Θ, or (ii) there is a minimal choice-set $\text{choice}_i(\{\Delta_1, \ldots, \Delta_n\})$ such that every derivation of ψ is based on a condition that is shown to be violated by $\text{choice}_i(\{\Delta_1, \ldots, \Delta_n\})$. We offer rules for this strategy in Fig. 19.11.

The first marking-rule $\boxtimes M$ reflects the following condition: if a formula ψ is derived at stage s under an adaptive condition ϕ which is part of all minimal choice-sets in $\Phi(\Gamma)$ at stage s', then at the next stage the formula ψ can be retracted.

The second marking-rule $\boxtimes M2$ reflects the following condition: if a formula ψ is derived always under an adaptive condition that is part of the same minimal choice-set in $\Phi(\Gamma)$, then at the next stage the formula ψ can be retracted. Here, the adaptive condition of the conclusion should be intended as saying that the retraction applies to all stages where the derivation of ψ was obtained under one of the listed conditions.

In Fig. 19.12, we illustrate with two trees the intended use of the marking-rules for Minimal Abnormality. In the first tree, it is shown how the derivation of a formula ψ

$$\dfrac{\Gamma;\Theta^-\vdash_s \psi \qquad \dfrac{\dfrac{\Pi}{\Gamma,\cdot\vdash_{s+1}\Phi(\Gamma)}\text{ Choice}}{\Gamma;\cdot\vdash_{s+2}\text{choice}_1(\{\Delta_1,\ldots,\Delta_n\})^{min}}\text{ MinChoice}\qquad \cdots \qquad \begin{matrix}\Pi'\\ \vdots\end{matrix}\qquad \Theta\cap\text{choice}_i\neq\emptyset}{\Gamma;\Theta^-\vdash_{s+n+1\boxtimes\text{MA}}\psi}\boxtimes\text{M1}$$

$$\dfrac{\Gamma;\Theta_1^-\vdash_s\psi \qquad \Gamma;\Theta_n^-\vdash_{s+n}\psi \qquad \dfrac{\dfrac{\Pi}{\Gamma,\cdot\vdash_{s'}\Phi(\Gamma)}\text{ Choice}}{\Gamma;\cdot\vdash_{s'+1}\text{choice}_1(\{\Delta_1,\ldots,\Delta_n\})^{min}}\text{ MinChoice}\qquad \Theta_1,\ldots,\Theta_n\cap\text{choice}_i\neq\emptyset}{\Gamma;\Theta_1^-\vdash_{\max(s+n,\,s'+1)+1\boxtimes\text{MA}}\psi}\,\ddagger\,\boxtimes\text{M2}$$

Fig. 19.12 Marking-trees for minimal abnormality

under an adaptive condition Θ is followed by a series of derivations Π for all possible choice-sets of minimally abnormal formulas; if the side-condition that requires Θ to occur in each such set holds, then marking can be applied. In the second (dual) tree, it is shown how a formula ψ derived under an adaptive condition Θ is marked if either (i) no alternative conditional derivation of ψ occurs in the proof, or (ii) every alternative conditional derivation of ψ that occurs in the proof depends on a condition Θ' that intersects with the same minimal choice-set $\text{choice}_1(\{\Delta_1,\ldots,\Delta_n\})^{min}$.

19.4.4 An Example with $\bigvee(\Delta^{min})$-Selection

The previous example is rather simple, in that it shows a formula that is first derived under an adaptive condition (referring to an abnormal formula assumed to be false), and then retracted after that condition is validated again.

Let us consider now a slightly more complex example. We want to show a situation in which a disjunction of two abnormalities can be derived: accordingly, there might be more than one formula to be marked. Let us start with a premise set $\Gamma = \{(p \vee r), \neg p, (p \vee q), \neg q, (\neg p \to q)\}$. Now consider the following derivation, dubbed $\mathbb{D}$:

$$\dfrac{\dfrac{\overline{\Gamma;\cdot\vdash_1 (p\vee r)}\text{ PREM}}{\Gamma;\cdot\vdash_2 p,r}\vee\text{E}\qquad \overline{\Gamma;\cdot\vdash_3 \neg p}\text{ PREM}}{\dfrac{\Gamma;\cdot\vdash_4 (p\wedge\neg p),r}{\Gamma;(p\wedge\neg p)^-\vdash_5 r}\quad (p\wedge\neg p)\in\Omega}\wedge\text{I}\quad\text{RC}$$

At stage 4 a disjunction of an abnormality with r is derived, and by RC at stage 6 the formula r is derived alone, assuming the relevant abnormality to be false. Consider now a second derivation, dubbed $\mathbb{D}'$:

$$\cfrac{\cfrac{\overline{\Gamma;\cdot \vdash_6 (p \vee q)}\;\text{PREM}}{\cfrac{\Gamma;\cdot \vdash_7 p, q}{\cfrac{\Gamma;\cdot \vdash_9 (p \wedge \neg p), q}{\cfrac{\Gamma;\cdot \vdash_{11} ((p \wedge \neg p), (q \wedge \neg q))^{min}}{\Gamma;\cdot \vdash_{12} \mathsf{UnRel}(\{p \wedge \neg p, q \wedge \neg q\})}\;\text{UnRel}}\;\text{RC}}}\;\vee\text{E}\quad \overline{\Gamma;\cdot \vdash_8 \neg p}\;\text{PREM}}{}}$$

Here, the abnormality $(p \wedge \neg p)$ that was previously assumed to be false is unconditionally derived in disjunctive form with a new abnormality $(q \wedge \neg q)$ at stage 11, where the latter is obtained by $\wedge$I from stages 7-9. If we join now the two branches $\mathbb{D}, \mathbb{D}'$ to form $\mathbb{E}$, we can apply the marking-rule (where $\mathsf{UnRel}(\Gamma)$ is the set $\{(p \wedge \neg p), (q \wedge \neg q)\}$):

$$\cfrac{\cfrac{\mathbb{D}}{\Gamma;(p \wedge \neg p)^{-} \vdash_5 r} \quad \cfrac{\mathbb{D}'}{\Gamma;\cdot \vdash_{12} \mathsf{UnRel}(\{p \wedge \neg p, q \wedge \neg q\})} \quad (p \wedge \neg p) \in \bigcup(\Delta(\Gamma))}{\Gamma;(p \wedge \neg p)^{-} \vdash_{13 \boxtimes \text{R}} r}\;\boxtimes\text{R}$$

At stage 13 the formula r is no longer valid, because its adaptive condition is in the set of unreliable formulas derived at stage 12. Now we can provide a further extension of this derivation dubbed $\mathbb{D}''$:

$$\cfrac{\cfrac{\overline{\Gamma;\cdot \vdash_{14} \neg p}\;\text{PREM} \quad \overline{\Gamma;\cdot \vdash_{15} \neg p \to q}\;\text{PREM}}{\Gamma;\cdot \vdash_{16} q}\;\to\text{I} \quad \overline{\Gamma;\cdot \vdash_{17} \neg q}\;\text{PREM}}{\Gamma;\cdot \vdash_{18} (q \wedge \neg q)}\;\wedge\text{I}$$

$\mathbb{D}''$ derives a single abnormality at stage 18, for which, as before, we skip the redundant step of deriving the singleton set of unreliable formulas. This also means that if we obtain a copy of derivation $\mathbb{D}$, where each step is renumbered consecutively, and join it to $\mathbb{D}''$ and $\mathbb{E}$, it is possible to deduce r anew with $(p \wedge \neg p)$ as its adaptive condition and accordingly leave this judgement unmarked at stage 20:

$$\cfrac{\cfrac{\mathbb{E}}{\Gamma;(p \wedge \neg p)^{-} \vdash_{13 \boxtimes \text{R}} r} \quad \cfrac{\mathbb{D}''}{\Gamma;\cdot \vdash_{18} (q \wedge \neg q)^{min}} \quad \cfrac{\mathbb{D}}{\Gamma;\cdot \vdash_{19} (p \wedge \neg p), r}}{\Gamma;(p \wedge \neg p)^{-} \vdash_{20} r}\;\text{RC*}$$

where $*$ is the side-condition that $(p \wedge \neg p) \in \Omega$.

19.4.5 Another Example

In the following example, the abnormalities are even more connected. For the sake of brevity, we do not derive a set of unreliable formulas when there is a single minimal disjunction of abnormalities. Let

$$\Gamma = \{p \vee q, \neg q,$$
$$(q \wedge \neg q) \vee (r \wedge \neg r), (q \wedge \neg q) \rightarrow (r \wedge \neg r),$$
$$(q \wedge \neg q) \vee (s \wedge \neg s), (q \wedge \neg q) \rightarrow (s \wedge \neg s)\}$$

Consider first the derivation $\mathbb{D}$:

$$\cfrac{\cfrac{\cfrac{\overline{\Gamma; \cdot \vdash_1 (p \vee q)}\text{ PREM}}{\Gamma; \cdot \vdash_2 p, q}\vee\text{E} \quad \overline{\Gamma; \cdot \vdash_3 \neg q}\text{ PREM}}{\Gamma; \cdot \vdash_4 (q \wedge \neg q), p}\wedge\text{I} \quad (q \wedge \neg q) \in \Omega}{\Gamma; (q \wedge \neg q)^- \vdash_5 p}\text{RC}$$

Which we extend as follows to form $\mathbb{E}$:

$$\cfrac{\mathbb{D} \qquad \cfrac{\overline{\Gamma \vdash_6 (q \wedge \neg q) \vee (r \wedge \neg r)}\text{ PREM}}{\Gamma; \cdot \vdash_7 ((q \wedge \neg q), (r \wedge \neg r))^{min}}\vee\text{E} \quad (q \wedge \neg q) \in \Delta^{min}}{\Gamma; (q \wedge \neg q)^- \vdash_{8\boxtimes\text{R}} p}\boxtimes\text{R}$$

We then construct $\mathbb{F}$ to show that a shorter disjunction of abnormalities can be derived:

$$\cfrac{\cfrac{\cfrac{\overline{\Gamma; \cdot \vdash_9 (q \wedge \neg q) \vee (r \wedge \neg r)}\text{ PREM}}{\Gamma; \cdot \vdash_{10} (q \wedge \neg q), (r \wedge \neg r)}\vee\text{E} \quad \overline{\Gamma; \cdot \vdash_{11} (q \wedge \neg q) \rightarrow (r \wedge \neg r)}\text{ PREM}}{\Gamma; \cdot \vdash_{12} (r \wedge \neg r), (r \wedge \neg r)}\rightarrow\text{E}}{\Gamma; \cdot \vdash_{13} r \wedge \neg r}\text{C}_R$$

As in the previous example, we put these branches together (and re-use a renumbered copy of $\mathbb{D}$) to obtain $\mathbb{G}$ and re-derive p on condition $(q \wedge \neg q)^-$:

$$\frac{\mathbb{E}}{\Gamma; (q \wedge \neg q)^- \vdash_{8 \boxtimes R} p} \quad \frac{\mathbb{F}}{\Gamma; \cdot \vdash_{13} (r \wedge \neg r)^{min}} \quad \frac{\mathbb{D}}{\Gamma; \cdot \vdash_{14} (q \wedge \neg q), p}$$
$$\frac{}{\Gamma; (q \wedge \neg q)^- \vdash_{15} p} \text{ RC*}$$

But consider then the following variant of $\mathbb{E}$, denoted by $\mathbb{E}^*$:

$$\frac{\mathbb{G}}{\Gamma; (q \wedge \neg q)^- \vdash_{15} p} \quad \frac{\overline{\Gamma \vdash_{16} (q \wedge \neg q) \vee (s \wedge \neg s)}^{\text{PREM}}}{\Gamma; \cdot \vdash_{17} ((q \wedge \neg q), (s \wedge \neg s))^{min}} \vee E \quad (q \wedge \neg q) \in \Delta^{min}$$
$$\frac{}{\Gamma; (q \wedge \neg q)^- \vdash_{18 \boxtimes R} p} \boxtimes R$$

and then a variant $\mathbb{F}^*$ of $\mathbb{F}$:

$$\frac{\overline{\Gamma; \cdot \vdash_{19} (q \wedge \neg q) \vee (s \wedge \neg s)}^{\text{PREM}}}{\Gamma; \cdot \vdash_{20} (q \wedge \neg q), (s \wedge \neg s)} \vee E \quad \overline{\Gamma; \cdot \vdash_{21} (q \wedge \neg q) \rightarrow (s \wedge \neg s)}^{\text{PREM}}$$
$$\frac{\Gamma; \cdot \vdash_{22} (s \wedge \neg s), (s \wedge \neg s)}{\Gamma; \cdot \vdash_{23} s \wedge \neg s} \text{ CR} \quad \rightarrow E$$

which can once more be used to re-derive p:

$$\frac{\mathbb{E}^*}{\Gamma; (q \wedge \neg q)^- \vdash_{18 \boxtimes R} p} \quad \frac{\mathbb{F}}{\Gamma; \cdot \vdash_{13} (r \wedge \neg r)} \quad \frac{\mathbb{F}^*}{\Gamma; \cdot \vdash_{23} (s \wedge \neg s)} \text{ UnRel} \quad \frac{\mathbb{D}}{\Gamma; \cdot \vdash_{14} (q \wedge \neg q), p}$$
$$\frac{\Gamma; \cdot \vdash_{24} \text{UnRel}(\{r \wedge \neg r, s \wedge \neg s\})}{\Gamma; (q \wedge \neg q)^- \vdash_{25} p} \text{ RC*}$$

19.4.6 An Example Based on Minimal Abnormality

Let $\Gamma = \{p \vee r, p \vee q, q \vee r, \neg p, \neg q\}$, and consider first a derivation $\mathbb{A}$ that ends with an application of RC.

$$\frac{\overline{\Gamma; \cdot \vdash_1 p \vee r}^{\text{PREM}}}{\Gamma; \cdot \vdash_2 p, r} \vee E \quad \overline{\Gamma; \cdot \vdash_3 \neg p}^{\text{PREM}}$$
$$\frac{\Gamma; \cdot \vdash_4 p \wedge \neg p, r}{\Gamma; (p \wedge \neg p)^- \vdash_5 r} \text{ RC} \quad \wedge I$$

Next, let $\mathbb{B}$ be the following derivation, where $*$ refers to the application of $\wedge$ I followed by the application of MINDAB, and $\star$ refers to the application of CHOICE followed by the application of MINCHOICE to single out $p \wedge \neg p$ from the only minimal disjunction of abnormalities:

$$
\cfrac{\cfrac{\cfrac{\cfrac{\Gamma; \cdot \vdash_6 p \vee q}{\Gamma; \cdot \vdash_7 p, q} \text{PREM}}{\Gamma; \cdot \vdash_9 p \wedge \neg p, q} \vee \text{E} \quad \cfrac{}{\Gamma; \cdot \vdash_8 \neg p} \text{PREM}}{\cfrac{\Gamma; \cdot \vdash_{11} (p \wedge \neg p, q \wedge \neg q)^{min}}{\Gamma; \cdot \vdash_{12} \text{choice}_1(\{\{p \wedge \neg p, q \wedge \neg q\}\})^{min}} \star} \wedge \text{I} \quad \cfrac{}{\Gamma; \cdot \vdash_{10} \neg q} \text{PREM}}{} *
$$

Combining both with an application of $\boxtimes$M2 gives us the derivation $\mathbb{C}$:

$$
\cfrac{\cfrac{\mathbb{A}}{\Gamma; (p \wedge \neg p)^- \vdash_5 r} \text{RC} \quad \cfrac{\mathbb{B}}{\Gamma; \cdot \vdash_{12} \text{choice}_1(\{\{p \wedge \neg p, q \wedge \neg q\}\})^{min}} \text{MINCHOICE}}{\Gamma; (p \wedge \neg p)^- \vdash_{13\boxtimes\text{MA}} r} \boxtimes\text{M2}
$$

The side-condition for the application of this marking-rule is satisfied because no other conditional derivation of r occurs earlier in the proof, and *a fortiori*, no deduction of r on a condition that does not intersect with the choice-set $p \wedge \neg p$ has been given.

Derivation $\mathbb{D}$ now provides a proof of r that relies on a different condition:

$$
\cfrac{\cfrac{\cfrac{\cfrac{\Gamma; \cdot \vdash_{14} q \vee r}{\Gamma; \cdot \vdash_{15} q, r} \text{PREM}}{\Gamma; \cdot \vdash_{17} q \wedge \neg q, r} \vee \text{E} \quad \cfrac{}{\Gamma; \cdot \vdash_{16} \neg q} \text{PREM}}{\Gamma; (q \wedge \neg q)^- \vdash_{18} r} \wedge \text{I}}{} \text{RC}
$$

At this point, we can see that neither of the following attempts to mark conditional deductions of r using the minimal abnormality marking-rule complies with the side-conditions for $\boxtimes$M2:

$$\frac{\mathbb{A}'}{\Gamma; (p \wedge \neg p)^- \vdash_{\mathbf{s}} r} \text{RC} \qquad \frac{\mathbb{B}'}{\Gamma; \cdot \vdash_{\mathbf{s}'} \text{choice}_1(\{\{p \wedge \neg p, q \wedge \neg q\}\})^{min}} \text{MinChoice}$$
$$\frac{}{?} \boxtimes \text{M2}$$

$$\frac{\mathbb{D}}{\Gamma; (q \wedge \neg q)^- \vdash_{18} r} \text{RC} \qquad \frac{\mathbb{B}''}{\Gamma; \cdot \vdash_{\mathbf{s}''} \text{choice}_2(\{\{p \wedge \neg p, q \wedge \neg q\}\})^{min}} \text{MinChoice}$$
$$\frac{}{?} \boxtimes \text{M2}$$

For the first attempt, it is the presence of judgement 18 elsewhere in the proof that shows that a derivation exists that does not depend on the underivability of $p \wedge \neg p$. For the second attempt, it is the presence of judgement 5 that (notwithstanding the marked judgement 12!) signals that a derivation of r that does not depend on the underivability of $q \wedge \neg q$ exists.

19.4.7 Reconstruction of Linear Adaptive Proofs

Before we consider the question of final derivability, which is needed to relate what is provable in dynamic proofs to a semantic consequence relation, we first show that every proof in **AdaptiveND** can, with the help of the numbering of the judgements, be mapped onto an (albeit somewhat redundant) adaptive proof in a linear format. This relates what can be derived at a stage in an **AdaptiveND**-proof to what can be derived at a stage in a linear proof.

We illustrate the procedure by reconstructing the linear proof that corresponds to the example form Sects. 19.3.1 and 19.4.2:

(1) $\neg p \vee q$ Prem $\emptyset$
(2) $\bigvee\{\neg p, q\}$ $\vee$E, (1) $\emptyset$
(3) p Prem $\emptyset$
(4) $\bigvee\{p \wedge \neg p, q\}$ $\wedge$I, (2, 3) $\emptyset$
(5) q RC, (4) $\{p\} \boxtimes^{10}$
(6) p Prem $\emptyset$
(7) $p \rightarrow \neg p$ Prem $\emptyset$
(8) $\neg p$ $\rightarrow$E, (6, 7) $\emptyset$
(9) p Prem $\emptyset$
(10) $p \wedge \neg p$ $\wedge$I, (8, 9) $\emptyset$

In this proof, the application of $\vee$E on line (2) is based on the representation of the disjunctive comma by a 'super-imposed' classical disjunction (a device that effectively plays the same role in adaptive logic, see Strasser (2014, §2.2, 2.7), whereas the application of $\wedge$I is valid in virtue of the **CLuN**-validity of $\neg p \vee q, p \vdash (p \wedge \neg p) \vee q$ which warrants the application of the unconditional rule (with empty conditions). The final marking is not added as a separate line, but is instead added in the fifth place on line 5 and labelled with the number of the line or stage at which the relevant abnormality was derived.

As illustrated in a second example based on the proof from Sect. 19.4.4, there is no guarantee that the translation of an **AdaptiveND** proof into a linear adaptive proof will contain all the markings required by the latter.

(1) $p \vee r$ Prem $\emptyset$
(2) $\bigvee\{p, r\}$ $\vee$E, (1) $\emptyset$
(3) $\neg p$ Prem $\emptyset$
(4) $\bigvee\{p \wedge \neg p, r\}$ $\wedge$ I, (2, 3) $\emptyset$
(5) r RC, (4) $\{p\}$ $\boxtimes^{??}$
(6) $p \vee q$ Prem $\emptyset$
(7) $\bigvee(p, q)$ $\vee$E, (6) $\emptyset$
(8) $\neg p$ Prem $\emptyset$
(9) $\bigvee(p \wedge \neg p, q)$ $\wedge$ I, (7, 8) $\emptyset$
(10) $\neg q$ Prem $\emptyset$
(11) $\bigvee(p \wedge \neg p, q \wedge \neg q)$ $\wedge$ I, (9, 10) $\emptyset$

Indeed, the above proof is a straightforward translation of the tree-form proof up to stage 11, but the marking of line 5, which should be added to the linear proof once $\bigvee(p \wedge \neg p, q \wedge \neg q)$ is derived, cannot be added by a mechanical translation procedure. This is because in the standard proof-format marking is governed by a definition, which simply stipulates when a line is marked, whereas in **AdaptiveND** marking is governed by rules and therefore requires the execution of additional inferential steps.[1] To show that proofs in **AdaptiveND** correctly capture the adaptive dynamics, we will therefore have to show that such 'missing markings' can always be obtained by a further extension of a proof.

In the next section, we complete our system with the required meta-theoretical analysis needed to define derivability at stage and final derivability.

19.5 Derivability

In the example from the previous section, we have illustrated how the marking condition establishes a dynamic derivability relation, which allows to derive formulas and retract them. Whenever a certain formula is derived on some $\phi \in \Delta^{min}$ adaptive condition, it might still be marked afterwards. Consequently, a judgement of the form $\Gamma; \Theta^- \vdash_\mathbf{s} \psi$ only expresses what is derived at a stage. This gives us the notion of derivability at a stage:

Definition 19.9 (*Derivability at stage*). A formula ψ is derived at stage $\mathbf{s}$ iff $\Gamma; \phi^- \vdash_{\mathbf{s}'} \psi$, where $\mathbf{s}' \leq \mathbf{s}$ and it is not the case that $\Gamma; \cdot \vdash_{\mathbf{s}''\boxtimes} \psi$ for some $\mathbf{s}' \leq \mathbf{s}'' \leq \mathbf{s}$.

---
[1] See also footnote 10 of Batens (2008) for a discussion of this distinction.

A more stable notion of derivability, called *final derivability*, holds when marking is no longer possible. This notion is customarily defined with a reference to possible extensions of a proof.[2] By only taking finite premise sets into consideration, we can pursue a more explicit characterisation of final derivability.

To this aim, one requires that the stage s at which a formula ψ is derived remains unmarked in all the extensions of the derivation tree, which can be obtained by using all *relevant* abnormalities as adaptive conditions. This relevance criterion is essential if one wants to guarantee finite surveyability of the proof-tree to establish whether a formula is never marked (again). We define, therefore, a set of *abnormalities relevant to* Γ. To do so we first identify the union set of all sub-formulas of the premise set Γ:

Definition 19.10 *(Subformulas of the premise set).* $\mathrm{Sf}(\Gamma) = \bigcup_{\phi \in \Gamma} \{\psi \mid \psi \text{ is a sub-formula of } \phi\}$.

From $\mathrm{Sf}(\Gamma)$, we then construe all the possible abnormalities that can be obtained from its members:

Definition 19.11 *(Abnormalities relevant to the premise set).* $\Omega(\Gamma) = \{\psi \wedge \neg\psi \in \Omega \mid \psi \in \mathrm{Sf}(\Gamma)\}$.

For **AdaptiveND**, **CLuN**^{r}, and **CLuN**^{m} the requirement that all $\psi \wedge \neg\psi$ should be in Ω is trivially satisfied. This condition becomes mandatory when Ω is based on a restricted logical form; e.g. when abnormalities are contradictions of the form $\psi \wedge \neg\psi$ with ψ atomic. In that case, our definition of $\Omega(\Gamma)$ is co-extensive with the more basic $\{\psi \wedge \neg\psi \mid \psi \in \mathrm{At}(\Gamma)\}$.

Theorem 19.2 *If $\bigvee(\Delta)$ is a minimal disjunction of abnormalities derivable from Γ, then $\Delta \subseteq \Omega(\Gamma)$.*

Proof We consider the possible ways of deriving a minimal disjunction of abnormalities by examining the structure of the proof-rules of minimalND.

1. Applying ($\vee$I) (or WR) can never result in a minimal disjunction of abnormalities. This excludes all proof-rules that can be used to deduce a judgement with a formula on the right that isn't yet a formula or sub-formula in one of the judgements it relies on.
2. A formula of the form $\phi \wedge \neg\phi$ can be derived on the right of the turn-style if it is already a sub-formula of some premise, or the result of ($\wedge$I).
3. If $\phi \wedge \neg\phi$ is the result of ($\wedge$I), each of its conjuncts should be derivable. We focus on the proof-paths to formulae of the form $\neg\phi$, of which there are four:

    (a) $\neg\phi$ is a premise;
    (b) $\neg\phi$ can be obtained by ($\wedge$E) from some $\neg\phi \wedge \psi$ on the right;
    (c) $\neg\phi$ can be obtained by ($\rightarrow$E) from some $\psi \rightarrow \neg\phi$ on the right;

---

[2] 'A is finally derived from Γ on line i of a proof at stage s iff (i) A is the second element of line i, (ii) line i is not marked at stage s, and (iii) every extension of the proof in which line i is marked may be further extended in such a way that line i is unmarked.' (Batens, 2007, 229).

(d) $\neg\phi$ can be obtained by $(\neg I)$ from ϕ on the left.

Cases (a–c) imply that $\neg\phi$ should be a positive part of a previously derived formula on the right, and hence ϕ should be a negative part of that formula. By induction over the length of proofs (with the rule **PREM** as the base-case), these three cases can be retraced to ϕ being a negative part of some premise.

Case (d) requires the presence or deduction of some ϕ on the left, either because the left-hand side is of the form Γ, ϕ and thus the result of applying (WL), or because it is of the form Γ with $\phi \in \Gamma$. In each of these cases, this can never lead to a judgement where (i) the left-side consists only of the premise set, and (ii) the right-side has no more formulae than before the application of $(\neg I)$. This implies that case (d) cannot lead to the deduction of a minimal disjunction of abnormalities.

Consequently, for every abnormality $\phi \wedge \neg\phi$ that occurs in a minimal disjunction of abnormalities, we can show that ϕ must occur as the negative part of some premise. A fortiori, this means that only abnormalities that can be formed from a member of $\Omega(\Gamma)$ can occur in a minimal disjunction of abnormalities. □

The focus on positive and negative parts of formulae goes back to Schutte (1960), and was previously used for the development of goal-directed proof-strategies for adaptive logics (Batens and Provijn 2001). The fact that we should pay attention to all negative parts of the premises should also be obvious in view of the semantics for **CLuN**, as the truth-value of negative formulae does not need to depend on the truth-values of its sub-formulae.

Theorem 19.2 helps us to characterize finite proof-trees to decide whether a formula is finally derived by identifying the abnormalities derivable in view of the syntactical form of the premises. But it can also be seen as a **CLuN**-specific variant of the *Derivability Adjustment Theorem* from Batens (2007). This result can be stated in multiple-conclusion form as follows:

$$\Gamma \vdash_{\textbf{ULL}} \phi \text{ iff } \Gamma \vdash_{\textbf{LLL}} \phi, \Delta \text{ for some finite } \Delta \subset \Omega$$

Or yet it can be seen as a **CLuN**-alternative of a result from Beall (2011) that relates **LP$^+$**, the multiple-conclusion extension of **LP**, and **CPL$^+$**, the multiple-conclusion extension of classical logic:

$$X \models^+_{\textbf{CPL}} Y \text{ iff } X \models^+_{\textbf{LP}} Y \cup \iota(X) \qquad \text{(LP/CPL)}$$

with $\iota(X) = \{p \wedge \neg p : p \in \text{At}(X)\}$.

Here, we do not use such connections to bridge different approaches to classical recapture, but instead rely on it to introduce the notion of a *complete proof-tree* with respect to derivable relevant disjunction of abnormalities:

Definition 19.12 *(Completeness relative to relevant abnormalities).* Let P be an AdaptiveND proof. We say that P is complete relative to $\Omega(\Gamma)$ at stage **s** if for every derivable $\bigvee(\Delta^{min})$ with $\Delta \subseteq \Omega(\Gamma)$ there is an **s**$'$ < **s** such that $\Gamma; \cdot \vdash_{\textbf{s}'} \bigvee(\Delta^{min})$.

Definition 19.13 *(Completeness relative to marking)*. Let P be an AdaptiveND proof. We say that P is complete relative to marking iff

Rel for every $\Gamma; \Theta^- \vdash_s \phi$ occurring in a tree $T_i \in P$, if P can be extended so that it includes a judgement $\Gamma; \cdot \vdash_{s'} \mathsf{UnRel}(\Gamma)$ with $\bigcup \Delta(\Gamma) \cap \Theta \neq \emptyset$, then there is a tree $T_{t>i} \in P$ that ends with $\Gamma; \Theta^- \vdash_{t\boxtimes} \phi$.

MinAb1 for every $\Gamma; \Theta^- \vdash_s \phi$, if every derivable $\mathsf{choice}_i(\{\Delta_1, \ldots, \Delta_n\})$ intersects with Θ, then there is a tree $T_{t>i} \in P$ that ends with $\Gamma; \Theta^- \vdash_{t\boxtimes} \phi$.

MinAb2 for every $\Gamma; \Theta^- \vdash_s \phi$, if for every other derivable $\Gamma; \Theta'^- \vdash_s \phi$, some derivable $\mathsf{choice}_i(\{\Delta_1, \ldots, \Delta_n\})$ intersects with each $\Theta, \ldots, \Theta'$, then there is a tree $T_{t>i} \in P$ that ends with $\Gamma; \Theta^- \vdash_{t\boxtimes} \phi$.

We can now formulate our notion of final derivability:

Definition 19.14 *(Final Derivability)*. A formula ψ is finally derived $\Gamma; \Theta^- \vdash_\checkmark \psi$ iff $\Gamma; \Theta^- \vdash_s \psi$ occurs in an abnormality and marking complete proof P, where $\Gamma; \Theta^- \vdash_{s'\boxtimes} \psi$ does not occur for any $s' \geq s$.

Theorem 19.3 $\Gamma; \Theta^- \vdash_\checkmark \psi$ *in* AdaptiveND *if and only if there is a final derivation of ψ from Γ in a standard linear adaptive proof.*

Proof $\rightarrow$ By assumption ψ is finally derived, therefore there is an abnormality and marking complete proof P in which it is derived. Then P, by definition, contains unconditional judgements for every deducible minimal disjunction of abnormalities in $\Omega(\Gamma)$ and ψ is not contained in any of those. Now consider the translation **P** of P in a linear adaptive proof: the same minimal disjunctions of abnormalities as in P are also unconditionally derived in **P**. By Theorem 19.2, this implies that all minimal disjunctions of abnormalities are unconditionally derived in **P**, and further extensions of the proof cannot lead to additional unconditionally derived minimal disjunctions of abnormalities. Because P contains all possible markings, every formula which is marked in P will be marked in **P**, and **P** will therefore be in accordance with the standard marking definitions. As ψ is not in any adaptive conditions of P which induces a marking, it will be derived in **P** as well. Moreover, ψ will be finally derived in **P** because further extensions would not lead to newly derived minimal disjunctions of abnormalities, and would not lead to additional marking either.

$\leftarrow$ Assume that ψ is finally derived in some linear adaptive proof **P**; then

for **CluN**^{R} $\Gamma \vdash_\mathsf{CluN} \psi \vee \phi$ with $\phi \cap U(\Gamma) = \emptyset$. By Theorem 19.1 there is a provable AdaptiveND judgement $\Gamma; \cdot \vdash_s \psi, \phi$. By applying RC, we obtain a judgement $\Gamma; \phi^- \vdash_{s+1} \psi$. If at some later stage s' an abnormality-complete proof is obtained, any $\bigcup \Delta(\Gamma)$ derived thereafter will, by Proposition 19.1 be a subset of $U(\Gamma)$. Consequently, any application of $\boxtimes$R to a judgement with condition ϕ^- would from that point also require $\phi \in \bigcup \Delta(\Gamma)$. Since this would contradict our assumption that $\phi \cap U(\Gamma) \neq \emptyset$, no such marking can be applied.

for **CluN**^{M} $\Gamma \vdash_\mathsf{CluN} \psi \vee \Theta$, with $\Theta \subseteq \Omega(\Gamma)$ and

1. Either for every $\mathsf{choice}_i(\{\Delta_1, \ldots, \Delta_n\}) \in \Phi(\Gamma)$ we have

$$\Theta \cap \mathsf{choice}_i(\{\Delta_1, \ldots, \Delta_n\}) = \emptyset.$$

By Theorem 19.1, there is a provable **AdaptiveND** judgement $\Gamma; \cdot \vdash_\mathbf{s} \psi, \Theta$, and by repeated applications of RC, we obtain the judgement $\Gamma; \Theta^- \vdash_{\mathbf{s+m}} \psi$. Let $t \geq s + m$ be a stage at which this proof is marking and abnormality-incomplete, and assume, *for reductio* that it ends with a judgement $\Gamma; \Theta^- \vdash_{t\boxtimes\mathsf{MA}} \psi$. Consequently, this judgement must occur as the final node of a marking-tree for minimal abnormality:

$$\cfrac{\Gamma;\Theta^- \vdash_s \psi \quad \cfrac{\Pi}{\Gamma, \cdot \vdash_{s+1} \Phi'(\Gamma)} \text{\scriptsize Choice} \quad \cfrac{}{\Gamma; \cdot \vdash_{s+2} \mathsf{choice}_1(\{\Delta'_1, \ldots, \Delta'_m\})^{min}} \text{\scriptsize MinChoice} \quad \ldots \quad \cfrac{\Pi'}{\Theta \cap \mathsf{choice}_i \neq \emptyset}}{\Gamma; \Theta^- \vdash_{t\boxtimes\mathsf{MA}} \psi} \text{\scriptsize{$\boxtimes$M}}$$

To complete our argument, we rely on the fact that every choice-set derivable from $\Phi'(\Gamma)$ is used as a premise for the application of the marking-rule, but we do not need to assume that $\Phi'(\Gamma)$ is identical to $\Phi(\Gamma)$. Because $\Phi'(\Gamma)$ is derived at a stage of the proof that is already abnormality-complete, we know that $\{\Delta'_1, \ldots, \Delta'_m\} \subseteq \{\Delta_1, \ldots, \Delta_n\}$ (each Δ'_i is equal to some Δ_i, but not vice-versa). Proposition 19.3 then guarantees that each $\mathsf{choice}_j(\{\Delta_1, \ldots, \Delta_n\})$ is a superset of some $\mathsf{choice}_i(\{\Delta'_1, \ldots, \Delta'_m\})$. Consequently, since the marking at stage t required that $\Theta \cap \mathsf{choice}_i(\{\Delta'_1, \ldots, \Delta'_m\}) \neq \emptyset$ for each choice-set in $\Phi'(\Gamma)$, it should also hold that $\Theta \cap \mathsf{choice}_i(\{\Delta_1, \ldots, \Delta_n\}) \neq \emptyset$ for each choice-set in $\Phi(\Gamma)$, which contradicts our initial assumption.

2. or for every $\mathsf{choice}_i(\{\Delta_1, \ldots, \Delta_n\}) \in \Phi(\Gamma)$, if $\mathsf{choice}_i(\{\Delta_1, \ldots, \Delta_n\}) \cap \Theta \neq \emptyset$, there is a Θ' such that $\Gamma \vdash_{\mathbf{CluN}} \psi \vee \Theta'$ with

$$\Theta' \cap \mathsf{choice}_i(\{\Delta_1, \ldots, \Delta_n\}) = \emptyset.$$

By Theorem 19.1, there are provable **AdaptiveND** judgements $\Gamma; \cdot \vdash_\mathbf{s} \psi, \Theta$ and $\Gamma; \cdot \vdash_{\mathbf{s'}} \psi, \Theta'$ for Θ and each such Θ'. As in the previous case, by repeated applications of RC we can derive corresponding $\Gamma; \Theta^- \vdash_{\mathbf{s+m}} \psi$ and $\Gamma; \Theta'^- \vdash_{\mathbf{s'+m'}} \psi$. Let t be a stage at which this proof is extended to an abnormality and marking complete proof, and assume, *for reductio*, that it ends with a judgement $\Gamma; \Theta^- \vdash_{t\boxtimes\mathsf{MA}} \psi$. Consequently, this judgement must occur as the final node of a marking-tree for minimal abnormality:

$$\cfrac{\Gamma; \Theta^- \vdash_s \psi \quad \ldots \quad \Gamma; \Theta'^- \vdash_{s'} \psi \quad \cfrac{\Pi}{\Gamma, \cdot \vdash_{s''} \Phi'(\Gamma)} \text{\scriptsize Choice} \quad \cfrac{}{\Gamma, \vdash_{s''+1} \mathsf{choice}_i(\{\Delta'_1, \ldots, \Delta'_m\})^{min}} \text{\scriptsize MinCh} \quad \forall(\Theta \ldots \Theta') \cap \mathsf{choice}_i \neq \emptyset}{\Gamma; \Theta^- \vdash_{t\boxtimes\mathsf{MA}} \psi} \text{\scriptsize{$\boxtimes$M}}$$

But then, by the same reasoning as above, Proposition 19.3 entails that if $\mathsf{choice}_i(\{\Delta'_1, \ldots, \Delta'_m\})$ intersects with every condition $\Theta, \ldots, \Theta'$, it must also intersect with some $\mathsf{choice}_j(\{\Delta_1, \ldots, \Delta_n\}) \in \Phi(\Gamma)$, which contradicts our initial assumption.

Therefore, $\Gamma; \Theta^- \vdash_{t\boxtimes\mathsf{MA}} \psi$ cannot occur at any stage t of a proof that is abnormality-complete at t. □

19.6 Concluding Remarks

To conclude, we would like to highlight certain distinctive features of the proposed calculus, and briefly discuss how these features can be used to reconsider the question of classical recapture. As we see it, the defeasible reasoning-forms formalized in our **AdaptiveND** system have three primary virtues:

1. They are formulated in a tree-format that forces one to state all information used in an inference-step explicitly, and this restricts the reliance on global features of a proof to a minimum (e.g. when checking that a disjunction of abnormalities is minimal);
2. The multiple-conclusion format leads to a transparent connection between the restricted inference-rules that are valid in **minimalND** (i.e. the lower limit logic) and their use as a premise of the conditional rule;
3. The explicit individuation of abnormalities as a subtype of the well-formed formulae.

The explicit connection between multiple conclusions and defeasible inferences brings a recent disagreement over the problem of classical recapture in the logic **LP** into focus.[3] In several papers, Graham Priest has explicitly endorsed the adaptive approach to classical recapture. To that effect, he has proposed his own *minimally inconsistent* **LP**: an adaptive logic based on a stronger paraconsistent logic (but without a detachable implication) and the minimal abnormality strategy (Priest 1991). This approach has been criticized by JC Beall, another prominent defender of the logic **LP**, on the ground that any all-purpose logic should at any cost prevent one to step from truth to falsehood (Beall 2012). This is a task that cannot in general be fulfilled by an adaptive logic, and indeed a task we shouldn't impute on adaptive logics in the first place (Priest 2012). By contrast, Beall's preferred take on classical recapture is that it should be handled with extra-logical means. The multiple-conclusion extensions of classical logic and **LP** already mentioned in the previous section provide formalisms in which this idea can be made precise, since (LP/CPL) can be seen as a minimalist expression of how paraconsistent logics like **LP** incorporate classical logic in a restricted form. Given the central role of similar multiple-conclusion judgements in **AdaptiveND**, results like (LP/CPL) should really be understood as agnostic between the different strategies for classical recapture. Indeed, whereas Beall advocates the view that **LP**$^+$ only presents us with logically viable options, these same options work as the motor behind any defeasible inference mechanism that allows one to favour one of them in the first place. Presentations of defeasible approaches to classical recapture based on the selection of minimally abnormal models bypass references to logical options: but their use in dynamic proofs relies implicitly or explicitly on the individuation of logical options that conform to a particular logical form. When adaptive logics are formulated according to the standard format, this type of connection is already made visible (Priest's minimally inconsistent **LP** is not formulated in this generic format) through the *Derivability Adjustment Theorem*

---

[3]See Allo (2016, 18ff) for a more detailed reconstruction of this debate.

mentioned in the previous section. **AdaptiveND** makes this connection even more explicit, by formulating its conditional rule with a multiple-conclusion judgement as a premise, and enforcing the condition that a logical option must have a particular logical form to be moved from the right-hand side where it is a logical possibility to the left-hand side where it is used as a negative condition.

The formal approach taken in the development of **AdaptiveND** signals another crucial departure from the terms in which the Priest/Beall debate is carried out, namely a departure concerning the individuation of abnormalities. Within the adaptive logic tradition, abnormalities are understood as formulae of a specific logical form, and the abnormality of models (e.g. how inconsistent they are) is measured relative to the abnormal formulas they verify. When compared to the road taken by minimally inconsistent **LP**, this has certain advantages (see Batens (2000) for a diagnosis of this problem in the first-order case). The same syntactic approach to abnormalities is integrated in **AdaptiveND** through the identification of a class of formulae of type Ω and the need to state membership of Ω when the conditional rule is applied. This approach is more general in the sense that it doesn't have to appeal to semantic concepts like *gluts* in its formulation, and it can explain how we step from logical options to defeasible inferences by only taking into account the logical form of the premises at hand. From a proof-theoretic viewpoint, this could be seen as a more explicit approach, whereas from the standpoint of the broader adaptive logic programme it is definitely more flexible.

Acknowledgements Patrick Allo—Supported by the European Union's Horizon 2020 research and innovation programme under the Marie Skłodowska-Curie Grant Agreement No. 657017.

References

Allo, P. (2016). Logic, reasoning and revision. *Theoria*, *82*(1), 3–31.
Batens, D., De Clerq, K., Verdée, P., & Meheus, J. (2008). Yes fellows, most human reasoning is complex. *Synthese*, *166*(1), 113–131.
Batens, D. (1980). Paraconsistent extensional propositional logics. *Logique & Analyse*, *23*(90–91), 195–234.
Batens, D. (1989). Dynamic dialectical logics. In P. Graham, R. Richard, & J. Norman (Eds.) *Paraconsistent logic—essays on the inconsistent* (pp. 187–217). Philosophia Verlag, München/Hamden/Wien.
Batens, D. (2000). Minimally abnormal models in some adaptive logics. *Synthese*, *125*(1), 5–18.
Batens, D. (2001). A general characterization of adaptive logics. *Logique & Analyse*, *173–175*, 45–68.
Batens, D. (2007). A universal logic approach to adaptive logics. *Logica Universalis*, *1*, 221–242.
Batens, D., & Provijn, D. (2001). Pushing the search paths in the proofs. a study in proof heuristics. *Logique & Analyse*, *44*(173–175), 113–134.
Beall, Jc. (2012). Why Priest's reassurance is not reassuring. *Analysis*, *72*(3), 517–525.
Beall, J. C. C. (2011). Multiple-conclusion LP and default classicality. *The Review of Symbolic Logic*, *4*(2), 326–336.
Pfenning, F. (2004). *Handout on automated theorem proving*. Technical Report, School of Computer Science, Carnegie Mellon University.

Priest, G. (1991). Minimally inconsistent LP. *Studia Logica*, *50*, 321–331.
Priest, G. (2006). *In contradiction* (2nd ed.). Oxford: Oxford University Press.
Priest, G. (2012). The sun may not, indeed, rise tomorrow: a reply to Beall. *Analysis*, *72*(4), 739–741.
Schütte, K. (1960). *Beweistheorie*. Berlin, Springer.
Strasser, C. (2014). Adaptive logics for defeasible reasoning. *Trends in Logic*. Springer.
Troelstra, A. S., & Schwichtenberg, H. (2000). *Basic Proof Theory* (2nd ed.). New York, NY, USA: Cambridge University Press.

Patrick Allo is a member of the Centre for Logic and Philosophy of Science at the Vrije Universiteit Brussel, and a research associate of the Digital Ethics Lab at the Oxford Internet Institute, University of Oxford. From October 2015 to September 2017 he was a Marie Skłodowska-Curie Fellow at the University of Oxford.

Giuseppe Primiero is Associate Professor of Logic at the University of Milan. Previously, he has held positions at the Department of Computer Science, Middlesex University London (UK), the Centre for Logic and Philosophy of Science, Ghent University (Belgium) and at the Philosophy Faculty, Leiden University (the Netherlands). His research interests are primarily in Logic and Computation, Philosophy of Computing and Information, Agent-based Modelling and Computer Simulations, History of Computing. Giuseppe currently acts as Secretary General for the Association Computability in Europe, is President of the Commission for the History and Philosophy of Computing, Member at Large of the Executive Board of the International Association for Computing and Philosophy and is Associate Editor for Philosophy of Computer Science & Technology for the Journal Philosophy and Technology (Springer).

Chapter 20
Denotation, Paradox and Multiple Meanings

Stephen Read

Abstract In line with the Principle of Uniform Solution, Graham Priest has challenged advocates like myself of the "multiple-meanings" solution to the paradoxes of truth and knowledge, due to the medieval logician Thomas Bradwardine, to extend this account to a similar solution to the paradoxes of denotation, such as Berry's, König's and Richard's. I here rise to this challenge by showing how to adapt Bradwardine's principles of truth and signification for propositions to corresponding principles of denotation and signification for descriptive phrases, applying them to give a "multiple-meanings" solution to the denotational paradoxes.

Keywords Multiple-meanings · Berry's paradox · Berkeley's paradox · Hilbert-Bernays' paradox · Epimenides' paradox · Heterologicality · Bradwardine

20.1 Paradox

Prior (1958) observed that Epimenides' claim that all Cretans are liars cannot have been the only claim made by a Cretan. Suppose it were and that it was true: then it would be a lie, and so false; but if false, then not a lie and if the only Cretan utterance, true. Prior inferred that Epimenides, or another Cretan, must have said something else, and indeed, something true. Thomas Bradwardine, writing 600 years before Prior, appreciated the point but drew a more radical and more plausible conclusion.

Graham Priest and I became firm friends over 40 years ago, when he came to St Andrews on a fixed-term Lectureship. We bounced ideas off one another from the start, and from time to time have agreed sufficiently on some topic to write joint papers. One thing he has never convinced me of, however (or at least, not long enough to survive his leaving the room), is dialetheism. I remain firmly committed to the law of Contravalence (that nothing can be both true and false) as much as to the laws of Non-Contradiction and Excluded Middle. But my ideas about logic, logical paradoxes and logical consequence, such as they are, would not have developed so fruitfully without his constructive criticisms.

S. Read (✉)
Arché Research Centre, University of St Andrews, St Andrews, Scotland
e-mail: slr@st-andrews.ac.uk

© Springer Nature Switzerland AG 2019
C. Başkent and T. M. Ferguson (eds.), *Graham Priest on Dialetheism and Paraconsistency*, Outstanding Contributions to Logic 18, https://doi.org/10.1007/978-3-030-25365-3_20

For it is in itself paradoxical to suppose that pure reflection on Epimenides' utterance could reveal to us that there was another Cretan utterance, let alone a true one. Rather, Bradwardine inferred, it is impossible for any utterance to mean only that that very utterance is a lie.[1] Similarly, no Cretan utterance can signify only that all Cretans are liars, if they all were liars. More generally, suppose that s signifies only that s is false, and suppose that s is false. It follows that something it signifies must fail to obtain, that is, it will not be false but true. So s must also signify that s is true, for Bradwardine claimed that signification is closed under consequence. Hence s cannot signify only that it is false.

The argument appeals to a principle that Bradwardine (2010, ¶6.3) cites as his second postulate:

> Every utterance signifies or means as a matter of fact or absolutely everything which follows from it as a matter of fact or absolutely.[2]

We can unpack this as claiming that the meaning of every utterance encompasses everything that follows from it either of necessity or even contingently, depending on how things are.[3] Thus Bradwardine allows both for inherently (absolutely) paradoxical utterances, and for (merely) contingent paradox, depending on contingent matters of fact, e.g., on who uttered it, or on the existence or nonexistence of other utterances and whether they are true or false.

We can formalize Bradwardine's postulate as a closure principle[4]:

$$(\forall p, q)((p \Rightarrow q) \rightarrow (\mathbf{Sig}(s, p) \rightarrow \mathbf{Sig}(s, q))) \tag{P2}$$

He also invokes his account of truth and falsity:

$$\mathbf{Tr}(s) =_{df} (\exists p)\mathbf{Sig}(s, p) \wedge (\forall p)(\mathbf{Sig}(s, p) \rightarrow p) \tag{TR}$$

that is, an utterance is true just when it is significative (there is something it signifies) and everything it signifies obtains, and

$$\mathbf{Fa}(s) =_{df} (\exists p)(\mathbf{Sig}(s, p) \wedge \neg p) \tag{FA}$$

---

[1] Bradwardine (2010, ¶ad A.4.3). See also Read (2009, §2). For the purposes of this paper, I will assume that the additional chapter contained in Appendix A to Bradwardine (2010) is indeed by Bradwardine. If not, it is certainly by an adherent and advocate of his views on the matter.

[2] Bradwardine (2010, ¶6.3): *Quelibet propositio significat sive denotat ut nunc vel simpliciter omne quod sequitur ad istam ut nunc vel simpliciter.*

[3] The externalist implications of this account of meaning or signification are explored in Cameron (2012). One might worry that the principle is too strong, just as many have objected to the unconstrained closure of knowledge under consequence. I explored ways to restrict Bradwardine's principle in Read (2015a).

[4] I've argued in a number of places, e.g., Read (2015b, pp. 399–400), that Bradwardine's second postulate should be interpreted as a closure principle, although Bradwardine's formal statement does not have that form. But he repeatedly applies it in that way.

that is, an utterance is false just when something it signifies fails to obtain.[5] (TR) and (FA) entail Bivalence, that every significative utterance is either true or false, and Contravalence, that none is both.[6]

From these principles, Bradwardine was able to show, as above, that every utterance which signifies its own falsity also signifies its own truth. Take Epimenides' claim that all Cretans are liars, for example. Since Epimenides was himself a Cretan, this claim entails that he is a liar, and hence that his own claim is a lie, so by (P2), his claim signifies that it is itself a lie. Now take any utterance, call it s, which signifies its own falsehood, and, in contrast to the earlier proof, suppose it signifies other things as well (as Epimenides' utterance does—it not only signifies that all Cretans are liars, but also that it itself is a lie, that Aenesidemus, also a Cretan, is a liar, that there are liars, and so on). If s is false, something it signifies must fail to obtain (by FA), so if it is not something else it signifies that fails, it must be that it is false that fails, that is, if it's false and whatever else it signifies obtains, it follows that it is true. But clearly, it signifies that it is false and whatever else it signifies, so by (P2) it signifies that it is true.[7] So any utterance which signifies that it is false also signifies that it is true. In particular, Epimenides' claim that all Cretans are liars also signifies, not only that it is itself false, but also that it is true (if all Cretans were liars). But it cannot be both true and false, so things cannot be wholly as it signifies, so by (FA) it is false. Moreover, we cannot infer from the fact that it is false that it is true, for (TR) requires for its truth that everything it signifies obtain, and that is impossible. Nothing can be both true and false.

This is Bradwardine's "multiple-meanings" solution to Epimenides' paradox. It can be extended to deal with Eubulides' Liar ("What I am saying is false"), the postcard (or "yes"–"no") paradox, the "no"–"no" paradox, Curry's paradox, the validity paradox and many others.[8] What, however, of the paradoxes of denotation?

---

[5]Restall (2008, p. 229–30) pointed out that Bradwardine's theory of signification collapses to triviality if "→" is taken to be material implication. In Read (2008, pp. 206–7) I observed that Bradwardine's argument works and his theory is nontrivial when "→" is taken as relevant implication.

[6]Whereas the variable "s" in (P2), (TR) and (FA) is a normal first-order variable ranging over utterances, the variables "p" and "q" are second-order propositional variables, that is, they should not be instantiated by *names* of propositions but by sentences expressing those propositions. For defence of the coherence of propositional quantification, see, e.g., Read (2006, 2007, 2008) or Rumfitt (2014).

[7]I noted in Read (2011, p. 231) that, strictly speaking, Bradwardine appeals here to a stronger principle than (P2):

$$(\forall p, q, r)((p \wedge q \Rightarrow r) \rightarrow (\mathbf{Sig}(s, p) \wedge \mathbf{Sig}(s, q) \rightarrow \mathbf{Sig}(s, r)))$$

in order to infer that if s signifies both that p and that q then s signifies that both p and q.

[8]See, e.g., Read (2006, 2010).

20.2 Denotational Paradox

Priest (2006b) observed that the denotational paradoxes are somewhat different from the usual semantic paradoxes, and the object of insufficient attention. Nonetheless, he thinks they share enough features with other paradoxes that they should yield to the same solution—"The Principle of Uniform Solution: same kind of paradox, same kind of solution."[9] Hence, any putative solution to the semantic paradoxes that cannot be adapted to deal with the denotational paradoxes is *ipso facto* inadequate. Of course, the slogan, "same paradox, same solution", is equivalent to "different solution, different paradox", threatening to undermine his point completely. The converse of Priest's principle is much more plausible: same solution, same kind of paradox. If they do yield to the same solution, so much the better for that solution, and the search for another solution can be called off; while if they do not, the possibility of a separate solution bringing out their different character is still open. The same point applies to the set-theoretic paradoxes. If they yield to the same solution, good, but if not, that in itself suggests they are different in kind.[10]

As a matter of fact, the Principle of Uniform Solution was invoked by Aristotle in Chap. 24 of *De Sophisticis Elenchis*. He there argues against solving the Hooded Man paradox by reference to the fallacy of the relative and the absolute on the ground that the Hooded Man is the same kind of puzzle as "Is this dog your father?", so they should have the same solution. The fallacious example is: "This dog is a father, this dog is yours, so this dog is your father." This puzzle does not yield to the fallacy of the relative and the absolute, so that cannot be right for the Hooded Man either—rather, they are both to be solved by reference to the fallacy of accident: just because the same thing is F and is G, it doesn't follow that it is an FG. Just because Coriscus is in a hood and is known to you, it doesn't follow that he is known to you in a hood; just because this dog is yours and a father, it doesn't follow that it is your father.

So the real question is whether Bradwardine's solution can be adapted to the denotational paradoxes, or whether a different solution is needed. If the latter, that in itself will suggest that the paradoxes are sufficiently different; while if Bradwardine's solution can be suitably adapted, that shows they are sufficiently similar and adds weight to the solution in further exploiting its explanatory character. So let us turn to that issue.

The simplest of the paradoxes of denotation is Berry's. Consider the description, "the least integer not denoted in fewer than 19 syllables".[11] Assuming that definite descriptions denote at most one thing, and given that there are only finitely many descriptions with fewer than 19 syllables, there must be a least integer not so denoted. But the above description denotes it in 18 syllables. Contradiction.

The paradox lends itself to many variations. König's paradox focuses on the description "the least indefinable ordinal," assuming that definability requires a

---

[9] Priest (2002, §11.5); see also Priest (2006b, p. 140).
[10] As Priest (2002, §17.2, p. 287, n. 39) himself notes.
[11] Russell (1908, p. 223), Priest (2002, §9.3, 2006a, p. 16).

unique description.[12] Given that there are uncountably many ordinals but only countably many descriptions,[13] there must be a least ordinal which is not definable (since the ordinals are well-ordered), which has just been defined by that very description. Richard's paradox considers definable real numbers between 0 and 1, of which again, there must be only countably many, hence they are listable.[14] Nonetheless, diagonalization defines the real number whose ith term differs (in some determinate way) from the ith term of the ith number on the list, which is not on the list. Contradiction.

Priest (2002, §4.9) identifies in Berkeley's Master Argument for idealism a paradox which he dubs "Berkeley's paradox". This paradox uses an indefinite description, "something I will never think about." It denotes something indefinitely, and enables us to think about an arbitrary one of the things we will never think about, which we just have. This transforms into a paradox of denotation: the expression "something not denoted" denotes something not denoted. Contradiction.

In his discussion of Bradwardine's solution to the liar paradox, Priest (2012) challenged its adherents to extend the "multiple-meanings" solution to solve these paradoxes of denotation, offering reasons why he thought this might be a challenge too far. I here rise to that challenge.

20.3 Denotation

As Priest (2012, p. 158) says, to deal with these paradoxes, a theory of descriptions and an account of denotation are needed. Each of the paradoxes in Sect. 20.2 uses a descriptive phrase of the form "$\nu x \phi x$", where "ν" is a variable-binding term operator (usually abbreviated to "vbto"), variously the definite description operator, ι (the ϕ), an indefinite description operator, ϵ (a/some ϕ), or the least number (or ordinal) operator, μ (the least ϕ).[15] There are various ways of dealing with these expressions, either as incomplete expressions, defining them away (as did Russell, for example), or as singular terms, whether taking them always to denote (so when there is no ϕ, taking them to denote some one and the same arbitrary object, such as 0, or perhaps different objects), or allowing them to be empty (so when there is no ϕ, taking them to denote nothing). The Russellian account is not appropriate for present purposes, since it denies that descriptive terms exist at all, defining them away existentially, whereas our task is to give an account of denoting which avoids the paradoxes of denotation. To be sure, denying they denote at all is one (path to a) solution, but a less radical path is to show how these phrases can denote non-paradoxically.

---

[12] König (1967) (in van Heijenoort 1967), Priest (2002, §9.3).

[13] Assuming a finite vocabulary for composing descriptions of finite length.

[14] Richard (1967), Priest(2002, p. 132, 2006b, p. 139).

[15] Here, and throughout the paper, "ϕ" ranges over properties, denoted by λ-terms, and ϕx represents both a formula ϕ of any complexity containing zero or more occurrences of the variable "x" free (and possibly other variables) and its β-transform $[(\lambda y)\phi y]x$. Here ϕy (in general, $\phi \tau$) results from ϕx by replacing all free occurrences of 'x' by "y", respectively, "τ", ensuring in the usual way that no variables are accidentally bound.

An example of the second, essentially Fregean, way of dealing with definite descriptions standardly takes the two axioms[16]:

$$\exists!x\phi x \rightarrow \phi(\iota x\phi x) \qquad (\iota\text{-}F1)$$

$$\neg\exists!x\phi x \rightarrow \iota x\phi x = \iota x\bot \qquad (\iota\text{-}F2)$$

where $\exists!x\phi x$ abbreviates $\exists x\forall y(\phi y \leftrightarrow x = y)$. So if there is a unique ϕ, $\iota x\phi x$ (that is, the ϕ) is indeed ϕ; if not, "$\iota x\phi x$" denotes some constant thing, the same for all empty or nonunique descriptions.[17] The consequence is that all terms are taken to denote, even empty terms such as "the greatest natural number," or incomplete singular terms like "the table". Of course, "the greatest natural number" doesn't denote the greatest natural number, since there isn't one. Nonetheless, it must denote something. "$\iota x\phi x$" always denotes on the Fregean account.

Priest's theory of denotation invokes both a Gödel-numbering operator, forming a Gödel-term $\langle \tau \rangle$ from each term τ, and a binary relation D of denotation, with the *Denotation Principle*[18]:

$$\forall x(D(\langle \tau \rangle, x) \leftrightarrow x = \tau) \qquad (\text{DEN})$$

Note that denotation is a function: a singular term "τ" denotes at most one object. What is missing in this account, however, is any mention of the content of the term "τ". Expressions like "the negative square root of 2" or "the smallest positive integer" denote what they do in virtue of their content. That is reflected in $(\iota\text{-}F1)$ and $(\iota\text{-}F2)$. If there is exactly one ϕ, then the ϕ is not only ϕ, but "the ϕ" denotes it in virtue of its being ϕ (and the only ϕ). If not, then "the ϕ" denotes $\iota x\bot$.

In particular, consider once again the term, "the greatest natural number". This cannot denote a natural number greater than all others, for there is no natural number greater than all others. But to realise that, we again need to examine the content of the term. It is because of the descriptive content of the term, and all it implies, that it fails to denote a natural number greater than all others, for nothing can have that property, which is inherently contradictory. Indeed, as we will see, we can sustain a theory on which it does denote—but not a natural number greater than all others.[19]

---

[16] See, e.g., da Costa (1980, p. 138), Read (1993, §5). In addition, for all vbtos ν, we have alpha-conversion: $\nu x\phi x = \nu y\phi y$, and extensionality: $\forall x(\phi x \leftrightarrow \psi x) \rightarrow \nu x\phi x = \nu x\psi x$. My justification for calling this a "Fregean" account is Frege's discussion in (2013, Part III 1 (a) 1, esp. §§63–64): "[A] more precise stipulation needs to be made here, so that for every object it is determined which object the half of it is; otherwise it is not permissible to use the expression, "the half of x", with the definite article ... [P]roper names are inadmissible that do not actually designate an object."

[17] This is a simplifying assumption, for the purposes of this paper. In line with footnote 5, not all contradictions are equivalent, and so different empty descriptions should really be allowed to denote different objects.

[18] See Priest (2006a, p. 25).

[19] *Pace* Priest (1997a), who shows that, with sufficient violence to logic, we can even construct a model containing the greatest natural number, which the description does denote.

20 Denotation, Paradox and Multiple Meanings

We also need to realise that an incomplete singular term, like "the table", say, also fails to denote what one might expect, since there is no unique thing which satisfies the description "table"—unless context adds further information to the description to fix a unique table.

Accordingly, what we need is a notion of the signification of a term. Where "τ" is a (meta-)variable over terms, let $\mathbf{Sig}(\langle\tau\rangle, \phi)$ express the fact that a term "τ" signifies (various) properties, ϕ. Then we require that the singular term "τ" denotes anything that has all the properties which it signifies, if there is one:

$$\exists! x(\forall\phi)(\mathbf{Sig}(\langle\tau\rangle, \phi) \to \phi x) \to \forall x(\forall\phi(\mathbf{Sig}(\langle\tau\rangle, \phi) \to \phi x) \to D(\langle\tau\rangle, x)) \tag{DEN-B1}$$

A second principle covers the case where nothing satisfies everything that "τ" signifies[20]:

$$\neg\exists x(\forall\phi)(\mathbf{Sig}(\langle\tau\rangle, \phi) \to \phi x) \to D(\langle\tau\rangle, \iota x\bot) \tag{DEN-B2}$$

But what if more than one thing has all the properties that "τ" signifies? "τ" is a singular term, so it should denote just one thing, if at all. The answer is to add a further principle regarding the signification of a term "τ". The same issue arises with Bradwardine's account of the signification of sentences. For example, Bradwardine clearly thinks, quite naturally, that "All Cretans are liars" signifies that all Cretans are liars, and that "What Socrates says is false" signifies that what Socrates says is false. Although not explicitly stated, the general principle he accepts is that $(\forall p)\mathbf{Sig}(\langle p\rangle, p)$.[21] In the same way, a term like "the Hooded Man" signifies being the Hooded Man, and "what Socrates says" signifies being what Socrates says. In general[22]:

$$\mathbf{Sig}(\langle\tau\rangle, (\lambda x)x = \tau). \tag{BUT}$$

(BUT) guarantees that if anything has all the properties that "τ" signifies, then a unique thing does, namely, τ. For suppose $\forall\phi(\mathbf{Sig}(\langle\tau\rangle, \phi) \to \phi x)$ and $\forall\phi(\mathbf{Sig}(\langle\tau\rangle, \phi) \to \phi y)$. Then $\mathbf{Sig}(\langle\tau\rangle, (\lambda x)x = \tau) \to x = \tau$ and $\mathbf{Sig}(\langle\tau\rangle, (\lambda x)x = \tau) \to y = \tau$. By (BUT), $\mathbf{Sig}(\langle\tau\rangle, (\lambda x)x = \tau)$. So $x = \tau = y$. Hence, if anything has all the properties "τ" signifies, then only one thing does.

In line with Bradwardine's observation that a proposition may signify more than may appear, and that signification is closed under consequence, we should also require that the signification of terms be closed under consequence[23]:

$$(\forall\phi, \psi)((\forall x)(\phi x \Rightarrow \psi x) \to (\mathbf{Sig}(\langle\tau\rangle, \phi) \to \mathbf{Sig}(\langle\tau\rangle, \psi)) \tag{CLO}$$

---

[20] We will see in Sect. 20.6 that the paradoxes can be strengthened to rule out the possibility of avoiding contradiction by supposing that the terms do not denote at all.

[21] See e.g., Read (2015b, p. 400).

[22] As Butler (1765, Preface, p. 37) wrote, "everything is what it is and not another thing".

[23] If necessary, we can generalize (CLO) in the same way (P2) was generalized in foonote 7.

Similarly, Bradwardinian versions of (ι-$F1$) and (ι-$F2$) must be predicated on the assumption that something uniquely satisfies everything that "$\iota x \phi x$" signifies:

$$\exists! x \forall \psi (\mathbf{Sig}(\langle \iota x \phi x \rangle, \psi) \to \psi x) \to \phi(\iota x \phi x) \qquad (\iota\text{-}B1)$$

and

$$\neg \exists! x \forall \psi (\mathbf{Sig}(\langle \iota x \phi x \rangle, \psi) \to \psi x) \to \iota x \phi x = \iota x \bot \qquad (\iota\text{-}B2)$$

We can augment (DEN-$B2$) with an exclusion principle, ensuring that "τ" denotes $\iota x \bot$ only when nothing satisfies everything that "τ" signifies:

$$\tau = \iota x \bot \leftrightarrow \neg(\exists! x)(\forall \phi)(\mathbf{Sig}(\langle \tau \rangle, \phi) \to \phi x) \qquad (\text{Exc})$$

The justification for adding (EXC) is twofold: first, by the thought that, despite Frege's practice, it is good to keep the denotation of terms that something satisfies distinct from those that nothing satisfies; secondly, that it allows us to preserve the *Denotation Principle* (DEN) and the requirement that all terms denote. However, (EXC) does lay down a stiff and puzzling requirement on the denotation of the contradictory term "$\iota x \bot$", namely, that nothing satisfies everything it signifies—that it cannot be exactly characterized. That requirement can be defended: after all, what could satisfy the characterization $(\lambda x) \bot$? However, it does suggest that whatever it is that contradictory terms denote, it does not, and could not, exist. In other words, it suggests that the theory is not so much Fregean, as Meinongian, or rather, noneist. Although Priest (2005, p. ix) remarks that "noneism is naturally committed to the idea that every term denotes something," there is more to noneism than that: it claims that everything is something, and some things don't exist (and may even qualify the claim that every term denotes something, to claim, e.g., that every term denotes something or things, as in Priest 2005, Chap. 8). Indeed, the present view is not Meinong's: Meinong was committed to the *Independence Principle*, of the independence of *Sosein* from *Sein*, and the *Characterization Principle*: that an object has the properties it is characterized as having, regardless of whether it exists.[24] That's false: in particular, $\iota x \bot$ does not have the property $(\lambda x) \bot$, nor many of the other properties signified by terms that denote it, as we will see.[25]

Given (EXC), we can show that "τ" always denotes τ. For suppose

$$\exists x (\forall \psi)(\mathbf{Sig}(\langle \tau \rangle, \psi) \to \psi x)$$

Then for some y,

$$(\forall \psi)(\mathbf{Sig}(\langle \tau \rangle, \psi) \to \psi y)$$

---

[24] See e.g., Priest (2005, p. vii).

[25] Priest (2005, p. 84) proposes that such descriptions characterize what they denote only at other, non-actual worlds. But, like me, he also believes that these other worlds don't exist. See Priest (2005, §7.3) and Read (2005).

in particular,

$$\mathbf{Sig}(\langle \tau \rangle, (\lambda x)x = \tau) \to y = \tau$$

But $\mathbf{Sig}(\langle \tau \rangle, (\lambda x)x = \tau)$, by (BUT). So $y = \tau$. Hence

$$(\forall \psi)(\mathbf{Sig}(\langle \tau \rangle, \psi) \to \psi \tau)$$

Moreover, suppose $(\forall \psi)(\mathbf{Sig}(\langle \tau \rangle, \psi) \to \psi z)$. Then by the same reasoning, $z = \tau$, so $y = z$, whence $\exists! x (\forall \psi)(\mathbf{Sig}(\langle \tau \rangle, \psi) \to \psi x)$, and $(\forall \psi)(\mathbf{Sig}(\langle \tau \rangle, \psi) \to \psi \tau)$, so by (DEN-B1), $D(\langle \tau \rangle, \tau)$. On the other hand, if

$$\neg \exists x (\forall \psi)(\mathbf{Sig}(\langle \tau \rangle, \psi) \to \psi x)$$

then by (DEN-B2), $D(\langle \tau \rangle, \iota x \bot)$ and by (EXC) $\tau = \iota x \bot$, so $D(\langle \tau \rangle, \tau)$. Either way, $D(\langle \tau \rangle, \tau)$, and so $(\exists x) D(\langle \tau \rangle, x)$, that is, all terms denote.

It follows that D is functional. For suppose that $D(\langle \tau \rangle, x)$ and $D(\langle \tau \rangle, y)$. Then either $\exists x (\forall \phi)(\mathbf{Sig}(\langle \tau \rangle, \phi) \to \phi x)$ or not. If the former, then as we showed above, $\exists! x (\forall \phi)(\mathbf{Sig}(\langle \tau \rangle, \phi) \to \phi x)$, whence $x = y$. But if $\neg \exists x (\forall \phi)(\mathbf{Sig}(\langle \tau \rangle, \phi) \to \phi x)$, then $D(\langle \tau \rangle, \iota x \bot)$, so by (EXC) $x = \iota x \bot = y$.

Finally, (DEN) immediately follows, given that D is functional: for if $x = \tau$, then since $D(\langle \tau \rangle, \tau)$, we have $D(\langle \tau \rangle, x)$; conversely, if $D(\langle \tau \rangle, x)$, then since $D(\langle \tau \rangle, \tau)$ and D is functional, $x = \tau$. So $\forall x (D(\langle \tau \rangle, x) \leftrightarrow x = \tau)$.

20.4 Berry's Paradox

Take the description, "the least natural number not denoted in English in fewer than 75 characters." Let "Bx" abbreviate $\neg \exists y (Nx \wedge (\ell y < 75) \wedge Dyx)$, where ℓy gives the number of characters in the term y. Call the members of the set $\{x : Bx\}$ the Berry numbers, that is, all the numbers not denoted in fewer than 75 characters. Since there are finitely many (alphanumeric) characters in English, say m, there are at most m^{75} names containing fewer than 75 characters, and so at most m^{75} natural numbers are denoted by such terms, a finite number. Hence there are a countable infinity of Berry numbers. The Berry numbers are natural numbers and thus well-ordered, and so $\{x : Bx\}$ has a least member, $\mu y By$, that is, $\iota y (By \wedge \forall x(Bx \to y \leq x))$. As a matter of fact, $\ell \langle \mu y By \rangle = 74$. Let π abbreviate $\mu y By$. Then we may be tempted to argue:

$$N\pi \wedge (\ell \langle \pi \rangle = 74) \wedge D(\langle \pi \rangle, \pi)$$

so $\exists y (N\pi \wedge (\ell y = 74) \wedge D(y, \pi))$

whence $\exists y (N\pi \wedge (\ell y < 75) \wedge D(y, \pi))$

But $B\pi$, so $\neg \exists y (N\pi \wedge (\ell y < 75) \wedge D(y, \pi))$. Contradiction.

The error lies in the claim that "π", that is, "the least natural number not denoted in English in fewer than 75 characters," denotes a Berry number smaller than all the others. What the paradox really shows is that the Berry term "π" is inherently contradictory. We can show that, although $\mathbf{Sig}(\langle\pi\rangle, \lambda x(Bx \wedge \forall y(By \to x \leq y)))$, "$\pi$" signifies more than just $\lambda x(Bx \wedge \forall y(By \to x \leq y))$, that is, being a Berry number smaller than all the others. First, suppose being such a Berry number were all "π" signified. Then by (DEN-B1), if n were that Berry number, "π" would denote n. But "π" has 74 characters, so n would be denoted by a description with fewer than 75 characters. Since "π" signifies being a Berry number smaller than all the others, it would follow from (CLO) that "π" signified being denoted by a description with fewer than 75 characters. So being a Berry number smaller than all the others would not be the only thing "π" signified. Consequently, being the least Berry number is not all that "π" signifies.

Now suppose "π" signifies being a Berry number smaller than all the others and ψ, where ψ encapsulates everything else that "π" signifies. Once again, if n were a Berry number smaller than all the others and ψ obtained, "π" would denote n, by (DEN-B1). But "π" has 74 characters, so n would be denoted by a description with 74 characters. Since "π" signifies being a Berry number smaller than all the others and being ψ, it follows from (CLO) that "π" signifies being denoted in fewer than 75 characters. So "π" signifies both being a Berry number smaller than all the others, that is, not being denoted with fewer than 75 characters, and being so denoted. So "π" is implicitly contradictory, and there is nothing which has all and only the properties signified by "π", and consequently "π" denotes the contradictory object: $\pi = \iota x \bot$.

What we have shown is that "π" does not denote something not denoted in fewer than 75 characters, for the description is implicitly contradictory, purporting to denote something which both is and is not denoted in fewer than 75 characters. The description denotes something which is denoted in fewer than 75 characters, for it is denoted by "the least number not denoted in fewer than 75 characters," that is, "π". Of course, whatever that object is, it is not a number not denoted in fewer than 75 characters smaller than all the others, for it is denoted by "π". Meinong's Characterization Principle must be denied.

One might be tempted to express this result as denying that "π" denotes the least number not denoted in fewer than 75 characters, i.e., π—that is, as showing that $\neg D(\langle\pi\rangle, \pi)$. Priest (2012, p. 157) rightly says that such a result would be "something of a *reductio* of the Bradwardine line." Actually, we have seen that the Bradwardinian theory is compatible with the universal truth of $D(\langle\tau\rangle, \tau)$. Nonetheless, the description, "the least natural number not denoted in English in fewer than 75 characters," does not denote a number not denoted in fewer than 75 characters. And in fact this paradoxical observation answers closely to the Principle of Uniform Solution. Recall that Bradwardine responds to the Liar by saying it is false. As Field (2006, p. 715) notes, this has a similarly puzzling and paradoxical air: the Liar says of itself that it is false, so Maudlin (on whom Field is commenting) and Bradwardine both say that the Liar is false and (since the Liar says that the Liar is false) that it is false that the Liar is false. But the Liar is false (according to Bradwardine) not because it isn't false, but because it isn't true—that is, something else it signifies fails to obtain. The

Liar sentence signifies not only that the Liar sentence is false (which is the case) but also that it is true (which is not so)—that is why it is false.

Priest (2012, p. 156) shows that Bradwardine is committed to a similar result about the heterologicality paradox. An object satisfies a predicate if it has all the properties that the predicate signifies, in symbols:

$$(\forall x)(x\$\langle\psi\rangle \leftrightarrow (\forall\phi)(\mathbf{Sig}(\langle\psi\rangle, \phi) \rightarrow \phi x))$$

Consider the predicate "$\neg x\$x$", and suppose $\neg\langle\neg x\$x\rangle\$\langle\neg x\$x\rangle$. Then for some ϕ,

$$\mathbf{Sig}(\langle\neg x\$x\rangle, \phi) \land \neg\phi(\langle\neg x\$x\rangle).$$

We may assume in line with (BUT) that $\mathbf{Sig}(\langle\neg x\$x\rangle, (\lambda x)\neg x\$x)$, and let ψ conjoin everything else that $\langle\neg x\$x\rangle$ signifies, so

$$\mathbf{Sig}(\langle\neg x\$x\rangle, (\lambda x)(\neg x\$x \land \psi x)).$$

Then, since something "$\neg x\$x$" signifies is not satisfied by '$\neg x\$x$', it follows that either $\langle\neg x\$x\rangle\$\langle\neg x\$x\rangle$ or $\neg\psi(\langle\neg x\$x\rangle)$. So if $\psi(\langle\neg x\$x\rangle)$ and $\neg\langle\neg x\$x\rangle\$\langle\neg x\$x\rangle$, then $\langle\neg x\$x\rangle\$\langle\neg x\$x\rangle$. But $\mathbf{Sig}(\langle\neg x\$x\rangle, (\lambda x)(\neg x\$x \land \psi x))$, so

$$\mathbf{Sig}(\langle\neg x\$x\rangle, (\lambda x)x\$x).$$

That is, '$\neg x\$x$' signifies not only that '$\neg x\$x$' does not signify itself, but also that it does, exactly parallel with the conclusion that any proposition signifying that it is itself not true also signifies that it itself is true. Consequently, as before, it follows that '$\neg x\$x$' cannot satisfy itself, since it signifies contradictory properties. And again as before, this does not suffice to infer that it does satisfy itself, since nothing can satisfy contradictory properties.

Thus not only does Bradwardine's theory deal with the heterological and Berry paradoxes, it does so in an entirely similar way to the other semantic paradoxes, such as the Liar, employing the closure principle to show that truth, satisfaction and denotation make contradictory demands of such paradoxical terms. $D(\langle\pi\rangle, \pi)$ and $\pi = \iota x\bot$, so $D(\langle\pi\rangle, \iota x\bot)$. Nonetheless, $\exists\phi(\mathbf{Sig}(\langle\pi\rangle, \phi) \land \neg\phi\pi)$, in particular, $\neg B\pi$, even though $\mathbf{Sig}(\langle\pi\rangle, \lambda x Bx)$.

20.5 Berkeley's Paradox

König's and Richard's paradoxes can be dealt with in much the same way as Berry's but require more technical apparatus from set theory. We should turn, therefore, to Berkeley's paradox. To recall, this is the paradox prompted by thinking about something not thought about, namely, what is denoted by the expression 'something not denoted'.

Priest extracts 'Berkeley's paradox' from what Gallois (1974) dubbed Berkeley's master argument for idealism. Berkeley challenges his realist opponent to conceive of things that are not conceived:

> That you conceive them unconceived or unthought of ... is a manifest repugnancy ... The mind ... is deluded to think it can and doth conceive bodies unthought of or without the mind, though at the same time they are apprehended by or exist in it self.[26]

Consequently, Berkeley claims, the idea of mind-independent objects, existing unperceived and unthought of, is incoherent.

Whatever the merits of this argument, Priest (2002, pp. 69–70) distils from it the paradox set out above, that the phrase 'something not denoted' denotes something not denoted, and hence something that is denoted. Note that 'something not denoted' is an indefinite description, whose logical behaviour is given by variants of $(\iota\text{-}B1)$ and $(\iota\text{-}B2)$[27]:

$$\exists x \forall \psi (\mathbf{Sig}(\langle \epsilon x \phi x \rangle, \psi) \to \psi x) \to \phi(\epsilon x \phi x) \quad (\text{-}B1)$$

$$\neg \exists x \forall \psi (\mathbf{Sig}(\langle \epsilon x \phi x \rangle, \psi) \to \psi x) \to \epsilon x \phi x = \iota x \bot \quad (\text{-}B2)$$

Note that "$\epsilon x \phi x$" is a singular term, denoting a single object, either one of the ϕs or $\iota x \bot$. Although more than one thing may satisfy ϕ (e.g., being a table), '$\epsilon x \phi x$' must signify more than that in order to pick out $\epsilon x \phi x$ from the ϕs uniquely—e.g., by (BUT), '$\epsilon x \phi x$' signifies $(\lambda x)(x = \epsilon x \phi x)$. As this shows, the signification of "$\epsilon x \phi x$" will not be purely descriptive. In particular, it will signify a choice function.[28]

We can formalize the indefinite description "something not denoted" with the existing notation as $(\epsilon x)\neg(\exists y) Dyx$—let us abbreviate this as ρ. Then it seems that $D(\langle \rho \rangle, \rho)$, while by definition, $\neg(\exists y) D(y, \rho)$, a contradiction.

The mistake, as before, is to think that nothing denotes ρ, just because $\mathbf{Sig}(\langle \rho \rangle, (\lambda x)\neg(\exists y) Dyx)$ and $D(\langle \rho \rangle, \rho)$. So we must ask, what properties, besides $(\lambda x)\neg(\exists y) Dyx$, does "$\rho$" signify? If that were all "ρ" signified, and if "ρ" denoted some object e (and not $\iota x \bot$), then by $(\epsilon\text{-}B1)$, $\neg(\exists y) D(y, e)$, but at the same time, $D(\langle \rho \rangle, e)$, a contradiction. So "$\rho$" must signify more than that—call it ψ. Then by (DEN-$B1$), if e were not denoted and $\psi(e)$, "ρ" would denote e. Since "ρ" signifies not being denoted and being ψ, it follows by (CLO) that "ρ" signifies being denoted by "ρ", and so being denoted by something. So "ρ" signifies contradictory properties—both being denoted by "ρ" and not being denoted at all, so by (DEN-$B2$), $D(\langle \rho \rangle, \iota x \bot)$.

---

[26] *Principles of Human Knowledge* §23, in Berkeley (1837, p. 12). See also his *Three Dialogues between Hylas and Philonous*, The First Dialogue (1837, p. 53).

[27] $(\epsilon\text{-}B1)$ and $(\epsilon\text{-}B2)$ adapt the usual Hilbertian axiom for indefinite descriptions, $(\exists x)\phi x \to \phi(\epsilon x \phi x)$ in the same way that $(\iota\text{-}B1)$ and $(\iota\text{-}B2)$ adapt the Fregean axioms for definite descriptions. See, e.g., Leisenring (1969, p. 40) and Priest (2002, §4.6). da Costa (1980, p. 139) points out that $(\epsilon\text{-}B2)$ follows from the extensionality axiom (see footnote 16).

[28] See e.g., Hilbert (1967, p. 466) and Corcoran and Herring (1971, p. 649).

20.6 Hilbert–Bernays' Paradox

We have developed our theory of descriptions on the Fregean basis that all descriptions denote—if not something satisfying the description, then some arbitrary object which serves as the denotation of all unsatisfiable descriptions. Priest (2005, §8.3) reminds us of a paradox due to Hilbert and Bernays which seems to undermine this assumption.[29] Consider the definite description, "the successor of the denotation of this description." Given that no number is its own successor, it seems that this description cannot possibly denote. For if it denoted some number n, then it would also denote $n + 1$. Given that the denotation of definite descriptions is unique (if it exists), it follows that $n = n + 1$. Contradiction. The only possible solutions seem to be that the description does not denote, or denotes more than one thing.

Priest (2005, §§8.5–8.6) explores the latter possibility, that the description "the successor of the denotation of this description" denotes more than one thing, namely, both what it denotes and its successor. This not only runs counter to the natural presumption in the theory of singular terms that their denotation is unique, it also leads swiftly to contradiction, contradictions of the laws of identity which Priest is willing to countenance.[30]

But the other option, that the description does not denote, is not open to us either, as Priest shows.[31] As before, let

$$(\mu x)\phi x = \iota x(\phi x \wedge \forall y(\phi y \rightarrow x \leq y))$$

and let

$$(\eta x)\phi x = \iota y(((\exists x)\phi x \rightarrow y = (\mu x)\phi x) \vee (\neg(\exists x)\phi x \rightarrow y = \iota x \bot))$$

Then if there is a ϕ, $(\eta x)\phi x$ will be the least ϕ, while if there isn't, $(\eta x)\phi x$ will be $(\iota x)\bot$. But that just identifies $(\eta x)\phi x$ with $(\mu x)\phi x$, for by definition, $(\mu x)\phi x$ is the least ϕ if there is a least ϕ and $(\iota x)\bot$ if there isn't. What this shows us, however, is that taking "$(\iota x)\phi x$" to denote even when there is no unique ϕ is not an unmotivated decision, but can be forced on us by careful choice of ϕ.

Priest (2006b, p. 147) also observes that mention of successor in the above paradox is only one special case.[32] One can develop the paradox for any number-theoretic function, $f : \mathbb{N} \rightarrow \mathbb{N}$, and show formally that any such function has a fixed point—though many functions, such as successor, clearly do not. Let δ be a denotation function from arithmetic terms to their denotations, that is, such that for all τ, $D(\langle\tau\rangle, \delta\langle\tau\rangle)$. So $\delta\langle\tau\rangle = \tau$. Then if δ is representable, the usual diagonalization lemma can be reworked to show that, for any $f : \mathbb{N} \rightarrow \mathbb{N}$, there is a term "$\chi$" for

---

[29] See also Priest (1997b, 2006b, §6).
[30] Priest (2005, §8.7) avoids the consequence that some number is its own successor, from which it would follow that $0 = 1$, by denying the substitutivity of identicals.
[31] Priest (1997b, §7, 2005, §8.4, 2006b, §6).
[32] See also Priest (1997b, p. 46, 2005, p. 158).

which
$$\chi = f(\delta\langle\chi\rangle) = f(\chi),$$

so every number-theoretic function has a fixed point. Contradiction.

Hilbert and Bernays' response is to conclude that the denotation function is not arithmetic, and so cannot be represented in arithmetic (if arithmetic is consistent), any more than arithmetic truth can, as recorded in Tarski's Theorem.[33] Priest (1997b, p. 47) observes, however, that, sound as this conclusion may be for formal arithmetic, it still leaves open the paradox in natural language, just as Tarski's Theorem cuts no ice with the Liar paradox. For natural language does appear to have a denotation function, instanced here by the description "the successor of the denotation of this description." Priest's own solutions to this paradox (that in Priest 2006b involving both truth-value gluts and truth-value gaps, and that in Priest 2005 necessitating violations of traditional laws of identity) differ from his dialetheic solution to the other paradoxes. The paradox does not seem readily to fit his common Inclosure schema.[34] If this is right, by the Principle of Uniform Solution the Hilbert–Bernays paradox is different in kind from the other paradoxes, of denotation and of truth. But we can already see that the description at the heart of the paradox is inherently contradictory, and so likely to submit to the standard Bradwardinian solution.

Let σ be short for the definite description "the successor of the denotation of "σ"," and suppose
$$\mathbf{Sig}(\langle\sigma\rangle, (\lambda x)(x = s(\sigma) \wedge \psi x))$$

gives the whole signification of "σ", where $s(x)$ is the successor function. Suppose $D(\langle\sigma\rangle, n)$. Then $\sigma = n$, by (DEN). But if $n = s(\sigma)$ and ψn, $n = s(\sigma) = s(n)$. So by (CLO), the signification of "σ" is contradictory. Consequently, since, as we saw, every description denotes, "σ" must denote something, namely, $\iota x \perp$.

20.7 Conclusion

Graham Priest challenged supporters of Thomas Bradwardine's "multiple-meanings" diagnosis of the logical paradoxes to show how Bradwardine's idea can be adapted to solve the paradoxes of denotation. He suggested (Priest 2012, p. 158) that Bradwardine's only option was to deny that the expressions in question denoted at all, since the idea that some expression "τ" did not denote τ seemed too far-fetched even for him. But properly understood, it is not at all far-fetched and entirely in keeping with Bradwardine's approach. The paradox is only apparent. Take "the greatest natural number". It can't denote a natural number greater than all the others, since there is no natural number greater than all the others. Instead, it denotes the contradictory

---

[33] Hilbert and Bernays (1939, pp. 268–9). On Tarski's Theorem, see, e.g., Boolos and Jeffrey (1980, p. 176).
[34] On the Inclosure schema as a uniform diagnosis of the paradoxes, see, e.g., Priest (2002, §9.4).

object, $\iota x \bot$. So, in a sense, $\iota x \bot$ is the greatest natural number, but of course, $\iota x \bot$ does not satisfy that description, for nothing does. The Characterization Principle is false. Just as Bradwardine says that the Liar sentence is false while also saying that "The Liar sentence is false" is false, so too his approach leads naturally to the conclusion that expressions like "something not denoted" denote something denoted, not something not denoted. For such paradoxical descriptive phrases cannot denote something possessing all the properties they signify, since they are implicitly contradictory and nothing has all the properties in question. In particular, $\iota x \bot$ does not have the "property" $(\lambda x)\bot$, on pain of contradiction.

The analysis above shows that a coherent account can be given of the apparently paradoxical descriptive phrases in Berry's, Berkeley's, and Hilbert–Bernay's paradoxes (and others) in keeping with Bradwardine's principles, resolving the paradoxes, and at the same time, just as Bradwardine maintained such standard logical principles as the laws of Bivalence and Contravalence, preserving the Fregean demand that all such phrases have denotation.

References

Berkeley, G. (1837). *Works*. London: Charles Daly.
Boolos, G., & Jeffrey, R. (1980). *Computability and logic* (2nd ed.). Cambridge: Cambridge UP.
Bradwardine, T. (2010). In S. Read (Ed. and Eng. Tr.), *Insolubilia*. Leuven: Peeters.
Butler, J. (1765). *Fifteen Sermons* (5th ed.). London: Robert Horsfield.
Cameron, M. (2012). Meaning: Foundational and semantic theories. In J. Marenbon (Ed.), *The Oxford handbook of medieval philosophy* (pp. 342–362). Oxford: Oxford UP.
Corcoran, J., & Herring, J. (1971). Notes on a semantic analysis of variable binding term operators. *Logique et Analyse, 55*, 644–57.
da Costa, N. (1980). A model-theoretical approach to variable binding term operators. In A. Arruda, et al. (Eds.), *Mathematical logic in Latin America* (pp. 133–62). Amsterdam: North-Holland.
Field, H. (2006). Maudlin's *Truth and Paradox*. *Philosophy and Phenomenological Research, 73*, 713–20.
Frege, G. (2013). In P. A. Ebert, M. Rossberg, & C. Wright (Trans., Eds.), *Basic laws of arithmetic: Derived using concept-script*. Oxford: Oxford University Press.
Gallois, A. (1974). Berkeley's master argument. *The Philosophical Review, 83*, 55–69.
Hilbert, D. (1967). In van Heijenoort (Ed.), *The foundations of mathematics* (464–479). Translation of Die Grundlagen der Mathematik, *Abhandlungen aus dem mathematischen Seminar der Hamburgischen Universität, 6*, 65–85 (1928).
Hilbert, D., & Bernays, P. (1939). *Grundlagen der Mathematik*. Berlin: Springer.
König, J. (1967). In van Heijenoort (Ed.), *On the foundations of set theory and the continuum problem* (pp. 145–149). Translation of Über die Grundlagen der Mengenlehre und das Kontinuumproblem, *Mathematische Annalen, 61*, 156–160 (1905).
Leisenring, A. C. (1969). Mathematical logic and Hilbert's ϵ-symbol. London: Macdonald.
Priest, G. (1997a). Inconsistent models of arithmetic, (i): The finite case. *Journal of Philosophical Logic, 26*, 223–245.
Priest, G. (1997b). On a paradox of Hilbert and Bernays. *Journal of Philosophical Logic, 26*, 45–56.
Priest, G. (2002). *Beyond the laws of thought* (2nd ed.). Oxford: Clarendon Press.
Priest, G. (2005). *Towards non-being*. Oxford: Clarendon Press.
Priest, G. (2006a). *In contradiction* (2nd ed.). Oxford: Clarendon Press.

Priest, G. (2006b). The paradoxes of denotation. In K. Bolander, V. F. Hendrix, & S. A. Pedersen (Eds.), *Self-reference* (pp. 137–150). Stanford, CA: CSLI.
Priest, G. (2012). Read on Bradwardine on the Liar. In C. D. Novaes & O. T. Hjortland (Eds.), *Insolubles and consequences: Essays in honour of Stephen Read* (pp. 155–161). London: College Publications.
Prior, A. N. (1958). Epimenides the Cretan. *Journal of Symbolic Logic*, *23*, 261–66.
Read, S. (1993). The slingshot argument. *Logique et Analyse*, *143–144*, 195–218.
Read, S. (2005). The unity of the fact. *Philosophy*, *80*, 317–342.
Read, S. (2006). Symmetry and paradox. *History and Philosophy of Logic*, *27*, 307–318.
Read, S. (2007). Bradwardine's revenge. In J. C. Beall (Ed.), *Revenge of the Liar* (pp. 250–261). Oxford: Oxford UP.
Read, S. (2008). Further thoughts on Tarski's T-scheme and the Liar. In S. Rahman, T. Tulenheimo, & E. Genot (Eds.), *Unity, truth and the Liar: The modern relevance of medieval solutions to the Liar paradox* (pp. 205–225). Berlin: Springer.
Read, S. (2009). Plural signification and the Liar paradox. *Philosophical Studies*, *145*, 363–375.
Read, S. (2010). The validity paradox. In M. Peliš (Ed.), *The logica yearbook 2009* (pp. 209–221). London: College Publications.
Read, S. (2011). Miller, Bradwardine and the truth. *Discusiones Filosóficas*, *12*(18), 229–235.
Read, S. (2015a). Paradox, closure and indirect speech reports. *Logica Universalis*, *9*, 237–251.
Read, S. (2015b). Truth, signification and paradox. In T. Achourioti, H. Galinon, J. Martínez Fernández, & K. Fujimoto (Eds.), *Unifying the philosophy of truth* (pp. 393–408). Berlin: Springer.
Restall, G. (2008). Modal models for Bradwardine's theory of truth. *Review of Symbolic Logic*, *1*, 225–240.
Richard, J. (1967). In van Heijenoort (Ed.), *The principles of mathematics and the problem of sets* (pp. 142–144). Translation of Les principes des mathématiques et le problème des ensembles, *Revue générale des sciences pures et appliquées*, *16*, 541 (1905).
Rumfitt, I. (2014). Truth and meaning. *Aristotelian Society Supplementary*, *88*, 21–55.
Russell, B. (1908). Mathematical logic as based on the theory of types. *American Journal of Mathematics*, *30*, 222–262.
van Heijenoort, J. (Ed.). (1967). *From Frege to Gödel: A source book in mathematical logic*. Cambridge, MA: Harvard University Press.

Stephen Read (D.Phil. Oxford) is Professor Emeritus of the History and Philosophy of Logic at the University of St. Andrews, Scotland and researcher at the Arché Research Centre for Logic, Language, Metaphysics and Epistemology. He is the author of Relevant Logic (Blackwells 1988) and Thinking About Logic (OUP 1995), and the editor and translator of Thomas Bradwardine's Treatise on Insolubles (Peeters 2010). He recently translated John Buridan's Treatise on Consequences (Fordham UP, 2015). With Catarina Dutilh Novaes, he edited The Cambridge Companion to Medieval Logic (CUP, 2016). He is the author of many articles both on contemporary philosophy of logic and language and on medieval logic. His current project, funded by the Leverhulme Trust, is on 'Theories of Paradox in Fourteenth-Century Logic'. Its aims are editions and English translations, from the manuscripts, of the treatises on 'Insolubles' (logical paradoxes) by Paul of Venice, Walter Sexgrave and John Dumbleton.

Chapter 21
Two Negations Are More than One

Greg Restall

Abstract In models for paraconsistent logics, the semantic values of sentences and their negations are less tightly connected than in classical logic. In 'American Plan' logics for negation, truth and falsity are, to some degree, independent. The truth of $\sim p$ is given by the falsity of p, and the falsity of $\sim p$ is given by the truth of p. Since truth and falsity are only loosely connected, p and $\sim p$ can both hold, or both fail to hold. In 'Australian Plan' logics for negation, negation is treated rather like a modal operator, where the truth of $\sim p$ in a situation amounts to p failing *in certain other situations*. Since those situations can be different from this one, p and $\sim p$ might both hold here, or might both fail here. So much is well known in the semantics for paraconsistent logics, and for first-degree entailment and logics like it, it is relatively easy to translate between the American Plan and the Australian Plan. It seems that the choice between them seems to be a matter of taste, or of preference for one kind of semantic treatment or another. This paper explores some of the differences between the American Plan and the Australian Plan by exploring the tools they have for modelling a language in which we have *two* negations.

$$\sim \quad \sim \quad \sim$$

I owe my start in logic and in philosophy to Graham Priest. From our first meetings, when he took me on, as an enthusiastic and naïve undergraduate Science student finding his way into philosophical logic from mathematics in 1989, through Ph.D. supervision in the early 1990s, and being a colleague and friend in the years since, my debt to Graham has been profound. My earliest work with Graham was on paraconsistent logics, and in particular, the first year of my Ph.D. was spent trying to work through the details of Australian Plan and American Plan semantics for substructural logics (Priest and Sylvan 1992; Restall 1993, 1995). It is a delight to return to some of these topics with the hindsight and experience of nearly 30 years, learning from Graham, being challenged and inspired by him, arguing with him and wrangling over these and many other issues. Thanks, Graham!

G. Restall (✉)
Philosophy Department, The University of Melbourne, Melbourne, Australia
e-mail: restall@unimelb.edu.au

© Springer Nature Switzerland AG 2019
C. Başkent and T. M. Ferguson (eds.), *Graham Priest on Dialetheism and Paraconsistency*, Outstanding Contributions to Logic 18,
https://doi.org/10.1007/978-3-030-25365-3_21

21.1 Negation on the American Plan

In a paraconsistent logic, the truth of the negation $\sim p$ doesn't necessarily *rule out* the truth of the thing negated, p. Truth and falsity can overlap. In many paraconsistent logics, models allow not only truth value 'gluts' like this, we also allow the dual case, of truth value gaps. The most natural, straightforward and elegant logic built on such a plan is now known as FDE, the logic of *first-degree entailment*.

It can be defined and understood in a number of different ways, but for our purposes it suits to introduce it as the generalisation of classical two-valued logic according to which evaluations are no longer functions assigning each sentence of a language a truth value from $\{0, 1\}$, but *relations* to those truth values. Relaxing the constraint that evaluations be Boolean functions means that sentences can be *neither* true nor false (the evaluation fails to relate the sentence to either 0 or 1) or *both* true and false (the evaluation relates the sentence to both truth values).[1]

Definition 21.1 A FDE-model for a propositional language consists of a relation ρ defined as follows: For each atomic sentence p, we posit by fiat whether it is true ($p\rho 1$) and whether it is false ($p\rho 0$). We then extend the relation ρ to a language including conjunction, disjunction and negation, as follows:

$(A \wedge B)\rho 1$ iff $A\rho 1$ and $B\rho 1$ $\qquad$ $(A \wedge B)\rho 0$ iff $A\rho 0$ or $B\rho 0$
$(A \vee B)\rho 1$ iff $A\rho 1$ or $B\rho 1$ $\qquad$ $(A \vee B)\rho 0$ iff $A\rho 0$ and $B\rho 0$
$\sim A\rho 1$ iff $A\rho 0$ $\qquad$ $\sim A\rho 0$ iff $A\rho 1$

The only deviation from classical propositional logic is that we allow for truth value gaps (ρ may fail to relate a given formula to a truth value) or gluts (ρ may relate a given formula to both truth values). Indeed, the possibilities of gaps and of gluts are, in a sense, *separable* or *modular*. It is quite straightforward to show that if a given interpretation ρ is a *partial function* on the basic vocabulary of a language—if it never over-assigns values to the extension of any predicate in that language—then it remains so over every sentence in that language. Sentences can be assigned gaps and not gluts. Similarly, if an interpretation is *decisive* over the basic vocabulary of some language—it never under-assigns values to the extensions of any predicate in that language—then it remains so over every sentence of that language. These sentences can be assigned gluts and not gaps. If an evaluation is *sharp* (if it allows for neither gaps nor gluts in the interpretation of any predicate), then it remains so over the whole language.

We can use FDE evaluations to analyse truth and consequence in the language of first order logic. One important notion goes like this:

Definition 21.2 An interpretation ρ is said to be a *counterexample* to the sequent $X \succ Y$ if and only if ρ relates each member of X to 1 while it relates no member

---

[1] The idea of a relational evaluation goes back to work by J. Michael Dunn, in the 1970s (Michael Dunn 1976), and has been made use of in some influential papers by Belnap (1977a, b). My presentation follows Graham Priest's treatment in *An Introduction to Non-Classical Logic* (Priest 2001, 2008).

of Y to 1.[2] In other words, an interpretation provides a counterexample to a sequent if it shows some way that the sequent fails to preserve truth. Given some set $\mathcal{M}$ of evaluations, a sequent is said to be $\mathcal{M}$-valid if it has no counterexamples in the set $\mathcal{M}$. We reserve the term 'FDE-valid' for those sequents which have no counterexamples at all.

All this is very well known in the literature on non-classical logics—see, for example, Priest's *An Introduction to Non-Classical Logic* (Priest 2008, Chap. 8) for more detail on the behaviour of these logical systems. The FDE-valid sequents include all of distributive lattice logic, with a de Morgan negation. For example, sequents such as these

$$A \vee (A \wedge B) \succ A \qquad A \succ A \wedge (A \vee B)$$
$$A \wedge (B \vee C) \succ (A \wedge B) \vee C$$
$$\sim(A \wedge B) \succ \sim A \vee \sim B \qquad \sim(A \vee B) \succ \sim A \wedge \sim B$$
$$\sim A \vee \sim B \succ \sim(A \wedge B) \qquad \sim A \wedge \sim B \succ \sim(A \vee B)$$
$$A \succ \sim\sim A \qquad \sim\sim A \succ A$$

are FDE-valid, while , $A, \sim A \succ B$ and $B \succ A, \sim A$ are not. (Any relational evaluation ρ where $A\rho 1$ and $A\rho 0$, while $B\not\rho 1$ is a counterexample to $A, \sim A \succ B$, and when $B\rho 1$ and $A\not\rho 1$ and $A\not\rho 0$, ρ is a counterexample to $B \succ A, \sim A$.) In addition, FDE-validity defined in this way satisfies the usual structural rules of identity, weakening (on the left and the right) and *Cut*:

$$A \succ A \qquad \frac{X \succ Y}{X, A \succ Y} \qquad \frac{X \succ Y}{X \succ A, Y} \qquad \frac{X \succ A, Y \quad X, A \succ Y}{X \succ Y}$$

Relational evaluations provide a very natural model for FDE. They show it to be an elementary generalisation of classical logic, allowing for gaps between truth values and over-assignment of those values. The interpretation of the connectives remains as classical as in two-valued logic, except for the generalisation to allow for gaps and gluts between the two semantic values. This is semantics on the 'American Plan'. It is interesting, substantial and philosophically salient. This understanding of negation is at the heart of Graham Priest's view of semantics (Graham Priest 1979; Priest 1987, 2001, 2008).

This semantics raises a question concerning the status of negation, and the interaction between truth and falsity. While truth and falsity are completely on a par when it comes to relational evaluations, they are not on a par when it comes to the definition of logical consequence. If we were to define logical consequence on the positive fragment of FDE—to restrict the language to $\wedge$ and $\vee$—we would not need to concern ourselves with falsity. We could just use ρ to keep track of whether a

---

[2] For this paper it suffices to take sequents to be pairs $X \succ Y$ of sets of formulas. We allow these sets to be infinite. I follow Humberstone (Lloyd Humberstone 2011) in using '$\succ$' as a sequent separator, as a sequent with a counterexample remains a sequent. I will reserve the assertion sign '⊢' for other uses.

formula is related to truth or not. Falsity, as far as validity is concerned, is a fifth wheel in the positive vocabulary language. It does enter into the semantics of logical consequence in order to give us the truth conditions for negation, but it does no other work. So, the question that is raised is this: what is so special about *negation* that means that we enrich our semantic statuses to keep track of this extra kind of information (negative information) as well as the information we already needed to track for evaluating validity?

To put this question another way: if our language were equipped with *two* kinds of negation instead of one, would this mean that we extend our evaluations to incorporate two different kinds of falsity? After all, that is the strategy of the American Plan.

This is a live question, because the negation of first-degree entailment, while a logical constant in the everyday sense, is not quite a logical *constant* in the same sense as the conjunction and disjunction of FDE. It is straightforward to see that conjunction and disjunction in FDE satisfy the following *defining rules*:

$$\frac{X, A, B \succ Y}{X, A \wedge B \succ Y} \qquad \frac{X \succ A, B, Y}{X \succ A \vee B, Y}$$

and that if any *other* connective satisfied one of the defining rule for conjunction, it would be logically equivalent to conjunction, and similarly for disjunction.[3] The same cannot be said for negation. There are no rules satisfied by negation in FDE-interpretations that force all FDE-negations to be equivalent. Negation, in FDE, is more like a $\square$ satisfying the conditions of the modal logic S5. There are bimodal logics with *two* S5 non-equivalent necessities, and there are 'bi-negation' logics with two non-equivalent FDE-negations.

This should not be surprising, for it is possible already to extend any FDE relational evaluation to model so-called 'Boolean negation', by adding the clauses:

$$-A\rho 1 \text{ iff } A\cancel{\rho} 1 \qquad -A\rho 0 \text{ iff } A\cancel{\rho} 0$$

and this is another FDE-negation. Now, to be sure, Boolean negation is not quite fully in the spirit of FDE, because it is the degenerate kind of negation with neither gaps nor gluts. We have a new kind of falsity (*untruth*) which stands to '−' as 0 stands to '∼'. Classical Boolean extensions of FDE are examples of logics with two non-equivalent FDE-negations.[4] It is possible, of course, for there to be *other* kinds of negations that satisfy the FDE conditions. One possible example might be found using the notion of polarity reversal for graded predicates. Consider the difference

---

[3] Suppose & also satisfied the defining rule for conjunction. By identity, $A \& B \succ A \& B$ holds, so by the defining rule for &, we have $A, B \succ A \& B$. But by the defining rule applied to ∧, we have $A \wedge B \succ A \& B$. Similarly, we can derive $A \& B \succ A \wedge B$, and then there is a short appeal through *Cut* to show that any derivable sequent in which $A \wedge B$ is used as a premise or a conclusion holds with $A \& B$ in its place, and *vice versa*. And the same holds for disjunction, too. Nothing like this holds for negation.

[4] I first learned of this sort of combination of negations in Meyer and Routley's work on Classical Relevant Logics (Robert 1973a,b).

between being *not* tall (and the boundary between tall and *not* tall, being vague, may perhaps permit gaps or overlap) and the difference between being tall and the *polar opposite* of tall—*short*. In a language where predicates have not only *complements* but *opposites*, there is scope for something that also satisfies the FDE conditions (for example, the opposite of the opposite takes you back to the original predicate), but care must be used in filling this out to the wider vocabulary. (A case could me made that the polar opposite of 'tall and reserved' should be 'short and outgoing' and not 'short or outgoing', so polarity reversal generalised to a sentential operator is not necessarily a de Morgan negation.) Regardless of how the details are filled out, there is a case to at least explore the behaviour of languages in which there are different notions of negation and opposition that satisfy the FDE constraints, because there is nothing in the logic that rules this out.

Nothing in what follows hangs on any particular choice of an interpretation for a language with two negations. My chief aim here is to understand the behaviour of negation in FDE by examining what happens when the language is extended to two negations. The aim is to not only shed light on FDE, relational evaluations and the American plan, but to also illuminate the distinctive features of that *other* semantics for FDE and negation, the *Australian* plan.

21.2 Negation on the Australian Plan

First-degree entailment, like all good logics, can be understood in more than one way. Another perspective on FDE is given when you follow the *Australian Plan*, and in particular, utilising the Routley star due to Routley and Routley (1972). On the Australian Plan (Robert 1986), negation is understood in the manner of a modal operator. The fact that the extension of a proposition is independent of its antiextension—or better, from the extension of its negation—is no more surprising than that there is independence between the extension of p and the extension of its necessitation $\Box p$. There are some statements p where p and $\Box p$ are both true. There are others where p is true and $\Box p$ is false. In the same way, the truth of p does not fix—on the Australian Plan—the truth of $\sim p$. Rather than taking falsity to be an independent and separate component of the semantic value of a statement from its truth, falsity is understood in terms of truth *elsewhere*.

Definition 21.3 A Routley frame is a set P of points endowed with a function $*$ where for each $x \in P$, $x^{**} = x$. A *Routley frame* $\langle P, * \rangle$ may be endowed with an interpretation $\Vdash$ for atomic sentences, determining whether $x \Vdash p$ or not at each point $x \in P$, and this is extended to the language of propositional logic by setting

$$x \Vdash A \wedge B \text{ iff } x \Vdash A \text{ and } x \Vdash B$$
$$x \Vdash A \vee B \text{ iff } x \Vdash A \text{ or } x \Vdash B$$
$$x \Vdash \sim A \text{ iff } x^* \not\Vdash A$$

and $\langle P, *, \Vdash \rangle$, so defined, is a *Routley model*.

We say that a sequent $X \succ Y$ holds on a model if and only there is no point $x \in P$ which serves as a counterexample to the sequent—i.e. where $x \Vdash A$ for each $A \in X$ and $x \not\Vdash B$ for each $B \in Y$.

A single Routley star $*$ on a frame P of points does not impose much structure on P. An operator determines a different partition of P into unordered pairs and singletons like so: $\{\{x, x^*\} : x \in P\}$. You can represent this in a diagram by connecting a point x and its star mate x^* with a link, like this:

A graph of this structure is special. Each node participates in one and only one link. They are very simple structures, with little complexity.

$$\sim \quad \sim \quad \sim$$

These models are rather unlike relational evaluations. However, it is straightforward to show that any sequents that hold in relational evaluations are exactly the sequents that hold in all Routley models.

Fact 21.4 *For each relational evaluation, there is a Routley model with a point satisfying exactly the formulas true at that evaluation. Conversely, for any point in a Routley model, there is a relational evaluation satisfying exactly the formulas true at that point.*

Proof Given each relational evaluation ρ, take the Routley model $\langle \{\rho, \rho^*\}, *, \Vdash \rangle$, where ρ^* is the relational evaluation that sets $p\rho^*1$ iff $p \not\rho 0$, and $p\rho^*0$ iff $p \not\rho 1$. Define $*$ in the obvious way (so, $\rho^{**} = \rho$) and set for atomic formulas p, $\rho \Vdash p$ iff $p\rho 1$ and $\rho* \Vdash p$ iff $p\rho^*1$, i.e. $p\not\rho 0$. It is an easy inductive argument to show that $\rho \Vdash A$ iff $A\rho 1$, and $\rho^* \Vdash A$ iff $A \not\rho 0$, and so, this is our Routley model that mimics the relational evaluation ρ.

Conversely, given a Routley model $\langle P, *, \Vdash \rangle$, and a point $x \in P$, define ρ by setting $p\rho 1$ iff $x \Vdash p$, and $p\rho 0$ iff $x^* \not\Vdash p$. This is a relational evaluation, and again, it is a straightforward inductive verification to show that $A\rho 1$ iff $x \Vdash A$, and $A\rho 0$ iff $x^* \not\Vdash A$, i.e. iff $x \Vdash \sim A$. ∎

Fact 21.5 *It is a consequence of this that* FDE *is not only the logic of arbitrary Routley models, it is also the logic of two point Routley models. No more than two points are needed to model* FDE.

So, we have two very different kinds of semantics for one logic. What is the difference between the American plan and the Australian plan? This is a real question, with philosophical bite.[5] These approaches are equivalent, in the sense that they model the same logic, but they do it in very different ways. These tools have very different affordances, as we will see when we attempt to take them beyond the very simple domain of modelling FDE.

We will not need to take them very *far* beyond that domain to address our question. We will see that these semantics are very different when we address the question of how they model *two* negations. Here, the difference between the American and Australian plans will come into sharper relief.

21.3 Two Kinds of Falsity

Given two negations and the American Plan, it is obvious how to *start* answering the question. We now have three semantic statuses: We have truth, falsity$_1$ and falsity$_2$. We will represent these as 1, 0 and F. That seems fair enough. We can extend the clauses for disjunction and conjunction:

$$(A \land B)\rho 1 \text{ iff } A\rho 1 \text{ and } B\rho 1$$
$$(A \land B)\rho 0 \text{ iff } A\rho 0 \text{ or } B\rho 0 \qquad (A \land B)\rho F \text{ iff } A\rho F \text{ or } B\rho F$$
$$(A \lor B)\rho 1 \text{ iff } A\rho 1 \text{ or } B\rho 1$$
$$(A \lor B)\rho 0 \text{ iff } A\rho 0 \text{ and } B\rho 0 \qquad (A \lor B)\rho F \text{ iff } A\rho F \text{ and } B\rho F$$

For our two negations, we know that we should *at least* enforce these conditions:

$$(\sim_1 A)\rho 1 \text{ iff } A\rho 0 \qquad (\sim_1 A)\rho 0 \text{ iff } A\rho 1$$
$$(\sim_2 A)\rho 1 \text{ iff } A\rho F \qquad (\sim_2 A)\rho F \text{ iff } A\rho 1$$

But what should we say about when $\sim_1 A$ is false$_2$, or when $\sim_2 A$ is false$_1$? There are many options, including adding semantic complexity to the values we already have at hand. If we aren't going to go beyond the three semantic statuses, however, the natural answer is this:

$$(\sim_1 A)\rho F \text{ iff } A\rho 1 \qquad (\sim_2 A)\rho 0 \text{ iff } A\rho 1$$

A negation's being *false* (in any sense of false) is, in some sense, a *positive* notion, and 1 is our only positive semantic status. In the American Plan, each semantic status of a complex formula is determined by the holding of a semantic status (positively) by the constituent formulas. We don't define truth or falsity in terms of the *failure* of the semantic status to subformulas.[6] Given the raw materials of 1, 0 and F, we don't

---

[5] De and Omori, for example, take it that the Australian Plan is philosophically defective, and the American Plan is the appropriate way to understand negation (Michael De and Hitoshi Omori 2017).

[6] This gives relational evaluations their significant power in terms of preservation and heridity results. If $\rho \subseteq \rho'$, in that ρ' assigns all the values that ρ does, and perhaps more, then ρ' makes true (and false) all the formulas made true (and false) by ρ. This result requires the positivity of the clauses for complex formulas. To break this is to lose one of the significant benefits of relational evaluations.

have much to work with. This clause is the natural one unless we wish to extend the modelling conditions further with other semantic statuses again such as keeping separate track not only of the two different kinds of falsity, but also the 1-falsity of 2-falsity (as a different kind of truth?), and more. But that way lies further complexity, and as far as I can see, makes the semantics look much less like relational evaluations and the purity of the American Plan.

$$\sim \quad \sim \quad \sim$$

So, what do our clauses for $\sim_1$ and $\sim_2$ tell us about the semantics for negation? Keeping our interpretation for validity as before, we know that both negations are FDE negations. The $\langle \wedge, \vee, \sim_1 \rangle$ and $\langle \wedge, \vee, \sim_2 \rangle$ fragments of the language are FDE, just as we would expect. Where things get interesting is the interaction between $\sim_1$ and $\sim_2$. One way to get a sense of just how interesting and different the logic is can be found by compiling 'truth tables' for the negations. Abusing notation a little, we can think of our evaluations as determining an eight-valued logic, with the values

$$\emptyset \quad \{1\} \quad \{0\} \quad \{F\} \quad \{1,0\} \quad \{1,F\} \quad \{0,F\} \quad \{1,0,F\}$$

where a formula is *assigned* the value S by a relational evaluation ρ where it is related to the values in S—and only those values—by ρ. Abusing notation a little more, we'll drop the set braces and think of the values as 0, $1F$, $10F$, etc. The truth tables for $\sim_1$, $\sim_2$, $\sim_1\sim_1$ and $\sim_2\sim_1$ are then:

p	$\sim_1 p$	$\sim_2 p$	$\sim_1\sim_1 p$	$\sim_2\sim_1 p$
$\emptyset$	$\emptyset$	$\emptyset$	$\emptyset$	$\emptyset$
1	$0F$	$0F$	1	1
0	1	$\emptyset$	$0F$	$0F$
F	$\emptyset$	1	$\emptyset$	$\emptyset$
10	$10F$	$0F$	$10F$	$10F$
$1F$	$0F$	$10F$	1	1
$0F$	1	1	$0F$	$0F$
$10F$	$10F$	$10F$	$10F$	$10F$

and a number of features stand out immediately. As in FDE, there are two fixed points for negation—for *both* negations here—the empty value $\emptyset$ and the total value $10F$. That is to be expected. The table shows clearly that the sequents $p \succ \sim_1\sim_1 p$ and $\sim_1\sim_1 p \succ p$ are valid (and similarly for $\sim_2$). The formulas p and $\sim_1\sim_1 p$ are designated in exactly the same rows. However, unlike in simple relational evaluations, p and $\sim_1\sim_1 p$ no longer receive the same semantic evaluations. When p is false$_1$ (only), $\sim_1\sim_1 p$ is not only false$_1$, it is also false$_2$. This double negation of p has gained a semantic status not had by p. But when p is false$_2$ only, its double negation $\sim_1\sim_1 p$ has value $\emptyset$. In this case, its double negation has *lost* a semantic status. In these bifurcated relational semantic valuations, A and $\sim_1\sim_1 A$ agree as far as 1 and 0, but they can disagree about F. A and $\sim_1\sim_1 A$ are no longer semantically equivalent in the strong sense: $\sim_2 A$ might fail (to be true) where $\sim_2\sim_1\sim_1 A$ might succeed (i.e. be true). The equivalence of A and $\sim_1\sim_1 A$ in relational evaluations is fragile. It can be disturbed by the addition of a new semantic status.

Another feature of this semantics for the two negations is the immediate collapse in the distinction between the two kinds of falsity, once a negation is involved. Notice that in the columns for $\sim_1 p$ and $\sim_2 p$, 0 and F never occur apart. This means that there are effectively only four different semantic statuses for negated formulas: $\emptyset$, 1, $0F$ and $10F$. Furthermore, on these values, $\sim_1$ and $\sim_2$ agree. The equivalence of $\sim_1\sim_1 p$ and $\sim_2\sim_1 p$ follows from a more general result: here, any negation of a negation if A is equivalent to any other negation of a negation of A. If we were to look at the sublanguage generated by all negations of atoms (rather than generated by the atoms), there is effectively only one negation, not two. So, the freedom of having two different semantic values for two different kinds of negations evaporates at the level of the negated formulas themselves if we use the natural falsity conditions for negations. To keep the level of freedom for negated sentences—while staying within the confines of the American Plan—we need to move to more complex clauses for the falsity conditions for negations. But what could they be? Perhaps the only way to be remain within the constraints of the American Plan, is to deny the presuppositions that $\sim_1$ has compositional evaluation conditions (at least with respect to F) and that $\sim_2$ has compositional evaluation conditions (at least with respect to 0).

$\sim\quad\sim\quad\sim$

This problem illustrates one well-understood issue with the American Plan. The bifurcated semantics, with separate conditions for truth and falsity, makes giving semantic clauses for *everything* else more complicated. The case of two 'negations' is just one simple illustration of this fact. It is one thing to give truth conditions for something (say, conditionals (Restall 1995), or in this case, another negative operator, another negation). Giving these truth conditions leaves open the thorny question of their *falsity* conditions. There is nothing in the *logic* that tells us, on the American Plan, how these falsity conditions are to relate to the truth conditions, and furthermore, there are ways to assign these conditions in such a way as to break the substitutivity of logical equivalents.

The Australian Plan has none of those difficulties. The extra work done to give 'modal' truth conditions for negation means that once truth conditions—at each point of the model structure—are given for a concept, the interaction between that concept and negation is already fixed. This is true, also, for the interaction between one negation and another, as we will see in the next section.

21.4 Two Routley Stars

On the Australian Plan, the prospect of having two FDE negations is immediate and affords no difficulty in giving truth conditions. The definition of the appropriate structure stares us in the face.

Definition 21.6 A *two-star frame* is a structure $\langle P, *, \star \rangle$ with two Routley stars. We endow a two-star frame with an interpretation $\Vdash$ in the usual way, adding clauses for two negations as follows:

$$x \Vdash \mathord{\sim}_1 A \text{ iff } x^* \nVdash A \qquad x \Vdash \mathord{\sim}_2 A \text{ iff } x^\star \nVdash A$$

and the definition of sequents holding on a two-star model are exactly the same as with Routley models.

There is no further question concerning how to evaluate $\mathord{\sim}_2\mathord{\sim}_1 p$. The model tells us that $\mathord{\sim}_2\mathord{\sim}_1 p$ holds at x if and only if p holds at $x^{\star *}$. Of course, where that *is* depends on the model, and on how the star operators, $*$ and $\star$, interact.

$$\sim \quad \sim \quad \sim$$

As to how $*$ and $\star$ interact, there is significant scope for complexity. Here is one way to get some sense of just how much complexity there is:

Fact 21.7 *Given any two-star frame $\langle P, *, \star \rangle$, the relation RB given by setting $xRBy$ iff $x^{\star *} = y$ is a permutation on P.*

Proof If $xRBy$ and $xRBz$ then $y = z$, since $y = x^{\star *} = z$. The converse relation RB^{-1} can be found by setting $xRB^{-1}y$ iff $x^{* \star} = y$, since if $x^{\star *} = y$ then $x^{\star * \star *} = y^{\star *}$ so $x^* = y^*$, so $x = x^{\star *} = y^{* \star}$. ∎

Here is an example of a two-star frame and its underlying RB permutation:

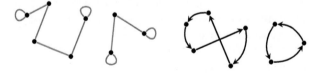

On any two-star frame, RB is a (single alternative) binary accessibility relation, and we have

$$x \Vdash \mathord{\sim}_2\mathord{\sim}_1 A \text{ iff for } all \text{ } y \text{ where } xRBy, \; y \Vdash A,$$
$$\text{iff for } some \text{ } y \text{ where } xRBy, \; y \Vdash A.$$

Let's abbreviate $\sim_2\sim_1 A$ by $\Box A$ (it could equally be $\Diamond A$, of course). This is the modal operator of single alternative modal logic, by the permutation RB on our set P of points. The converse modality $\Box^{-1}$, modelled by RB^{-1}, is given by setting $\Box^{-1} A = \sim_1\sim_2 A$. Clearly $\Box\Box^{-1} A$ is equivalent to $\Box\Box^{-1} A$ is equivalent to A on all of our two-star frames.

So, we have an interesting logical structure on our frames, and a recognisable normal modal logic—the logic of permutations—is embedded in two-star frames. In fact, we can say more. The structure of two Routley stars is enough to generate *any* permutation.

Fact 21.8 *Given any permutation σ on a set P, there is some two-star frame $\langle P, *, \star\rangle$ such that the RB relation on that frame is the permutation σ.*

Proof The permutation σ on P decomposes into its cycles in the following way: we say that $x, y \in P$ are in the same cycle if and only if some finite number (including zero) of applications of σ or its converse sends x to y. It is easy to see this is an equivalence relation on P, by design. Each cycle has some finite non-zero size (is an n-cycle for some n), or is infinite, in which case it has of order type $\omega^* + \omega$. The action of the permutation on P is determined by the structure of its cycles.

So, for any cycle, use the following rubric to define $*$ and $\star$ in order to generate such a cycle out of the action of two Routley stars:

For a cycle of even length, on $\{1, \ldots, 2n\}$ (where $n \geq 1$), set

$$1^* = 1, (2k)^* = 2k + 1 \; (k = 1, \ldots, n-1), \; 2n^* = n$$
$$(2l+1)^\star = 2l + 2 (l = 0, \ldots, n-1)$$

The structure looks like this

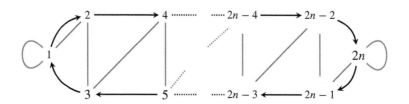

For a cycle of odd length, on $\{1, \ldots, 2n+1\}$ ($n \geq 0$), assign

$$1^* = 1, (2k)^* = 2k + 1 (k = 1, \ldots, n)$$
$$(2l+1)^\star = 2l + 2 (l = 0, \ldots, n-1), (2n+1)^\star = 2n+1$$

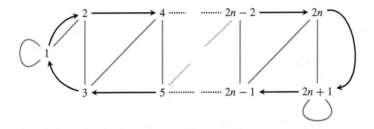

And finally, for an infinite $\omega^* + \omega$ sequence on $\mathbb{Z}$ assign

$$n^* = -n \text{ for each } n \qquad n^* = 1 - n \text{ for each } n$$

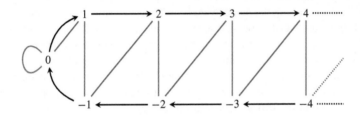

In this way, we can assign $*$ and $\star$ on each cycle of the permutation σ, such that the RB permutation generated by them agrees with σ. Each permutation is given by some two-star model. ∎

So, we have a rich family of structures. The logic of such structures can vary significantly. Here is one example of how the logic of a structure varies with the kinds of cycles present in its RB permutation.

Fact 21.9 *The following sequents hold:*

$$\Box^{n_1} p \wedge \cdots \wedge \Box^{n_k} p \succ p \qquad p \succ \Box^{n_1} p \vee \cdots \vee \Box^{n_k} p$$

on any frame if and only if its RB permutations consists of cycles where each cycle has some length l where $l \mid n_i$ for some i.

Proof Suppose $\Box^{n_1} p \wedge \cdots \wedge \Box^{n_k} p$ holds at a point x, where x is on some l cycle where $l \mid n_i$. Since $\Box^{n_i} p$ holds at x, and x is n_i RB-steps away from itself, p holds at x too. Suppose that p holds at a point x, where x is on some l cycle where $l \mid n_i$. The only point n_i steps away from x is x itself, so $\Box^{n_i} p$ holds at x also, and so does $\Box^{n_1} p \vee \cdots \vee \Box^{n_k} p$.

Conversely, suppose the frame has a component which is *not* an l cycle for any $l \mid n_i$. That is, it has a component which is an $\omega^* + \omega$ chain, or a cycle of some length k where $k \nmid n_i$ for each i. In any case, choose a point x in this component, and let p be true at each of the points n_i steps from x for each i. This does not include x itself. So, $\Box^{n_1} p \wedge \cdots \wedge \Box^{n_k} p$ holds at x but p doesn't. For the other sequent, do the reverse:

p holds at x but at none of the points n_i steps from x. This is a counterexample to the other sequent. ∎

So, for example, $\Box^4 p \wedge \Box^5 p \succ p$ holds in frames consisting of 4-cycles and 5-cycles. This sequent *also* holds in frames consisting of 2-cycles, 4-cycles and 5-cycles, because 2 divides 4.

These differences make a difference in the logic of negation. Frames consisting solely of 73-cycles are such that iterated $\sim_2 \sim_1$ double negations of length 73 collapse into nothingness, *but not before that*. Can sense be made of that?

So, constructions like these show us that we have infinitely many different logics, corresponding to different classes of frames, each with subtly different negation interaction conditions.

$$\sim \quad \sim \quad \sim$$

The variations among the logics of two-star frames do not end there. We have seen just a little of what you can do with different RB permutations. However, there is more to a frame than this. (You cannot, in general, recover the two-star frame $\langle P, *, \star \rangle$ from its RB permutation. For example, these two different two-star frames on $\{a, b\}$, one in which the stars $*$ and $\star$ keep a and b fixed, and the other, in which both stars swap a and b, have the same RB permutation, the identity function on $\{a, b\}$.) Exactly what other rich variety can be found in the class of two-star frames is left for further research.

21.5 On Semantics for Negations

So, while the Australian Plan and the American Plan are equivalent when it comes to the semantics of negation by itself, they differ significantly when we add negation to something else—even if that something else is just another negation. There are complexities on both sides. Following the American Plan, and extending relational evaluations to include another semantic status beyond truth and (the original) falsity, leads to questions about how this new status and falsity are to interact—questions that are not easy to answer. Furthermore, when you answer them, you may find that desirable features which you formerly had (such as the substitutivity of logical equivalents) no longer hold. However, there is little risk of semantic hyperinflation, of rich structures of possible propositional values generated by your semantic primitives.

On the Australian Plan, the situation is reversed. There is no need to ask difficult questions about how different semantic statuses are to interact. Once the modelling condition for each negation is set up, the interactions between those negations are fixed. Giving the truth conditions for each negation in each point in the model structure *fixes* the semantic values—and hence the interactions with other connectives, including the other negation. However, the potential for interactions between the devices used to interpret each negation means that we move beyond a simple four-valued or eight-valued logic into the complexity and variety of many different algebras of potential semantic values, many of them infinite and intricate.

Once we move beyond the simple language of first-degree entailment itself, we discover that the American Plan and the Australian Plan are very different. We are only beginning to understand the rich possibilities for modelling non-classical logics.

Acknowledgements Thanks to Jc Beall, Rohan French, Lloyd Humberstone, Dave Ripley and Shawn Standefer for discussions on these topics, and to two referees who gave suggestions to improve the paper. Thanks *especially* to Graham Priest, whose encouragement, example, guidance and challenge have given me more—over nearly 30 years!—than I can express in words. Whereof one cannot speak, ... ¶ This research is supported by the Australian Research Council, through Grant DP150103801, and Elephant9 and Reine Fiske's album, *Atlantis*. ¶ A version of this paper is available at http://consequently.org/writing/two-negations/.

References

Belnap, N. D. (1977a). How a computer should think. In G. Ryle (Ed.), *Contemporary aspects of philosophy*. Oriel Press.

Belnap, N. D. (1977b). A useful four-valued logic. In J. Michael Dunn & G. Epstein (Eds.), *Modern uses of multiple-valued logics* (pp. 8–37). Dordrecht: Reidel.

De, M., & Omori, H. (2017). There is more to negation than modality. *Journal of Philosophical Logic*, 1–19.

Humberstone, L. (2011). *The connectives*. The MIT Press.

Meyer, R. K., & Martin, E. P. (1986). Logic on the Australian Plan. *Journal of Philosophical Logic*, *15*(3), 305–332.

Meyer, R. K., & Routley, R. (1973a). Classical relevant logics I. *Studia Logica*, *32*, 51–66.

Meyer, R. K., & Routley, R. (1973b). Classical relevant logics II. *Studia Logica*, *33*, 183–194.

Michael Dunn, J. (1976). Intuitive semantics for first-degree entailments and "coupled trees". *Philosophical Studies*, *29*(3), 149–168.

Priest, G. (1979). The logic of paradox. *Journal of Philosophical Logic*, *8*(1), 219–241.

Priest, G. (1987). *In contradiction: A study of the transconsistent*. The Hague: Martinus Nijhoff.

Priest, G. (2001). *An introduction to non-classical logic*. Cambridge University Press.

Priest, G. (2008). *An introduction to non-classical logic: From if to is*. Cambridge: Cambridge University Press.

Priest, G., & Sylvan, R. (1992). Simplified semantics for basic relevant logics. *Journal of Philosophical Logic*, *21*(2), 217–232.

Restall, G. (1993). Simplified semantics for relevant logics (and some of their rivals). *Journal of Philosophical Logic*, *22*(5), 481–511.

Restall, G. (1995). Four valued semantics for relevant logics (and some of their rivals). *Journal of Philosophical Logic*, *24*(2), 139–160.

Routley, R., & Routley, V. (1972). Semantics of first degree entailment. *Noûs*, *6*(4), 335–359.

Greg Restall is Professor of Philosophy at the University of Melbourne. He received his Ph.D. from the University of Queensland in 1994, and has held positions at the Australian National University and Macquarie University, before moving to Melbourne in 2002. His research focuses on formal logic, philosophy of logic, metaphysics and philosophy of language, and even some philosophy of religion. He is the author of three books, An Introduction to Substructural Logics (Routledge, 2000), Logic (Routledge, 2006) and Logical Pluralism (Oxford University Press, 2006; with Jc Beall), and his research papers (and other things) can be seen at his website: http://consequently.org. His research has been funded by the Australian Research Council, and he is a Fellow of the Australian Academy of the Humanities.

Chapter 22
Inconsistency and Incompleteness, Revisited

Stewart Shapiro

Abstract Graham Priest (*In contradiction*) introduces an informal but presumably rigorous and sharp 'provability predicate'. He argues that this predicate yields inconsistencies, along the lines of the paradox of the Knower. One long-standing claim of Priest's is that a dialetheist can have a complete, decidable, and yet sufficiently rich mathematical theory. After all, the incompleteness theorem is, in effect, that for any recursive theory A, *if A is consistent*, then A is incomplete. If the antecedent fails, as it might for a dialetheist, then the consequent may also fail to hold. One somewhat friendly purpose of my 'Incompleteness and inconsistency' was to improve the technical situation for the dialetheist, eschewing reliance on an informal provability predicate. Another, less friendly purpose was to bring out what I took to be some untoward consequences of the situation. It seems that Priest accepted at least some of the improvements that I attempted. In the second edition of *In contradiction*, he responded to the alleged untoward consequences. One purpose of this note is to revisit the technical and philosophical situation. There were some errors in my original presentation, brought out by discussion with Priest and by Hartry Field's analysis of the second incompleteness theorem in such contexts. A second task here is to present a sort of Curry version of the Gödel incompleteness situation. I tentatively conclude that even for a dialetheist, an interesting and complete theory is not as easy to come by as it may look—at least not for theories of arithmetic that are plausible for a dialetheist.

Keywords Soundness · Beall Jc · Proof · Church's thesis · Curry · Incompleteness

Graham Priest's *In contradiction* introduces an informal but presumably rigorous and sharp 'provability predicate'. In the 2006 edition (Priest 2006), it appears semi-formally as $\beta(x)$. The idea is that for each natural number n, $\beta(n)$ holds just in case there is a rationally compelling proof in, say, mathematized English, of the sentence (also in mathematized English) whose code is n.

S. Shapiro
The Ohio State University, Columbus, USA
e-mail: shapiro.4@osu.edu

© Springer Nature Switzerland AG 2019
C. Başkent and T. M. Ferguson (eds.), *Graham Priest on Dialetheism and Paraconsistency*, Outstanding Contributions to Logic 18,
https://doi.org/10.1007/978-3-030-25365-3_22

Priest argues that β yields inconsistencies, along the lines of the paradox of the Knower. In line with his celebrated dialetheism, we thus have true contradictions involving this predicate, sentences that are both true and not true. Moreover, Priest argues, the predicate meets the conditions for Gödel's incompleteness theorem, and thus will yield Gödel-sentences.

One long-standing claim of Priest's is that the dialetheist can have a complete, decidable, and yet sufficiently rich mathematical theory. After all, the incompleteness theorem is, in effect, that for any recursive theory A, *if A is consistent*, then A is incomplete. If the antecedent fails, as it might for a dialetheist, then the consequent may also fail to hold.

By itself, of course, this does not entail that we *can* have a complete and decidable sufficiently rich mathematical theory. It only shows that the incompleteness theorem does not rule this out.

Priest shows how to construct models of an inconsistent arithmetic which are complete, in the sense that, for any sentence A in the language, either A or $\neg A$ (or both) is true. Moreover, the sets of truths in some of these models is recursive. Indeed, in one such model, *every* sentence is true (and every sentence is false). So the set of 'truths' in this trivial model is clearly recursive.

Given these trivial models, it seems that for the dialetheist, completeness, in this usual sense, is not the gold standard. Say that a theory, with an intended interpretation, is *super-sound* if it proves all and only the true sentences in the language. Perhaps super-soundess is the ultimate goal, for a dialetheist. Most dialetheists, including Priest, hold that every sentence is either true or false (or both). So super-soundness entails completeness.[1]

One somewhat friendly purpose of my 'Incompleteness and inconsistency' (Shapiro 2002) was to improve the technical situation for the dialetheist, eschewing reliance on an informal provability predicate. Another, less friendly purpose was to bring out what I took to be some untoward consequences of the situation.

It seems that Priest accepted at least some of the improvements that I attempted. In the second edition of *In contradiction* (Priest 2006), he responded to the alleged untoward consequences. One purpose of this note is to revisit the technical and philosophical situation. There were some errors in my original presentation, brought out by discussion with Priest and by Field's (2006) analysis of the second incompleteness theorem in such contexts.

A second task here is to present a sort of Curry version of the Gödel incompleteness situation. I tentatively conclude that even for a dialetheist, super-soundness is not as easy to come by as it may look—at least not for theories of arithmetic that are plausible or otherwise interesting.

---

[1] For a gap theorist—one who says that certain sentences are neither true nor false—super-soundness is weaker than completeness. Presumably, a gap theorist does not want a complete theory.

22.1 The Play with Informal, or Naive Proof

Let $Proof(x, y)$ be a predicate that says that x is the code of a proof, in mathematized English, whose conclusion is the sentence coded by y. This is a derivative notion related to Priest's β-predicate. We would have

$$\beta(y) \equiv \exists x (Proof(x, y)).$$

Priest argues that $Proof$ is a *recursive* predicate of natural numbers, and is thus expressible in ordinary Dedekind–Peano arithmetic. His argument for this goes via Church's thesis, together with a premise to the effect that mathematicians typically reach consensus on the correctness (or incorrectness) of purported proofs.

I do not think this use of Church's thesis is warranted, for a number of reasons. I list them here, more or less in increasing order of importance. First, Priest's argument is made in the context of a foundationalist epistemology for mathematics. The idea is that Proof is an absolute notion, based on derivation from self-evident premises, using self-evident rules of inference. Presumably, normal, competent humans do have the ability to tell whether an axiom or inference rule is self-evident. I do not think the foundationalism here is warranted (see Shapiro 2009), and the argument for consensus (over correct proofs) loses its plausibility with respect to a more holistic epistemology. And the history of mathematics has seen some furious arguments over the correctness of purported proofs. To be sure, the issue of foundationalism, even for mathematics, remains controversial, and I do not put much weight on it here, one way or the other.

Second, to even ask if a given set of strings is recursive, we need a single language, with a single alphabet. If not, then we don't really have a well-defined set of strings, and so the question of whether it is recursive is ill-formed. English, of course, does has a single alphabet, but the target language here is *mathematized* English, the language used by professional mathematicians, in the English speaking world, in the course of their work. As mathematics evolves, so do its languages. If we focus on a given time-slice of English, say the language spoken by mathematicians in 2017, then the $Proof$ relation is not comprehensive enough for its purposes. It would not apply to future languages, with undreamt of vocabulary.[2]

One might claim that the language in question is that of formalized set theory, with its single non-logical term for membership. But in that case, consensus is not always reached easily. Proofs translated into the language of set theory are notoriously tedious and difficult to evaluate. There is no 'data' concerning how easily consensus is reached on those proofs.

---

[2] A referee pointed out that the new mathematical concepts introduced in a given era are typically explained in terms of the language of the preceeding era. Suppose that this goes on for several iterations. We explain the mathematics of, say, four or five generations of new concepts in a single language. I would think that we could not count on consensus as to what is correct in the original language.

Third, the alleged consensus on the correctness of proofs, presented informally, is based on a relatively small and finite sample of mathematicians. Typical appeals to Church's thesis are not like this. I know of no other serious claim that a given set is recursive just because a group of people seem to be able to detect membership in it in typical (and, of necessity, a finite number of) instances.

Fourth, the cases where there is consensus among mathematicians all involve full classical logic (or perhaps intuitionistic logic), and take place in mathematical theories in which there seems to be no threat of inconsistency. *That* is the arena in which consensus over the correctness of a given proof is quickly determined. However, it is crucial to Priest's purposes that we can invoke the notion of informal proof in the more general arenas, which *do* have inconsistencies (e.g. to get the paradoxes of the Liar and the Knower on board). In such cases, there simply is no consensus, not even on the cases actually encountered. Most mathematicians do not tolerate inconsistencies at all (at least not true ones). And those mathematicians who tolerate true contradictions do not agree on what the correct logic is. So, in such cases, there really is not much in the way of consensus over whether alleged proofs are correct or not. Moreover, with these weak, paraconsistent logics, it is proving very difficult to adjudicate long, informal chains of reasoning. Again, no consensus.

Fifth, and most important, to invoke Church's thesis in an argument for recursiveness, it is not enough to note that people (trained mathematicians or otherwise) seem to reach consensus on certain verdicts most of the time. To invoke Church's thesis, one must specify an *algorithm*, a step by step procedure for computing a value, or deciding a question, a procedure that involves no creativity or use of intuition. And that we clearly do not have.

The early discussion against Church's thesis contains arguments where it is alleged that humans have such and such an ability, say to prove a certain kind of theorem. Those arguments were almost universally rejected at the time (see, for example, Kalmár 1959 and Mendelson 1963). Again, when it comes to the truth or falsity of Church's thesis, the cases that matter are those where we have something that is generally *agreed to be an algorithm*. All such cases have proved to be recursive. An instruction to 'check whether such and such a string is a good proof' does not count as a step in an algorithm. And there seems to be no (non-question-begging) way to specify an algorithm that will tell whether an alleged informal proof is correct or not.

22.2 Getting More Formal

Fortunately, the appeal to Church's thesis is not necessary for most of the relevant themes. In my 2002 paper (Shapiro 2002), I tried to sharpen Priest's argument, avoiding the appeal to Church's thesis, by working in a formal language.

Begin with the theorems of ordinary PA. Add a truth predicate T to the language, with the full T-rules (formulated with a detachable conditional). And include formulas that contain T in the induction scheme.

Of course, inconsistency results, in the usual manner. The underlying logic of this enterprise is paraconsistent, presumably LP together with rules for a detachable conditional.

Call the resulting theory PA*. Whether one is a dialetheist or not, this is (a gesture towards) a rigorous deductive system. Anyone—of any philosophical or logical persuasion—can study its deductive features.

Let $PRFPA^*(x, y)$ be a predicate, in the language of ordinary Dedekind–Peano arithmetic, that says that y is the code of a PA* derivation whose last line is coded by x. *This* predicate is primitive recursive, and thus expressible in the language of PA.

A Gödel sentence for PA* is a fixed point for

$$\neg(\exists y)PRFPA^*(x, y).$$

That is, we have, in the language of ordinary PA, a sentence G such that

$$G \leftrightarrow \neg(\exists y)PRFPA^*(\ulcorner G \urcorner, y)$$

is provable in ordinary PA.

I took it as straightforward that PA* proves its own soundness, as formulated with the detachable conditional (but see below):

$$(\forall x)[(\exists y)PRFPA^*(x, y) \to T(x)].$$

The 'proof' would go by induction. The axioms are true and the rules of inference preserve truth.

Given all of the above, we have that

$$\text{PA*} \vdash G \text{ and } \text{PA*} \vdash \neg G.$$

See Shapiro (2002) for details.

Notice that the sentence G is purely arithmetic; it has no occurrences of the truth predicate. So PA* proves contradictions *in the language of arithmetic*.

Up to this point, I was trying to be helpful. But Shapiro (2002) went on to sketch what I took to be some untoward consequences of all of this. A thoroughgoing dialetheist (like Priest) will hold that PA* is, in fact sound, namely all of its theorems are true. And our thoroughgoing dialetheist will maintain the tight equivalence between strings and natural numbers. So our thoroughgoing dialetheist will hold that there are true contradictions concerning what is and what is not derivable in certain deductive systems.[3] In particular, G both is and is not a theorem of PA*.

---

[3] A referee suggested that a dialetheist might resist the connection between the formalized proof predicate, which goes via the coding, and actual proof. This possibility was raised in Shapiro (2002). Priest seemed to reject this, accepting the usual tight connection between statements of what is provable and their coded arithmetic counterparts.

I balked at this. Priest takes this balking at this as a version of the incredulous stare. In retrospect, I admit that it very well might have been that. I admit that I have trouble understanding what it means to say that a given sentence both is and is not provable in a given deductive system. Having a derivation seems to preclude that there is no such derivation. But statements of my own psychology are not all that helpful in philosophical dialectic.

Priest (2006) shows how the above results play themselves out in models of inconsistent arithmetic. Any such model makes at least one atomic sentence both true and false. Since, the only primitive arithmetic predicate needed in arithmetic is identity, all models of inconsistent arithmetic have inconsistent identities, sentences of the form:

$$(\exists x)(\exists y)(x = y \,\&\, x \neq y), \text{ or perhaps better}$$

$$(\exists x)\, x \neq x.$$

That is, every model of inconsistent arithmetic will sport a number that is not identical to itself (and, of course, also identical to itself). The relevant thought is that the code of the proof of G is one such number. Given that, Priest is able to explain what is going on with the contradictions concerning derivability. There is a number, n that is the code of a derivation of G, and yet *every* number (including n) is *distinct from* any code of a derivation of G. So G is and is not derivable in PA*.

Given that our Gödel sentence G really is derivable in PA*, the code of its proof is a *standard* natural number. So PA* entails that there are true contradictory statements concerning the standard natural numbers. Indeed, there is an ordinary numeral g such that PA* proves that

$$g = g \,\&\, g \neq g.$$

I admit to having trouble understanding how this can be true (as would follow from the soundness of PA*), but this seems to be just another instance of the incredulous stare. And it is not polite to stare.

'Incompleteness and inconsistency' (Shapiro 2002) also claimed that the result shows that, for Priest, ordinary PA is inconsistent, since PA is sound and complete for primitive recursive relations. That was a mistake on my part. The usual way to show that PA is complete for primitive recursive relations presupposes the consistency of the facts concerning primitive recursive relations. Unfortunately, this was not the only flaw in Shapiro (2002).

22.3 Correction Involving Detachment

In what follows, let '⊃' be the ordinary, material conditional, so that $A \supset B$ is equivalent to $\neg A \vee B$, and let '→' be a detachable conditional, properly axiomatized (as in, for example, both editions of *In contradiction* (Priest 2006)). The idea is that modus ponens (or →-elimination) holds for this connective.

The recipe described in Shapiro (2002) was to first replace all of the conditionals in theorems of PA with detachable conditionals. The plan was to begin with the theorems of PA, so modified, and *then* add a naive truth predicate. Otherwise, I suggested, something is lost in the process. We end up with a weaker arithmetic than the one we started with, something that PA establishes but PA* does not.

Unfortunately, this makes PA* trivial—everything is provable in it. So something purely arithmetic *is* lost in the process. Let me explain.

Recall our Gödel sentence G, where the following is provable in PA:

$$G \equiv \neg(\exists y)PRFPA^*(\ulcorner G \urcorner, y).$$

So

$$G \supset ((\exists y)PRFPA^*(\ulcorner G \urcorner, y) \supset 0=1)$$

is also provable in PA. Replacing the material conditionals with detachable conditionals, we get

$$G \to ((\exists y)PRFPA^*(\ulcorner G \urcorner, y) \to 0=1).$$

However, PA* had better *not* prove this. Suppose, as above, that PA* proves both G and $\neg G$. The latter is (equivalent to) $(\exists y)PRFPA^*(\ulcorner G \urcorner, y)$. So PA* proves $0=1$. A disaster.

So, it seems, the full power of some purely arithmetic theorems has to be given up in the resulting inconsistent theory. I take it to be open just how strong an arithmetic we can start with—just how many arithmetic theorems can be stated with the detachable conditional, and end up with an interesting inconsistent theory that satisfies the naive truth rules and both proves and refutes a standard Gödel sentence for that theory (and may in fact be super-sound).

22.4 Correction Involving Soundness

As noted above, I stated in Shapiro (2002) that it should be straightforward to prove, in PA*, that PA* is sound (as formulated with a detachable conditional):

$$(\forall x)[(\exists y)PRFPA^*(x, y) \to T(x)]. \quad \text{(Sound)}$$

The proof, I suggested, would go by induction. The axioms are true, and the rules of inference preserve truth.

Field (2006) has shown that this reasoning is problematic, even in dialetheic contexts. With the exception of a few sub-structural and non-transitive logics, one cannot have it that even modus ponens preserves truth—thanks to the Curry paradox. Of course, LP and its like are neither sub-structural nor non-transitive.

It does not follow, of course, that PA* does *not* prove its own soundness. Only that the obvious way to show soundness does not work.

There is a sort of dilemma here for the dialetheist. Let us assume (for the sake of argument) that PA* is sound—that only truths are provable (even though some falsehoods are also provable). If PA* does not prove its own soundness, then we have not overcome incompleteness. There is one truth, namely soundness, that PA* fails to prove (assuming that it does not prove and fail to prove soundness, and assuming it does not prove the negation of soundness ...). We can be more precise.

22.5 Gödel Meets Curry

Let us continue to use 'PA*' for whatever (recursive) theory is put forward by our dialetheist, in order to overcome incompleteness. One might then check that the resulting theory PA* is itself sound and complete (but, of course, not consistent). It seems doubtful that this can be accomplished, for a reasonable theory.

To explain. Presumably, our dialetheist does not want every sentence to be provable in the preferred theory. That would be most uninteresting. Let $\perp$ be any sentence that the dialetheist definitely does not want to be provable. One candidate is '$0 = 1$'. Another is '$\forall x T x$', a statement that everything is true. Let C be a fixed point for the following formula:

$$(\exists y) PRFPA^*(x, y) \to \perp.$$

where, again, we use a detachable conditional That is, we have the following:

$$C \leftrightarrow [(\exists y) PRFPA^*(\ulcorner C \urcorner, y) \to \perp]. \tag{*}$$

If $\perp$ is arithmetic, then (*) is provable in ordinary PA, but, in any case, (*) should be provable in PA*.

Suppose that C is provable in PA*. Let n be the code of a proof. Then the following will be provable:

$$PRFPA^*(\ulcorner C \urcorner, n).$$

An instance of existential introduction and two instances of detachment, and we have that $\perp$ is provable in PA*, and thus, PA* is a disaster. Indeed, if $\perp$ just is $\forall x T x$, then our conclusion is that PA* is trivial, it proves everything. If $\perp$ is $0 = 1$, then we have that PA* proves that, which is bad enough.

So we can conclude that C is not provable in PA* (or else PA* is not all that interesting for our dialetheist, as it would be a disaster).

But it seems that C is *true*. After all, C 'says' that if C is provable, then (in effect) all hell breaks loose. But the above, informal argument, establishes just that. We started by assuming that C is provable in PA* and concluded, on the basis of that assumption, that PA* proves $\bot$. So, *if* C is provable in PA*, then so is $\bot$. But this is what C 'says'.[4]

So there is at least one truth that PA* fails to prove. Of course, it is still open that PA* is, technically, complete, if it proves $\neg C$. Without more detail on how the detachable conditional '$\rightarrow$' works, I don't know how to rule this out. But this is not a particularly welcome situation for our dialetheist.

At the outset, we defined a theory to be *super-sound* if it proves all and only the truths of the theory. The foregoing conclusion is that our candidate PA* is not super-sound. There is at least one truth, C, that PA* fails to prove (assuming, again, that in PA*, all hell does not break loose).[5]

Addendum

Let us briefly examine the perspective of another dialetheist, Jc Beall (e.g. Beall 2009). Beall argues that dialetheias—true contradictions—arise *only* in semantic contexts, when we add an unrestricted truth predicate T, subject to the 'transparency' rule that A and $T\ulcorner A\urcorner$ can always be substituted for each other. So, in particular, there are no true contradictions in the language of arithmetic.[6]

Beall has no need for a detachable conditional. The only conditional in the formal language is the 'material' one, '$\supset$', where $A \supset B$ is equivalent to $\neg A \vee B$. The underlying logic is LP. According to Beall, modus ponens (arrow elimination) is truth-preserving in contexts in which there is no danger of contradiction, but, even there, it is not logically valid.

Start with the theorems of ordinary PA, as formulated with a material conditional. Add a truth predicate, with the usual transparency rule that any formula A can be substituted with $T\ulcorner A\urcorner$. And include sentences with T in the induction scheme. Call the resulting theory PA**.

Say that a finite sequence P of sentences in the expanded language is a SPOOF if every line in P is one of the following three things:

1. a theorem of PA**.
2. follows from previous lines by a rule of the logic (i.e. LP) or a truth-rule (transparency).
3. follows from two previous lines via detachment and *both of those lines are purely arithmetic* (i.e. they contain no occurrences of the truth predicate T).

---

[4]I don't claim that these remarks constitute a formal proof of C. An attempt to formalize the informal argument presented here would presumably be a conditional proof, and it is open to the dialetheist to reject that for the conditional in C. Other attempts to reconstruct Curry-type reasoning in the formal system invoke contraction, which is rejected for the detachable conditional. Thanks to a referee for pointing this out.

[5]To follow the previous note, it is perhaps open to a dialetheist to claim that C is not true, in which case $\neg C$ is true. So if PA* is super-sound, there should be a proof, in PA*, of $\neg C$.

[6]A referee suggested that it is open for someone to claim that true contradictions arise only in semantic contexts, but that such contexts do (or at least might) result in true contradictions in the language of arithmetic. But this is not Beall's view. As noted, he holds that there are no true contradictions in the language of arithmetic. As Beall (2009, p. 16) puts it, 'The spandrels of ttruth bring gluts into our language, but they do not "spill" gluts back into our otherwise classical base language. In this way, the resulting dialetheism is very limited'. So for Beall, it is acceptable to reason with classical logic in arithmetic theories and in theories of syntax.

To be sure, for Beall, SPOOFs are not proofs, and their conclusions are not logical consequences of the axioms of PA**.

We can show (from the outside, so to speak) that, for Beall, every line in a SPOOF is true. It would go by induction on the length of the SPOOF. The axioms are all true and the rules of inference preserve truth. Moreover, for Beall, modus ponens, when restricted to purely arithmetic sentences, also preserves truth.[7] As noted, he holds that there are no true contradictions in the language of arithmetic.

Moreover, it is easy to check that SPOOF is a recursive predicate on (codes of) sequences of sentences in the formal language. So it can be represented in ordinary PA.

Let $SPF(x, y)$ be a (purely arithmetic) formula that says that y is the code of a SPOOF and x is the code of the last line of that SPOOF.

Suppose that there is a SPOOF whose last line expresses soundness (using the non-detachable conditional). The last line in question would be:

$$(\forall x)[(\exists y)SPF(x, y) \supset Tx]$$

Notice that this assumption is not ruled out by the results of Field (2006). The theory does not have a detachable conditional, and so the Curry-type reasoning does not go through. For this reason, this statement of soundness may not be all that interesting.

Now let G be a fixed point for

$$(\forall y)\neg SPF(x, y).$$

So we have

$$G \equiv (\forall y)\neg SPF(\ulcorner G \urcorner, y).$$

This is a theorem of ordinary PA and thus of PA**. And, again, the '$\equiv$' is a material (i.e. not detachable) biconditional. But this equivalence is also purely arithmetic.

By the foregoing assumption, there is a SPOOF of soundness:

$$(\forall x)[(\exists y)SPF(x, y) \supset Tx].$$

Instantiating, we have a SPOOF of

$$(\exists y)SPF(\ulcorner G \urcorner, y) \supset T\ulcorner G \urcorner.$$

And, by transparency:

$$(\exists y)SPF(\ulcorner G \urcorner, y) \supset G. \qquad (**)$$

---

[7] This claim does not run afoul of the conclusions in Field (2006). The indicated argument takes place in a meta-language (one whose logic is classical).

But notice that this is purely arithmetic. There is no occurrence of the truth predicate T. So we can proceed as in ordinary PA. As usual, $\neg G$ is logically equivalent to $(\exists y)SPF(\ulcorner G\urcorner, y)$. Substituting into (**), we have a SPOOF of $\neg G \supset G$. And that is logically equivalent to G.

So, formalizing all of the above, there is a SPOOF of G. Let m be the code of such a SPOOF. So, in ordinary PA, we have

$$SPF(\ulcorner G\urcorner, m)$$

So, in ordinary PA, we have

$$(\exists y)SPF(\ulcorner G\urcorner, y)$$

This last is equivalent to $\neg G$.

So we have a SPOOF of G and a SPOOF (actually, an ordinary PA-proof) of $\neg G$. Since G is arithmetic, this is, by Beall's lights, a BAD thing, as he is committed to arithmetic being consistent and to the truth of all lines in all SPOOFs.

The culprit here, of course, is the above assumption that there is a SPOOF of soundness. I take the foregoing to be a refutation of that assumption. For Beall, there can be no SPOOF of soundness, even formulated with the non-detachable material conditional. To be sure, this is not a direct criticism of Beall (2009), since, so far as I know, he never argued that there should be such a SPOOF.

References

Beall, J. (2009). *Spandrels of truth*. Oxford: Oxford University Press.
Field, H. (2006). Truth and the unprovability of consistency. *Mind*, *115*, 567–605.
Kalmár, L. (1959). An argument against the plausibility of Church's thesis. In *Constructivity in mathematics* (pp. 72–80). Amsterdam: North Holland.
Mendelson, E. (1963). On some recent criticisms of Church's thesis. *Notre Dame Journal of Formal Logic*, *4*, 201–205.
Priest, G. (2006). *In contradiction: A study of the transconsistent* (2nd, revised ed.). Oxford: Oxford University Press (1st ed., Dordrecht: Martinus Nijhoff Publishers, 1987).
Shapiro, S. (2002). Incompleteness and inconsistency. *Mind*, *111*, 817–832.
Shapiro, S. (2009). We hold these truths to be self evident: But what do we mean by that? *Review of Symbolic Logic*, 175–207.

Stewart Shapiro is currently the O'Donnell Professor of Philosophy at The Ohio State University, and he serves as a visiting Professor at the University of Connecticut and The Hebrew University of Jerusalem. He specializes in philosophy of mathematics, philosophy of language, logic and philosophy of logic, recently developing an interest in semantics. Professor Shapiro has taught courses in logic, philosophy of mathematics, philosophy of science, philosophy of religion, Marxism, aesthetics and medical ethics. He has an interest in Jewish Philosophy and in Jewish ethics, and occasionally teaches courses on those topics.

Chapter 23
GP's LP

Neil Tennant

Abstract This study takes a careful inferentialist look at Graham Priest's Logic of Paradox (LP). I conclude that it is sorely in need of a proof-system that could furnish formal proofs that would regiment faithfully the "naïve logical" reasoning that could be undertaken by a rational thinker *within* LP (if indeed such reasoning could ever take place).

23.1 Introduction

Graham Priest first put forward his Logic of Paradox, now known by the acronym LP, in Priest (1979). Subsequently, LP was one of many paraconsistent logics that he considered in his survey article (Priest 2002). (Unsourced page references will be to the latter.) By definition, a paraconsistent logic is one that does not admit (hence certainly does not allow one to derive, let alone have as a primitive rule) the inference known as *Explosion*:

$$A, \neg A : B.$$

Priest's LP was a creature of formal semantics. At its inception, it had no proof-system. All it had was a logical consequence relation $\models_{LP}$ defined in the usual way— the usual way, that is, for a *many*-valued logic. In such logics, there are more than two truth-values, and one is concerned with preserving "designatedness" of truth-value. Many-valued logics, therefore, evince a vestige of classical, two-valued thinking in

---

I am grateful to Graham for a helpful email exchange, and for his comments on an earlier draft. Thanks are owed to Matthew Souba for a careful reading of an earlier draft. I am grateful to Julia Jorati and Florian Steinberger for vetting my translation of the German passage quoted in footnote 8.

N. Tennant
Department of Philosophy, The Ohio State University, Columbus, OH 43210, USA
e-mail: tennant.9@osu.edu

© Springer Nature Switzerland AG 2019
C. Başkent and T. M. Ferguson (eds.), *Graham Priest on Dialetheism and Paraconsistency*, Outstanding Contributions to Logic 18,
https://doi.org/10.1007/978-3-030-25365-3_23

the way they partition their many values into two classes—the designated and the undesignated. LP employs certain 3-valued truth tables (or perhaps one should say: *truth-value* tables) for the connectives ¬, ∧, ∨ and →.

At p. 224 of Priest (1979), we read that "the logic of naïve proofs [which is what Priest is after—NT] is not classical". Nevertheless, it turns out that LP is "classical enough" for negation and conjunction to suffice for the definition of disjunction and the conditional. We read at p. 227 *loc. cit.* that Priest is content to define $A \vee B$ de Morgan-wise as $\neg(\neg A \wedge \neg B)$; and thereafter to define $A \to B$ as the usual disjunction $\neg A \vee B$.

The truth-values that (Priest 1979) proposes for LP are *True, False,* and *both-True-and-False*. *True*, i.e., *True only,* is abbreviated as t; *False*, i.e., *False only,* is abbreviated as f; and *both-True-and-False* is equated with *Paradoxical,* and abbreviated accordingly as p. The row-by-row truth-value tables for the connectives ¬ and ∧, as well as the (primitive or defined) ∨ and →, are as follows. We state them so that the rows can be read "from left to right", i.e., from values of immediate constituents to values of the respective compounds.

A	$\neg A$
t	f
p	p
f	t

A	B	$A \wedge B$	$A \vee B$	$A \to B$
t	t	t	t	t
t	p	p	t	p
t	f	f	t	f
p	t	p	t	t
p	p	p	p	p
p	f	f	p	p
f	t	f	t	t
f	p	f	p	t
f	f	f	f	t

In this form, significant patterns are difficult to discern. This can be remedied by resorting to stating a 3-by-3 matrix for each binary connective and using some font changes to highlight the patterns.

∧	t	p	f
t	t	p	f
p	p	p	f
f	f	f	f

∨	t	p	f
t	t	t	t
p	t	p	p
f	t	p	f

→	t	p	f
t	t	p	f
p	t	p	p
f	t	t	t

As Priest notes, these are the truth-value tables of Kleene's (strong) 3-valued logic in Kleene (1952, pp. 332) *ff.*[1] The main difference is that in LP the designated values are *t* and *p*, whereas in Kleene's logic only *t* is designated.

Observation 23.1 Priest (1979), *at p. 227, describes the truth-value table for* → *as resulting from defining* $A \to B$ *as* $\neg A \lor B$; *but there should be no objection in principle to anyone who opts to treat* → *as a* primitive *connective in LP, furnished with that truth-value table. It would then be very natural to pose the question: What rules of inference govern "*→*"?*

By the end of this study, it will emerge, perhaps surprisingly, that there appears to be no straightforward answer to this question.

23.2 The Formal Semantics of LP

Definition 23.1 An *assignment* τ is a function assigning values in $\{t, p, f\}$ to certain atoms. (Which atoms these are will depend on the context.) The value $\tau(\varphi)$ is defined in the familiar way, by appeal to the truth-value tables of LP. When $\tau(\varphi)$ is a designated value, we shall say that τ designates φ, and abbreviate this as $\tau \Vdash \varphi$. When τ designates every member of Δ, we shall say that τ designates Δ, and write $\tau \Vdash \Delta$.

Logical implication (the semantic relation) in LP can now be defined as follows.

Definition 23.2

$$\Delta \models_{LP} \psi \text{ if and only if } \forall \tau (\tau \Vdash \Delta \Rightarrow \tau \Vdash \psi),$$

i.e., for every truth-value assignment τ to the atoms involved, if τ designates every member of Δ, then τ designates ψ.

We shall read $\Delta \models_{LP} \psi$ as "the inference $\Delta : \varphi$ is LP-valid"; or "Δ LP-implies φ"; or "φ is an LP-consequence of Δ".

Remember that truth-value assignments are single-valued.

Definition 23.3 φ and ψ are (semantically) *equivalent* just in case they have the same truth(-value) table.

For example, $\neg A \lor B$ is equivalent to $A \to B$:

---

[1] Kleene had *u* as his third value, standing for 'undefined' (or, at p. 335, "unknown"). He was concerned with partial recursiveness of compounds, not at all with paradoxes or "truth-value gluts".

A	B	$\neg A$	$\neg A \vee B$	$A \to B$
t	t	f	t	t
t	p	f	p	p
t	f	f	f	f
p	t	p	t	t
p	p	p	p	p
p	f	p	p	p
f	t	t	t	t
f	p	t	t	t
f	f	t	t	t

Observation 23.2 *That $\varphi \models_{LP} \psi$ and $\psi \models_{LP} \varphi$ is not in general sufficient to ensure that φ and ψ are semantically equivalent. This is because mutual LP-implication is compatible with one of the two sentences being t, and the other p, in some same row.*

Definition 23.4 φ is a *daudology* just in case φ takes a designated value on every LP-assignment; i.e., just in case $\emptyset \models_{LP} \varphi$.

N.B. This is a neologism, not a typo. "Daudology" takes its initial letter from "designated". It is the LP-analog of a tautology in the two-valued case.

23.2.1 Some Low-Level Explorations of LP-Consequence

All the results of this subsection are obtained by using only Core Logic in the metalanguage.

Observation 23.3 *A good example of a daudology is $A \vee \neg A$:*

A	$\neg A$	$A \vee \neg A$
t	f	t
p	p	p
f	t	t

Note that we do not need a solid column of t-entries; all we need is the absence of any f-entries. This is because both t and p are designated.

For $\Delta \cup \{\varphi\}$, we shall write Δ, φ and typically understand the latter to mean also that $\varphi \notin \Delta$.

Lemma 23.1 *If $\tau \not\Vdash \neg \varphi$ then $\tau \Vdash \varphi$.*

Proof Suppose $\tau \not\Vdash \neg\varphi$. Then by definition of $\Vdash$ we have $\tau(\neg\varphi) = f$. Hence, by the truth-value table for $\neg$ we have $\tau(\varphi) = t$. So by definition of $\Vdash$ again we have $\tau \Vdash \varphi$. □

Note that the proof of Lemma 23.1 implicitly relies on the following assumption, which could be called the Law of Excluded Fourth: *an assignment τ assigns, to every sentence whose atoms it deals with, a determinate one of the three values t, p or f.* The Law of Excluded Fourth underlies Priest's whole three-valued semantical approach to LP-consequence. The core meta-logician is entitled, therefore, to appeal to that law when investigating the properties of LP that are revealed in this subsection.

Lemma 23.2 (LP-consequence satisfies Dilution)
If $\Delta \models_{LP} \psi$, then $\Delta, \varphi \models_{LP} \psi$.

Proof Suppose $\Delta \models_{LP} \psi$. Suppose μ is an arbitrary assignment dealing with the atoms involved in Δ, φ, ψ. Suppose $\mu \Vdash \Delta, \varphi$. Then $\mu \Vdash \Delta$. Hence by main supposition, we have $\mu \Vdash \psi$. But μ was arbitrary. Thus $\Delta, \varphi \models_{LP} \psi$. □

When one is working with sequents of the form $\Delta : \Gamma$, where Δ and Γ are finite sets of sentences, the natural reading of "$\Delta \models \Gamma$" (for any many-valued logic) is that no assignment designates every sentence Δ but none in Γ. We are adopting that reading here, but treating only of sequents whose succedents are at most a singleton. (We write "$\Delta \vdash \varphi$", however, rather than "$\Delta \vdash \{\varphi\}$".) We shall follow the convention of rendering "$\Delta \models \emptyset$" as "$\Delta \models \bot$". This means, on the reading adopted, that no assignment designates every member of Δ.

Lemma 23.3 (Classical Reductio is LP-valid)
If $\Delta, \neg\varphi \models_{LP} \bot$, then $\Delta \models_{LP} \varphi$.

Proof Suppose $\Delta, \neg\varphi \models_{LP} \bot$. It follows that for no τ do we have both $\tau \Vdash \Delta$ and $\tau \Vdash \neg\varphi$. Suppose μ is an arbitrary assignment dealing with the atoms involved in Δ, φ. Suppose $\mu \Vdash \Delta$. Then $\mu \not\Vdash \neg\varphi$. Hence by Lemma 23.1 $\mu \Vdash \varphi$. But μ was arbitrary. Thus, $\Delta \models_{LP} \varphi$. □

Observation 23.4 *We have used constructive reasoning in the metalanguage to show that* Classical *Reductio is LP-valid. This should strike the unsuspecting reader as surprising. It provokes the question "How does LP pull this off?" (i.e., reveal itself, 'constructively' at the metalevel, to validate a strictly classical, nonconstructive, rule of reasoning). The answer must be that an awful lot is packed into the Law of Excluded Fourth. That principle is so powerful that merely constructive reasoning can draw out from it the LP-validity of Classical Reductio.*

Lemma 23.4 (Double-Negation Elimination is LP-valid)
$\neg\neg\varphi \models_{LP} \varphi$.

Proof Suppose μ is an arbitrary assignment dealing with the atoms in φ. Suppose $\mu \Vdash \neg\neg\varphi$. Then either $\mu(\neg\neg\varphi) = t$ or $\mu(\neg\neg\varphi) = p$.

Suppose first that $\mu(\neg\neg\varphi) = t$. Then $\mu(\neg\varphi) = f$; whence $\mu(\varphi) = t$.
Suppose second that $\mu(\neg\neg\varphi) = p$. Then $\mu(\neg\varphi) = p$; whence $\mu(\varphi) = p$.
Either way, φ is assigned a designated value under μ. That is, $\mu \Vdash \varphi$.
But μ was arbitrary. Hence, $\neg\neg\varphi \models_{LP} \varphi$. □

Lemma 23.5 (Dilemma is LP-valid)
If $\Delta, \varphi \models_{LP} \psi$ and $\Gamma, \neg\varphi \models_{LP} \psi$, then $\Delta, \Gamma \models_{LP} \psi$.

Proof Suppose that
$$\Delta, \varphi \models_{LP} \psi \text{ and } \Gamma, \neg\varphi \models_{LP} \psi.$$

We shall show that
$$\Delta, \Gamma \models_{LP} \psi.$$

Let μ be an arbitrary assignment dealing with the atoms involved in $\Delta, \Gamma, \varphi, \psi$. Suppose that
$$\mu \Vdash \Delta, \Gamma.$$

It follows immediately that
$$\mu \Vdash \Delta \text{ and } \mu \Vdash \Gamma.$$

We shall now show that $\mu \Vdash \psi$.

We reason by using proof by cases (Disjunction Elimination) in the metalanguage. We know that μ assigns one of the three values t, p, or f to any sentence. So, in particular,
$$(\mu(\varphi) = t \text{ or } \mu(\varphi) = p) \text{ or } \mu(\varphi) = f.$$

Suppose on the one hand that either $\mu(\varphi) = t$ or $\mu(\varphi) = p$. Then $\mu \Vdash \varphi$. It follows that
$$\mu \Vdash \Delta, \varphi.$$

By main supposition, we have
$$\Delta, \varphi \models_{LP} \psi.$$

So
$$\mu \Vdash \psi.$$

Suppose on the other hand that $\mu(\varphi) = f$. Thus $\mu(\neg\varphi) = t$, whence
$$\mu \Vdash \neg\varphi.$$

It follows that
$$\mu \Vdash \Gamma, \neg\varphi.$$

By main supposition we have
$$\Gamma, \neg\varphi \models_{LP} \psi.$$
So
$$\mu \Vdash \psi.$$

Either way, we have $\mu \Vdash \psi$.

But μ was an arbitrary assignment such that $\mu \Vdash \Delta, \Gamma$. It follows that
$$\Delta, \Gamma \models_{LP} \psi.$$

□

Lemma 23.6 *Suppose* $\Delta, \varphi \models_{LP} \bot$. *Then* $\Delta \models_{LP} \varphi \to \psi$.

Proof The main supposition amounts to the following:
$$\forall \tau \ \neg(\tau \Vdash \Delta \wedge \tau \Vdash \varphi).$$

Let μ be an arbitrary assignment to the atoms involved in Δ, φ, ψ. Suppose that $\mu \Vdash \Delta$. We shall show that $\mu \Vdash \varphi \to \psi$; and this will establish the lemma. This we do by using Disjunctive Syllogism in the metalanguage. We know that μ assigns one of the three values t, p, or f to any sentence. So, in particular,
$$(\mu(\varphi \to \psi) = t \ \text{ or } \ \mu(\varphi \to \psi) = p) \ \text{ or } \ \mu(\varphi \to \psi) = f.$$

We are seeking to show that μ designates $\varphi \to \psi$, i.e.,
$$\mu(\varphi \to \psi) = t \ \text{ or } \ \mu(\varphi \to \psi) = p.$$

So all we have to do is rule out the possibility that $\mu(\varphi \to \psi) = f$. Assume for (constructive) *reductio*, then, that $\mu(\varphi \to \psi) = f$. By the 3-valued table for $\to$, we have
$$\mu(\varphi) = t \ \text{ and } \ \mu(\psi) = f.$$

Thus $\mu \Vdash \Delta \wedge \mu \Vdash \varphi$, contradicting the main supposition. □

Lemma 23.7 *Suppose* $\Delta, \varphi \models_{LP} \psi$. *Then,* $\Delta \models_{LP} \varphi \to \psi$.

Proof The main supposition amounts to the following:
$$\forall \tau (\tau \Vdash \Delta, \varphi \Rightarrow \tau \Vdash \psi).$$

Let μ be an arbitrary assignment to the atoms involved in Δ, φ, ψ. Suppose that $\mu \Vdash \Delta$. We shall show that $\mu \Vdash \varphi \to \psi$; and this will establish the lemma. This we do by using Disjunctive Syllogism in the metalanguage. We know that μ assigns one of the three values t, p, or f to any sentence. So, in particular,

$$(\mu(\varphi \to \psi) = t \text{ or } \mu(\varphi \to \psi) = p) \text{ or } \mu(\varphi \to \psi) = f.$$

We are seeking to show that μ designates $\varphi \to \psi$, i.e.,

$$\mu(\varphi \to \psi) = t \text{ or } \mu(\varphi \to \psi) = p.$$

So all we have to do is rule out the possibility that $\mu(\varphi \to \psi) = f$. Assume for (constructive) *reductio*, then, that $\mu(\varphi \to \psi) = f$. By the 3-valued table for $\to$, we have

$$\mu(\varphi) = t \text{ and } \mu(\psi) = f.$$

Thus $\mu \Vdash \Delta, \varphi$. Hence by main supposition we have

$$\mu \Vdash \psi \,;$$

but this contradicts

$$\mu(\psi) = f.$$

$\square$

We are using Modus Ponens here in the metalanguage, even though it is not LP-valid.[2] But the discrepancy, once registered, is innocuous. This is because it is our prerogative as investigators of LP at the metalevel to be able to reason in normal mathematical fashion about the system, since it is mathematically well defined. Both Modus Ponens and Disjunctive Syllogism are part of (even the constructive) mathematician's unrelinquishable tool kit. Otherwise, how could any useful consequences be drawn out from the various definitions that the LP-theorist provides of such notions as $\models_{LP}$? The "naïve logician" would not proceed any differently. For we are talking here about the *inferential* part of the "logic of naïve *proof*", which *obviously* includes all the usual logical inferences employed by mathematicians. This prescinds from the *axiomatic* basis of the logic of naïve proof in, say, arithmetic. Priest's comments about naïve reasoning in arithmetic, in Sects. II.2 and II.6 of Priest (1979), in no way commit the formalizer to *not* incorporate inferential rules such as Modus Ponens and Disjunctive Syllogism in whatever formal system of proof will result from one's investigations. See also Observation 23.6.

Corollary 23.1 *Suppose* $\Delta \models_{LP} \psi$. *Then* $\Delta \models_{LP} \varphi \to \psi$.

Proof Suppose $\Delta \models_{LP} \psi$. Then by Lemma 23.2, we have $\Delta, \varphi \models_{LP} \psi$. By Lemma 23.7 it follows that $\Delta \models_{LP} \varphi \to \psi$. $\square$

Corollary 23.1 also has this direct proof: Suppose $\Delta \models_{LP} \psi$. Let μ be an arbitrary assignment dealing with Δ, φ, ψ. Suppose $\mu \Vdash \Delta$. Then $\mu \Vdash \psi$. Hence $\mu \Vdash \varphi \to \psi$.

---

[2]This point is owed to Matthew Souba.

23.2.2 Considerations of Natural Deduction

In general, a natural deduction is a proof of some conclusion θ from some (finite) set Δ of undischarged assumptions (premises). For any system $\mathcal{S}$ of formal proof, we write
$$\mathcal{P}_{\mathcal{S}}(\Pi, \theta, \Delta)$$
for "Π is an $\mathcal{S}$-proof of the conclusion θ from the set Δ of undischarged assumptions". Note that "from the set" means, here, "using exactly the set".

Suppose there *is* a natural deduction system to be had for LP, in which all LP-proofs are LP-sound. Expressed as a single metalinguistic sentence, we have
$$\forall \Pi \, \forall \Delta \, \forall \theta (\mathcal{P}_{\mathrm{LP}}(\Pi, \theta, \Delta) \Rightarrow \Delta \models_{\mathrm{LP}} \theta)$$

Equivalently, we have the metalinguistic rule of inference
$$\frac{\mathcal{P}_{\mathrm{LP}}(\Pi, \theta, \Delta)}{\Delta \models_{\mathrm{LP}} \theta}$$

Lemma 23.6 forces the following reflection. This rule is admissible in LP:

$$\begin{array}{c} \overline{\Delta, \varphi}^{(i)} \\ \vdots \\ \underline{\bot}_{(i)} \\ \varphi \to \psi \end{array}$$

Lemma 23.7 and Corollary 23.1 together force a similar reflection. The conventional Rule of Conditional Proof (a.k.a. $\to$-Introduction), which permits vacuous discharge, is also admissible in LP:

$$\begin{array}{c} \overline{\Delta, \varphi}^{(i)} \\ \vdots \\ \underline{\psi}_{(i)} \\ \varphi \to \psi \end{array}$$

Note that these last two rules are the two parts of the rule of $\to$-Introduction in Core Logic. (See, for example, Tennant 2015b; and Appendix 1 below.)

Lemma 23.8 *Explosion* $(A, \neg A : B)$ *is not LP-valid. That is,*
$$A, \neg A \not\models_{\mathrm{LP}} B.$$

Proof It is clear how to invalidate Explosion. Let A be assigned p. Then $\neg A$ too takes the value p. Now assign B the value f. This yields a counterexample to Explosion: each of the premises A and $\neg A$ enjoys a designated value, whereas the conclusion B does not. □

The form of definition of $\models_{LP}$ guarantees that it is unrestrictedly transitive:

Lemma 23.9 (Unrestricted Cut)
If $\Delta \models_{LP} \varphi$ and $\Gamma, \varphi \models_{LP} \psi$, then $\Delta, \Gamma \models_{LP} \psi$.

Proof Suppose that
$$\Delta \models_{LP} \varphi \tag{23.1}$$

and
$$\Gamma, \varphi \models_{LP} \tag{23.2}$$

Suppose μ is an arbitrary assignment dealing with the atoms involved in $\Delta, \Gamma, \varphi, \psi$. Suppose that
$$\mu \Vdash (\Delta \cup \Gamma) \tag{23.3}$$

We shall show that $\mu \Vdash \psi$. From (23.3), it follows that
$$\mu \Vdash \Delta \tag{23.4}$$

and
$$\mu \Vdash \Gamma \tag{23.5}$$

From (23.1) and (23.4), we have
$$\mu \Vdash \varphi \tag{23.6}$$

From (23.2), (23.5) and (23.6), we have $\mu \Vdash \psi$. □

Corollary 23.2 *As a special case of Lemma 23.9, we have the following:*
If $\models_{LP} \varphi$ and $\Gamma, \varphi \models_{LP} \psi$, then $\Gamma \models_{LP} \psi$.

Proof Set $\Delta = \emptyset$ in Lemma 23.9. □

Observation 23.5 *Corollary 23.2 says one can suppress daudologies as premises.*

Now it is well known, from Lewis's famous argument, that Explosion follows from the combination of

1. having $A \vee B$ implied by A (and implied by B);
2. having Disjunctive Syllogism: $\neg A, A \vee B : B$; and
3. having unrestricted transitivity.

23 GP's LP

This stares one in the face upon arranging the following little bits of proof in an inviting pattern:

$$\frac{A}{A \vee B}$$

$$\frac{A \vee B \quad \neg A}{B}$$

Since LP validates $\vee$I (hence may have it as a rule) and enjoys unrestricted transitivity, but invalidates Explosion, it follows that LP must invalidate Disjunctive Syllogism. And so it does.

Lemma 23.10 *Disjunctive Syllogism ($\neg A, A \vee B : B$) is not LP-valid. That is,*

$$\neg A, A \vee B \not\models_{LP} B.$$

Proof It is clear how to invalidate Disjunctive Syllogism. We adduce once again the assignment

$$\tau(A) = p \quad \tau(B) = f$$

By LP's truth-value tables, we have

$$\tau(\neg A) = p, \quad \tau(A \vee B) = p \,;\, \text{but } \tau(B) = f.$$

The value p is designated but the value f is not. So Disjunctive Syllogism is not LP-valid. □

Observation 23.6 *The proofs of Lemma 23.6 and of Lemma 23.7 use Disjunctive Syllogism in the metalogic. But Lemma 23.10 says Disjunctive Syllogism is not LP-valid. So the advocate of LP cannot undertake the foregoing metalogical reasoning that yields the insights of Lemma 23.6 and of Lemma 23.7. This reveals a reflexive instability in the position of the LP-advocate—unless some* alternative *passage of metalogical reasoning can be furnished, which is formalizable in a proof-system for LP.*

I cannot, however, find any ready alternative. I would be happy to be instructed in this regard by an LP-er.

Disjunctive Syllogism as stated above is an inference from the premises $A \vee B$, $\neg A$ to the conclusion B. Metalinguistically, one would state

$$A \vee B, \neg A \models_S B$$

for any logical system S validating Disjunctive Syllogism. For LP, however, we have, on the one hand,

$$A \vee B, \neg A \not\models_{LP} B,$$

whence also (by $\wedge$I and unrestricted transitivity)

$$(A \vee B) \wedge \neg A \not\models_{LP} B.$$

On the other hand, the THEOREM in Sect. III.8 on p. 228 of (Priest 1979) tells us that the classical tautologies are exactly the daudologies. Among the latter, we have

$$\models_{LP} ((A \vee B) \wedge \neg A) \to B$$

Note that this is a *formal semantic* claim, using the double turnstile $\models_{LP}$.

23.2.3 The Trouble with Modus Ponens

If Priest is correct in thinking that there is a sound and complete system of natural deduction for LP—giving rise to a *single* turnstile $\vdash_{LP}$ coextensive with $\models_{LP}$—then we know that to match the last semantic claim (the second of the two results stated at the end of Sect. 23.2.2) there must be some proof in this system of the form

$$\begin{array}{c} \emptyset \\ \Pi \\ ((A \vee B) \wedge \neg A) \to B \end{array}$$

The first of the two results stated at the end of Sect. 23.2.2—

$$(A \vee B) \wedge \neg A \not\models_{LP} B$$

—shows that the usual rule of detachment (i.e., Modus Ponens):

$$\begin{array}{cc} \Delta & \Gamma \\ \vdots & \vdots \\ \varphi & \varphi \to \psi \\ \hline & \psi \end{array}$$

fails rather conspicuously in LP. Indeed, it can fail even in a context where the newly accumulated premises $\Delta \cup \Gamma$ are jointly *consistent*:

$$\begin{array}{c} \emptyset \\ \Pi \\ \dfrac{(A \vee B) \wedge \neg A \quad ((A \vee B) \wedge \neg A) \to B}{B} \end{array}$$

23 GP's LP

So one *cannot* have, in LP, the following rule, either primitive or derived[3]:

$$\frac{\varphi \quad \varphi \to \psi}{\psi}$$

There is of course a deep motivating reason for the LP-er to invalidate Modus Ponens. If he *were* entitled to the instance

$$\frac{(A \lor B) \land \neg A \quad ((A \lor B) \land \neg A) \to B}{B}$$

then he would be able to construct a proof of Explosion as follows:

$$\frac{\dfrac{\dfrac{A}{A \lor B} \quad \neg A}{(A \lor B) \land \neg A} \quad \dfrac{\emptyset}{\Pi} \atop ((A \lor B) \land \neg A) \to B}{B}$$

And this, by Lemma 23.8, would mean that the would-be natural-deduction system for LP is unsound with respect to $\models_{\text{LP}}$.

Note that this problem for Modus Ponens arises even without inquiring into the nature of any LP-proof Π that would have to be vouchsafed by any LP-proof-system establishing all daudologies as theorems. We are entitled to assume that LP-proofs, however they might be constructed, are LP-sound. That granted, we know that the final step of the "proof" just given must be LP-invalid by the following considerations:

1. the left immediate subproof establishes the following true statement of LP-consequence:

$$A, \neg A \models_{\text{LP}} (A \lor B) \land \neg A;$$

---

[3]Priest, of course, is aware of this; see Priest (1979, p. 228) *infra*, where he points out that $A, A \to B \not\models B$. (He does not use the subscript LP with $\models$. His $\models$ is our $\models_{\text{LP}}$.)

Note also that the generalized (or parallelized) elimination rule

$$\frac{\varphi \to \psi \quad \varphi \quad \begin{array}{c} \overline{\psi}^{(i)} \\ \vdots \\ \theta \end{array}}{\theta}{}_{(i)}$$

fails a fortiori, since it has Modus Ponens as a degenerate instance:

$$\frac{\varphi \to \psi \quad \varphi \quad \overline{\psi}^{(i)}}{\psi}{}_{(i)}.$$

2. the right immediate subproof Π establishes the following true statement of daudologousness:
$$\models_{LP} ((A \vee B) \wedge \neg A) \to B \, ;$$

3. $\models_{LP}$ is transitive;
4. therefore, if the final step (of what looks like Modus Ponens) were LP-valid, it would follow that
$$A, \neg A \models_{LP} B \, .$$

But this we already know is impossible.

This trouble with Modus Ponens is hardly surprising. Priest does, after all, define $A \to B$ as $\neg A \vee B$. Thus, the primitive-looking would-be rule of Modus Ponens:

$$\frac{A \quad A \to B}{B}$$

is really only a form of Disjunctive Syllogism:

$$\frac{A \quad \neg A \vee B}{B},$$

which we have already seen to be LP-invalid.

On further reflection, the puzzle with $\to$ deepens. Suppose one were seeking to have it as a *primitive* connective in LP. Consider how (in a classical system, at least) the two sentences $A \to B$ and $\neg A \vee B$ are interdeducible. Here, for example, are the two proofs establishing their interdeducibility in Classical Core Logic. Each of these proofs contains a subproof that establishes an LP-invalid sequent.

$$\cfrac{\cfrac{\cfrac{A \quad A \to B}{B}}{\neg A \vee B} \quad \cfrac{\overline{\neg A}^{(1)}}{\neg A \vee B}}{\neg A \vee B}^{(1)} \qquad \cfrac{\neg A \vee B \quad \cfrac{\cfrac{\overline{\neg A}^{(1)} \quad \overline{A}^{(2)}}{\bot} \quad \overline{B}^{(1)}}{B}}{\cfrac{B}{A \to B}^{(2)}}^{(1)}$$

Under our supposition, the LP-advocate would not be able to avail himself of either of these proofs. For the proof on the left uses Modus Ponens; while the proof on the right contains a subproof that establishes Disjunctive Syllogism (in the form $\neg A \vee B, A : B$). The latter emerges therein as derived, courtesy of Core Logic's liberalized rule of $\vee$-Elimination. This rule allows one to bring down as the main conclusion of a proof by cases the conclusion of either of the two case-proofs should the other case-proof's conclusion be absurdity ($\bot$). If one were (like the intuitionistic logician) to eschew Core Logic's liberalized rule of $\vee$-Elimination, then one would obtain equiform conclusions for the two case proofs by insinuating into the first case proof an application of the rule Ex Falso Quodlibet:

$$
\begin{array}{c}
(1)\underline{} \quad \underline{}(2) \\
\neg A \quad A \\
\underline{} \\
\bot \quad \underline{}(1) \\
\neg A \vee B \quad B \quad B \\
\underline{}(1) \\
\underline{B}\,\underline{}(2) \\
A \to B
\end{array}
$$

The offending appeal to, in effect, Disjunctive Syllogism, remains within this intuitionistic version of the right-hand core proof given above. But the advocate of LP would now cast a jaundiced glare at the invocation of EFQ, which is not allowed by his lights. It would be a mistake, however to lay the "blame" for Disjunctive Syllogism at the door of EFQ. EFQ is reprehensible all on its own; while, in the view of the core logician (who *rejects* EFQ), *Disjunctive Syllogism is perfectly acceptable.* Indeed, Disjunctive Syllogism is *perfectly* valid!—it is valid, and has no valid proper subsequent.

We have seen, then, that even though $A \to B$ and $\neg A \vee B$ *have the same truth-value table* (as we saw above), nevertheless *neither* of them can be deduced from the other in the usual way in any "standard" system of natural deduction for LP that might seek to subject $\to$ to anything like its "own" rules as a primitive. (A logical operator \$ has rules of "its own" just in case the rules in question are stated in such a way that \$ is the only operator explicitly occurring in them.) It is highly unusual for (semantically) equivalent sentences to fail thus to be interdeducible. More familiar are certain systems in which interdeducibility fails to secure synonymy—see Smiley (1962). Smiley proposed, as necessary and sufficient for the synonymy of any two sentences within a logical calculus, that they be inter-replaceable, *salva veritate*, in all statements of deducibility. Smiley's exigent definition of synonymy makes it trivial, therefore, that synonymous sentences are interdeducible. We have seen that what we defined above as LP-equivalence—possession of the same three-valued LP-truth-value table—cannot *even* secure LP-interdeducibility in any natural sense; and therefore certainly cannot guarantee synonymy in Smiley's yet more demanding sense. Despite Observation 23.1, Priest appears to be denied the option of adopting $A \to B$ as a *primitive* conditional, and making it behave (in a system of natural deduction) exactly the way it ought to behave if it were understood as being captured also as "merely" an abbreviation for $\neg A \vee B$. With the latter, "abbreviatory" option, of course, the question of *interdeducibility* would not arise; or, at least, it would be settled trivially in the affirmative: from $\neg A \vee B$ one can deduce $\neg A \vee B$, just by writing it down. And one can construe this either way—that one has deduced the (defined) $A \to B$ from $\neg A \vee B$, or *vice versa*.

Modus Ponens, for the would-be LP-proof-theorist, is *verboten*. But *no* proof-theorist for any system S can refuse to accept a rule of inference (as a means of constructing S-proofs) that happens to be S-valid. So the would-be LP-proof-theorist cannot refuse "to be given" any of the four classical rules of negation respectively known as the Law of Excluded Middle, Dilemma, Double-Negation Elimination and Classical Reductio ad Absurdum. By the same token, in light of Lemma 23.7 and

Corollary 23.1, the would-be LP-proof-theorist also cannot refuse "to be given" the Rule of Conditional Proof.

This makes the situation with LP rather puzzling for the inferentialist. For Lemma 23.7 and Corollary 23.1 together showed that the conventional Rule of Conditional Proof—which is part of the rule of →-Introduction in Core Logic—is LP-admissible (i.e., LP-valid). The remaining part of the rule of →-Introduction that is found in Core Logic is that which allows one to infer the conditional upon refuting its antecedent. And *this* part is *also* LP-admissible, as Lemma 23.6 shows.

So the conventional Rule of Conditional Proof—indeed, even the full rule of →-Introduction of Core Logic—might as well be adopted as the rule of →-Introduction for LP. In unrestrictedly transitive systems, the usual "harmoniously balancing" companion rule, the rule of →-Elimination, is taken to be Modus Ponens. But the latter *is not allowed* as a rule of inference in LP. Conclusion: LP cannot regiment the logical behavior of → by means of introduction and elimination rules that are in harmony with one another.[4]

So: what is the inferentialist to do, when confronted with the strange relation $\models_{LP}$? What might be left in the inferentialist's toolbox to make the conditional arrow intelligible to one who wishes to learn *how to reason deductively* within, or by means of, LP? Or, if not in LP itself, then at least in some other system with just as good a claim as LP to capturing the "logic of naïve proofs"?

23.3 A Different Route for the Inferentialist

Tennant (1979) put forward a different paraconsistent logic, namely Core Logic.[5] (For its natural deduction rules, see Appendix 1 below; for the corresponding sequent rules, see Appendix 2.) One main aim of Core Logic is to track the ways in which the premises of a proof are relevant to its conclusion. This motivating concern was not necessary to handle the paradoxes, let alone to lay any logical foundations for such an arresting philosophical claim as Priest's, that some *contradictions* are *true*. The interest was only in how logicians might avail themselves of techniques of proof-normalization in order to establish conclusions only from genuinely relevant premises.[6] Not long after our papers appeared, Priest made the humorous remark to me that whereas I was countenancing truth-value *gaps*, he was "merely" countenancing truth-value *gluts*. But the contrast of course, between our respective systems runs far deeper than that. The contrast is in both output—the field of the consequence or deducibility relation—and the *methodology*—formal-semantical (in Priest's case)

---

[4]I am grateful to a referee for referring me to Priest (2008), at p. 125, where Priest points out that changing just one entry in LP's 3-valued truth table for the conditional turns it into the conditional of RM_3, for which Modus Ponens holds. But, as Priest then points out, this change turns the whole *system* into RM_3. And this would take us right off our titular topic.

[5]The system was originally called Intuitionistic Relevant Logic.

[6]The project was accomplished for the propositional part of Core Logic in Tennant (1992), and reached its full fruition for first-order logic in Tennant (2015c).

and proof-theoretical (in mine). This study dwells on this contrast, and seeks to draw some lessons from it.

It turned out (from the perspective of this Core logician) that by getting the logical rules right—in a perfectly tweaked form, differing slightly but crucially from their form in Gentzen (1935) and Prawitz (1965)—it was possible to hold that the resulting logical system forged *analytical* logical connections among sentences in two important and complementary regards. Such connections obtained *both* because of the meanings conferred on the logical operators by the logical rules governing them; *and also* because those rules were so formulated that when their applications were arranged so as to form proofs, relevant connections were preserved among the *extra*-logical expressions occurring in the premises and the conclusion of any core proof. This phenomenon of relevant connection is preserved also in Classical Core Logic, whose strictly classical rules for negation are carefully tweaked in similar fashion.[7] So, to the extent that one is after a "logic of naïve proof", Core Logic and Classical Core Logic are perfect vehicles for the rigorous formalization of reasoning in constructive and classical mathematics respectively.[8]

At its inception *this* particular logical system (Core Logic) was a system of *proof*. The mode of investigation was thoroughly *inferentialist*. It was only later that it was discovered that the proof-theoretic methods had fruitful application in the diagnosis and defusing of the logico-semantic paradoxes.[9]

The inferentialist logician who inquires after how a particular logical system is characterized wants to be given a *system of proof*—ideally, a set of *primitive rules*

---

[7]For details, see Tennant (2015c).

[8]That the contrast between Gentzen's system of natural deduction and the artificial Frege–Hilbert systems reflects favorably on the former appears to be tacitly conceded at p. 25 in Hilbert and Ackermann (1938)—the second, "improved" (*verbesserte*) 1938 edition of *Grundzüge der theoretischen Logik*, whose first edition appeared in 1928. There Hilbert and Ackermann write

> Wir erwähnen endlich noch als eine Sonderstellung einnehmend den von G. GENTZEN aufgestellten „Kalkül des natürlichen Schließens"[fn], der aus dem Bestreben hervorgegangen ist, das formale Ableiten von Formeln mehr als bisher dem inhaltlichen Beweisverfahren, wie es z. B. in der Mathematik üblich ist, anzugleichen. Der Kalkül enthält keine logischen Axiome, sondern nur Schlußfiguren, die angeben, welche Folgerungen aus gegebenen Annahmen gezogen werden können, sowie solche, die Formeln liefern, bei denen die Abhängigkeit von den Annahmen beseitigt ist.

I would translate this as follows (rather than using the translation provided in Hilbert and Ackermann (1950) at p. 30):

> We mention finally one more system, one occupying a special position, the "calculus of natural deduction" set up by G. Gentzen, which emerged from the endeavor to make the formal derivation of formulae resemble more closely than it has until now the contentful procedure of proof that is customary, for example, in mathematics. The calculus contains no logical axioms, but only rules of inference, which specify which consequences can be drawn from given assumptions, as well as rules that deliver formulae while rendering them independent of [certain] assumptions.

[9]See Tennant (1982, 1995, 2015a).

of inference by means of which proofs can be constructed in the now familiar way. Moreover, it is a special mark in favor of a particular kind of proof-system if the proofs it furnishes are *homologous to*, or *smoothly regiment*, the more informal proofs that they formalize and that carry conviction for expert reasoners in areas like mathematics. The preferred format for such description is that of natural deduction and/or the sequent calculus, familiar from the *loci classici* Gentzen (1935) and Prawitz (1965). Logical rules in systems of natural deduction deal with single dominant occurrences of logical operators—in conclusions of introduction rules, and in major premises of elimination rules. The introduction and elimination rules in natural deduction are mirrored by the Right and Left rules, respectively, in the sequent calculus.

If thinkers *could* reason in an ordinary, but appropriately adjusted, sort of way within a framework like Priest's once they had liberated themselves from the dogma of consistency, then one would expect their patterns of reasoning—and the primitive steps that are sanctioned within them—to be formalizable "naturally". This is probably why the assurance was given, in Priest (1979) at p. 241, that "It is not difficult to give an axiom or rule system for LP …".

It is difficult, however, to find in the literature either a natural-deduction presentation, or a sequent-calculus presentation, of LP.

In an email exchange in April and May 2016, I received Priest's helpful answer to my question "[W]here, in your estimation, is the best or fullest treatment of a proof system (natural deduction or sequent calculus) for your Logic of Paradox?" He directed me to the rules given in Sect. 4.6 of Priest (2002). These are the rules of the system of First Degree Entailment, plus Excluded Middle:

$$\overline{\alpha \vee \neg \alpha}$$

The quantifier rules are in Sect. 6.4, and are the usual ones. (LP confines all its significant differences from Classical Logic to the propositional level. The same is true of Core Logic.)

LP of course cannot contain Explosion:

$$\frac{\alpha \quad \neg \alpha}{\beta}$$

because LP is paraconsistent. FDE plus Explosion is Kleene's system K3. Another "rule" that needs to be mentioned at this stage is Implosion, the dual of Explosion (and which a relevantist would regard as *similarly irrelevant!*):

$$\frac{\beta}{\alpha \vee \neg \alpha}$$

Priest commented

> I think you need LEM, not Implosion. Normally these would be equivalent, but without either you have FDE, which has no logical truths, and so Implosion does not deliver LEM. [FDE plus LEM] is sound and complete [for LP].

So the list of officially approved rules for propositional LP is the one given (in Sect. 23.4 below), these rules being taken from pp. 302, 303 and 309 in (Priest 2002).

There emerged from the aforementioned email exchange the following further important points about the rules of inference for LP.

1. There are no rules just for negation, in isolation. The de Morgan-like rules (see Sect. 23.4 below) plus LEM are "the rules" for negation.
2. Likewise, there are no rules just for the conditional, in isolation.
3. There is no straightforward way to identify the intuitionistic fragment of LP (if there is such a thing) by simply dropping certain primitive rules of (classical) LP.
4. Implosion—despite its inherent irrelevance—is semantically valid in LP.
5. There is no "normal form" proof of Implosion by means of the rules for LP. (This will be shown at the end of Sect. 23.4.)

Another comment is in order. Among the rules of FDE is the classical rule of Double-Negation Elimination. But as Priest observed, FDE has no logical truths; hence by means of its rules, one can prove no theorems. Hence despite containing Double-Negation Elimination, FDE does not prove the Law of Excluded Middle. This is rather peculiar. For, against the background of Intuitionistic Logic, the four strictly classical rules of Double-Negation Elimination, Classical Reductio, Dilemma, and the Law of Excluded Middle are interderivable. Moreover, in order to derive DNE using LEM and the standard rule of Proof by Cases, one has to use EFQ. In the absence of EFQ, then (as in any paraconsistent logic) LEM is *prima facie* weaker than DNE. Yet in LP it seems to be the other way round. It appears that DNE is weaker than LEM; for the system containing DNE needs to have LEM *added* in order to yield the full classical system of LP. This terrain, for the logician sensitive to the usual marks of nonconstructivity, appears to be full of potholes.

23.4 Rules for Propositional LP

The following set of rules for propositional LP were gleaned, then, from Priest (2002), with helpful guidance from its author.[10]

From p. 302:

$\wedge$I $\quad\quad\quad \dfrac{\alpha \quad \beta}{\alpha \wedge \beta}$

$\wedge$E $\quad\quad\quad \dfrac{\alpha \wedge \beta}{\alpha} \quad\quad\quad \dfrac{\alpha \wedge \beta}{\beta}$

$\vee$I $\quad\quad\quad \dfrac{\alpha}{\alpha \vee \beta} \quad\quad\quad \dfrac{\beta}{\alpha \vee \beta}$

---

[10]The rule $\wedge$I was given the mistaken label $\vee$I on p. 302.

From p. 303:

∨E
$$\frac{\alpha \vee \beta \quad \overset{\overline{\alpha}}{\underset{\gamma}{\vdots}} \quad \overset{\overline{\beta}}{\underset{\gamma}{\vdots}}}{\gamma}$$

So far, so good ... for this is straight out of Gentzen (1935) and Prawitz (1965). These are the familiar Introduction and Elimination rules for ∧ and ∨. But *where are the rules for ¬?* ... All we learn is that the behavior of negation is constrained only *in relation to* conjunction, disjunction and itself, as follows. Such multi-operator constraining of how any single operator behaves is what makes it so difficult to identify an *intuitionistic* subsystem of LP.

From p. 309:

$$\frac{\neg(\alpha \wedge \beta)}{\neg\alpha \vee \neg\beta} \qquad \frac{\neg\alpha \vee \neg\beta}{\neg(\alpha \wedge \beta)}$$

$$\frac{\neg(\alpha \vee \beta)}{\neg\alpha \wedge \neg\beta} \qquad \frac{\neg\alpha \wedge \neg\beta}{\neg(\alpha \vee \beta)}$$

$$\frac{\alpha}{\neg\neg\alpha} \qquad \frac{\neg\neg\alpha}{\alpha}$$

LEM
$$\overline{\alpha \vee \neg\alpha}$$

Conspicuously missing here, for the Gentzen–Prawitz proof-theorist, is any account of negation *on its own*, or *in its own right*. LP appears not to provide so much as a rule of constructive *reductio ad absurdum* (i.e., negation introduction), let alone a rule of negation elimination.

Bear in mind that Priest is happy with the following *abnormal* proof of Implosion from LEM, using the time-honored trick of applying ∧E immediately after ∧I:

$$\frac{\dfrac{\overline{A \vee \neg A} \quad B}{(A \vee \neg A) \wedge B}}{A \vee \neg A}$$

This makes B *spuriously* relevant to the ultimate conclusion $A \vee \neg A$. A simple $\wedge$-Reduction transforms the foregoing "proof" into a single invocation of LEM, which is all that is "really going on":

$$\dfrac{\dfrac{\overline{A \vee \neg A} \quad B}{(A \vee \neg A) \wedge B}}{A \vee \neg A} \quad \stackrel{\wedge\text{-Redn.}}{\rightsquigarrow} \quad \overline{A \vee \neg A}$$

The conclusion $A \vee \neg A$ does not *really* follow *from* an arbitrary, thematically unconnected "premise" B after all. Implosion implodes.

But not for LP! According to Priest, the would-be proof of Implosion exploiting the abnormality trick really does count as a proof in his "natural deduction" system for LP. So that system will be devoid of any meaningful normalization theorem concerning its proofs, which is one of the main motivations for using natural deduction as one's format for fully regimented proofs.

23.5 Upshot; and a Proposed Agenda for the LP-er

The upshot of all this, for the inferentialist logician, is a sense of both bewilderment and bafflement as to *how one is permitted to reason "within"* LP. There is no easily recognizable "body of deductive reasoning" produced by LP-experts that lends itself to faithful and homologous formalization by means of such LP-proofs as are available in the rather sketchy system that this inquiry has uncovered. My own experience is that logicians who think that they understand "what it would be to reason like an LP-er" are really only reasoning *about* LP (in particular: about its semantic consequence relation $\models_{\text{LP}}$). Their reasoning does not, on any particular occasion when it is directed towards some conclusion, lend itself to formalization as a proof *in* a proof-system for LP, a system that could generate a deducibility relation $\vdash_{\text{LP}}$ that might be shown (at the metalevel) to coincide with $\models_{\text{LP}}$. Rather, they "reason LP-ishly" at arm's remove, "one level up", as metalogicians using ordinary mathematical reasoning (as I have above!) to work out whether particular sequents lie in the extension of $\models_{\text{LP}}$. They don't work "within LP itself", in accordance with *primitive deductive rules* of LP, which could then in turn be studied by a proof-theorist. As we have pointed out above, there is a reflexive instability in making moves at the metalevel, when reasoning about $\models_{\text{LP}}$, that are not themselves catered for within $\models_{\text{LP}}$ itself.

If anything positive is to emerge from these modest investigations, one hopes it will be this: LP-ers ought to devote some of their ingenious energies to the devising of a natural-deduction system, or sequent calculus, for LP that could be offered as "the" organon of inference in "naïve logic". It's just not good enough to stick with $\models_{\text{LP}}$ "down there", and reason about it like any classical or constructive logician "up here". Rather, one must get down there with the requisite logical goggles on, in order to appreciate or better understand the murky currents of thought that the LP-er, by his own lights, can allow himself to be guided by when setting out from certain premises and seeking to arrive at some conclusion LP-implied by them. One should also be mindful of the fact that some naïve reasoners might wish their reasoning to be strictly *constructive*. That means we ought to be prepared to identify a strict subset of the primitive rules of LP that would enable and allow all the inferential moves that the naïve *constructive* reasoner wishes to make.

If such an investigation were to be undertaken by LP-devotees intent on furnishing us at long last with a proof-system for LP worthy of the title "(regimentation of) naïve logic", then the next item on the collective agenda would be to find points of contrast or similarity with the proof-system of Core Logic. For the latter system also lays claim to the title of 'naïve logic', on at least three scores: (i) it avoids altogether EFQ and its ilk (so it is a nonexplosive, or paraconsistent, logic); but (ii) it respects every ordinary logical inference (such as Disjunctive Syllogism) which naïve beginners find (correctly!) to be intuitively correct; while (iii) it can be deployed to regiment the reasoning that is actually involved in generating the various logico-semantical paradoxes, and at the same time allows the theorist to point out what it is about (the formalizations of) those passages of reasoning that reveal that one is dealing thereby with the paradoxical, rather than with the straightforwardly inconsistent.[11]

The sort of thing I am asking the LP-er to provide is: some set of logical rules, each, ideally, focusing on just one logical operator at a time, plus—*perhaps*—some structural rules (as in some sequent calculi) that do not mention any operators; something, indeed, like the systems $\mathbb{C}$ of Core Logic, and $\mathbb{C}^+$ of Classical Core Logic. These are laid out in the Appendix, and are furnished here as a guide. Note, however, that the core systems have no structural rules other than the Rule of Initial Sequents (REFLEXIVITY). Both CUT and THINNING are *admissible* only, not derivable (and a fortiori not primitive) rules of the systems.

Both $\mathbb{C}$ and its classicized extension $\mathbb{C}^+$ are paraconsistent, and are *transitive* in the following important sense:

> Any pair of proofs of the sequents $\Delta : \varphi$ and $\Gamma, \varphi : \psi$ respectively can be effectively transformed into a proof of a sequent either of the form $\Delta', \Gamma' : \varphi$ or of the form $\Delta', \Gamma' : \bot$, for some $\Delta' \subseteq \Delta$ and $\Gamma' \subseteq \Gamma$.

That is, CUT is admissible.[12]

---

[11] In support of these closing claims, see Tennant (2015a, 2016, 2017).

[12] We mean here that the following metalinguistic inference holds:

$$\frac{\Delta \vdash \{\varphi\} \quad \Gamma, \varphi \vdash \{\psi\}}{\Delta, \Gamma : \{\psi\}}$$

where $\Theta \vdash \Xi$ means that there is a proof of some subsequent of $\Theta : \Xi$, i.e., a proof of some sequent of the form $\Theta' : \Xi'$, where $\Theta' \subseteq \Theta$ and $\Xi' \subseteq \Xi$.

Appendix 1: Natural Deduction Rules for Core Logic $\mathbb{C}$

A box annotating a discharge stroke indicates that we must have used at least one of the indicated undischarged assumptions in the subordinate proof in question. This is called *non-vacuous* discharge of assumptions. *Vacuous* discharge, by contrast, is indicated by a diamond. Note that the absence of any vertically descending dots above a major premise for elimination indicates that it must "stand proud", with no proof work above it.

$(\neg I)$
$$\boxed{}\text{---}(i)$$
$$\underbrace{\varphi, \Delta}$$
$$\Pi$$
$$\underline{\bot}\text{---}(i)$$
$$\neg\varphi$$

$(\neg E)$
$$\Delta$$
$$\Pi$$
$$\underline{\neg\varphi \quad \varphi}$$
$$\bot$$

$(\wedge I)$
$$\Delta_1 \quad \Delta_2$$
$$\Pi_1 \quad \Pi_2$$
$$\underline{\varphi_1 \quad \varphi_2}$$
$$\varphi_1 \wedge \varphi_2$$

$(\wedge E)$
$$(i)\text{---}\boxed{}\text{---}(i)$$
$$\underbrace{\varphi_1, \varphi_2, \Delta}$$
$$\vdots$$
$$\underline{\varphi_1 \wedge \varphi_2 \quad \theta}\text{---}(i)$$
$$\theta$$

$(\vee I)$
$$\Delta \quad \Delta$$
$$\Pi \quad \Pi$$
$$\underline{\varphi_1} \quad \underline{\varphi_2}$$
$$\varphi_1 \vee \varphi_2 \quad \varphi_1 \vee \varphi_2$$

$(\vee E_{\theta\bot})$
$$\boxed{}\text{---}(i) \quad \boxed{}\text{---}(i)$$
$$\underbrace{\varphi_1, \Delta_1} \quad \underbrace{\varphi_2, \Delta_2}$$
$$\Pi_1 \quad \Pi_2$$
$$\underline{\varphi_1 \vee \varphi_2 \quad \theta \quad \bot}\text{---}(i)$$
$$\theta$$

$(\vee E_{\bot\theta})$
$$\boxed{}\text{---}(i) \quad \boxed{}\text{---}(i)$$
$$\underbrace{\varphi_1, \Delta_1} \quad \underbrace{\varphi_2, \Delta_2}$$
$$\Pi_1 \quad \Pi_2$$
$$\underline{\varphi_1 \vee \varphi_2 \quad \bot \quad \theta}\text{---}(i)$$
$$\theta$$

$(\vee E_{\theta\theta})$
$$\boxed{}\text{---}(i) \quad \boxed{}\text{---}(i)$$
$$\underbrace{\varphi_1, \Delta_1} \quad \underbrace{\varphi_2, \Delta_2}$$
$$\Pi_1 \quad \Pi_2$$
$$\underline{\varphi_1 \vee \varphi_2 \quad \theta \quad \theta}\text{---}(i)$$
$$\theta$$

$(\vee E_{\bot\bot})$
$$\dfrac{\varphi_1 \vee \varphi_2 \quad \begin{array}{c}\overbrace{\varphi_1\,,\,\Delta_1}^{\Box\text{---}(i)} \\ \Pi_1 \\ \bot\end{array} \quad \begin{array}{c}\overbrace{\varphi_2\,,\,\Delta_2}^{\Box\text{---}(i)} \\ \Pi_2 \\ \bot\end{array}}{\bot}\text{---}(i)$$

$(\to I)(a)$
$$\dfrac{\begin{array}{c}\overbrace{\varphi_1\,,\,\Delta}^{\Box\text{---}(i)} \\ \Pi \\ \bot \end{array}}{\varphi_1 \to \varphi_2}\text{---}(i)$$

$(\to I)(b)$
$$\dfrac{\begin{array}{c}\overbrace{\varphi_1\,,\,\Delta}^{\Diamond\text{---}(i)} \\ \Pi \\ \varphi_2 \end{array}}{\varphi_1 \to \varphi_2}\text{---}(i)$$

$(\to E)$
$$\dfrac{\varphi_1 \to \varphi_2 \quad \begin{array}{c}\Delta_1 \\ \Pi_1 \\ \varphi_1\end{array} \quad \begin{array}{c}\overbrace{\varphi_2\,,\,\Delta_2}^{\Box\text{---}(i)} \\ \Pi_2 \\ \theta \end{array}}{\theta}\text{---}(i)$$

$(\exists I)$
$$\dfrac{\begin{array}{c}\Delta \\ \Pi \\ \psi_t^x\end{array}}{\exists x\psi}$$

$(\exists E)$
$$\dfrac{\exists x\psi(x) \quad \begin{array}{c}\overbrace{\psi(a)\,,\,\Delta}^{\Box\text{---}(i)} \\ \Pi \\ \theta\end{array}}{\theta}\text{---}(i)$$
where a does not occur in any sentence in Δ or in $\exists x\psi(x)$ or in θ

$(\forall I)$
$$\dfrac{\begin{array}{c}\Delta \\ \Pi \\ \psi(a)\end{array}}{\forall x\psi(x)}$$
where a does not occur in any sentence in Δ or in $\forall x\psi(x)$

$(\forall E)$
$$\dfrac{\forall x\psi \quad \begin{array}{c}\overbrace{\psi^x_{t_1}\,,\,\ldots\,,\,\psi^x_{t_n}\,,\,\Delta}^{(i)\text{---}\,\ldots\,\Box\,\ldots\,\text{---}(i)} \\ \Pi \\ \theta\end{array}}{\theta}\text{---}(i)$$

The classicized extension $\mathbb{C}^+$ of Core Logic may be obtained by adding one or other of the following two rules (Classical Reductio or Dilemma).

(CR)
$$\begin{array}{c} \square\underline{\quad}(i) \\ \neg\varphi \\ \vdots \\ \underline{\bot}\,(i) \\ \varphi \end{array}$$

(Dil)
$$\begin{array}{cccc} \square\underline{\quad}(i) & \square\underline{\quad}(i) & \square\underline{\quad}(i) & \square\underline{\quad}(i) \\ \varphi & \neg\varphi & \varphi & \neg\varphi \\ \vdots & \vdots & \vdots & \vdots \\ \underline{\psi \qquad \psi}\,(i) & & \underline{\psi \qquad \bot}\,(i) \\ \psi & & \psi \end{array}$$

Appendix 2: Sequent Calculus Rules for Core Logic $\mathbb{C}$

The only structural rule is REFLEXIVITY, or the Rule of Initial Sequents:

$$\overline{\varphi:\varphi}$$

The logical rules are as follows:

$(:\neg)$ $\quad \dfrac{\Delta,\varphi:}{\Delta:\neg\varphi}$

$(\neg:)$ $\quad \dfrac{\Delta:\varphi}{\Delta,\neg\varphi:}$

$(:\wedge)$ $\quad \dfrac{\Delta_1:\varphi_1 \quad \Delta_2:\varphi_2}{\Delta_1,\Delta_2:\varphi_1\wedge\varphi_2}$

$(\wedge:)$ $\quad \dfrac{\Delta,\varphi_i:\Gamma}{\Delta,\varphi_1\wedge\varphi_2:\Gamma}$ for $i = 1, 2$;

$\left(\text{which affords also the more economical } \dfrac{\Delta,\varphi,\psi:\Gamma}{\Delta,\varphi\wedge\psi:\Gamma}\right)$

$(:\vee)$ $\quad \dfrac{\Delta:\varphi_i}{\Delta:\varphi_1\vee\varphi_2}$ for $i = 1, 2$

$(\vee:)$ $\quad \dfrac{\Delta_1,\varphi_1:\Gamma_1 \quad \Delta_2,\varphi_2:\Gamma_2}{\Delta_1,\Delta_2,\varphi_1\vee\varphi_2:\Gamma_1\cup\Gamma_2}$ where $\Gamma_1\cup\Gamma_2$ is at most a singleton

$(:\rightarrow)$ $\quad \dfrac{\Delta,\varphi_1:}{\Delta:\varphi_1\rightarrow\varphi_2} \quad \dfrac{\Delta:\varphi_2}{\Delta\setminus\{\varphi_1\}:\varphi_1\rightarrow\varphi_2}$

$(\rightarrow:)$ $\quad \dfrac{\Delta_1:\varphi_1 \quad \Delta_2,\varphi_2:\Gamma}{\Delta_1,\Delta_2,\varphi_1\rightarrow\varphi_2:\Gamma}$

$(:\exists)$ $\quad \dfrac{\Delta:\psi_t^x}{\Delta:\exists x\psi}$

$$(\exists :) \quad \frac{\Delta, \psi_a^x : \Gamma}{\Delta, \exists x\psi : \Gamma} \text{ where the conclusion sequent has no occurrences of } a$$

$$(: \forall) \quad \frac{\Delta : \psi}{\Delta : \forall x \psi_x^a} \text{ where } a \text{ occurs in } \psi \text{ but in no member of } \Delta$$

$$(\forall :) \quad \frac{\Delta, \psi_{t_1}^x, \ldots, \psi_{t_n}^x : \Gamma}{\Delta, \forall x\psi : \Gamma}$$

$$(CR) \quad \frac{\neg\varphi, \Delta : \emptyset}{\Delta : \varphi}$$

$$(Dil) \quad \frac{\varphi, \Delta : \psi \quad \neg\varphi, \Gamma : \psi}{\Delta, \Gamma : \psi} \quad \frac{\varphi, \Delta : \psi \quad \neg\varphi, \Gamma : \emptyset}{\Delta, \Gamma : \psi}$$

References

Gentzen, G., & Untersuchungen über das logische Schliessen. (1934–5). Mathematische Zeitschrift, I, II: 176–210, 405–431 Translated as 'Investigations into Logical Deduction', in The Collected Papers of Gerhard Gentzen, edited by M. E. *Szabo, North-Holland, Amsterdam, 1969*, 68–131.

Hilbert, D., & Ackermann, W. (1938). *Grundzüge der Theoretischen Logik* (2nd ed.). Berlin: Springer.

Hilbert, D., & Ackermann, W. (1950). *Principles of Mathematical Logic*. New York: Chelsea Publishing Co.

Kleene, S. C. (1952). *Introduction to Metamathematics*. Princeton, New Jersey: Van Nostrand.

Prawitz, D. (1965). *Natural Deduction: A Proof-Theoretical Study*. Stockholm: Almqvist & Wiksell.

Priest, G. (1979). The logic of paradox. *Journal of Philosophical Logic, 8*(1), 219–241.

Priest, G. (2002). Paraconsistent logic. In D. Gabbay & F. Guenthner (Eds.), *Handbook of Philosophical Logic* (pp. 287–393)., Volume 6 Dordrecht: Kluwer Academic Publishers.

Priest, G. (2008). *An Introduction to Non-Classical Logic: from If to Is*. Cambridge: Cambridge University Press.

Smiley, T. (1962). The independence of connectives. *The Journal of Symbolic Logic, 27*, 426–436.

Tennant, N. (1979). Entailment and proofs. In *Proceedings of the Aristotelian Society* (Vol. LXXIX, pp. 167–189).

Tennant, N. (1982). Proof and paradox. *Dialectica, 36*, 265–296.

Tennant, N. (1992). *Autologic*. Edinburgh: Edinburgh University Press.

Tennant, N. (1995). On paradox without self-reference. *Analysis, 55*, 199–207.

Tennant, N. (2015a). A new unified account of truth and paradox. *Mind, 123*, 571–605.

Tennant, N. (2015b). Cut for classical core logic. *Review of Symbolic Logic, 8*(2), 236–256.

Tennant, N. (2015c). The relevance of premises to conclusions of core proofs. *Review of Symbolic Logic, 8*(4), 743–784.

Tennant, N. (2016). Normalizability, cut eliminability and paradox. *Synthese, special issue Substructural Approaches to Paradox* (pp. 1–20). https://doi.org/10.1007/s11229-016-1119-8.

Tennant, N. (2017). *Core Logic*. Oxford: Oxford University Press.

Neil Tennant is Arts & Sciences Distinguished Professor in Philosophy at Ohio State University. His most recent book, *Core Logic*, is an exhaustive study of the philosophical, proof-theoretic, and computational properties of core logic, a constructive and relevant deductive calculus.

Chapter 24
Expanding the Logic of Paradox with a Difference-Making Relevant Implication

Peter Verdée

Abstract In this paper, we aim to devise a logic that can deal with both the paradoxes that motivate dialetheism and the paradoxes related to the irrelevance of material implication. We propose the semantics and the sequent calculus of a relevant logic inspired by difference-making accounts of causation and arguably true to Graham Priest's Logic of Paradox **LP**: a relevant logic that validates those and only those **LP**-consequences that are not irrelevant. The new logic's relevant implication is then added to **LP** to obtain a logic with the logical vocabulary required to cope with the mentioned paradoxes.

Keywords Relevance logic · Paradoxes · Paraconsistent logic · Dialetheism · Logic of Paradox · Non-transitivity · Interventionism · Causality

24.1 Introduction

In this paper, we present the logic **LPR** and its sequent calculus. **LPR** formalizes a relevant implication the behaviour of which is faithful to the ideas behind Graham Priest's Logic of Paradox (henceforth **LP**). First, we argue why one would want a logic that coherently tackles both the issues of irrelevance and the problems Priest's logic is supposed to overcome. Then we develop a specific account of relevance inspired by 'difference making' theories of causation. For each given Tarskian consequence relation **L** (for example, **LP**-consequence), this account aims to define a notion of relevant consequence, here called *logical cause*,[1] true to important aspects of **L**, in the sense that the new consequence relation will validate all consequences of **L**, except the blatantly irrelevant ones. **LPR** is **LP**'s logic of logical cause and is thus a

---

[1] We would like to emphasize that this is merely a name. We nowhere want to claim that there is any deep link between our 'logical cause' notion and causation in the empirical sciences. There are at most some formal resemblances.

P. Verdée (✉)
University Catholique de Louvain, Louvain-la-Neuve, Belgium
e-mail: peter.verdee@uclouvain.be

© Springer Nature Switzerland AG 2019
C. Başkent and T. M. Ferguson (eds.), *Graham Priest on Dialetheism and Paraconsistency*, Outstanding Contributions to Logic 18, https://doi.org/10.1007/978-3-030-25365-3_24

formalization of what might be called **LP**'s relevant consequence relation. When the object language implication corresponding to the meta-linguistic **LPR**-consequence relation is added to **LP**, one obtains a logic that could be used to cope with both the paradoxes of irrelevance and the paradoxes for which **LP** was originally devised.

But first, we motivate why one would want to add a relevant implication faithful to **LP** to **LP**. **LP** was never supposed to be a relevant logic and it aims to solve other, fundamentally different problems than logic of relevance. We argue that it makes sense to overcome the two kinds of problems in one logic.

It is well known that both the classical characterization of negation and the classical characterization of implication are incoherent in combination with certain intuitively plausible principles. We will call such incoherences *paradoxes*: incompatibilities of the formalization of a linguistic phenomenon with intuitively plausible principles concerning that phenomenon.

Paradoxes of classical negation. Graham Priest has convincingly argued that there are sentences such that the sentence and its negation are both true given some intuitively plausible semantic principles. Obviously, this is incoherent with classical logic (henceforth **CL**). Let us call a sentence *false* iff its negation is true.

A first sentence that could be seen as both true and false is the liar sentence

$L: L$ is false,

which is proven to be both true and false by means of reductio (when a sentence is proven to be true on the hypothesis that it is false, it is in fact proven to be true) and the intuitive T-scheme: A is true iff A. Suppose that L is false, then L is true by the T-scheme. Consequently, by reductio (ad absurdum), L is proven to be true. By the T-scheme then, we can conclude that L is also false. Following the intuitive T-scheme, many other sentences with a truth predicate should be both true and false, among which specific instances of the Curry paradox, which we will discuss further on.

A second example is the Gödel sentence for the notion of informal mathematical proof

G: there is no informally provable mathematical proof for G,

whose truth and falsity is proven by means of reductio, double negation and soundness of informal proof. The argument goes as follows. If we assume that G is false, then, by the definition of G, there is an informal mathematical proof for G. By the soundness of informal proof, one can conclude from this that G is true. By reductio, this entails that G is true also without the hypothesis. We have thus provided an informal proof for G. And, by the definition of G, this entails that G is false (next to also being true).

A third example is the self-membership for the Russell collection in naive collection theory

$R : r$ is a member of r, where $r = \{x \mid x$ is not a member of $x\}$,

which can be proven by means of comprehension and reductio.

If one accepts the construction of sentences L, G and R, some intuitive principles and some basic rules of reasoning, the dialetheist's conclusion that there are sentences that are both true and false is hard to escape. However, this position is untenable given the **CL**-characterization of negation, for, given this characterization, as soon as one sentence is both true and false, ex falso quodlibet allows us to conclude that all sentences are true and false. In that case truth and logic lose their discriminatory force, and become useless and meaningless. These are the *paradoxes of classical negation*.

A logic that (1) does not forbid the formalization of sentences like L, G and R and (2) accepts intuitively plausible principles like the T-scheme, comprehension and reductio should be able to deal with sentences that are both true and false. Graham Priest's dialetheist logic **LP** (cf. Priest 1979) is the best-known logic that allows for this (moreover it validates excluded middle, De Morgan laws and double negation laws).

Paradoxes for CL's implication. Most people who have not followed formal logic courses would intuitively assume that an implication or conditional can only be true if there is a connection between premises and conclusion. The latter is obviously not the case in **CL**. All of the following are valid in it.

If A is true, $B \to A$ is true
If B is false, $B \to A$ is true
$A \to B$ is true or $B \to C$ is true[2]
If $A \to B$ is false, then A is true and B is false

To see that the classical formalization $A \to B$ of 'A implies B' is far from natural, substitute in the expressions above A, B and C by, respectively, 'there is life on earth', Fermat's last theorem, and the consistency of Peano arithmetic, and $\to$ by 'implies'. Most rational agents who have not been introduced to **CL**'s implication will consider the obtained expressions as false or meaningless (cf. for example, Priest 2008).

So **CL** is unable to account for some pre-theoretic ideas concerning truth, collections, proof *and implication*. Its negation is unable to deal with contradictory sentences and its implication is unable to deal with requirements of the relevance of the antecedent of an implication for its consequent and vice versa. Unless the widespread pre-theoretic conceptions mentioned above are internally incoherent, it is a useful project to formulate a formal system which is able to deal with all of them together, both the negation and the implication related problems. Devising such a system will be the purpose of this paper.

More concretely, we will discuss the addition of an appropriate relevant implication to the logic **LP**. We define **LP** semantically as follows. The language of **LP** is constructed with binary connectives $\vee$ (disjunction), $\wedge$ (conjunction) and unary connectives $\neg$ (paraconsistent negation), and $\sim$ (classical negation).[3]

---

[2]The version in which $C = A$ is much better known, but this one is more general and even more absurd.

[3]Classical negation is only added in this paper's treatment of **LP** to increase the expressiveness of the logic; it can easily be removed without harm to any of the results in this paper.

The logic of paradox. An interpretation is an **LP**-*interpretation* iff the following conditions are met[4] (suppose there are no restrictions on truth and falsity, prior to this definition).

A is true or A is false.
$\neg A$ is true iff A is false.
$\neg A$ is false iff A is true.
$A \wedge B$ is true iff A is true and B true.
$A \wedge B$ is false iff A is false or B is false.
$A \vee B$ is true iff A is true or B true.
$A \vee B$ is false iff A is false and B is false.
$\sim A$ is true iff A is (classically) not true.

One can define the following conditional-like and biconditional-like connectives.

$A \supset B =_{df} \neg A \vee B$.
$A \sqsupset B =_{df} \sim A \vee B$.
$A \equiv B =_{df} (A \supset B) \wedge (B \supset A)$ (A and B are both true or both false).

A formula is *verified by an* **LP**-*interpretation* iff it is true in the interpretation (but not that it may be false at the same time). The set of *tautologies of* **LP** is the set of formulas true in all **LP**-interpretations and $\Gamma \vDash_{\mathbf{LP}} \Delta$ (multiple conclusion consequence—defined in Beall (2011) as **LP**$^+$) iff each **LP**-interpretation that verifies all members of Γ also verifies at least one of the members of Δ.

CL's multiple conclusion consequence is obtained by replacing '**LP**-interpretations' by '**CL**-interpretations', i.e. those **LP**-interpretations in which no formulas are both true and false. **LP**'s tautologies are the same as **CL**'s. The difference with **CL** is only seen in the consequence relation $\vDash_{\mathbf{LP}}$.

Conditionals in the logic of paradox. The project to formalize conditionals (if A then B) or implications (A implies/entails B) in the context of dialetheic negation has given rise to many existing proposals. Graham Priest has discussed both the addition of a relevance logic implication to **LP** and the qualities of **LP**'s own material implication $A \supset B$ (cf. Priest 2017). Weber (2012) has based his dialetheic set theory on a relevance logic extension of **LP**. Priest, Goodship (1996) and Omori (2015) have proposed ways to treat the paradoxes by means of **LP**'s material implication. Batens's logic **CLuNs** (cf. Batens and De Clercq 2004) adds a $\sqsupset$-like connective to **LP**.

The most important reason to consider a non-classical account of implication in this context has been the attempt to avoid Curry's paradox.

C: the truth of C implies B.

If 'C is true' is true, then, by the T-scheme, C is true. Hence, by its definition, the truth of C implies B is true. By Modus Ponens, we obtain B. We have thus proven that 'C is true' implies B. This means, by definition, that C is true. By the T-scheme

---

[4] The meta-language in this paper ('not', 'iff', etc.) is completely classical.

we know this implies that 'C is true' is true. But we have proven this to imply B. So B is true, while B could have been any formula.

So, as long as the definition of sentences like the Curry sentence is allowed, we can prove whatever formula A from the existence of the appropriate Curry sentence where B is replaced by A. Every formula is provable and so logic loses its discriminatory power all over again.

This paradox seems independent of considerations about negation. The only specific inference pattern we have (implicitly) used here is Contraction (next to Modus Ponens and transitivity): Conclude that A implies B, from the fact that 'A implies B' follows from A. It seems obvious to diagnose the problem here as a result of the way implication is characterized (viz. allowing contraction). One has, therefore, often judged conditionals/implications in **LP** on their ability to avoid the destructive disasters of the Curry paradox for the dialetheist. This is very natural: introducing a new logical symbol should not destroy the original project of avoiding the paradoxes of truth and self-reference. Some relevant logics are indeed non-contractive, and so introducing such a relevance conditional could prima facie solve two problems at once: problems of irrelevance and the Curry paradox.

However, an interesting observation made in several recent works (often called the Goodship project, see e.g. Omori (2015)) is that the Curry paradox can be prevented if the T-scheme, comprehension, and other similar paradoxical schemas are formalized with material equivalence $\equiv$ of **LP**, for the simple reason that this equivalence connective does not detach: the truth of both $A \equiv B$ and A does not imply the truth of B. If A is also false, $A \equiv B$ is also true when B is false only. This is very interesting in a dialetheic context: formulated with $\equiv$, detaching paradoxical schemas is not a valid inference whenever a dialetheia is involved. For example, the so-obtained T-scheme[5] $A \equiv \text{IsTrue}('A')$ then comes down to the following.

' 'A' is true' and 'A' are both true or are both false.

Given this version of the scheme, from ' 'A' is true' one can only conclude that 'A' is true if ' 'A' is true' is not also false. This version of the T-scheme will never lead to triviality, because there will always be an acceptable interpretation (it suffices to make ' 'A' is true' both true and false to make the specific instance of the scheme true for that A). This way this solution provably prevents all of the ways in which the mentioned paradoxes, also the Curry paradox, can lead to triviality independently of how negation or implication are interpreted inside the instances of the paradox prone schemes. This means that in the weaker version of the T-scheme mentioned above, the 'A' can contain all kinds of implication-like connectives ($\supset$, $\sqsupset$, relevant implications, strict implication, etc.) and even classical negation $\sim$, without risking that paradoxes would trivialize the theory.

Once one chooses this solution for the self-reference paradoxes for notions such as truth, informal proof, and collection mentioned above (henceforth the *dialetheic paradoxes*), the choice of a proper implication can be made independently of the

---

[5]The IsTrue('A') is intentionally left unspecified, as the formal specification does not matter here.

solution for those paradoxes. In this case, the implication does not need to be non-contractive or otherwise weakened for reasons that have nothing to do with what an implication is supposed to be. It then makes sense to study the paradoxes of relevance (in **LP**) independently from the dialetheic paradoxes, *in such a way that the solution respects the logical road taken to solve the dialetheic paradoxes*. The question of how to solve both the dialetheic and the relevance paradoxes of **CL** can then be further specified in the following way.

> How to develop and formalize an implication that is not susceptible to the paradoxes of relevance and is in some sense faithful to the essence of **LP**'s consequence relation?

To solve this, we will use an approach to relevance that is fundamentally pluralistic: one bases the relevant consequence relation (here **LPR**) on a given (irrelevant) underlying consequence relation (here **LP**) and makes the latter relevant by getting rid of *only* its obviously irrelevant consequences, keeping the rest intact.[6] In the next section, the concept of relevant consequence we have in mind will be defined loosely based on the difference-making account of causation. Subsequently, we will develop a simple, adequate sequent calculus for **LP**'s relevant consequence relation **LPR**. Then a relevant object language implication → is defined that projects **LP**'s relevant consequence relation **LPR** into the object language. The obtained logic will be called **LPR**$^\rightarrow$.

Finally, to obtain a logic that can indeed deal with the dialetheic paradoxes in the right way (i.e. just like **LP** handles them) and has an adequate (i.e. allowing for the **LP**-consequences that do not pose an irrelevance thread) implication that is not suffering from the irrelevance paradoxes, we need to add the obtained relevant object language implication → (conservatively) to **LP**. This needs to be done in such a way that the resulting consequence relation is Tarskian. The exact technical elaboration of such a logic **LP**$^\rightarrow$ is left to the appendix of the paper, because no interesting new element is added to the logic **LPR**$^\rightarrow$.

24.2 Logical Cause for LP

We will build our pluralistic account of relevance on some ideas inspired by the difference-making account of causation from philosophy of science. Causal consequence is linguistically and intuitively very close to logical consequence. Still, the

---

[6]This approach is not new. In (1991), Gehrard Schurz has defined the following account of relevant consequence. Let's call it Schurz-consequence.

$\Gamma \vdash_L^{cr} A$ iff $\Gamma \vdash_L A$ and there exists no non-empty $\mu \subseteq \mathbb{N}$ such that $\Gamma \vdash \pi^\mu A$ hold for every π.

For our purposes, it suffices to know that π^μ is the result of substituting in A predicate π by an arbitrary predicate. So an A is a Schurz-consequence of a set of premises iff A is a logical consequence and substituting one predicate by an arbitrary predicate in A breaks the relation of logical consequence. For Schurz there are two coherent notions of consequence given one logic **L**: a Tarskian $\vdash_L$ notion and a relevant one $\vdash_L^{cr}$. We also make this distinction, but Schurz' specific formalization of consequence is very different from the one we will propose here.

contemporary philosophical literature on the two kinds of consequence are completely separated. The presence of some linguistic or intuitive resemblances obviously does not entail anything about their actually being similar, but it could be useful to check in how far the literature in the one domain could be useful for the other (as a source of inspiration).

Specifically, for the topic of relevance, some notions of causation come in very handy. Irrelevant causal consequences simply do not exist. There being a link between cause and effect (possibly only a statistical correlation) is probably the only common feature of most theories of causation in no matter which field. It is entirely self-evident that, if the effect is always produced, independently of the candidate cause, there is no causal relation between candidate cause and effect. Similarly with impossible candidate causes: it would be very weird to claim that an impossible event can cause an effect.

The case of logical consequence could not be more different. Most of the formal accounts consider (necessarily) true propositions to be consequences of no matter which set of premises. They moreover allow arbitrary formulas to be consequences of (necessarily) false premises (that is, if the logic considers anything necessarily false at all—this is not the case for **LP**). Irrelevance is by and large accepted in logical consequence, while it is by and large considered problematic in causal consequence.

Let us look at a currently prominent account of causation: the difference-making account by James Woodward (cf. Woodward 2003). Here is a slight simplification of his definition of direct cause:

> In the context of a variable set, a variable A is a direct cause of B iff there is an intervention on A that manipulates A w.r.t. B, keeps all other variables in the variable set unchanged, and changes the value of B.

This can intuitively be understood by means of an example of an old-fashioned portable radio with some knobs. Given this definition, the statement 'knob 1 directly causes (an increase in) the volume of the sound' comes down to the existence of a possible intervention that manipulates knob 1 such that all other knobs and buttons of the radio are left unchanged and the volume of the sound changes. Put otherwise: knob 1 directly causes the increase in volume iff it is possible to change *only* this button in such a way that the volume increases.

We want to define a notion of relevant consequence inspired by this. We will say that (the truth of) a premise logically causes a conclusion (to be true) in the context of other premises whenever there is a way to manipulate only the causing premise (switch its truth value from false to true) in such a way that thereby also the conclusion is switched from false to true. A causing premise is like the right knob to bring about a certain effect (e.g. the volume): it makes a difference for the conclusion. The reader understands that this is intuitively close to relevance: a premise is relevant to the conclusion can be understood as the premise makes a difference for the conclusion.

If we try to modify this definition so that it becomes useful to define (relevant) logical consequence, we need to keep a number of important aspects in mind. First of all, Woodward only considers variables and his account does not specify whether a causal influence is positive or negative. Given his definition, saying that (the variable) 'smoking' causes (the variable) 'cancer' is, for example, equivalent to 'non-smoking' causes 'cancer'. In the case of logical consequence, it seems essential that we speak of positive influence of the cause/premise on the effect/conclusion. That's why one should in any case substitute 'manipulates' and 'changes' by 'manipulates positively' and 'changes positively' resp. Second, there can be multiple premises. They could maybe correspond to the 'other variables in the variable set'.

> Given a set of formulas Γ, a formula A is, in the context of Γ, a logical cause of B iff there is an intervention I on the sequent $\Gamma, A \vdash B$ s.t. if I positively manipulates A and keeps the truth value of all members of Γ unchanged, then necessarily the value of B is changed positively.

Prima facie this seems to make sense, but the problem lies in the way in which one is to interpret 'intervention' in a logical context. Woodward has a detailed definition of what an intervention is, but that is of little use to us here.

To indicate what kind of intervention he has in mind, Woodward speaks of a *surgical* intervention, i.e. an intervention that only manipulates the candidate cause, no other variables. In the case of the question, whether B is a logical consequence of A in the context of Γ, we only want to manipulate the formula A and leave the members of Γ intact. We want to install a switch on A that can force it to be false or true, without interfering in the truth of the members of Γ. Installing such a switch is far from evident. A can be a tautology (is true in all interpretations) or a logical falsehood (is not true in any interpretation), or it can simply have subformulas in common with Γ, so that it may be that changing its truth value does not go without also making some members of Γ false.

As a solution, we propose that a surgical intervention for a logical sequent be nothing but the idea of what we call an *abstraction* of the sequent: a way to see subformulas of a sequent as black boxes. Those black boxes can be modelled by primitive sentences. As an example, consider the sequent $p \wedge \sim p \vdash p \wedge (\sim p \vee q)$. We need to install a switch on $p \wedge \neg p$. Consider some abstractions that are able to do that:

$$r_{p \wedge \sim p} \vdash p \wedge (\sim p \vee q)$$

$$p \wedge r_{\sim p} \vdash p \wedge (\sim p \vee q)$$

$$p \wedge r_{\sim p} \vdash p \wedge (r_{\sim p} \vee q)$$

The subscript notation is used to indicate which formulas the new primitive formulas are supposed to be a black box for. In all three cases, the premise can be made both false and true. In the first two cases, however, flipping the switch on the premise does not necessitate that the conclusion also becomes true. One could say that these surgical interventions do therefore not justify that $p \wedge (\sim p \vee q)$ is a relevant consequence of $p \wedge \sim p$. Flipping the switch on the premise defined by the intervention in the last expression, however, does necessitate the conclusion to become true. This last fact could be seen as a sufficient justification for the relevant validity of the original sequent.

This concept of 'abstraction' is technically definable as follows:

Definition 24.1 An *abstraction* of a triple $\langle \Gamma, A, B \rangle$ is a triple $\langle \Gamma', A', B' \rangle$ iff there is a series of subsequent applications of Uniform Substitution[7] starting from $\langle \Gamma', A', B' \rangle$ and resulting in $\langle \Gamma, A, B \rangle$.

We are now ready to adapt Woodward's concept of direct cause to the case of relevant logical consequence.[8] We start by a semi-precise version of logical cause.

> Given a set of formulas Γ, a formula A is, in the context of Γ, a *logical cause* of B iff there is an intervention (i.e. an abstraction) on $\langle \Gamma, A, B \rangle$ that allows for manipulating the truth (i.e. the being verified) of A and B without affecting the truth of Γ s.t. a positive manipulation of A verifying all of Γ by such an intervention is always accompanied by a positive change in B.

This can be made precise as follows.

Definition 24.2 Given a set of formulas Γ, a formula A is, *in the context of* Γ, a *logical cause* of B iff there is an abstraction $\langle \Gamma', A', B' \rangle$ of $\langle \Gamma, A, B \rangle$ s.t.

(i) there is a model that verifies the members of Γ' but unverifies[9] B',
(ii) there is a model that verifies the members of Γ' and A', and
(iii) $\Gamma', A' \vdash B'$ is valid.

Now that the definition is established, we can give some examples. For starters, let $\Gamma = \emptyset$. Again we use **LP** with classical negation $\sim$ as our underlying logic.

Example 24.1 Does p logically cause $p \vee q$? The answer is yes because the trivial intervention $\langle \emptyset, p, p \vee q \rangle$ is such that (i) there is a model that unverifies $p \vee q$ (ii) there is a model that verifies p and (iii) in every model that verifies p also $p \vee q$ is verified, because $p \vdash p \vee q$. This generalizes to all cases for which (i) there is a model that verifies the premise, (ii) one that does not verify the consequent and (iii) the consequent is a logical consequence of the premise. In multiple consequence sequent notation: whenever $\not\vdash B$ and $A \not\vdash$, B is a logical cause of A iff $A \vdash B$. But there are also cases in which we do have $\vdash B$ or $A \not\vdash$ but A nevertheless is a logical cause of B, as will be clear from the following examples.

---

[7] *Uniform Substitution* is the rule that enables the substitution of *every* occurrence of a sentential letter (usually in one formula, here in a triple of a set of formulas and two formulas) by one and the same formula.

[8] To avoid confusion: what we are defining here is an asymmetric logical consequence, not logical equivalence.

[9] We use 'unverifying' as a synonym for 'not verifying', where 'not' is a classical negation. Note that, in **LP**, this is something else than 'falsifying'. Had we chosen for 'falsifying', this condition would have been trivially satisfied in **LP**. There is always a model that verifies an arbitrary set of sentences and falsifies another arbitrary sentence.

Example 24.2 Does p logically cause $p \vee \neg p$? The answer is yes because there is an intervention $\langle \emptyset, p, p \vee q \rangle$ such that (i) there is a model that unverifies $p \vee q$ and (ii) there is a model that verifies p, and (iii) in every model that verifies p also $p \vee q$ is verified, because $p \vdash p \vee q$. Example 24.2 is a result of Uniform Substitution on Example 24.1, and its answer follows therefore from the answer to Example 24.1.

Example 24.3 Does $p \wedge \sim p$ logically cause p? The answer is yes because there is an intervention $\langle \emptyset, p \wedge q, p \rangle$ such that (i) there is a model that unverifies p, (ii) there is a model that verifies $p \wedge q$, and (iii) in every model that verifies $p \wedge q$ also p is verified, because $p \wedge q \vdash p$.

Example 24.4 Does $p \wedge \sim p$ logically cause $q \vee \neg q$? No it does not, because every (minimally destructive) intervention that is able to verify $p \wedge \sim p$ and is also able to unverify $q \vee \neg q$ will need to abstract respectively $\sim p$ and $\neg q$ into two different sentential letters (also different from p and q). Take r and s as examples (but it works with any two sentential letters different from p and q). For the third condition to be satisfied, $p \wedge r$ should have $q \vee s$ as a logical consequence. This is clearly not the case.

Now let us consider cases where $\Gamma \neq \emptyset$.

The cases where $\Gamma, A \nvdash$ and $\Gamma \nvdash B$ reduce again to usual logical consequence: A logically causes B in the context of Γ iff $\Gamma, A \vdash B$.

Example 24.5 Let $\Gamma = \{p \wedge \sim p\}$. Does, in the context of Γ, r logically cause $p \wedge r$? Yes it does, because the intervention $\langle \{p \wedge q\}, r, p \wedge r \rangle$ is such that (i) $p \wedge q, r \nvdash$, (ii) $p \wedge q \nvdash p \wedge r$, and (iii) $p \wedge q, r \vdash p \wedge q$.

Example 24.6 Does, in the context of $\{p \wedge r\}$, $p \wedge q$ logically cause $p \wedge r$? Maybe it comes as a surprise that this is indeed a correct logically causal claim. How is that possible given that the alleged cause is redundant for the effect in that context? There is an intervention $\langle \{s \wedge r\}, p \wedge q, p \wedge r \rangle$ such that: (i) $p \wedge r, p \wedge q \nvdash$ (ii) $s \wedge r \nvdash p \wedge r$, and (iii) $s \wedge r, p \wedge q \vdash p \wedge r$. Remark that this is nothing else but an instance of the much less surprising fact that in the context of $\{A \wedge B\}$, $C \wedge D$ logically causes $C \wedge B$.

We now extend logical causation to also take into account negative contexts. A positive context is a set of formulas that are supposed to be true, and a negative context is a set of formulas that are supposed not to be true. In order to do that, first naturally extend the definition of abstraction and intervention to quadruples containing two sets of formulas and two formulas.

Definition 24.3 Given two sets of formulas Γ and Δ, a formula A is, *in the positive context* of Γ and *in the negative context* of Δ, a *logical cause* of B iff there is an intervention $\langle \Gamma', \Delta', A', B' \rangle$ on $\langle \Gamma, \Delta, A, B \rangle$ s.t.

  (i) there is a model that verifies the members of Γ', unverifies the members of Δ' and unverifies B' (in other words $\Gamma' \nvdash B', \Delta'$),

(ii) there is a model that verifies A' and the members of Γ', but unverifies the members of Δ' (in other words $\Gamma', A' \nvdash \Delta'$) and
(iii) each model that verifies each of the members of $\Gamma \cup \{A\}$ also verifies one of the members of $\Delta \cup \{B\}$ (in other words $\Gamma', A' \vdash B', \Delta'$).

To evaluate whether NULL causes B, condition (ii) and '$\cup\{A\}$' should be removed from the definition. To evaluate whether A causes NULL, condition (i) and '$\cup\{B\}$' should be omitted.

We can now define multiple conclusions sequents for logical causation.

Definition 24.4 Where Γ and Δ are sets of formulas, we say that Γ *logically causes* Δ, abbreviated by $\Gamma \Vdash \Delta$, iff, for every $A \in \Gamma$ and every $B \in \Delta$, A logically causes B in the positive context $\Gamma - \{A\}$ and in the negative context $\Delta - \{B\}$.

Some further specification are required for the cases in which Γ or Δ or empty sets. The empty set logically causes Δ ($\Vdash \Delta$) iff, for each $B \in \Delta$, NULL causes B in the negative context of $\Delta - \{B\}$. Γ logically causes the empty set ($\Gamma \Vdash$) iff, for each $A \in \Delta$, A causes NULL in the positive context of $\Gamma - \{A\}$.

Fortunately, we can heavily simplify Definition 24.3. To do this, we first introduce another definition: *strictly relevant consequence*, viz. valid consequence that is no longer valid if one of its elements is removed.

Definition 24.5 Γ is said to *strictly relevantly* have Δ as a *consequence* iff $\Gamma \vdash \Delta$ and, for all $A \in \Gamma$, $\Gamma - \{A\} \nvdash \Delta$ and, for all $B \in \Delta$, $\Gamma \nvdash \Delta - \{B\}$

The following theorem presents the simplification of the notion of local cause. It is easy to prove: condition (iii) in Definition 24.3 corresponds to the validity of $\langle \Gamma', \Delta' \rangle$ and (i) and (ii) to the strict relevance of this validity.

Theorem 24.1 $\Gamma \Vdash \Delta$ *iff there exists an intervention $\langle \Gamma', \Delta' \rangle$ on $\langle \Gamma, \Delta \rangle$ such that Δ' is a strictly relevant consequence of Γ'.*

Suppose again the underlying logic is **LP** with classical negation $\sim$. We then get the following behaviour of $\Vdash_{\text{LPR}}$. Note that **LPR** validates all consequences that do not contain completely unrelated premises or conclusions, for if it is completely unrelated, there will be no abstraction of the consequence in such a way that it becomes strictly relevant.

1. $A \Vdash_{\text{LPR}} A$
2. $p, q \Vdash_{\text{LPR}} p \wedge q$ but $p, q, r \nVdash_{\text{LPR}} p \wedge q$ and $p, q \nVdash_{\text{LPR}} p \wedge q, r$
3. $p \vee q \Vdash_{\text{LPR}} p, q$ but $p \vee q, r \nVdash_{\text{LPR}} p, q$ and $p \vee q \nVdash_{\text{LPR}} p, q, r$
4. $\Vdash_{\text{LPR}} p \vee \neg p$ but $r \nVdash_{\text{LPR}} p \vee \neg p$
5. $p \Vdash_{\text{LPR}} p \vee \neg p$ and $\neg p \Vdash_{\text{LPR}} p \vee \neg p$ but $p, \neg p \nVdash_{\text{LPR}} p \vee \neg p$
6. $q \vee \neg q \nVdash_{\text{LPR}} p \vee \neg p$
7. $p \Vdash_{\text{LPR}} p \vee q$ but $p \nVdash_{\text{LPR}} p, q$
8. Where A is a tautology, $\Vdash_{\text{LPR}} A$, but $q \nVdash_{\text{LPR}} A$ (if q does not occur in A)
9. $\Vdash_{\text{LPR}} p, \neg p$ but $p, \neg p \nVdash_{\text{LPR}}$ ($\neg$ is paraconsistent, but not paracomplete)

10. $q \not\Vdash_{LPR} p, \neg p$
11. $\Vdash_{LPR} p, \sim p$ and $p, \sim p \Vdash_{LPR}$, but $p, \sim p \not\Vdash_{LPR} q$
12. $p \Vdash_{LPR} p \wedge (q \vee \neg q)$ and $p \Vdash_{LPR} (p \wedge q) \vee \neg q$
13. $p, \sim p \vee q \Vdash_{LPR} q$ but $p, \neg p \vee q \not\Vdash_{LPR} q$
14. $p \wedge \sim p \Vdash_{LPR} p \wedge (\sim p \vee q)$ and $p \wedge (\sim p \vee q) \Vdash_{LPR} q$, but $p \wedge \sim p \not\Vdash_{LPR} q$
15. $q \Vdash_{LPR} p \vee (q \wedge \neg p)$ and $p \vee (q \wedge \neg p) \Vdash_{LPR} p \vee \neg p$, but $q \not\Vdash_{LPR} p \vee \neg p$
16. $p, q \Vdash_{LPR} p \wedge q$ and $p \wedge q \Vdash_{LPR} p$, but $p, q \not\Vdash_{LPR} p$
17. $p \Vdash_{LPR} p \vee q$ and $p \vee q \Vdash_{LPR} p, q$, but $p \not\Vdash_{LPR} p, q$
18. $p \wedge p \Vdash_{LPR} p, p \wedge \neg p$

To see how one can prove the occurrences of $\not\Vdash_{LPR}$ in these examples, when in fact $\Vdash_{LP}$ holds, just consider that in all these cases there is a conclusion or premise that is not just redundant, but redundant in such a way that the sequent cannot be abstracted in such a way that it becomes non-redundant. Usually (but not necessarily) this is the case when the propositional letters used in the redundant formula have nothing to do with the other formulas. Consider e.g. item 14: q is completely redundant in $p \wedge \sim p \Vdash_{LP} q$, as we already have $p \wedge \sim p \vDash_{LP}$, and, because q is primitive and has nothing to do with $p \wedge \sim p$ there is no use in further abstracting the sequent to overcome redundancy.

In all the examples, one observes that the relation is neither left nor right monotonic: adding a formula left or right of a valid $\Vdash_{LPR}$ does not warrant (relevant) validity. In some sense, only the formulas are included in a sequent that really do some work in making the sequent valid. On the other hand, the relation is obviously formal/structural, in the sense that Uniform Substitution always preserves (relevant) validity.[10] Tautologies and logical impossibilities are derivable at the right-hand side resp. left-hand side of the turnstile with the other side of the turnstile empty, as seen in items 9 and 11. Except for formulas that do no work at all in the validity of a logical consequence, all **LP**-consequences are accepted, even if this can, in combination with other principles lead to strict irrelevances. This is illustrated in items 14 and 15. The first two are in both cases correct logically causal consequences, while the third is not. Nevertheless the third is just the result of the subsequent application of the first two principles. This is an essential feature of $\Vdash_{LPR}$: it is not transitive in the following sense.

$\Vdash_{LPR}$ is not transitive: That $\Gamma \Vdash_{LPR} A$ and $A \Vdash_{LPR} \Delta$ does not always entail $\Gamma \Vdash_{LPR} \Delta$

---

[10]To avoid confusion: while we are, in this paper, certainly not defining a traditional relevant logic in the relevance logic tradition (due to the lack of transitivity), **LPR** is just as non-monotonic and formal/structural as traditional relevance logics. Just like in e.g. the logic **R**, one gets monotonicity when premises of an implication are conjoined (as in $(A \wedge B) \to C$) but one gets non-monotonicity if premises are fused (as in $A \to (B \to C)$). Here we choose for the technically much more interesting fusion of premises, but conjoining them is also perfectly possible in **LPR**, as a simpler one-premise case. In that case there is some kind of monotonicity, also for **LPR**.

This lack of transitivity is also illustrated (proven) by items 16 and 17. This may seem very weird at first but, while items 14 and 15 can maybe still be seen as illustrations that mainly suggest that $\Vdash_{\mathbf{LPR}}$ is ill-behaved and problematic, items 16 and 17 are more intuitive properties of a relevant consequence relation that does not validate consequences with redundant formulas (it is **LP**-valid and none of premises or conclusions can be dropped). The relevant consequence in our sense should essentially neither be (left or right) monotonic nor transitive (but as is clear from item 1, the relation is reflexive). We would like to stress that we do not see the lack of transitivity as a disadvantage. It is a simple fact that, given the way we approach relevance, the relation of being a relevant consequence can never be transitive. This is not at all an undesired side effect of technical choices. We believe that it is even perfectly compatible with intuitions about the relevant consequence. A mathematician who writes in her paper 'From Theorem A and Theorem B together, it follows that A-and-B' is true' seems perfectly reasonable. Also writing that A-and-B entails A seems unproblematic. Yet, writing that Theorems A and B together entail A would seem rather odd, exactly because the second premise is completely redundant. For a more concrete example, instantiate this A and B as follows: let n be some specific natural number, A be 'n is a prime number', and B be 'there is a natural number m such that $n = 2^m + 1$'. Of course, these intuitions are not completely uncontroversial, but they are at least convincing enough to be open to the possibility that relevant consequence might turn out to be non-transitive.

Of course, it is completely sensible to demand that the notion of deduction or of derivation is indeed transitive. The logic that determines which theorems can be deduced from the axioms or postulates of theories should probably be transitive (and monotonic) no matter in which context. Otherwise one would not be able to build further on knowledge already obtained through deduction. Every single derivation would need to start again from the basic axioms or postulates. But our approach here does not at all go against the transitivity of the logic of deduction, derivation or theory structuring. For all these purposes we accept the validity of **LP**'s consequence relation. We merely claim that 'relevant implication/consequence' (corresponding to **LP**-consequence) is a separate notion that does not need to have this property. Incidentally, remark that most contemporary accounts of conditionals (e.g. so-called *conditional logics*) are not transitive either.

Item 18 may seem a bit bizarre. $p \wedge \neg p$ seems completely redundant. However, consider that $p \wedge q \Vdash_{\mathbf{LPR}} r, p \wedge \neg r$ is unproblematic and that the sequent in item 18 is the result of applying Uniform Substitution to the former sequent (substituting r and q both by p).

In Sect. 24.4, we will use $\Vdash_{\mathbf{LPR}}$ to add a relevant implication to **LP**, but first we will develop a proof theory for $\Vdash_{\mathbf{LPR}}$.

24.3 A Sequent Calculus for Logical LP-Cause

24.3.1 The Axioms and Rules of LPR

In the previous section, we have semantically defined the logically causal consequence relation $\Vdash$ for any Tarskian consequence relation. We now develop an elegant and simple proof theory for the logically causal consequence relation $\Vdash_{\mathbf{LPR}}$ of **LP**.

The calculus derives sequents of the form $\Gamma \vartriangleright \Delta$ from other sequents of that form, where Γ and Δ are possibly empty sets of formulas of **LP** with classical negation[11] $\sim$.

Definition 24.6 An **LPR**-*proof* for $\Gamma \vartriangleright \Delta$ is a tree with root $\Gamma \vartriangleright \Delta$ s.t. for each node, the sequent at that node is derived from the sequents of its children justified by one of the rules of **LPR** and all leaves contain axiom sequents.

The only axiom schema:

$$A \vartriangleright A$$

The only structural rule:

$$\frac{\Gamma, A, A \vartriangleright \Delta}{\Gamma, A \vartriangleright \Delta} \text{ CON}$$

The rules for $\neg$:

$$\frac{\Gamma, A \vartriangleright \Delta}{\Gamma \vartriangleright \neg A, \Delta} \text{ R}\neg \qquad \frac{\Gamma, A \vartriangleright \Delta}{\Delta, \neg\neg A \vartriangleright \Delta} \text{ L}\neg\neg \qquad \frac{\Gamma \vartriangleright A, \Delta}{\Delta \vartriangleright \neg\neg A, \Delta} \text{ R}\neg\neg$$

$$\frac{\Gamma \vartriangleright \neg A, \Delta}{\Gamma \vartriangleright \neg(A \wedge B), \Delta} \text{ R}\neg\wedge 1 \qquad \frac{\Gamma \vartriangleright \neg B, \Delta}{\Gamma \vartriangleright \neg(A \wedge B), \Delta} \text{ R}\neg\wedge 2$$

$$\frac{\Gamma, \neg A \vartriangleright \Delta}{\Gamma, \neg(A \vee B) \vartriangleright \Delta} \text{ L}\neg\vee 1 \qquad \frac{\Gamma, \neg B \vartriangleright \Delta}{\Gamma, \neg(A \vee B) \vartriangleright \Delta} \text{ L}\neg\vee 2$$

$$\frac{\Gamma_1, \neg A \vartriangleright \Delta_1 \quad \Gamma_2, \neg B \vartriangleright \Delta_2}{\Gamma_1, \Gamma_2, \neg(A \wedge B) \vartriangleright \Delta_1, \Delta_2} \text{ L}\neg\wedge \qquad \frac{\Gamma_1 \vartriangleright \neg A, \Delta_1 \quad \Gamma_2 \vartriangleright \neg B, \Delta_2}{\Gamma_1, \Gamma_2 \vartriangleright \neg(A \vee B), \Delta_1, \Delta_2} \text{ R}\neg\vee$$

The rules for $\sim$:

$$\frac{\Gamma, A \vartriangleright \Delta}{\Gamma \vartriangleright \sim A, \Delta} \text{ R}\sim \qquad \frac{\Gamma \vartriangleright A, \Delta}{\Delta, \sim A \vartriangleright \Delta} \text{ L}\sim$$

The rules for $\vee$:

$$\frac{\Gamma \vartriangleright A, \Delta}{\Gamma \vartriangleright A \vee B, \Delta} \text{ R}\vee 1 \qquad \frac{\Gamma \vartriangleright B, \Delta}{\Gamma \vartriangleright A \vee B, \Delta} \text{ R}\vee 2 \qquad \frac{\Gamma_1, A \vartriangleright \Delta_1 \quad \Gamma_2, B \vartriangleright \Delta_2}{\Gamma_1, \Gamma_2, A \vee B \vartriangleright \Delta_1, \Delta_2} \text{ L}\vee$$

The rules for $\wedge$:

---

[11] We add this negation merely for illustrative reasons; those who object against it can leave it out without any harm.

$$\frac{\Gamma, A \triangleright \Delta}{\Gamma, A \wedge B \triangleright \Delta} \, L\wedge 1 \qquad \frac{\Gamma, B \triangleright \Delta}{\Gamma, A \wedge B \triangleright \Delta} \, L\wedge 2 \qquad \frac{\Gamma_1 \triangleright A, \Delta_1 \quad \Gamma_2 \triangleright B, \Delta_2}{\Gamma_1, \Gamma_2 \triangleright A \wedge B, \Delta_1, \Delta_2} \, R\wedge$$

24.3.2 Examples of Proofs

LPR-proof for $r \wedge q \triangleright \neg p \vee (p \wedge q)$

$$\cfrac{\cfrac{q \triangleright q \quad \cfrac{\cfrac{p \triangleright p}{\triangleright \neg p, p} \, R\neg}{q \triangleright \neg p, p \wedge q} \, R\wedge}{\cfrac{\cfrac{q \triangleright \neg p, \neg p \vee (p \wedge q)}{q \triangleright \neg p \vee (p \wedge q), \neg p \vee (p \wedge q)} \, R\vee}{\cfrac{q \triangleright \neg p \vee (p \wedge q)}{r \wedge q \triangleright \neg p \vee (p \wedge q)} \, L\wedge} \, \text{CON}}}{}$$

LPR-proof for $(p \wedge \sim p) \vee q \triangleright q \wedge (\neg r \vee r)$

$$\cfrac{\cfrac{\cfrac{\cfrac{\cfrac{p \triangleright p}{\sim p, p \triangleright} \, L\sim}{p \wedge \sim p, p \triangleright} \, L\wedge}{\cfrac{p \wedge \sim p, p \wedge \sim p \triangleright}{p \wedge \sim p \triangleright} \, \text{CON}} \, L\wedge}{(p \wedge \sim p) \vee q \triangleright q} \quad q \triangleright q \quad \cfrac{\cfrac{\cfrac{\cfrac{r \triangleright r}{\triangleright \neg r, r} \, R\neg}{\triangleright \neg r \vee r, r} \, R\vee}{\triangleright \neg r \vee r, \neg r \vee r} \, R\vee}{\triangleright \neg r \vee r} \, \text{CON}}{(p \wedge \sim p) \vee q \triangleright q \wedge (\neg r \vee r)} \, R\wedge \, Lv$$

LPR-proof for $\neg(p \wedge (q \vee r)) \triangleright \neg(p \wedge r)$

$$\cfrac{\cfrac{\cfrac{\cfrac{\neg p \triangleright \neg p \quad \cfrac{\neg r \triangleright \neg r}{\neg(q \vee r) \triangleright \neg r} \, L\neg\vee}{\neg(p \wedge (q \vee r)) \triangleright \neg p, \neg r} \, L\neg\wedge}{\neg(p \wedge (q \vee r)) \triangleright \neg(p \wedge r), \neg r} \, R\neg\wedge}{\cfrac{\neg(p \wedge (q \vee r)) \triangleright \neg(p \wedge r), \neg(p \wedge r)}{\neg(p \wedge (q \vee r)) \triangleright \neg(p \wedge r)} \, \text{CON}} \, R\neg\wedge}{}$$

LPR-proof for $p \wedge p \triangleright p, p \wedge \neg p$

$$\cfrac{\cfrac{p \triangleright p \quad \cfrac{\cfrac{p \triangleright p}{\triangleright \neg p, p} \, R\neg}{p \triangleright p, p \wedge \neg p} \, R\wedge}{p \wedge p \triangleright p, p \wedge \neg p} \, L\wedge}{}$$

24.3.3 Soundness and Completeness

Theorem 24.2 $\Gamma \Vdash_{\mathbf{LPR}} \Delta$ *if there is an* **LPR**-*proof for* $\Gamma \triangleright \Delta$.

Theorem 24.3 *If* $\Gamma \Vdash_{\mathbf{LPR}} \Delta$, *then there is an* **LPR**-*proof for* $\Gamma \triangleright \Delta$.

We omit the proofs, because they can easily be obtained by the adequacy proofs for **NTR** presented in Peter Verdée and De Bal (2019), which is basically the logical cause for classical logic. The only differences with those results are the following.

- **NTR** is defined as a relation between multisets of formulas, while **LPR** is defined as a relation between sets. This is nothing but a simplification. Wherever a multiset is used in the adequacy proofs of **NTR**, one can replace this by the set of all members of that multiset. One can also slightly modify the present paper in such a way that all premises and conclusions become multisets. Although a multiset characterization is more desirable to avoid consequences like $(A \to B) \to (A \to (A \to B))$ in the logic with the explicit object language implication defined below, we want to avoid the technical clutter in this paper to focus on the mere possibility of conceiving a relevant implication that is true to the ideas behind **LP**.
- In **NTR**, there is of course no paraconsistent negation $\neg$. However, the proof for this extended language goes along very similar lines as the original proof, where the proof for the paraconsistent negation can be adapted from the proof of the classical negation in that result.

24.4 Logical LP-Cause and Relevant Implication

24.4.1 From Relevant Consequence to Relevant Implication

In the previous sections, we looked into the idea of logical cause as a notion of relevant logical consequence inspired by one prominent way to make precise what causal consequence amounts to. We defined an interesting consequence relation $\Vdash_{\mathbf{LPR}}$ of causal logical consequence based on a Tarskian consequence relation $\vdash$, such as $\vDash_{\mathbf{LP}}$. The latter results in the logic **LPR** for which we also gave a sequent calculus.

By means of this relation $\Vdash_{\mathbf{LPR}}$ one can build a relevant implication that could then be added to **LP**. We project the consequence relation of **LPR** exactly into the object language connective $\to$, as follows.

Definition 24.7 $\mathbf{LPR}^{\to}$ is the logic where $\to$ is added to the language of **LP** in such a way that no $\to$ occurs in the scope of another logical symbols (in other words: there are only relevant implications at the outside of formulas) and the consequence relation $\vDash_{\mathbf{LPR}^{\to}}$ is recursively defined as follows.

(i) $\Gamma \Vdash_{\mathbf{LPR}^\to} \Delta$ iff $\Gamma \Vdash_{\mathbf{LPR}} \Delta$, when Γ and Δ do not contain $\to$,
(ii) $\Gamma \Vdash_{\mathbf{LPR}^\to} A \to B, \Delta$ iff $\Gamma, A \Vdash_{\mathbf{LPR}^\to} \Delta, B$, and
(iii) $\Gamma, A \to B \Vdash_{\mathbf{LPR}^\to} \Delta$ iff there are $\Gamma_1, \Gamma_2, \Delta_1$, and Δ_2 such that $\Gamma_1 \cup \Gamma_2 = \Gamma$, $\Delta_1 \cup \Delta_2 = \Delta$, $\Gamma_1 \Vdash_{\mathbf{LPR}^\to} A, \Delta_1$ and $\Gamma_2, B \Vdash_{\mathbf{LPR}^\to} \Delta_2$.

It does not come as a surprise that, in order to obtain a proof theory for **LPR**$^\to$, one merely needs to add the following rules to the sequent calculus of **LPR**.

$$\frac{\Gamma, A \triangleright B, \Delta}{\Gamma \triangleright A \to B, \Delta} \, \text{R}{\to} \qquad \frac{\Gamma_1 \triangleright A, \Delta_1 \quad \Gamma_2, B \triangleright \Delta_2}{\Gamma_1, \Gamma_2, A \to B \triangleright \Delta_1, \Delta_2} \, \text{L}{\to}$$

This is not a fully characterized relevant implication, because we did not specify how negation, conjunction and disjunction behave when $\to$ occurs in their scope, let alone how one would then again apply implications to formulas that contain, e.g. disjunctions of implications. We will not get into that problem here, because it is far from obvious how this should go. There are in general few coherent intuitions about the right interpretation of, for example, negations of logical implications. And a fortiori in paraconsistent logics like **LP**. The negation of the implication could be interpreted as affirming the possibility of a counterexample to the implication, not necessarily in this world. One of the issues is: should one have an interpretation that can verify both $A \to B$ and the existence of a possible counterexample for $A \to B$? In a dialetheic framework, one could expect the answer to this question to be positive, but further philosophical investigation is required to find out whether this makes sense and in which context.

Nevertheless, remark that, technically speaking, it is easy to extend **LPR**$^\to$ to the unrestricted language with unlimited nesting of $\neg, \vee, \wedge$, and $\to$: simply let the rules of the sequent calculus of **LPR**$^\to$ be unrestrictedly valid. But it is not because one can technically define it, that it has any value philosophically speaking, so we would like to be prudent here.

24.4.2 No Irrelevances

The logic **LPR**$^\to$ is a relevant logic. To illustrate how relevance is achieved in the sequent calculus, have a look at the following *failed* proto-proofs for some of the well-known paradoxes of irrelevance.

$$\frac{\dfrac{A, B \triangleright B}{B \triangleright (A \to B)} \, \text{R}{\to}}{\triangleright B \to (A \to B)} \, \text{R}{\to}$$

This is not a full proof (the proof does not start with an axiom) and we cannot complete it either, because we cannot get rid of the redundant A.

$$\frac{\dfrac{A, \neg A \triangleright B}{\neg A \triangleright A \to B} \text{R}\to}{\triangleright \neg A \to (A \to B)} \text{R}\to$$

In this proto-proof, we cannot get rid of the redundant B and hence will not find a full proof. But note that this is not even irrelevantly valid as there is an **LP**-model making A and $\neg A$ true but B false.

$$\frac{\dfrac{\dfrac{\dfrac{\dfrac{A, B \triangleright B, C}{A \triangleright B, B \to C} \text{R}\to}{\triangleright A \to B, B \to C} \text{R}\to}{\triangleright A \to B, (A \to B) \vee (B \to C)} \text{R}\vee}{\triangleright (A \to B) \vee (B \to C), (A \to B) \vee (B \to C)} \text{R}\vee}{\triangleright (A \to B) \vee (B \to C)} \text{CON}$$

We cannot get rid of the redundant A and C and hence will not find a full proof. However, we should be careful here: $(A \to B) \vee (B \to C)$ is not even a well-formed formula in our restricted language.

24.4.3 Adding a Relevant Implication to LP

While the implication in the obtained logic **LPR**$^\to$ arguably captures the relevant fragment of **LP** (because it corresponds to the relevant consequence relation **LPR**), it is not an extension of **LP**, except in an unusual sense[12]:

$$\vdash_{\textbf{LPR}} \bigwedge \Gamma \sqsupset A \text{ iff } \Gamma \vDash_{\textbf{LP}} A.$$

A Tarskian conservative extension of **LP** with a relevant implication would make it possible to formulate theories in a mixed language with usual **LP**-connectives PLUS the relevant implication. It is useful to notice that **LPR** is not a deductive logic (non-transitive, non-monotonic), so it does not even make sense to close theories under the **LPR**-consequence relation. However, if we want to use the presented results for the original purpose of this paper—coping with both dialetheic and relevance paradoxes—we do need a deductive logic that is able to overcome the paradoxes. It is perfectly possible to develop a deductive logic that is a conservative extension of **LP** and contains the **LPR**$^\to$-implication. We define this logic **LP**$^\to$ in Appendix A. The there defined logic involves some technical clutter on top of **LPR**$^\to$, but introduces no new ideas or unexpected outcomes.

Given this last detour, we have obtained a way to add a relevant implication to **LP**, true to **LP**-consequence itself.

---

[12] $\sqsupset$ is defined on Sect. 24.1.

24.5 Difference Between LPR$^\rightarrow$ and R

We have defined a new relevant logic based on difference-making causation. But what makes this logic differ from traditional relevance logics such as **R** (cf. Anderson and Belnap 1975)? The answer is simple: while they are also paraconsistent to some extent, they are not tailored for the logic **LP** and, for this reason, validate *too many and too few* implications.[13] Because the details of our notion of 'logical cause' depend entirely on the underlying Tarskian consequence relation, and we chose **LP** as the underlying logic, $\rightarrow$ is almost by definition in line with **LP**. Logical cause was created to validate all the consequences of the underlying Tarskian logic, in this case **LP**, except the irrelevant ones.

R has too many valid implications. Specifically, the relevance logic **R** validates arrow versions of Modus Tollens, i.e.

$$\vdash_\mathbf{R} \neg B \rightarrow ((A \rightarrow B) \rightarrow \neg A),$$

and Contraposition, i.e.

$$\vdash_\mathbf{R} (A \rightarrow B) \rightarrow (\neg B \rightarrow \neg A)$$

and

$$\vdash_\mathbf{R} (\neg A \rightarrow \neg B) \rightarrow (B \rightarrow A).$$

None of those rules are valid for $\supset$, $\sqsupset$, $\vDash_\mathbf{LP}$ or in a proof theory for **LP**. They can also be falsified by some simple semantic modal reasoning, where $\rightarrow$ is reduced to the strict implication. For Modus Tollens: suppose p is both true and false, moreover, let the q-true worlds be a subset of p-true worlds. This says nothing about the truth of q: it can easily be false only, and so $\neg p$ and $q \rightarrow p$ do not necessarily entail $\neg q$. For Contraposition: suppose that there is some link between the falsity of p and the falisity of q, in the sense that all the p-true-only worlds are a subset of all the q-true-only worlds (and hence $\neg q \rightarrow \neg p$). There is nothing that forbids us to construct the p-true worlds and q-true worlds as supersets of both the q-true-only worlds and the p-true-only in such a way that neither is a subset of the other. Then there are p-worlds that are not q-worlds and so $p \rightarrow q$ is not true. The other direction of contraposition can be falsified in exactly the same way (the negation is involutive).

So the arrow versions of Modus Tollens and Contraposition should not be valid in a formalization of relevant **LP**-implication. Traditional relevant logic **R** violates this maxim while in our logic **LPR**$^\rightarrow$, they are indeed not valid.

---

[13]The quasi-relevant logic **RM3** (cf. Avron 1991) is somewhat closer, but it validates all 'bad' consequences of **R** mentioned below and more, such as $(p \wedge \neg p) \rightarrow (q \vee \neg q)$.

$$\cfrac{\cfrac{\cfrac{B, \neg B \vartriangleright \cfrac{A \vartriangleright A}{\vartriangleright A, \neg A} \text{R}\neg}{A \to B, \neg B \vartriangleright \neg A} \text{L}\to}{\neg B \vartriangleright (A \to B) \to \neg A} \text{R}\to}{\vartriangleright \neg B \to ((A \to B) \to \neg A)} \text{CON} \qquad \cfrac{\cfrac{\cfrac{B, \neg B \vartriangleright \cfrac{A \vartriangleright A}{\vartriangleright A, \neg A} \text{R}\neg}{A \to B, \neg B \vartriangleright \neg A} \text{L}\to}{A \to B \vartriangleright \neg B \to \neg A} \text{R}\to}{\vartriangleright (A \to B) \to (\neg B \to \neg A)} \text{R}\to$$

The sequent $B, \neg B \vartriangleright$ cannot be proven, for the simple reason that B and $\neg B$ can be verified together in **LP**. Depending on what sequent calculus one devises for **R**, $B, \neg B \vartriangleright$ will probably be provable in it.

R has too few valid implications. There are also some arguably relevant **LP**-implications that are not valid in the relevance logic **R**:

$$\not\vdash_{\mathbf{R}} A \to (B \to (A \wedge B)),$$

$$\not\vdash_{\mathbf{R}} B \to (B \wedge (A \vee \neg A),$$

$$\not\vdash_{\mathbf{R}} B \to ((B \wedge A) \vee \neg A).$$

In each of the cases the premises **LP**-imply the conclusions, in such a way that they are indispensable (and therefore relevant) for the validity of the implication. They are unproblematically provable in the presented logic, as the following proofs show.

$$\cfrac{\cfrac{\cfrac{A \vartriangleright A \quad B \vartriangleright B}{A, B \vartriangleright A \wedge B} \text{R}\wedge}{A \vartriangleright B \to (A \wedge B)} \text{R}\to}{\vartriangleright A \to (B \to (A \wedge B))} \text{R}\to \qquad \cfrac{\cfrac{B \vartriangleright B \quad \cfrac{\cfrac{A \vartriangleright A}{\vartriangleright A, \neg A} \text{R}\neg}{\vartriangleright A \vee \neg A} \text{R}\vee}{B \vartriangleright B \wedge (A \vee \neg A)} \text{R}\wedge}{\vartriangleright B \to (B \wedge (A \vee \neg A))} \text{R}\to$$

$$\cfrac{\cfrac{\cfrac{\cfrac{B \vartriangleright B \quad \cfrac{A \vartriangleright A}{\vartriangleright A, \neg A} \text{R}\neg}{B \vartriangleright B \wedge A, \neg A} \text{R}\wedge}{B \vartriangleright B \wedge A, (B \wedge A) \vee \neg A} \text{R}\vee}{B \vartriangleright (B \wedge A) \vee \neg A, (B \wedge A) \vee \neg A} \text{R}\vee}{\cfrac{B \vartriangleright (B \wedge A) \vee \neg A}{\vartriangleright B \to ((B \wedge A) \vee \neg A)} \text{R}\to} \text{CON}$$

Decidability. Another important difference is the fact that there is an analytic sequent calculus for **LPR**$^\to$. This shows that the logic here presented is decidable while Urquhart (1984) has shown that **R** is undecidable, already at the propositional level. **New paradoxes**? Some valid implications in **LPR**$^\to$ will look suspicious to traditional relevance logicians. Here is the easiest one, albeit one with a classical negation $\sim$:

$$\Vdash_{\mathbf{LPR}} (\sim A \vee B) \to (A \to B).$$

Prima facie this seems paradoxical: how can one derive a relevant implication (strong link between sets of possible worlds) from a merely true disjunction (in this world)?

For example, from 'I go to work tomorrow or on Mars it is on average colder than on earth' it does not (seem to) follow that, 'I don't go to work tomorrow' implies that 'it is colder on Mars and than on earth'. However, we argue that this is not a problem, because we propose the following reading of that schema: if we already have the information that $\sim A \vee B$, where $\sim$ is classical, then new information A allows us to derive B. Our approach is not so much about modality or relations between possible worlds but rather about pieces of information (relevantly) entailing other pieces of information (thinking about relevant implication in terms of information is, by the way, rather common, also in traditional relevance logics, for example based on the semantics developed in Urquhart (1972)).

24.6 Conclusion

In conclusion, we first have analysed some paradoxes of self-reference and paradoxes of relevance. We have argued that it reasonable to look for independent solutions for the two kinds of paradoxes, combined in one logic that overcomes both problems of **CL**. For the first kind, **LP** with a material equivalence connective is a solution, but there are no straight forward ways to add or define implications in **LP** that solve the paradoxes of relevance and that are true to the other features of **LP**.

As a proposal that fulfils these requirements, we have developed an account of logical causation in which Woodward's difference-making definition of cause is applied to logical consequence. The result is an interesting kind of relevant consequence relation which could be applied to any given Tarskian consequence relation, also to **LP**, which results in **LPR**. The relation is non-monotonic and non-transitive, but it is formal/structural. We moreover have developed a sequent calculus for this relevant consequence relation based on **LP**. This consequence relation is then projected into the object language by extending the language with a relevant implication $\rightarrow$, resulting in the logic **LPR**$^{\rightarrow}$. Finally, this relevant implication is added to the traditional logic **LP** resulting in **LP**$^{\rightarrow}$.

This extension of **LP** with a relevant implication, based on **LP**, could be used to solve both the dialetheic paradoxes and those concerning irrelevance.

Appendix

A Adding the Relevant LPR-Implication to LP: The System LP$^{\rightarrow}$

Here we define, by means of a sequent calculus that is very similar to the one for **LPR**, a consequence relation that is deductive (monotonic, transitive and reflexive), defines the **LPR**-relevant implication, and is a conservative extension of **LP**. The most important challenge here is to devise a consequence relation that is fully transitive without also making the defined implication transitive (otherwise it does not define logical cause as we defined it above).

The reason why $\vdash_{\mathbf{LPR}}$ as defined above is not transitive or, in other words, why cut is not admissible in it, is that the connectives on the left side of the turnstile have a much stronger meaning than the corresponding connectives on the right side. The same holds for the implication connective that exactly reflects the turnstile in $\mathbf{LPR}^{\rightarrow}$. We need to keep this instability for the implication, but remove it for the turnstile. We do this by adding two auxiliary symbols to the language, $+$ and $-$, to distinguish 3 interpretations of symbols : the normal (formulas not preceded by $+$ or $-$), the stronger (formulas preceded by $+$) and the weaker interpretation (formulas preceded by $-$). The turnstile becomes transitive by being explicit about the difference in strength left and right of the implication-arrow and requiring the same interpretations on both sides of the turnstile.

In relevant logic, it is common to distinguish two kinds of disjunctions (the extensional $A \vee B$ and the intensional $\sim A \rightarrow B$) and two kinds of conjunctions (the extensional $A \wedge B$ and the intensional $\neg(A \rightarrow \neg B)$). In the logic we have presented above, we have used the intensional disjunction and the extensional conjunction for the strong interpretations on the left side of the turnstile, and the extensional disjunction and the intensional conjunction for the weak interpretations on the right side of the turnstile. In what follows we return to the traditional symmetric turnstile: standard we use the extensional interpretations and whenever there is a $+$ resp. $-$, we use the strong resp. weak interpretations, independently of the side of the turnstile. As there is no longer a difference in interpretation on the left and the right, the logic will be transitive.

We thus need three introduction rules, instead of one, per connective and side of the turnstile: a normal, a strong ($+$) and a weak ($-$) version. This perfectly restores the transitivity of the turnstile (see proof below). The arrow is kept non-transitive by having introduction rules for it that introduce it as having a strong interpretation ($+$) for the antecedent and a weak interpretation ($-$) for the implication. This way all desired properties of the original implication are maintained in the transitive calculus: an implicational formula is tautological in **LPR** iff it is in the new transitive logic.

The second issue is monotonicity. Of course, we cannot simply add a general weakening rule to **LPR**. This would make its implication obviously irrelevant, as we would have $A, B \triangleright B$ (via $B \triangleright B$ and weakening), which would allow the derivation of $\triangleright A \rightarrow (B \rightarrow B)$, which is obviously an irrelevant implication. As a solution, we can allow weakening with formulas that will never be used as antecedent or consequents of the implication. We do this by making a distinction between two ways for formulas to occur in sequents: as a possibly irrelevant bystander and as a guaranteed relevant part of the sequent. The first kind we denote by adding a superscript N to the formulas, the second by adding a superscript R. Weakening is then only possible with superscript ^{N} and introduction of $\rightarrow$ is only possible based on parts of sequents with superscript ^{R}. It is important to mention that, unlike $+$ and $-$, these super and subscripts are *not* part of the object language, but belong to the proof theory's formalism.

24 Expanding the Logic of Paradox with a Difference-Making ...

In what follows, X means either R or N.
The two axiom schemata:

$$A^R \triangleright A^R \quad +A^R \triangleright -A^R$$

Structural rules:

$$\frac{\Gamma, A, A \triangleright \Delta}{\Gamma, A \triangleright \Delta} \text{ CON}$$

$$\frac{\Gamma \triangleright \Delta}{\Gamma, A^N \triangleright \Delta} \text{ LWeak} \qquad \frac{\Gamma \triangleright \Delta}{\Gamma \triangleright A^N, \Delta} \text{ RWeak}$$

$$\frac{\Gamma, A^R \triangleright \Delta}{\Gamma, A^N \triangleright \Delta} \text{ LRtoN} \qquad \frac{\Gamma \triangleright A^R, \Delta}{\Gamma \triangleright A^N, \Delta} \text{ RRtoN}$$

The rules for $\neg$:

$$\frac{\Gamma, A^X \triangleright \Delta}{\Gamma \triangleright \neg A^X, \Delta} \text{ R}\neg \qquad \frac{\Gamma, A^X \triangleright \Delta}{\Delta, \neg\neg A^X \triangleright \Delta} \text{ L}\neg\neg \qquad \frac{\Gamma \triangleright A^X, \Delta}{\Delta \triangleright \neg\neg A^X, \Delta} \text{ R}\neg\neg$$

$$\frac{\Gamma, \mp A^X \triangleright \Delta}{\Gamma \triangleright \pm(\neg A)^X, \Delta} \text{ R}\pm\neg \qquad \frac{\Gamma, \pm A^X \triangleright \Delta}{\Delta, \pm(\neg\neg A^X) \triangleright \Delta} \text{ L}\pm\neg\neg \qquad \frac{\Gamma \triangleright \pm A^X, \Delta}{\Delta \triangleright \pm(\neg\neg A)^X, \Delta} \text{ R}\pm\neg\neg$$

$$\frac{\Gamma \triangleright \neg A^X, \Delta}{\Gamma \triangleright \neg(A \wedge B)^X, \Delta} \text{ R}\neg\wedge 1 \qquad \frac{\Gamma \triangleright \neg B^X, \Delta}{\Gamma \triangleright \neg(A \wedge B)^X, \Delta} \text{ R}\neg\wedge 2$$

$$\frac{\Gamma \triangleright -(\neg A)^X, \Delta}{\Gamma \triangleright -(\neg(A \wedge B))^X, \Delta} \text{ R}-\neg\wedge 1 \qquad \frac{\Gamma \triangleright -(\neg B)^X, \Delta}{\Gamma \triangleright -(\neg(A \wedge B))^X, \Delta} \text{ R}-\neg\wedge 2$$

$$\frac{\Gamma \triangleright +(\neg A)^X, +(\neg B)^X, \Delta}{\Gamma \triangleright +(\neg(A \wedge B))^X, \Delta} \text{ R}+\neg\wedge$$

$$\frac{\Gamma, \neg A^X \triangleright \Delta}{\Gamma, \neg(A \vee B)^X \triangleright \Delta} \text{ L}\neg\vee 1 \qquad \frac{\Gamma, \neg B^X \triangleright \Delta}{\Gamma, \neg(A \vee B)^X \triangleright \Delta} \text{ L}\neg\vee 2$$

$$\frac{\Gamma, +\neg A^X \triangleright \Delta}{\Gamma, +\neg(A \vee B)^X \triangleright \Delta} \text{ L}+\neg\vee 1 \qquad \frac{\Gamma, +\neg B^X \triangleright \Delta}{\Gamma, +\neg(A \vee B)^X \triangleright \Delta} \text{ L}+\neg\vee 2$$

$$\frac{\Gamma, -\neg A^X, -\neg B^X \triangleright \Delta}{\Gamma, -\neg(A \vee B)^X \triangleright \Delta} \text{ L}-\neg\vee$$

$$\frac{\Gamma, \neg A^X \triangleright \Delta \quad \Gamma, \neg B^X \triangleright \Delta}{\Gamma, \neg(A \wedge B)^X \triangleright \Delta} \text{ L}\neg\wedge \qquad \frac{\Gamma \triangleright \neg A^X, \Delta \quad \Gamma \triangleright \neg B^X, \Delta}{\Gamma \triangleright \neg(A \vee B)^X, \Delta} \text{ R}\neg\vee$$

$$\frac{\Gamma_1, +\neg A^X \triangleright \Delta_1 \quad \Gamma_2, +\neg B^X \triangleright \Delta_2}{\Gamma_1, \Gamma_2, +\neg(A \wedge B)^X \triangleright \Delta_1, \Delta_2} \text{ L}+\neg\wedge \qquad \frac{\Gamma_1 \triangleright -\neg A^X, \Delta_1 \quad \Gamma_2 \triangleright -\neg B^X, \Delta_2}{\Gamma_1, \Gamma_2 \triangleright -\neg(A \vee B)^X, \Delta_1, \Delta_2} \text{ R}-\neg\vee$$

$$\frac{\Gamma, -\neg A^X \triangleright \Delta \quad \Gamma, -\neg B^X \triangleright \Delta}{\Gamma, -\neg(A \wedge B)^X \triangleright \Delta} \text{ L}-\neg\wedge \qquad \frac{\Gamma \triangleright +\neg A^X, \Delta \quad \Gamma \triangleright +\neg B^X, \Delta}{\Gamma \triangleright +\neg(A \vee B)^X, \Delta} \text{ R}+\neg\vee$$

The rules for $\sim$:

$$\frac{\Gamma, A^X \triangleright \Delta}{\Gamma \triangleright \sim A^X, \Delta} \text{ R}\sim \qquad \frac{\Gamma \triangleright A^X, \Delta}{\Delta, \sim A^X \triangleright \Delta} \text{ L}\sim$$

$$\frac{\Gamma, \mp A^X \triangleright \Delta}{\Gamma \triangleright \pm\sim A^X, \Delta} \text{ R}\pm\sim \qquad \frac{\Gamma \triangleright \mp A^X, \Delta}{\Delta, \pm\sim A^X \triangleright \Delta} \text{ L}\pm\sim$$

The rules for $\vee$:

$$\dfrac{\Gamma \triangleright A^\times, \Delta}{\Gamma \triangleright A \vee B^\times, \Delta} \text{R}\vee 1 \qquad \dfrac{\Gamma \triangleright B^\times, \Delta}{\Gamma \triangleright A \vee B^\times, \Delta} \text{R}\vee 2 \qquad \dfrac{\Gamma, A^\times \triangleright \Delta \quad \Gamma, B^\times \triangleright \Delta}{\Gamma, A \vee B^\times \triangleright \Delta} \text{L}\vee$$

$$\dfrac{\Gamma \triangleright -A \vee -B^\times, \Delta}{\Gamma \triangleright -(A \vee B)^\times, \Delta} \text{R}-\vee \qquad \dfrac{\Gamma, -A \vee -B^\times \triangleright \Delta}{\Gamma, -(A \vee B)^\times \triangleright \Delta} \text{L}-\vee$$

$$\dfrac{\Gamma_1, +A^\times \triangleright \Delta_1 \quad \Gamma_2, +B^\times \triangleright \Delta_2}{\Gamma_1, \Gamma_2, +(A \vee B)^\times \triangleright \Delta_1, \Delta_2} \text{L}+\vee \qquad \dfrac{\Gamma \triangleright +A^\times, +B^\times, \Delta}{\Gamma \triangleright +(A \vee B)^\times, \Delta} \text{R}+\vee$$

The rules for $\wedge$:

$$\dfrac{\Gamma \triangleright A^\times, \Delta \quad \Gamma \triangleright B^\times, \Delta}{\Gamma \triangleright A \wedge B^\times, \Delta} \text{R}\wedge \qquad \dfrac{\Gamma, A^\times \triangleright \Delta}{\Gamma, A \wedge B^\times \triangleright \Delta} \text{L}\wedge 1 \qquad \dfrac{\Gamma, B^\times \triangleright \Delta}{\Gamma, A \wedge B^\times \triangleright \Delta} \text{L}\wedge 2$$

$$\dfrac{\Gamma \triangleright +A \wedge +B^\times, \Delta}{\Gamma \triangleright +(A \wedge B)^\times, \Delta} \text{R}+\wedge \qquad \dfrac{\Gamma, +A \wedge +B^\times \triangleright \Delta}{\Gamma, +(A \wedge B)^\times \triangleright \Delta} \text{L}+\wedge$$

$$\dfrac{\Gamma_1 \triangleright -A^\times, \Delta_1 \quad \Gamma_2 \triangleright -B^\times, \Delta_2}{\Gamma_1, \Gamma_2 \triangleright -(A \wedge B)^\times, \Delta_1, \Delta_2} \text{R}-\wedge \qquad \dfrac{\Gamma, -A^\times, -B^\times \triangleright \Delta}{\Gamma, -(A \wedge B)^\times \triangleright \Delta} \text{L}-\wedge$$

The rules for $\rightarrow$:

$$\dfrac{\Gamma, +A^\text{R} \triangleright -B^\text{R}, \Delta}{\Gamma \triangleright A \rightarrow B^\text{R}, \Delta} \text{R}\rightarrow \qquad \dfrac{\Gamma_1 \triangleright +A^\text{R}, \Delta_1 \quad \Gamma_2, -B^\text{R} \triangleright \Delta_2}{\Gamma_1, \Gamma_2, A \rightarrow B^\text{R} \triangleright \Delta_1, \Delta_2} \text{L}\rightarrow$$

$$\dfrac{\Gamma, -A^\text{R} \triangleright +B^\text{R}, \Delta}{\Gamma \triangleright +(A \rightarrow B)^\text{R}, \Delta} \text{R}\rightarrow \qquad \dfrac{\Gamma_1 \triangleright -A^\text{R}, \Delta_1 \quad \Gamma_2, +B^\text{R} \triangleright \Delta_2}{\Gamma_1, \Gamma_2, +(A \rightarrow B)^\text{R} \triangleright \Delta_1, \Delta_2} \text{L}\rightarrow$$

$$\dfrac{\Gamma, A \rightarrow B^\times \triangleright, \Delta}{\Gamma, -(A \rightarrow B)^\times \triangleright \Delta} \text{L}-\rightarrow \qquad \dfrac{\Gamma \triangleright A \rightarrow B^\times, \Delta}{\Gamma \triangleright -(A \rightarrow B)^\times, \Delta} \text{R}-\rightarrow$$

Definition 24.8 $\Gamma \vdash_{\text{LP}\rightarrow} \Delta$ iff $\Gamma^\text{N} \triangleright \Delta^\text{N}$ is derivable in the calculus defined by the axioms and rules introduced above.

Theorem 24.4 *Whenever no $\rightarrow$ or $\sim$ occurs in $\Gamma \cup \Delta$, $\Gamma \vdash_{\text{LP}} \Delta$ iff $\Gamma \vdash_{\text{LP}\rightarrow} \Delta$ (in other words: $\text{LP}^\rightarrow$ is a conservative extension of LP).*

This can be proven by using the fact that one can readily define a sound and complete tableaux method for **LP**, based on the rules of $\text{LP}^\rightarrow$.

Theorem 24.5 $\text{LP}^\rightarrow$ *is a structural (formal), monotonic, reflexive and transitive conservative extension of* **LP**.

Proof $\text{LP}^\rightarrow$ is structural because $\text{LPR}^\rightarrow$ has only formal rules and axioms. In view of the fact that **LPR** has the same theorems as **LP**, we have that, if there is a proof tree for $\triangleright \bigwedge \Gamma_1 \sqsupset \bigvee(\Delta_1 \cup \{A\})$ and for $\triangleright \bigwedge(\Gamma_2 \cup \{A\}) \sqsupset \bigvee \Delta_2$, then there is one for $\triangleright \bigwedge(\Gamma_1 \cup \Gamma_2) \sqsupset \bigvee(\Delta_1 \cup \Delta_2)$. This proves transitivity of $\text{LP}^\rightarrow$ for formulas in which $\rightarrow$ does not occur. In view of the structurality of $\text{LP}^\rightarrow$, this generalizes to all the formulas. ∎

Let us look at some illustrations of proofs in the logic $\mathbf{LP}^{\rightarrow}$.

$$\cfrac{\cfrac{\cfrac{p^R \rhd p^R}{\rhd p^R, \neg p^R} R\neg}{\rhd p \lor \neg p^R} R\lor}{\cfrac{q^N \rhd p \lor \neg p^R}{q^N \rhd p \lor \neg p^N} \text{LRtoN}} \text{LWeak}$$

$$\cfrac{\cfrac{\cfrac{\cfrac{\cfrac{\cfrac{p^R \rhd p^R}{\rhd p^R, \neg p^R} R\neg}{\rhd p^N, \neg p^R} \text{RRtoN}}{\rhd p^N, \neg p^N} \text{RRtoN}}{\rhd p^N, \neg p^N, q^N} \text{RWeak}}{\cfrac{\rhd \neg p \land \neg q^N, p^N, q^N}{\rhd (\neg p \land \neg q) \lor p^N, q^N} R\lor}{\rhd ((\neg p \land \neg q) \lor p) \lor q^N} R\lor$$

$$\cfrac{\cfrac{\cfrac{\cfrac{q^R \rhd q^R}{\rhd q^R, \neg q^R} R\neg}{\rhd q^N, \neg q^R} \text{RRtoN}}{\rhd q^N, \neg q^N} \text{RRtoN}}{\cfrac{\rhd q^N, \neg q^N, p^N}{\rhd q^N, \neg q^N, p^N} \text{RWeak}} R\land$$

This is a proof for $\rhd ((\neg p \land \neg q) \lor p) \lor q^N$, but there is no proof for $\rhd ((\neg p \land \neg q) \lor p) \lor q^R$. However there is one for $\rhd - (((\neg p \land \neg q) \lor p) \lor q)^R$.

$$\cfrac{\cfrac{+p^R \rhd -p^R}{\rhd -p^R, --\neg p^R} R-\neg \quad \cfrac{+q^R \rhd -q^R}{\rhd -q^R, --\neg q^R} R-\neg}{\cfrac{\rhd -(\neg p \land \neg q)^R, -p^R, -q^R}{\rhd -(\neg p \land \neg q) \lor -p^R, -q^R} R\lor}{\cfrac{\rhd -((\neg p \land \neg q) \lor p)^R, -q^R}{\rhd -((\neg p \land \neg q) \lor p) \lor -q^R} R\lor}{\rhd -(((\neg p \land \neg q) \lor p) \lor q)^R} R-\lor$$

$$\cfrac{\cfrac{\cfrac{\cfrac{p^R \rhd p^R}{\rhd p^R, \neg p^R} R\neg}{p^N \rhd p^R} \text{RRtoN}}{p^N \rhd p^N} \text{RRtoN} \quad \cfrac{\cfrac{\cfrac{\cfrac{q^R \rhd q^R}{\rhd q^R, \neg q^R} R\neg}{\rhd q \lor \neg q^R} R\lor}{\rhd q \lor \neg q^N} \text{RRtoN}}{p^N \rhd q \lor \neg q^N} \text{RWeak}}{p^N \rhd (q \lor \neg q) \land p^N} R\land$$

Similarly, there is no proof for $p^R \rhd (q \lor \neg q) \land p^R$, but there is one for $+p^R \rhd -((q \lor \neg q) \land p)^R$ whence also $\rhd p \rightarrow (q \lor \neg q) \land p^N$. In general, we can say that the $+$ and $-$ vocabulary is required to get to transitivity, taking away the imbalance between left and right side of $\rhd$. But this seriously weakens the logic. From there we need the ^{R} and ^{N} meta-linguistic notation to re-obtain full $\mathbf{LP}$ and to make the consequence relation monotonic. All this is done without loss of expressiveness: adding ^{R} to formulas, $+$ before premises, and $-$ before conclusions restores $\mathbf{LPR}$.

Theorem 24.6 $\Gamma \vdash_{\mathbf{LPR}} \Delta$ *iff there is a proof for* $\{+A \mid A \in \Gamma\}^R \rhd \{-A \mid A \in \Delta\}^R$ *in an* $\mathbf{LP}^{\rightarrow}$*-proof.*

Moreover, to some extent, the new logic $\mathbf{LP}^{\rightarrow}$ is even more expressive than $\mathbf{LPR}$ if relevant implication is concerned: it can express $\mathbf{LPR}$'s inference rules in the object language. The following proof shows that, form $A \vdash_{\mathbf{LPR}} B$ and $C \vdash_{\mathbf{LPR}} D$, one can conclude $A, B \vdash_{\mathbf{LPR}} C \wedge D$ (i.e. a sub-rule of **LPR**'s R∧).

$$\cfrac{\cfrac{\cfrac{+C^R \rhd +C^R \qquad -D^R \rhd -D^R}{C \to D^N, +C^R \rhd -D^R}\text{RRtoN} \qquad \cfrac{+A^R \rhd +A^R \qquad -B^R \rhd -B^R}{A \to B^N, +A^R \rhd -B^R}\text{RRtoN}}{\cfrac{A \to B^N, C \to D, +A^R, +C^R \rhd -(B \wedge D)^R}{\cfrac{A \to B^N, C \to D, +A^R \rhd -(C \to (B \wedge D))^R}{A \to B^N, C \to D^N \rhd A \to (C \to (B \wedge D))^N}\text{R∧}}\text{R∧}}}{}\text{R∧}$$

It should not come as a surprise that object language transitivity of the relevant implication is not provable: $A \to B, B \to C \nvdash_{\mathbf{LP}^{\to}} A \to C$ although it is (maybe problematically?) provable in **LPR**, as the following proof shows.

$$\cfrac{\cfrac{\cfrac{+A^R \rhd -A^R \qquad +B^R \rhd -B^R}{+(A \to B)^R, +A^R \rhd -B^R}\text{RRtoN} \qquad +C^R \rhd -C^R}{\cfrac{+(A \to B)^R, +(C \to A)^R, +C^R \rhd -B^R}{+(A \to B)^R, +(C \to A)^R \rhd -(C \to B)^R}\text{RRtoN}}}{}\text{RRtoN}$$

The pure $\mathbf{LP}^{\to}$-version of transitivity of $\to$ does not work, as one can observe here.

$$\cfrac{\cfrac{\cfrac{-A^R \rhd +A^R \qquad -B^R \rhd -B^R}{+(A \to B)^N, -A^R \rhd -B^R}\text{RRtoN} \qquad +C^R \rhd -C^R}{\cfrac{C \to A^N, A \to B^N, +C^R \rhd -B^R}{C \to A^N, A \to B^N \rhd C \to B^N}\text{RRtoN}}}{}\text{RRtoN}$$

The latter is not a proof because $-A^R \rhd +A^R$ is not an axiom.

References

Anderson, A. R., & Belnap, N. D. (1975). *Entailment: The logic of relevance and necessity* (Vol. 1). Princeton: Princeton University Press.
Avron, A. (1991). Natural 3-valued logics–characterization and proof theory. *Journal of Symbolic Logic, 56*(1), 276–294.
Batens, D., & De Clercq, K. (2004). A rich paraconsistent extension of full positive logic. *Logique et Analyse, 47*, 185–188, 227–257. Appeared 2005.
Beall, J. C. (2011). Multiple-conclusion lp and default classicality. *The Review of Symbolic Logic, 4*(2), 326–336.
Goodship, L. (1996). On dialethism. *Australasian Journal of Philosophy, 74*(1), 153–161.
Omori, H. (2015). Remarks on naive set theory based on lp. *The Review of Symbolic Logic, 8*(02), 279–295.
Verdée, A. S. P., & De Bal, I. (Forthcoming). A non-transitive relevant implication for classical logic consequence.
Priest, G. (1979). The logic of paradox. *Journal of Philosophical Logic, 8*(1), 219–241.
Priest, G. (2008). *An introduction to non-classical logic: From if to is*. Cambridge introductions to philosophy. Cambridge: Cambridge University Press.
Priest, G. (2017). What if? the exploration of an idea. *The Australasian Journal of Logic, 14*(1).
Schurz, G. (1991). Relevant deduction. *Erkenntnis, 35*(1–3), 391–437.
Urquhart, A. (1972). Semantics for relevant logics. *The Journal of Symbolic Logic, 37*(1), 159–169.

Urquhart, A. (1984). The undecidability of entailment and relevant implication. *The Journal of Symbolic Logic, 49*(4), 1059–1073.
Weber, Z. (2012). Transfinite cardinals in paraconsistent set theory. *Review of Symbolic Logic, 5*(2), 269–293.
Woodward, J. (2003). *Making things happen: A theory of causal explanation.* Oxford studies in philosophy of science. USA: Oxford University Press.

Peter Verdée is a professor of logic and philosophy of language at the Institut Supérieur de Philosophy at the University Catholique de Louvain in Belgium. He obtained his Ph.D. in 2008 with a dissertation on dynamic proof methods and goal-directed reasoning under the supervision of Diderik Batens. Before he started his present job in 2014, he did postdocs in the CLPS group in Ghent, Belgium and in the CLE group in Campinas, Brazil and a long research stay at the ILLC in Amsterdam. During his research in Ghent, he became an expert in adaptive logics, their metatheory, computational properties and applications to mathematics. Currently, his main long-term research project is on non-transitive approaches to relevant logic and their applications in philosophy, for the study of counterfactuals, grounding, hyperintensionality, and explanation. Next to this, he works on non-monotonic pragmatic foundations of mathematics, putting defeasible reasoning and its dialectics at the very core of the foundational philosophy of mathematics. Since he works at the UCL, he also started studying the formal properties of manipulation accounts of causation and their application to problems of mental causation. Finally, throughout his whole academic life, he has had a special interest in paraconsistent logic. On this topic, he recently co-edited a collective book with Holger Andreas in Springer's Trends in Logic series.

Chapter 25
On Non-transitive "Identity"

Heinrich Wansing and Daniel Skurt

Abstract Graham Priest takes the relation of identity to be non-transitive. In this paper, we are going to discuss several consequences of identity as a non-transitive relation. We will consider the Henkin-style completeness proof for classical first-order logic with a non-transitive "identity" predicate, Leibniz-identity in Priest's second-order minimal logic of paradox, and the question whether or not identity of individuals should be defined as Leibniz-identity.

Keywords Identity · Transitivity · Ship of Theseus · Leibniz's Law · Henkin-style completeness proofs · Minimal second-order LP

25.1 Introduction

Graham Priest is known for advocating various highly controversial philosophical theses, the most prominent being, perhaps, the claim known as dialetheism, namely, the contention that there are dialetheas, propositions that are both true and false. Another striking violation of orthodoxy is his view that identity is a non-transitive relation.[1] Conceiving of identity as a non-transitive relation opens paths to solving a number of problems.

---

[1] In Bader (2012, p. 141), Ralf Bader argues that "the occasional identity relation and the contingent identity relation are both non-transitive". These relations restrict identity to particular worlds or times. Bader, however, concludes that due to failure of transitivity, the restricted notions do not classify as identity relations.

H. Wansing (✉) · D. Skurt
Ruhr University Bochum, Bochum, Germany
e-mail: Heinrich.Wansing@rub.de

D. Skurt
e-mail: Daniel.Skurt@rub.de

It

- may be utilized to avoid the Ship of Theseus problem (Priest 2009, 2010);
- can be used to evade the slingshot argument, see, e.g., Shramko and Wansing (2009);
- prevents Priest's gluon theory (Priest 2014) from infinite regress.

In the present paper, we shall discuss and comment upon Priest's nonstandard conception of identity.

There are several logical, metaphysical, and meaning-theoretic issues related to identity (or sameness, or equality) that have been discussed in the literature, see Deutsch (2008), Forrest (2016), French (2016), Mackie and Jago (2013), Noonan and Curtis (2014). One issue pertinent to the question of transitivity or non-transitivity of identity is the problem of sameness of physical objects over time. This problem is surveyed in Gallois (2016), where the puzzle of identity through time is presented as follows:

1. If a changing thing really changes, there can't literally be one and the same thing before and after the change.
2. However, if there isn't literally one and the same thing before and after the change, then nothing has really undergone any change.

A very fundamental question concerning identity is whether identity is a relation at all, or rather a property. Following some remarks by Ludwig Wittgenstein, who in Wittgenstein (1922, 5.5301) maintained that it is obvious "[t]hat identity is not a relation between objects", Wehmeier (2004, 2012) rejects identity understood as a binary relation that every object bears to itself and to no other objects. He introduces a Wittgensteinian predicate logic of which the characteristic feature is "that, from the range of a bound variable 'x', all objects are excluded that are values of variables occurring free within the scope of the quantifier that binds 'x'," and points out that, for a language without individual constants, Wittgensteinian predicate logic and first-order logic with identity are expressively equivalent. If the restriction to constant-free languages is lifted, the expressive power of first-order logic with identity can be restored by introducing a predicate expressing co-referentiality of individual constants (Wehmeier 2008); see also the critical discussion in Trueman (2014), Wehmeier (2014).

Another basic question concerning identity is whether it is understood as an extensional or an intensional notion. Intuitionistically, intensional identity is a mind-related notion. Following Troelstra (1975, p. 308) "in standard intuitionistic terminology 'a is intensionally equal to b' is equivalent to saying that 'a and b are the same mental construction'." According to Troelstra, there is a range of non-extensional identity relations between extensional and intensional equality, but all of them are equivalence relations, so that none of them coincides with Priest's non-transitive identity relation.

The literature on the notion of identity either as primitive or as defined in second-order logic is vast, and we cannot even briefly touch all major philosophical aspects of identity. We shall, for example, neither discuss necessary versus contingent identity, i.e., the semantics of the identity predicate in intensional contexts of alethic modalities, nor the semantics of the identity predicate in hyperintensional contexts such as belief ascriptions. Doing so would require considering alethic modal, doxastic, and other extensions of various first- and second-order logics, which is beyond the scope of the present paper.[2] In this paper, we shall raise and to some extent discuss the following issues:

- Is the question whether the notion of identity is inconsistent or even gives rise to dialetheas conceptually related to the transitivity or non-transitivity of identity?
- If identity is deprived of transitivity, what does that mean for the Henkin-style completeness proof for classical first-order logic with the impoverished notion?
- Does second-order minimally inconsistent LP, the non-monotonic, minimal models version of the first-order Logic of Paradox enjoy the properties of classical recapture and reassurance?
- Are there methodological or other reasons for treating identity as a primitive notion rather than one explicitly defined in second-order logic?

These questions will be addressed against the background of several papers in which Graham Priest has presented his views on identity.

25.2 Identity in First-Order LP

In Priest (1987), Graham Priest added a primitive identity predicate "=" to the first-order Logic of Paradox, LP.[3] The interpretation $d(=)$ of "=" in a model $\langle D, d \rangle$ for first-order LP is the pair $\langle d^+(=), d^-(=) \rangle$, where $d^+(=)$ (the extension of "=") is defined as the diagonal relation on the domain D: $\{\langle x, x \rangle \mid x \in D\}$, and $d^-(=)$ (the anti-extension of "=") is arbitrary, except that $(d^+(=) \cup d^-(=)) = D^2$, where the third occurrence of "=" in the latter equation may be assumed to denote the diagonal relation in the metalanguage's set theory.

Given a model $\mathcal{M} = \langle D, d \rangle$ and a variable assignment $s: VAR \longrightarrow D$, the denotation den(t) of a term t in $\mathcal{M}$ under s is defined as usual. Priest takes as the set of truth values the set $\{\{0\}, \{1\}, \{0, 1\}\}$ of non-empty subsets of the set of classical truth values 1 (*true*) and 0 (*false*). An evaluation function v maps every pair consisting of a formula and a variable assignment s to a truth value from $\{\{0\}, \{1\}, \{0, 1\}\}$, and the defining clauses for identity statements are as follows:

$1 \in v(t = t', s)$ iff $\langle \text{den}(t), \text{den}(t') \rangle \in d^+(=)$;
$0 \in v(t = t', s)$ iff $\langle \text{den}(t), \text{den}(t') \rangle \in d^-(=)$.

---

[2] Another topic we will not discuss here is the view that a whole is identical to its parts. The "hybrid identity" suggested by Wallace (1999), "is transitive, reflexive, symmetric, it obeys Leibniz's Law, etc.; the exception is that the hybrid identity relation allows us to claim that many things can be identical to a singular thing" (Wallace 1999, p. 4). Since hybrid identity is transitive, it should be of little interest to Priest.

[3] Note that sometimes we will use the symbol "=" metalinguistically (implicitly or explicitly).

A formula $t = t'$ is then said to be true (false) in $\langle D, d \rangle$ iff $1 \in v(t = t', s)$ ($0 \in v(t = t', s)$) for any variable assignment s. Single-conclusion semantical consequence is then defined as preservation of designated elements, which in the relational formulation of the semantics gives one

$\Gamma \models \alpha$ iff for all models, $\mathcal{M}$, and all assignments, $s \colon VAR \longrightarrow D$,
it is true that if $1 \in v(\beta, s)$ for every $\beta \in \Gamma$, then $1 \in v(\alpha, s)$.

With reference to that semantics, in Priest (2010), the paper which is devoted to expounding the non-transitivity of identity, Priest points out that "True sentences of the form $a = a$ and $a \neq a$ are standard fare in paraconsistent theories of identity". In the second edition of Priest (1987), we can find some remarks concerning the identity predicate in a language containing an intensional, relevant conditional, $\rightarrow$. The semantics used there makes use of a non-empty set P of logically possible worlds, relative to which validity is defined, as well as a set of non-normal, "logically impossible" worlds (Priest 1987, p. 273):

> The identity predicate, like all predicates, has an extension and an anti-extension at every world. For it to function as identity, it needs to satisfy the condition
>
> For all $w \in P$, $d_w^+(=) = \{\langle a, a \rangle : a \in D\}$
> This ensures that $x = x$ is a logical truth. (Clearly, one should not expect this constraint to extend to impossible worlds as well. In such worlds, $x = x$ may fail.) It is also not difficult to check that $\{x = y\} \models \beta \leftrightarrow \beta(x/y)$ (where y is free when substituted for x).

A form of substitutivity of identity thus holds when a primitive identity predicate is added to first-order $LP_\rightarrow$, first-order LP with a relevant conditional.[4]

Whilst the interpretation of identity in first-order LP clearly allows sentences of the form $a = a$ and $a \neq a$ to be true in a model $\langle D, d \rangle$, in Priest (2010) Priest does not explicitly say that he himself indeed takes the identity predicate to give rise to dialetheas. He presents a version of the famous Ship of Theseus problem in terms of his motorbike and points out that "Dialecticians, such as Hegel, have delighted in such considerations, since they appear to show that the bike both is and is not the same" (Priest 2010, p. 406) and, after having presented some further considerations, he remarks that "various properties standardly taken to be possessed by identity (consistency, transitivity, substitutivity) are not to be taken for granted philosophically" (Priest 2010, p. 407).

In his short and much earlier note "Sorites and identity" (Priest 1991a, p. 294) Priest emphasizes the structural similarity between the Sorites paradox and the Ship of Theseus problem and notes that the latter can be solved by giving up substitutivity of identity. Moreover, in his later paper (Priest 1998) (see also Priest 2003, 2008, Chap. 25), Priest defines a notion of fuzzy identity for which reflexivity and symmetry are valid, whereas transitivity is not. Since substitutivity of identity entails transitivity,

---

[4]Priest points out that due to working with a semantics that makes use of both normal and non-normal worlds, implications of the form $(x = y \wedge \beta) \rightarrow \beta(x/y)$ are not valid but can be validated by requiring that at non-normal worlds, w, $d_w^+(=) \subseteq \{\langle a, a \rangle : a \in D\}$. When non-normal worlds are not assumed, as in Chap. 6 of Priest (1987), $(x = y \wedge \beta) \rightarrow \beta(x/y)$ emerges as valid.

"substitutivity of identity, *real identity*, fails" (our emphasis) (Priest 1998, p. 337). In both papers, Priest does not mention failure of consistency for the primitive identity predicate. It is, however, clear that Priest takes the notion of identity to be inconsistent. According to Priest et al. (1989, p. 7) "in change ... there is at each stage a moment when the changing item is both in a given state, because it has just reached that state, but also not in that state, because it is not stationary but moving through and beyond that state", and in Priest (2014, p. 69), Priest explains with respect to the Ship of Theseus problem in terms of his motor bike changing over time that

> [a]t the two ends we have clear cases of truth and falsity. Being neither true nor false is indeed symmetrical with respect to these, but so is being both true and false; and, as far as that goes, just as good. However, there are reasons to suppose that the latter is a better answer. Assuming this to be the case, then in the borderline cases there are going to be a_js which are both identical with, and distinct from, a_i and a_k

where a_i is a motorbike in the sorites sequence, a_k is a motorbike far away from a_i in the sequence, and a_j is a borderline case in between a_i and a_k.

Edwin Mares (2004, p. 272) explicitly maintains that "our notion of identity is overdefined" and motivates this claim by a version of the Ship of Theseus problem. Let us refer to Theseus' ship as t. Then t is rebuilt over time by successively replacing its planks by new planks so as to eventually obtain ship t' that has no original planks. Moreover, the original parts and pieces are used to reassemble another ship t''. Mares claims that the theory of identity that is applied to the Ship of Theseus problem is inconsistent. Let us call that theory $T_=$. The theory $T_=$ is not defined but it is assumed to support the following principles:

(a) If two objects are made of exactly the same materials, then they are identical.
(b) If the material an object is made of changes slowly over time, it remains the same thing.
(c) If two material things are both complete and are wholly spatially separated at a given time, then they are distinct.
(d) Identity is transitive.

By (a), t is identical to t'', by (b) t' is identical to t, and by transitivity, (d), the ship t' is identical to t''. The latter, however, contradicts (c). The continuity criterion (b) is contentious, and the usual sorites version of the problem builds on the intuition that t' is not identical to t.

In any case, we may notice that whilst transitivity of identity can be derived from substitutivity of identity, consistency of identity is a separate matter. There are theories of identity, in which the set $\{a = b, a \neq b\}$ is satisfiable, that is, there are models in which both $a = b$ and $a \neq b$ receive a value that is preserved in valid inferences, and identity is transitive, such as Priest's first-order LP with a primitive identity predicate. The inconsistency of the identity predicate thus does not entail the non-transitivity of identity. Conversely, giving up transitivity of identity, say in classical first-order logic with identity, which is the topic of the following section, does not entail the satisfiability of $\{a = b, a \neq b\}$.

25.3 Henkin-Style Completeness for First-Order Logic with Non-transitive "Identity"

In first-order logic, real identity, now understood as the diagonal relation on a given domain, is proof-theoretically underdetermined by its standard axiomatization. The diagonal of a domain is not implicitly definable by a set of first-order sentences, see Hodges (1983, p. 69) or Ketland (2011, Theorem 3.15).

Van Dalen (2004, p. 82) explains that "from the axioms alone, we cannot determine the precise nature of the interpreting relation. We explicitly adopt the convention that '=' will always be interpreted by real identity." The standard axioms for identity are called "a second best" by Hodges (1983); they are true in all first-order structures with real identity and they entail every sentence in a first-order language $\mathcal{L}$ with identity that is true in every $\mathcal{L}$-structure:[5]

$\forall x(x = x)$
all closed formulas of the form $\forall xy((x = y) \to (\varphi \to \varphi[y/x]))$

For the enterprise of giving up transitivity, a less compact axiomatization is better suited; the following is taken from van Dalen (2004, p. 81):

I_1 $\forall x(x = x)$
I_2 $\forall xy((x = y) \to (y = x))$
I_3 $\forall xyz(((x = y) \land (y = z)) \to (x = z))$
I_4 $\forall x_1 \ldots x_n y_1 \ldots y_n (\bigwedge_{i \leq n}(x_i = y_i) \to (t(x_1, \ldots, x_n) = t(y_1, \ldots, y_n)))$

$\forall x_1 \ldots x_n y_1 \ldots y_n (\bigwedge_{i \leq n}(x_i = y_i) \to (\varphi(x_1, \ldots, x_n) \to \varphi(y_1, \ldots, y_n)))$

where all formulas in I_4 are closed.

Assume now that we drop I_3 and thus give up transitivity of the identity predicate. We then also lose substitutivity, i.e., I_4. A two-place relation that is reflexive and symmetric is called a *similarity relation*. Let us use "≈" as the impoverished, non-transitive "identity" predicate. We may wonder whether there is a canonical similarity relation that stands to "≈" as real identity stands to "=". Whatever that relation might be, we may expect that real similarity for individuals is proof-theoretically underdetermined. Thus, we merely assume that the extension of "≈" in any model is a reflexive and transitive relation on the non-empty domain of that model.

In the Model Existence Lemma for classical first-order logic with identity, it must be shown that any consistent set Γ of first-order formulas has a model. One considers

---

[5]Priest (in 2009 and elsewhere) considers the reflexivity and substitutivity of identity not in terms of the standard axioms for identity, but semantically as

Law of Identity $\models a = a$
Schema of the Substitutivity of Identicals $a = b, A_x(a) \models A_x(b)$

a maximally consistent Henkin extension T_m of the theory of Γ in a language $\mathcal{L}_m$. The final step is to construe the quotient structure of the term model. In particular, two closed terms a, b are defined to be equivalent iff their identity is provable in T_m:

$$a \sim b := T_m \vdash a = b$$

and the individuals of the Henkin model $\mathfrak{A}$ then are the equivalence classes of closed terms under the relation $\sim$, i.e., $D^{\mathfrak{A}}$ the domain of $\mathfrak{A}$ is the set $\{[t] \mid t \text{ is a closed term of } \mathcal{L}_m\}$.

Let $Term_m^c$ be the set of closed terms of $\mathcal{L}_m$ and let $t^{\mathfrak{A}}$ be the denotation of t in $\mathfrak{A}$. We may now ask what is to replace the question marks in the following equivalence statement that parallels the one for real identity:

$\mathfrak{A} \models a = b$ iff $[a] = [b]$ where $[a] := \{d \in Term_m^c \mid T_m \vdash a = d\}$ and $a^{\mathfrak{A}} = [a]$ and
$\qquad\qquad\qquad\qquad\qquad\qquad [b] := \{d \in Term_m^c \mid T_m \vdash b = d\}$ and $b^{\mathfrak{A}} = [b]$
real identity of the
equivalence classes
of the closed terms
a and b

$\mathfrak{A} \models a \approx b$ iff $\mathbf{X}_a \; ? \; \mathbf{X}_b$ where $\mathbf{X}_a := ?, \mathbf{X}_b := ?, a^{\mathfrak{A}} = \mathbf{X}_a$ and $b^{\mathfrak{A}} = \mathbf{X}_b$
real similarity of
$\mathbf{X}_a$ *and* $\mathbf{X}_b$ *of the*
closed terms a and b

The problem is to answer these questions simultaneously, and in a first step, the idea is to define $\mathbf{X}_a$ and $\mathbf{X}_b$ as non-empty sets such that for these sets, "real similarity" can be understood as *overlap* in the sense of non-empty intersection of non-empty sets:

$\mathfrak{A} \models a \approx b$ iff $\mathbf{X}_a \cap \mathbf{X}_b \neq \emptyset$ where $\mathbf{X}_a := ?, \mathbf{X}_b := ?, a^{\mathfrak{A}} = \mathbf{X}_a$ and $b^{\mathfrak{A}} = \mathbf{X}_b$
non-empty intersection
of $\mathbf{X}_a$ *and* $\mathbf{X}_b$ *of the*
closed terms a and b

If we look at the Ship of Theseus problem in its sorites version, the successive removal of planks reaches a point at which the transitivity of the "identity" relation is broken. It is not clear when exactly that point is reached, but intuitively it is definitely reached when *all* planks have been exchanged. What can be said about the point at which transitivity breaks? At that point, we have a ship which is not within the

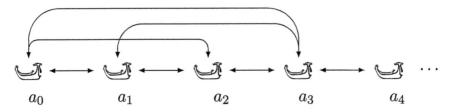

Fig. 25.1 The Ship of Theseus, ⌣ℐ, (plank exchanges not visible); a_i is the ship at moment t_i $(0 \leq i \leq 4)$

outreach of the original ship in a certain sense of "outreach". The set $\mathbf{X}_a$ will then be defined as the outreach of "a" under the binary relation $\vdash \_ \approx \_$ (see Fig. 25.1).

Definition 25.1 Let S be any similarity relation on a non-empty set D, and let $a \in D$. The *outreach of a under S*, S_a, is defined as follows:

$$S_a := \{\langle c, c' \rangle \in S \mid c = a \text{ or } c' = a\}.$$

Observation 25.1 Let S be any similarity relation on a non-empty set D, and let $a, b \in D$. Then $\langle a, b \rangle \in S$ iff $(S_a \cap S_b) \neq \emptyset$.

Proof The direction from left to right is obvious. Since S is reflexive, $S \neq \emptyset$. If $\langle a, b \rangle \in S$, then, by symmetry of S, $\langle a, b \rangle$ and $\langle b, a \rangle$ belong to $(S_a \cap S_b)$. If $\langle e, d \rangle \in (S_a \cap S_b)$, then $\langle e, d \rangle \in S_a$ and thus $e = a$ or $d = a$. (i) If $\langle a, d \rangle \in (S_a \cap S_b)$, then $\langle a, d \rangle \in S_b$. If $a = b$, then $\langle a, b \rangle \in S$, and if $a \neq b$, then $d = b$ and $\langle a, b \rangle \in S$, too. The case (ii), $\langle e, a \rangle \in (S_a \cap S_b)$, is analogous. q.e.d.

We can thus define $\mathbf{X}_a$ as $\{\langle c, c' \rangle \in \textit{Term}_m^c \times \textit{Term}_m^c \mid T_m \vdash c \approx a \text{ or } T_m \vdash c' \approx a\}$. For the Henkin model $\mathfrak{A}$, it must be shown that the definition of individuals as equivalence classes is such that the compound terms and the predicates are well defined. By using the definition of individuals as outreaches, the problem does not arise because the function that maps a to S_a is injective.

Real similarity for sets may be understood as the relation of non-empty intersection, and real similarity is thus the existential version of the "Leibniz-identity" for sets x, y, defined in a first-order language without a primitive identity predicate, which we will call "Leibniz-similarity":

$$\begin{aligned}&\text{Leibniz-identity of sets} \quad x = y \text{ iff } \forall z (z \in x \leftrightarrow z \in y)\\ &\text{Leibniz-similarity of sets } x \approx y \text{ iff } \exists z (z \in x \wedge z \in y).\end{aligned} \quad (25.1)$$

The Henkin-style completeness proof for classical first-order logic with identity can thus be modified to obtain completeness for the result of dropping Axiom I_4 from van Dalen's axiomatization.

25.4 Identity in Second-Order Minimal LP

For Graham Priest, there is no doubt that defining identity of individuals by means of Leibniz's Law is a natural thought (Priest 2009, 2010, 2014): $x = y := \forall X(X(x) \leftrightarrow X(y))$. To this end he makes use of second-order LP.

The language of second-order LP is exactly the same as that of first-order LP, with the exception that quantifiers now can range over relations as well. But for the sake of simplicity, Priest considers only monadic predicates and no function symbols. A model (or interpretation), I, for the second-order language is a triple $\langle D_1, D_2, d \rangle$, where D_1 is the non-empty domain of first-order quantification and D_2 is the non-empty domain of second-order quantification. Members, D, of D_2 are of the form $\langle D^+, D^- \rangle$, where $D^+, D^- \subseteq D_1$ and $D^+ \cup D^- = D_1$. Now, to ensure that every member of D_1 and D_2 has a name, we enrich the language in the following way. If $c \in D_1$, we add an individual constant k_c to the language. If $D \in D_2$, we add a predicate constant P_D to the language. Furthermore, the function d assigns to every individual constant a member of D_1 and to every predicate a member of D_2. If P is a predicate, we will write $d(P)$ as $\langle d^+(P), d^-(P) \rangle$, where $d^+(P)$ and $d^-(P)$ are the extension and the anti-extension of P, respectively. The separate truth- and falsity conditions for closed formulas are straightforward and can be found for example in Priest (2014, p. 29).

But, there are choices to be made. In the *standard* semantics for classical second-order logic the predicates range over *all* subsets of the domain of individuals. Therefore it seems natural to allow the predicate variables in LP to range over all extension/anti-extension pairs, and in Priest (2002), Priest did not impose any restrictions on the domain of second-order quantification except for being non-empty and to contain every extension/anti-extension pair. But then the definition of identity as Leibniz-identity is dubious, to say the least, as has been pointed out by Hazen and Pelletier (2016). One element of D_2 would be the pair $\langle D_1, D_1 \rangle$, and assigning this pair to all predicates P will make $P(k_c)$ and also $P(k_c) \to P(k_c)$ both true and false, for every $c \in D_1$. But this will make $k_c = k_c$ both, true and false, for every $c \in D_1$. However, not even the most frenetic dialetheist would claim that *every* object is and is not identical with itself. The discussion of gluon identity in Priest (2014, p. 24) suggests that this problem might be a reason why Priest in (2010, p. 408) uses a nonstandard semantics. He stipulates that for every $A \subseteq D_1$ there is a $B \subseteq D_1$ such that $\langle A, B \rangle \in D_2$ and emphasizes that he does not assume that every pair of the form $\langle A, B \rangle$, where $A \cup B = D_1$, belongs to D_2. Nevertheless, it is unclear if the problem just mentioned does not carry over to this semantics. A discussion of several versions of second-order LP can be found in Hazen and Pelletier (2016).

Like LP, second-order LP is very weak insofar as *modus ponens* and *disjunctive syllogism* are not valid. But, since LP is a sublogic of classical logic, there are models of LP that are *classical* (Priest 2010). In such classical models we have for every $\langle D^+, D^- \rangle \in D_2$, $D^+ \cap D^- = \emptyset$. As Priest claims, those classical models are just the models of second-order classical logic. In second-order LP, Leibniz-identity does not satisfy transitivity nor substitutivity of identicals but, if we restrict ourselves to classical models, then the identity predicate would behave in the same way as in classical logic. Exhibiting the same patterns of inference as classical logic is known as classical recapture.

Priest turned this idea into a formal non-monotonic logic, minimally inconsistent LP. The first-order version was introduced in Priest (1991b), while the second-order versions can be found in Priest (2010, 2014). However, to show that first-order minimal LP behaves in the same way as classical logic in consistent situations, is not trivial. A proof of classical recapture for languages without identity was presented in Crabbé (2012), and so far no proof is known of classical recapture in second-order minimally inconsistent LP. In this section we will not fill this gap, but give some considerations to the effect that hopefully with the right definition of minimality one can indeed get classical reasoning in classically consistent situations, so that second-order minimally inconsistent LP would indeed emerge as a logic that has consistency as a default assumption and that allows identity to satisfy transitivity and substitutivity of identicals in consistent situations.

Definition 25.2 Let I be an interpretation, then $I! = \{P_D k_d | d \in D_1, D \in D_2$ and $d \in D^+ \cap D^-\}$. $I!$ is the set of inconsistent atomic sentences of interpretation I. We define a preorder, $\prec$, on interpretations. If I_1 and I_2 are interpretations, we stipulate $I_1 \prec I_2$ iff

1. $I_1! \subseteq I_2!$ and,
2. $D_1^{I_1} \supseteq D_1^{I_2}$[6]

Let Σ be a set of sentences. A model I of Σ is then minimal iff for all models of Σ, J, if $J \prec I$, then $I! \subseteq J!$. Furthermore,

$$\Sigma \vDash_m H \text{ iff all minimal models of } \Sigma \text{ are models of } H.$$

Intuitively, it should hold that if Σ is classically consistent, then it has only minimal models I, in which $I! = \varnothing$. But this is not a trivial consequence of the definition. In fact, Crabbé showed in (2011) that in first-order minimal LP classical recapture is only valid if the domains of the bigger models are supersets of the domains of the smaller models and that there are counterexamples for other possible relations between the domains of the models. It seems that this problem carries over to second-order minimal LP. To make things worse, if one is willing to use a primitive notion of identity, the result would be a minimal logic that never satisfies classical recapture and other meta-properties (cf Crabbe 2011, 2012). Therefore, the step from a primitive notion to a defined one in second-order LP might be necessary from a technical point, if one wants to have LP as a base system.

Priest notes in (2014, p. 74) that classical recapture may not hold for his definition of second-order LP,[7] since there is no reason to suppose that for every $X \subseteq D_1$ there

---

[6]Priest briefly discusses this restriction of the first-order domains in Priest (2014) as a way of guaranteeing reassurance and classical recapture, but he did not define it formally.

[7]However, Priest (2014) emphasizes that classical recapture holds in languages without primitive identity for the first-order fragment.

is a $\langle X, D_1\setminus X\rangle \in D_2$. A way out of this misery might be to stipulate the following requirement for the set D_2: for every $X \subseteq D_1$, $\langle X, D_1\setminus X\rangle \in D_2$.

To prove classical recapture in Crabbé (2012), Crabbé uses a unique definition of minimality, for which classical recapture holds, and then shows that it is equivalent to the definition of minimality for LP. We are confident that one can adapt Crabbé's definition in order to prove classical recapture for second-order minimal LP, but we will leave this for future investigation.[8]

However, assuming second-order minimal LP satisfies classical recapture, what would be the implications for a second-order account of identity? As Priest notices, if objects a, b, c behave consistently, i.e., for no $D \in D_2$, $a, b, c \in D^+ \cap D^-$, we would get transitivity back and, furthermore, substitutivity for consistent objects. In the end, this was one of the reasons for using a non-monotonic logic. However, for this result, there is a price to pay. One of the appeals of LP is its simplicity as a natural generalization of classical logic, but to fulfill Priest's program, one needs to use an unintuitive definition of minimality in order to guarantee classical recapture.

There is another issue that is neither mentioned in Priest (2010) nor in Priest (2014), namely a property named "reassurance", which is the following condition: if a set of sentences Σ is not LP-trivial, i.e., $\Sigma \nvDash_{LP} H$ for some H, then we have $\Sigma \nvDash_m H$ for some H. Even though this property does not concern classically consistent premise sets, if reassurance does not hold, we could have contradictions that lead to explosion for the minimal consequence relation. In personal communication Priest noted that he does not know if reassurance holds, but this property seems crucial to the whole program of dialetheism itself, if one wants to make use of a non-monotonic version of second-order LP. So far it is not clear if Crabbé's results[9] can be used to obtain a second-order version of reassurance in the language discussed above. We will leave this topic to future investigation as well.

However, in second-order languages that contain function symbols, reassurance does not seem to hold if function symbols in LP are interpreted as classical functions. Consider the following set of sentences Σ,[10] where b, u_1 and u_2 are binary and unary function symbols, respectively:

1. $\forall x \forall y \forall X (X(u_1(b(x, y))) \leftrightarrow X(x))$
2. $\forall x \forall y \forall X (X(u_2(b(x, y))) \leftrightarrow X(y))$
3. $\forall x \forall y \forall X (X(b(u_1(x), u_2(x)) \leftrightarrow X(x))$
3. $\forall x \neg \forall X (X(b(x, x)) \leftrightarrow X(b(x, x)))$

First, observe that all finite models of Σ are trivial and have a singleton first-order domain, since b (by 1., 2., 3.) is a bijection between $D_1 \times D_1$ and D_1. Now, let I be

---

[8] We note just in passing that there are other definitions of minimality. As Christian Straßer mentioned in personal communication, at least for first-order LP, but not in languages with first-order identity, a different definition is necessary to obtain other properties such as strong reassurance, which minimal LP does not satisfy in its current version.

[9] The proof of reassurance in Crabbé (2012) is incomplete; see Skurt (2017) for a completed proof.

[10] See Crabbé (2011, p. 483) for a structurally similar counterexample to reassurance for first-order minimal LP.

a nontrivial infinite model of Σ and O be the following set $\{b_I(o, o) | o \in D_1\}$, where b_I is the interpretation of b under I. Furthermore, let F be a permutation of D_1^I, such that $F[O]$ is a proper subset of O. We now define a model J in the following way:[11] $D_1^J = D_1^I$, for every pair $\langle A_I, B_I \rangle \in D_2^I$ we define $\langle A_J, B_J \rangle \in D_2^J$ as follows, $A_J = A_I$ and $B_J = \{o \in D_1^J | o \notin A_J \text{ or } o \in F[O]\}$ and $b_J(o_1, o_2) = F(b_I(o_1, o_2))$ and $u_{iJ}(o) = u_{iI}(F^{-1}(o))$, with $i = 1, 2$ and $o, o_1, o_2 \in D_1^I$. It is now easy to observe that J is also a model of Σ and that $J! \subseteq I!$. Further checking makes it clear that $J \prec I$. Since we made no assumptions about I, it follows that we cannot find a minimal model for Σ. As a referee pointed out, in this particular example a restriction on the second-order domain, in the same way as with the first-order domain, could have avoided the problem. But it is not clear at all whether the restriction $D_2^{I_1} \supseteq D_2^{I_2}$ guarantees reassurance in general.

The above remarks should not be seen as a decisive argument against defining identity of individuals by means of Leibniz's Law in second-order minimal LP. However, they show that there are open questions that need to be answered before one can seriously consider using this definition of identity in, for example, paraconsistent set theory or for solving paradoxes like the sorites paradox.

25.5 Substitutivity Without Transitivity?

Transitivity of identity can be derived from substitutivity of identity.[12] In Cobreros et al. (2014, p. 78), Pablo Cobreros et al. present the somewhat surprising plan to "develop two notions of identity built on ideas close to Priest's", where the "first notion of identity is non-transitive but substitutivity works", whereas the second notion of identity is transitive but "sensitive to expressions of (in)definiteness." The first announcement in particular is irritating, so let us see what is going on.

The first and nonstandard notion of identity is defined semantically in the context of a semantically introduced logic called second-order ST. The language of second-order ST is that of second-order LP, and the semantics of second-order ST differs from

---

[11] Note that all occurrences of "=" are understood as metalinguistic real identity.

[12] Transitivity of identity is a special case of substitutivity of identity if either the above axiom I_3 is replaced by the classically equivalent

I_3' $\forall xyz((x = y) \to ((y = z) \to (x = z)))$

or axiom I_4 is replaced by the classically equivalent

I_4' $\forall x_1 \ldots x_n y_1 \ldots y_n (\bigwedge_{i \leq n}(x_i = y_i) \to (t(x_1, \ldots, x_n) = t(y_1, \ldots, y_n)))$

$\forall x_1 \ldots x_n y_1 \ldots y_n ((\bigwedge_{i \leq n}(x_i = y_i) \land (\varphi(x_1, \ldots, x_n)) \to \varphi(y_1, \ldots, y_n)))$

where all formulas in I_4 are closed.

that of second-order LP only but crucially in the notion of semantical consequence (entailment).

Definition 25.3 Let $\Gamma \cup \{\alpha\}$ be a set of second-order LP formulas. Then Γ entails α in ST ($\Gamma \models_2^{ST} \alpha$) iff there is no model for second-order LP, $\mathcal{M}$, such that (i) for every $\beta \in \Gamma$, $1 \in v(\beta)$ but $0 \notin v(\beta)$ and (ii) $0 \in v(\alpha)$ but $1 \notin v(\beta)$.

This definition results in a so-called *p*-entailment (plausibility entailment) relation, a kind of consequence relation introduced and investigated by Frankowski (2004a, b), see also Shramko and Wansing (2011). If $\{\{1\}\}$ is seen as the set of designated values, $\{\{0\}\}$ as the set of anti-designated values, and $\{\{0, 1\}\}$ as the set of values that are neither designated nor anti-designated, then the relation $\models_2^{ST}$ leads from designated premises to not anti-designated single conclusions, which is characteristic of *p*-entailment relations. Although in general *p*-entailment relations fail to be Tarskian consequence relations because they are not always transitive, for the classical vocabulary without real identity, $\models_2^{ST}$ coincides with consequence in classical second-order logic.

The semantical definition of the first "tolerant identity" predicate from Cobreros et al. (2014) leads to a non-transitive similarity relation, $\approx$, that satisfies substitutivity in the following sense:

$$\models_2^{ST} \forall x \forall y \forall P(\neg(Px \land (x \approx y)) \lor Py)$$
$$Px, (x \approx y) \models_2^{ST} Py$$

Cobreros et al. (2014) use a functional semantics, and the monadic predicates are interpreted by elements from a function space $D_2 \subseteq \{1, \frac{1}{2}, 0\}^{D_1}$, where D_1 is the non-empty domain of individuals and Priest's values $\{0\}$, $\{1\}$, and $\{0, 1\}$ are denoted by "0", "1", and "$\frac{1}{2}$", respectively. The relation $\approx$ is defined by stipulating that a formula $(a \approx b)$ receives the designated value 1 just in case for every $f \in D_2$, $|f(a) - f(b)| < 1$.

The second notion of identity considered in Cobreros et al. (2014) is Leibniz-identity in second-order ST. Since the conditional of second-order ST is classical implication, Leibniz-identity in second-order ST is transitive and, therefore, does not capture Priest's non-transitive notion of identity. Nevertheless, Cobreros et al. suggest to use Leibniz-identity in second-order ST for an analysis of the Ship of Theseus problem in terms of Priest's motorbike, PM, namely to use it in combination with the earlier similarity relation $\approx$. They emphasize that they obtain a principle of tolerance for the defined identity predicate, $=$, in the following sense:

$$\models_2^{ST} \forall x \forall y(\neg((\text{PM} = x) \land (x \approx y)) \lor (\text{PM} = y))$$
$$(a \approx b), a = \text{PM} \models_2^{ST} b = \text{PM}$$

Whatever the merits of this approach are, a general and serious drawback of working with $\models_2^{ST}$ for defenders of a paraconsistent logic is that the negation of second-order ST fails to be paraconsistent. The relation $\models_2^{ST}$ should therefore not have much appeal to dialetheists such as Priest. Also, the failure of transitivity of $\models_2^{ST}$ certainly is another not entirely unproblematic aspect of second-order ST.

25.6 Should Identity Be Defined by Leibniz's Law or Not?

Van Dalen (2004, p. 149) refers to Leibniz's Law as a *"natural* means to define identity for individuals" (our emphasis), namely as Leibniz-identity. Using a detachable biconditional, van Dalen points out that in case there is already a primitive identity symbol "$\stackrel{\circ}{=}$" in the language, both primitive identity and Leibniz-identity are provably equivalent, so that Leibniz-identity seems to be unproblematic.[13] For one direction we may just use the substitutivity of the primitive first-order identity, for the other direction we have:

$$\frac{x \stackrel{\circ}{=} x \quad \frac{\forall X(X(x) \leftrightarrow X(y))}{x \stackrel{\circ}{=} x \leftrightarrow x \stackrel{\circ}{=} y}}{x \stackrel{\circ}{=} y}$$

Priest, however, rejects the substitutivity of identity as a primitive first-order notion and, moreover, the failure of *modus ponens* in second-order LP breaks the above inference. With respect to his motorbike changing over time, Priest (2010, p. 406) remarks:

> But the categorical distinction between the thing itself and its properties is one which is difficult to sustain; to suppose that the bike is something over and above all of its properties is simply to make it a mysterious *ding an sich.*

Does this mean that Priest prefers a second-order definition of identity over identity as a primitive notion in first-order logic? This is maybe not so clear. In Priest (2010, p. 414), Priest remarks that in comparison to the theory of the non-transitive primitive fuzzy identity predicate from Priest (2003), the theory of the non-transitive Leibniz-identity in second-order minimally inconsistent LP "would do just as well". He also points out in Priest (2010, p. 414 f. and 2014, p. 72 f.) that he could have avoided using second-order logic by deploying as a metatheory a naive set theory based on LP in which identity of individuals is defined as Leibniz-identity of sets. Nevertheless in Priest (2014, p. 70) Priest emphasizes that Leibniz-identity in second-order minimal LP is a theory of identity that "handles the paradoxes of identity (at least, the ones that are not relatively superficial) in a simple and natural way." Moreover, as to Leibniz-identity in general he writes that "[t]here can be no doubt that the Leibnizian definition of identity produces something naturally thought of as identity (Priest 2014, p. 22) and rejects two putative arguments against Leibniz-identity (Priest 2014, p. 23 f.).

But Leibniz-identity is exposed to a number of problems. Timothy (Williamson 2006, p. 375) explains that

---

[13]Nevertheless, van Dalen (2004, p. 150 f.) discusses second-order arithmetic in a second-order language with *first-order* identity.

For practical purposes, one could even use the open formula $\forall P(Px \to Py)$ simply to define identity, although it is unlikely that second-order quantification is conceptually more basic than identity in any deep sense. But the appeal to second-order quantification may not satisfy those who are seriously worried about the problem of interpreting the identity predicate. For how do we know, or what makes it the case, that the second-order quantifier $\forall P$ should be interpreted in the standard way?

and then points to the nonstandard models for which Henkin showed classical second-order logic to be complete and for which the definition does not guarantee to capture real identity.

In her book on extensions of first-order logic, Maria Manzano prefers to treat identity as primitive in second-order logic and gives four reasons for that preference (Manzano 2005, p. 56):

1. On nonstandard structures, $\forall P(Px \to Py)$ does not guarantee to define the "prototypical" (i.e., real) identity.
2. In subsystems of calculi for second-order logic and second-order logic with abstraction that lack comprehension, not only the substitutivity of identity for individuals but also the reflexivity of identity of individuals is no longer derivable.
3. Even in standard structures the equality for *relations* could be different from real identity. (Adding equality for relations as primitive avoids that problem.)[14]
4. With identity as primitive, the combination of the comprehension and the extensionality axioms allows the definition by abstraction.

We have already seen that the consistency or inconsistency of identity as a primitive predicate is independent from the transitivity or non-transitivity of identity. The same holds true for Leibniz-identity. We may, for instance, consider David Nelson's paraconsistent constructive logic N4, see, for example, Almukdad and Nelson (1984), Kamide and Wansing (2015), Odintsov (2008), Wansing (2008). In any reasonable second-order version of N4 (or its extension by a falsity constant, $\bot$, N4$^\bot$), Leibniz-identity will be an equivalence relation, and identity will be inconsistent in the sense that the set $\{a = b, \sim(a = b)\}$ is satisfiable. If one prefers to (i) use a paraconsistent logic, (ii) keep transitivity of identity, and (iii) give up substitutivity of identity, Leibniz-identity in second-order N4 or N4$^\bot$ might seem to be attractive because in these systems both interderivability and provable equivalence fail to be congruence relations: substitutivity fails. However, working with Leibniz-identity in second-order N4 or N4$^\bot$ does not do justice to the separate treatment of truth and falsity as independent and equally important dimensions of semantic evaluation in Nelson's logics. In second-order N4 or N4$^\bot$, it would be appropriate to work with *strong Leibniz-identity*

$$x = y := \forall P((P(x) \leftrightarrow P(y)) \wedge (\sim P(x) \leftrightarrow \sim P(y)))$$

---

[14] van Dalen (2004, p. 151) introduces equality for relations in addition to equality for individuals in order to formulate the following Axiom of Extensionality (notation slightly modified): $\forall x_1 \ldots x_n(Xx_1 \ldots x_n \leftrightarrow Yx_1 \ldots x_n) \leftrightarrow X = Y$.

which is an equivalence relation and does satisfy substitutivity.[15]

One might see it as an advantage of defining identity in second-order LP that thereby the non-transitivity of identity is not postulated but follows from the failure of *modus ponens* for material implication, that is, the connective defined by $\alpha \supset \beta := \neg \alpha \vee \beta$, in LP. If one, however, considers the failure of *modus ponens* as a serious drawback of material implication in LP, this putative advantage vanishes. We believe that identity defined in second-order LP has further drawbacks even for Priest. It may be seen as a drawback for Priest that as a consequence of being explicitly defined, the satisfiability of $\{a = b, a \neq b\}$ for the explicitly defined identity predicate in second-order LP logic is inherited from the satisfiability of other pairs $\{\alpha, \neg \alpha\}$. Concerning an object that changes its properties, Priest (2010, p. 412) explains:

> Consider some object, a; and suppose, for the sake of illustration, that its properties at some time are consistent. Let P be one of these properties. Suppose that at some later time it comes to acquire, *in addition*, the property $\neg P$, all other properties remaining constant. Call the object that results b. Then even after this time, $Qa \equiv Qb$ for every Q. (Recall that $Pa \equiv Pb \models \neg Pa \equiv \neg Pb$.) Hence, $\forall X(Xa \equiv Xb)$, that is, $a = b$. But since Pa and $\neg Pb$, $\neg(Pa \equiv Pb)$; thus $\neg \forall X(Xa \equiv Xb)$. So $a \neq b$. Thus, a and b are both identical with each other and distinct from each other.

Priest here recalls the earlier observation (Priest 2010, p. 409) that $\alpha \equiv \beta \models \neg \alpha \equiv \neg \beta$ in second-order LP.[16] The set of second-order formulas $\{\forall X(Xa \equiv Xb), \neg \forall X(Xa \equiv Xb)\}$, i.e., $\{a = b, a \neq b\}$, *is* satisfiable in second-order LP. However, it is satisfiable only in models in which $(1 \in v(Pa)$ and $1 \in v(\neg Pa))$ or $(1 \in v(Pb)$ and $1 \in v(\neg Pb))$ for at least one predicate P. It thus seems that in second-order LP with Leibniz-identity, instead of identity giving rise to satisfiable contradictions, contradictions give rise to identity being an inconsistent property. The problem would be avoided if, as suggested by Hazen and Pelletier (2016), one would include in second-order LP a primitive identity of individuals in addition to the defined Leibniz-identity.

---

[15]First-order N4 extends first-order first-degree entailment logic, FDE, by intuitionistic implication and is complete with respect to a Kripke-style semantics that draws a distinction between support of truth and support of falsity at information states, see Odintsov and Wansing (2003). Adding a primitive identity predicate to N4 or N4$^\perp$ leads to a problem regarding the construction of the quotient structure of the term model. If it is assumed that for every information state, the support of truth of $a = b$ entails the real identity of a and b at that state, the model validates $\forall x \forall y((x = y) \vee \neg(x = y))$. The identity predicate of the quotient structure therefore cannot be (really) identified with real identity, which, according to van Dalen (2004, p. 171) "is definitely embarrassing for a language with function symbols." For intuitionistic logic, the problem is dealt with in van Dalen (2004) by defining a certain modification of intuitionistic Kripke models.

[16]Note that propositional as well as quantified LP is not self-extensional, i.e., mutual derivability of formulas is not a congruence relation. For atomic formulas α, β, the classical validities $\alpha \vee \neg \alpha$ and $\beta \vee \neg \beta$ are provably equivalent in LP, but their negations $\neg(\alpha \vee \neg \alpha)$ and $\neg(\beta \vee \neg \beta)$ are not.

25.7 Conclusion

We have seen that Graham Priest's thinking about identity has undergone some changes over the years. Priest's use of a primitive identity predicate for the diagonal relation in first-order LP led him to an equivalence relation that does not exclude the satisfiability of $\{a = b, a \neq b\}$. The consideration of the sameness of physical objects over time (and other phenomena) led him to the rejection of substitutivity of primitive identity of individuals and hence to giving up transitivity of identity, and in order to account for the Ship of Theseus problem, he introduced a notion of non-transitive fuzzy identity. Eventually, Priest suggested defining identity of individuals as Leibniz-identity in second-order minimal LP. This choice, though clearly motivated, is not unproblematic, in particular if it is driven to the point that second-order minimal LP is to be used as metatheory, thereby expulsing the standard notion of real identity entirely. Clearly, non-transitive "identity" is not identical with what is usually taken to be real identity, the diagonal relation. What we have been emphasizing in the present paper can be summarized as follows:

- Giving up transitivity of identity is independent of the consistency or inconsistency of identity.
- Giving up transitivity of identity is independent of whether identity is primitive or defined in second-order logic.
- The result of weakening the identity predicate by giving up transitivity is a predicate that denotes a similarity relation; set-theoretically, similarity can be understood as the relation of non-empty intersection.
- There are several varied reasons for treating identity as a primitive notion; taken together these reasons have some weight.

We prefer to use a primitive concept of identity and not to dispel the linguistic resources for referring to the diagonal relation from our object- or metalanguages.

Acknowledgements We would like to thank the audience at the workshop *Logic in Bochum II*, June 27–29 2016, and especially Graham Priest, for helpful comments on a presentation of an embryonic version of the present paper.

References

Almukdad, A., & Nelson, D. (1984). Constructible falsity and inexact predicates. *Journal of Symbolic Logic, 49*, 231–233.

Bader, R. M. (2012). The non-transitivity of the contingent and occasional identity relations. *Philosophical Studies, 157*, 141–152.

Cobreros, P., Egré, P., Ripley, D., & van Rooij, R. (2014). Priest's Motorbike and Tolerant Identity. In R. Ciuni, H. Wansing, & C. Willkommen (Eds.), *Recent trends in philosophical logic* (pp. 75–83). Dordrecht: Springer.

Crabbé, M. (2011). Reassurance for the logic of paradox. *Review of Symbolic Logic, 4*, 479–485.

Crabbé, M. (2012). Reassurance via translation. *Logique et Analyse, 55*, 281–293.

Deutsch, H. (2008). Relative identity. In E. N. Zalta (Ed.) *The Stanford encyclopedia of philosophy* (Winter 2008 Edition). https://plato.stanford.edu/archives/win2008/entries/identity-relative/.

Forrest, P. (2016). The identity of indiscernibles. In E. N. Zalta (Ed.) *The Stanford encyclopedia of philosophy* (Winter 2016 Edition). https://plato.stanford.edu/archives/win2016/entries/identity-relative/.

Frankowski, S. (2004a). P-consequence versus q-consequence operations. *Bulletin of the Section of Logic, 33*, 197–207.

Frankowski, S. (2004b). Formalization of a plausible inference. *Bulletin of the Section of Logic, 33*, 41–52.

French, S. (2016). Identity and individuality in quantum theory. In E. D. Zalta (Ed.) *The Stanford encyclopedia of philosophy* (Winter 2016 Edition). http://plato.stanford.edu/entries/qt-idind/

Gallois, A. (2016). Identity over time. In E. N. Zalta (Ed.) *The Stanford encyclopedia of philosophy* (Winter 2016 Edition). https://plato.stanford.edu/archives/win2016/entries/identity-time/.

Hazen, A. P., & Pelletier, F. J. (2006). Second-order logic of paradox. *Notre Dame Journal of Formal Logic*.

Hodges, W. (1983). Elementary Predicate Logic. In D. Gabbay & F. Guenthner (Eds.), *Handbook of philosophical logic* (Vol. 1, pp. 1–131). Dordrecht: Reidel.

Kamide, N., & Wansing, H. (2015). *Proof theory of N4-related paraconsistent logics, studies in logic* (Vol. 54). London: College Publications.

Ketland, J. (2011). Identity and indiscernibility. *Review of Symbolic Logic, 4*, 171–185.

Mackie, P., & Jago, M. (2013). Transworld identity. In E. N. Zalta (Ed.) *The Stanford encyclopedia of philosophy* (Fall 2013 Edition), https://plato.stanford.edu/archives/fall2013/entries/identity-transworld/.

Manzano, M. (2005). *Extensions of first-order logic.* Cambridge: Cambridge UP.

Mares, E. (2004). Semantic dialetheism. In G. Priest, J. C. Beall, & B. Armour-Garb (Eds.), *The law of non-contradiction: new philosophical essays* (pp. 264–275). Oxford UP, Oxford.

Noonan, H., & Curtis, B. (2014). Identity. In E. N. Zalta (Ed.) *The Stanford encyclopedia of philosophy* (Summer 2014 Edition). https://plato.stanford.edu/archives/sum2014/entries/identity/.

Odintsov, S. P. (2008). *Constructive negations and paraconsistency.* Trends in Logic (Vol. 26). Dordrecht: Springer.

Odintsov, S. P., & Wansing, H. (2003). Inconsistency-tolerant description logic. Motivation and basic systems. In V. Hendricks, & J. Malinowski (Eds.), *Trends in Logic. 50 Years of Studia Logica* (pp. 301–335). Kluwer Academic Publishers, Dordrecht.

Priest, G. (2007). *In contradiction*, 2nd ed., Oxford UP, Oxford, 2007 (first edition 1987).

Priest, G. (2014). *One. Being an investigation into the unity of reality and of its parts, including the singular object which is nothingness*, Oxford UP, Oxford.

Priest, G., Routley, R., & Norman, J. (Eds.). (1989). Paraconsistent logic. Essays on the inconsistent, Philosophia, Munich.

Priest, G. (1991a). Sorites and identity. *Logique et Analyse, 135–136*, 293–296.

Priest, G. (1991b). Minimally inconsistent LP. *Studia Logica, 50*, 321–331.

Priest, G. (1998). Fuzzy identity and local validity. *The Monist, 81*, 331–342.

Priest, G. (2002). Paraconsistent Logic. In D. Gabbay & F. Guenthner (Eds.), *Handbook of philosophical logic* (2nd ed., pp. 287–393). Dordrecht: Kluwer Academic Publishers.

Priest, G. (2003). A site for sorites. In J. C. Beall (Ed.), *Liars and heaps: New essays on paradox* (pp. 9–23). Oxford: Oxford UP.

Priest, G. (2008). *An introduction to non-classical logic.* Cambridge UP, Cambridge: From If to Is.

Priest, G. (2009). A Case of Mistaken Identity. In J. Lear & A. Oliver (Eds.), *The force of argument: Essays in honor of timothy smiley* (pp. 205–222). London: Routledge.

Priest, G. (2010). Non-transitive identity. In R. Dietz & S. Moruzzi (Eds.), *Cuts and clouds: Vaguenesss. Its Nature and Its Logic*: Oxford UP, Oxford.

Shramko, Y., & Wansing, H. (2011). *Truth and falsehood: An inquiry into generalized logical values*, Trends in Logic (Vol. 36) Dordrecht: Springer.

Shramko, Y., & Wansing, H. (2009). The slingshot argument and sentential identity. *Studia Logica, 91*, 429–455.
Skurt, D. (2017). FDE circumscription. *Australasian Journal of Logic, 14*, 326–355.
Troelstra, A. (1975). Non-extensional equality. *Fundamenta Mathematicae, 82*, 307–322.
Trueman, R. (2014). Eliminating identity: A reply to Wehmeier. *Australasian Journal of Philosophy, 92*, 165–172.
van Dalen, D. (2004). *Logic and structure* (4th ed.). Berlin: Springer.
Wallace, M. (1999). *Composition as identity*, Ph.D. thesis, Department of Philosophy, University of North Carolina at Chapel Hill, 1999. https://cdr.lib.unc.edu/indexablecontent/uuid:f4c4d877-0b88-4a72-8e7b-13a8c17c2317.
Wansing, H. (2008). Constructive negation, implication, and co-implication. *Journal of Applied Non-Classical Logics, 18*, 341–364.
Wehmeier, K. (2004). Wittgensteinian predicate logic. *Notre Dame Journal of Formal Logic, 45*, 1–11.
Wehmeier, K. (2008). Wittgensteinian tableaux, identity, and co-denotation. *Erkenntnis, 69*, 363–76.
Wehmeier, K. (2012). How to live without identity-and why. *Australasian Journal of Philosophy, 90*, 761–777.
Wehmeier, K. (2014). Still living without identity: Reply to trueman. *Australasian Journal of Philosophy, 92*, 173–175.
Williamson, T. (2006). Absolute Identity and Absolute Generality. In A. Rayo & G. Uzquiano (Eds.), *Absolute generality* (pp. 369–389). Oxford: Clarendon Press.
Wittgenstein, L. (1961). *Tractatus Logico-Philosophicus* (1922), In D. Pears & B. McGuinness (trans.), New York, Humanities Press.

Heinrich Wansing is a professor of Logic and Epistemology at the Ruhr-University Bochum, Germany. Before that he was a professor of Philosophy of Science and Logic at Dresden University of Technology (1999–2010). He took his M.A. and his Ph.D. in Philosophy at the Free University of Berlin and his Habilitation in logic and analytical philosophy at the University of Leipzig. He is the author of *The Logic of Information Structures* (Springer 1993), *Displaying Modal Logic* (Kluwer 1998), *Truth and Falsehood. An Inquiry into Generalized Logical Values* (with Y. Shramko, Springer 2011), *Proof Theory of N4-related Paraconsistent Logics* (with N. Kamide, College Publication 2015), and numerous articles in professional journals. Heinrich Wansing has been working mainly on philosophical logic, including the semantics and proof theory of modal, constructive, paraconsistent, many-valued, and other non-classical logics. Moreover, he is the editor-in-chief of the book series *Trends in Logic* (Springer), a managing editor of the journal *Studia Logica*, and a member of a number of other editorial boards of logic and philosophy journals.

Daniel Skurt is currently a postdoc working with Heinrich Wansing at the Ruhr-University Bochum, Germany. In 2017, he took his Ph.D. in Logic at the Ruhr-University in Bochum, in 2012 his M.A. in Logic at the University of Leipzig and in 2007 his Dipl.-Ing. in Electrical Engineering at the Technical University of Chemnitz. His research interests include many-valued logics, modal logics, and paraconsistent logics.

Chapter 26
At the Limits of Thought

Zach Weber

> *How wonderful that we have met with a paradox.*
> *Now we have some hope of making progress.*
>
> – Niels Bohr
>
> *Before I studied dialetheism, mountains were mountains and rivers were rivers.*
> *After studying dialetheism for some time,*
> *mountains were no longer mountains and rivers were no longer rivers.*
> *But now that I have found rest, I see*
> *mountains are mountains,*
> *and rivers are rivers.*
>
> – adapted from *Compendium of the Five Lamps 1252; cf.* (Garfield and Priest 2009)

Abstract The inclosure schema has been proposed by Priest as the structure of many paradoxes (Priest in Beyond the limits of thought. Oxford University Press, 2002). The inclosure analysis has many virtues, especially as a step toward a uniform solution to the paradoxes. Inclosure suggests that paradoxes arise at the limits of thought because the limits can be surpassed, and also not; and so dialetheism is true. I explore the consequences of accepting Priest's proposal. From a thoroughly dialetheic perspective, then, I find that the import of inclosure changes: (i) some limit phenomena cannot be contradictory, on pain of absurdity, and (ii) true contradictions are better thought of as local, not "limit" phenomena. Dialetheism leads back from the edge of thought, to the inconsistent in the every day.

Keywords Inclosure schema · Dialetheism and paraconsistent logic · Paradoxes

---

Z. Weber (✉)
University of Otago, Dunedin, New Zealand
e-mail: zach.weber@otago.ac.nz

© Springer Nature Switzerland AG 2019
C. Başkent and T. M. Ferguson (eds.), *Graham Priest on Dialetheism and Paraconsistency*, Outstanding Contributions to Logic 18, https://doi.org/10.1007/978-3-030-25365-3_26

26.1 Does a Limit Always Have the Other Side?

Are there thoughts we cannot think? Wittgenstein observes in the preface to the *Tractatus* that

> ...in order to draw a limit to thinking we should have to be able to think both sides of this limit (we should, therefore, have to be able to think what cannot be thought).

Wittgenstein is noticing that the very existence of any limit already suggests the other side. This is plausible to our spatial intuition based on common experience: for any wall or fence, there is always a far side, which the wall is blocking access to. Surely even if you came to the edge of the universe, you could stick out your arm![1] And what goes for fences, goes for thoughts. Wittgenstein intends his claim as a solemn modus tollens, but *dialetheism* takes the thrilling modus ponens direction. There *are* thoughts we cannot think—and we are thinking one of them *right now*.[2]

The central thesis of Priest's *Beyond the Limits of Thought* is that the limits to thought can be transcended—indeed, by virtue of recognizing one as a limit, we have already in some sense transcended it.[3] In this essay, I will look at some of the virtues of Priest's analysis of the limits of thought. I will then consider what happens to the inclosure analysis if one follows it out as Priest suggests, committing to dialetheic paraconsistency. I find that the philosophical direction of the inclosure schema begins to point too far outwards—past inconsistency, into absurdity. At the same time, considerations from vagueness and the sorites paradox, along with the paraconsistent mathematics of inclosure, begin to point *inwards*, at a more localized conception of contradiction. Therefore, while a common conception of dialetheias has them as rare and far away,[4] my considered dialethetic view is that contradictions cannot occur at extreme limits, and may occur well before any limit. Inclosure-based dialetheism is not dialetheic *enough*.

On a personal note, I read Priest's' *Beyond the Limits* when I was finishing my undergraduate degree, and it is no exaggeration to say that it changed my life. The basic insight that barriers must always, at least conceptually, have a far side, seemed immediately true. The possibility that naive set theory itself bears this all out, and that paraconsistent logic makes it tractable, seemed worth wholeheartedly dedicating my research to. I am glad I did.

---

[1] As Simplicius reports Archytas arguing in the *Physics* (sixth century BCE).

[2] For the thinkability (or not) of contradiction, see Routley and Routley (1985, p. 210); cf. Priest (2016).

[3] "The thesis of the book is that such limits are dialetheic ... the limits of thought are boundaries which cannot be crossed, but yet which are crossed" (Priest 2002, p. 3).

[4] Not only in more modest forms of dialetheism such as found in Beall (2009), but also Priest (2006b, Chap. 8, Sect. 4). See Sect. 26.6 below.

26.2 Paradoxes: In Search of a Uniform Solution

Now many years since Russell wrote to Frege,[5] and several steps further down the research track, we have moved from observing that there *are* paradoxes, to seeking a deeper theory *of* the paradoxes. A better understanding of the paradoxes would come, ideally, through some kind of "grand unified theory" of the paradoxes, or at least an abstract characterization of a wide class of paradoxes.[6] The value of such a characterization is attached to what Priest calls the

Principle of uniform solution If two paradoxes are of the same kind, they should be solved in the same way.

Unifying the paradoxes would make it possible to deal with them all at once. Whatever works for one, will work for all. And if a proposed solution only works for some paradoxes, but the family is unified, then that solution is not adequate, "bound to appear somewhat one-eyed, and as not having come to grips with the fundamental issue" (Priest 2002, p. 166).

A unified theory would not only have predictive power, in the sense of isolating the essential conditions for a paradox to arise. Unification would provide a satisfactory *explanation*. This fits with Kitcher's *unificationist account* of explanation, that places a premium on making connections between apparently disparate phenomena.

> Science advances our understanding of nature by showing us how to derive descriptions of many phenomena, using the same pattern of derivation again and again, and in demonstrating this, it teaches us how to reduce the number of facts we have to accept as ultimate (Kitcher 1989, p. 423).

Kitcher's views are mainly about (empirical) science, but it works for mathematics too (Kitcher 1989, p. 437); cf. (Mancosu 2011, Sect. 6.2). Along these lines, a formal schematic for the paradoxes of self-reference would advance our understanding by making sense of what is otherwise a bewildering, stinging swarm. Once a pattern emerges, there is beauty in the swarm.

26.3 Inclosures

26.3.1 The Schema...

The Inclosure Schema has been proposed as the structure of the paradoxes of self-reference.[7] The basic picture is that of a space with an operation that can escape

---

[5] "I find myself in complete agreement with you on all essentials. ... There is just one point where I have encountered a difficulty...." Russell to Frege, 1902 (van Heijenoort 1967, p. 124).

[6] One can see Russell (1905) already attempting this. A more recent proposal is (Lawvere 1969), very readably exposited in Yanofsky (2003).

[7] First in Priest (1991) following Russell: "There are some properties such that, given any class of terms all having such a property, we can always define a new term also having the property

any subregion of the space, but always stays in the overall space. More formally, an *inclosure* is a pair $\mathcal{I} = \langle W, \partial \rangle$, with W a set and a function $\partial : \mathscr{P}(W) \longrightarrow W$ from the powerset of W to W such that

Existence $W = \{x : \varphi(x)\}$ exists, and $\psi(W)$;
and for all $X \subseteq W$ such that $\psi(X)$,
Transcendence $\partial(X) \notin X$
Closure $\partial(X) \in W$

When all the conditions are satisfied, a contradiction is immediate, since W is a subset of itself: thereby $\partial(W) \in W$ and $\partial(W) \notin W$.

In the canonical case, Russell's paradox, the totality is the set of all sets V and the diagonalizer is $r(X) = \{x \in X : x \notin x\}$; the contradiction is that $r(V) \in V$ and $r(V) \notin V$; details in Sect. 26.4.2.1 below. It is no surprise that Russell's paradox is an inclosure. Russell abstracted the core of the inclosure schema from his paradox. Priest generalizes it in a natural way, whereby, contra Ramsey, inclosure applies both to the set-theoretic paradoxes and to semantic paradoxes like the liar.[8] And the inclosure schema is apt for modeling all sorts of phenomena, not only within mathematics and formal semantics, as the vast sweep of topics in *Beyond the Limits* suggests.

The idea that the *sorites paradox* fits the inclosure schema came well after even the second edition of *Limits*.[9] The sorites is not a paradox of self-reference, so it fitting the schema expands the meaning of being an inclosure paradox. But it is a natural addition.[10] If φ is a vague predicate, like being a heap of sand, then two quantities of sand that differ by only a grain do not differ with respect to being a heap. Vagueness means that

if one satisfies φ, so does the other—the principle of tolerance (Priest 2010, p. 70).

That is how closure is satisfied. The principle of tolerance does look very much like a closure principle. It has the right "feel" for an inclosure: novel objects outside (but nearby to) a totality of φ things get pulled back in, exactly because the predicate tolerates some stretching. We will look more closely at the details of sorites in Sect. 26.4.2.4 below.

Inclosure is a diagnostic tool. It shows how a paradox arises, not what to do about it (Weber 2010a). In the case of the set-theoretic and semantic paradoxes, though, seeing is believing: once you watch the inclosure form, you begin to see that the result—a true contradiction—is inevitable.

---

in question" (Russell 1905, p. 142). Cf. (Landini 2009). Priest further generalizes what he calls "Russell's schema" by adding the condition that $\psi(W)$.

[8] In the case of the liar, the carrier set is $T = \{p : p \text{ is true}\}$. The diagonal ℓ takes subsets X of truths to the sentence "this sentence is not in X." Either $\ell(X) \in X$ or not. If $\ell(X) \in X$, what $\ell(X)$ says would be false—but every sentence of X is true, since X is a subset of T. So $\ell(X) \notin X$ by reductio. (Here and throughout, reductio is the rule: if p implies $\neg p$ then $\neg p$.) Thus $\ell(X)$ is true. Therefore, it is a member of T. Contradiction: $\ell(T)$ is the liar sentence. It is both in T and not.

[9] I think it was suggested first by Colyvan at the Australasian Association of Philosophy meeting in Armidale 2007; see Colyvan (2008).

[10] I presuppose familiarity with the sorites paradox; see Weber (2010c).

26.3.2 ...and Its Virtues

The inclosure analysis of the paradoxes is both wide and deep. If it succeeds, inclosure is explanatory, across two dimensions. It unifies a wide range of otherwise disparate entities, and provides a clear story about what they have in common. With respect to Kitcher's unificationist account of explanation, the inclosure explanation provides a basic pattern for the underlying mechanics of contradiction. If inclosure is right, the paradoxes arise as a result of the way diagonalizers behave in a closed space; the inconsistency is the collision at the boundary.[11] By comparison, other descriptions of the paradoxes tend to be in terms of symptoms, or effects (e.g., self-reference), without getting at the root cause. Priest compares it to understanding a volcano simply as the eruption at the top, versus the geothermal activity occurring deep below (Priest 2002, p. 279).

> Once one understands *how* it is that a diagonaliser manages to operate on a totality of objects of a certain kind to produce a novel object of the same kind, it becomes clear *why* a contradiction occurs at the limit (Priest 2002, p. 136).

As ever, better understanding helps avoid overreactions. In contrast to orthodox set theory, inclosure points away from a blanket ban on all quantitatively "overlarge" sets, toward a more qualitative and subtle understanding; cf. (Priest 2013). Paradoxes are about "shape" rather than "size". Perhaps most pleasingly, inclosure works at a higher order of explanation. It shows what is *compulsive* about true contradictions, what makes them so hard to *solve*. That is, inclosure models the much-remarked on notion of 'revenge'. For example (see Priest 2002, Sect. 9.6), Russell's solution to the paradoxes is captured in the slogan: "Whatever involves *all* of a collection must not be one of that collection." This solution, like most that deny closure, is self-undermining in a simple way.

> By his own theory, Russell's solution cannot be expressed [Transcendence]; but he does express it [Closure]. ...In trying to solve the problem, Russell just succeeds in reproducing it. His theory is, therefore, less of a solution to the contradiction at the limits of thought than an illustration of it (Priest 2002, p. 140).

Inclosures are models not just of pathological sentences or the like, but of the dynamics of those pathologies propagating in our *solutions* to the paradoxes.

The story is attractive.[12] All the semantic and set-theoretic paradoxes of self-reference are inclosure contradictions. Coupled with the principle of uniform solution, this makes dialetheic paraconsistency highly explanatory and leaves most other solutions looking hopelessly incomplete. More expansively, inclosure looks (or at least, at one point looked to me) like an expression of Nietzsche's revelation: "And life itself confided this secret to me: 'Behold,' it said, 'I am that which must always overcome itself'" [On Self-Overcoming, in *Thus Spoke Zarathustra* (1883)].

---

[11]"An immovable force meets an irresistible object; and contradiction, in the shape of an inclosure, is the result" (Priest 2002, p. 233).

[12]Of course, not without its critics, e.g., (Abad 2008; Badici 2008; Dümont and Mau 1998; Zhong 2012).

26.4 Problems with Inclosure

Except. Two aspects of inclosure arguments jump out.[13] First, they are *informal* arguments using familiar logical steps, such as the law of excluded middle, reductio, and perhaps some other controversial moves like contraposition. The arguments look valid, at least classically, but they are not laid out as formal proofs. Second, it is not clear that the contradictions established really are the right targets. In the case of the Russell paradox, the inclosure is that the Russell set is both a set and not—meaning that the contradiction is really about V, the set of all sets, and the fact that $\in$ is inconsistent over V. But this is a rather sophisticated way of seeing things. One might have thought that the basic paradox is that the Russell set is a member of *itself* and not, that the contradiction is essentially only about the Russell set, and the fact that its membership is internally inconsistent. The inclosure analysis makes the paradoxes "external" to the diagonal objects, as a core aspect of the explanation.

Let us take these issues—the validity of inclosure arguments, and what they argue—in turn.

26.4.1 Informal Arguments

As Priest notes, with a little finesse almost anything can be made to fit the inclosure schema (Priest 2002, Sect. 9.5). How do we determine which paradoxes really do fit?

To get a better grip on the problem, take two examples that are the right shape for inclosures, but are not genuine paradoxes. Consider all the actions that can be performed by an omnipotent being. Famously, He can create a stone (closure) so heavy He cannot lift (transcendence), and then can both lift it and not. This would be a paradox, except that the preponderance of evidence is that no such being exists. Or for a mathematical example, take the natural numbers $\mathbb{N}$; for any subset X of numbers, take as diagonalizer the least natural number n greater than any member of X; then the least number greater than any number would be a number and not—except that there is no reason to think that $\mathbb{N}$ is closed under any such proposed diagonalizer.[14] In both cases, we would need independent, sound arguments to make these real inclosure paradoxes.

According to Priest, satisfaction of the inclosure conditions needs to be *prima facie* (or perhaps *a priori*) plausible (Priest 2002, Sect. 17.2); cf. (Weber et al. 2014). Prima-facie-validity will vary from reasoner to reasoner, though, depending on their logical training. And more importantly, "seeming" valid is not a reliable guide to truth. Once one gets past initial diagnostic devices, more precise and reliable instruments than mere prima facie plausibility are needed. The way to distinguish a genuine

---

[13] Both are mentioned at (Priest 2002, Sect. 17.2). Some of the points I make below may be related to criticisms that Priest rebuts in that section.

[14] As opposed to the ordinals more generally, where the existence of such a diagonalizer is essential for the existence of the ordinals—so the result, Burali-Forti's paradox, is a genuine dialetheia (Priest 2002, Chap. 8).

contradiction from a contradiction-shaped joke is to show that the conditions of inclosure are *genuinely* satisfied.

What does genuine satisfaction mean? Let us suppose the following, as a hypothesis.[15] Supposing that the inclosure analysis of the paradoxes is correct, then the best (though not only) response is dialetheic paraconsistency. Some contradictions really are true, and logical consequence really does not validate explosion (*ex contradictione quodlibet*) or its correlates. Moreover, if logic is content neutral, the same logic should always be in force. This means we should be using a fully paraconsistent metatheory, all the way up, with only paraconsistently valid inferences. Any appeal to non-paraconsistent inferences in some "metalanguage" is anathema, since after all

> ...the whole *point* of the dialetheic solution to the semantic paradoxes is to get rid of the distinction between object language and meta-language (Priest 1990, p. 208, emphasis original).

If we are trying to explain paradoxes, *all reasoning must be paraconsistently valid*. That is, we suppose that "genuine satisfaction" of the inclosure schema reduces to the inclosure conditions being provable (deductively justified and true) in a paraconsistent setting. If this can be done, the inclosure is genuine—a true contradiction!—and if not, not.[16] The details of different paraconsistent logics differ but a large family of them share certain core features: the target contradiction must be established without using rules like disjunctive syllogism or invalid contrapositions;[17] and it must be established without at the same time proving *triviality*. The logic I have in mind has an LP extensional base plus a conditional that obeys modus ponens (e.g., see Priest 2006b, Chap. 18). The framework would include full naive set theory, too, in the vicinity of Routley (1977) or Weber et al. (2016).

Once this program is taken on board, we return to Russell's paradox, an inclosure if there ever was one, and consider how the argument works.

26.4.2 Formal Arguments

26.4.2.1 Russell's Paradox as Inclosure over "The Universe" Is Invalid

Let us check the details of how the argument that Russell's paradox fits inclosure is meant to go (Priest 2002, Sect. 9.1). The existence condition is satisfied by the set of all sets

---

[15]For an argument, see Weber et al. (2016).

[16]This is complicated by issues in paraconsistent model theory. (Thanks to a referee for complicating it.) As discussed in Weber et al. (2016), there is a sense in which all arguments are invalid, because of the existence of a "trivial" counterexample. So the connection between *absence of proof* and *existence of counterexamples* via a completeness theorem is lost, and it is harder to say what "*p* is not provable" amounts to. For the purposes of this paper, we take a down-in-the-dirt approach. "Provable" means the existence of a sound argument (without consideration of what classically inclined people sometimes ask for—that the argument is "only" valid, not also invalid); and "not provable" means provisionally that all attempted proofs so far use some invalid step.

[17]Cf. Priest (1989) in Priest et al. (1989).

$$V = \{x : x \text{ is a set}\}$$

The property ψ is vacuously filled by $V = V$. Consider the function

$$r(X) = \{x : x \in X \text{ and } x \notin x\}$$

that picks out all the non-self-membered members of any $X \subseteq V$. Either $r(X) \in r(X)$ or not. If $r(X) \in r(X)$ then $r(X) \in X$ and $r(X) \notin r(X)$; so $r(X) \notin r(X)$, by contraposition. Then for all $X \subseteq V$,

$$r(X) \notin X$$

because otherwise $r(X) \in r(X)$, which we've just showed isn't so. This establishes transcendence. But also

$$r(X) \in V$$

just because everything is in V. And that's a contradiction: when the diagonal hits $r(V) = \{x : x \notin x\}$, which is *the* Russell set, we have $r(V) \in V$ and $r(V) \notin V$. The Russell set is not a set.

But when we review the reasoning, checking for satisfaction rather than prima facie plausibility, the argument breaks down. If $r(X) \in \{x \in X : x \notin x\}$, then $r(X) \notin r(X)$; so $r(X) \notin r(X)$ because the law of excluded middle and argument by cases are valid. That validates reductio. This is fine, but not enough to show that $r(X) \notin X$. For suppose otherwise, that $r(X) \in X$. Then $r(X) \in X$ and $r(X) \notin r(X)$ both hold, so $r(X) \in r(X)$ by definition. That would mean $r(X)$ is inconsistent. But that is just a contradiction about $r(X)$, not proof that $r(X) \notin X$. Even if contraposition is permitted (and we ignore any concerns about contraction), we are only working with the implication.

If $r(X) \in X$ and $r(X) \notin r(X)$, then $r(X) \in r(X)$ and $r(X) \notin r(X)$.

All contradictions are at least false, so falsity-preservation backwards does mean that one of the premises is false—but we can't isolate which one without using disjunctive syllogism.[18] The argument to transcendence over X fails paraconsistently.[19]

---

[18] Cf. (Priest 2013, p. 1274), where working with contrapositions is treated semantically.

[19] Anticipating this, Priest says that "in case the inferences used to establish Transcendence and Closure are not always dialetheically valid, we may define the conditions more cautiously" (Priest 2002, p. 130, footnote 7). The idea is that disjunctive syllogism *is* paraconsistently valid, in the form

$$p, \neg p \vee q \therefore (p \& \neg p) \vee q$$

And from there, transcendence can be (in words)
  either the diagonal of x is not in x, or some (other) contradiction is true
and closure can be
  either the diagonal of x is in the totality W, or some (other) contradiction is true

26.4.2.2 Russell's Paradox Is an Inclosure *over Itself*

A simple version of the Russell paradox *can* be made to fit the inclosure schema, in a way that is paraconsistently valid (Priest 2002, p. 130, footnote 8). Let r itself be the totality W and the diagonalizer just be the *identity*, $\partial(x) = x$. With this rendering, a contradiction follows by paraconsistently valid reasoning: for any $x \subseteq r$,

- if $x \in x$ then $x \in r$ and so $x \notin x$; so $x \notin x$ by reductio, so $\partial(x) \notin x$ (transcendence);
- but then $x \in r$ by definition, so $\partial(x) \in r$ (closure).

This is, effectively, the version of the Russell contradiction that I think is true, and the "shape" of true contradictions I endorse below. Whether this is a meaningful instance of inclosure though is dubious. The diagonalizer is doing nothing; all the energy is coming from the totality, top-down, rather than an inevitable contradiction surging *toward* the totality. All the features that make an inclosure distinctive are absent. The very reasons Priest gives for the importance of the underlying mechanics of inclosure (Priest 2002, p. 279) would suggest that this instance is degenerate, or at least not explanatory in the right sort of way. If an inclosure is supposed to be like a volcano, with the diagonalizing magma erupting at the dialetheic crater at the top, then this instance of inclosure is like the extensionless "volcano" that is the geometric point at the center of the earth.

26.4.2.3 Russell's Paradox as Inclosure over the Universe Is Absurd

Consider the *universe*[20]

$$\mathcal{V} = \{x : \top\}$$

where $\top$ is the True: everything entails $\top$. Set theoretically, $\top$ can be played by the formula $\exists y (x \in y)$, collecting everything that has at least one property. And consider again the diagonal $r(X)$. Closure is automatic, $r(X) \in \mathcal{V}$, because $\mathcal{V}$ is the universe. Transcendence, though, would be

$$r(\mathcal{V}) \notin \mathcal{V}$$

And the target contradiction will then be that the diagonal of W is both in W and not, or else some other contradiction is true. (In the case of the Russell paradox, the "other" contradiction is that $r(X) \in r(X)$ and not, for arbitrary subset X.) This is not far off from what I will suggest in the next sections—that the true contradiction may be located somewhere other than the "edge" of the totality W. But this is not merely a more cautious formulation of the inclosure schema. It amounts to saying: *either* the paradoxes lead to contradictions at the limits of an inclosure ... *or* maybe they lead to some other contradictions somewhere else, for some other reason.

[20] The "universe" V and the universe $\mathcal{V}$ have the same extension, but not the same anti-extension. They are different sets. See Weber (2010b) for further details.

which entails $\neg\top$, or

$$\bot$$

But from $\bot$, everything follows. It can be played by the formula $\forall y(x \in y)$, stating that x has every property, and thus the property φ for any sentence φ whatsoever. $\bot$ is absurdity: "that than which no sillier can be thought" (Slaney 1989, p. 476).

There cannot be an inclosure on the (true) universe.[21]

26.4.2.4 The Sorites Paradox as an Inclosure Is Invalid

Looking at the sorites makes it easier to understand what is going on with the Russell paradox. I just present the problem here, following (Priest 2010, p. 70), and comment on it in Sect. 26.6.

Take a sequence A of objects $a_0, \ldots, a_n$ where $\varphi(a_0)$ and $\neg\varphi(a_n)$, for a vague predicate φ. Then the carrier set $W = \{x \in A : \varphi(x)\}$ of the inclosure is just a subset of this sequence, and a proper subset since $a_n \notin W$. That satisfies existence. Transcendence is immediate, since we are dealing with a finite sequence of objects: for any $X \subseteq W$, there is a first A not in X, call it $\partial(X)$. But how to get closure? For a nonempty subset X, its diagonal is the first thing after some $a_i \in W$. Then because φ is vague, $\partial(X) \in W$, too, by the principle of tolerance.

But how to formulate tolerance? If it is with a *conditional*

$$\varphi(a_m) \Rightarrow \varphi(a_{m+1})$$

that satisfies modus ponens, then the satisfaction of the sorites paradox is a disaster: by mathematical induction $\forall m \varphi(a_m)$. *Everything* is φ. Since no one is prepared to accept that everything is a heap of sand, the (conditional) sorites paradox cannot be solved in the same dialetheic way as the other paradoxes. If it fits the inclosure schema, this violates the principle of uniform solution and undermines the inclosure schema as a diagnostic tool. Compare this with the Russell argument in Sect. 26.4.2.3, where fitting the inclosure points to absurdity.

If tolerance is formulated materially, meantime,

$$\neg\varphi(a_m) \vee \varphi(a_{m+1})$$

then crisis is averted, just because disjunctive syllogism is invalid. Now, though, the reasoning for closure in the inclosure schema breaks down. Take a_X to be the last member of some $X \subseteq W$. Then $\neg\varphi(a_X) \vee \varphi(\partial(X))$, by material tolerance. And $\varphi(a_X)$. But this just gives us

$$(\varphi(a_X) \wedge \neg\varphi(a_X)) \vee \varphi(\partial(X))$$

---

[21] This is closely related to Curry's paradox, which is a long-running problem for dialetheism, and which would take us too far afield to address here. Priest's position is that Curry paradoxes "have nothing to do with contradictions at the limit of thought" (Priest 2002, p. 169); for recent debate, see Weber et al. (2014), Beall (2014), Priest (2017).

and in particular, no way to conclude (without disjunctive syllogism) that $\varphi(\partial(W))$. The limit contradiction does not obtain, and the sorites does not fit the inclosure. The sorites *does* give rise to a contradiction, but not necessarily the inclosure-intended one. We have that, for *some* φ-thing, it is both φ and not—but that thing need not be the boundary. Compare this with the Russell argument in Sect. 26.4.2.1, where local contradictions do not suffice for a global one.

26.4.3 Whither Inclosure?

These cases show that the possibility of local contradictions—before reaching the inclosure limit—either blocks inclosure arguments, or renders them too general to be explanatory. And the utter impossibility of absurdity means transcendence needs to fail against some totalities. I'll spend the rest of the paper unpacking these claims, first considering the meaning of absurdity, and then the transition to locality.

26.5 Absurdity

According to *non*-paraconsistent logics, there is no difference between inconsistency and absurdity. All contradictory limits are equally unapproachable. A dialetheic paraconsistent view is, therefore, uniquely positioned to discern the following: some inconsistent limits can be surpassed, but some cannot. There can be inclosures on sets that contain everything, sets with the same extension as the universe, e.g., $\{x : x = x\}$. The one true universe, though, $\{x : \top\}$, cannot be escaped, on pain of absurdity. This claim is only *interesting* from a dialetheic paraconsistent point of view. There are totalities for which even the unstoppable force of diagonalization must stop.

The result in Sect. 26.4.2.3 means that the driving theme—the philosophical import—of Priest's inclosure analysis is wrong.[22] In spirit, at least, *Beyond the Limits* suggests that *every* limit of thought can be overcome—that objects no matter how immovable may still be moved—and crucially that thoughts on the "other side" of the limit (like "now I am thinking the other side of the limit") are *true*. This is okay when the limit is given by consistency. One can think the other side of the limit, whereby one is in contradiction, but still coherent. For *coherence*, though, if the set $\{p : p$ is not absurd$\}$ has inconsistent membership, then absurdity follows. The sentence "this sentence is absurd" is simply false.[23]

---

[22] Or at least, my longstanding interpretation thereof: "…it has seemed to people that though there be no greater than the infinite; yet there be a greater. This is, in fact, the leitmotif of the book" (Priest 2002, Sect. 2.0). Perhaps I got this impression most from the conclusion of the book, where Priest quotes a martial arts teacher (p. 225): "Whatever is your maximum, kick two inches above that."

[23] By the law of excluded middle, everything is either absurd or not. But if p is not absurd, that does not make it *true*.

Fig. 26.1 *17th C. woodcut, German school [modified]*

Some limits do not, as it turns out, have a far side.[24] A two-dimensional Euclidean shape has only one side. It has no backside, as Borges used to great effect in his short story "The Disc" (1975): "...the Euclidean circle, which has but one face". You may *think* about the other side of a two-dimensional disc, of course, just as you may think about the superluminal spaceship Santa is bringing for Christmas, or the koan "everything is true". But to think a thought does not mean that thought is true. Thank goodness (Fig. 26.1).

In reply, someone convinced of the "ability to *break through every barrier*"[25] can offer a simple answer to the problems outlined above. It is to deny that there is a $\bot$ particle or equivalent in our language.[26] In the truth-functional logic of LP, there is no sentence that is absurd (false on every valuation). Perhaps logics without $\bot$ are friendlier to dialetheism?

If we drop absurdity from the language entirely, there may be some interesting directions to pursue, but I suspect the problem does not go away. Notice that $\bot$ need not be an out-and-out absurdity; it just needs to be something that is in no way (actually) true.[27] There must be something that is false-and-in-no-way-true, even just contingently, or there was no reason to adopt a paraconsistent logic: some things are

---

[24] More recently, Priest discusses **everything** and its complement, **nothing** (Priest 2014, Sects. 4.6, 6.13, boldface original). Priest says that we should expect them to be (self) contradictory. I think it is clear here that **everything** is a collection that, while including absolutely everything, also can coherently not include some things, so it is not the true totality of $\top$; by the same token, **nothing** is the fusion of an empty set with no members that may also include some members, so it is not the true nothingness of $\bot$.

[25] Priest (2002, p. 117) interpreting Cantor via Michael Hallet.

[26] Cf. Casati and Fujikawa (201x).

[27] Like this not true sentence: "It was the eldritch scurrying of those fiend-born rats, always questing for new horrors, and determined to lead me on even unto those grinning caverns of earth's centre where Nyarlathotep, the mad faceless god, howls blindly in the darkness to the piping of two amorphous idiot flute-players" [H.P. Lovecraft, "The Rats in the Walls" (1924)].

not (actually) true, even if some contradictions are. And whatever that false-only sentence is demarcates a line that we will not cross. The negation of that sentence defines our universe.

Priest points out that rationality does not rely on consistency (Priest 2006a, Chaps. 6, 7); inconsistency is not the only reason why we might deny something. The moon is not a frog, and is in no way also a frog—not because of logic but because of the way the moon is, and the way frogs are. The main evidence that no moons are also frogs is *not* that it would be a logical contradiction for an astronomical body to be a short-bodied amphibian. Someone who needed deductive logic to make this observation is making a kind of mistake. By the same token, the reason that absurdity is beyond the ultimate horizon of thought is not that it would be *inconsistent* to cross that boundary. You can go up to the edge of absurdity and stick out your head, but you won't find enlightenment; you'll get your head cut off.[28]

Some limits to thought cannot be transcended, or even approached too closely. Absurdity is not a limit with a far side.

26.6 Locality

The ways in which paraconsistent logic recasts the inclosure reasoning tell us something deeper about the paradoxes. In Sect. 26.4 I remarked that the Russell paradox should *prima facie* be about the Russell set, not about the universe of all sets. The paradox should not be that the Russell set transcends V, but rather that it is self-inconsistent. Now we have formal evidence that this is correct. The argument in Sect. 26.4.2.1 establishes that Russell contradictions can occur locally. In particular, the Russell set simpliciter, $r = \{x : x \notin x\}$, generates its own simple contradiction, $r \in r$ and $r \notin r$. And *then*, as a corollary, when $V = \{x : x = x\}$ we get $r \in V$ and $r \notin V$, since by the axiom of extensionality r is a set that differs from itself with respect to membership (Restall 1992, p. 427). So the inclosure schema catches this downstream contradiction, but not at the source. We can take the source itself to be an inclosure, as in Sect. 26.4.2.2, but doing so is not very illuminating.

Similarly with the sorites, granting that the conditional form of tolerance must be false, then we are left with a nice version of the 'line drawing' form of the paradox (Hyde 2008). There must be a pair of objects that are very nearby each other with respect to φ, but also not, because one of them is φ and the other not. Then either the first member of the pair is both φ and not, or the second one is. A contradiction

---

[28] Maybe you can stick out just your finger:

> Gutei raised his finger whenever he was asked a question about Zen. A boy attendant began to imitate him in this way. When anyone asked the boy what his master had preached about, the boy would raise his finger. Gutei heard about the boy's mischief. He seized him and cut off his finger. The boy cried and ran away. Gutei called and stopped him. When the boy turned his head to Gutei, Gutei raised up his own finger. In that instant the boy was enlightened [from *The Gateless Barrier* [Mumonkan] (13th c.), trans. Senzaki and Reps].

obtains, but not at any particular extremity of the sorites sequence (Weber 201x). The contradiction is intrinsic to the local pair.

What do these examples show? A common presentation of dialetheias has them as rare and far away: spandrels, or singularities that occur only at the edge of the universe. In everyday situations, we can assume consistency, reason with classical logic, and carry on as if the paradoxes never happened.[29] Dialetheism and paraconsistency kick in only during extreme and unusual circumstances, like at the top of the ordinals, standing before Cantor's Absolute. But the maths suggests that contradictions may occur well before any limit, and cannot occur at extreme limits.

I submit, then, that any story about dialetheias as 'singularities' be revised. Paradoxes certainly are found in dramatic limiting cases—but not only there. There are contradictions much lower down, as Priest points out (Priest 2013, p. 1272). The famous paradoxes are only the noisiest of many. With vagueness a cite for dialetheias, and the ubiquity of vagueness, most paradoxes are usually just too innocuous to notice as paradoxes. (Anyone who has ever lived in a place with snow can tell you that it is white, but also that snow is often not all that white.) Contradictions already have a bad reputation, and thinking of them as the serpents where HERE BE DRAGONS only reinforces the bias.[30] Dialetheias-as-limit-cases perpetuates the idea that contradictions are marginal and therefore easy to ignore.

The reconsideration of inclosure in light of a paraconsistent metatheory shows a different way. For any data, we simply want to give the best description of it we can; dialetheism councils that sometimes, that description will be inconsistent. If this is right, then we shouldn't encourage the sentiment that an inconsistent description will be defective or bizarre, that it will involve impossible Escher towers or a psychadellic ouroboros. Dialetheic descriptions will be as mundane as bald men, heaps of sand, crossing a threshold, or comprehending a set. Dialetheism isn't a special theory for anomalies. It is the consequence of accepting very boring theses, like "collections are sets," "truth is what is the case," or even "sometimes it is raining and sometimes it is not." Contradictions may occur anywhere, as local and rather banal properties of everyday objects.[31]

Here is an objection to my suggestion. Doesn't this "local" view, in part motivated by a dialetheic account of vagueness, mean that everything is inconsistent in some sense? Almost every predicate is vague. That's a lot of inconsistency! In particular, it undermines the claim that most ordinary reasoning is untouched by paraconsistent considerations, if inconsistencies are statistically the norm. Priest rejects the presupposition (Priest 2017):

> Take a vague predicate, such as "red". The vast majority of objects are not red, and consistently so. It follows that the collection of objects which are red and not red are a very small proportion of the total.

---

[29] This is argued by Priest in (1979, Sect. 4), (1989), and (2006b, Chap. 8). An even more restricted place for dialetheias is outlined in Beall (2009).

[30] "My money's still on the dragon" (Priest 2008, p. 140).

[31] As hinted at in Beall and Colyvan (2001).

True. But take almost any object; it is a borderline case of *some* vague predicate. Any red thing is a borderline case of some sub-collection within the red things—e.g., the set of "red but also a bit yellowish orange" things. So the collection of objects x for which there is *some* predicate φ such that φx and not φx is the *entire* total. If motion and change are also sites for dialethia (Priest 2006b, Chaps. 11, 12), then similar comments apply: we are in contradiction when standing in a doorway—both in the room and not—but also every time we pass through the two-dimensional plane at the boundary of any subregion of the room. (See Weber and Cotnoir 2015.) We are (almost literally) swimming in contradictions.

To make the point more directly, there was already the following set, out there in aussersein[32]:

$$V^r = \{x : x = x \text{ and } r \in r\}$$

the universe to the degree that the Russell set is a member of itself. (Any true contradiction would do.) You are a member of this set, because you are self-identical, and $r \in r$. You are not a member of this set, because $r \notin r$, so the conjunction of "$r \in r$" with anything is false. Everything has at least one inconsistent property already, from naive comprehension alone.[33]

I take away from all this that deductive logical reasoning should simply *never* presume consistency. Dialetheism shows that non-paraconsistent logic is invalid, tout court.[34] Dialetheism offers a radical insight into the nature of the world. It confuses the position to say that one of the most basic assumptions in most of Western thought is false, but then to say that usually classical logic is just fine, because dialetheism is mostly irrelevant to the ordinary world.[35] Priest is correct that "counterexamples to inferences such as the disjunctive syllogism occur only in the transconsistent" (Priest 2006b, p. 222). But the transconsistent is when you meet your friend at "noonish", when you think about all the times you wax nostalgic, when you remark that nothing your uncle says is true. The transconsistent is *everywhere*.

---

[32] A phrase from Routley (1977), where this set also comes from. Ripley calls it the weber set (Ripley 2015, p. 559) but I think it should pretty clearly be the *routley set*.

[33] This does not, however, make everything inconsistent. It makes the non-identity of discernibles false: a and b may differ in some way without being non-identical (Weber 2010b).

[34] Minimally, we cannot presume consistency when reasoning about notoriously paradoxical objects in set theory or semantics. I might agree with Priest's *methodological maxim* (Priest 2006b, p. 116) about 'quasi-valid' principles like disjunctive syllogism and contraposition, that

> Unless we have specific grounds for believing that the crucial contradictions in a piece of quasi-valid reasoning are dialetheias, we may accept the reasoning.

But I think Priest has showed that there *are* grounds for believing that there are true contradictions everywhere, and especially in applying diagonalizers to subsets of the universe. So the methodological maxim has the modus tollens, not the intended modus ponens, effect.

[35] In a different context (Priest 2006b, p. 243), Priest sounds like he would agree: "Shapiro's objections stem from being half-hearted about dialetheism. If one endorses an inconsistent arithmetic, but tries to hang on to either a consistent computational theory or a consistent metamathematics of proof, one is in for trouble. The solution to Shapiro's problems is, therefore, not to be half-hearted, and to accept that these other things are inconsistent too."

26.7 Within the Limits of Thought

Priest's inclosure is an illuminating view of paradoxes. But under inclosure, we are only seeing contradictions as the end-result—the terminal node of a long process. If these limits can always be crossed, inclosure threatens that the diagonal will eventually drive out of even the inconsistent universe and into the endless void of noise that lies beyond. Paraconsistent mathematics directs us otherwise. A thoroughgoing dialetheic account finds contradictions to be localized.

The inclosure analysis of paradoxes, especially as it offers a unified theory, and especially when read from a classical starting point, is a compelling argument for dialetheism. To the extent that it is a *key* premise in the argument for dialetheism, that premise would appear to be (self-)undermined. However, as pointed out in Sect. 26.3 (and in Weber 2010a), that diagnosis is independent of dialetheism. Most simply, dialetheism follows from extremely simple proofs that start from the naive set comprehension sechema and end in the Russell paradox or liar paradox. Nothing I've said puts pressure on those simple proofs; the paradoxes are theorems, with or without the inclosure story. Rather what I've said is that the inclosure analysis is too classical a way of thinking about the paradoxes—that if a contradiction is going to happen, it will be as some eschatological apocalypse at the end of time. Inclosure points to contradictions, but it points *away*. Once we have climbed up over the inclosure, the direction is reversed. The paradoxes point to contradictions *inward*.

I do think the need for an *explanation* of the paradoxes is freshly opened, or perhaps the meta-question of whether there even should be such a thing as the 'explanation for the paradoxes', since after all according to dialetheism *there are no true contradictions* (Priest 1979)—but one thing at a time. With a fully paraconsistent place to stand, we are better positioned to proceed. We can stop looking for the explanation at the edge of the universe. Once we see that there are contradictions at the liminal edges of coffee cups, and that they are still just coffee cups, that is when we can stop transcending these propositions and see the world aright.[36]

Acknowledgements Versions of this paper (some with entirely different conclusions) were presented at the University of Melbourne, the University of Canterbury, the University of Otago, the University of Auckland, and the University of Kyoto—thanks to participants and collaborators at those places. Thanks to the editors of this volume, and two anonymous referees who made detailed and helpful suggestions. Research supported by the Marsden Fund, Royal Society of New Zealand.

Appendix: Double Inclosures

There is fun to be had with inclosure structures.

A *double inclosure* is a pair $\langle \mathcal{I}_0, \mathcal{I}_1 \rangle$ such that

$$\mathcal{I}_0 = \langle W_0, \partial_{W_0} \rangle, \quad \mathcal{I}_1 = \langle W_1, \partial_{W_1} \rangle$$

---

[36]Following the Pears and McGuinness translation of the *Tractatus* 6.54.

are inclosures, where each carrier set is the *complement* of the other:

$$\overline{W_0} = \{x : x \notin W_0\} =: W_1$$

If $W_0 = \{x : \varphi(x)\}$ then $W_1 = \{x : \neg\varphi(x)\}$, and $\overline{W_1} = W_0$ by double negation elimination. Basic calculations show that the members of the double inclosure pair share their limit contradictions:

$$\partial(W_0) \in (W_0 \cap \overline{W_0}) = (W_0 \cap W_1) = (W_1 \cap \overline{W_1}) \ni \partial(W_1)$$

So there is an overlapping and non-empty boundary between the two complements. Each of these inclosures can be said to be dual to the other.

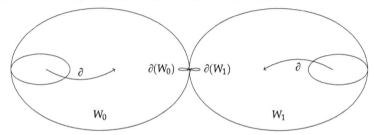

Not all inclosures have a dual. Mirimanoff's paradox, for instance, concerns the set of all well-founded sets; but rather like the truth teller (this sentence is true) the set of all non-well-founded sets does not support an inclosure; see Barwise and Moss (1996). Nevertheless, double inclosures exist. If we take, for example, W_0 to be the set of all true sentences, then W_1 is the set of all falsities. For any $X \subseteq W_1$, the sentence $\partial_{W_1}(X) =$ "this sentence is in X" is not in X. (If it were, then it would be false, so it is not, by reductio.) And then ipso facto it is in W_1. But then $\partial_{W_1}(W_1)$ is "this sentence is false" which is in W_1 by usual reasoning. So the dual structure to the liar inclosure is also an inclosure.

What else might this model? Priest has argued that the sorites paradox fits the inclosure schema (Priest 2010) (Sect. 26.4.2.4). Sorites paradoxes concern vague predicates, and if φ is vague, then $\neg\varphi$ is vague, too. Insofar as the inclosure schema matches the sorites, a double inclosure might express it nicely.

Or take an *inclosure chain* to be a (decreasing) sequence of inclosures

$$\langle \mathcal{I}_0, \mathcal{I}_1, \mathcal{I}_2, \ldots \rangle$$

such that

$$W_0 = \{x : \varphi_0(x)\}$$
$$W_1 = \{x : \varphi_0(x) \& \varphi_1(x)\}$$
$$\vdots$$
$$W_n = \{x : \varphi_0(x) \& \ldots \& \varphi_n(x)\}$$
$$\vdots$$

where each W_i carries an inclosure. Then $W_0 \supseteq W_1 \supseteq W_2 \ldots$ by conjunction–elimination, and the diagonalizer ∂_{n+1} on $\mathcal{I}_{n+1}$ is the diagonalizer ∂_n on $\mathcal{I}_n$ restricted to the carrier set W_{n+1}. So each $\mathcal{I}_{n+1}$ is a *sub-inclosure* of $\mathcal{I}_n$. There will be a corresponding sequence of contradictions at each boundary,

$$\langle \partial(W_0), \partial(W_1), \ldots \rangle$$

What might this model? Priest has argued that motion and change involve inconsistency (Priest 2006b, Chaps. 11, 12). An inclosure chain could capture Hegel's claim, that "Contradiction is the very moving principle of the world" [*Lesser Logic* (1812), par 119]—to which he adds: "and it is ridiculous to say that a contradiction is unthinkable."

Exercise for the reader on an idle Sunday afternoon. Put these ideas together for a *double inclosure chain*. [Hint: Such a chain models what happens if you think about inclosures for too long.]

References

Abad, J. (2008). The inclosure schema and the solution to the paradoxes of self-reference. *Synthese*, *160*(2), 183–202.

Badici, E. (2008). The liar paradox and the inclosure schema. *Australasian Journal of Philosophy*, *86*(4), 583–596.

Barwise, J., & Moss, L. (1996). *Vicious circles*. CSLI Publications.

Beall, J. (2009). *Spandrels of truth*. Oxford University Press.

Beall, J. (2014). The end of inclosure. *Mind*, *123*(491), 829–849.

Beall, J., & Colyvan, M. (2001). Looking for contradictions. *Australasian Journal of Philosophy*, *79*, 564–9.

Casati, F., & Fujikawa, N. (201x). Nothingness, meinongianism, and inconsistent mereology. *Synthese*. forthcoming.

Colyvan, M. (2008). Vagueness and truth. In H. Dyke (Ed.), *From truth to reality: New essays in logic and metaphysics*. Routledge.

Dümont, J., & Mau, F. (1998). Are there true contradictions? A critical discussion of Graham Priest's beyond the limits of thought. *Journal for General Philosophy of Science*, *29*(2), 289–299.

Garfield, J., & Priest, G. (2009). Mountains are just mountains. In Garfield and Damato (Eds.), *Pointing at the moon: Buddhism. Logic and analytic philosophy* (pp. 71–82). Oxford University Press.

Hyde, D. (2008). Sorites paradox. In E. Zalta (Ed.), *Stanford encyclopedia of philosophy*. http://plato.stanford.edu/entries/sorites-paradox/.

Kitcher, P. (1989). Explanatory unification and the causal structure of the world. In P. Kitcher & W. Salmon (Eds.), *Scientific explanation* (pp. 410–505). Minneapolis: University of Minnesota Press.

Landini, G. (2009). Russell's schema, not priest's inclosure. *History and Philosophy of Logic*, *30*(2), 105–139.

Lawvere, W. (1969). Diagonal arguments and cartesian closed categories. *Lecture Notes in Mathematics*, *92*, 134–145.

Mancosu, P. (2011). Explanation in mathematics. In E. Zalta (Ed.), *Stanford encyclopedia of philosophy*.

Priest, G. (1979). The logic of paradox. *Journal of Philosophical Logic*, *8*, 219–241.

Priest, G. (1989). Reductio ad absurdum et modus tollendo ponens (pp. 613–626).
Priest, G. (1990). Boolean negation and all that. *Journal of Philosophical Logic, 19*, 201–215.
Priest, G. (1991). The limits of thought-and beyond. *Mind, 100*(3), 361–370.
Priest, G. (2002). *Beyond the limits of thought*. Oxford University Press.
Priest, G. (2006a). *Doubt truth be a liar*. Oxford.
Priest, G. (2006b). *In contradiction: A study of the transconsistent* (2nd ed.). Oxford: Oxford University Press.
Priest, G. (2008). Hopes fade for saving paradox. *Philosophy, 85*, 109–140.
Priest, G. (2010). Inclosures, vagueness, and self-reference. *Notre Dame Journal of Formal Logic, 51*(1), 69–84.
Priest, G. (2013). Indefinite extensibility—dialetheic style. *Studia Logica, 101*, 1263–1275. Special issue: Advances in philosophical logic, edited by Heinrich Wansing, Roberto Ciuni and Caroline Willkommen.
Priest, G. (2014). *One*. Oxford University Press.
Priest, G. (2016). Thinking the impossible. *Philosophical Studies, 173*, 2649–2662.
Priest, G. (2017). What if? The exploration of an idea. *Australasian Journal of Logic*. Special issue: Non-classicality: Logic, philosophy, mathematics, edited by Z. Weber, P. Girard, and M. McKubre-Jordens.
Priest, G., Routley, R., & Norman, J. (Eds.). (1989). *Paraconsistent logic: Essays on the inconsistent*. Munich: Philosophia.
Restall, G. (1992). A note on naïve set theory in *LP*. *Notre Dame Journal of Formal Logic, 33*, 422–432.
Ripley, D. (2015). Naive set theory and nontransitive logic. *Review of Symbolic Logic, 8*(3), 553–571.
Routley, R. (1977). Ultralogic as universal? *Relevance Logic Newsletter, 2*, 51–89. In new edition of *Exploring Meinong's Jungle and Beyond* as volume 3, edited by Zach Weber, with contributions. Synthese Library Springer, forthcoming 2018.
Routley, R., & Routley, V. (1985). Negation and contradiction. *Revista Colombiana de Matemáticas, 19*, 201–231.
Russell, B. (1905). On some difficulties in the theory of transfinite numbers and order types. *Proceedings of the London Mathematical Society, 4*, 29–53.
Slaney, J. K. (1989). *Rwx is not curry-paraconsistent* (pp. 472–480).
van Heijenoort, J. (Ed.). (1967). *From Frege to Gödel: A a source book in mathematical logic, 1879–1931*. Cambridge, MA: Harvard University Press.
Weber, Z. (2010a). Explanation and solution in the inclosure argument. *Australasian Journal of Philosophy, 88*(2), 353–357.
Weber, Z. (2010b). Extensionality and restriction in naive set theory. *Studia Logica, 94*(1), 87–104.
Weber, Z. (2010c). A paraconsistent model of vagueness. *Mind, 119*(476), 1025–1045.
Weber, Z. (201x). Sorites in paraconsistent mathematics. In O. Bueno & A. Abasnezhad (Eds.), *On the sorites paradox*. Springer. forthcoming.
Weber, Z., Badia, G., & Girard, P. (2016). What is an inconsistent truth table? *Australasian Journal of Philosophy, 94*(3).
Weber, Z., & Cotnoir, A. (2015). Inconsistent boundaries. *Synthese, 192*(5), 1267–1294.
Weber, Z., Ripley, D., Priest, G., Hyde, D., & Colyvan, M. (2014). Tolerating gluts. *Mind, 123*(491), 791–811.
Yanofsky, N. S. (2003). A universal approach to self-referential paradoxes, incompleteness and fixed points. *Bulletin of Symbolic Logic, 9*(3), 362–386.
Zhong, H. (2012). Definability and the structure of logical paradoxes. *Australasian Journal of Philosophy, 90*(4), 779–788.

Zach Weber is a Senior Lecturer in the Department of Philosophy, University of Otago. His research is about dialetheic paraconsistency, especially with regard to mathematics and paradoxes. His Ph.D. is from the University of Melbourne. He was a postdoctoral research fellow at the University of Sydney and has been a visiting researcher at the University of Connecticut, the Czech Academy of Sciences, and the University of Kyoto. Since 2013 he has been co-organizing a series of events on non-classical mathematics, and in 2017 co-edited a special issue of the Australasian Journal of Logic, Non-Classicality: logic, mathematics, philosophy.

Chapter 27
Some Comments and Replies

Graham Priest

Abstract In this chapter I comment on and give a number of replies to matters raised in the papers on my work on paraconsistency and dialetheism in this volume.

27.1 Introduction: An Orientation

First of all, let me say a warm thank you to all those who have contributed papers to this volume. Many of them are old friends and/or colleagues and/or ex-students and/or coauthors. I feel honoured that they should have considered it worth spending the time and thought required to contribute to the volume; and I thank them warmly for the kind remarks they make about me.[1]

In what follows I shall comment on each of the papers. There is much more to be said about nearly all of them, but given the context, I shall have to restrict myself to what I take to be the central points; and generally speaking, I do not think this is the place to enter into detailed technical issues. As one might expect, I shall have more to say about some of the papers than others. I shall take the papers in alphabetical order of the authors. I have tried to avoid cross-references, though I have used these sometimes where not to do so would have resulted in large chunks of repetition.[2] Occasionally, I have permitted myself a few remarks of a more personal nature.

In what follows, I will have to refer to some of my books a number of times. To avoid prolix referencing, I will refer to them as follows: IC, *In Contradiction*

---

[1] Many thanks, too, go to all those who sent me comments on earlier drafts of the following sections, which certainly helped improve them. Since I have no better place to say it, let me also say a heartfelt 'thank you' to Can Başkent and Tom Ferguson for their initiative and all the hard work involved in producing this volume.
[2] I shall frequently refer to what is said, sometimes quoting. At the time of writing, however, I do not have the appropriate page numbers. So I will reference by section numbers.

G. Priest (✉)
Department of Philosophy, The Graduate Center, City University of New York, New York, USA
e-mail: priest.graham@gmail.com

Department of Philosophy, The University of Melbourne, Melbourne, Australia

© Springer Nature Switzerland AG 2019
C. Başkent and T. M. Ferguson (eds.), *Graham Priest on Dialetheism and Paraconsistency*, Outstanding Contributions to Logic 18,
https://doi.org/10.1007/978-3-030-25365-3_27

(Priest 1987); BLoT, *Beyond the Limits of Thought* (Priest 1995a); TNB, *Towards Non-Being* (Priest 2005a); DTBL, *Doubt Truth to be a Liar* (Priest 2006a); INCL, *Introduction to Non-Classical Logic* (Priest 2008); ONE, *One* (Priest 2014d). A '2' following the acronym will indicate the second edition. Page references are to second editions, where they exist.

27.2 Allo and Primiero (Chap. 19): Adapting Adaptive Logic

Patrick Allo and Giuseppe Primiero deliver a propositional multiple-conclusion versions of Batens' adaptive logic based on the lower limit logic *CLuN*. This is clearly an interesting technical construction. I am content to leave an analysis of it, its strengths and weaknesses, to those who understand adaptive logic better than I do. They locate the construction in the context of classical recapture. Let me say a few things about that.

Paraconsistent logics are normally proper sublogics of classical logic. In particular, the disjunctive syllogism fails: $A, \neg A \vee B \nvdash B$. Yet classical logic seems to work very well in many contexts—for example, classical mathematics, where (one might hope) there are no inconsistencies. It, therefore, behoves a paraconsistent logician to explain how.[3] To a certain extent, this may be done by an account of the conditional which does not identify it with the material conditional, so that the disjunctive syllogism is not necessary for applying *modus ponens*. However, this will not deal with reasoning where the disjunctive syllogism proper is used.

Adaptive logic is a beautiful and quite general strategy for classical recapture. Not only does it deliver this, but it does so in such a way as to take account of the fact that inconsistencies can have a *local* role in which inferences they invalidate. I learned the idea of adaptive logic from Batens many years ago, and used it to fashion my own adaptive system, *LPm*.[4]

LPm is a simple construction for showing how the adaptive trick may be turned. Batens' own construction is much more general, and may well have other advantages over *LPm*.[5] If so, all well and good. We are now discussing the best way in which the classical recapture can be carried out, not *that* it can be.

Allo and Primiero locate their construction (Sect. 19.6) in the context of Beall's recent thoughts about classical recapture.[6] Here, I agree with them entirely. Beall suggests doing away with the disjunctive syllogism altogether. Where one might normally think to employ it, what one has is simply a choice between the conclusion

---

[3] The methodological significance of this was pointed out as long ago as IC, 8.5.
[4] See Priest (1991).
[5] As I point out in Priest (2017), 6.3.
[6] See Beall (2011, 2012).

and a contradiction. General considerations of rationality can determine which of these to accept.[7]

Beall is, of course, right that quite general conditions of rationality, and not just what follows from what, play a role in rational belief. However, adaptive logics show how a default assumption of consistency results in elegant non-monotonic notions of consequence, and how classical recapture can be obtained formally. I can really see no objection to this.[8] In particular, I find Beall's rejection of any notion of non-monotonic reasoning very strange. In real life, most of our reasoning is non-monotonic—or to give it a more traditional name, inductive.[9]

27.3 Batens (Chap. 8): Logic and Metatheory, a Beligian Take

Diderik Batens and I have been friends, for most of the years I have worked on paraconsistency. Many times, I have visited him and the impressive school of logicians he built up in Gent. And when it was decided to hold the first world congress on paraconsistency, Gent was the natural place for it, and Diderik organised the historic conference. Of course, we have discussed paraconsistency and related issues for all these years. I have learned much from these discussions,[10] and I'm delighted to be able to carry on our conversation here. As he says, we tend to come at things from rather different perspectives; and that difference is certainly too big an issue to take on here. We also disagree about some more particular things, though I often think that there is much more agreement between our views than he does. Often, it seems to me, it is just a matter of reorientation. We may disagree about that too!

Anyway, Batens' paper here is rich with ideas and arguments. I can take up only a few of the issues he raises. I will concentrate on what I take to be the two main topics: logical pluralism and paraconsistent metatheory.

27.3.1 Logical Pluralism and Related Issues

As far as deductive logic goes, I am a logical monist.[11] That means that given some inference there is, in principle, one correct answer to the question, when suitably understood, of whether or not it is deductively valid. Call this *the Question*. I am not against logical pluralism (the denial of monism); if it turns out to be the case, so be it. I don't think that this affects the nature of, or arguments for, dialetheism at all. It

---

[7] The strategy goes back to IC, 8.5.
[8] See Priest (2017), Sect. 6.2.
[9] See Priest (2012a).
[10] For some of the discussion, see Priest (2014a).
[11] See Priest (2001).

is just that, so far as I can see, an adequate case for pluralism has not been made; and methodologically, monism is the simpler view.[12]

Of course, I am well aware that there are many pure logics (classical, intuitionist, paraconsistent, etc.). And as pieces of pure mathematics, these are all equally good. Moreover, suitably understood, they can all be thought of as providing answers to the Question. That, of course, does not imply that there is more than one correct answer. Neither do I think that what theory logicians take to be right cannot change over time; of course it can: it has. Nor do I wish to suggest that I or anybody else has the logic which does always answer the Question correctly—though of course, we have reasons to suppose that some theories are better than others. In all these ways, logic is much the same as physics.[13]

What does actually follow is that there is a unique deductive *logica ens*—something that may be distinct from both our theories and our practices. Batens says (Sect. 8.2) that he does not understand this. I find this rather surprising. It is, after all, what logicians have been trying to characterise since Aristotle and the Stoics. Perhaps it is impossible to *achieve* what they have been trying to do; however, *what* they are trying to achieve sees perfectly intelligible.

Naturally, inferences are expressed in language, and, as Batens notes, the languages we actually speak (as opposed to some formal language), are highly idiomatic and ambiguous. It may well be that the correct response to the Question is, 'It depends what you mean'. And this meaning may need to be clarified before a sensible answer is possible. A standard logicians' assumption is that this has already been done before the Question is addressed.[14] But once it is done, one may fairly expect a straight answer to the Question. To say that the inference is valid in classical logic, but not intuitionist logic, would reasonably be seen as avoiding the question.[15]

Of course, again as Batens points out, the meanings of our words can change over time, as, then, may answers to the question 'What do you mean?' That does not imply that the answer to the Question changes. Nor does it imply that there is no such thing as determinate meaning—Derrida notwithstanding. It just means that the results of clarification may change with time. To put it in very traditional terms, an answer to the Question requires us to determine what propositions are being expressed by the sentences involved. And of course, in the process of change, new concepts may be introduced. A new word is coined, or the meaning of an old word is revised, to express a concept not hitherto expressible. This is a way in which the *logica ens*

---

[12] I guess I've always been disposed to accept monism, just because in the early days of paraconsistency, many paraconsistent logicians, such as da Costa, endorsed pluralism as a way of attempting to legitimate what they were doing. This always struck me as a failure of nerve.

[13] For a more general discussion, see Priest (2014b).

[14] It is no accident that much philosophical discussion concerns clarification of meanings. It is a sensible thing to do before any kind of debate—and not just in philosophy.

[15] Imagine that a logician is called as an expert witness because one of the arguments used by the defence is particularly complicated. The judge asks the logician whether the conclusion of the argument actually follows.

may change. It can be augmented, simply because new propositions are coming into play.[16]

Batens ties his view to a patchwork view of knowledge (perhaps better, rational belief). Our knowledge about the world is no unified whole—and maybe never will be; it is a patchwork of sometimes inconsistent views. With this, I am completely in agreement. This is one reason why Bryson Brown and I invented the methodology of Chunk and Permeate.[17] As far as I can see, all this is perfectly compatible with what I have said above, which does not mention epistemology at all. Of course, as our theories change, new words may be coined, or old words may come to have new meanings. But as I have already said, this is quite compatible with logical monism.

Which brings me to adaptive logic, a subject for which I have enormous admiration. Adaptive logic is a species of default reasoning, or, to give it a more traditional name, inductive reasoning. This is non-monontonic, in the sense that from certain information we may correctly infer a conclusion because, in some sense, that would be the case in a normal situation. The conclusion can be rescinded, though, given further information, to the effect that we are not in a normal situation. The hackneyed example: Tweety is a bird; so Tweety flies—which is fine until and unless we learn that Tweety is a penguin.

Developments in non-monotonic logic are a great development in twentieth- century logic. Such logics can be handled semantically by having a normality order on interpretations—the valid inferences being the ones where the conclusion holds in all models of the premises which are as normal as possible, given those premises.[18]

Non-monotonic logic is not a rival to a deductive logic. Such a logic standardly has a deductive logic as its basis. (Batens calls this the *lower limit logic*.) A non-monotonic logic is built atop of this, and *extends* the valid inferences, by adding a normality ordering. The most usual non-monotonic systems take classical logic as the lower limit logic, and use various empirical default assumptions to generate the notion of normality involved.

One of Batens' great achievements was to take, instated of classical logic as the lower limit logic, a weaker (often paraconsistent) logic, to which might be added, not empirical default assumptions, but default assumptions about what one might call logical normality. Prime amongst such assumptions is one of consistency. Adaptive logic is, hence, a very general approach to non-monotonicity; and over the years, Batens and his school have clearly shown how such logics can be used for many kinds of default assumptions, including such things as monosemy—as mentioned in his paper.[19]

---

[16] Whether it can change in other ways may depend on what, exactly it is that determines what follows from what. (See Priest 2014b, Sect. 4.1.) If one takes validity to be a relation between abstract entities (such as propositions or mathematical structures of a certain kind—as in Priest 1999), then presumably not.

[17] Brown and Priest (2004).

[18] See Priest (1999).

[19] For a deductive logic which handles ambiguity of denotation, see Priest (1995b).

I have described the basis of adaptive logics in some detail so that, as I hope becomes clear, there is nothing in such a project with which I feel the need to disagree. On the contrary, I have used it in my own limited way, to construct the adaptive logic *LPm*. (See Sect. 27.2 above.) In particular, there is nothing in this project which in any way challenges monism about deductive logic—any more than inductive logic challenges deductive logic; and there is certainly nothing in this view which is anti-formal-logic: adaptive logics *are* formal logics; nor need there be any tension between adaptive logics and paraconsistent logics. Adaptive logics often *are* paraconsistent (in that, according to such logics, contradictory premises do not imply everything). Whether adopting an adaptive logic delivers a defensible solution to the paradoxes of self-reference is far too large a topic to take on here.

27.3.2 Metatheory

Let me now turn to the question of the semantics of deductive paraconsistent logic, and particularly the semantics of *LP*. This is the topic to which Batens turns in the second half of his paper.

To address matters here, there is a crucial prior issue. Batens is concerned with the model-theoretic definition of validity. This is formulated in set-theoretic terms, so one has to determine what account of set theory to endorse. A dialetheic account of the set-theoretic paradoxes accepts the naive comprehension schema:

- $\exists x \forall y (y \in x \text{ iff } A)$

where A can is arbitrary.[20] The most crucial question is how to interpret the 'iff'. There are currently two possible views on the table.[21]

One option is to take the underlying logic of the theory to be a relevant logic, and take the 'iff' to be a relevant biconditional. This approach was pioneered by Routley.[22] I have certainly contributed to this project. But undoubtedly the idea has been developed in its most sophisticated form by Weber.[23] Quite how much metatheory one can construct in this approach is still not known. However, for such an approach, the consequences spelled out by Batens in the first part of Sect. 4 do, indeed, seem to follow (if the relevant biconditional contraposes). Indeed, related issues were pointed out by Weber.[24] How damaging these consequences are would require a substantial discussion. However, I forego this here, because I am inclined to a different approach to naive set theory.

---

[20] One normally insists that x not occur free in A, but in a relevant logic, this actually implies the more general condition. The argument for this based on an underlying substructural logic can be found in Cantini (2003), Theorem 3.20. The same argument applies to a relevant logic.

[21] Maybe more if one goes substructural.

[22] Routley (1977).

[23] Weber (2010, 2012).

[24] Weber (2016a). However, Weber endorses the axiom that truth and falsity in an interpretation are exclusive, which seems to exacerbate the matter.

This is to take the underlying logic of the theory to be *LP*, and to take the 'iff' to be its material biconditional.[25] The conditional of *LP* does not detach. Hence, there is no hope of trying to prove set-theoretic theorems in this theory. One must sail on a different tack. This is itself model-theoretic.

One can prove that there are models of this theory which verify, not only the naive comprehension schema, but also *all* the theorems of *ZFC*. If one assumes that the universe(s) of set theory is (are) represented by such a model (models),[26] then one can simply take over all the result of *ZFC*, including standard metatheory—including that of *LP*.

And for such an approach, the results about validity set out by Batens do not go through. To establish the results about the relation R he uses would appear to use invalid arguments. Thus, to show the existence of an inconsistent R, one would have to reason as follows. Suppose that k is the Russell set, $\{x : x \notin x\}$, and let K be $k \in k$. Then $K \wedge \neg K$. By naive comprehension, we may define a propositional interpretation, R, such that:

- $\langle x, y \rangle \in R \equiv (x \in \mathbb{P} \wedge (y = 0 \vee y = 1) \wedge K)$

where $\mathbb{P}$ is the set of propositional parameters. One cannot, however, infer that $R(p, 1)$, $R(p, 0)$, $\neg R(p, 1)$, or $\neg R(p, 0)$, since the material biconditional does not detach.

This does not mean that logical consequence is a consistent notion, however. The argument for this was given by Young,[27] and is gestured at by Batens in the final pages of the section in question. The argument requires that the set theory should be able to establish the existence of a standard model; that is, a model $\mathfrak{M}$, such that for any formula A[28]:

- $[\mathfrak{M} \Vdash A] \equiv A$

For then, by self-referential techniques, one can find a sentence, D_0, such that:

- $\mathfrak{M} \Vdash D_0 \equiv \mathfrak{M} \nVdash D_0$

It follows in *LP* that $\mathfrak{M} \Vdash D_0 \wedge \mathfrak{M} \nVdash D_0$, and so $\neg(\mathfrak{M} \Vdash D_0 \supset \mathfrak{M} \Vdash D_0)$. Given that the validity of an inference from A to B is defined as $\forall x(x \Vdash A \supset x \Vdash B)$ (sticking to the one-premise case, for simplicity), it follows that $p \not\models p$ (though $p \models p$ as well), since the inference has a substitution instance which is a counter-model. The obvious generalisation of this argument to other forms of inference does not go through. Whether the inconsistency of the validity relation spreads further still requires investigation.

---

[25] This approach is explained in IC2, Chap. 18, and at greater length, in Priest (2017), Sects. 10–12.

[26] Or at least, such models of a relatively low degree of inconsistency—to rule out, for example, the trivial model.

[27] Young (2005).

[28] Of course, this cannot be done in *ZFC*—assuming it to be consistent. But one can show that there are models of naive set theory and *ZFC* in which this is the case. (See Priest 2017, Sect. 11.)

Batens' objection at this point is simply that the validity relation is inconsistent, contrary to my view.[29] However, as far as I can recall, I have never suggested that it was. Indeed, given that we have a standard model—and so, in effect, a truth predicate—and techniques of self-reference, this is exactly what should be expected. Neither does this seem to me to be particularly problematic. After all, the inference in question is still valid.[30]

Finally, Batens points out that worries about the inconsistency of logical consequence might spill over into worries about non-triviality—and specifically that one might be able to prove (rather trivially!) that every theory is non-trivial; that is, for any T, there is some A such that $T \nvdash A$. Given that we are allowed to assume standard results about soundness and completeness, this means that for some $\mathfrak{A}$, $\mathfrak{A} \Vdash T \wedge \mathfrak{A} \nVdash A$. Now, given Young's argument, it is true that if the language of T is that of set theory, then $\mathfrak{M} \nVdash D_0$; but in general it will not be the case that $\mathfrak{M} \Vdash T$, since $\mathfrak{M}$ is a very particular model.[31]

27.4 Berto (Chap. 2): Impossible Conceiving

Franz Berto and I have worked together on the version of noneism that he and I favour for some years now, and it has been a very fruitful collaboration. He has now started to think more about imagination, and I'm very happy to able to push that project forward. What is at issue in his paper is whether one can imagine the impossible. I certainly think you can; so does he. He thinks that things might be a bit more complicated than I have suggested, though.

I think that conceiving a state of affairs and imagining it are much the same thing.[32] To conceive of something is simply to bring a representation of that state before the mind. (As Berto notes, this does not imply that one can conceive of every impossibility, or even every possibility. There may be some states of affairs—possible and impossible—that transcend anything I can represent to myself, perhaps because they are too complex.[33]) Berto thinks that imagining is not quite the same as conceiving: it is a special sort of conceiving: conceiving as imagining.

In conceiving as imagining, the kind of representation one brings before the mind is, in some sense, a pictorial image. A paradigm example of this is when I imagine Socrates drinking hemlock. When I do this, I have a visual image of an old guy with a beard and a snub nose sitting on a couch with a cup in his hand. The image

---

[29] He also asks how results about the semantics can have been established in *ZF*, if they are inconsistent. This they cannot, since they use unrestricted comprehension, and so go beyond *ZF*.

[30] For a further discussion of the above these matters, see Priest (201+a).

[31] Moreover, even if such 'cheap' proofs of non-triviality *were* available, this does not undercut the value of more substantial proofs. For such proofs normally establish not only non-triviality, but also limitations on the class of sentences in the language which T can show to be contradictory.

[32] At least in the relevant sense of 'imagine'. There is a sense of the word in which you imagining something implies that you are not certain of it. That is not the sense in question here.

[33] On all these things, see TNB2, 9.6.

involved in conceiving as imagining does not have to be visual, however. It could be an auditory image—for example, of an orchestra playing Beethoven's *Ode to Joy*—or a kinesthetic image—for example, of my performing a karate kata. One might say that in such imagining, one runs a sensory system in the brain, 'but offline'. By contrast, in conceiving-as-not-imagining the representation is a linguistic one. Thus, I might conceive its being the case that intuitionistic logic is correct. No sensory image is involved. The representations, in this case, are linguistic statements such as the sentence 'Excluded Middle is not valid', and so on.

Now I am quite happy to say that conceiving of the latter kind *is* imagining. This sort of imagining is exactly what I do when I imagine intuitionist logic to be correct. But I am happy to agree that the representations involved in imagining can be linguistic or sensory—though I think that the distinction may be vague. Diagrams and maps, for example, seem to have elements which are both pictorial and linguistic. (Thus, when I imagine that intuitionist logic is correct, I might imagine a diagram of Kripke counter-model to Excluded Middle. There is certainly a visual image involved; but the failure of Excluded Middle is not something that one can literally see.)

Set all this aside, however. Berto thinks that conceiving as imagining puts up stiffer resistance to the thought that one can imagine the impossible. Thus, suppose that it is a necessary truth that water is H_2O, and that I imagine that it is not. I might have a mental image of this wet stuff, such that once one zooms in, the molecules have some other constitution. One might simply aver, as some have, that it was not *water* that I imagined, but some other wet stuff. Berto's reply is that even in pictorial representation, there is more to matters than the phenomenology. Thus, if I imagine that Hillary Clinton won the 2016 US election, I have a visual image of her celebrating, surrounded by streamers, etc. But how do I know that it is her, and not just someone who looks like her? Well, *ex hypothesi*, it is *her* I am imagining, not a *doppelgänger*. Similarly (as Berto notes), when I imagine that water is not H_2O, it is *water* that I am imagining, and not some *doppelgänger*.[34] As Berto puts it, even in the pictorial case, it is not like looking at the situation through a telescope; there is an element of fiat about who or what it is that are the objects in the picture.

I think that Berto is quite right about this. However, it seems to me that even without this element of fiat, one can have this kind of imagination of the impossible. We are all familiar with visual illusions. One of the most interesting for present purposes is the waterfall effect. A perceptual—usually visual—system is conditioned by constant motion in one direction. Once this is taken away, one gets a negative after-image, of whatever one is looking at moving in the other direction. But if one focuses one's attention at a point in the visual field it appears to be moving in that direction *and* be stationary as well.[35] That is exactly how subjects of the illusion describe what they see, and you will too if you undertake the experiment. Now, something's being stationary and being in motion (in the sense in question) is an impossibility. Yet I can perfectly well picture what it is like for something to be stationary and in motion,

---

[34] See TNB2, p. 195.
[35] For discussion and references, see DTBL, Sect. 3.3.

because I have seen what it is like: I have done the experiment. Yet, no element of fiat is involved in this: that's just how it looks—as through a telescope.

For good measure, I note that there are auditory phenomena of the same kind. One can generate the sound of a note that appears to be perpetually rising, but stays constant.[36] That is of course, impossible. Now, at the end of the Beatles' 'Day in the Life' the orchestra plays a note that actually does gradually ascend. But I can imagine listening to the Beatles 'Day in the Life' when the note which the orchestra plays at the end perpetually rises but remains constant. There is an element of fiat involved in this. *Ex hypothesi*, it is the *Beatles*' 'Day in the Life' that is the subject of my imagining. I am not imagining a situation in which the Rolling Stones composed 'Day in the Life'. But there is no fiat involved in the notes' doing what it does. I know exactly what it sounds like.

A final comment on granularity. I agree with Berto that representations will have a granularity. Some will be more detailed than others—though any representation is likely to be partial, and so silent on some details. However, I do think that I can 'imagine building step by step a perfectly valid proof' of Goldbach's Conjecture (Sect. 2.7). This is all in my imagination. So the validity itself may be imaginary. What I cannot do is imagine building a step by step perfectly valid proof, such that it is actually valid. Were I able to do this, then I would indeed have solved the Goldbach problem.[37] Imagining a state of affairs, as Berto notes, cannot (perhaps sadly) make it obtain.

27.5 Brady (Chap. 3): True or False (Only) Strikes Back

Ross Brady has proved many impressive results in relevant/paraconsistent logic. I still regard his proof of the non-triviality of naive set theory as one of the milestones in the development of the subject.[38] Perhaps one of the main things he and I have argued about over the years is whether the paradoxes of self-reference are best handled via truth value gaps or truth value gluts.[39] His essay in the collection puts the issue in the more general context of how many semantic values there actually are.

First, let me get some distracting matters out of the way. The paper he refers to as the one I gave in Istanbul in 2015 is, I think, Priest (2015). In that, I did not defend the idea that logic should have four—or five—values. I argued that the best way to make formal sense of the Buddhist *catuṣkoṭi* is to do so in a 4-valued logic, and specifically, *FDE*. The paper also discusses adding a fifth value, e, ineffability— though, as it briefly indicates, if one does so, one has to think of the bearers of values

---

[36] See Shepard (1964), Tenny (1969).

[37] Van Inwagen's objection fails none the less because, *ex hypothesi*, it is [building a valid detailed proof of the Conjecture] which I am imagining.

[38] This appeared as Brady (1989), but the result is much earlier. I remember that we had a whole mini-conference around it in Canberra in 1979.

[39] See IC, Chap. 1.

as states of affairs, not as sentences.[40] Given that, we are not considering *semantic* values. In particular, this has nothing to do with the fifth value as a value for linguistic nonsense of any kind, for which I hold no brief.[41] Indeed, I have never argued that we need more than four semantic values. I quite agree with Brady that we do not.

Next, Brady argues that validity is about deducibility. For me, validity is about truth- (or, more precisely, satisfaction-) preservation. This is too big an issue to take on here.[42] I note briefly only two things. First, Brady's claim (Sect. 3.2) that truth preservation is about propositions, which are, by definition, either true or false, and not both, is about as tendentious a definition as I think one might find (whether or not it is standard). Propositions (as opposed to questions and commands) are the *kind* of thing that can be true or false; but this says nothing about whether any achieve neither or both of these statuses.) Second, on a truth-theoretic semantics there is absolutely no problem about disjunction or the particular quantifier. $A \vee B$ is true iff A and B is true. Even if we know that $A \vee B$ is true, it does not follow that we should know which of these it is, much less that it be proved.[43]

Anyway, set these matters aside. Given any theory, for any A, there are, as Brady says, four possibilities:

1. A is provable and $\neg A$ is not
2. $\neg A$ is provable and A is not
3. Neither is provable
4. Both are provable

And there are theories which deliver each of these possibilities for various As. This brings me to the most significant point of disagreement concerning what Brady says in his essay. He wishes to eliminate possibility 4. I certainly do not.

He says (Sect. 3.1) that:

> the case for dropping the contradictory value [4 ...] will depend on ideal formal systems that represent conceivable concepts and would involve reconceptualising any concept or concepts that lead to contradiction.

In other words, the value is to be eliminated in an ideal situation. I suppose the most obvious thing to say is that even if the value is not present in an ideal situation, we are not normally in one, and hence we have such values. I will return to this matter in due course; but the crucial question is why one should suppose that eliminating case 4 is an ideal, that is, something for which one should strive.

The main considerations for this are marshalled in Sect. 3.3. The first is that Hilbert intended that consistency was what formal systems were meant to achieve.

---

[40] The matter is discussed at much greater length in Priest (2018a), Chap. 5.

[41] Indeed, I am happy to take (atomic) such sentences, if they occur in the language at all, simply to be false. See IC 4.7.

[42] I have discussed the matter in DTBL, Chap. 11.

[43] If anything, it is the proof-theoretic semantics which is problematic. For given such an account of validity (of the kind one finds in, for example, intuitionist logic), $A \vee B$ is provable iff A is provable or B is provable; and this may well not be the case in the relevant logics of the kind that Brady favours (as he, himself, points out).

Well, yes. That was his agenda in the philosophy of mathematics; but that was only ever one such agenda; it is now gone; and in any case, Hilbert knew nothing about paraconsistent logics.

The next observation is that ideal logical systems are meant to capture logical notions such as conjunction and negation with conceptual clarity. We may certainly agree with this; but there is nothing unclear about a dialetheic theory of negation, or about something and its negation both being provable.[44]

Next, we are told that when we have a logical system in which something and its negation are both provable:

> in order to avoid inconsistency, people would be inclined to re-examine the concepts to see if such a system can be made consistent by fixing up the axiomatisation. It would be thought that a conceptual clash between concepts would have taken place or a particular concept would have been overdetermined.

I am not sure what people Brady has in mind here—wise people, dogmatic people, twentieth-century people? Certainly, not everyone has this tendency in all cases. Even scientists feel no need to render their concepts consistent if those concepts do their job properly. The seventeenth and eighteenthth centuries infinitesimal calculus was known to operate on the basis of an inconsistent notion of infinitesimal; but no one at the time felt the need to revise this.[45]

But in any case, since this is about what should be the case ideally, the point is not about what people *do* do, but about what they *ought* to do. Why should such revisions be made? We are told that such a contradiction shows that a concept has been overdetermined—implying, I presume, that it has been incorrectly charactised. But why should it not be in the very nature of a concept to be contradictory? *Prima facie*, the liar paradox shows exactly that the notion of truth is overdetermined in this sense.

Perhaps, for some inconsistent concepts, there would be a good reason to revise them. Thus, consider the concept:

- *x has priority of way at a road junction*

If this were inconsistent, in such a way that different drivers both had priority, it would be sensible to revise it. The point of traffic laws is a practical one; and the law embedded in this concept would certainly have impractical consequences. But such considerations do not generalise. Grant that the concept of truth is inconsistent. Why should we revise it? The concept of truth serves us perfectly well as it is. It does not cause death on the roads.[46]

---

[44] In Sect. 3.4 Brady says that Boolean negation is the 'intended negation'. Intended by whom? Certainly not be me. Boolean negation is a theory of how negation behaves—just a false one. (See DTBL, Chaps. 4, 5.) Brady tells me in discussion that what he meant was that Boolean negation has 'conceptual completeness', in the sense that the conditions under which something is true also determine the conditions under which its negation is true. (They are simply the complement.) This is certainly so; but that negation behaves like this is simply part of the same false theory.

[45] It is true that the notion was abolished in the nineteenth century with the consistent notion of a limit; but this was done for quite different reasons. See Lakatos (1978).

[46] See IC, 13.6.

Later on in the same section, Brady appears to offer another argument. It is better if our concepts are conceivable, and inconsistent concepts—such as *round square*—are not conceivable. So inconsistent concepts should be revised. However, inconsistent concepts may well be quite conceivable. Something round and square may not be *visualisable*. But such is true with many quite consistent concepts, such as that of being a chiliagon. Indeed, the concept *round square* is quite conceivable. We conceive it in saying that anything that satisfied it would have inconsistent properties—we must understand it to know that this is so; and if this is not to conceive it, I have no idea what conception is. Similarly, the notion of truth, if it is inconsistent, is still quite conceivable.

In Sect. 3.5 of his paper Brady adduces one further argument for the special case of contradictions in a metatheory.[47] There can be no contradictions to the effect that something is both provable and not provable. First, I note that many statements in a metatheory, as the term is usually used, are not about provability at all. Hence, even if the point about provability were true, it would not rule out contradictions in a metatheory in general. However, I find Brady's argument for his conclusion unpersuasive. If I understand it right, it goes like this. To show that something is provable one gives the proof. One cannot show that something is unprovable in this way. To do so, one needs some quite independent kind of proof procedure (algebraic, model theoretic, etc.), one that is 'outside of the recursive proof process'. Hence 'non-proof cannot overlap with proof'. It seems to me that this does not follow at all. Indeed, the fact that proof and non-proof are to be established by independent mechanisms opens up the very possibility that they may give conflicting results.[48]

So much for case 4 of our four cases. Let me say a final word about what Brady says about case 3: neither true nor false. In Sect. 3.4 he argues that in ideal cases—albeit ones that may not be realisable—there should be nothing in this category either.

Now, first, I entirely agree with him—and not just in an ideal case, but in the actual case! Though one may make a case for truth value gaps, I have never accepted this.[49] Next, I must say that this part of Brady's paper particularly surprised me. For years, Brady has been arguing that the correct solution to the paradoxes of self-reference is to recognise the paradoxical sentences as neither true nor false.[50] Given what he is now arguing, if it be the case that a dialetheic solution to the paradoxes of self-reference does not hold in the ideal case, it is also the case that neither does his! Third, given his background assumptions, to realise the ideal situation for case 3, the theory in question has to be decidable. For many concepts, such as validity in first-order logic, that is entirely impossible. Given so, how can one be sure that

---

[47] In Sect. 3.3 he also avers that 'metatheory is currently assumed to be two-valued'. Well, it is not so assumed by intuitionists or by those who hold that metatheory can be carried out in a paraconsistent metalanguage. See, e.g. Dummett (1977), Chap. 5, and IC2, Chap. 18. But in any case, have we never heard of current assumptions being wrong?

[48] A quite different argument to the effect that something cannot be provable and not provable is given by Shapiro (2002). I have replied to this in IC2, 17.8.

[49] See IC, Chap. 4.

[50] A view which, he tells me, he still endorses.

the same is not true of case 4? Maybe in many cases, it is impossible to eliminate contradiction.[51] Indeed, this would appear to be exactly what extended paradoxes of self-reference have taught us concerning truth.[52]

27.6 Carnielli and Rodrigues (Chap. 9): Expressing Consistency (Consistently)

In their paper, Walter Carnielli and Abilio Rodrigues present natural deduction systems for the logics they call *BLE* and *LET$_J$*. Essentially, the first is Nelson's Logic N_4, and the second augments this with a classicality operator to give an *LFI*. They claim that these deliver a proof-theoretic account of meaning, and can be motivated in terms of evidence preservation. Much of the material, together with more technical details, appear in Carnielli and Rodrigues (2017). I shall leave to those who are concerned with proof-theoretic semantics the question of whether these systems are adequate for that purpose. Here, I will just comment on a few other philosophical issues they raise, end especially those connected with dialetheism.

First a preliminary issue. Carnielli and Rodrigues (hereafter, C&R) note that the use of a paraconsistent logic (such as N_4) does not commit one to dialetheism. Indeed not. This is something that I have pointed out many times.[53] Nor does using *LP* imply a commitment to dialetheism either. One may hold that the actual world is consistent, and that the interpretations in which contradictions hold represent impossible situations. Indeed, there are interpretations of *LP* which dispense with the middle 'contradictory' value altogether.[54] The interpretation of paraconsistent logics where the semantic values are given informational interpretations ('told true' and 'told false') are also well known.[55] Having said that, the mere fact that there are interpretations of a paraconsistent logic, such as theirs, which does not endorse dialetheism is not an argument against it. There may be interpretations of it which do; or, alternatively, if dialetheism is correct, this may simply be the wrong logical system. C&R do not engage with the arguments for dialetheism at all.

Next, they attribute to me views about dialetheism which I do not hold.[56] A dialetheia is a pair of sentences (or their conjunction, if you like) of the form A and $\neg A$, such that both are true. If we take 'false' to mean having a true negation,

---

[51] Further on that matter, see Priest (2014a).

[52] In Sect. 3.6 of his paper Brady briefly addresses the matter of extended paradoxes. His position, if I understand it correctly, is to resort to the Tarskian object/metalanguage distinction. But this move, apart from being problematic in its own right, as shown by Kripke (1975), does not eliminate extended paradoxes. See IC, 1.5.

[53] See, e.g. Priest (2002), Sect. 2.2.

[54] See Brown (1999).

[55] See, e.g. Belnap (1977).

[56] In fairness to them, these are views I hear not infrequently. I have no idea why this is so, since the views are without textual support; indeed, they are against textual support.

this means that A is both true and false. Dialetheism is the claim that some As are dialetheias.[57] The definition is not committed to any particular view of truth. That dialetheism presupposes no particular theory of truth is spelled out at length in DTBL, Chap. 2. There, I point out (among other things) that one can have a verificationist view of truth, that is, one that is 'epistemically constrained'—where truth is warranted assertibility—of the kind espoused by anti-realists.[58]

It is therefore mistaken to claim that 'the dialetheist claims that some contradictions are ontological in the sense that they are due to some 'inner contradictory essence of reality'.[59] Indeed, as IC2, 20.6 points out, it is not even clear that the claim that there are contradictions, in reality, makes sense.[60] For it to do so, one has to endorse some kind of correspondence theory of truth, holding reality to comprise facts or fact-like entities.[61] I have never endorsed such a view. Indeed, the only theory or truth I have ever advocated (IC, 4.5—the 'teleological theory of truth'), is anything but such a realist theory.[62] So let me say it one more time: it is not clear that a philosophically substantial claim to the effect that there are contradictions in reality makes sense; and even if it does, I am not committed to it.

I now turn to matters concerning meaning. Broadly speaking, there are two contemporary approaches to sentence meaning. One is that the meaning of a sentence is determined by its truth conditions.[63] The other is that it is determined by rules of use, and especially proof. The first line was endorsed, famously, by Frege, and later taken up by Davidson. The second was pioneered by Gentzen. I have endorsed a version of the first view.[64] C&R endorse a version of the second. (Though what view of truth they endorse is not clear to me; nor is the connection they envisage between truth and meaning. It is certainly not of the intuitionist kind, since they explicitly distinguish between warranted assertibility and truth.) That particular difference is too big an issue to take on here, though I note that nothing about dialetheism presupposes a truth-conditional account of meaning either. Indeed, given a proof-theoretic account of meaning, if our rules for the use of words (or our rules of proof)—maybe those concerning the T-Schema—are such as to establish A and $\neg A$ then dialethe-

---

[57]This is how matters are defined in IC, p. 4.

[58]On anti-realism and truth, see Glanzberg (2013), Sect. 4.

[59]Carnielli and Rodrigues (2017), p. 1. I have no idea from where they are drawing the quotation. This is certainly not something I would say.

[60]Except in the entirely banal one that for a sentence to be true it requires the cooperation both of words and of the world. Thus, 'Brisbane is in Queensland' is true because of the meanings of 'Brisbane' and 'Queensland', and of Australian geography.

[61]Matters are made worse by the fact that C&R mis-state this view. They say (Sect. 9.8): 'let us recall that a true contradiction would be made true by an object a and a property P such that a does and does not have the property P'. No. This is to import classical truth conditions where they are obviously not apt. a should be in the extension of P and the anti-extension of P, which is quite different from not being in the extension of P.

[62]And I point out that the account is compatible with different kinds of sentences having different kinds of truth-makers (IC, p. 57).

[63]Or perhaps truth-in-a-possible-world conditions, if we are dealing with modal notions.

[64]See IC, 9.4.

ism holds, since these things are true in virtue of meaning, whatever theory of truth one endorses. C&R's claim that their logic does not support dialetheism (Sect. 9.8) because one cannot have (non-trivially) $\neg A \wedge oA$ and $\neg A \wedge oA$. But one can have A and $\neg A$, so this observation seems to miss the point.[65]

Let me make one further remark about truth-theoretic views of meaning.[66] C&R, as do many people, appear to conflate truth *simpliciter* (*ts*, from now on) with truth-in-an-interpretation (*tii*, from now on). (The fact that Tarksi wrote seminal papers on both tends to abet this confusion.) *ts* is a non-relational notion that satisfies the *T*-Schema (or at least some close cousin.) *tii* is a set-theoretic relation, and it has one major function: to deliver a model-theoretic notion of validity. Of course, one might hope that there is one interpretation such that truth in it coincides with *ts*, but this may not be the case. Witness the fact that the set theory *ZFC* can give a definition of validity for its own language, but on pain of inconsistency, it cannot establish the existence of such an interpretation, by the Tarski indefinability proof.

Now, the notion of truth (and coordinately, of reference) that is at issue in a truth-conditional account of meaning is *ts*, not *tii*. C&R criticise the idea that model theory has anything to do with meaning. Quite rightly. But it was never supposed to do so. The appropriate notion of truth for this job is *ts*.

A final comment on what C&R say about evidence. They argue that the notion of validity in their preferred logic can be motivated in terms of the preservation of evidential support, by analogy with the way that validity in intuitionist logic can be motivated in terms of preserving provability. There are reasons to suppose that this motivation does not succeed. For evidence, unlike proof, is defeasible. A can provide evidence for C, but B can override this. (So 'Tweety is a bird' is evidence that Tweety flies. But 'Tweety is a bird and Tweedy is a penguin' is not evidence that Tweety flies—quite the opposite.) In other words, evidence is non-monotonic. However, in C&R's logic, if $A \vdash C$ then $A \wedge B \vdash C$. For good measure, the logic validates adjunction $A, B \vdash A \wedge B$; but there appear to be perfectly good situations where we can have evidence for A and evidence for $\neg A$, which does not provide evidence for $A \wedge \neg A$. The preface paradox is one such example.[67]

27.7 Coniglio and Figallo-Orellano (Chap. 10): Categorically Non-deterministic

In their paper, Marcelo Coniglio and Aldo Figallo-Orellano show us many of the details of the model theory of the paraconsistent logic *mbC*, a paraconsistent logic with a negation operator and classicality operator both with non-deterministic seman-

---

[65] Unless they are identifying truth/falsity with classical truth/falsity. That would clearly be entirely question begging in the context of any argument against dialetheism.

[66] Also, a minor comment. C&R claim (Sect. 9.5) that classical logic cannot be given a proof-theoretic semantics. This is false. See Read (2000). One merely has to find an appropriate notion of proof-theoretic harmony.

[67] See IC, 7.4.

tics. There are, naturally, many such systems, including the original da Costa C systems. Coniglio and Figallo-Orellano hereafter C&FO) cite plurivalent logics as another example of logics with non-deterministic semantics. I don't think that this is the right way to look at them. Plurivalent logics are logics in which sentences can take a plurality of semantic values, and if one constructs one on top of a many-valued logic, then the connectives are deterministic, in the sense that the collection of values of the inputs determines the collection of the collection of the values of the output uniquely.[68] However, this is tangential to the main concern of C&FO's paper.

This is not the place to comment on the details of their paper, so I will just say this. In the last 60 years or so, most mathematical work in model theory has been on classical logic. C&FO's paper shows that model-theoretic investigations of non-classical logics, such as paraconsistent logics, can be every bit as sophisticated as that of the model theory of classical logic. There is clearly much interesting mathematics to be undertaken here.

Let me, however, make a few philosophical comments on something central to their paper: the use of a classicality operator. Dialetheism is the view that some statements are dialetheias. As such, it is not committed to any particular view about which statements are dialetheic: that will be the concern of particular applications of the view. The major application investigated over recent years has been to the paradoxes of self-reference—so much so, that many people, I believe, think of it simply as a view concerning these paradoxes. It is not: there are many possible applications of the view; the one concerning the paradoxes of self-reference is a very important one, but it is by no means the only one. It has never even seemed to me to be the most ungainsayable. That honour surely belongs to the application concerning legal contexts, where the ability of legislatures to make things true by fiat is transparent.[69] In the end, it may not be the most profound application either. Applications concerning the limits of thought/language[70] are perhaps more so.

So suppose that one does not endorse a dialetheic solution to the paradoxes of self-reference. Then there is no reason why an appropriate paraconsistent logic should not have a classicality operator, ∘. The major objection to this is that, assuming that negation, ¬, behaves is a reasonable fashion, one may define the operator $\dagger A$ as $\neg A \wedge A\circ$, and $\dagger A$ will behave as does Boolean negation.[71] This invites the thought that $\neg A$ is not *really* negation, but some other strange operator; consequently, things of the form $A \wedge \neg A$ are not really contradictions, but something else. If the negation symbol has non-deterministic semantics, there is weight to this thought. For the semantics of natural language is compositional: semantic values of wholes are determined by semantic values of parts. This is how we are able to understand complex sentences that we have never heard before. A non-deterministic negation symbol is a perfectly

---

[68] See Priest (2014c).
[69] See IC, Chap. 13.
[70] As found in BLoT and elsewhere.
[71] For some investigation of the matter, see Omori (2018).

fine technical device, but just because of its non-compositional nature, it cannot be an adequate account of negation as it is used in a natural language.[72]

If the underlying logic is *LP*, a logic in which negation has a deterministic semantics, however, this thought is unfounded. In this logic, the truth table for negation may be written as:

A	$\neg A$
t	f
b	b
f	t

Where t is *true only*, f is *false only*, and b is *true and false*. The truth table for † is then:

A	$\dagger A$
t	f
b	f
f	t

Thus, the very semantics of † is *predicated* on the assumption that some things may be true and false, and so of dialetheism. Moreover, $\neg$ toggles between truth and falsity, just as one should require of negation; † does not. So, it is † that is not really negation, and $A \wedge \dagger A$ that is not really a contradiction. The fact that many logicians have taken classical logic to provide the correct theory of negation does not, of course, make it so—any more than the fact that many physicists took classical mechanics to be the correct theory of motion made it so. Both theories could be thought of as true only in virtue of an inadequate diet of examples, as Wittgenstein put it.[73]

If one does endorse a dialetheic solution to the paradoxes of self-reference, matters are less straightforward. For one can then formulate a liar sentence, L, of the form $\dagger T \langle L \rangle$; and if one takes the *T*-Schema to be formulated with a detachable conditional, this generates triviality.[74] One then has to hold that Boolean negation, and so †, is semantically defective in an appropriate sense. In that way, the view agrees with that of intuitionists, who also take Boolean negation to be semantically defective—though for quite different reasons. Nor does the classical logician have any reason to feel smug about this. For, together, a truth predicate which satisfies the *T*-Schema and a negation which satisfies the conditions of Boolean negation, deliver triviality. One cannot have both. And there is no doubt which of the *T*-Schema and Explosion is the more counter-intuitive. Moreover, to insist that satisfying the rules of Boolean negation is sufficient to *guarantee* a logical operator a meaning manifests a naivety about meaning which *tonk* should have disposed of once and for all years ago.[75]

---

[72] See Priest and Routley (1989a), Sect. 2.2.

[73] For example, *Philosophical Investigations*, Sect. 593.

[74] I note, however, that if it is formulated with a non-detachable conditional, such as the material conditional of *LP*, this is not the case, See Priest (2017), Sect. 8.

[75] For a fuller discussion, see DTBL, Chap. 5.

27.8 Cotnoir (Chap. 11): How to Have Your Doughnut and Eat It

Aaron Cotnoir's paper poses a problem concerning points of topological discontinuity. I entirely agree that the problem is a puzzling one, and for the reasons he gives.

He considers a consistent solution to the problem, according to which multiple points can be co-located, and a solution according to which the point of tear is a gluon of the whole which is being torn. He argues that the consistent story is preferable: the gluon story has too many moving parts. I am inclined to agree with him on this. Gluon theory was not designed, after all, to solve this problem. It seems to me, however, that there is a solution which is preferable to both. To see what this is, let us consider the most simple case.

Take a one-dimensional continuum, C, and mark a point, p, on it. Now, tear the continuum at p into a left part, L, and a right part R. Let l be the right-hand end of L, and r be the left-hand end of R. After the tear, where is p? Then there four possibilities:

- p is l and not r
- p is r and not l
- p is l and r
- p is neither l nor r

As Cotnoir points out, the first two cases are implausible. Intuitively, the situation is symmetric, and these answers are not. So we are left with the third and fourth cases.

Solutions along both lines are possible. In Case 3, $l = p = r$. Since, patently, l is not r, we require a non-transitive account of identity. Such is entirely possible, and it does not have to be connected with gluon theory.[76] However, such a solution seems to have an insuperable difficulty, as Cotnoir, in effect, points out. After the tear, L and R are disjoint. But this cannot be if p belongs to both.[77] So we are left with Case 4. It would seem bizarre to suppose that p has gone somewhere else. So it must have ceased to exist.[78] One might, I suppose, ask where it has gone. But that would be a rather silly question. It hasn't gone anywhere. The tearing just destroyed it.

Cotnoir's solution is rather different.[79] He suggests that p was, in fact, the two points, l and r, all along. They were just co-located before the tearing. The fact that

---

[76] If p is an inconsistent object (see below), we can apply the appropriate paraconsistent machinery. See Priest (2010a).

[77] In this way the fission involved here seems to be unlike the fission of an amoeba, where there appear to be no similar considerations.

[78] This does not mean that it has become a non-existent object, though that is a theoretical option. While we are on the issue of non-existent objects, a minor comment about what Cotnoir says about them in the context of gluon theory. He says (Sect. 11.1.2) that if $u \neq u$ then u does not exist, since $\neg \exists x\, x = u$. No. What follows is that $\neg \mathfrak{S} x\, x = u$. The particular quantifier is not existentially loaded. That u does not exist would be $\neg Eu$, where E is the one-place existence predicate. See ONE, §P7.

[79] Actually, he doesn't address this case. I extrapolate from what he says about the others.

two point-sized things can be located at the same point obviously poses problems, akin to those of the medieval problem of how many angels there can be on the head of a pin. If they are in the same place, how can they be distinguished? Indeed, the solution seems to reproduce the element of arbitrariness. How come l went to the left, and r went to the right, instead of vice versa? Indeed, how come they didn't both go the same way? And why suppose there were two if one will do the job?

One might, I guess, ask where l and r came from if they weren't there originally. But that question seems to make no more sense than the question of where p went to. They didn't come *from* anywhere: the tearing simply created them. And in any case, one may ask of Cotnoir's solution the reverse question: how come l and r were there in the first place?[80]

Let us call the above scenario the *fission case*. We can run this backwards. We start with L and R, and then join l and r at p, which unites L and R into C. Let us call this the *fusion case*. Since this is simply the fission case running backwards, the situation is the same in reverse. In particular, at the joining, l and r go out of existence, and the distinct point p comes into existence.

Let us now turn to the first major case that Cotnoir considers. This is where a sphere is progressively deformed until the point at the top, a, reaches the point at the bottom, b, that being the only place where the two halves of the deformed sphere are joined. The two halves are then torn apart at this location.

The phase of the progression up to the meeting of a and b is essentially the fusion case. The material to which a and b are attached is different from the linear continuum case, since it is three dimensional. However, we still have two distinct points, a and b, which move together and coalesce. Hence, a and b then go out of existence and p comes into existence, being the single point that joins the two halves of the deformed sphere. The next phase of the progression is essentially the same as the fission case. A tear is made at p generating two distinct objects. Thus, p goes out of existence, and two new points come into existence, one on each half, marking the sites of the tear. Note that there is no reason why these two points, a' and b', should be the same as a and b. They can be entirely new. We do not have to worry, as in Cotnoir's case, as to why a went one way, and b went the other. a' *just is* the point on, say, the left-hand half, and b' just is the point on the other half.

Let us now turn to Cotnoir's second major example, the sphere deforming into a torus. The first phase of the progression is the same as in the previous case. a and b move together, and go out of existence when the point of tear-to-be, p, comes into existence. The difference comes at the next stage. p goes out of existence, but instead of two points coming into existence, a continuum of points come into existence—those forming the inner circumference of the torus. The fact that more points come into existence than before changes nothing. It still cannot be the case that p is identical to each of these, or the torus would be joined at the middle. And we do not have

---

[80]Cotnoir's answer (Sect. 11.2.2), if I understand it, is that the number of co-located points is determined by the potential boundaries the location could be involved in. If this is so, it might seem more plausible to say that the points are there potentially, and are actualised (come into existence) only when the boundary is formed.

to suppose, as for Cotnoir, that the point of singularity actually housed a continuum of points before the tear (exacerbating the angel problem). Of course, one can ask why so many points came into existence in this case. The answer is simple—and is essentially the same as in the two-point case. A continuum of points is required to deliver the integrity of the post-tear structure. Since, *ex hypothesi*, that is the structure that comes into being, those are the points that come into being. Or, to put it another way: if some of the points did not come into being, we would not have a torus, but a torus missing some bits. If it is a torus that comes into being, so, therefore, must those points.

I have now explained what I take to be a better solution to Cotnoir's puzzle. It seems to me to have none of the problematic features of either of the solutions he considers.

The observant will have noticed that paraconsistency and dialetheism have nothing to do with this solution. Does this mean that they are absent from the scenario? Not necessarily. Come back to a pre-tear structure, say the linear continuum, C. L and R were still there, just united. One may, therefore, ask whether p belonged to L or R before the split. p is a boundary between the two, and boundaries are—almost by definition—contradictory objects, both separating and joining the things of which they are the boundary. It therefore seems natural to suppose that p is symmetrically poised, in both L and R, but in neither. It joins them because it is in both; and it separates them because it is in neither. So, if $x < p$ then x is consistently in L; if $x > p$ then x is consistently in R; and p is inconsistently in both.[81]

27.9 Dicher and Paoli (Chap. 18): *ST*, *LP*, and All That

Bogdan Dicher and Francesco Paoli take us into the world of Cut-free logics. Before I turn to their paper, some preliminary comments. In standard many-valued logics, there is a single set of designated values. The technique of having a different set of values for the premises and conclusion(s) is an interesting one, and allows for a logic which can invalidate principles such as Identity and Cut, which standard many-valued logics validate.

The use of such semantics by Cobreros, Egré, Ripley, and van Rooij to deliver a logic *ST*, which may provide a solution to paradoxes such as the liar, is an intriguing one. Proof theoretically, one may obtain a consequence relation which delivers the same valid inferences as classical logic. However, when there are axioms or rules of inference for distinguished predicates, such as the *T*-schema, the failure of Cut blocks the argument to triviality, since one may have things of the form $\Rightarrow A, \Rightarrow \neg A$, $A, \neg A \Rightarrow B$, but not $\Rightarrow B$.

Now, first, it has always seemed to me that this is not a way of avoiding dialetheism with respect to the liar paradox. Given self-reference and the rules endorsed for

---

[81] For a formal model, see IC, 11.3. For Cotnoir's own take on inconsistent boundaries, see Cotnoir and Weber (2015).

truth, one can construct a liar sentence, λ, in the usual way, and establish that $\Rightarrow \lambda$ and $\Rightarrow \neg\lambda$ (and so $\Rightarrow \lambda \wedge \neg\lambda$). In other words, the principles for truth establish contradictions; so we have dialetheism. From this perspective, the logical machinery is just a different way of endorsing dialetheism and blocking to Explosion.

Next, since in this logic $A, \neg A \Rightarrow B$, it is not paraconsistent, at least according to the standard definition. But since the argument from a contradiction to triviality is blocked, the conclusion one might well draw from this is that the standard definition is wrong (or at least, works only when Cut is present). A more adequate definition might be that a consequence relation, $\Rightarrow$, is paraconsistent if there are theories, $\mathcal{T}$, and sentences, A, B such that $\mathcal{T} \Rightarrow A, \mathcal{T} \Rightarrow \neg A$, but not $\mathcal{T} \Rightarrow B$.[82]

Third, is this logical machinery the best one for a dialetheist to endorse? Perhaps. The machinery obviously has its attractions. As far as the liar paradox goes,[83] it is very neat. Whether it works so well for other arguably dialetheic areas, such as law, motion, the limits of thought/language is an issue that still needs to be addressed. The main problem is, of course, that the logic is very weak. All the classically valid inferences might be available, but without Cut, much apparently unproblematic reasoning (for theories with non-logical axioms/rules) is unavailable. The same is true of the logic *LP*, of course. (I pointed this out in the very first paper I wrote on the subject.[84]) Over the years, many ways of overcoming this weakness have been investigated, such as augmenting the logic with a detachable conditional, and employing default reasoning.[85] Whether the Cut-free approach can do as well or better in the matter is an issue so far unaddressed, and far too large an issue to take on here.

This brings us to Dicher and Paoli (hereafter, D&P).[86] In axiomatic presentations of logic, there is a standard distinction between rules that are truth-preserving (like *modus ponens*) and rules that are merely validity-preserving (such as Necessitation). The latter are fine if one is just trying to generate the logical truths, but if one is using logic as an organon of proof in general, the former are required. D&P point out, correctly, that exactly the same distinction, with exactly the same point, can be made with respect to the rules of a sequent calculus. In the light of this, they show that the sequent calculus given by the authors of *ST* is incomplete with respect to the semantics, and provide an extended sequent calculus, LK^-_{INV}, which is complete. This is an insightful piece of work.

---

[82] The alternative definition of 'paraconsistent', together with the relevance of Cut (transitivity) was already pointed out in Sect. 1 of Priest and Routley (1989a).

[83] And maybe the set-theoretic paradoxes as well, though this puts the question of 'classical recapture' at centre stage.

[84] Together with ways of addressing the matter. See Priest (1979), Sect. 4.

[85] More recently, I have explored the matter of classical recapture in mathematics in terms of mathematical pluralism. See, e.g. Priest (201+b).

[86] One small side comment. D&P mention the connection between Gödel's incompleteness theorems and semantic closure (and specifically the deployment of a naive truth predicate). This is certainly a connection I have stressed before (e.g. IC, Chap. 3). However, the points made concerning the import of the theorems for dialetheism can be made without appeal to a truth predicate. See IC2, Chap. 17.

Turning to its philosophical consequences, D&P note that, modulo a very natural notion of when distinct formulations deliver the same logic, LK^-_{INV} and LP are the same logic.[87] Given this, any rivalry between ST and LP would seem to disappear. As they point out, one might contest the notion in question. Whether it is appropriate to do so, and, if so, what the upshot is, I will leave for the authors of ST to determine.

However, D&P have another card in their hand. Even supposing that ST and LP are distinct, it remains the case that the language of ST is incomplete. Just as one might expect there to be logical constants to represent the values 1 and 0, namely $\top$ and $\bot$, one should expect there to be one that represents the value 0.5. A constant behaving as does λ fits the bill nicely. In other words, one does not need the liar paradox to deliver dialetheism: it is built into logic itself.

I take all this to be an elegant way of making much more precise my initial reaction to ST—that it is very much dialetheism-friendly.

27.10 Dunn and Kiefer (Chap. 12): Heeding Firefighters

Michael Dunn is a veteran warrior of relevant logic. Our interest in relevant logic brought us together in the early 1980s, and we have been friends since then. In their paper, he and Nicholas Kiefer present a problem about conflicting information in the context of a paraconsistent logic. Before we turn to this, a couple of preliminary comments.

Dialetheism is one application of paraconsistent logic; but of course it is not the only one, as I have often pointed out. Handling information for which one cannot guarantee consistency is another.[88] Such an application might well suggest that, in the semantics of a paraconsistent logic, the values *true/false* should be thought of as *informed as (told) true*, and *informed as (told) false*. Such an interpretation has appealed to many of those, such as Nuel Belnap and Dunn himself, hailing from the US Pittsburg relevant logic group (as opposed to the Australian Canberra group).

This interpretation has well-known problems, though. The first concerns conjunction. One is often given the information that A and the information that $\neg A$, where one would not want to claim to have been given the conjoined information, as is required by the semantics of relevant logic. Indeed, the situation given by Dunn and Kiefer (hereafter, D&K) is exactly an example of such a thing.[89] The poor person fleeing from the burning hotel room is not at all inclined to infer that there is one exit which is both to the left *and* to the right. The informational understanding of semantic values motivates much more naturally a discussive paraconsistent logic, where conjunction introduction fails.[90] A second problem besets disjunction: one is

---

[87] The result was proved in Pynko (2010). The connection between LP and LK minus Cut was, in fact, already noted in *IC*, p. 78.
[88] See, e.g. Priest and Routley (1989b), Sect. 4ζ, Priest (2002), Sect. 2.2.
[89] As is the paradox of the preface. See IC, Sect. 7.4.
[90] See Priest (2002), Sect. 4.2.

often informed that $A \vee B$ without being informed either that A or that B. But on the favoured Belnap/Dunn interpretation, if $A \vee B$ is 'told true' so is either A or B.

Moreover, there is nothing about the more standard reading of the semantic values as *true* and *false* which commits one to dialetheism. These are values that a sentence has in an interpretation (or even a world of an interpretation). There is nothing to imply that interpretations in which things are both true and false represent *actual* situations. In logic, one reasons about many sorts of situation: actual, possible, and maybe impossible too. A non-dialetheic advocate of relevant logic may hold that the interpretations in which things are both true and false are of the latter kind.[91]

With these preliminary remarks out of the way, let us turn to D&K's Paradox of the Two Firefighters. Actually, I wouldn't call this a paradox. There is nothing paradoxical about the situation in which the escaper-to-be finds themself. It could be all too real. Still, there is certainly a puzzle here. What should the person do in the face of the contradictory information, and why? Intuitively, without information, the person should choose from the three directions at random. If only one firefighter speaks, they should go that way. If two speak, they should choose between the two directions at random. The question is, then, why? D&K suggest two possible solutions to the problem, one of which uses a paraconsistent logic, and one of which uses classical probability theory.

Although D&K do not set things up this way, this is really a problem of decision theory ('there is nothing like the pressure of needing to decide which hallway to take to avoid being burned to death', Sect. 12.3). So let me give what seems to me to be the most natural decision-theoretic solution to the puzzle. (To what extent this is preferable to D&K's two solutions, I leave the reader to consider.)

Let G_L, G_R, and G_S be, respectively: go left, right, and straight on. And let F_L, F_R, and F_S be, respectively: find an exit left, right, and straight on. Let $pr(A)$ be the probability of A; let $val(A)$ be the value of A; and let $\mathcal{E}(A)$ be the expectation of A. Then:

- $\mathcal{E}(G_L) = pr(F_L).val(F_L) + pr(\neg F_L).val(\neg F_L)$
- $\mathcal{E}(G_R) = pr(F_R).val(F_R) + pr(\neg F_R).val(\neg F_R)$
- $\mathcal{E}(G_S) = pr(F_S).val(F_S) + pr(\neg F_S).val(\neg F_S)$

Let the value of finding a door be v, and the value of not finding one be $-v$, where v is some positive real.[92] And prior to any information, we can assume that the probabilities of each of the Fs is the same, namely, $p > 0$ (or else prepare to die!). Thus:

- $\mathcal{E}(G_L) = \mathcal{E}(G_R) = \mathcal{E}(G_S) = pv + (1-p).-v = v(2p-1)$

Since these actions all have the same expectation, one should choose between them at random.

Now, suppose that we have only the information given by the *go-left* firefighter, then:

---

[91] See Priest (1998), p. 414.

[92] Perhaps one should not assume one is the negative of the other; but nothing much hangs on this assumption, and it keeps matters simple.

- $\mathcal{E}(G_L) = 1v + 0. - v = v$
- $\mathcal{E}(G_R) = \mathcal{E}(G_S) = 0v + 1. - v = -v$

Clearly, going left has the highest expectation, and so one should do that.[93] Of course, given only the information given by the *go-right* firefighter, the situation is symmetric, so:

- $\mathcal{E}(G_R) = 1v + 0. - v = v$
- $\mathcal{E}(G_L) = \mathcal{E}(G_S) = 0v + 1. - v = -v$

So we come to the case with inconsistent information. One of the possibilities that D&K consider is that of aggregating probabilities; but we can aggregate expectations directly instead. We may suppose that each of the two bits of information is just as good as the other, and so average them out. This gives:

- $\mathcal{E}(G_R) = \frac{1}{2}(v - v) = 0$
- $\mathcal{E}(G_L) = \frac{1}{2}(v - v) = 0$
- $\mathcal{E}(G_S) = \frac{1}{2}(-v - v) = -v$

Clearly, going left and right have equal expectations, and a better expectation than going straight on. So one should choose between these two at random. Hence, all our pre-theoretic intuitive judgments have been vindicated.[94]

Finally, I note that this solution has nothing to do with paraconsistent logic, which is not, then, required to solve the problem.

27.11 Égré (Chap. 4): Respectfully Yours

There is a time-honoured strategy for resolving an apparent contradiction: draw a distinction. And this is clearly exactly the right thing to do in many cases. I say truly, 'it is 5pm and it is noon'. That sounds like a contradiction, but in fact, what I mean is that it is 5pm in London, and noon in New York. In this case, the family of parameters required to disambiguate is discrete. (There is a finite number of time zones.) But the family can equally be continuous. Thus, one might say that something is red to degree 0.6, and not red to degree 0.4, the parameters being the real numbers. In such a case, one may fairly speak of the property as coming by degrees.

---

[93] Note that provided that $p < 1$, $v > v(2p - 1)$, so the information has increased the expectation of going left. An increase of information does not have to guarantee that an expectation goes up. Thus, suppose that the firefighter says that there is one exit, and it is to the left, but that it is probably closed. Then the probability of F_L becomes 1, but the value of finding it, and so the expectation of going left, drops. One would, none the less, prefer to have this information than none.

[94] This analysis treats the information given as dependable. We may take into account the possibility that it is not so with a simple dominance argument. If it is dependable then, as shown, one should go left. If it is not, each direction is equally good. So going left is the dominant strategy.

This raises the question of whether all apparent dialetheias may be resolved by the strategy of paramaterisation. According to Paul Égré, they can[95]:

> I argue that this relation [equivocation more subtle than lexical ambiguity]... underlies the logical form of true contradictions. The generalisation appears to be that all true contradictions really mean 'x is P is some respects/to some extent, but not in all respects/not to all extent'.

This raises two questions: [α] Are all apparent dialetheias produced by predicates subject to parameterisation? [β] In cases where they are, does this resolve the contradiction involved? Let me address these questions in turn.

27.11.1 The Variety of Dialetheias

Égré offers no argument for a positive answer to [α], and one is necessary. For this certainly does not appear to be the case. The putative examples of dialetheias that have been offered include:

1. Paradoxes of semantic self-reference, such as the liar.[96]
2. Set-theoretic paradoxes, such as Russell's.[97]
3. Paradoxes concerning Gödel's theorems (involving the sentence 'this sentence is not provable').[98]
4. Contradictions concerning the instant of change, and, more generally, motion.[99]
5. Legal contradictions concerning inconsistent legislation.[100]
6. Paradoxes concerning the limits of language (or thought), in which one appears to have to describe the ineffable.[101]
7. Statements concerning the unity of objects.[102]
8. Statements connected with multi-criterial terms.[103]
9. Statements concerning the borderline area of vague predicates.[104]

---

[95] *Abstract*. In one place, he appears more circumspect, saying that he is concerned only with contradictions with respect to vague predicates. ('In this paper I propose to [address]... the semantic treatment of contradictory sentences involving vague predicates' (*Introduction*).) But the rest of the paper eschews this qualification. So I presume that he thinks that all *prima facie* dialetheias involve vague predicates. At the very least, nothing in the paper shows any awareness of other possibilities.

[96] For example, IC, Chap. 1.

[97] For example, IC, Chap. 2.

[98] For example, IC, Chap. 3.

[99] For example, IC, Chaps. 11, 12.

[100] For example, IC, Chap. 13.

[101] For example, BLoT.

[102] For example, ONE, Part 1.

[103] For example, IC, Sect. 4.8.

[104] For example, Priest (2010b).

8 and 9 relate to question [β]; and Égré argues that 1 is a special case of 9. I will return to these matters in a moment. This is hardly the place to comment at length on the others, but let us review them briefly.

2 seems to have little to do with matters of respect or degree. The membership of a set does not come by degrees: it is an all-or-nothing matter. Note that the sets involved in the paradoxes in question are not fuzzy sets, as the set of all males might be. They are perfectly crisp mathematical objects deploying only pure mathematical predicates. Similar points apply to 3. This concerns what is mathematically provable. Mathematical proof is deductive. The strength of a deductive argument does not come by degrees, as does that of an inductive argument.

For 4, just consider the situation at an instant of change, e.g. when I am leaving a room, and symmetrically poised between being in and not in it. Actually, this particular case could well be thought of as involving vagueness, and so degrees, since both I and the room are extended (and vague) objects. But this is not essential to the example, which applies just as much to point particles crossing a two-dimensional boundary. (See IC, p. 161.) As another example, consider the point of midnight between Monday 1st and Tuesday 2nd. There is no vagueness in either Monday, Tuesday, or the instant of midnight. All are, or are bounded by, precise points.

Examples of 5 occur when there is a law of the form: people in category X may do Z; people in category Y may not do Z. Someone, a, then turns up who is in both categories X and Y. Now, the categories may themselves be vague, but the membership of a in X and Y can be as determinately true as one might wish. Note, also, that in legal matters, there is room for a qualification of respects. Thus, something may hold in one jurisdiction but not in another. But in the case at hand, both clauses may be taken as parts of a single law in a single jurisdiction.

Examples of kind 6 depend on arguments to the effect that something or other is ineffable. There are many such arguments, and different philosophers have endorsed different ones. In some of these, one might try to invoke a distinction of respect, though rarely does this seem to succeed. Consider just one example: orthodox Christianity says that God is so different from his creatures that it is almost impiety to suggest that human categories apply to him. Yet theologians say much about God—with human categories. (What else do they have?) In response to this, theologians of a *via negativa* persuasion have suggested that one can say nothing *positive* of God. One can say only that he is *not* this and *not* that. But theologians seem to say an awful lot of positive things about God (such as that he is omnipotent, omniscient, etc.). Indeed, even the claim that (only) negative things can be predicated of God is positive. But in any case, if we take to heart the claim that God is literally ineffable, then one can say *nothing* about God—not even that. There is no room for respects or degrees. Even to say *a little bit* is to say something.

Finally, 7. This arises when we are forced to recognise that something (a gluon) both is and is not an object. As such, this can be seen as a special case of 6. For if something is not an object, one can say nothing of it, so is it is ineffable: to say 'a is such and such' is to treat a as an object. But for the same reason, when one says something, this requires the object of predication to be just that—an object.

27.11.2 Dialetheism and Differences of Degree/Extent

So let us turn to question [β]. But first, a preliminary comment about this kind of case. To make their point, Égré, and many of those like him who approach philosophical issues via empirical linguistics, appeal to what (many/most/some) people are wont to say. Such a form of argument must be treated with great care (as Égré, in effect, notes), since people will say all kinds of false things: 'the sun rises in the morning', 'the sky is blue', 'white people have often oppressed black people' (the white people are sort of pink, and the black people are various shades of brown). What people say might provide some *prima facie* evidence for the truth of what is said. But in the end, this cannot replace looking at the evidence/arguments for its truth.

That said, let us start with examples of case 9. These arise when one has a vague predicate, say *red*, and something is on the borderline between red and not red. Égré suggests that a vague predicate, P, depends on a property that comes by degrees. Let us write the degree of a's being P as $|a|_P$, where this is a member of some closed interval of non-negative reals, X. Égré holds that there are $\theta_0, \theta_1 \in X$ (determined by context), with $\theta_0 < \theta_1$, such that a is completely P if $\theta_1 \leq |a|_P$; a is partially P (or P in some, but not all, respects) if $\theta_0 < |a|_P < \theta_1$; and a is not at all P if $|a|_P \leq \theta_0$. Borderline cases are those *a*s in the middle class.

There are already issues here to do with higher order vagueness, but let us set such worries aside.[105] The picture so far does not avoid dialetheism, for the simple reason that we have so far said nothing about $\neg P$, which is equally a vague predicate. Given $|a|_P$, where is $|a|_{\neg P}$? Could both be $\geq \theta_1$? We need a theory of negation.

Égré does not address the matter in the paper, but he refers us to his work with Cobreros, Ripley, and van Rooij, where they endorse the logic *ST*. Semantically, this is a three-valued logic, with values, 1, 1/2, and 0. Negation maps 1 to 0, vice versa, and 1/2 to 1/2. We may take Pa to have the value 1 if $\theta_1 \leq |a|_P$; the value 1/2 if $\theta_0 < |a|_P < \theta_1$; and the value 0 if $|a|_P \leq \theta_0$. So Pa is partially true iff both Pa and $\neg Pa$ are 'half true'.

Now, there are a number of things to be said about this. First, there are three-valued logics for negation in which both Pa and $\neg Pa$ have the value 1. For example, there are fuzzy relevant logics where both can have the value 1 (at a world).[106] Theories of negation are many. Hence, we need an argument for Égré's theory.

Next, note that there will, in general, be differences of degree *within* the three categories. Pa and Qa may take the value 1, even though $|a|_P > |a|_Q$. It would seem that for every a such that $|a|_P$ is less than the maximum value (if, indeed, there is one) Pa is true in some respects, but not all. To call only those in the middle category *partially* true, or true to some extent, therefore seems to misrepresent the situation:

---

[105] All accounts of vagueness have issues with higher order vagueness. See the introduction the Keefe and Smith (1999). I have had my say on the matter in Priest (2010b).

[106] See INCL, Chap. 11.

It is more plausible to interpret the trichotomy as *just true*, *true and false*, and *just false* (as in *LP*).[107] The question is whether the value of $|a|_P$ is such as to make both Pa and $\neg Pa$ assertible—which would seem to be the case if one can take the empirical data as indicative of this.[108] We are back with unvarnished dialetheism.

Let us now turn to examples of case 8. These arise where a predicate can have different criteria of application, which can fall apart. Thus, consider the predicate 'has a temperature of $x°$ absolute'. There are different ways of verifying a statement of this form. We may measure temperature with a mercury thermometer, an electrochemical thermometer, the frequency of black-body radiation emitted, and so on. The method used may depend on both the object in question and on how hot it is. But in some cases, different measuring devices may be used to measure a temperature of the same thing. (Thus, the temperature of seawater can be measured by both a—correctly functioning—mercury thermometer and a—correctly functioning—electrochemical thermometer. And normally in such cases, these will give the same result (to within experimental error). But such determinations way well come apart. In such a case an object can have a temperature of $x°$ and not of $x°$. Indeed, such cases are not unknown in the history of science. A plausible historical concerns the notion of the angular size of bodies in seventeenth-century optics.[109]

Now, it might well be thought that predicates of the form 'has a temperature of $x°$' are examples of what Égré calls polar opposites, and so may be handled by some form of aggregation (Égré's f), described in Sect. 4.4. But this would be to misunderstand how such predicates work. If we found ourselves in this situation of conflict, and assuming that this was not due to experimental error, but a quite systematic phenomenon, we would most certainly not compute the degree to which an object has a temperature of $x°$ by taking a weighted average. What would happen is that the notion of temperature would bifurcate into two, one determined by each of the criteria.[110]

Nor should the fact that such conceptual fission takes place show that there was no dialetheia. The new concepts are (one would hope) consistent. But it remains the case that statements involving the old concept were inconsistent. The contradictions are verifiable, and so true—even if the verifications are never actually performed.[111]

---

[107] Alternatively, if one is really serious about degrees, it makes more sense to move to a fuzzy logic—which raises quite different issues.

[108] Cobreros et al. (2012, 2013), call things in category 1 strongly assertible, and things in category 1/2 weakly assertible. These are terms of art. The question is simply whether $|a|_P$ is such as to make both Pa and $\neg Pa$ true enough to be assertible (in the context).

[109] See Maund (1981), pp. 317f.

[110] Conceptual fission of this kind is well known in the history of science. Thus, the notion of *mass* in Newtonian dynamics bifurcated into *rest mass* and *inertial mass* in the dynamics of Special Relativity. See Field (1973).

[111] One might take such predicates to have multiple referents, which can be used to parameterise the predicate. However, be the referents many, the sense is one; and each verification is sufficient to apply the predicate. (See Priest and Routley 1989c, Sect. 2IIi.) Moreover, speakers who used the term *mass* most certainly did not *mean* 'rest mass or inertial mass'.

And of course, we may well have concepts of this kind for which we never discover the inconsistency, and so which never undergo conceptual fission.

Finally, let us come to examples of kind 1, the liar paradox, concerning a sentence, L, or the form $\neg T \langle L \rangle$. Égré argues that the truth predicate is a vague predicate, and so the contradiction of the liar is defused by the general considerations concerning vagueness. There are a couple of things to be said about this matter.

The first is that it is not clear that the truth predicate is a vague predicate. Why not take T to be a perfectly crisp predicate? Even if P is itself a vague predicate, we may hold that $T \langle Pa \rangle$ takes the value 1 or 0 depending on whether the degree to which a has the property P is greater or less than some (contextually determined) cut-off point. In other words, T is a crisp but indexical predicate.

But suppose that the value of Pa is identical with that of $T \langle Pa \rangle$. Then if A has a classical value, so does $T \langle A \rangle$; and if A has a non-classical value, so does $T \langle A \rangle$. But what of $T \langle L \rangle$? To argue that this has a non-classical value because L does is clearly question begging. *Any* putative solution to the liar paradox may say that things must be *thus and such*, or a contradiction will arise. What is necessary is an independent argument that things are *thus and such*, or the move is entirely *ad hoc*.

Next, and most importantly, the semantic value of the liar sentence is, in a sense, neither here nor there. The liar paradox is generated by an argument. Here is one way of putting it, in sequent calculus form:

$$
\begin{array}{rcll}
T \langle L \rangle & \vdash & T \langle L \rangle & 1 \\
T \langle L \rangle, \neg T \langle L \rangle & \vdash & & 2 \\
\neg T \langle L \rangle & \vdash & \neg T \langle L \rangle & 3 \\
L & \vdash & L & 4 \\
T \langle L \rangle & \vdash & L & 5 \\
& \vdash & \neg T \langle L \rangle, L & 6 \\
& \vdash & L, L & 7 \\
& \vdash & L & \\
\end{array}
$$

1 is the identity sequent. 2 and 3 are applications of the rule for negation. 4 holds since $\neg T \langle L \rangle$ just is L. 5 is an application of the T-schema. 6 is the rule for negation again. 7 is just the identity of L and $\neg T \langle L \rangle$ again; and 8 is contraction. Similarly:

$$
\begin{array}{rcll}
T \langle L \rangle & \vdash & T \langle L \rangle & 1 \\
T \langle L \rangle, \neg T \langle L \rangle & \vdash & & 2 \\
\neg T \langle L \rangle & \vdash & \neg T \langle L \rangle & 3 \\
L & \vdash & L & 4 \\
L & \vdash & T \langle L \rangle & 5 \\
L, \neg T \langle L \rangle & \vdash & & 6 \\
\neg T \langle L \rangle, \neg T \langle L \rangle & \vdash & & 7 \\
\neg T \langle L \rangle & \vdash & & 8 \\
L & \vdash & & 9 \\
& \vdash & \neg L & 10 \\
\end{array}
$$

But Égré endorses the T-schema (Sect. 4.5.1), and all the other steps are valid in the logic ST, which is his preferred logic, given the analysis in his paper. In the logic one can even conjoin the conclusions to infer $\vdash L \wedge \neg L$.[112] So the liar contradiction follows from premises and rules that Égré endorses. One should therefore accept it. Dialetheism.

His analysis will not, then, defuse the contradiction of the liar paradox.

27.12 Ferguson (Chap. 13): Collapsing in Unusual Places

I discovered the Collapsing Lemma in the late 1980s, when working on a technical problem concerning minimally inconsistent models.[113] Later, I discovered that it had quite different applications in the construction of inconsistent models of arithmetic and set theory. The models were both technically interesting in their own right, and had important philosophical ramifications—for example, connected with Gödel's Theorems, and various aspects of set theory.[114] In his paper, Thomas Ferguson notes these things, and goes on to establish a number of interesting results concerning the possibility or otherwise of extending the Collapsing Lemma to machinery beyond that of *LP*.

Speaking generally, the Collapsing Lemma depends on the truth functions and quantifiers involved behaving monotonically, in a certain sense. Namely, in Ferguson's notation, whenever an input changes from b to t or f, the output is never changed from t to f, or vice versa. That is, consistentising an input cannot change a classical output. The result can therefore be extended beyond the bounds of first-order logic to other logical machinery which is monotonic—for example, second-order quantifiers and modal operators.[115] However, the Lemma will fail once one is dealing with non-monotonic connectives/quantifiers. In particular, there is no version of the Lemma which will work for a many-valued truth-functional logic with a detachable conditional, as Ferguson nicely shows; and I have no idea of how one might turn the trick for a logic of some other kind.

Nice as it would be if this were possible, I don't see that it is a serious problem if it is not. The Lemma was always part of an investigation to see what could be done with classical machinery in a paraconsistent context. Classical logic elects to work within the framework of conjunction, negation, the particular quantifier, and the things that can be defined in terms of these—and in *LP*, these are all monotonic in the appropriate sense. In particular, classical mathematics works within this framework. Hence, a limitation of the Collapsing Lemma implies no limitation of the investigation of

---

[112] It is also the case that $A, \neg A \vdash B$, for arbitrary A and B; but one cannot infer $\vdash B$, since Cut in not valid.

[113] See Priest (1991). I subsequently found out that a similar result had been proved by Mike Dunn (1979) some years earlier.

[114] See IC2, Chaps. 17, 18, and Priest (2017), Sect. 11.

[115] See Priest (2002), Sects. 7.2, 7.3. On many-valued modal logic in general, see INCL2, Chap. 11a.

paraconsistent versions of classical theories. The mantra of paraconsistent logic with regard to classical logic/mathematics has always been: we can do anything you can do—and a lot more besides![116]

27.13 Ficara (Chap. 5): Priest as a Hegelian (or Hegel as a Priestian)

It has always seemed to me that the most salient and ungainsayable dialetheist in the history of Western philosophy, between Aristotle's wildly influential but fatally flawed attack on the view in the *Metaphysics*[117] and contemporary times, is Hegel. I know that suggesting that Hegel was a dialetheist is wont to provoke fits of apoplexy in a number of Hegel scholars, who can see no further than so-called 'classical' logic.[118] However, I have defended the view elsewhere,[119] and I shall not revisit the issue here.

Elena Ficara's paper concerns a different (though related) possible similarity between my view and Hegel's. Now, the world *logic* gets used in many ways, and I think it is silly to argue about what the right use is. One just needs to be clear about how someone uses the word. Hegel and I use it in somewhat different ways, as will be clear to anyone who compares Hegel's *Logic* and my *Introduction to Non-Classical Logic*. I use the word in the way that most contemporary logicians use it, as being about validity—that is, what follows from what, and why. For Hegel, logic is what is covered in his *Logic*—roughly, human conceptual thought and its rational evolution. I think that what I mean by logic is pretty close to what Hegel calls *subjective logic* (*Verstandeslogik*). For him, this is the Aristotelian syllogistic of his day.

That point notwithstanding, and assuming that Ficara's exegesis of Hegel is right,[120] there does seem to be a similarity between, pairs of our terms: (a) *logica utens, natürliche Logik*; (b) *logica docens, die Logik*; (c) *logica ens, das Logische*. The first comprises the norms of a reasoning practice. The second comprises logical theories, as found in logic textbooks. The third is what it is that such theories aim to capture. And as Ficara points out, according to both Hegel and myself, (b) can be revised. In particular, it can be changed in order the better to bring it into line with (c).[121]

---

[116] On that matter, see IC, Chap. 8, especially 8.5.
[117] See DTBL, Chap. 1
[118] Though not Ficara, as I know from many illuminating discussions.
[119] See Priest (1990, 2019a).
[120] I am no Hegel scholar, and am usually happy to leave the nitty-gritty of Hegel-exegesis to those, like Ficara, who knows his work better than I do.
[121] Concerning (a), Ficara says (Sect. 5.2) says that 'one can/also admit that reasoning practices and norms are grounded in metaphysical views about what there is and its nature. Priest implicitly assumes this insofar as he states that inferential norms are based on our view about the meanings of the connectives'. Now there certainly can be metaphysical assumptions inherent in a reasoning

The major difference that Ficara points out concerns this process of revision. For me, this is to be done by a very general process of theory choice.[122] For Hegel, it happens in the process of dialectical thought (undertaken by *Geist*). This is his *Vernunftlogik*.[123]

Concerning this matter, Ficara says (Sect. 5.3):

> While Hegel postulates the idea of a rational logic, i.e. embeds his view of logical revision in a conception of logic as conceptual and philosophical (i.e. self-revising and truth-oriented) analysis of natural language, Priest sticks to the idea of an external operation, which follows the model of rational theory choice among rival logical theories. Logic revision is for Hegel, as we have seen, the result of Hegel's very idea of logic as the analysis of *das Logische*, i.e. of logic as analysis and individuation of forms of truth. Revision, intended as the procedure of adjustment between theories and data, is an operation actuated by logic itself. In this sense Hegel's idea of logic does not admit the distinction between pure and applied logic. In non-Hegelian terms, Hegel's logic as rational logic would involve both the construction of a model and reflection on the adequacy of the model.

Now, I think that this matter is, to a large extent, terminological. Hegel uses the word *logic* to encompass rationality in general. I use it for just one aspect of rationality. We both agree that there is a process of rational revision going on; and (assuming that Ficara is right about Hegel) we both agree that this involves a dialectic between theory and data.[124]

However, it is certainly true that Hegel and I think of the mechanism of revision quite differently. Hegel's view, I take it, can be accepted only if one endorses Hegel's idealism. This has always struck me as somewhat whimsical view; but I certainly don't intend to defend that claim here. So let me just defend my own view against the problems that Ficara sees.

Ficara notes essentially two problems. The first concerns the fallibility of the data of theory choice (Sect. 5.3):

> why should the data be a criterion of rationality if they are our intuitions about validity and our intuitions can be wrong?

The second is that the method needs an explanation of the connection between the data of logical theorising and what actually follows from what (Sect. 5.3):

---

practice—think of Dummett on classical logic *vs* intuitionist logic—but I wouldn't say that disputes about the meanings of the connectives necessarily have metaphysical ramifications.

[122] See Priest (2016a). This can be thought of as a sort of reflective equilibrium as Ficara (fn 19) notes. This does not mean that it does not 'admit the metaphysical meaning of logic'. *Logic ens* is the very subject of the deliberative equilibrium. I note also that theories of *validity* are liable to engage with theories of other notions, such as *meaning* and *truth*. In that way, theorising about validity is entangled with a number of the more general issues which play an important role in matters for Hegel.

[123] For a formal account of this, see Priest (201+c).

[124] And for what it is worth, it seems to me that the distinction between (in my terms) pure and applied logic makes just as much sense for Hegel as it does for me. A pure logic is simply a bit of pure mathematics, which may be applied for many purposes, or never applied at all—though Hegel is working before the impact of non-Euclidean geometry on mathematics, which finally brought home the distinction between pure and applied mathematics.

Priest's theory about logic revision needs to be completed by a conception about the nature of the data, and their relation to truth, or *logica ens*.

Take the second point first. Theorising is an important rational activity. We are aware of some phenomenon, be it motion, time, ethics, language. We wish to understand the whats and whys of it. To do so we construct a theory to account for the phenomenon. Since we are aware of the phenomenon, there are already things we believe about it, or can ascertain directly. These provide the initial data to which the theory answers.

Now, giving reasons is something we all do naturally, though perhaps badly. It is certainly a skill that can be improved if one studies mathematics, law, and doubtless many other subjects. We come to see that sometimes a reason offered really does support a conclusion; and sometimes that it does not. In other words, there seem to be facts about what follows from what. *Logica ens* is the truth of this matter; *logica docens* is a theory about what the truth is. (And the best theory we have at any time is our best guide to what the reality it aims to describe is like.) The initial views we have about what follows from what are data to which the theory must answer. These are the views which characterise—at least initially—the very phenomenon at issue.

Which brings us to the first question. Adequacy to the data is a criterion (though not the only one) of what it is that makes a theory rationally acceptable. (To account for these is, after all, a large part of what the theory was produced *for*.) In that sense, this is a 'criterion of rationality'. However, this does not mean that the data are infallible. In theorisation about any complex issue, data are fallible. That this is so even for empirical data is one of the hard lessons of the twentieth-century philosophy of science.[125] Inadequacy to a piece of data is certainly a rational black mark against a theory; but a theory that is strong in other respects can overturn the datum—especially if one can give an independent reason as to why one was mistaken about its truth. This does not undercut the rational use of data: the dialectic between theory and data, in which views about the truth of both can be revised, just is a feature of rationality.[126]

---

[125] See, e.g. Chalmers (2014), Chap. 2.

[126] Ficara writes (Sect. 5.3) that the criteria for the revision of logical theory:

> are, for Priest, the standard ones of rational theory choice: adequacy to the data, fruitfulness, non-ad hocness, etc. For Hegel, the only criterion that orients the critique of logic as theory is *das Logische* as conceptual truth, as the way the connectives *are* and validity *is*. It is a realistic meaning of truth, the correspondence of the logical theories with the logical fact. Interestingly, a correspondence that is already given in the empirical data logic deals with (our intuitions about validity).

> Now, I think that there is a certain confusion here. The aim of both Hegel and myself it to get logic as theory right, or at least, better. In that sense, we both want to get the theory to correspond to reality. Where we might disagree is on what method will achieve this, and why. In neither case is this correspondence already guaranteed by our intuitions, which, we can both agree, are sometimes wrong, or at least, inadequate.

27.14 Field (Chap. 6): Out in Left Field

Which brings us to Hartry Field. Field and I have been discussing logic, and especially solutions to the semantic paradoxes, for many years now; and the discussions have caused me to think much harder about many issues. I have enjoyed the discussions, and learned a great deal from them. I very much value Field's open-mindedness and intellectual honesty. We agree about much, and especially on the fact that 'solutions' to the paradoxes which employ 'classical' logic are not viable.

The solutions to the paradoxes that he and I endorse also have much in common. But there is one crucial disagreement: his solution to the paradoxes is to reject Principle of Excluded Middle (PEM); mine is to reject Explosion. His present paper is the latest written contribution to our discussions.[127] As ever, it is rich and insightful— and there is absolutely no way that I do justice to everything it contains here. (To do this would require a piece at least as long!) So I will restrict myself to the most important things. I will take these up in more or less the order in which he raises them, though I collect together a bunch of more minor points in a section at the end.

27.14.1 K_3 and LP: Duality

Field points out (Sect. 6.1) that there is a substantial duality between K_3 ('middle value' non-designated) and *LP* ('middle value' designated). He concludes on the basis of this that, setting aside issues of conditionality, the dispute between advocates of the two logics is simply notational. In particular, the *LP* theorist's acceptance of *A* is the K_3 theorist's rejection of ¬*A*, and vice versa. Perhaps, as far as formal matters go, this is correct. But enforcing this duality appears to have implausible consequences in a wider context, since acceptance and rejection have essential roles to play in other areas, such as action and its rationality. Someone who accepts that it will rain has grounds to take an umbrella. Someone who merely rejects the claim that it will not rain does not have the same ground; for this fact, classical logic having gone, gives them no reason to suppose that it will rain! Similarly, someone who is a dialetheist about the liar paradox, *L* (such as me), has grounds to write a book advocating *L* and ¬*L*. Someone who merely rejects both (such as Field) does not. To interpret Kripke's 'Outline of a Theory of Truth' (1975) as advocating both *L* and ¬*L*, would seem to be an act of gross perversity!

---

[127] The previous exchanges were Field (2005, 2008), Priest (2005b), especially Part V, and Priest (2010c).

27.14.2 Conditionals

Let us now turn to the subject of conditionals. I'm happy to note that Field's views and mine on the matter seem to be converging, though they certainly don't coincide yet![128]

Let us set aside for the moment the matter of restricted quantification—I'll come back to this—and just talk about the ordinary conditional.[129] My current thinking on this can be found in Priest (2009, 2018b). In these places, I give a semantics for a relevant conditional logic of a very standard kind, using a formula-indexed binary relation. Technically the semantics endorsed in spelled out in INCL, 10.7.[130]

Field gives his preferred account in Sect. 6.2. It deploys a world-indexed ordering relation, rather than a sentence-indexed binary relation. But these techniques are well known to be different ways of doing much the same thing. Both of us are clear that this is the semantics for a basic logic, and might well be strengthened by constraints on the relation/ordering. Both of us admit impossible worlds. Field allows for the possibility of both moderately impossible worlds and anarchic worlds.[131] In INCL Chaps. 9 and 10, I give semantics with each of these—and of course, if you have anarchic worlds, some of these will behave in the same way as moderately impossible worlds. The structure of impossible worlds is an important philosophical question, though neither of us thinks that it is really important in this context.[132] So we are very much on the same page here.

Perhaps the main difference between us at this point is this. The frameworks in question can be used to deliver both relevant logics and irrelevant logics. Field is happy with an irrelevant logic. For example, he takes it that if B is any logical truth, so is $A > B$. Thus, since $B > B$ is a logical truth, so is $A > (B > B)$. I prefer a relevant

---

[128] Field's semantics for conditionals have gone through many iterations. (See Field 2008, 2014, 2016, 201+.) When, in the past, I have pressed him on the subject of ordinary language conditionals, he has always said that he didn't care much about these. He is happy simply to replace natural language. I see now that he takes ordinary language conditionals more seriously. For the evolution of my own views, see IC, Chap. 6, IC2, 19.8, Priest (2009, 2018b). I have always taken ordinary language conditionals seriously. I have endorsed both relevant logic and conditional logic, separately—though perhaps I have never brought these two things together in the context of the paradoxes of self-reference.

[129] Field writes this as ▷. I will stick to the notation of INCL, and use >.

[130] On the relevance (in the technical sense) of the conditional in question, see 10.11, Ex 12.

[131] In the text to fn. 11, Field says 'Priest also tends, after allowing for anarchic worlds, to ignore them, since allowing them would invalidate almost every law of conditionals. In the footnote, he refers to Sects. 9.4.6, 9.4.7 of INCL2. A warning: INCL is a textbook, and should not be taken as expressing my own views. Also, as Sect. 9.4.5 makes clear, only conditionals are at issue in this discussion of logical anarchy.

[132] I think that the importance of anarchic worlds really kicks in when one is dealing with intensional operators (TNB, Chap. 1), but they also have a use in the semantics of counter-logical conditionals. Field thinks that for such conditionals one does not need all of them at once, and different contexts will determine different semantics. I think that it is simpler to have a uniform semantics, and allow context to pick out which are the worlds relevant to a conditional. (See Priest 2016b, 3.3.) These matters are of little import here.

logic. For example, the following does not seem true, let alone logically true: If every instance of the law of identity fails, then if snow is green, snow is green.) However, so far, nothing crucial seems to hang on this.[133] He also insists that the account of the conditional in 'non-classical contexts'—those which contain the truth predicate—should reduce to this account in classical contexts. For me, this constraint is trivially realised, since I take the logic to be the same whether or not sentences contain the truth predicate.[134]

Let us now turn to restricted quantification. There is still a central agreement here. The problem with restricted quantification, once one foregoes the material conditional, is how to express restricted universal quantification. He and I both think that 'All As are Bs' should be understood as of the form $\forall x(A(x) \mapsto B(x))$, where $\mapsto$ is not the conditional $>$.[135] However, we give different accounts of what $\mapsto$ is. This is probably the main disagreement between us, as far as conditionals go. Field gives a somewhat complex account of the semantics of his conditional. I prefer a simpler approach.

In fact, there are several ways of doing much the same thing. In any standard relevant logic, there is a conditional operator, $\rightarrow$.[136] If the semantics is of the Routley/Meyer kind, there is a ternary relation, R, such that:

- $A \rightarrow B$ is true at world w iff for all a and b such that $Rwab$, whenever A is true at a, B is true at b.

There may also be a binary (heredity) relation on worlds, $\sqsubseteq$, such that for all A:

- if $x \sqsubseteq y$ then if A is true at x, A is true at y

One way of defining $\mapsto$ is by giving it the following truth conditions:[137]

- $A \mapsto B$ is true at world w iff for all a and b such that $Rwab$ and $a \sqsubseteq b$, whenever A is true at a, B is true at b.

Another is by simply defining it thus:

- $A \mapsto B := (A \wedge t) \rightarrow B$

---

[133] Since Field has a semantics with impossible worlds, I did wonder why he did not endorse a relevant logic, when one would come at no apparent cost. He says that the usual relevant logics are not adequate to account for the normal English conditional (Sect. 6.4.2). Agreed. But the semantics at issue now are the semantics of a relevant conditional logic. And an irrelevant logic seems to do no justice to such conditionals, as I have just observed.

[134] Why does Field not do the same—that is, take his generalisation of these semantics to be the correct logic right from the start? After all, the truth predicate is a part of natural language. I presume that there is a good answer, though it isn't clear to me. Perhaps, it is because it would make the semantics of the natural language conditional far too complex to be grasped by lesser mortals than logicians!

[135] Field writes the conditional as $\rightarrow$. I think that this is too confusing when relevant conditionals may come into play. So I will use a different notation.

[136] I note that the presence of this operator is not necessary for $>$ being a relevant conditional.

[137] As is done in Beall et al. (2006)—hereafter BBHPR.

where t is the logical constant of relevant logic, which is, intuitively, the conjunction of all truths.[138] An even simpler way is to define it thus:

- $A \mapsto B := (A \to B) \vee B$

These are not all exactly the same,[139] but they all share the crucial property, that $B \models A \mapsto B$.[140]

The crucial question at this point is whether $\mapsto$, as either Field or I define it, has the appropriate properties. So, what, then, are the appropriate properties? Field (Sect. 6.3.1) gives a list of inferences he 'takes to be compelling'; BBHPR (Sects. 2, 3) give an overlapping, more systematic, but incompatible list of desiderata. Now, some things really do seem to be necessary for an account of restricted universal quantification. For the rest, it seems to me, these are a legitimate matter of theoretical 'give and take'.[141]

To start with the former: both Field and I agree that one should have[142]:

[1] $A(a), \forall x(A(x) \mapsto B(x)) \models B(a)$

and

[2] $\forall x B(x) \models \forall x(A(x) \mapsto B(x))$

Both his account and mine validate these inferences.

Turning to the latter, it would be tedious to hammer through all the other examples from the lists, especially the more marginal ones. So let me just comment on the most significant disagreement here. This concerns negation. Field endorses (1_c), that is: $\forall x(A(x) \mapsto B(x)), \neg B(a) \models \neg A(a)$. BBHPR explicitly reject this (desideratum

---

[138] This is done in IC2, pp. 254f.

[139] For example, the third, but not the first two, satisfies the inferential version of (2^*) on Field's list: $\neg \forall x(A(x) \mapsto B(x)) \models \neg \forall x B(x)$. Whilst the first two, but not the third, satisfy the inferential version of (4a) on Field's list: $\forall x(A(x) \mapsto B(x)), \forall x(A(x) \mapsto C(x)) \models \forall x(A(x) \mapsto (B(x) \wedge C(x)))$.

[140] Actually, one might even consider the possibility of defining $\mapsto$ as $(A > B) \vee B$. But this raises novel complexities. For example, $\mapsto$ is not, then, transitive. It might be thought absurd that restricted universal quantification is not transitive. However, there are standard counterexamples to transitivity in conditional logic which can easily be modified to apply to quantification. Thus, suppose that there are just two candidates for election. Then the following inference seems to fail:

- Anyone whose competitor dies before the election will win.
- Anyone who wins will have a disappointed competitor.
- So anyone whose competitor dies before the election will have a disappointed competitor.

However, this is not the place to discuss these matters.

[141] Field seems to endorse the policy of 'the stronger the better' (Sect. 6.3.1). Now, I have never been persuaded by arguments of this kind. I have heard them all too often in the defence of the material conditional. But even granting that strength is a desideratum, it has to be modulo other things, such as a solution to the paradoxes of self-reference. Thus, one can certainly strengthen the logics that both he and I favour by adding pseudo *modus ponens* $((A \wedge (A \mapsto B)) \mapsto B)$. We both reject this, for reasons connected with the Curry Paradox.

[142] In what follows I shall discuss the inferential versions of principles. I will return to the matter of the conditional versions towards the end of this section.

B1),[143] since, given [2], the principle delivers Explosion. This is fine for Field, for whom Explosion is valid anyway. It is not fine if one is to endorse a paraconsistent logic, as I do. Similarly, Field endorses (2_c): $\forall x \neg A(x) \models \forall x(A(x) \mapsto B(x))$, which delivers Explosion even faster.[144] Indeed, one really should not expect this inference in a dialetheic context. Suppose that everything satisfies $\neg A(x)$. It may yet be the case that some a is such that $A(a)$, as well; and there is absolutely no reason to suppose that $B(a)$. Field says that he can see no independent reason for giving up principles of this kind which does not ascribe to the restricted universal quantifier a modal character (Sect. 6.4.1). The independent reason is exactly paraconsistency/dialetheism. And to reject this as a ground in the present context is clearly to beg the question.

Two further points. Field objects (Sect. 6.4.1) that, in a dialetheic context, there are certain valid principles of inference about the restricted universal quantifier, and which he finds compelling, for which one cannot endorse the logical truth of the corresponding conditional. Certainly, but this seems no real problem. First, there are many valid inferences for which the corresponding conditional is not a logical truth, such as *modus ponens*—as both Field and I agree. Next, in most reasoning, it is the inference that does all the hard work, not the conditional. Third, the failure of the rules of inference in question is a consequence of paraconsistency itself. And, as the saying goes, one person's *modus tollens* in another person's *modus ponens*.

Finally (Sect. 6.4.1), Field is worried by the fact that $\mapsto$ does not reduce to $\supset$ 'in classical contexts'. Never mind whether or not it *should*; mine does. A classical context—i.e. one encompassed by the semantics of classical logic—is one where there is just one world, and every sentence has the value 1 or 0, but not both (and not just one where there are no contradictions). In such a context, whichever of the three definitions of $\mapsto$ is used, $A \mapsto B$ is true at the world iff $A \supset B$ is.

27.14.3 Curry Paradoxes and Quasi-Naivety

Field points out (Sect. 6.5) that, for any conditional, there will be a corresponding Curry paradox. For the material conditional, $A \supset \bot$, is just $\neg A$. The corresponding Curry paradox is, hence, just the Liar, and so will have the same solution.—The failure of PEM (aka, $\supset$-Introduction) for Field; the failure of disjunctive syllogism (aka, Explosion) for me.—But what of the Curry paradox for other conditionals, and particularly the conditional involved in the T-Schema? Some have suggested that this ought to have the same solution as that for $\supset$. There is really no *a priori* reason why all conditionals should behave in the same way with respect to their

---

[143] To be precise, they reject the inference of contraposition for $\mapsto$, but the reason given applies equally to this principle.

[144] Ditto for 5*, as Field notes. He says that he finds 5* 'totally compelling' (Sect. 6.4.1). From a paraconsistent perspective, which has independent virtues—even if one is not a dialetheist—there is total uncompellingness.

Curry paradoxes; but Field points out that, in any case, the point has no force against his solution, which is to reject conditional introduction for this conditional, too.

The same point does not apply to a paraconsistent solution, since this rejects conditional introduction (that is, Contraction) for the conditional of the T-Scheme, but not for $\supset$.[145] The question then becomes whether the Liar and the Curry in question are of the same kind. This is an exceptionally vexed issue. I think that they are not, but this is not the place to go into that matter.[146]

In Sect. 6.6 Field turns to the contraposibility of the T-Schema. We both hold that $T \langle A \rangle \leftrightarrow A$, for the appropriate $\rightarrow$. He also holds that $\neg T \langle A \rangle \leftrightarrow \neg A$ (naivety). I do not (semi-naivety). Indeed, Field holds that for any A, A and $T \langle A \rangle$ are intersubstitutable—at least in non-intensional contexts. Thus, take a conditional that satisfies $\neg A \leftrightarrow \neg A$. Intersubstitutivity gives. $\neg T \langle A \rangle \leftrightarrow \neg A$. For me, take any dialetheia, A. Then A is true and false; so $T \langle A \rangle$ is certainly true. But, generally speaking, there is no reason why A should not be just true.

Of course, if one is a deflationist about truth, and holds that A and $T \langle A \rangle$ have exactly the same content, then intersubstitutivity follows. I have never been a deflationist about truth, however.[147] And if A and $T \langle A \rangle$ really do have the same content, it follows that if someone believes A, they believe $T \langle A \rangle$, and vice versa. But it seems that someone can believe one without the other—if, for example, they have slightly odd views about truth.

Notwithstanding, it is open to a dialetheist about the semantic paradoxes to endorse naivety. Beall, for example, does.[148] I am inclined against this. For a start, it spreads contradictions beyond necessity, turning *any* dialetheia into a dialetheia about truth.[149] Moreover, semi-naivety permits one to draw useful distinctions. Thus, one can express the thought that A is *true only* by saying that $T \langle A \rangle$ and $\neg T \langle \neg A \rangle$. Given naivety, these two are equivalent (modulo double negation): there is no distinction between something's being true and its being true only. Of course, this does not articulate the distinction in a way that enforces the consistency of truth-only. There will be sentences which are false and true only. (Such as the liar in the form $\neg T \langle A \rangle$, though perhaps not in the form $T \langle \neg A \rangle$, as Field notes.) The distinction is expressed, none the less.[150] More of this later.

A standard fixed-point construction shows that a model for a language without T can be extended conservatively with a naive truth predicate. Since semi-naivety is weaker than naivety, the addition of a semi-naive truth predicate is also conservative. In a previous essay[151] Field called this 'uninteresting', since it does not show how the contraposed truth predicate can fail. It was not meant to: it was simply a proof of

---

[145] I assume here that the conditional of the T-Schema is not $\supset$, though this is not obvious. See Priest (2017).

[146] On this, see Priest (2017), Sect. 15.

[147] See, e.g. IC, Chap. 4.

[148] Beall (2009).

[149] See IC, 4.8.

[150] Again as Field notes, dual considerations apply to his approach.

[151] Field (2008), p. 371.

conservative extension. But in response to Field, I gave a non-triviality proof which shows how it may do so.[152] In particular, there are A's such that $\neg A$ holds in the model, but $\neg T \langle A \rangle$ does not. The contraposed T-Scheme is therefore invalid, since it has invalid instances.

In the model, the contraposed T-Schema fails only for T-free sentences. I do not take this to show that it fails only under such conditions. That was not the point of the construction, which was just to show that one may have a model of the T-Schema in which its contraposed form is not valid. So it is a fair question to ask when one may have $\neg A$ without $\neg T \langle A \rangle$. This will happen when $A \wedge \neg A$ holds, but $T \langle A \rangle \wedge \neg T \langle A \rangle$ does not. For the second of these to be true, something (else) must force us to suppose so. (Contradictions should not be multiplied beyond necessity.) The model shows that if A is T-free nothing so forces us. Sometimes, as we have seen, we are so forced; for example, when A is $\neg T \langle A \rangle$. But there seem to be T-ful sentences where this does not appear to be the case; for example when A is $T \langle \neg A \rangle$. So when are we forced to accept the T-ful contradiction? The answer, I think, will be given by models of the T-Schema which are, in an appropriate sense, minimally inconsistent. However, I have nothing useful to say on that matter at present.

Field notes that someone might object that a non-classical theory of truth is not really about *truth*, since its saves (Sect. 6.6):

> the truth schema in name only. The charge is that the connective $\leftrightarrow$ in the non-classical logician's preferred version of '$True(\langle A \rangle) \leftrightarrow A$' is some contrived connective, far from what motivates the idea that $True(\langle A \rangle)$ should be equivalent to A.

A naiveist may reply that the equivalence is best understood as intersubstitutivity. But a semi-naiveist may reply, instead, that $\leftrightarrow$ is not at all contrived. It is exactly what we mean when we say 'if and only if', in the context of the T-Schema. Of course, what we do mean is contentious. But to assume it is not what a non-classical logician says is just to beg the question against them. I note also that if the relevant biconditional is the ordinary English (bi-)conditional, $>$, then this does not contrapose. $True \langle A \rangle <> A$ does not entail $\neg True \langle A \rangle <> \neg A$; and we have semi-naiveism.

27.14.4 L'Affaire Gödel

In Sect. 6.7 Field raises matters to do with Gödel's Theorems.[153] This takes us back to where the dialetheic journey started for me.[154] The idea that the theorems might motivate dialetheism was a provocative but simple one. But, like all philosophical ideas of any interest, matters have turned out to be more complex. Let me say how things now appear to me, especially *vis à vis* Field's comments. Let us start with the relatively uncontentious matters, and work our way up to the most contentious.

---

[152] Priest (2010c), Sect. 11.

[153] More on this topic, see the discussion of Shapiro, Sect. 27.21 below.

[154] See Priest (1979, 1984).

Behind Gödel's proof of his first incompleteness theorem, there is an obvious paradox of self-reference. Let us write $Prov\ x$ for 'x is provable', and angle brackets as a name-forming device. Anything provable is so. That is:

- $Prov\ \langle A \rangle \to A$

Let us call this, for want of a better name, *Löb's Principle*.[155]

One can, of course, ask for the justification of Löb's Principle. And if a proof is proof in some particular formal systems of arithmetic, the schema is known to be unprovable, on pain of contradiction (Löb's Theorem). But here we are not yet dealing with proof in some formal system, but proof *simpliciter*—what we might call the naive notion of proof. To be provable in this sense is simply to be established as true. Löb's Principle seems, then, to be a plain *a priori* truth.

Now, consider the sentence 'this sentence is not provable'; that is, a sentence, G, of the form $\neg Prov\ \langle G \rangle$.[156] Substituting this in Löb's Principle gives us $Prov\ \langle G \rangle \to \neg Prov\ \langle G \rangle$. Hence, we have proved $\neg Prov\ \langle G \rangle$. But this is just G. So we have demonstrated $Prov\ \langle G \rangle$. We have then established both the Gödel sentence and its negation. Let us call this *Gödel's Paradox*. The paradox is clearly a paradox very similar to the 'Knower Paradox', and is in the same family as the Liar.[157] If one subscribes to a dialetheic solution to the Liar, then one should equally subscribe to a dialetheic solution to this.

It is clear that one can avoid the dialetheic conclusion if one rejects the PEM, and so the inference to $\neg Prov\ \langle G \rangle$, as I presume Field would. However, such a move requires a justification—and one not simply of the form 'if one does not reject this, a contradiction will arise'. Note that $Prov\ \langle A \rangle$ entails $T\ \langle A \rangle$, but not vice versa. So $Prov\ \langle A \rangle \vee \neg Prov\ \langle A \rangle$ does not entail $T\ \langle A \rangle \vee \neg T\ \langle A \rangle$. A justification is therefore required independent of the failure of the PEM for T-ful sentences. The case for the failure of the PEM for T-ful sentences is, I take it, something like Kripke's: truth and falsity are determined 'from the ground up', and there is nothing to determine the truth of ungrounded sentences. Hence, they are neither truth nor false. There is, as far as I can see, no similar argument to be made for provability. So at this point, Field seems to have offered no solution to this paradox.

The notion of proof I have been talking of till now is an informal notion. Let us now turn to how matters stand if we are dealing with some formalisation of the naive notion of proof, i.e. representing it as proof within some formal axiom system. In IC, 3.5 (pp. 49f.) I gave an argument to the effect that, given an intuitively sound notion of formal proof, $Prov$, one can give an equally intuitively sound argument for $Prov\ \langle A \rangle \to T\ \langle A \rangle$. Löb's Principle, of course, follows from this and the T-Schema. Field has convinced me that, intuitive as this proof may be, one is not entitled to it, for

---

[155] And *Prov* means provable from things *including* Löb's Principle. So it is not provability in some other system, as Field moots in his last paragraph of the section.

[156] The construction of such a sentence in formal arithmetics requires that the primitive recursive function of diagonalisation be representable by a function symbol. If we have only the usual successor, addition, and multiplication function symbols at our disposal, we can construct only a G materially equivalent to $\neg Prov\ \langle G \rangle$. This fact has no material effect on the considerations to follow.

[157] See BLoT, 10.2.

reasons to do with Curry's paradox, at least if the system of proof is an axiom system which uses *modus ponens*[158]—though I do think that this sort of argument is what underlies claims that the Gödel sentence, G, for, say, Peano Arithmetic (hereafter, PA) is true. However, for the paradoxical argument to run, one does not have to bring truth into it, as I have shown above. And Löb's Principle itself strikes me as something one should have in any formal system which attempts to capture our intuitive notion of proof—just as much as one should have the T-Schema in any formal system which attempts to capture our intuitive notion of truth.

None of this assumes that the axiom system we are talking about is one in which the theorems are recursively enumerable (re), or even arithmetic.[159] (Though this is sometimes built into the definition of a formal system, there is no technical necessity to do this—the usual definition of a formal proof works for any set of axioms.) In this case, there is no reason to suppose that the proof predicate for the system can be defined in purely arithmetic terms.

There are, however, arguments to the effect that for a formal system adequate to our naive notion of proof (for, say, arithmetic), the theorems are re.[160] I don't claim that these are definitive, but they have a certain force. And if what they show is correct, the proof predicate for the system, and the corresponding Gödel sentence, is expressible in purely arithmetic vocabulary—where, presumably, the PEM is not an issue. It follows that the set of true purely arithmetic sentences is inconsistent. Nor is there anything technically unfeasible about this. We know that there are perfectly sensible re theories in the language of arithmetic which are complete (in the sense of containing everything true in the standard model), but inconsistent. Unsurprisingly, each validates both its Gödel sentence the negation thereof.

If the above is correct, then the set of sentences true in the language of arithmetic is inconsistent. It does not follow that PA is inconsistent; nor have I ever claimed that it is. Shapiro (2002) claims that I am, none the less, committed to this. However, his argument fails. It invokes the claim that all recursive sets/predicates are representable in PA. Now, a binary relation, Θ, is representable in a theory, $\mathfrak{T}$, if there is some formula of two free variables, $\theta(x, y)$, such that:

- if $\langle n, m \rangle \in \Theta$ then $\theta(\langle n \rangle, \langle m \rangle) \in \mathfrak{T}$
- if $\langle n, m \rangle \notin \Theta$ then $\neg\theta(\langle n \rangle, \langle m \rangle) \in \mathfrak{T}$

where $\langle k \rangle$ is the gödel number of the numeral of k. Thus, if Θ is the proof relation then, in the present scenario, it is recursive. So if PA is consistent then, clearly, Θ cannot be represented in PA—though it may be representable in an inconsistent arithmetic.[161] Of course, there is a standard proof that all decidable sets are representable in PA. One may, therefore, ask where that breaks down. I point this out in IC2, 17.7, where all this is discussed.

---

[158] See Priest (2010c), Sect. 5.

[159] See IC, 17.5.

[160] These are given in Priest (1984), Sect. 6, and IC, 3.2.

[161] I note also that this opens up intriguing new possibilities in computation theory. See, e.g. Weber (2016b).

This takes us to an argument, which Field's text displays in Sect. 6.6. To the extent that this concerns provability in PA (or similar system), then, whatever else there is to be said about things, the matter is the same: the argument simply assumes that all recursive relations can be represented in PA. They are not, if those relations are inconsistent and PA is not.

However, to the extent that the argument is taken to concern, not provability in an axiomatic arithmetic, but truth in the language of arithmetic, the matter is different. Field sketches an argument of mine, and then says (for *reductio*), that 'we might equally argue' in terms of the reasoning he then gives. Now, first, what should one make of his argument?

This takes a dialetheic sentence from outside the language of arithmetic, Q, to show that a sentence within the language is dialetheic. The sentence is $\langle Q \rangle = \langle Q \rangle$, and it is perhaps not so surprising that that sentence is dialetheic. After all, we know that in any of the inconsistent models of arithmetic there are dialethic identities. But the argument can be generalised. Let s be any non-empty decidable set of natural numbers. (In Field's case, this is $\{\langle Q \rangle\}$.) This is defined in the language of arithmetic (that is, defined in the true theory—whatever that is) by a formula $A_s(x)$. We now consider the set $s' = \{x : Q \wedge x \in S\}$.[162] Since Q, $s' = s$; and since $\neg Q$, $s = \emptyset$. Hence s' is defined in the language of arithmetic by $A_s(x)$ (and $x \neq x$; the matter there is the same). Take any $n \in s$. Then $n \in s'$, so it follows that $A_s(\mathbf{n})$ (where $\mathbf{n}$ is the numeral of n); but $n \notin s'$, so $\neg A_s(\mathbf{n})$.

The main problem with this argument concerns the claim that $s' = s$. This takes us into issues of paraconsistent set theory. It has to be shown that $\forall x(x \in s \text{ iff } x \in s')$, that is, $\forall x(x \in s \text{ iff } Q \wedge x \in s)$. What the 'iff' is here depends on how one understands paraconsistent set theory. There are two main possibilities.[163] The first is to formulate set theory in an appropriate relevant logic, and to take the 'iff' to be the relevant biconditional. But in that case, the sentence fails from left to right. The other possibility is to formulate set theory in *LP* and take the biconditional to be its material biconditional. In this case, the sentence holds. But if $n \in s$, then since $\neg(Q \wedge n \in s)$, so does its negation. In this approach, extensionality tells us that $\forall x(x \in s \equiv x \in s') \supset s = s'$. And since this is a material conditional, we cannot detach the identity—even in a default form, since the antecedent is contradictory. Hence, the argument breaks in this case also.

Finally, it is clear from these considerations that this is nothing like the argument that Field extracts from my texts, and of which has argument is supposed to be a *reductio ad absurdum*. That argument is to the effect that the naive provability predicate can be expressed in the language of arithmetic, and so that the paradoxical argument concerning it can be reproduced in the language of arithmetic. This is clearly quite different. In particular, it does not depend on constructing a *doppelgänger* set using a dialetheia.

Finally, we come to the most contentious matter. Field says that the claim that PA is inconsistent, or even just that there are true contradictory Σ_1 sentences in the

---

[162]*Doppelgänger* of this kind are well known in paraconsistent set theory. See, e.g. Weber (2012).
[163]See IC2, Chap. 18, and also the discussion of Batens, Sect. 27.3.2 above.

language of PA 'strikes him as totally incredible'. I don't for a moment doubt Field's judgments concerning his own mental states, but I do question the rationality of his certitude.

Why might one be so certain that there are no dialetheias amongst the Σ_1 sentences of the language of arithmetic? One cannot, of course, claim that PA is consistent and that every true Σ_1 and Π_1 sentence can be proved in PA (the negation of a Σ_1 sentence being Π_1). We know from Gödel's theorem itself that we cannot prove all Π_1 statements true in the standard model in PA. This is true in spades if the correct model is one of the inconsistent arithmetics. PA is radically incomplete with respect to this.[164]

A more hopeful suggestion is to the effect that it is unclear how a true statement asserting the existence of something with a Δ_0 property could also be false. The point is made by Shapiro (2002), and is answered in IC2, 17.8. The answer is to the effect that in an inconsistent arithmetic the identity relation (which is of course Δ_0) is itself inconsistent, and this can 'spread inconsistency' higher up the arithmetic hierarchy.

The grounds for certitude about the consistency of PA strike me as equally dubious. Gödel's paradox fails to be representable in PA only by a whisker, and almost by luck. And who is to say that there are not other paradoxes of this or a similar kind lurking in the area? So what is Field's certitude based on? Hardly the fact that the axioms are self-evident. The fate of Frege's axioms taught us a lesson never to be forgotten about that. Certainly not the fact that we have a consistency proof for the axioms: the proof is in a system stronger than the axioms themselves. Perhaps that we have a very clear intuitive model of the axioms? But the consistency of the picture is no better than the consistency of the axioms themselves. The fact that we have not found an inconsistency so far? Given the infinity of possible proofs, this is not a very good induction. And I note also that there are at least some mathematicians who take the possibility of such inconsistency very seriously; for example, the (non-Hartry) Fields Medal-winning Vladimir Voevodsky.[165]

27.14.5 *Paradoxes of Denotation*

Next (Sect. 6.9), we turn to a crucial matter where Field and I disagree: the paradoxes of denotation, and specifically Berry's paradox. Denotation can be defined in terms of satisfaction. So any model-theoretic construction that accommodates satisfaction accommodates denotation. Except that... the paradoxes of denotation have

---

[164] Assuming that it is consistent. And if it is not, then because it is based on classical logic, it is radically unsound!

[165] Voevodsky (2012); and his Princeton Colleague, Edward Nelson (2015). I should make it clear that I am not endorsing the work of either of these people. I merely cite them to show that some very good mathematicians do not share Field's incredulity. And just in case anyone is tempted to misunderstand what I have said here: I am not arguing against the consistency of PA; merely against our certitude that it is consistent.

peculiarities all of their own.[166] One is that they use some sort of description operator essentially; and once such is in the language, the proofs of standard fixed-point constructions (such as Field's) break down.

Now, IC 1.8 formalises the argument to contradiction in Berry's paradox in a logic which does not contain the PEM. So, Field's solution to the Liar paradox appears not to apply to it. The formalisation in IC uses a least number operator satisfying the principle:

[Mu] $\exists x A(x) \vdash A(\mu x A(x))$

the quantifiers ranging over natural numbers, appropriate precautions being taken to prevent the clash of bound variables. Note that this principle says nothing about what the denotation—if any—of '$\mu x A(x)$' is when nothing satisfies $A(x)$. Nor is there anything implausible about it, even when there are cases of denotation failure.

In his analysis of the argument, Field, taking it to use definite descriptions, objects to the inference from 'there is a n such that Fn' to 'there is a least n such that Fn' on the ground that it presupposes the PEM. However, there is no such step in the argument as formalised, so the objection is beside the point.

Of course, one might think that an *analogous* objection applies to the use of [Mu] directly. It does not. When one applies Field's argument directly to [Mu], one obtains the following. Let B be an arbitrary sentence, and let $A(x)$ be $x = 1 \vee (x = 0 \wedge B)$. Now clearly, $A(1)$, and so $\exists x A(x)$. Let τ be $\mu x A(x)$. [Mu] gives $A(\tau)$; that is, $\tau = 1 \vee (\tau = 0 \wedge B)$. This entails that $\tau = 1 \vee \tau = 0$. If $\tau = 0$ then we can rule out the first disjunct, and so B follows. If $\tau = 1$, then $\neg A(0)$, since 1 is the least n such that $A(n)$. That is, $\neg(0 = 1 \vee (0 = 0 \wedge B))$, which, given that $0 = 0$, entails $\neg B$.

But if one cannot assume the PEM, this argument fails. For that 1 is the least n such that $A(n)$ does not imply $\neg A(0)$. ($A(0)$ may be 'neither true nor false'.) Indeed, assuming that the extensional connectives in this context work in the standard way (say of K_3), if B is neither true nor false, so is $A(0)$. Hence, the argument for $B \vee \neg B$ begs the question.

Indeed, given an interpretation of the language of arithmetic which allows for the possibility that the PEM fails, this can be extended to an interpretation for the least number operator. '$\mu x A(x)$' denotes the least number, n, such that $A(n)$ (is true), if there is such (and whatever one wants to say about the matter if this condition fails). This verifies [Mu], and is a conservative extension, which does not, therefore, deliver the *PEM*.

Now, as I noted, Field's discussion is predicated on the assumption that the least number operator can be defined as a definite description: $\iota x(A(x) \wedge \forall y(y < x \rightarrow \neg A(y))$. This is not the μ-operator of the last paragraph. Of course, in a classical context, the two operators coincide. This just shows that an equivalence that works in a classical context can fail is a non-classical context of the kinds that Field and I endorse. That is a lesson which has been learned in non-classical logic many times over. Field's assumption of the equivalence would, therefore, seem to be vitiated by distinctions drawn in his own framework.

---

[166] See Priest (2006b).

27 Some Comments and Replies

Finally, as Field notes, the argument for Berry's paradox can be run equally well with an indefinite description operator, ε, instead of a least number operator, μ. Such operators are well known to deliver conservative extensions of the underlying logic—whether or not the underlying logic is classical. (One merely augments the semantics with a choice function, to be employed in the denotation conditions for ε-terms.) Field avers 'I don't think it's in the least clear that there there's much cost to regarding the ε-operator ... as illegitimate'. The thought, presumably, is that the notion is incoherent in some way. But it is quite coherent, both intuitively, and on all the standard semantics. So, there had better be an independent argument for this claim, or this is simple *ad hoc*ery.

27.14.6 Expressibility and Revenge

In Sects. 6.11, 6.12 Field turns to the topic of expressibility and revenge. He considers a predicate, M,[167] satisfying (for me) the condition (I quote):

[M1] It should be legitimate to accept $M \langle A \rangle$ iff it is legitimate to reject $\neg A$

and (for him):

[M2] It should be legitimate to accept $\neg M \langle A \rangle$ iff it is legitimate to reject A

I am not entirely clear what 'legitimate' means here, or, for that matter, the sort of 'should' that is in question. But I don't think that that is a crucial matter at the moment.

He notes that neither of us can accept the existence of a predicate satisfying the respective conditions, or All Hell breaks loose. He says that we should each take the existence of such a predicate to be 'an illusory ideal': there is no such notion. I agree with him completely. Of course, a lot more should be said about this, if the thought is not simply to be of the disappointing kind: 'if there is such an M, I'm in trouble'. For me, at least, the existence of M is delivered by Boolean negation, †. One may define $M \langle A \rangle$ as $T \langle A \rangle \land †F \langle A \rangle$. M can, in turn, be used to state the truth conditions for Boolean negation: $†A$ is true iff $M \langle \neg A \rangle$. I have said what I have to say about Boolean negation elsewhere (DTBL, Chap. 5), and there is no need to repeat it here.

Field notes that there is a way of obtaining part of what would be required by using an appropriate conditional. Thus, for me, $T \langle A \rangle \rightarrow \bot$ will do some of the job. But as he points out, for reasons to do with the Curry paradox this can be no more than partial. Indeed, asserting $T \langle A \rangle \rightarrow \bot$ will not even count as a rejection of A for all speakers. A trivialist will assert it, and reject nothing.[168] Field offers a hierarchy of predicates which approximate more and more closely the (illusory) ideal. I am not

---

[167] M for 'true with no monkey business'.
[168] On trivialism, see DTBL, Chap. 3.

inclined to follow him down this path. (Over the years, I have developed an antipathy to hierarchies—of all kinds.) I think that there are better ways of proceeding.[169]

Both Field and I think that there are some things that are true in a mundane sense ('Field is a person'); some things are false in a mundane sense ('Priest is a frog'), and some other things which are neither of these ('this sentence is not true'). The pressure for the existence of the predicate M comes from a certain take on talking about things that are or are not in the third category.

I think that there is a perfectly good way of doing this.[170] Things that are in the first category are those A such that $T \langle A \rangle \land \neg T \langle \neg A \rangle$; things that are in the second category are those A such that $T \langle \neg A \rangle \land \neg T \langle A \rangle$; things that are in the third category are those A such that $T \langle A \rangle \land T \langle \neg A \rangle$. Of course, these predicates do not express matters consistently. In particular, there will be some things that are in more than one category, such as the liar, $\neg T \langle L \rangle$. They express them none the less.[171]

Field says 'the absence of general [by which he means 'consistent'] notions of non-paradoxical truth and non-paradoxical falsehood ... makes life awkward: Priest frequently uses such notions in informal statements of his position'. It makes life awkward only for those who assume that I intend to be consistent. *Caveat emptor.*

Indeed, the demand for consistency is one that cannot be met: by me or by anyone else. There is nothing that can be asserted which forces consistency. A classical logician's assertion of †A does not do this. It merely guarantees that any inconsistency collapses into triviality. (In that way, it is like an assertion of $A \to \bot$.)[172]

Note that Field cannot do the dual thing. Thus, to say that something is in the middle category cannot be expressed—even allowing for a failure of PEM—by $\neg T \langle A \rangle \land \neg T \langle \neg A \rangle$, since this collapses into the contradiction $A \land \neg A$. This is a difference between Field and myself—and an important one.

Using the speech act of rejection, Field can express his attitude to statements in the third class. He can reject both A and $\neg A$. The trouble is that statements prefixed by force operators do not embed in propositional contexts. Indeed, if there were a predicate, M, which applied to just those statements in the third category, contradiction would arise.[173] For we could then construct a sentence, F, of the form $\neg M \langle F \rangle$. Suppose that $M \langle F \rangle$. Then F follows; that is $\neg M \langle F \rangle$. So we appear to have proved $\neg M \langle F \rangle$, that is F, and so $M \langle F \rangle$. This argument assumes the PEM, in the form $M \langle F \rangle \lor \neg M \langle F \rangle$. Hence, this must be rejected, but then we must reject the equivalent $F \lor \neg F$. So $\neg M \langle F \rangle$, and we are back with the contradiction anyway.

---

[169] Indeed, it is not at all clear that the hierarchy does avoid revenge problems, as Welch (2008, 2011) has shown.

[170] For more on what follows, see IC2, 20.4.

[171] As already noted in Sect. 27.14.3, above.

[172] Field say 'at first blush ... a dialetheist must assert that a sentence is non-paradoxically true in order to preclude a hearer from thinking that while he believes the sentence, he also believes ... its negation'. First blush indeed. That matter is taken care of by the Gricean conversational implicature of stating the whole relevant truth. *Exactly* the same point applies to the classical logician who asserts something. How is one supposed to know that they are not a dialetheist?—a point that I have heard Field himself make (in my defence) in seminars.

[173] See Priest (2005b), §3.

At this point, one can say that M is a meaningless predicate. Indeed, elsewhere Field does say something like this: it is 'ultimately unintelligible'.[174] Yet one needs a predicate of this kind if one is going to make generalisations about our three categories, such as 'Not everything is in the first or second category'. (If $\dashv$ is the force operator of rejection, $\exists A(\dashv A \land \dashv \neg A)$ makes no sense.)

Indeed, to add insult to injury, Field himself seems to give us such a predicate. Come back to [M1] and [M2]. Field uses the expression 'it is legitimate to reject'. There is no suggestion that this is meaningless—on the contrary. So A's being in the third category can be characterised by the predicate 'it is legitimate to reject A and it is legitimate to reject $\neg A$'.

Field and I have a long-running dispute about whether his account of the paradoxes avoids revenge paradoxes. It still seems to me that it does not.

27.14.7 Other Matters

I turn now to a few miscellaneous matters.

In Sect. 6.8 Field takes up the topics of set theory and model theory. Most of what I have to say about set theory and model theory I have said in IC2, Chap. 18. So only a few extra comments are required here.

I have endorsed a model-theoretic account of meaning, as opposed to an inferentialist account.[175] I don't have anything to add here to what I said about the matter in DTBL, Chap. 11. I might note that if the chips ultimately fall in the other direction, I don't have a problem with this.

The importance for me of a standard model is not that it delivers a compositional account of meaning: a model-theoretic account does this anyway. A model explains why one is justified in applying the logic to the subject of the model. A standard model of set theory explains why one is so justified with respect to set theory. Without that, one has to attempt some dodge, such as the Kreisel squeeze argument.[176]

In Sect. 6.12 Field raises the question of an appropriate paraconsistent set theory for model theory, and whether there can be a standard model in it. I have addressed this in my reply to Batens (27.3.2). So I need say no more about it here.[177]

In Sect. 6.10 Field takes up the issue of paradoxes of validity, and what he calls epistemic paradoxes. I agree with him that there is nothing much about the former

---

[174] Field (2008), p. 356. In Sect. 6.11.2 of the present essay, Field complains that I misrepresent him in Priest (2010c), p. 137, since I claim that he cannot define determinacy/indeterminacy. This is unfair. Of course I know Field's definition of D. In the passage, he cites (as I would have hoped that context makes clear) I am talking not about D, but what he calls 'super-determinacy', that is M.

[175] So it's not true to say (as Field does Sect. 6.11) that the dialetheist has no 'general notion of validity'. Validity is preservation of truth-in-an-interpretation, for every interpretation.

[176] See Priest (2010c), Sect. 10.

[177] I note the application of the Collapsing Lemma invoked to deliver models of set theory does not deliver a model of inconsistent arithmetic, but tweaking the collapse will.

which is relevant to his views about the paradoxes versus mine. The latter is a little different.

Field formulates a number of sentences concerning belief and cognate notions that might be thought paradoxical, and refers to Caie (2012). Caie discusses how a rational and perfectly introspective agent should respond to the sentence:

- I do not believe this very sentence to be true.

and advocates a solution involving a failure of the PEM.

Field does not discuss the relation of the paradoxical sentences he cites to Caie's; nor does he say much concerning his own thoughts about a solution to such paradoxical sentences—beyond pointing out that perfect introspectibility is implausible, and that the paradoxes 'put pressure on the coherence of the epistemic notions they employ'. So I will not discuss these examples further. Nor is this the place to discuss Caie's paradox and his detailed solution. So let me just note the following.

There is a paradox in the family concerning the sentence[178]:

- It is not rationally permissible to accept this sentence.

A dialetheist can simply accept the contradictory conclusion. Field cannot. The paradox does require one idealising assumption, namely that: if A entails B, then if it is rationally permissible to accept A, it is rationally permissible to accept B. And one might certainly doubt this in general. Thus, someone might not realise that A entails B. However, it is hard to reject the particular instance of the principle used in the argument, since the inference from A to B in question is not only very short, but is actually presented in the paradoxical argument.

The paradox also uses the premise that:

- It is not rationally permissible to accept that (A and it is not rationally permissible to accept that A).

Again, one might contest this, but since Field has not said how he would do this, there is not much more to say about the matter. At this point, Field has offered no solution to the problem. Neither, I note, does Caie's solution of rejecting the PEM appear relevant, since this is not used in the paradoxical argument. So whatever solution to the paradox Field envisages will be quite different from his solution to the Liar paradox—unlike a dialetheist's.

27.14.8 Wrapping Field Up

In summing up in Sect. 6.13, Field hangs the preference for his account of the paradoxes over mine on the matter of restricted quantification (the rest is merely aesthetic!). I have argued that that matter is not as decisive as he thinks. Much more important are the facts that his account cannot handle paradoxes such as Berry's and

---

[178] See Priest (2010c), Sect. 7 and DTBL, Sect. 6.6.

that it does not escape revenge paradoxes. Of course, I do not expect that Field will agree with me on these matters, and knowing him, there will be more to be said about these things!

27.15 Girard (Chap. 14): Possibly Impossible

In his essay, Patrick Girard takes us into the world of dialetheic conditional and modal logic, an important area. In the first part of his paper, he deploys a variable accessibility relation to accommodate impossible worlds. In the second, he takes the much more daring step of a semantics which is itself dialetheic. Let me comment on these two parts separately.

To accommodate conditionals with impossible antecedents in a sensible fashion, given a worlds semantics for these, one needs some sort of impossible worlds.[179] In standard modal logic, it is normal to think of the accessibility relation, xRy, as saying that y is possible relative to x—in whatever sense of possibility is in question. If it is not the case that xRy then, in that sense, y is impossible relative to x. This notion of impossibility can then be deployed for conditionals with impossible antecedents.

Such a notion of impossibility will do for many kinds of impossiblilty, such as physical and—contradictory obligations aside—moral. Girard wants to extend this to other kinds of impossibility. This is certainly fine for some of them. Thus, false mathematical statements are normally taken to be impossible. Yet there is nothing in the semantics of standard modal/conditional logic which requires them to be true at all worlds of an interpretation. So we can, as Girard does, take these things to fail at worlds outside the R-equivalence class of worlds containing the actual world.

Exactly the same is generally true of metaphysical impossibilities.[180] There is an issue here about identity statements, however. Many people, following Kripke, hold that true statements of identity are (metaphysically) necessarily so. Thus, it is necessarily true that Hesperus is Phosphorous. Now, Girard gives the semantics for only a propositional logic, and so he does not say what semantics he is using for identity. However, if we stick with classical semantics (as Girard says that he does in the first part of his paper), since 'Hesperus is Phosphorus' is true at the actual world, it will be true at all worlds, possible and (Girard's) impossible. Given this, conditionals with false identity statements may well come out with the wrong value, such as if Hesperus were not Phospherus, modern political philosophy would be badly mistaken. (Compare: if Hesperus were not Phospherus, modern astronomy would be badly mistaken.)[181]

---

[179] See Priest (2016b), 3.3, Berto et al. (2018), and the references therein.

[180] Assuming there to be such. For a skeptical view, see Priest (201+d).

[181] To accommodate such conditionals, the truth conditions for identity need to be non-classical at impossible worlds. See INCL, 23.6, 24.6.

Perhaps most importantly, the construction cannot handle counter-logicals. Thus, as Girard points out, suppose that the liar sentence, L, is both true and false, as for dialetheism about the semantic paradoxes. Then the following is true:

[L] If $L \land \neg L$, then a consistent solution to the semantic paradoxes is correct

This is intuitively false, though it is true in the semantics, since the antecedent is true at no worlds.

The problem with counter-logicals does not end there, though. For example, consider the conditional:

- If intuitionist logic and philosophy are correct, then $G \lor \neg G$

where G is a statement of Goldbach's conjecture. This is presumably false. Yet, given the classical semantics for disjunction and negation, the conclusion holds in all worlds, so the conditional is true. Or again, whatever the relevant notion of conditionality at issue, let the *Law of Identity* refer to the logical truth of the schema 'If A then A'. Now consider:

- If every instance of the Law of Identity fails, then if G then G

This certainly appears false, though the consequent is true in all Girard worlds. To handle counter-logicals generally, one needs, precisely, worlds where any kind of logic can hold (or fail).

In the second part of his paper, Girard advocates moving from a classical logic to a paraconsistent logic to handle dialetheism about sentences such as [L]. Naturally, I am happy to go along with this. However he goes further, giving the semantics of the logic in a paraconsistent metatheory, indeed, a metatheory based on naive set theory.

He gives two reasons for this. One is the need to handle paradoxes about worlds that may turn up in the metatheory. I am on board with this too.[182] The second is a problem noted by Martin (2014). Suppose that we are working in the logic *LP*. Then in all the worlds, including the actual world, @, for any A, $\neg(A \land \neg A)$ holds. Hence $\Box \neg(A \land \neg A)$, that is $\neg \Diamond(A \land \neg A)$ holds at @. But for some Bs, $B \land \neg B$ holds at @. Hence something impossible happens at @. So @ is an impossible world.

I'm not persuaded by this, since the last step doesn't follow. What follows is simply that something contradictory holds at @. We are assuming that contradictions may hold at @. Given that the accessibility relation is reflexive, $\Diamond(B \land \neg B)$ holds at @, and so $\Diamond(B \land \neg B) \land \neg \Diamond(B \land \neg B)$ is just one of them; some of the contradictions that hold at @ concern what is possible itself. @ is a possible world none the less. For me, a possible world is a world where the laws of logic are the same as those of the actual world, so an impossible world is a world where they are different. (See INCL, 9.7.) @ is then possible, by definition. But even for Girard, the possible worlds are, *ex hypothesi*, those accessible from @ via R. To make @ impossible, it would have to be the case that @R@ does not hold (and @ would then be both possible and impossible).

---

[182] See Priest (2018c).

To implement his semantics, Girard formulates this is a naive set theory based on a relevant logic (where the conditional does not contrapose, and so Martin's argument breaks down).[183] However, the most distinctive feature of the semantics is that it dispenses with an independent notion of falsity, and uses homophonic truth conditions for negation:

- $\neg A$ is true at a world, w (in an interpretation) iff it is not the case that A is true at w

The thought that one might do this is a very natural one, and has occurred to many people.[184] A cost is that it makes the semantics themselves inconsistent, but if its underlying logic is paraconsistent, where is the problem in this?[185]

By way of reply, note that, though a homophonic semantics is very natural, it is not mandatory. A non-homophonic semantics is perfectly appropriate for many purposes, such as relating the semantics to metaphysical concerns. Indeed, I note that Girard himself uses non-homophonic clauses for the modal operators.[186]

Next, note that the homophonic truth conditions spread contradictions. Thus, any contradiction, $A \wedge \neg A$, in a world of an interpretation delivers a contradiction *about* it: A is and is not true at the world. Contradictions should not be multiplied beyond necessity (IC, 4.9). Moreover, the contradictions appear to spread where they are really not wanted. PEM ($A \vee \neg A$) is a logically valid schema. But take a world of an interpretation, w, where for some B, $B \wedge \neg B$ holds. Since $\neg B$ is true at w, B is not true at w. And since B is true at w, $\neg\neg B$ is true at w, so it is not the case that $\neg B$ is true at w. So $B \vee \neg B$ is not true at w. Hence PEM is not a logically valid (as well). More generally, let w be the trivial world. Then for any A, $\neg A$ is true at w, so A is not true at w. Hence there are no logical truths.[187]

I note, finally, that moving to a paraconsistent logic certainly removes the problem concerning the conditional [L]. However, it does nothing to rectify the shortcomings of the earlier semantics with respect to counter-logical conditionals in general.[188] Thus, since $G \vee \neg G$ is true in every world, the conditional:

- if intuitionist logic and philosophy are correct, then $G \vee \neg G$

is still true.

---

[183] I prefer to formulate the semantics in naive set theory in a rather different way. See my reply to Batens, Sect. 27.3.2 above

[184] And I note that this move is quite independent Girard's strategy concerning impossible worlds.

[185] Girard says (Sect. 14.3) that his semantics is not recursive, and the clauses must be thought of simply as semantic axioms. This seems to me to be a mistake. The clauses are perfectly well-grounded, and if the set theory has the means to turn a recursive definition into an explicit definition, the axiomatic clauses can be turned into such a definition.

[186] A homophonic clause for $\Box$ would be: $\Box A$ is true at a world w (of an interpretation) iff $\Box(A$ is true at $w)$.

[187] Weber, Badia, and Girard (himself) (2016), point out a number of the fraught model-theoretic consequences of what amount to homophonic truth conditions for negation.

[188] Or conditionals with false identity statements as antecedents, unless one modifies the usual semantics for identity.

Hence, I am not inclined to go down the homophonic path along which Girard beckons.

One final and more tangential comment on Girard's paper. He endorses the principle POS: if the antecedent of a true conditional is possible, so is the consequent. I did think this true at one time, but Dave Ripley and Yale Weiss persuaded me that I was mistaken. When evaluating a conditional, one looks at worlds where the antecedent is true, and where certain information bleeds across from the world of evaluation.[189] Normally, if the antecedent of the conditional is possible, the bleeding will not take us to an impossible world. Thus, to evaluate the conditional, 'if I jump out of a 17th floor the window, I will get hurt', the information that bleeds across concerns the laws of gravity, human biology, etc.; and we evaluate the consequent at worlds where these and the antecedent hold. Nothing forces such worlds to be impossible. However, suppose that we have a correctly programmed computer that searches for a proof in Peano Arithmetic of its Gödel sentence; a light will go on iff it finds it. Now, consider the conditional: if the light were to go on, something impossible would have happened. One can hear this as true, and there is nothing impossible about the antecedent. Yet if one hears it as true, the information that bleeds across from the base world includes the claim that the machine is working correctly; and if this and the antecedent are realised, we must be in an impossible world.[190] Of course, without this piece of information bleeding across, it would be more plausible to accept the truth of the following: if the light were to go on, the machine would not be working correctly. In general, what information bleeds across in the evaluation of a conditional is contextually dependent. POS is not, then, an appropriate *logical* constraint, though it may well hold for many (most?, normal?) contexts.

27.16 Humberstone (Chap. 15): Everything You Wanted to Know About *LP* Negation...

Let us now turn to Lloyd Humberstone's paper. Humberstone and I have known each other for many years, and our intellectual paths have often crossed—over the last 18 years at meetings of the Melbourne Logic Group, but many times before that as well. His logical interests and mine certainly do not coincide, but there is a significant overlap, and I have often benefitted from his insights. The present paper shows the scholarly knowledge, logical acumen, ingenuity, and thoughtful care which characterise his work, such as his magisterial book on the logical connectives.[191]

The present paper is rich in insight, and one thing it shows clearly is that the characterisation of a number of notions is a very sensitive matter. Humberstone is clear about distinctions which many others slide over. In what follows I will restrict myself largely to addressing two issues concerning *LP* which the paper raises:

---

[189] See Priest (2018b).
[190] This particular example is due to Vander Laan (2004).
[191] Humberstoneone (2011).

27 Some Comments and Replies

the uniqueness of the characterisation of negation; and negation as a contradictory forming operator. I will deal with these in the next two sections. In the third, I will comment on some specific remarks by Humberstone, which I was unable to integrate into the more general discussion without disrupting it too much.

27.16.1 LP Negation: Uniqueness

I take it that the fundamental property of negation is that it toggles between truth and falsity. That this holds for negation can be agreed by partisans of many different logics. I am, of course, aware that there are those who would demur: there are, after all, many theories of negation. Still, the aim here is not to defend a theory of negation, but to discuss its characteristics. Given this property, we have:

- $\neg A \in T$ iff $A \in F$
- $\neg A \in F$ iff $A \in T$

where T is the set of truths, and F is the set of falsehoods. If one thinks that $\neg A$ is true iff A is not true, then one can simply give truth conditions for negation, and the falsity conditions will go along for the ride. However, once one is contemplating logical gaps or gluts, truth and falsity have to be treated even-handedly.

Suppose that we generalise these conditions from truth *simpliciter* to truth in an interpretation. If the logic is one in which truth and falsity are the only semantic considerations then, in every interpretation:

- $\neg A$ is true iff A is false
- $\neg A$ is false iff A is true

These conditions are those of many logics. If there are no further constraints on truth and falsity, and conjunction and disjunction behave in a standard way, we have the logic *FDE*. If one insists that nothing can be both true and false (in an interpretation), we have K_3. If one insists that everything is either true or false, we have *LP*. If one insists on both, we have classical logic. (I assume that validity is defined as truth preservation in all interpretations.) Which, if any, of these constraints is correct is a separate semantic/metaphysical issue. I endorse those of *LP*, though this is not the place to go into the matter.[192] As I have just said, the point here is not to defend a theory but to explore its properties.

These conditions then define $\neg$ uniquely, in at least a couple of different senses. Thus, suppose there is another operator, $\neg'$, which satisfies the same conditions. That is, in any interpretation:

- $\neg' A$ is true iff A is false
- $\neg' A$ is false iff A is true

---

[192] That position is defended in IC, Chap. 4.

then in any interpretation:

- $\neg A$ is true iff $\neg' A$ is true
- $\neg A$ is false iff $\neg' A$ is false

It follows that:

- $\neg A \dashv \vDash \neg' A$

There is also a stronger sense of uniqueness. In classical logic, an operator is called a truth function if whenever the inputs have the same truth value, the output has the same value. In the present context, we have to say, instead, that whenever the inputs have the same truth and falsity values, the output has the same truth and falsity values. In the language of *LP*, all the connectives (including $\neg'$ if this is added) have this property. And provided all the connective have this property, we have:

- $C(\neg A) \dashv \vDash C(\neg' A)$

If our logics are extended to ones with a world semantics, matters are slightly more complicated. Let us say that a logical operator is *truth functional in a modal sense* if whenever all the inputs have the same truth/falsity value at every world, so does the output. This condition is satisfied by the usual extension of the language of *LP* by modal operators, and/or a conditional connective.[193] Then the previous two bullet points still hold.[194]

Nothing in this section gainsays any of Humberstone's results. As he makes clear, e.g. in his discussion of sequents in Sect. 15.5, he is concerned with a notion of unique characterisation in proof-theoretic terms. By contrast, here I have given a notion of unique characterisation in semantic terms.[195]

27.16.2 LP Negation: Contradiction Formation

Let us now turn to the question of whether the negation of *LP* is a contradiction-forming operator—hereafter cdfo.[196] First, some context. I have claimed that negation is a cdfo, and have argued that the negations of many logics do not satisfy this

---

[193] See INCL, Chaps. 9, 10, and 11a.4.

[194] There are semantics with impossible worlds in which (some) sentences are assigned arbitrary truth/falsity values at these. If the truth/falsity values of $\neg$ and $\neg'$ can come apart at these worlds, then one will still have $\neg A \dashv \vDash \neg' A$, validity being defined as truth preservation at all possible worlds. However, one may no longer have $C(\neg A) \dashv \vDash C(\neg' A)$. The same is true, of course, of $\wedge$ and $\wedge'$, $\vee$ and $\vee'$, etc.

[195] In particular, Humberstone's function which is the same as the $\neg$ of *LP*, except that it maps b to f does not satisfy the condition that if A is false (i.e. b or f), $\neg A$ is true (i.e. b or t).

[196] Since Humberstone objects to 'cfo'.

condition.[197] Slater agreed that negation is a cdfo, but charged that the negation of *LP* is itself not a cdfo.[198]

To look at the matter, we first need to address the question of what contradictories are. Traditional logic is pretty clear on this matter. Consider the following conditions. Let us call these the *naive* conditions:

- it can't be the case that both A and B
- it must be the case that (at least) one of A and B

If A and B satisfy the first but not the second, they are contraries. If they satisfy the second but not the first, they are subcontraries. If they satisfy both, they are contradictories. Note that this is what Humberstone terms 'mere' contraries/subcontraries.

But how should one understand the modal terms deployed here? Any way of making matters precise will deliver an understanding of what it is to be a contrary/subcontrary/contradictory. And as Humberstone's paper makes abundantly clear, there may be many ways one might do this, delivering different non-equivalent notions.

One way to understand these two conditions (respectively) is as follows:

- there is no interpretation in which both A and B are true
- in any interpretation at least one of A and B is true

This is pretty much what Slater means in his critique of paraconsistent negation. However, it is not a happy way of cashing out an understanding of the notions, at least if that understanding is the standard one in the history of Western logic. For a start, the notion of an interpretation is a creature of the last 100 or so years, and is alien to the thought of Aristotle, Buridan, Leibniz, Kant, etc. Worse, a standard example of contraries is: 'x is red' and 'x is green'. In modern logic, one can but write these as Rx and Gx. But there are interpretations where something can satisfy both these predicates. So they are not contraries on this account. Similarly, a standard example of subcontraries is: 'part of x is the Northern Hemisphere', and 'part of x is in the Southern Hemisphere', as applied to geographical features on the Earth. Writing these as Nx and Sx, respectively, there are interpretations where nothing satisfies both, and so they are not subcontraries according to this account.[199]

A better understanding of these conditions is to take the modal operators deployed at face value. The conditions then become:

Con $\quad \neg \Diamond (A \wedge B)$, i.e. $\Box \neg (A \wedge B)$
SubCon $\quad \Box (A \vee B)$

---

[197] DTBL, Chap. 4, and Priest (2007).

[198] Slater (1995, 2007).

[199] Another possible understanding of what it is for A and B to be contraries is mooted by Humberstone (Sect. 15.3): $A, B \vdash$. (Similarly, with $\vdash A, B$ for subcontraries, and with both for contradictories.) Clearly, this will not do in general, for similar reasons. One would not expect it to be a fact of logic that: $Rx, Gx \vdash$. Nor is it appropriate when B is $\neg A$. For this simply builds Explosion into the definition of a cdfo, and so begs the question against a paraconsistentist, as Humberstone notes later in the section.

And what is the negation sign here? It is whatever those who deployed the informal locutions intended. That is, it is the negation of ordinary language. How that behaves is, of course, a contentious matter. But those of an *LP* persuasion, such as myself, will take this to be *LP* negation.

For negation to be a cdfo in this sense, we then have:

- $\Box \neg (A \wedge \neg A)$
- $\Box (A \vee \neg A)$

And since we are dealing with negation, which is a matter of logic, and not red/green, north/south, we should expect these to be logical truths. Moreover, if our modal operator satisfies standard conditions, and in particular Necessitation (if $\models A$ then $\models \Box A$) and veridicality ($\Box A \models A$) then these are equivalent to:

- $\models \neg (A \wedge \neg A)$
- $\models A \vee \neg A$

Now neither nor *FDE* nor K_3 satisfies both of these conditions; but *LP* does. So in this sense, the negation of *LP* is a cdfo.[200]

There other logics where these conditions are not met. Thus in intuitionist logic one has the first, but not the second; and in da Costa's *C*-systems one has the second but not the first. Now, the whole issue about cdfos arose originally because Richard Routley (as he then was) and I argued on the basis of this observation that the negation of the *C*-systems is not a cdfo, but a (mere) subcontrary-forming operator.[201] This point, then, stands.

Of course, we may, in the end, have to conclude that, according to our best theory of negation, negation is not a cdfo in this sense. But the fact that an account of negation does not deliver an operator which is a cdfo in the naive sense, is a black mark against it. So much of the history of logic would, then, have to be written off as mistaken.[202]

Let us now turn to the matter of the uniqueness of contradictories. DTBL, p. 78, says that contradictories, unlike contraries and subcontraries, are unique. How may one best understand this claim?

As I have suggested, A and B are contradictories iff $\Box(A \vee B)$ and $\Box \neg (A \wedge B)$. Now, if $\bot$ is a logical constant such that $\bot \models C$, for all C. Then in classical logic B is logically equivalent to $B \vee \bot$. So if A and B are contradictories, so are A and

---

[200] Humberstone (Sect. 15.1) quotes a passage from DTBL and says, 'I read this as saying that for a one-place connective, #, to count as a c[d]fo in logic S it is necessary and sufficient that we have, for an arbitrary formula, A, of the language of that logic ... $\vdash_S A \vee \#A$ and $\vdash_S \#(A \wedge \#A)$', and points out that these conditions are satisfied when $\neg A$ is $\top$. Whether or not this is a reasonable interpretation of the passage, it is not what I intended, as I hope is now clear. That the wide scope # is negation is presupposed.

Yet another way of interpreting negation as a cdfo is Wansing's as discussed by Humberstone in his Appendix B. This is that $A \vdash \neg\neg A$ and $\vdash \neg(A \wedge \neg\neg A)$. These conditions are satisfied in intuitionist logic. The failure of Double Negation in intuitionist logic makes these a particularly unsuitable way of cashing out the naive (and traditional) understanding of the notion.

[201] Priest and Routley (1989a), p. 165.

[202] For further discussion, see DTBL, especially 4.4.

$B \vee \bot$. Hence, different formulas can be contradictories of A. Moreover in LP, B is also logically equivalent to $B \vee \bot$. So the same point applies.

However, in classical logic:

(*) $A \vee B, \neg(A \wedge B), A \vee C, \neg(A \wedge C) \models C \equiv B$

Hence, $\Box(A \vee B), \Box\neg(A \wedge B), \Box(A \vee C), \Box\neg(A \wedge C) \models \Box(C \equiv B)$. So contradictories are unique up to necessary material equivalence. One might think of $\Box(B \equiv C)$ as saying that B and C express the same proposition. In this case, there is a unique proposition which is the contradictory of A.

This is not true in LP, where (*) does not hold. (Make A both true and false.) However, we do have:

(**) $A \vee B, \neg(A \wedge B), A \vee C, \neg(A \wedge C) \models (C \equiv B) \vee A!$

where $A!$ is $A \wedge \neg A$. Hence, we have $\Box(A \vee B), \Box\neg(A \wedge B), \Box(A \vee C), \Box\neg(A \wedge C) \models \Box((C \equiv B) \vee A!)$. Let us say that B and C are *materially equivalent relative to A* if $(C \equiv B) \vee A!$ Contradictories of A are then unique up to necessary material equivalence relative A. If one thinks of $\Box((B \equiv C) \vee A!)$ as saying that B and C express the same proposition relative to A, then the contradictory of A is a unique proposition relative to A.

Note that this is not the same with respect to subcontraries, mere or not (or contraries—the dual considerations apply). Classically:

- $\Box(A \vee B), \Box(A \vee C) \not\models \Box(B \equiv C)$

So subcontraries of A are not unique up to necessary material equivalence. And in LP:

- $\Box(A \vee B), \Box(A \vee C) \not\models \Box((B \equiv C) \vee A!)$

Hence, subcontraries of A are not unique up to necessary material equivalence relative A.

27.16.3 Some Further Comments

In this final section, I will comment on a number of further claims in Humberstone's essay, which I think merit a mention.

(i) In Sect. 15.2 he quotes a passage from DTBL and says that it is rather hard to understand. Let me see if I can do better. Let us suppose that some things are neither true nor false; that is, for some As neither A nor $\neg A$. One might object that $\neg$ is not really a cdfo, since the PEM is violated. (So $\neg A$ should hold in all the cases that A fails.) Whether or not this is a good objection, the parallel objection to someone who holds that some things are both true and false—that is, for some As, both A and $\neg A$—does not work, simply because logical gluts do not violate the PNC, as we have seen.

(ii) In a footnote to the same passage, Humberstone says that he does not understand the notion of surplus content. I agree that its explanation in Priest (2007) is not very clear. What I had in mind was this. Classically, $\neg(A \wedge \neg A)$ rules out $A \wedge \neg A$. In a paraconsistent context, it does not, so there can be a 'surplus'.[203] Of course, classically a contradiction entails everything, and paraconsistently, it does not. So in that sense, it is a classical contradiction that has a surplus.

(iii) A little later in the section, Humberstone notes that defending the view that $\neg \Diamond (A \wedge \neg A)$ is a statement of the PNC by pointing out that it allows for $\Diamond (A \wedge \neg A)$, is, in effect, question begging against Slater. But of course, to say that it does not do so because of this, is equally question begging against me. Questions of onus of proof are tricky. However, in this context, it was Slater who initiated the argument by claiming that the negation of *LP* is not a cdfo; and so the onus of proof must fall on him.

(iv) In fn 22, Humberstone says that I follow the motto: when the going gets tough, go homophonic. I don't think that's right, if by 'homophonic' one means giving the truth conditions of some notion using that very notion. Though this may be a perfectly sensible strategy, it is not one that I fall back on when 'the going gets tough'. He references a passage in DTBL, but the point here is not about homophony but about using the same logic in the object and metalanguage—a quite different point, and one which I do not fall back on: it's right up front.

(v) At the end of Sect. 15.3, Humberstone comments on a passage in DTBL, where I argue that Con and Subcon deliver Double Negation. He points out, correctly, that this presupposes the uniqueness of contradictories, and objects that Con and Subcon do not deliver such, on the ground that they hold when negation is interpreted as $\top$. As I intend them, however, Con and Subcon require that we are dealing with the correct negation—that of *LP*—as I have already said. It may fairly be replied that since Double Negation holds in *LP*, the argument does, in a certain sense, presuppose this.

(vi) In a digression at the start of Sect. 15.6 Humberstone comments on a passage in DTBL where I discuss a certain notion of logical pluralism. I claim that classical negation and intuitionistic negation have different meanings,[204] but that one cannot put them together in a single language. He says that he infers from the latter fact that they do not have different meanings. I infer that it may not be possible to combine independently meaningful things. After all, Boolean negation—assuming it to be meaningful—cannot be combined with the intuitionistic conditional without this collapsing into the classical conditional; similarly, Boolean negation cannot be combined with a naive truth predicate without total collapse.[205] Again, as they say, one person's *modus ponens* in another's *modus tollens*.

---

[203] I note also, as, in effect, does Humberstone in Sect. 15.4, that $A \wedge \neg A$ entails $\neg A \wedge \neg\neg A$. So, in this sense, a dialetheia is neither true nor false. Even more surplus!

[204] Incidentally, in a footnote, Humberstone asks what truth conditions I had in mind for intuitionistic negation. It was those in Kripke semantics.

[205] See DTBL, 5.5.

27.17 Istre and McKubre-Jordens (Chap. 17): Relevant Conditionals and Naive Sets

With Erik Istre and Maarten McKubre-Jordens, we venture into the world of relevant naive set theory.

I have always thought that the correct approach to the paradoxes of set theory is to accept the naive principle of set abstraction and the consequent paradoxical contradictions, but to use a paraconsistent logic to prevent the spread of contradiction.[206] A natural thought is to use an appropriate relevant logic as the paraconsistent logic in question. In the late 1970s, Brady's proof showed that this could be done without triviality.[207]

Of course, showing what can't be proved is not sufficient. One also needs to show what *can* be proved. In particular, one needs to show how one can carry out natural set-theoretic reasoning, such as that pursued by Cantor. The aim was never to be revisionary, in the way that intuitionism is. And this is not a trivial problem. The set theoretic axioms are strong, but to avoid triviality the underlying logic must be much weaker than 'classical' logic. Simple reasoning about basic set-theoretic operations, such as unions and compliments, is routine,[208] but after that, the going gets tough.[209]

This problem exercised many of those in the Canberra logic group in the late 70s and early 80s, including Routley, Meyer, Brady, Mortensen, Slaney, and myself. The paper by Istre and McKubre Jordens (hereafter I&MJ) well explains the sort of problems we hit. I am not persuaded by all the ones they note. Thus, they claim (Sect. 17.3) that there is a problem with meta-rules, since these 'make a claim about what proofs can be constructed'. Not so; they are constitutive of what *counts* as a proof.[210] However, problems of the kind they document are very real. So much so that, at the time, all of us, I think, gave up trying to solve them. I certainly did. I always hoped that someone might be able to do it; but after that, I started to think about other ways of achieving the 'classical recapture' in set theory, and especially how to use a model-theoretic approach to achieve this end.[211]

A highly significant advance on the problem was made by Weber in his Melbourne PhD thesis of 2009, where he showed how to obtain most of the standard results concerning transfinite ordinals and cardinals.[212] His trick was not to try to reconstruct the classical proofs—as we had been trying to do—but to develop quite new proofs which deploy inconsistency essentially. (And whether or not one is persuaded by the legitimacy of Weber's techniques, they are certainly fascinating.)

---

[206] See IC, Chap. 3.

[207] Brady (1989).

[208] This was all spelled out in Routley (1977).

[209] And without the PEM one seems completely hamstrung, as I&MJ note (fn 1).

[210] A very minor comment on their paper. They say in Sect. 17.2 that $A \to (A \wedge A)$ is not provable in *DKQ*. This is surely a slip. It follows from their A1 and A6.

[211] See IC2, Chap. 18, Priest (2017), Sects. 27.10, 27.11, and 27.3.2 above.

[212] See Weber (2010, 2012).

That work, however, still leaves much to be done. In particular, one needs to show that things like model theory can be done in this relevant setting. Weber has continued to work on this, sometimes tweaking the underlying relevant logic in the process;[213] and maybe something like this can be done. But at the moment, problems of the kind that I&MJ document still stand in the way. Without solutions to these, I think that the best approach to paraconsistent set theory is the alternative one I indicated above.

27.18 Mares (Chap. 16): They Are, Are They?

For a long time, relevant logic was something of an outlier in the family of non-classical logic, and something of an ugly duckling. That is started life simply in axiomatic form didn't help. And when a robust semantics appeared, in the shape of Routley/Meyer semantics, matters seemed to become worse. First, the semantics appeared complex and somewhat awkward. Next, it became clear that the family of relevant logics was enormous. That, in itself, is not a serious problem. The same is true, of course, of modal logics. There, one just has to say which notion of modality one has in mind, and choose the most appropriate system for it. The same for relevant logic: one just has to say which notion of conditionality one has in mind, and choose the most appropriate system for that. The problem is that conditionality is a much more vexed issue than modality, and so it is much less clear how to choose. And the semantics were, in a sense, too good: one could do nearly anything with them! Not only that, but the machinery deployed, notably the ternary R and the Routley $*$-operator seemed to want for a plausible philosophical interpretation. Hence their properties could provide no guidance in this matter.

The situation is now much improved. Philosophical issues are slowly coming into focus.[214] The formal machinery has also been shown to have interesting technical properties.[215] Moreover, intimate connections have emerged with the family of substructural logics, with information and its flow, and with other accounts of conditionality.[216]

It can now be said that, at least as far as propositional logic goes, the machinery is well under control, both philosophically and technically. The same is not true of its extension to a first-order context, however. Questions concerning predication, quantification, identity, are still vexed. Ed Mares, another stalwart of the band of Australasian logicians, has been at the forefront of work in this area, trying to bring order into the complexity.[217] His paper here continues his investigations of identity in the context of relevant logic. Drawing on an analogy between identity, as a binary

---

[213] See, e.g. Weber (2016a).
[214] See, e.g. Restall (1999), Beall et al. (2012).
[215] See, e.g. Urquhart (1984).
[216] On which see, respectively, Restall (2000), Mares (1996), and Beall et al. (2012) again.
[217] As well as in investigations of the philosophical basis of relevant logic, as in Mares (2004).

relation between objects, and biconditionality, as a connective between sentences, he provides two semantics for identity, and discusses their properties.

As he says, there is a clear analogy between identity and the biconditional. It certainly does not mandate the view that the two behave in an isomorphic fashion, and, as Mares notes, analogous principles may well be semantically independent.[218] However, the analogy is, at the very least, highly suggestive. Mares' first semantics is simple and clean[219]; the second, somewhat less so. In part, this is due to the presence of the ternary R; but only in part. The semantics has a feeling, often voiced concerning the original Routley/Meyer semantics, that they are being 'rigged to get what you want'. The feeling is exacerbated by the presence of the predicate *Ind*. Given the need for this, one will naturally ask: why not give the truth conditions of identity using this, and have done with it?

Still, the semantics are clever and interesting. Rather than discuss the details further, I will just comment on one of the more general matters Mares raises. That is, the schema he calls *FullSub*:

- $(a = b \wedge A_x(a)) \rightarrow A_x(b)$

This is a very natural principle concerning identity; and one which, for the appropriate As, one would need a good reason to hold to be incorrect.

Of course, one should not expect this principle—or even more restricted versions of substitutivty—when A may contain intensional verbs, such as 'believes', 'admires', etc. One can believe that Richard Routley was Richard Routley without believing that Richard Routley was Richard Sylvan. Or one can admire Routley without admiring Sylvan, believing this to be a different person. But there appear to be counterexamples even in the more limited context of the vocabulary of first-order logic. In particular, substitution into the scope of a conditional appears to have counterexamples. Thus,[220] it is clearly true that:

- If the Morning Star is (were) not Venus, modern astronomy is (would be) badly mistaken.

But the Morning Star is Venus. However, it does not follow that:

- If Venus is (were) not Venus, modern astronomy is (would be) badly mistaken

It would not be modern astronomy that is mistaken, but modern logic. Or again, consider:

- If I am (were) not Graham Priest then, I am not (would not be) the author of *In Contradiction*.

But I am Graham Priest. Yet it does not follow that

---

[218] In particular, there might well be good reasons to endorse what Mares calls FullSub, but not Pseudo *Modus Ponens*.
[219] A small comment. He says (Sect. 16.4) that the normal worlds of N_* are complete. This is not so. See INCL 9.6.
[220] See INCL, 19.5.

- If Graham Priest is (were) not Graham Priest then, he is not (would not be) the author of *In Contradiction*.

Whatever the consequences of the failure of the Principle of Identity are, the consequent doesn't seem to be one of them.[221]

Next, in the semantics of INCL2, 24.6 (though not in Mares' semantics) it is the Subset Constraint that validates FullSub. The constraint is to the effect that the extension of $=$ at a non-normal world is a subset of its extension at normal worlds (the set of pairs $\langle d, d \rangle$, where d is a member of the domain). Concerning this constraint, Mares says (Sect. 16.8):

> There is a problem with the subset constraint. Consider a model in which 'Hesperus is Mercury', does not hold in any normal situation. Then, in no non-normal point is Hesperus identical to Mercury. So we have valid on the model
>
> > Hesperus is Mercury implies that the moon is made of green cheese.

Let us disentangle what is going on here.[222]

For a start, the subset constraint does not give rise to 'fallacies of relevance', at least in one sense. It can be shown, for example, that in the quantified relevant logic B, with constant domain, and identity satisfying the subset constraint, for any logical truth of the form $A \to B$, A and B have a predicate (maybe identity) in common.[223] In particular, then, with obvious notation $h = m \to M$ is not a logical truth.

Moreover, it is not a problem that this conditional is true in some models. After all, one can make any non-logical truth hold in some model—e.g. 'Hesperus is Mercury' and 'Some red things are not coloured'. Irrelevant conditionals could simply be like this.

However, there is a genuine worry here. Consider the model in which truth values get assigned correctly. Then, assuming the necessity of true identities, in that model $h = m$ does not hold at any normal (possible) world, and so, given the Subset Constraint, at any world. It follows that $h = m \to M$ is true at the base world of the model, and so true *simpliciter*. This certainly does not seem to be right.

One can, of course, have a restricted form of FullSub, where one does not substitute into a conditional context (and more generally, a context whose evaluation requires a world shift). To do this, one requires a contingent identity logic (as noted in INCL, 19.5.8). Mares indicates that this is the way he is inclined to go. He motivates the position as follows. Suppose that in our model's true identities are necessarily so. That

---

[221] Substitutivity into a conditional hold in Mares' systems. However, the conditionals in these systems are essentially entailment conditionals, and the conditionals in the counterexamples are hardly of this kind. One might not, therefore, be too worried by them.

[222] I note that Mares uses the word 'implies' for the conditional. I think it best to avoid this. In my experience it is at the root of much confusion in students between the conditional and the validity relation.

[223] The proof extends that of INCL, p. 220, ex. 11. For details, see (Priest, 2019b) Sect. 3.

is, if an identity statement is true at the base world, it is true in all normal (possible) worlds. And suppose that 'Hesperus in Venus' is thus true. Then (Sect. 16.8):

> the following is valid on the frame:
>
> Hesperus is Mercury implies that Mercury is Venus.
>
> This sentence is, I think, true... But I do not think that the counterfactual version is true:
>
> If Hesperus were Mercury, then Mercury would be Venus.
>
> In stating the counterfactual one is asking the audience to imagine what the world would be like like if Hesperus were not Mercury. They might imagine that a bright star someone pointed to at in the evening turned out to be the closest planet to the sun, and that there is another planet between the Earth and the sun which is Venus.

Now, I am not persuaded that there is a difference of kind between indicative and subjunctive conditionals. In particular, there does not seem to me to be much difference between conditionals with antecedents in the present indicative and the present subjunctive.[224] Moreover, someone who holds that conditionals with necessarily false antecedents are vacuously true[225] might well say that, in the scenario envisaged by Mares, a person is not imagining that Hesperus, *that very planet*, is not Mercury; merely that the name 'Hesperus' had been pinned on some other planet.

Yet I certainly agree that in evaluating conditions such as the one in question, one does need to consider worlds where Hesperus (that very planet) is not Venus (that very planet).[226] This may be done by deploying a contingent identity semantics.[227] Objects have parts, avatars, or whatever,[228] which may vary from world to world, and the truth conditions of predicates (including the identity predicate) make reference to these. This does not require true identities to fail at normal worlds (though it is compatible with this).[229] The worlds where true identities fail may just be impossible worlds. As long as there are some worlds where they fail, FullSub will hold when substituting into non-conditional contexts, but fail otherwise. Indeed, substitutivity in the even stronger[230] form $a = b, A_x(a) \models A_x(b)$ will fail if substituting into a conditional context.[231]

---

[224] For further discussion, see Priest (2018b).

[225] Such as Williamson (201+).

[226] See Berto et al. (2018).

[227] See INCL, Chap. 17.

[228] Mares pointed out to me that the notion of an avatar is in some ways similar to Castañeda's (1989) notion of a guise.

[229] See TNB, 2.9.

[230] Or weaker, depending on which way you think is up!

[231] INCL, 24.7.10.

27.19 Read (Chap. 20): Bradwardine Comes Back from the Dead

Stephen Read is my oldest philosophical colleague, friend, and coauthor. We have shared thoughts and debated ideas, both in my two periods in St Andrews (1974–1976, 2000–2013), as well as between those times and after them. I have learned much from him in the process. We share many interests, most prominently logical paradoxes, relevant logic, and medieval logic—though I, unlike Read, am very much an amateur in the last of these. Indeed, I owe my interest in it entirely to him. In my first year in St. Andrews, he opened my eyes to the richness of logic in the period.

One of Read's very significant achievement over recent years is to have resuscitated the solution to paradoxes in the family of the liar advanced by Thomas Bradwardine (1300–1349), articulating and defending it with all the resources of modern logic.[232] Behind this solution is a theory of truth and signification. A sentence may signify many things. Indeed, according to Bradwardine, it signifies everything that follows from it. A sentence is true if *everything* that it signifies is the case. So using $\mathbf{Sig}(s, p)$ to express that s signifies that p[233]:

- $T \langle A \rangle \leftrightarrow \forall p(\mathbf{Sig}(\langle A \rangle, p) \to p)$

Given that $\mathbf{Sig}(\langle A \rangle, A)$, the left-to-right direction of the T-Schema is forthcoming. But given that A will signify other things as well, the right-to-left direction is not, so the paradoxical argument is blocked. Bradwardine argues that the liar sentence ('this sentence is false') signifies not only that it is false but that it is true. Hence, signifying a contradiction, it is false. So something it signifies (viz., that it is true) is not the case. Since the right-hand side of the biconditional is false, one cannot infer the left.

In 'Read on Bradwardine on the Liar' (hereafter RBL),[234] I raised a couple of problems for the Bradwardine/Read account. One was that the account seemed unable to account for paradoxes of denotation, such as Berry's paradox. In his paper here, Read cleverly takes up the challenge to show that it can.[235]

The paradoxes of denotation require descriptive terms of some kind. For present purposes, and since we are dealing principally with Berry's paradox, which concerns natural numbers, let us suppose that these are formed with the least number operator, μ (*the least number such that*). (Read uses a definite description operator, ι. Nothing

---

[232] His many papers on the topic are referenced in his paper here, but not a bad place to start is with Spade and Read (2017).

[233] One might also add the clause $\exists p \mathbf{Sig}(\langle A \rangle, p)$ to the right-hand side, but since $\mathbf{Sig}(\langle A \rangle, A)$, this is redundant.

[234] Priest (2012b).

[235] The other problem was that Read's account uses propositional quantification; and if this is legitimate, there are versions of the Liar paradox which finesse the machinery of the Bradwardine solution. Nothing in Read's present paper addresses this issue. I note that in (2008) Read addresses a version of the Liar paradox given by Tarski, which uses propositional quantification and the operator 'says that'. His comments there do not apply to the version I give, which does not use such an operator.

hangs on this fact: exactly the same considerations apply to each.) The argument for Berry's paradox[236] then depends on two standard principles concerning such terms and their denotations:

- $D(\langle t \rangle, x) \leftrightarrow t = x$
- $\exists x A \to A_x(\mu x A)$

In the first of these, $D(x, y)$ is the denotation predicate (x denotes y), and t is any closed term of the language. This is the D-Schema. In the second, we assume a relabelling of bound variable to avoid any clash when the μ-term is substituted. This is the Least Number Principle (LNP): if something satisfies A then the least thing that satisfies A is such a thing. We need to make no assumption about how the μ-term behaves if nothing satisfies A (though, in his account, Read in fact does).

Since the D-Schema is the analogue of the T-Schema, I had assumed in RBL that it was this which should be Bradwardinised for a corresponding approach to the paradoxes of denotation. This would give something like:

- $D(\langle t \rangle, x) \leftrightarrow \forall y(\mathbf{Sig}(\langle t \rangle, y) \to t = y)$

though what **Sig** means in this context is somewhat unclear; and even given this, the modification does not appear to do what is required.

In the present paper, Read endorses the unmodified D-Schema (though, interestingly, he deduces this from considerations concerning signification). What then is supposed to break the paradox is the Bradwardinised LNP. This now becomes:[237]

- $\exists x \forall P(\mathbf{Sig}(\langle \mu y A \rangle, P) \to Px) \to A_y(\mu y A)$

Here, the upper case variables range over properties, and $\mathbf{Sig}(\langle t \rangle, P)$ means that the term t signifies the property P. There is already a notable movement from the Bradwardinian theory here, since we are no longer modifying semantic principles, but the behaviour of a much more general piece of logical machinery: descriptions. Moreover, there is another significant departure. In the Bradwardinian theory of truth the entities signified by a sentence are of the same syntactic category. The same is true of a Bradwardinian account of the Heterological Paradox, where open sentences signify properties.[238] In the present case, terms signify properties—something of a different syntactic category.

Anyway, given this machinery, if '$\mu y A$' signifies (were to signify) just one property, which it possesses—any property whatever; it need have nothing to do with the property A intuitively specifies—then one can infer that $A_y(\mu y A)$. However, if it may signify other things, we cannot; so the paradoxical argument is blocked.

However, why should we suppose that terms have multiple significations? Indeed, why should we suppose that they have a signification at all? Read points out that a descriptive term has a sense. Thus 'the most ignorant and stupid president the United States has ever had', has a sense which picks out its bearer. (No prizes.) We may

---

[236] As given in IC, 1.8.
[237] I have changed Read's notation to bring it in line with the conventions used in the present essay.
[238] See both Read's paper and mine.

therefore think of the signification of a descriptive term as its sense, that is, in effect, the property specified by the open sentence used in its construction. But signification is supposed to be a feature of all terms, not just descriptive ones. What does 'Aristotle' signify? Assuming that 'Aristotle' is a rigid designator in the sense of Kripke, it has no sense.[239] Read tells us that this includes at least the property $\lambda x \, x =$ Aristotle. Quite generally, Read tells us, any term t signifies the property $\lambda x \, x = t$.

But given that this is part of the signification, why do we need to suppose that the signification of a descriptive term is something else *as well*, given by its sense? Read postulates that if a term signifies some property, X, it signifies any property, Y, such that anything that is X is Y (CLO). If this were the case, the term t would signify the property $\lambda x(x = t \lor x =$ Donald Trump), which is certainly different. CLO is the analogue of what Bradwardine assumes about whole sentences (that if B follows from A, and a sentence signifies A, it signifies B). But even given that closure under signification is the case for closed sentences—and why should one suppose that? As far as I understand it, Bradwardine just postulates this—why suppose that it holds for other grammatical categories?

And the answer had better not be that otherwise Berry's paradox would give us a contradiction. One can turn *any* principle involved in the generation of a paradox into a conditional with some antecedent condition, and then use the paradox argument to infer that the condition is not satisfied. This is cheap. What we need for a genuine solution is an independent argument that the condition is not satisfied—in this case, that a term has multiple significations.

But there are other problems with the solution. Grant that this approach blocks the Berry argument. It also blocks every other argument in which we use the least number operator—and all other kinds of descriptions.[240] That's too much: we reason correctly using descriptions all the time. Given Read's approach, to deploy any description, $\mu x A$, we need to establish that the modified description principle can be applied. How do we do this? Proving that $\exists x A$ will not do, as Berry's paradox shows us. What we have to prove instead is that $\exists x \forall P(\mathbf{Sig}(\langle \mu y A \rangle, P) \to Px)$. The fact that this is not the case when the properties that '$\mu x A$' signifies are not mutually consistent might suggest that we can take it to hold if they *are*. But proving consistency is, we know, hard, and in general highly non-effective. Moreover, whatever it takes to show that this condition holds, we cannot even make a start on it till we know what properties the term $\mu x A$ signifies. Nothing in the story so far tells us this.[241] Without a resolution of these issues, the job is at best half done.[242] In truth, this was already an issue with the Bradwardine solution to the liar paradox. We frequently use the

---

[239] Kripke's arguments in *Naming and Necessity* to the effect that proper names are not descriptions of any kind are well known, and need no rehearsal here.

[240] Read notes that the strategy blocks many other paradoxes: König's paradox, Berkeley's paradox, Hilbert and Bernays' paradox. In a sense, this is not, therefore, surprising.

[241] Read tell us that $\mu x A$ signifies $\lambda x A$, and anything that this property entails. But that cannot be all. $\mu x A$ signifies $\lambda x A$ and $\lambda x \, x = \mu x A$, and these do not entail each other if we are not entitled to assume that $A_x(\mu x A)$.

[242] A dialetheic approach to the paradoxes, of course, faces a similar issue. This is the problem of 'classical recapture'. Thus, applications of the disjunctive syllogism are not valid; yet we frequently

right-to-left direction of the *T*-Scheme. (Everything true is found in the *Bible*. We should take an eye for an eye and a tooth for a tooth. So this is found in the *Bible*.) But, in the present context, the matter is much more acute, simply because we are dealing with a piece of logical machinery that is topic neutral.

Finally, and in any case, the Bradwardinian machinery would, in the end, seem to be beside the point. Grant that the least number operator works in the way that Read says that it does. There appears to be a perfectly intelligible neighbouring operator which delivers paradox. Let me illustrate using the Berry case. The simple combinatorial argument in Berry's argument assures us that, for a certain condition, *B*, with one free variable x (x is number not definable in such and such a way), $\exists xB$. By the properties of natural numbers, there is a least such. Now, never mind what the term μxB signifies, fix on this number, and give it a name. If you like, call this $\mu^* xB$. Then, if *B* specifies a purely extensional context (as it does in the Berry case)—that is, one where the only thing relevant to whether an object satisfies it, is the object itself, nothing to do with the way that it is specified—then by the very construction, $B_x(\mu^* xBx)$. We may simply run Berry's argument for this.

Let me now turn to another matter raised by Read: the Principle of Uniform Solution (PUS) and the Inclosure Schema. Berry's Paradox is a touchstone for proffered solutions to the semantic paradoxes, such as Read's.[243] The Principle of Uniform Solution (PUS) says: same kind of paradox, same kind of solution (BLoT, 11.5, 11.6). So if two paradoxes which are of the same kind are given different solutions, these cannot (both) be correct.

Since the Liar Paradox and Berry's Paradox would clearly seem to be of the same kind, it should be the case that solutions such as Read gives are of the same kind. Whether Read's solution to the two paradoxes satisfies this criterion is somewhat moot, as I have noted. Certainly, the machinery deployed in both cases is the same. However, the natural analogue of Read's solution to the Liar paradox qualifies the *D*-Schema in the way that the *T*-Schema is modified; and this, his approach does not do. However, what is at issue here is what counts as the same kind of solution, and this is a somewhat murky depth we need not plumb here.[244]

Read is, in any case, not persuaded by the PUS. He claims (Sect. 20.2) that the converse principle (same kind of solution, same kind of paradox) is 'much more plausible'. Moreover, one can contrapose the PUS: different kind of solution, different kind of paradox. Whilst (perhaps) logically equivalent, this suggests applying the principle differently. We take solutions to provide a criterion for individuating

---

use the syllogism unproblematically. How so? There is a substantial literature on this (starting with IC, Chap. 8, and IC2, Chap. 16).

[243] But also Field's. See Sect. 27.14.5 above.

[244] Read notes (Sect. 20.6) that my solution to the Hilbert and Bernays paradox involves faulting some traditional laws of identity, and so is of a kind different from my solution to the other paradoxes of denotation. This is not quite right. The solution is not to amend the laws of identity: that is just a consequence. The dialetheic solution to the usual paradoxes of self-reference is to suppose that a sentence may have more than one truth value (that is, Fregean reference); the solution I give to the Hilbert and Bernays paradox is to suppose, analogously, that a term may have more than one referent. Again, this takes us into the question of what counts as the same kind of solution.

paradox kinds. Thus, the mere fact that two paradoxes have different kinds of solution shows, *ipso facto*, that they are of different kinds, 'the possibility of a separate solution bringing out their different character'.

Now, first, I take the converse of the PUS to have very little plausibility. The fact that two paradoxes have the same kind of solution most certainly does not show that they are of the same kind. Thus, a dialetheist may hold that the liar paradox and paradoxes about the instant of change have a dialetheic solution.[245] This hardly shows that they are the same kind of paradox: self-reference has nothing to do with the instant of change.

Second, the methodology of individuating paradoxes in terms of their solutions is seriously flawed. Thus, consider the liar paradox. The self-reference required for this can be obtained in many ways: with a demonstrative (*this*); with a definite description (*the first sentence...*), gödel coding, and so on. Suppose that one were to give a solution to the version where demonstratives are used, deploying a theory according to which a demonstrative that is self-referential is ungrammatical; but that one were to give a solution to the version where a description is used, deploying a theory according to which the T-Schema fails for sentences containing descriptions. This would be bizarre. We have a sense that it is the same thing that is going on in these different formulations of the liar paradox, however self-reference is achieved; and focussing on the mode of self-reference is missing the nerve of the paradox.

Of course, how to articulate this sense of 'the same thing' is no easy matter. That is exactly the point of the Inclosure Schema.[246] This is a schema involving an operator ('the diagonaliser') which, when applied to a bunch of objects of a certain kind delivers a novel object of the same kind. And when one sees how it does this, one understands how it is that a contradiction will be generated at the limit, when the operator is applied to the totality of all such objects. The mechanism which produces the contradiction is, as it were, revealed, and one understands *why* the paradox arises. Moreover, it is not just I who think that the Inclosure Schema does this. Recall that the schema is just a tweak of one proposed by Russell (1908), where he explains that it exposes the mechanism which generates the paradoxes of self-reference. Of course, the diagnosis of why this kind of paradox arises does not determine a solution. That is another matter entirely. For Russell, it was an enforcement of the Vicious Circle Principle; for a dialetheist, it is accepting the contradiction at the limit. But as the PUS says, the same kind of solution is required, whatever that is.

Finally, in Sect. 20.6 Read raises the interesting question—which, oddly, had not occurred to me before—of whether Hilbert and Bernays' paradox fits the Inclosure Schema. At present, I have found no way to show that it does so. The problem is that the things that fit the Inclosure Schema are limit paradoxes. We have a totality, and an operation on subsets of that totality, which gives rise to a contradiction when things are pushed as far as (im)possible. Hilbert and Bernays' paradox doesn't seem

---

[245] Paradoxes of the instant of change were well known to the medievals under the title of 'incipit and desinit'. On a dialetheic solution to these, see IC, Chap. 11; and on medieval connections, see Priest (201+e).

[246] BLoT, 11.5, 11.6 and BLoT2 17.2.

to be a limit phenomenon of this kind. The contradiction involving the fixed point just doesn't appear to be something that happens at the limit of some totality. Until now, the major example of a paradox of self-reference which does not fit the Schema is Curry's paradox. Whether it is in the same family as the other standard paradoxes of self-reference is a vexed question.[247] But at any rate, Curry's paradox does not seem to be a limit paradox either. So maybe now we have two examples of this kind.

27.20 Restall (Chap. 21): A Tale of Two Negations

Greg Restall and I have also been friends and colleagues for many years. Over these years, I have probably discussed logic with him more than with anybody else. And though his views on logic differ from mine in many ways, I have learned much from our discussions, and his technical and philosophical insights.

Restall's paper takes us into the world of negation. The system of First Degree Entailment (*FDE*) is, as he says (Sect. 21.2), the most natural, straightforward, and elegant way of handling truth value gaps and gluts. It also provides a stable basis for extensions to theories of modality, conditionality, and other topics. The system is now some 50 years old, and is very well understood. As Restall's paper shows, however, there are still novel things to be learned about it.

As he notes, there are two well-known semantics for *FDE*.[248] According to one—the relational semantics—there are two truth values (*true* and *false*), and sentences may have two, one, or none, of these.[249] The other semantics is a modal semantics, and uses the Routley $*$ operator.[250] I have always preferred to the first approach, because of its transparent conceptual simplicity.[251] Whatever its technical versatility, the philosophical meaning of the Routley $*$ has, however, always been somewhat opaque, as is the matter of why a world-shift should poke its nose into the truth conditions of negation.[252]

As Restall shows, both semantics can be augmented to produce a second negation satisfying the same inferential rules as the usual *FDE* negation. The two negations interact in quite different ways in the two semantics, though. The matters concerning

---

[247] On which, see Priest (2017), Sect. 15.

[248] See INCL, Chap. 8.

[249] Equivalently, one may formulate this as a four-valued logic, with values *true* (only), *false* (only), *both*, or *neither*.

[250] These have come to be known as the 'American plan', and the 'Australian plan', respectively. As far as I know, the terms first appeared in print in Routley (1984). I have never cared particularly for the terminology. It is true that it bespeaks *something* of the origins of the ideas. But it seems to me as inappropriate as calling Frege's and Russell's ideas, the German plan and the English plan for classical logic.

[251] See IC, Chaps. 5, 6, and IC2, 19.8. True, I do not accept the *neither* possibility philosophically; but that does not bear on the present issue.

[252] Restall himself goes some way towards addressing these matters in his (1999).

how to interpret the machinery which I noted in the last paragraph are on display in the constructions.

Truth and falsity are pre-theoretical notions; one might call them 'folk notions'—though doubtlessly the folk don't pay too much attention to their details. The thought that truth and falsity are exclusive and exhaustive is natural enough; but so are the thoughts that they might not be. It surfaces in many metaphysicians. Thus, Aristotle argued in *De Interpretatione*, Chap. 9, that contingent statements about the future are neither true nor false; Hegel argued in his *Logic* that motion—amongst many other things—realises contradictions; then there are the many theorists of vagueness who take the borderline zones of vague predicates to deliver truth value gaps or gluts.[253] It also surfaces in the thought of the folk themselves, as XPhi studies make clear.[254] It is exactly, these possibilities concerning truth and falsity that the relational semantics make manifest.[255]

In Restall's extension of the relational semantics for *FDE* (Sect. 21.3), we still have 'good old fashioned' falsity; but he adds *another* notion of falsity, which behaves in a parallel fashion.[256] One might naturally think (as I did): what on earth is that supposed to be? The philosophical meaning of this second notion is opaque. One might, of course, raise a skeptical question. How does one know that it is the first falsity that represents the usual notion, and not the new notion? But that point can be set aside. For each negation, on its own, behaves in exactly the same way. In a sense, as Restall notes, they collapse into each other. Differences emerge only with their interaction. We are still faced with the question of why one should have *two* structurally parallel notions of falsity at all.

Compare this with Restall's extension of the ∗ semantics (Sect. 21.4)—which is clearly the more interesting extension mathematically. One is not at all inclined to ask what the second ∗ operator means, precisely because it was not clear what the original one was supposed to mean in the first place. There seems to be no particularly good reason why the machinery of stars should not be multiplied *ad lib*.

None of this bears on Restall's interesting technical investigations and results, of course. I am just using his construction to bring out philosophical issues behind the two technologies for negation—a matter that Restall does not broach in his paper.

---

[253] On these, see, e.g. INCL2, 7.9, 11a.7, Priest (1990, 2010b), and Fine (1975), respectively.

[254] See, e.g. Ripley (2001), Alxatib and Pelletier (2011).

[255] Restall observes (Sect. 21.1) that though both truth and falsity are involved in the semantic conditions of the connectives, validity is usually defined in terms of only truth preservation forwards. It might be natural to add a clause requiring falsity preservation backwards. In fact, for *FDE*, this would not change the validity relation at all (INCL, 8.10, ex. 8). It would for *LP*, since we would then no longer have $\models A \vee \neg A$, though we would have $B \wedge \neg B \models A \vee \neg A$. However, if one requires a consequence relation for which falsity preservation backwards is important, one can always define one, $\models^{\#}$, in the obvious way: $A \models^{\#} B$ iff $A \models B$ and $\neg B \models \neg A$ (with its natural generalisation to the multiple premise and conclusion case).

[256] Though there are other possibilities for adding a second notion of falsity, as he notes.

27.21 Shapiro (Chap. 22): *L'Affaire Gödel*

27.21.1 Background

Stewart Shapiro and I have been friends for many years. For over a decade we were both Arché Professorial Fellows at the University of St. Andrews, and we had many fruitful and enjoyable discussions over these years.[257] His paper here revisits *l'Affaire Gödel*.[258] I think it will help the discussion of this to put it into the context of the history of the whole matter.

In 1971 I attended the Bertrand Russell Memorial Logic Conference in Uldum.[259] During this, there was a talk on Gödel's Incompleteness Theorems by Moshé Machover. A central point of his discussion was how it could be possible that there are things which are unprovable, but which we could yet know to be true. The thing in question was, of course, the sentence that says of itself that it is not provable, that is, a sentence, G, of the form $\neg Prov \langle G \rangle$, where $Prov$ is the proof predicate for the axiomatic arithmetic in question.[260] For the rest of these comments I will refer to this—perhaps somewhat inappropriately, given the context—as the 'undecidable sentence'. This sentence changes from theory to theory, of course. The theory in question will, I hope, be clear from the context in what follows.

Anyway, Machover's problem piqued my curiosity. The problem was not, of course, that there are things that can not be proved in some system, but proved in another. The point was that we can recognise the truth of the undecidable sentence for, say, PA by means which are in some sense implicit in what we can already prove. One way one might make the point is this. PA uses an axiom schema of induction. But whatever intuition supports the schema supports, equally, the second-order version of induction, where one merely replaces the schematic variables with a second-order variable. Yet in second-order arithmetic one can prove the first-order undecidable sentence. Similarly, Dummett suggests that reasoning by mathematical induction is constitutive of our concept of natural number, but[261]:

> once a system has been formulated, we can, by reference to it, define new properties, not expressible in it, such as a true statement of the system: hence, by applying induction to such new properties, we can arrive at a conclusions not provable in it.

What is going on here?

---

[257] He is fond of telling the following story. In our visits to St. Andrews, we frequently shared an apartment. Often we would start a discussion at supper, and take it up again at breakfast. He is an evening person, and I am a morning person. He would get the better of the discussion in the evening, and I would get the better of it the following morning.

[258] For more on the matter see the comments on Field, Sect. 27.14.4 above.

[259] Actually, this was the first conference I ever attended. The results of the conference were subsequently published as Bell et al. (1973).

[260] Standard proofs of the fixed-point theorem deliver only a G equivalent to $\neg Prov \langle G \rangle$. But if the function symbols available include one for diagonalisation, we can obtain a literal identity. (See IC 3.5.)

[261] Dummett (1978a), p. 195. Dummett's view is discussed in IC, 3.2.

As I thought more about the phenomenon it seemed to me that the essence of the matter was how we show that the undecidable sentence is true in the standard model of arithmetic. (As is evident to anyone who has ever thought about soundness proofs, showing that the axioms of PA hold in the standard model invokes the very claims that are stated in the axioms.) Since we are dealing with the standard model, we might just as well talk about truth *simpliciter*. Hence, to carry out such reasoning, we need a language with a truth predicate. And of course, if the undecidable sentence is indeed provable, the theory is inconsistent. So we have a contradiction on our hands. Paraconsistency was therefore required. Trying to tie down this thought was what lead to the discussion of Gödel's theorem in Sect. 2 of the 'Logic of Paradox' (Priest 1979) and two later places.[262] In what follows, let me call the formulation of the argument in these places, the *original formulation*.

So, *pace* the introduction to Shapiro's paper, the aim was never to have a 'complete, decidable, yet sufficiently rich mathematical theory'. I did (and do) not find a problem with the thought that there are true mathematical claims that cannot be proved. What worried me was the thought that there was a particular unprovable claim that we could know to be true. And there was never a suggestion that arithmetic should be decidable—welcome as it might be if this were the case.[263]

At any rate, I now think the original formulation overshot the mark. For there to be an undecidable sentence in the first place, the theory in question had to be axiomatisable (or at least, arithmetic). I argued that naive mathematical proof (all of it) was, in principle, axiomatic. This was unnecessary. All that needed to be axiomatisable was the fragment of it in which the argument for the truth of the undecidable sentence was carried out—a much more modest claim.

Anyway, and to return to the history: The original formulation of the argument made no attempt to spell out the details of what an inconsistent arithmetic which can prove its own undecidable sentence would be like.[264] Things changed when I became aware of the potential for applying the work of Meyer on relevant arithmetic,[265] and especially once the techniques of the Collapsing Lemma because available in Priest

---

[262] I revisited the matter in Priest (1984), Sects. 5–7. IC, Chap. 3, is a slightly topped up distillation of these two discussions.

[263] I note that the notion of completeness is ambiguous in this context. Completeness might mean that, for every A, either A or $\neg A$ is provable; or it might mean that everything true is provable. The inconsistent arithmetics constructed in INCL2, Chap. 17, are complete in the first sense; they may not be complete in the second; that depends on what actually *is* true. The axiomatic theories constructed there are finite, and so decidable. (Later, Paris and Sirokofskich (2008) showed that there were infinite decidable models.) But decidability was simply a route to showing axiomatisability. I presume that there are complete (in the first sense) axiomatisable but non-decidable inconsistent arithmetics, though I know no proof of this.

[264] That was one of the tasks that Shapiro (2002) determined to undertake. It was a good shot, but turned out to have problems, as Shapiro notes in Sects. 22.3, 22.4 of the present paper.

[265] As spelled out in Meyer and Mortensen (1984). (The first edition of IC was essentially finished in 1983. It did not appear until 1987 because of the difficulty of finding a publisher, as the preface to IC2 explains.)

(1991). The resulting implications were spelled out in Priest (1994a, 2003).[266] These formed the basis of the material in IC2, Chap. 17.

Two things are notable about these inconsistent arithmetics. First, the working assumption about inconsistent arithmetics in the original formulation was that the underlying logic of the theory was a logic with a detachable conditional. The inconsistent models of arithmetic showed how the result—reasoning to the truth of the undecidable sentence—could be achieved when the underlying logic was *LP*, and so doesn't need a detachable conditional.[267] This means, of course, that reasoning within the theory, one does not use *modus ponens*. Given whatever axioms there are, the rules of *LP* deduction suffice. Since the discovery of these inconsistent arithmetics, I have tended to the view that the phenomenon at issue is best understood in the language of *LP*. This is, after all, the standard classical assumption as well—and if it is correct, issues about a detachable conditional fall by the wayside.

The second notable thing is that the language of the theories constructed does not, in a certain sense, contain a truth predicate. In one sense, *Prov* is a truth predicate for the theory. For any A, either A is in the theory or it is not. Since *Prov* represents proof in the theory, in the first case *Prov* $\langle A \rangle$ is in the theory; in the second case, both $\neg A$ and $\neg Prov \langle A \rangle$ are in the theory. In both cases, $A \equiv Prov \langle A \rangle$ is in the theory. But of course, this does not give us bi-deducibility for the T-Schema. However, there is a well-known construction that does deliver such a truth predicate. One can take an inconsistent model of arithmetic, add a truth predicate to the language, and conservatively extend the theory to one which verifies the T-Schema in a bi-deducibile form (as IC2, 17.3 notes). If one starts with an axiomatic theory of arithmetic and adds the T-Schema, one obtains an axiomatic theory. So the extended theory has an arithmetic proof predicate, *Prov*. One can show that $Prov \langle A \rangle \supset A$ is true in the theory (as in IC2, 17.4), as, then, is $Prov \langle A \rangle \supset T \langle A \rangle$. It does not follow that one can prove that $\forall x(Prov\, x \supset Tx)$, however. If the numbers in the model are all standard (and *Prov* is defined in such a way that if n is not the code of a formula then n satisfies $\neg Prov\, x$) then this is true in the extended model. However, it does not follow that it is provable.

Given these things, I think that the original formulation of matters overshot the mark in another way. First, the proof of the undecidable sentence does not require the quantified form of soundness: $\forall x(Prov\, x \supset Tx)$. Since a particular sentence is at issue, the schema $Prov \langle A \rangle \supset T \langle A \rangle$ will do just as well. Indeed, even the truth predicate itself is not necessary. The plain $Prov \langle A \rangle \supset A$ will do. Second, and perhaps more importantly, it is not necessary that soundness—in any of these forms—is provable from other things.[268] Even if it is a simple axiom schema, this will suffice

---

[266] I note that the first of these deploys what is, in effect, the Hilbert and Bernays paradox about denotation. I no longer accept that argument. See TNB, Chap. 8.

[267] Interestingly, an axiomatisation for some of these inconsistent arithmetics using the logic A_3, which conservatively extends *LP* with a detachable conditional, was given by Tedder (2015).

[268] Happily. Since, as Field later stressed (see Sect. 27.14.4 above), and as Shapiro notes in Sect. 22.4 of his paper, natural as the argument it is, Curry paradox considerations rule this out—at least if the logic of the theory uses *modus ponens*.

for the proof of the undecidable sentence. The question of proving its truth from other things also, therefore, falls by the wayside.

27.21.2 Church's Thesis and Its Application

With this background, let us now turn to the contents of Shapiro's paper. Sections 22.2, 22.3, and 22.4 provide a discussion of infelicities in Shapiro's earlier proposal of an appropriate formal arithmetic. There is little for me to comment on here.

Sect. 1 provides a number of arguments against the appeal to Church's Thesis in the original formulation of the argument (though some of them raise matters of more general import). The relevant question here is the extent to which the considerations he raises apply to a version of the argument to dialetheism which does not overshoot the mark, as the original formulation did. Shapiro raises five concerns. Let us take them in order.

Point 1: This is an objection to mathematical foundationalism—and I clearly did appeal to this in the original formulation. (This was, after all, the 1970s, and foundationalism was still the dominant view.) Mathematical foundationalism is, in fact, a view I have subsequently come to reject.[269] The foundationalism of the original formulation was used to justify the thought that naive mathematical proof is axiomatic. This, in turn, served two guarantee two things necessary for the argument:

[E] That there *is* an undecidable sentence.
[S] That what is proved is true.

Throw in the claim that the undecidable sentence is provable, and one has the sought result.

Now, granted that foundationalism in general is wrong, and that there is no reason to suppose that mathematics, *in toto*, is axiomatic, what of the fragment of it involved in the argument for the undecidable sentence? This fragment contains little more than the basic facts of arithmetic (e.g. those available in PA), plus the statement of soundness. Let us call this fragment of naive proof the *modest fragment*. The modest fragment would certainly seem to be axiomatic. That is, [E].

What about [S]? It is hard to cast doubt on the basic facts of arithmetic. So the main suspect for doubt is soundness itself. However, if we understand *proved* simply as *established as true*, this just seems analytic. True, soundness is perhaps not something that mathematicians would normally concern themselves with explicitly, but no mathematician is going to contest the claim that things that have been proved are true. (Recall that most mathematical proof is not proof in a formal system. Formal regimentation comes later—if at all.) Moreover, even if soundness is false for the

---

[269] Indeed, Shapiro's views on the philosophy of mathematics, which he was developing in St Andrews, and which were subsequently to appear in Shapiro (2014) helped to push me in that direction, though the move was well underway before that. (See Priest and Thomason 2007.) Indeed, rereading what I wrote many years ago, I was surprised to find that the move was even prefigured in my early papers on the undecidable sentence. (Priest 1979, Sect. IV.11, and Priest 1984, p. 171.)

system in question, we can, presumably, revise it to make it so—and the basic facts of arithmetic are unlikely to be junked in the process.[270]

Point 2: This concerns [E]. Shapiro points out that for the notion of recursiveness to make sense, we need a fixed language, and the language of mathematics is both somewhat indeterminate and changes. Again I agree; but the point seems to have no force against the modest fragment of naive proof. For the language of Peano Arithmetic will do, perhaps augmented by a truth predicate. And considerations of temporal change are irrelevant: we are dealing with how we reason here and now.

Point 3: This concerns the notion of consensus (or agreement). Shapiro points out that in many areas of mathematics, consensus is sometimes hard to achieve, and is based on a 'relatively small and finite sample of mathematicians'. First, since I have not mentioned the topic till now, why is consensus relevant? It is invoked twice in IC, Chap. 3.[271] In both cases consensus is used to support the idea that naive proof is axiomatic. This is no longer an issue, it seems to me, for the modest fragment. (See Point 1.) And in any case, it seems reasonable to suppose that there would be consensus concerning that particular fragment.

Point 4: This also concerns consensus. Shapiro says that where there is consensus, it is based on the assumption of classical (or at least intuitionist) logic—which I am obviously in no position to endorse. Now, for a start, mathematical proof is carried out informally. (No one argues *à la Principia Mathematica*.) I doubt that most mathematicians who work outside of the foundations of mathematics (which is most of them) know (or care!) much about formal logic—classical or otherwise. They just reason in a way that seems right to them. Moreover, as far as the modest fragment goes, this concerns only the inferences of *LP* (and maybe some inferences concerning a detachable conditional and truth). There is nothing very contentious about any of these for most mathematicians.

Point 5: Perhaps the preceding observations serve to circumvent many of the disagreements between Shapiro and myself; but this may not be the case with respect to the last point, which concerns the application of Church's Thesis itself. In my original formulation of matters, this was invoked to support [E]. Shapiro objects that appeal to the thesis is not appropriate since to do so 'one must specify an algorithm, a step by step procedure for computing a value, for deciding a question, a procedure that invokes no creativity or use of intuition' (Sect. 22.1). Now, Church's Thesis can be put in different ways; but one standard version of it is to the effect that if a function is effectively computable, then it is recursive in the technical sense. Indeed, this is Church's own formulation of the Thesis[272]:

> We now define the notion, already discussed, of an *effectively calculable* function of positive integers by identifying it with the notion of recursive function of positive integers.

The previous discussion in question consists simply of examples. The function at issue in the present case is that which maps (the code of) a sequence of sentences

---

[270] See IC, p. 46.
[271] Pp. 40 (fn 3), 41.
[272] Church (1936), p. 356.

to 1 or 0, depending on whether it is sound argument. And, as Church himself says elsewhere[273]:

> ...consider the situation which arises if the notion of proof is non-effective. There is then no certain means by which, when a sequence of formulas has been put forward as a proof, the auditor may determine whether it is in fact a proof. Therefore he may fairly demand a proof, in any given case, that the sequence of formulas put forward is a proof; and until the supplementary proof is provided, he may refuse to be convinced that the alleged theorem is proved. This supplementary proof ought to be regarded, it seems, as part of the whole proof of the theorem...

Indeed, that proof is effectively recognisable and truth is not is a central way in which the two differ, as Dummett, for example, has argued.[274] Now, it may well be that this appeal to effective recognisability breaks down where consensus breaks down, but I have already dealt with this matter. (Points 3 and 4.) So I still see no problem with invoking Church's Thesis in the present context.

In any case, and as Shapiro notes at the beginning of Sect. 22.2, arguments for [E] are finessed if one can actually produce such a system. This was one of the purposes of the system Shapiro unsuccessfully produced in (2002). But the inconsistent arithmetics of IC2, Chap. 17 do this—though perhaps they do not provide entirely what is required, since there are many axiomatic inconsistent arithmetics, so this does not determine the system uniquely. Different considerations are required to do so—if, indeed, there is a uniquely correct such system: it could that there is an indeterminacy in the matter. Equations may go inconsistent for suitably large (*very* large) numbers; but where, exactly, we may not know.[275]

27.21.3 Curried Undecidability

This brings us, finally, to the interesting matter of the Curried undecidable sentence, which Shapiro discusses in Sect. 22.5. Shapiro claims that it shows that even in an axiomatic framework which can prove its own Gödel undecidable sentence, there are still true but unprovable sentences. That, *per se*, would not be a problem: as I have already said, the argument was never aimed at showing that all truths are provable. The problem was with a sentence we could know to be true, but which was not provable. How do things stand with the Curried undecidable sentence?

We assume that we have an inconsistent axiomatic arithmetic. Then there is a sentence, C, of the form $Prov \langle C \rangle \Rightarrow \bot$. The first question is how one is to understand $\Rightarrow$. If the arithmetic is one where the only conditional available is $\supset$, then C is $Prov \langle C \rangle \supset \bot$. This is logically equivalent to $\neg Prov \langle C \rangle$. So this is simply the Gödel undecidable sentence, and there is nothing new here.

---

[273] Church Church (1956), p. 53.
[274] Dummett (1978b).
[275] The matter is discussed in Priest (1994b).

Suppose, however, that we are contemplating an arithmetic that has a detachable conditional, $\rightarrow$, and that this $\Rightarrow$. If C is provable, then so are $Prov \langle C \rangle \rightarrow \bot$ and (since $Prov$ represents provability) $Prov \langle C \rangle$. Hence, the theory is trivial. If it is not, then C cannot be proved. So, $\neg Prov \langle C \rangle$ is true. All this is as Shapiro notes.

What of the truth of C? Given that we are not dealing with a material conditional, we cannot simply infer that $Prov \langle C \rangle \rightarrow \bot$, i.e. that C is true. But, as Shapiro notes, there is another argument. Suppose that $Prov \langle C \rangle$, then by soundness, $Prov \langle C \rangle \rightarrow \bot$, and so $\bot$. Hence, by conditional proof, $Prov \langle C \rangle \rightarrow \bot$, i.e. C. This is essentially the $Prov$ form of the Curry paradox, and the argument must fail for similar reasons. The reason given in IC, Chap. 7 is that $\rightarrow$ does not satisfy Absorption (Contraction). As is well known, in a natural deduction context, this means that in applications of the rule of conditional proof one cannot discharge more than one occurrence of an assumption, which this argument does.[276] So is C true or not? Without a theory of the conditional and its interaction with the other machinery (and maybe even with it), one cannot say.

It remains the case that we have shown $\neg Prov \langle C \rangle$ to be true. Just as for the Gödel undecidable sentence, this should therefore be provable our formal arithmetic. Whether or not it is, will also depend on the theory and, crucially, how it handles $\rightarrow$.[277]

27.22 Tanaka (Chap. 7): Taking Exception to Candrakīrti

Koji Tanaka is one of the few logicians who shares my interests in the Asian philosophical traditions, and he brings this interest to bear in his discussion of so-called logical exceptionalism. I think that the best thing I can do to address his comments it to start by saying what I take logical exceptionalism to mean—or at least, what it means in as far as it applies to me.[278]

In matters of any complexity, we have to theorise. Thus, we do this in physics and biology, but also in linguistics, history, ethics, and metaphysics. In each case, there is something we wish to explain and understand (and maybe, for certain kinds of topics, make predictions). We construct theories which do this, and then accept whichever does the best job of the matter. Of course, we may change our mind as to which one does so, as new theories are discovered, or we come to understand old theories better.

Certainly, the theories must do justice to the data concerning whatever it is we are trying to explain, though data is just as liable to be fallible as theory. But adequacy to

---

[276] As Shapiro observes in fn 4. Note that, for the same reason, the proof of Löb's Theorem, which is a version of Curry-reasoning, breaks down.

[277] We need to be able to encode the following reasoning. Suppose that $Prov \langle C \rangle$. Then $Prov \langle Prov \langle C \rangle \rightarrow \bot \rangle$. But $Prov \langle A \rangle \rightarrow A$, so $Prov \langle C \rangle \rightarrow \bot$, and so $\bot$. Hence we have shown that $Prov \langle C \rangle \vdash \bot$. (Note that we have not proved the corresponding conditional.) Thus, $\neg \bot \vdash \neg Prov \langle C \rangle$, and so $\neg Prov \langle C \rangle$.

[278] As explained and defended in Priest (2014b, 2016a).

the data alone is not enough. There may be theories that do equal justice to the data—or more likely, no (extant) theory may account for all the data. So other criteria, such as simplicity, unifying ability, extent of *ad hoc* auxiliary assumptions, etc., come into play. The theory it is rational to accept is the one which does best overall—in some way of cashing out this idea.

The general picture is familiar enough. Anti-exceptionalism about logic is the view that logic, in one sense of that word, is no exception to the picture.[279] Logic, in this sense, is a theory about what follows from what, and why. It should be stressed that this is not, in any simple sense, a descriptive theory about how people *actually* reason. That is a matter for cognitive psychology. It is about the norms that govern correct reasoning.[280]

Logicians have been constructing such theories—which disagree on many matters—as long as there have been logicians. The theories are contestable, fallible, and modern accounts, at least, apply quite sophisticated mathematical ideas. Seen in this way, the epistemology of logic that emerges is quite distinct from a traditional *a priori* account of the Kantian kind. In this, the laws of logic are available simply to reflection, are certain, and not rationally contestable

An obvious question that arises in this context is what kind of data it is which is relevant in logical theorising. I take this to be those simple inferences (and perhaps inference schemas) that strike us as correct; and I suppose that these are *a priori* in one sense of that term. One does not have to go and look or listen to determine their apparent acceptability; one just has to think.[281] But, as ever in theorising, what appears to be so may not be so.

With this background, let me now comment on Tanaka's paper. First, he asks at the end of Sect. 7.3 what, exactly it is that makes my view different from a traditional *a priori* view of logic. I hope that the answer to that question is now clear.

Next, Tanaka suggests that my account is, or can best be seen as, a form of the *lokaprasiddha* (common sense) account of Madhyamaka Buddhism. Now, how, exactly, to understand this view is somewhat contentious, but this is not the place to enter into scholarly disputes about this matter; so let us just accept Tanaka's interpretation. According to this, truth is simply what people accept. ('Truth is nothing more than what people on the street assent to and knowledge is nothing more than what they think' Sect. 7.4.)[282] In particular, a claim about validity is true if people

---

[279] Just to be clear, this doesn't have anything much to do with Quine's 'change of logic, change of subject' argument, as Tanaka suggests (Sect. 7.1). That's another matter. See DTBL, Chap. 10, especially 10.9.

[280] In one sense of that word. Thus, Harman (1986) prefers to use the word 'reasoning' for what would now be called belief revision. Naturally, there are connections between these two senses. When juggling one's beliefs, if one is aware that B follows from A, then one shouldn't accept A but not B. Exactly how one cashes out this idea is no easy matter, however, as MacFarlane (2004) shows.

[281] I suppose that in some sense this is an empirical survey—albeit one with a single subject! I don't need to find out what others think, any more than I, as a native speaker of English, have to consult others to determine whether 'the cat sat on the mat' is grammatical.

[282] For a rather different interpretation of truth in Madhyamaka, see Priest et al. (2010).

accept it to be so. Now, it is certainly the case that data against which we judge our logical theories is determined by the inferences that strike us as valid. But that is about where what is in common between this view and mine ends. In particular, it is not the case that 'whether or not an inference is valid is ... just a matter of what people in the street would accept' (Sect. 7.4). The results of logical theorising may well, and frequently do, overturn the views of 'people in the street'. Nor is it true that 'we may not have any sustained reason for why we believe [a given inference] to be valid' (Sect. 7.4). The theories logicians develop—be they model-theoretic, proof-theoretic, or whatever—provide just such reasons. And it would just be a travesty of my account to say of it that (Sect. 7.4):

> there is no need to analyse logical concepts, the notion of validity or anything. All there is to logic is what can be expressed by things like 'This inference looks good to me', or 'that inference strikes me as valid'.

Even if one must start from 'common sense' the aim of theorising is always, as it were, to get behind appearances. This is so in all forms of theorising, in logic as elsewhere.

Indeed, the *lokaprasiddha* account is in dire danger of collapsing into relativism about truth, since it 'reduces truth and knowledge to mere opinions and beliefs' (Sect. 7.5). I would reject such a relativism entirely. I take it that there are objective truths about validity, just as much as there are about physics. Indeed the whole point about theorising is to delve into the matter of what *is* true in this sense, and to deliver us our best (current) view about what this is.

Let me, finally, turn to the connection between these methodological issues and paraconsistency. Tanaka says that my methodological views on logic are 'a part of [my] argument for paraconsistent logic' (Sect. 7.2). In a sense this is true. The sense is that they mount a case against those historically benighted people who think that so-called classical logic, *qua* theory—or any other theory, for that matter—is God-given and uncontestable. It is not true that I think that applying this method *must* deliver the conclusion that a paraconsistent logic is (currently) the most rational theory to accept (though as a matter of fact, I think it does). What comes out of applying this method can be determined only by, well, applying the method.[283] One has to look at how well each of the relevant theories performs on each of the theoretical desiderata, and on their aggregation.

A couple of issues that Tanaka mentions here are relevant. First, I think it quite true, as he says, that most people who have never studied any logic find bizarre claims to the effect that instances of Explosion are valid. This is surely a black mark against classical logic; but it is one that it does not have much difficulty digesting. The validity of the inference follows from a theory that is simple and powerful. The inference turns out to be vacuously valid; and the intuition that it is invalid can simply be explained by saying that such a kind of validity does not normally occur to people.

More telling are examples where the premises are thought to be true, and the conclusion is not. If a person can be brought to agree that there are such situations

---

[283] A case study in applying this can be found in Priest (201+f). No one has yet undertaken the daunting task of applying the method to logic quite generally.

then they will surely judge that Explosion is not deductively valid. After all, not everything is true. And truth preservation appears to be a rather minimal necessary condition for deductive (as opposed to non-deductive) validity. So any inference where the premises are taken to be true, and the conclusion is taken to be untrue will strike one as (deductively) invalid.

The point of contention here will (of course) be whether there *are* situations in which a contradiction is true. The example of visual illusions that Tanaka cites is not of this kind (*contra* the view he attributes to me in Sect. 7.6). These are, after all, *illusions*.[284] A contradiction may *appear visually* to hold, but it does not really do so. Thus, in the waterfall illusion, the object on which one is focussing is not really both moving and not moving; and we know this to be so.

Examples which *are* of this kind are those which dialetheists standardly cite.[285] Of course, such examples may well be contentious; and at issue will be how one might explain away the apparently contradictory nature of the situations. One gets a glimpse here of the enormity of the task of evaluating logical theories. One cannot, in the last instance, disentangle such evaluation from the evaluation of theories in semantics, metaphysics, the theory of norms, and numerous other areas.[286]

27.23 Tennant (Chap. 23): Through an Inferentialist Telescope

Neil Tennant takes us into the world of what he nowadays calls Core Logic, and casts an inferentialist eye on *LP*. Our interest in each other's work goes back to the 1970s, when we both developed ideas in paraconsistency and in logics which are, in some sense, relevant. I developed *LP*; he developed his cut-free logic. Philosophical interest in cut-free logics, and more generally substructural logics, is now reasonably commonplace.[287] It was not in those days. And over the years, Tennant has articulated his view, throwing into the mixture anti-realism, inferentialism, and paradox.

Behind his paper, there is an important difference between us concerning meaning. He subscribes to an inferentialist account of this; I subscribe to a truth-conditional account. For an inferentialist, the meanings of the logical constants are determined by rules of inference. Matters can be set up equivalently in terms of either natural deduction or a Gentzen-style sequent calculus. For the sake of definiteness in what follows, let me talk in terms of the latter. Each logical operator comes with a pair of rules, one of which shows how to introduce sentences containing the operator on the left-hand side of the sequent connector, and the other of which shows how to introduce them on the right-hand side. Moreover, the rules have a certain balance (or harmony, as it is often called), which allows for an appropriate cut theorem (or,

---

[284]DTBL, 3.3.

[285]For a short catalogue of these, see Sect. 27.11.1 above.

[286]For some further discussion, see Priest (201+g).

[287]For example, with the work of Cobreros et al. See Sects. 27.9, 27.11 above.

in the case of natural deduction, a normal form theorem). This account of meaning goes naturally (though not invariably) with a proof-theoretic notion of validity, and a view of truth as warranted assertibility. For someone who subscribes to a truth-conditional account, the meanings of the connectives are given by their truth conditions (generalised to truth-in-an-interpretation-conditions, or, for some logics, truth-at-a-world-in-an-interpretation-conditions). This account of meaning goes naturally (though not invariably) with a model-theoretic notion of validity, and a more robust notion of truth.

That difference is a deep philosophical one, and this is not the place to discuss it. (I have had my say on the matter in DTBL, Chap. 11, and I shall not repeat it here.[288]) I note only the following. Many of Tennant's critical remarks concerning *LP* turn on the fact that its system of proof does not satisfy the sort of desiderata one would want if one is an inferentialist. Indeed it does not; but for those who prefer a model-theoretic account of validity, such constraints are of no import. The proof system is just a way of characterising the notion of validity in a combinatorial fashion.[289]

With these background comments, let me turn to four more specific matters.

First, as Tennant points out, and as is well known, the material conditional in *LP*, $A \supset B$, that is $\neg A \lor B$,[290] does not satisfy detachment. Ever since the early years of work on *LP*, it has been a standard thought that the language needs to be augmented by a conditional that does,[291] and so the semantics need to be extended to accommodate this. The working assumption has usually been that the conditional is one of an appropriate relevant logic. Proof-theoretically, much of the material pertaining to relevant logic has now been subsumed under the investigation of substructural logics, which contains many results concerning a proof theory of the kind that inferentialists like. However, this is not the place to go into that matter.[292]

Next, if someone thinks that some particular logic, *L*, is the correct one, it clearly behoves them to give the metatheory for *L* (in whatever form that takes) in a way that is logically kosher. In particular, if moves are made in the metatheory which are not valid in *L*, there had better be an appropriate story to tell about this. As Tennant

---

[288] Though let me say that I would not go to the wall over this matter. If an inferentialist account turns out to be correct, then so be it. There are certainly proof-theoretically defined logics that are appropriate for dialetheism. Indeed, since Core Logic (or Classical Core Logic) is paraconsistent, then, depending on many other moving parts, that might even turn out to be the most appropriate one.

[289] Thus, the puzzles (Sect. 23.2.3) and potholes (Sect. 23.3) that Tennant perceives are ones that appear only if one is driving on the inferentialist side of the road.

[290] I note that one can simply define $\supset$ in this way. If one wishes to take it as an undefined symbol, it can be characterised by the two-way natural deduction rules (or the equivalent for a sequent calculus):

$$\frac{\neg A \lor B}{A \supset B} \quad \frac{A \land \neg B}{\neg(A \supset B)}$$

These rules will not, of course, satisfy an inferentialist of Tennant's kind. But there is nothing wrong with them from a model-theoretic perspective of validity.

[291] A possibility that, oddly, Tennant does not mention.

[292] For an excellent exposition of the area, see Restall (2000).

notes, the standard model-theoretic semantics for *LP* uses things like the disjunctive syllogism,[293] which is not valid in *LP*. This has motivated the many discussions of 'classical recapture'.[294]

In fact, the situation is even more acute for someone such as myself. For a model-theoretic semantics is carried out in set theory, and this would standardly be taken to be *ZF*; but such a set theory is not correct if one subscribes to a paraconsistent set theory. So, again, it behoves someone like me to show either how the standard model-theoretic reasoning can be accomplished in a paraconsistent set theory (using a conditional which detaches, since the material conditional is of no use in this regard), or explain how the *ZF* reasoning itself is acceptable. Both approaches are to be found in the literature. I prefer the second. Since I have discussed this in Sect. 27.3.2 above, I will not go into the matter again here.

Third, Tennant claims that a virtue of his account is that it provides a 'faithful and homologous' formalisation of informal arguments, which a suitable proof-theoretic account, such as his, does, and which *LP*—and more generally, a model-theoretic account of validity—does not do. Now, neither of us thinks that a theory of logic should be a theory of the way that people actually reason. That is a matter for the cognitive psychology. Rather, a logic should undergird good reasoning of this kind. For contemporary logicians, the paradigm of good reasoning has always been mathematics.

Next, I note that when people reason—and reason correctly—they reason informally, even in mathematics.[295] Arguments are not laid out in some natural deduction system or sequent calculus. Such would be far too prolix. Those for whom this is important take it that the arguments *could* be regimented in an appropriate fashion. Tenant claims that the regimentation should be given in terms of the rules beloved by an inferentialist proof theorist. However, I see no reason for this unless one supposes in advance that such is the correct understanding of validity. Indeed, the original and most famous regimenters of mathematical reasoning were Frege, in *Grundgesetze*, and Russell and Whitehead, in *Principia*; and their regimentations were clearly nothing like the kind Tennant endorses.

Of course, we will often want to put a piece of reasoning under the microscope. This is highly useful in assessing the correctness of a complex piece of reasoning. However, what is at issue here is whether the steps in the argument are valid. This is the business of a theory of validity, and it does not have to be an inferentialist one. A model-theoretic one will do just as well.

Finally, though Tennant subscribes to a paraconsistent logic, he is no dialetheist. In particular, he is not a dialetheist about the semantic paradoxes.[296] Though he endorses the *T*-Schema in its bi-deducibility form, he argues that the failure of Cut (or normalisability) avoids the dialetheic conclusion. He points out that one can establish

---

[293] At least if the conditional of the metatheory is supposed to be $\supset$.

[294] Starting with Priest (1979), Sect. IV, and IC, Chap. 8.

[295] Tennant invokes the fact (Sect. 23.1) that in Priest (1979) I refer to naive proofs. As the context, I would hope, makes clear, this was the sort of proof I was talking about.

[296] For this and what follows, see Tennant (2015).

that $\vdash \neg Tl$ and that $\neg Tl \vdash \bot$, where l is the liar sentence; but one cannot deploy Cut to infer $\vdash \bot$. However, in Core Logic, the second fact delivers $\vdash \neg\neg Tl$. Even if we cannot infer $\vdash Tl$ for reasons of an intuitionist kind, it remains the case that we have established a contradiction. If the deductions are kosher, we have dialetheism. Moreover, the failure of the inference to $\bot$ is just what the dialetheist needs to prevent the spread of contradiction.[297]

27.24 Verdée (Chap. 24): Making *LP* Relevant

Let us now turn to the paper by Peter Verdée, and the issues of conditionality which it raises.

The material conditional of *LP*, $\supset$, does not detach. It is therefore an obvious idea that the language needs to be augmented with one that does. Such a thought has been pursued since the inception of *LP*.[298] A plausible thought is that the conditional should be that of an appropriate relevant logic. If the conditional is to be the one used to formulate principles which generate the paradoxes of self-reference, such as the *T*-Schema and the naive comprehension schema of set theory, then not all such conditionals are appropriate. Thus, the Absorption principle, $A \to (A \to B) \vdash A \to B$, will deliver triviality, in the shape of the Curry paradox. Relevant logics which contain this or related principles, such as *R* and Verdée's own conditional, are not, therefore, appropriate. If, however, dialetheism is not to be deployed to solve these paradoxes; or if it is, but the conditional of the naive principles is a non-detachable conditional such as $\supset$,[299] there is no problem with a relevant logic containing such principles.

As something of an aside, let me note the following. It is sometimes objected to a dialetheic solution to the paradoxes of self-reference that it is not uniform: the solution to the Liar paradox rejects Explosion; the solution to Curry's paradox rejects Absorption. Since these paradoxes appear to be of the same kind, they should have the same kind of solution (the Principle of Uniform Solution). However, for a start, if the conditional of the Curry sentence is $\supset$, then the Curry paradox is solved by rejecting the disjunctive syllogism, $A, \neg A \vee B \vdash B$. In *LP* this is a very simple equivalent of Explosion. Hence the solutions to the two paradoxes are exactly the same. If, however, the condition of the Curry sentence is a detachable conditional, $\to$, this is no longer the case. But there is no *a priori* reason why *different* sorts conditionals should require the *same* sorts of solution.[300] And there are, in fact, good

---

[297] One may make a similar point about the cut-free ST system of Cobreros et al. See Sect. 27.11.2 above.

[298] Priest (1979), Sect. 4, IC, Chap. 6.

[299] As suggested by Goodship (1996), and discussed in Priest (2017).

[300] Thus, one can formulate the sorites paradox with different sorts of conditional, and different solutions may be appropriate for different conditionals. For some, it may be appropriate to say

reasons as to why the Liar paradox and Curry's paradox with a detachable conditional are *not* of the same kind.[301]

Anyway, and to return to Verdée's paper, he suggests a novel kind of relevant conditional. The strategy used can be applied to any consequence relation, but Verdée's main aim is to apply it to *LP*. The strategy comes in two stages. The first is to take a consequence relation and use it to define a sub-relation which is relevant (presumably in usual variable-sharing terms, though Verdée never specifies what it is for logic to be relevant). The second is to augment the language of the original consequence relation with a conditional which mirrors the relevant consequence relation.[302]

This is not the place to discuss the details of the construction, so I will just note a couple of points. The strategy of starting with a consequence relation and filtering out the irrelevant instances is a well-known one. The logics produced are often termed 'filter logics' for obvious reasons.[303] Typically, though not invariably, filter logics are non-transitive, as is Verdée's logic. Second, the strategy employed by Verdée is very close to that which generates Tennant's Core Logic.[304] Both function by taking a consequence relation and filtering out those inferences where a premise or conclusion is redundant.[305] Given that the classical inference $A, \neg A \vdash B$ already falls by the wayside on this approach, one might wonder what the benefits are of moving away from classical logic to apply the strategy. I will leave that matter for Verdée to address.

Finally, I note that Verdée's conditional is distinct from that of all the usual relevant logics. It is not contained in any of these, since it does not contain transitivity. And it does not contain any of these, since, as Verdée notes (Sect. 24.5), it verifies things such as $A \to (B \to (A \land B))$, which does not hold in the usual systems of relevant logic. So there is an issue as to which of the two approaches gives the best sort relevant conditional—however one might understand that question. That, however, is too big an issue to take on here.

---

that the conditional premises are untrue; for some, it may be appropriate to say that detachment is invalid. (See Priest (2010b), Sect. 7.)

[301] See Priest (2017), Sect. 15.

[302] On a small matter: Verdée uses the word 'implication' for both a conditional connective and a validity relation. Of course, he is not confused about the distinction, and there is a venerable tradition using the word for the conditional. However, I think that the terminology is unfortunate. 'Implies' is not a connective; it is a relation. And in my experience, the terminology encourages a confusion in students between the connective and the consequence relation.

[303] For a discussion and references, see Priest (2002), Sect. 4.1.

[304] See Sect. 27.23 above.

[305] The easiest way to see this is to compare Verdée's Definition 5 and Theorem 1 with the presentation of Tennant's approach in Priest (2002), Sect. 4.1.

27.25 Wansing and Skurt (Chap. 25): If a is b, and b is c, a is c, Isn't It?

My training in logic was very much in classical mathematical logic. I suppose that as the years have gone on, I have come to see how the mathematical techniques I learned could be applied to produce many other logics—non-classical logics. These, in turn, opened up new avenues of approach to philosophical problems, both traditional and contemporary. This is very much true concerning the topic of identity. I started by assuming, like most contemporary logicians, that the standard textbook account of identity gets it right. Nothing made me challenge this assumption for many years. However, I started to think otherwise in the 1990s, when I came to realise how an account according to which the transitivity of identity (TI) fails could be applied to solve a number of interesting philosophical problems, and how, plausibly, my thinking had been channelled by what Wittgenstein called an inadequate diet of examples (mainly from mathematics).[306] The first ideas in this direction concerned the failure of transitivity in continuum-valued logics, but in ONE the failure of transitivity was obtained by defining $x = y$ as $\mathfrak{A}P(Px \equiv Py)$ in second-order *LP*.

27.25.1 Identity Itself

In their paper, Heinrich Wansing and Daniel Skurt (hereafter W&S)—whom I came to know very well in a brief but happy sojourn at the Ruhr University of Bochum in 2013—take us into this world of non-transitive identity. They make a number of interesting technical observations. However, in keeping with my policy in writing these comments, I shall not go into technical matters. What I will do is to consider the objections they raise to the account of identity in ONE—which, to eliminate any uncertainty about my view (of the kind that W&S find in their Sect. 25.6), is my current view.[307]

First, W&S insist on calling identity according to the standard account *real* identity (e.g. Sects. 25.3, 25.7). Indeed, even the title of their essay uses scare quotes when referring to non-transitive identity.[308] Now that is an entirely tendentious way of putting the matter. I take it that, orthodox as the account may be, it in fact mischaracterises identity—though the characterising conditions may well hold in 'nor-

---

[306] 'A main cause of philosophical disease—a one-sided diet: one nourishes one's thinking with only one kind of example'. *Philosophical Investigations*, Sect. 593. Wittgenstein (1968), p. 144^e.

[307] I note, as they do (Sect. 25.6), that the non-transitivity of the identity relation and its inconsistency are, in principle, quite different issues. There are theories of identity which verify each of these but not the other.

[308] They say 'non-transitive "identity" is not identical with what is usually taken to be real identity' (Sect. 25.7). That, at least, is true. But note, again, the scare quotes around 'identity'.

mal' circumstances. Whether the standard account of identity gets the real identity relation right is exactly what is at issue in this matter.[309]

It remains the case that one may, in the metalanguage, have a notion, $\asymp$, specified as in the standard account, and so where its extension is $\{\langle x, x \rangle : x \in \mathcal{D}_1\}$, $\mathcal{D}_1$ being the first-order domain).[310] However, it does not follow that this should be used to state the truth conditions of $=$ in the object language. Why not? Because it is not identity (or at least one can assume so only by begging the question)! What, then is the relation $\asymp$? The simple answer is that it is co-substitutivity, a relation that is, of course, language-dependent.

W&S's Sect. 25.3 raises the question of similarity relations. Similarly, relations are reflexive and symmetric, but not transitive. So, it may occur to the reader to wonder why $=$, as I have defined it, is no more than a similarity relation. The answer is simple. A similarity relation may not satisfy the condition $\mathfrak{A}P(Px \equiv Py)$. Consider being similar with respect to colour; and let a and b be distinct objects that are similar in this respect. a, we may suppose, emits light of frequency ν_a, and b emits light of frequency ν_b. Let Q be the property of emitting light of frequency ν_a (and suppose this to be a consistent condition). Then Qa is true (and true only) while Qb is false (and false only). It follows that $\mathfrak{A}P(Pa \equiv Pa)$ is not true.

27.25.2 Second-Order LP

Section 25.4 of W&S's paper raises questions concerning the range of the second-order quantifiers, second-order minimal inconsistency, and reassurance.

Let $\mathcal{D}_2$ be the second-order domain. If this contains every extension/anti-extension pair of the form $\langle X, Y \rangle$ such that $X \cup Y = \mathcal{D}_1$—call this the *full* $\mathcal{D}_2$—or even just the pair $\langle \mathcal{D}_1, \mathcal{D}_1 \rangle$, then every sentence of the form $a = b$ is true and false. As W&S note, this is an undesirable consequence. But as W&S also note, ONE puts no constraints on $\mathcal{D}_2$ other than that it be non-empty. I left it open what other constraints $\mathcal{D}_2$ should satisfy, though I was very clear (2.7) that it should not be full, to rule out exactly the sort of consequences that W&S point to. The denizens of the domain need to be metaphysically real properties.[311]

---

[309] In Sect. 25.2 they quote me as saying that TI for real identity fails. But as I would hope the context makes clear, by 'real identity' I am referring to identity as it really is, and not how some mistaken theory takes it to be.

[310] The following points are made in ONE, 5.11.

[311] In their paper, W&S refer to Hazen and Pelletier (2018), who prove a number of interesting results about second-order *LP* and the Leibniz-defined identity in this. Hazen and Pelletier refer in their Abstract to the characterisation of $\mathcal{D}_2$ (if I understand the allusion correctly), saying that 'it will be extremely difficult to appeal to second-order *LP* for the purposes that its proponents advocate, until some deep, intricate, and hitherto unarticulated metaphysical advances are made'. They surely overplay their hand here. One does not have to determine every parameter of a piece of logical machinery before one can reasonably deploy it—especially since the applications can provide a constraint on fixing those parameters.

Next, W&S raise the question of classical recapture with respect to second-order minimal *LP*, *LPm*. Classical recapture is a statement to the effect that for consistent Σ, if *A* follows from Σ in classical logic, it follows in *LPm*. In their definition (Definition 2) of the consistency ordering, ≺, they impose a constraint to the effect that if $\mathcal{I} \prec \mathcal{J}$ then the first-order domain of $\mathcal{I}$ is a superset of the first-order domain of $\mathcal{J}$.[312] They then note that classical recapture is problematic. Perhaps so. However in the discussion of first-order *LPm* in IC, 16.4–16.5, no such constraint is imposed; and that being so, classical recapture is immediate. Neither is it imposed in the discussion of the second-order case in ONE, 5.13. And as I note there, this gives recapture for classical first-order logic *with* (standard) *identity*; though what happens with second-order inferences may depend on how the details of the second-order domain are filled out. The first-order inferences would seem to be good enough for most practical purposes, however. After all, it's all you have on the standard account favoured by W&S.

So much for classical recapture. Reassurance is another matter. Reassurance is the claim that if Σ is non-trivial under *LP*, it is non-trivial under *LPm*. This holds for propositional logic, but as W&S note, extending the result to first-order logic—and *a fortiori* second-order logic—has turned out to be a very tricky matter,[313] and domain-restriction conditions play an important role in the matter. Much work remains to be done to sort things out. Here I note only two things.

First, at one time I did take reassurance to be a necessary condition for a suitable notion of minimal inconsistency. I no longer think so. If it holds 'for the most part', that will be fine. If I may quote myself from Priest (2017), Sect. 6.3:

> Now, even without Reassurance, *LPm* would seem to do everything that one would like: it delivers a more generous notion of consequence than *LP*, where irrelevant contradictions do not invalidate classical inferences, and which delivers all classical consequences given consistent premises. [Footnote: The reason given in *IC*, 16.6, for the desirability of Reassurance is as follows. Taking triviality to be a mark of incoherence, Reassurance guarantees that a coherent situation will never be turned into an incoherent one under *LPm*. This may be more than is required, though. It might be quite sufficient if mostly, or normally, *LPm* does not turn a non-trivial situation into a trivial one. If there are some exceptions, and *LPm* is otherwise robust, we might take the triviality exposed to speak against the coherence of the original situation.]

As I go on to note there, there is, in any case, a very easy way to obtain reassurance. One simply redefines the *LPm* consequence relation as follows. $\Sigma \vDash_m A$ iff:

- (some minimally inconsistent model of Σ is non-trivial, and every minimally inconsistent model of Σ is a model of *A*) or (every minimally inconsistent model of Σ is trivial—in particular, if there are none—and $\Sigma \vDash_{LP} A$).

Reassurance follows simply, and $\vDash_m$ still has all the crucial properties.[314]

---

[312] They go on to call this clause 'unintuitive'. My intuitions—for or against—do not stretch this far.

[313] As Crabbé (2011, 2012) has made clear.

[314] As I also note there, Batens works with a different notion of minimal inconsistency, which guarantees Reassurance.

At the end of the section, W&S state that the open questions they note on these matters 'need to be answered before one can seriously consider using [Priest's] definition of identity in, for example, paraconsistent set theory or for solving paradoxes like the sorites paradox'. To the extent that these questions need answering, I take it that I have now answered them.

27.25.3 The Leibniz Definition

Finally, in Sect. 25.6 W&S turn to the matter of whether identity should be defined *via* the Leibniz condition. They quote a passage from Williamson (2006) saying that:

> it is unlikely that second-order quantification is conceptually more basic than identity in any deep sense. But the appeal to second-order quantification may not satisfy those who are seriously worried about the problem of interpreting the identity predicate. For how do we know, or what makes it the case, that the second-order quantifier $\forall P$ should be interpreted in the standard way?

The standard way referred to in the last sentence is where the second-order domain is the full power set of the first-order domain. (Williamson's discussion is in the context of classical logic.)

We may ignore the point about being conceptually more basic. If people subscribe to some kind of methodology of conceptual foundationalism, I leave it to them to figure out what is more fundamental than what. For my part, I take all such programmes of conceptual analysis to be flawed.

As for the second point, the context of Williamson's quotation is a concern with how one can be sure that someone who characterises identity by the usual first-order conditions really means identity (as usually understood), since there are non-standard interpretations of the language in which it is not so interpreted. He is pointing out that appealing to a characterisation of identity by the Leibniz condition in second-order logic is of no help here, since there are non-standard interpretations of the second-order quantifiers as well. Now, what, exactly, the intended interpretation of the second-order logic is in the present case is as yet undetermined, pending a specification of the range of the second-order variables. But in any case, Williamson's comments are beside the point in the present context. True, a syntactic characterisation of the second-order semantics in question is likely to underdetermine; but as the context makes clear, this is equally the case with the standard first-order theory of identity. As we know, any axiomatic theory in the language of first- or second-order logic with an infinite model will have non-standard interpretations.[315] The non-standard are always with us. What, then, about usage ensures that we have a standard interpretation is an important question; but it is not one on the agenda here.

W&S then refer to a passage from Manzano (2005). In the context, she is addressing the question of why, though one may define standard identity in (full) classical

---

[315] Actually, in the case of *LP*, the restriction to having an infinite model can be dropped, due to the Collapsing Lemma.

second-order logic, it is a good idea to take it as primitive anyway. The first three reasons she gives are to the effect that in less than full second-order logic the Leibniz definition may not deliver standard first-order identity. As hardly needs to be said, this is not a reason for not characterising identity with the Leibniz condition if the aim is not to recapture the usual account of identity—she even points out that the substitutivity of identicals may fail!—and, moreover, one is not concerned with subsystems of classical logic anyway. Such considerations are simply, therefore, beside the point.[316]

In the last paragraph of the section W&S, cite two further problems they see with defining identity by the Leibniz condition in *LP*. The first is the failure of *modus ponens* for the material conditional in *LP*. As the saying goes, this is not a bug, it is a feature. It is this which undergirds the account of unity in ONE! It might be suggested that since most contemporary logicians hold identity to be transitive, its failure requires some 'independent motivation'; and perhaps the gluon theory of ONE is unusual enough not to count. But as ONE, 5.7, 5.8 notes, there are well-known (if frequently ignored) apparent counterexamples to TI. And as 5.3 points out, in this context, standard arguments for TI are bankrupt.

Of course, it might be suggested that the *LP* biconditional is the wrong one to use in the Leibniz condition, and that a detachable conditional should be used. That matter is taken up in ONE, 2.6.

The second of W&S's supposed problems is that the possibility of contradictory identities is inherited from other contradictions. I simply fail to see this as a problem. Why should the contradictory nature of identity not follow from other contradictions? As ONE, Chap. 1, argues, one should expect gluons to be inconsistent objects. One is, hence, committed to a paraconsistent logic. As explained in Chap. 2, the Leibniz definition of identity then gives rise to a non-transitive identity. This explains *why* identity is not transitive, when one might have expected otherwise. It also explains why gluons do not generate a Bradley-style regress, solving the problem of unity. Dialetheism about gluons is, therefore, fundamental to the whole story.

In Sect. 25.7 W&S add a final complaint:

> [Priest defines] identity of individuals as Leibniz identity in second-order minimal *LP*. The choice, though clearly motivated, is not unproblematic, in particular, if it is driven to the point that second-order minimal *LP* is to be used as a metatheory, thereby expulsing the standard notion of real identity entirely.

Now, as I have explained in Sect. 22.3.2 above, the semantic metatheory is best thought of in terms of *ZF*, which one can make sense of in *LP*-based (not, *nota bene*, *LPm*-based) naive set theory. And matters concerning identity are irrelevant. For as is well known, *ZF* can be expressed in a language with a single predicate, $\in$, the principle of extensionality being formulated as $\forall z(x \in z \equiv x \in y) \supset (x \in w \equiv y \in w)$.

---

[316]Manzano's fourth point is slightly different. This is that (classical) primitive identity, comprehension, and extensionality allows us to introduce set/property abstracts. Outwith this context, abstracts have to be taken as primitive. I see no reason why this is problematic. If abstracts are required, it seems no worse to take these as primitive than to take identity itself as primitive.

In their final sentence, W&S say that they prefer the standard theory of identity to my account. They are, of course, entitled to their preferences. However, they do not say how they would solve the philosophical problems that a non-transitive identity allows. Nor, as I have indicated, do I find their problems with non-transitive identity very persuasive.

27.26 Weber (Chap. 26): Contradictions Before the Limit

Weber's results on relevant naive set theory are clearly the most significant advances in the topic since, and on a par with, Brady's groundbreaking work in the late 1970s. Brady showed what *could not* be proved: suitably formulated, relevant naive set theory was non-trivial. The other side of the coin is what *could* be proved. Could one establish standard results of the theory of transfinite cardinals and ordinals in the theory? Many people in the Canberra group of the late 1970s (including myself) struggled with this problem, and gave up. Weber showed how it could be done some 30 years later.[317]

Over recent years, Weber has also been the most ardent advocate of dialetheism, as well as one who has significantly stretched the dialetheic frontiers. In his essay in this collection, he returns to what first attracted him to dialetheism: the Inclosure Schema (IS). There is much I agree with in the essay, and some things with which I do not. I think the best way to put the matter into focus is, again, to provide an historical perspective.

IC was a full frontal assault on the Principle of Non-Contradiction. The paradoxes of self-reference loomed large in this. I used them as a battering ram, as it were, on what I took to be the weakest part of the defences, eroded, as they had been for decades—indeed, for centuries, for those who know their history of Western logic—by failure to reach consensus. But these were not the only applications of dialetheism there, as the third part of the first edition makes clear.[318]

Thanks to discussions with Uwe Petersen in the mid-1980s I came to see the connection between the paradoxes of self-reference and the limit phenomena important to the thought of Kant and Hegel. The result was BLoT. The main thesis of this was that there are certain limits that are dialetheic (the limit of what can be expressed; the limit of what can be described or conceived; the limit of what can be known; the limit of iteration of some operation or other, the infinite in its mathematical sense[319]). These are all such that one can go no further; yet one can.

In the process of writing BLoT I formulated the Inclosure Schema. An inclosure arises where there is a totality, Ω, and an operator, δ, such that it appears to be the case that when δ is applied to any subcollection of Ω of an appropriate kind, it

---

[317] This was in his doctoral thesis, much of which appeared in Weber (2010, 2012).
[318] I give a category of some possible applications of dialetheism in Sect. 27.11.1 above. 1–5 and 8 all appear in the first edition of IC.
[319] BLoT, p. 3.

produces an object not in that subcollection (Transcendence) but still in Ω (Closure). These conditions obviously produce contradiction when applying δ to Ω itself. I took the idea of the IS essentially from Russell, though I tweaked it to accommodated paradoxes such as the Liar cleanly.

The IS was never meant to be an argument for dialetheism; that was not its point[320]—though careless expression sometimes might have suggested otherwise. It was a diagnostic schema for characterising a class of paradoxes (hence unifying them), and showing why they arise. What to do about them is another matter. Russell, after all, was no dialetheist. Of course, if the conditions of the IS are true, then so is the contradiction they generate; and the dialetheic solution is simply to accept this.

The main application of the IS in BLoT is as an argument against orthodoxy (of the time and contemporary) which espouses quite different solutions to the set-theoretic paradoxes and the semantic paradoxes. These both fit the IS, and so, being the same kind of paradox, should have the same kind of solution (the Principle of Uniform Solution).[321]

It had always seemed to me that sorites paradoxes were of a quite different kind from paradoxes of self-reference, and I was not at all inclined to a dialetheic solution for them. But in discussion with Mark Colyvan in 2007, it struck me that the sorites paradoxes were inclosure paradoxes (that is, paradoxes fitting the IS) as well. Because of the Principle of Uniform Solution, I therefore came to accept a dialetheic solution.

27.26.1 Limits and Dialetheism

With these preliminary considerations, we can now turn to Weber's paper.

First, and most straightforwardly, Weber shows (Sects. 26.4.2, 26.5) that, on pain of triviality, not all limits can be transcended. True; but the aim of BLoT was not to show that they could be.[322] (I don't think it ever occurred to me that this might be the case.) So to the extent that Weber interpreted me as saying so (fn. 22), that *is* a misinterpretation.[323] To transcend a boundary, there must be something that takes you to the other side. Such, though contradictory, is the case for the limits of thought with which BLoT deals. The limit of the non-trivial rules this out—on pain of triviality.

The next topic (the rest of Sect. 26.4) concerns problems Weber finds with the IS itself. He makes two points in this context. The first (Sect. 26.4.1) is that there are arguments that have the right shape for an inclosure, but are not paradoxes. One

---

[320] As Weber notes at the end of Sect. 26.3.1.

[321] BLoT, 11.5.

[322] 'Limits of *this kind* provide boundaries beyond which certain processes... cannot go; a sort of conceptual *ne plus ultra*. The thesis of this book is that such limits are dialetheic...' BLoT, p. 3. (First italics added here.)

[323] Weber also quotes me as quoting Cantor talking about breaking through every barrier. The 'every' here concerns just ordinals.

example concerns an omnipotent god; a second concerns diagonalising out of the natural numbers.[324] Some have made a similar point using the 'barber paradox'.[325]

As Weber notes (and BLoT 17.2 explains) for the IS conditions to create paradoxes we have to have good reasons, or at least a *prima facie* kind, to suppose that the conditions are true. There are none in these cases. Weber concurs, but objects:

> Prima facie validity will vary from reasoner to reasoner, though, depending on their logical training. And more importantly, 'seeming' validity is not a reliable guide to truth. Once one gets past initial diagnostic devices, more precise and reliable instruments than mere prima facie plausibility are needed. The way to distinguish a genuine contradiction from a contradiction-shaped joke is to show that the conditions of the inclosure are genuinely satisfied.

Now the first point is true to a certain extent, but should not be over-played. The naive principles that generate the paradoxes of self-reference, and the reasoning from them, have appeared *prima facie* obvious to all those who have thought about them. That is why, after all, we call these things paradoxes, and not simply *reductio* arguments.

The rest of Weber's quote is also agreed; but the IS was never meant to be an argument for dialetheism, as I have explained. It just characterises a class of paradoxes. It is an independent question as to what to do about them. For the contradictions they generate to be true, the IS conditions have to be true. That will require separate arguments. A major one such is that no consistent solution to the paradoxes works, simply because the principles it invokes merely succeed in relocating the inclosure.[326]

27.26.2 Reasoning to the IS Conditions

Weber's next point is more substantial. The IS conditions are generated by naive principles about truth, sethood, and related matters. But to establish the IS conditions we need to reason from these principles. Weber points out that some of the reasoning involved may not be dialetheically valid. Point, again, taken.[327] In BLoT (p. 130, fn 7) drawing on the fact that if Σ entails A in classical logic then, for some B, Σ entails $A \vee B!$ in *LP* (where $B!$ is $B \wedge \neg B$), I noted that even if we cannot establish that $\delta(\Omega) \in \Omega!$ in a kosher fashion, we still have a contradiction in the form $\delta(\Omega) \in \Omega! \vee B!$. Thus, the IS conditions still deliver contradiction.

Weber notes this, but observes, correctly, that, in this case, the contradiction may no longer concern the limit (Ω). True; though most of the arguments for instances of the IS conditions given in BLoT *are* dialethically valid (for example, those concerning truth, knowledge, expressibility), as may be checked on a case by case basis.

---

[324] Actually, I don't how the first of these is supposed to fit into the IS. What, for example, is the diagonaliser? But that is of no great moment here.

[325] See BLoT2, 17.2

[326] BLoT, pp. 228 ff.

[327] I agree with Weber that the logic of an acceptable metatheory should be paraconsistent, though, despite what Weber says, the point appears to me to be irrelevant here. For the arguments to work, they have to be valid; one does not have to prove them to be so in a metatheory. For a discussion of paraconsistent metatheory, see the reply to Batens, 27.3.2 above.

27 Some Comments and Replies

Weber analyses a couple of cases where they are not. It will pay to look at these more closely. One is a version of Russell's paradox where Ω is V (Sect. 26.4.2.1). He points out that in proving Transcendence, given $X \subseteq \Omega$, and having shown that $\delta(X) \notin \delta(X)$, the best we can do is to establish that:

- $\delta(X) \in X \wedge \delta(X) \notin \delta(X) \vdash \delta(X) \in \delta(X)!$

(changing his r to δ). From this we cannot get $\delta(X) \notin X$ by valid means. But given that $\delta(X) \notin X$ or $\delta(X) \in X$, we have either $\delta(X) \notin X$ or $\delta(X) \in \delta(X)!$, so if Transcendence fails, this is because there are already inconsistent boundaries (within Ω), as BLoT claims.

The second case that Weber analyses is the sorites paradox (which does not feature in BLoT). Given a sorites sequence of objects, $\Omega = \{a_0, ..., a_n\}$ is the set of things satisfying some vague predicate, P; and if $X \subseteq \Omega$, $\delta(X)$ is a_{i+1}, where i is the largest j such that $a_j \in X$. As Weber points out (Sect. 26.4.2.4), given that the conditionals in the sorites argument are material, one cannot establish Closure. If $X \subseteq \Omega$, it needs to be shown that Pa_{i+1}. The closest we can get is $Pa_i! \vee Pa_{i+1}$. What the conditions of the sorites entail is that $\bigvee_{0 \leq i \leq n} Pa_i!$.[328] This does not show where the contradiction lies.

However, if one consults the dialetheic models of a sorites transition, in which borderline elements are contradictory, we have something of the form[329]:

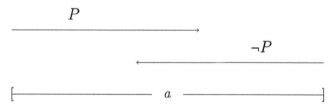

Hence, if a is in the borderline area, $Pa!$ holds. There is a clear sense in which such as constitute the boundary of Ω. Hence, again, the boundary is indeed contradictory. Similarly, such as constitute the boundary of $\overline{\Omega} = \{x : \neg Px\}$, so the boundary of this is also contradictory.[330]

27.26.3 Local Contradictions

Having said these things, it remains the case, as Weber points out, that inclosures may be within a larger totality. He establishes this with respect to the IS and another version of Russell's paradox, where Ω is the Russell set, R, itself (not V). As demonstrated, this is an inclosure contradiction. However, he says that:

---

[328] Priest (2010b), Sect. 5.
[329] Priest (2010b), Sect. 4.
[330] This is an example of the 'double inclosure' in Weber's appendix.

whether this is a meaningful instance of the inclosure schema is dubious. The diagonalizer is doing nothing; all the energy is coming from the totality, top-down, rather than a contradiction surging towards the totality.

I fail to see this. The diagonaser is still transcending lesser collections. It's just that the eruption occurs only part way up the mountain (of the absolute), to pick up the metaphor. Note also that the domain of the inclosure in this case is R, so the contradiction does occur at the boundary of the relevant inclosure.

More generally, it was never an aim of BLoT to show that dialetheias occur *only* at absolute infinities. Even in BLoT, there are many inclosure contradictions which are not of this kind: the liar (where Ω is the set of truths; Berry's paradox, where Ω is a subset of the natural numbers; König's paradox, where Ω is the set of definable ordinals). These contradictions arise a long way from absolute infinity, and so are hardly contradictions at 'the edge of the universe' (as Weber puts it Sect. 26.6). And as BLoT, pp. 170 f. shows, there can indeed be inclosures within inclosures.

More generally, as I have noted, IC argues for the existence of many 'local' dialetheias—concerning motion and law, for example. These are not only nothing to do with absolute infinity; they are not even inclosure contradictions. So I quite agree with Weber's comment (Sect. 26.1) that 'inclosure-based dialetheism is not enough'.

IC, Chap. 8, argues that since dialetheias are relatively rare, we may use classical logic as a default inference-engine until one is shown that one cannot.[331] Weber worries (Sect. 26.6) that this rash of local contradictions will make the methodology vacuous. I don't think so. The question is how to understand rarity. Even if there is just one dialetheia, p_0, there are as many dialetheias as there are formulas. (If q is any true statement, $p_0 \wedge q$ is another dialetheia.)[332] The point is that dialetheias are statistically infrequent in our reasoning.[333] Grant that dialetheias occur at instantaneous states of change. We rarely reason about the truly instantaneous states. Grant that dialetheias occur in the border areas of vague predicates. We rarely reason about such areas. (Even if the predicate 'red' is vague, most things are simply not red.) Similar remarks apply to paradoxical sets and truths.

To summarise my agreements and disagreements with Weber: Dialetheias are many. Some of these are inclosure contradictors; some are not. And even where they are inclosure contradictions, the domain of the inclosure may not be absolutely infinite, but some much smaller totality. It remains the case that inclosure contradictions show there are limits of certain notions (the true, the expressible, and so on), where the notions go no further... though they do. All as BLoT says.

---

[331] I agree with Weber (Sect. 26.6) that 'deductive reasoning should never presume consistency'. The inference engine is one of default reasoning, and so non-deductive, as shown most clearly when the strategy is implemented in *LPm*.

[332] So, again, I agree with Weber (Sect. 26.6) that all objects will satisfy inconsistent conditions. Let a be any object, and P any condition that it satisfies; then $p_0 \wedge Pa$ and $\neg(p_0 \wedge Pa)$.

[333] See IC, 8.4.

27.27 Conclusion: Looking Back Over My Shoulder

As I have been writing these comments over the last six months, I have had to go back and read—or reread—a number of things I have written on matters paraconsistent and dialetheic over the last 40 years or more. In doing so, I have come face to face with the way in which my own thinking on these matters has evolved during the course of that journey. I was struck by how impossible it would have been to predict how things would evolve—and continue to evolve: in writing these comments I have had to think through a number of matters afresh.

Be that as it may, the journey I have made, I have not made on my own. Friends, colleagues, students, and many whom I hardly know—some now, sadly, dead—have been fellow travellers. Their investigations and critical insights, positive and negative, have all helped to enrich the road. I am indeed fortunate to have been accompanied by such good friends.

References

Alxatib, S., & Pelletier, F. (2011). The psychology of vagueness: Borderline cases and contradictions. *Mind and Language, 26,* 287–326.
Beall, J. C. (2009). *Spandrels of truth.* Oxford: Oxford University Press.
Beall, J. C. (2011). Multiple-conclusion *LP* and default classicality. *Review of Symbolic Logic, 4,* 326–336.
Beall, J. C. (2012). Why Priest's reassurance is not so reassuring. *Analysis, 72,* 517–525.
Beall, J. C., & Armour-Garb, B. (Eds.). (2005). *Truth and deflationism.* Oxford: Oxford University Press.
Beall, J. C., Brady, R., Dunn, J. M., Hazen, A., Mares, E., Meyer, R., et al. (2012). On the ternary relation and conditionality. *Journal of Philosophical Logic, 41,* 595–612.
Beall, J. C., Brady, R., Hazen, A., Priest, G., & Restall, G. (2006). Relevant restricted quantification. *Journal of Philosophical Logic, 35,* 587–598.
Bell, J., Cole, J., Priest, G., & Slomson, A. (1973). *Proceedings of the Bertrand Russell memorial logic conference.* Leeds: University of Leeds.
Belnap, N. (1977). A useful four-valued logic. In J. M. Dunn & G. Epstein (Eds.), *Modern uses of multiple-valued logic* (pp. 5–37). Dordrecht: Reidel.
Berto, F., French, R., Priest, G., & Ripley, D. (2018). Williamson on couterpossibles. *Journal of Philosophical Logic, 47,* 693–713
Brady, R. (1989). The non-triviality of dialectical set theory, Chap. 16 of Priest, Routley, & Norman (1989).
Brown, B. (1999). Yes, Virginia, there really are paraconsistent logics. *Journal of Philosophical Logic, 28,* 489–500.
Brown, B., & Priest, G. (2004). Chunk and permeate I: The infinitesimal calculus. *Journal of Philosophical Logic, 33,* 379–388.
Caie, M. (2012). Belief and indeterminacy. *Philosophical Review, 121,* 1–54.
Cantini, A. (2003). The undecidability of Grišin's set theory. *Studia Logica, 74,* 345–368.
Carnielli, W., & Rodrigues, A. (2017). An epistemic approach to paraconsistency: A logic of evidence and truth. *Synthese, 751.* https://doi.org/10.1007/s11229-017-1621-7.
Castañeda's, H. N. (1989). *Thinking, language, and experience.* Minneapolis, MN: University of Minnesota Press.

Chalmers, A. (2014). *What is this thing called science?* (4th ed.). Brisbane: University of Queensland Press.
Church, A. (1936). An unsolvable problem of elementary number theory. *American Journal of Mathematics*, *58*, 345–363.
Church, A. (1956). *Introduction to mathematical logic*. Princeton, NJ: Princeton University Press.
Cobreros, P., Égré, P., Ripley, D., & van Rooij, R. (2012). Tolerant, classical, strict. *Journal of Philosophical Logic*, *41*, 347–385.
Cobreros, P., Égré, P., Ripley, D., & van Rooij, R. (2013). Reaching transparent truth. *Mind*, *122*, 841–866.
Cotnoir, A., & Weber, Z. (2015). Inconsistent boundaries. *Synthese*, *192*, 1267–1294.
Crabbé, M. (2011). Reassurance for the logic of paradox. *Review of Symbolic Logic*, *4*, 479–485.
Crabbé, M. (2012). Reassurance via translation. *Logique et Analyse*, *55*, 281–293.
Dummett, M. (1977). *Elements of intuitionism*. Oxford: Oxford University Press.
Dummett, M. (1978a). The philosophical significance of Gödel's theorem. Chap. 12 of *Truth and other enigmas*. London: Duckworth.
Dummett, M. (1978b). The philosophical basis of intuitionist logic. Chap. 14 of *Truth and other enigmas*. London: Duckworth.
Dunn, J. M. (1979). A theorem in 3-valued model theory, with connections to number theory type theory, and relevant logic. *Studia Logica*, *38*, 149–169.
Field, H. (1973). Theory change and indeterminate reference. *Journal of Philosophy*, *70*, 462–481.
Field, H. (2005). Is the liar both true and false? Chap. 2 of Beall and Armour-Garb.
Field, H. (2008). *Saving truth from paradox*. Oxford: Oxford University Press.
Field, H. (2014). Naive truth and restricted quantification. *Review of Symbolic Logic*, *7*, 147–191.
Field, H. (2016). Indicative conditionals restricted quantification and naive truth. *Review of Symbolic Logic*, *9*, 181–208.
Field, H. (201+). Properties, propositions and conditionals. *Australian Philosophical Review* (forthcoming).
Fine, K. (1975). Vagueness, truth, and logic. *Synthese*, *30*, 265–300.
Glanzberg. M. (2013). Truth. In E. Zalta (Ed.), *Stanford encyclopedia of philosophy*. https://plato.stanford.edu/entries/truth/.
Goodship, L. (1996). On dialethism. *Australasian Journal of Philosophy*, *74*, 153–161.
Harman, G. (1986). *Changes in view*. Cambridge, MA: MIT Press.
Hazen, A., & Pelletier, J. (2018). Second-order logic of paradox. *Notre Dame Journal of Formal Logic*, *59*, 547–558.
Humberstoneone, L. (2011). *The connectives*. Cambridge, MA: MIT Press.
Keefe, R., & Smith, P. (Eds.). (1999). *Vagueness: A reader*. Boston: MIT Press.
Kripke, S. (1975). Outline of a theory of truth. *Journal of Philosophy*, *72*, 690–716.
Lakatos, I. (1978). Cauchy and the continuum: The significance of non-standard analysis for the history and philosophy of mathematics. *The Mathematical Intelligencer*, *1*, 151–161; reprinted as pp. 43–60, vol. 2, of Lakatos' *Collected Papers*, eds. J. Worrall & G. Currie. Cambridge: Cambridge University Press, 1978.
MacFarlane, J. (2004). *In what sense (if any) is logic normative?* A talk presented to the Central Division of the American Philosophical Association.
Manzano, M. (2005). *Extensions of first-order logic*. Cambridge: Cambridge University Press.
Martin, B. (2014). Dialetheism and the impossibility of the world. *Australasian Journal of Philosophy*, *93*, 61–75.
Mares, E. (1996). Relevant logic and the theory of information. *Synthese*, *109*, 345–360.
Mares, E. (2004). *Relevant logic: A philosophical interpretation*. Cambridge: Cambridge University Press.
Maund, J. B. (1981). Colour: A case of conceptual fission. *Australasian Journal of Philosophy*, *59*, 308–322.
Meyer, R. K., & Mortensen, C. (1984). Inconsistent models for relevant arithmetic. *Journal of Symbolic Logic*, *49*, 917–929.

Nelson, E. (2015). *Inconsistency of primitive recursive arithmetic*. arXiv: abs/1509.09209
Omori, H. (2018). From logics of formal inconsistency to logics of formal classicality. *Logic Journal of the IGPL*. https://doi.org/10.1093/jigpal/jzy056.
Paris, J., & Sirokofskich, A. (2008). On *LP*-models of arithmetic. *Journal of Symbolic Logic, 73*, 212–226.
Priest, G. (1979). The logic of paradox. *Journal of Philosophical Logic, 8*, 219–241.
Priest, G. (1984). Logic of paradox revisited. *Journal of Philosophical Logic, 13*, 159–173.
Priest, G. (1987). *In contradiction*. Dordrecht: Martinus Nijhoff; 2nd ed. Oxford: Oxford University Press, 2006.
Priest, G. (1990). Dialectic and dialetheic. *Science and Society, 53*, 388–415.
Priest, G. (1991). Minimally inconsistent *LP*. *Studia Logica, 50*, 321–331.
Priest, G. (1994a). Is arithmetic consistent? *Mind, 103*, 337–349.
Priest, G. (1994b). What could the least inconsistent number be? *Logique et Analyse, 37*, 3–12.
Priest, G. (1995a). *Beyond the limits of thought*. Cambridge: Cambridge University Press; 2nd ed. Oxford: Oxford University Press, 2002.
Priest, G. (1995b). Multiple denotation ambiguity, and the strange case of the missing amoeba. *Logique et Analyse, 38*, 361–373.
Priest, G. (1998). What's so bad about contradictions? *Journal of Philosophy, 95*, 410–426; reprinted as Chap. 1 of G. Priest, J. C. Beall, & B. Armour-Garb, *New essays on the law of non-contradiction*. Oxford: Oxford University Press.
Priest, G. (1999). Validity. In A. Varzi (Ed.), *The nature of logic (European Review of Philosophy)*, pp. 183–206, vol. 4. Stanford, CA: CSLI Publications; reprinted as Chap. 11 of Priest (2006a).
Priest, G. (2001). Logic: one or many? In J. Woods & B. Brown (Eds.), *Logical consequences: Rival approaches*, pp. 23–38. Oxford: Hermes Scientific Publishers; reprinted as Chap. 12 of Priest (2006a).
Priest, G. (2002). Paraconsistent logic. In D. Gabbay & F. Guenthner (Eds.), *Handbook of philosophical logic* (2nd ed.). Dordrecht: Kluwer Academic Publishers.
Priest, G. (2003). Inconsistent arithmetic: Issues technical and philosophical. In V. F. Hendricks & J. Malinowski (Eds.), *Trends in logic: 50 years of Studia Logica* (pp. 273–99). Dordrecht: Kluwer Academic Publishers.
Priest, G. (2005a). *Towards non-being*, Oxford: Oxford University Press; 2nd ed. 2016.
Priest, G. (2005b). Spiking the field artillery, Chap. 3 of Beall and Armour-Garb (2005).
Priest, G. (2006a). *Doubt truth to be a liar*. Oxford: Oxford University Press.
Priest, G. (2006b). The paradoxes of denotation. Chap. 7 of T. Bolander, V. F. Hendricks, & S. A. Pedersen (Eds.), *Self-reference*. Stanford, CA: CSLI Lecture Notes.
Priest, G. (2007). Reply to Slater. In J. Y. Béziau, W. Carnielli, & D. Gabbay (Eds.), *Handbook of paraconsistency* (pp. 467–474). London: College Publications.
Priest, G. (2008). *Introduction to non-classical logic: From if to is*. Cambridge: Cambridge University Press.
Priest, G. (2009). Conditionals: A debate with Jackson. Chap. 13 of I. Ravenscroft (Ed.), *Minds, worlds and conditionals: Themes from the philosophy of Frank Jackson*. Oxford: Oxford University Press.
Priest, G. (2010a). Non-transitive identity. In Chap. 23 of R. Dietz & S. Moruzzi (Eds.), *Cuts and clouds: Vagueness, its nature and its logic*. Oxford: Oxford University Press.
Priest, G. (2010b). Inclosures, vagueness, and self-reference. *Notre Dame Journal of Formal Logic, 51*, 69–84.
Priest, G. (2010c). Hopes fade for saving truth. *Philosophy, 85*, 109–140.
Priest, G. (2012a). The sun may, indeed, not rise tomorrow: A reply to Beall. *Analysis, 72*, 739–741.
Priest, G. (2012b). Read on Bradwardine on the Liar. In C. D. Novaes & O. Hjortland (Eds.), *Insolubles and consequences: Essays in honour of Stephen Read* (pp. 155–161). London: College Publications.
Priest, G. (2014a). Contradictory concepts. Chap. 10 of E. Weber, D. Wouters, & J. Meheus (Eds.), *Logic, reasoning and rationality*. Dordrecht: Springer.

Priest, G. (2014b). Revising logic. Chap. 12 of P. Rush (Ed.), *The metaphysics of logic*. Cambridge: Cambridge University Press.
Priest, G (2014c). Plurivalent logic. *Australasian Journal of Logic, 11*, Article 1. http://ojs.victoria. ac.nz/ajl/article/view/1830.
Priest, G. (2014d). *One*. Oxford: Oxford University Press.
Priest, G. (2015). None of the above: The Catuṣkoṭi in Indian Buddhist logic. Chap. 24 of J.-Y. Beziau, M. Chakraborty, & S. Dutta (Eds.). *New directions in paraconsistent logic*. Dordrecht: Springer.
Priest, G. (2016a). Logical disputes and the a Priori. *Logique et Analyse, 59*, 347–366; and *Principios: Rivista de Filosofia, 23*, 29–57.
Priest, G. (2016b). Thinking the impossible. *Philosophicl Studies, 173*, 2649–2662, and *Argumenta, 2*, 181–94 (2017).
Priest, G. (2017). What if: The exploration of an idea. *Australasian Journal of Logic 14*. https://ojs. victoria.ac.nz/ajl/article/view/4028/3574.
Priest, G. (2018a). *The fifth corner of four*. Oxford: Oxford University Press.
Priest, G. (2018b). Some new thoughts on conditionals. *Topoi, 305*, 369–377.
Priest, G. (2018c). Paradoxical propositions. *Philosophical Isues, 28*, 300–307.
Priest, G. (2019a). Kant's excessive tenderness for things in the world, and Hegel's dialetheism. Ch. 2 of S. Lapointe (Ed.), *Logic from kant to russell: Laying the foundations for analytic philosophy*. London: Routledge.
Priest, G. (2019b). Berry's paradox...again. *Australasian Journal of Logic, 16*, 41–48.
Priest, G. (201+a). Dialetheic metatheory (to appear).
Priest, G. (201+b). A note on mathematical pluralism and logical pluralism (to appear).
Priest, G. (201+c). The structure of logical dialectic (to appear).
Priest, G. (201+d). Metaphysical necessity: A skeptical perspective. *Synthese* (to appear).
Priest, G. (201+e). Some contemporary solutions to some medieval problems about the instant of change (to appear).
Priest, G. (201+f). Logical theory choice: The case of vacuous counterfactuals (to appear).
Priest, G. (201+g). Metaphysics and logic: An observation in metametaphysics (to appear).
Priest, G., & Routley, R. (1989a). Systems of paraconsistent logic, Chap. 5 of Priest, Routley, & Norman (1989).
Priest G., & Routley, R. (1989b). Applications of paraconsistent logic, Chap. 13 of Priest, Routley, & Norman (1989).
Priest, G., & Routley, R. (1989c). The philosophical signficance and inevitability of paraconsistency, Chap. 18 of Priest, Routley, & Norman (1989).
Priest, G., Routley, R., & Norman, J. (Eds.). (1989). *Paraconsistent logic: Essays on the inconsistent*. Munich: Philosophia Verlag.
Priest, G., Siderits, M., & Tillemans, T. (2010). The truth(s) about the two truths. Chap. 8 of the Cowherds, *Moonshadows*. New York, NY: Oxford University Press.
Priest, G., & Thomason, N. (2007). 60% proof: Proof, lakatos and paraconsistency. *Australasian Journal of Logic, 5*, 89–100.
Pynko, A. (2010). Gentzen's cut-free calculus versus the logic of paradox. *Bulletin of the Section of Logic, 39*, 35–42.
Read, S. (2000). Harmony and autonomy in classical logic. *Journal of Philosophical Logic, 29*, 123–154.
Read, S. (2008). The truth schema and the liar. In S. Rahman, T. Tulenheimo, & E. Genot (Eds.), *Unity, truth and the liar: The modern relevance of medieval solutions to the liar paradox* (pp. 205–225). Berlin: Springer.
Restall, G. (1999). Negation in relevant logics (how I stopped worrying and learned to love the Routley star). In D. Gabbay & H. Wansing (Eds.), *What is negation?* (pp. 53–76). Dordrecht: Kluwer Academic Publishers.
Restall, G. (2000). *Introduction to substructural logic*. London: Routledge.

Ripley, D. (2001). Contradictions at the borders. In R. Nouwen, R. van Rooij, U. Sauerland, & H. C. Schmitz (Eds.), *Vagueness in communication* (pp. 169–188). Berlin: Springer.
Routley, R. (1977). Ultralogic as universal. *Relevant Logic Newsletter, 2*, 50–89, 138–75; reprinted as an appendix to Exploring Meinong's Jungle and Beyond. Canberra: Research School of Social Sciences, ANU.
Routley, R. (1984). The American plan completed: Alternative classical style semantics without stars. *Studia Logica, 43*, 131–158.
Russell, B. (1908). Mathematical logic as based on the theory of types. *American Journal of Mathematics, 30*, 222–262.
Shapiro, S. (2002). Incompleteness and inconsistency. *Mind, 111*, 817–832.
Shapiro, S. (2014). *Varieties of logic*. Oxford: Oxford University Press.
Shepard, R. (1964). Circularity in judgments of relative pitch. *Journal of the Acoustical Society of America, 36*, 2346–2353.
Slater, B. H. (1995). Paraconsistent logic? *Journal of Philosophical Logic, 24*, 451–454.
Slater, B. H. (2007). Dialetheias are mental confusions. In J. Y. Béziau, W. Carnielli, & D. Gabbay (Eds.), *Handbook of paraconsistency* (pp. 459–466). London: College Publications.
Spade, P. V., & Read, S. (2017). Insolubles. In E. Zalta (Ed.), *Stanford encyclopedia of philosophy*. https://plato.stanford.edu/entries/insolubles/.
Tedder, A. (2015). Axioms for finite collapse models of arithmetic. *Review of Symbolic Logic, 8*, 529–539.
Tennant, N. (2015). A new unified account of truth and paradox. *Mind, 124*, 571–605.
Tenny, J. (1969). *For Ann (rising)*. https://www.youtube.com/watch?v=bbKbE8y95sg.
Urquhart, A. (1984). The undecidability of entailment and relevant implication. *Journal of Symbolic Logic, 49*, 1059–1073.
Vander Laan, D. (2004). Counterpossibles and similarity. Chap. 20 of F. Jackson & G. Priest (Eds.), *Lewisian themes: The philosophy of David K. Lewis*. Oxford: Oxford University Press.
Voevodsky, V. (2012). *What if current foundations of mathematics are inconsistent?* Lecture. Institute of Advanced Studies, Princeton University. https://www.youtube.com/watch?v=O45LaFsaqMA&t=2687s.
Weber, Z. (2010). Transfinite numbers in paraconsistent set theory. *Review of Symbolic Logic, 3*, 71–92.
Weber, Z. (2012). Transfinite cardinals in paraconsistent set theory. *Review of Symbolic Logic, 5*, 269–293.
Weber, Z. (2016a). What is an inconsistent truth table? *Australasian Journal of Philosophy, 94*, 533–548.
Weber, Z. (2016b). Paraconsistent computation and dialetheic machines. In H. Andreas & P. Verdee (Eds.), *Logical studies of paraconsistent reasoning in science and mathematics* (pp. 205–221). Dordrecht: Springer.
Weber, Z., Badia, G., & Girard, P. (2016). What is an inconsistent truth table? *Australiasian Journal of Philosophy, 94*, 533–548.
Welch, P. (2008). Ultimate truth via stable truth. *Review of Symbolic Logic, 1*, 126–142.
Welch, P. (2011). Truth, logical validity, and determinateness. *Review of Symbolic Logic, 4*, 348–359.
Williamson, T. (2006). Absolute identity and absolute generality. In A. Rayo & G. Uzquiano (Eds.), *Absolute generality* (pp. 369–389). Oxford: Oxford University Press.
Williamson, T. (201+). Counterpossibles in metaphysics. In B. Armour-Garb & F. Kroon (Eds.), *Fictionalism in philosophy* (to appear).
Wittgenstein, L. (1968). *Philosophical investigations*. Oxford: Basil Blackwell.
Young, G. (2005). *Revenge: dialetheism and its expressive limitations*. PhD thesis, University of Glasgow.

Chapter 28
Crossing Boundaries

Graham Priest

Abstract In this paper I review some changes in philosophy and logic, and the way that these relate to my work over the years.

28.1 From Mathematics to Philosophy

I suppose that my professional life started by crossing a boundary: that between mathematics and philosophy. Most of my undergraduate studies, and all my graduate studies, were in mathematics—all be the latter in mathematical logic. But by the time I had finished that, I knew two things: first, that philosophy was a lot more fun than mathematics; second, that I would only ever be, at best, a mediocre mathematician. So when St. Andrews offered me a job in the Department of Logic and Metaphysics, I jumped at it. (In those days, as was the tradition in Scottish Universities, there were two philosophy departments: a Department of Logic and Metaphysics and a Department of Moral Philosophy.) To this day, I have no idea why they offered me a job: they hired me to teach the philosophy of science. But I am forever grateful that they did.

Thus, when I became a professional philosopher, I knew virtually no philosophy. I have had to learn nearly everything I know about it by reading, talking to colleagues and teaching it. (Something that students never know is that the teacher always learns more about the subject than do the students.) And I have loved every minute of this. In retrospect, I regard the fact that I came to philosophy without the blinkers imposed by an undergraduate education in the subject an enormous advantage. It has meant that I was free to wander the length and breadth of the land of philosophy, as its various regions took my interest.

It is not my aim to discuss this journey in detail here, though. For philosophy itself has made a journey in the past 100 years, crossing many boundaries of its own;

G. Priest (✉)
Department of Philosophy, The Graduate Center, City University of New York, New York, USA

Department of Philosophy, University of Melbourne, Melbourne, Australia
e-mail: priest.graham@gmail.com

© Springer Nature Switzerland AG 2019
C. Başkent and T. M. Ferguson (eds.), *Graham Priest*
on Dialetheism and Paraconsistency, Outstanding Contributions to Logic 18,
https://doi.org/10.1007/978-3-030-25365-3_28

and I am as much a product of the *Zeitgeist* as an agent in it. In what follows, I will talk about some of the boundaries it has crossed, and, in the process, make a few comments about myself.

28.2 From Classical Logic to Non-classical Logic

Formal logic is a distinctive part of philosophy, since it tends to deal in technicalities, which can appear rather alien to those in other areas of philosophy. It is an integral part of philosophy nonetheless. It has close relationships with epistemology and metaphysics—though at some times, those metaphysical connections have been denied.

Formal logic has undergone significant developments in Western philosophy for over 2,000 years, theories of what follows from what and why, coming and going. Perhaps the most significant development in the history of the subject occurred around the turn of the twentieth century, when sophisticated mathematical tools (such as formal semantics, algebraicisation, axiomatics) were applied to the subject for the first time. What emerged was so called classical logic—though how inappropriate this terminology is is evident to anyone who knows the history of the subject. This was the account of logic delivered by Frege and Russell, and polished by succeeding generations of logicians, including Hilbert, Tarski, and Gentzen. So successful was this account that it soon became the orthodox logical theory of its day; and it is still the account of logic that one will learn now if one takes the first course in logic.

From its origins, it has never been free of problems, however. Indeed, the problems concerning truth, conditionals, vagueness, were manifest. In the initial flush of success, these problems could be swept under the carpet. The paradigm of reasoning for logicians of this period was the mathematics of their day, and classical logic did a good job of accounting for the reasoning in this. Other things could be taken as fringe concerns.

Matters started to change in seriousness around the 1960s and 1970s, when it became clear that the mathematical techniques that delivered classical logic could be applied to deliver a whole host of other logics. And many of these non-classical logics appeared to provide a much more successful account of reasoning than did classical logic once one moves it away from the platonic mathematical heaven.

Unsurprisingly, then, we have since seen the rise of a sophisticated study of non-classical logics, their properties and applications. This, in turn, has generated intense philosophical debates about whether classical logic can be endorsed as the 'one true logic', whether some other system warrants this label, or whether there is any such thing. These investigations and debates now continue apace. Thus is was that logic crossed over the boundary from classical logic to non-classical logic.[1]

Though my training was in classical mathematical logic, I had acquired an interest in non-classical logic as an undergraduate at Cambridge, when being supervised by

---

[1] For more discussion of the revisability of logic, see Priest (2014).

Sue Haack. (She was then a Ph.D. student, writing a thesis which would eventually appear as her *Deviant Logic*.) I found these logics technically fascinating, but could also see many of the philosophical advantages such logics possessed over classical logic.

For many years as a professional philosopher, I taught logic courses, including many on non-classical logic. I had never intended to write a textbook, but by the late 1990s, I got fed up with the fact that there was really no textbook on non-classical logic: one had to select many little bits of the research literature that were simple enough for undergraduates to understand. So, it was that over one summer I wrote up my lecture notes. *Introduction to Non-Classical Logic* was the result.

The original *Introduction* covered only propositional logic. Several friends who used the book to teach told me that they had to augment the material with notes on quantified non-classical logic. I came to agree that leaving the matter at propositional logic was leaving the job half done. So, I wrote to Cambridge University Press, suggesting a second volume on quantifiers and identity. Within the space of a few days, I had a contract from CUP for this. (I have never had a book accepted so fast!) So (not without some regrets about the amount of time it was taking me to write the material from scratch) I wrote the second volume. CUP eventually decided to publish the two volumes as one, producing the current *Introduction*.

It now seems to me that, for the most part, the undergraduate logic curriculum in philosophy departments has not caught up with developments in the subject. Many departments, because of their size, cannot afford to appoint a specialist logician, so some poor individual is assigned to teach the subject. They pull a textbook off the shelf, and simply teach students to fill in the 1s and 0s. This can be not only rather dull, it can give the impression that logic was brought down by Frege on tablets of *Begriffschrift*, from Mt. Jena, leaving the misleading impression that there is nothing more to be said. The first course in logic being on classical logic, the second course, if there is one, is on metatheory. Now, of course, one has to know metatheory if one wants to be a card-carrying logician; but most philosophy students do not, and they do not need to know how to prove soundness and completeness results. What they need to know are the techniques of non-classical logic (such as those of modality, counterfactuals, truth value gaps and gluts, intuitionism), since these have now become integral to many areas metaphysics, epistemology and the philosophy of language.

It still seems to me that the first course on logic should cover classical logic. That, after all, is where contemporary logical techniques are at their simplest. But the course should also highlight the problem areas of classical logic, to avoid giving the impression that that is the end of the story. The second course on logic should be on non-classical logic. A more technical course, if one can be taught, should come later for those with the appropriate mathematical inclinations. I hope that *Introduction to Non-Classical Logic* is helping to move the curriculum towards something more appropriate to philosophy at the start of the twenty-first century.

28.3 From the PNC to Dialetheism

The thought that contradictions cannot be rationally accepted has been high orthodoxy in Western philosophy. Aristotle succeeded in persuading philosophers, by means of arguments that can only be described as both convoluted and lame, that the Principle of Non-Contradiction (PNC) was not only true but obviously so. (What this says about philosophers' rationality, I leave the reader to contemplate.) True, there have been some who have balked against the orthodoxy. The obvious example is Hegel. But such exceptions serve to underline the orthodoxy of the view, not to undermine it—particularly since most Hegel commentators since have tried to interpret Hegel as a friend of consistency. It is a striking fact that in Western philosophy every aspect of Aristotle's philosophy has been rejected, or at least seriously problematised, since his death—with one exception: the PNC.

In the first part of the twentieth century, some progressive thinkers did start to challenge the orthodoxy, however. The most systematic was Łukasiewicz. But one also finds thoughts, which challenge the idea in Meinong and Wittgenstien (after the period of the *Tractatus*). However, I think it fair to say that philosophy crossed the bridge to contemplate the possibility of holding that some contradictions might be true only with the contemporary dialetheic movement.

I do not intend to imply by this that a rejection of the PNC is now orthodox; it most certainly is not. Rather, the point is this. In the 1970s and 1980s, dialetheism was viewed by most philosophers as so absurd as to be entirely ignorable.[2] This, I think, is no longer the case. As people have come to see how hard it is to defend the PNC, and as the possibilities of plausible applications of dialetheism to a number of different areas have grown, most philosophers have at least been forced to recognise that it is an option in logical space that has to be at least acknowledged.[3]

I first started to countenance the view that some contradictions might be true when I was writing my doctorate (which had nothing to do with the matter). The idea struck me as absurd as everyone else. But considerations to do with Gödel's Incompleteness Theorem, which quickly moved to closely connected issues concerning paradoxes of self-reference, eventually persuaded me otherwise. I started to work out the idea of how this might be possible in formal logical terms by formulating the logic which has now become known as the Logic of Paradox, *LP*. The name was given to me by Alan Slomson, when I gave a talk in Leeds. I don't now think it's a great name, since lots of dialetheias don't seem to have much to do with paradoxes of any kind. But the name has stuck.

---

[2]Though dialetheism was soon on the philosophical agenda in Australasia in the 1970s and 80s, it took a long time for it to be taken seriously in the rest of the English-speaking philosophical world. In fact, Hugh Mellor once told me that he didn't realise that *I* was serious till my debate with Timothy Smiley at the Joint Session in 1993! The fact that such good philosophers as Hartry Field, Stewart Shapiro, and Mark Sainsbury (in his book *Paradoxes*) were prepared to engage with dialetheism helped to change matters. I owe them a debt of gratitude.

[3]For more on the history of PNC and dialetheism, see Priest (2007).

Over the next few years, I started to think through matters much more carefully. What other applications of dialetheism might there be? How might one defend the PNC? How does rationality work if contradictions are rationally tolerable? It still seemed to me that there must be something obviously wrong with dialetheism, which I was simply missing. Every time I gave a talk on the matter, I expected someone to put up their hand and show me what I was missing. But this never happened, so I started to think that the PNC was, after all, a bit of outdated dogma.

Anyway, I put those early reflections together in *In Contradiction*, which has since become something like a dialetheist's manifesto. It took me 4 years to find a publisher for this. I had rejection after rejection. Some were simply polite refusals. Some were on the basic of caustic reviewer's reports. My favourite (in retrospect!) was from a referee who said that the whole idea was so absurd that there was nothing more to be said about the matter, and then spent the next five pages engaging with the ideas. (Maybe the person was a secret dialetheist...)

After this book, because of discussions with Uwe Petersen, I came to see the connection between the paradoxes of self-reference and the limit phenomena central to the thought of Kant and Hegel. That prompted *Beyond the Limits of Thought* (a book which is, in its own way, about crossing boundaries). Because of the reading that I was doing for that, I developed a strong interest in the history of Western philosophy, which has never disappeared. It was easier to find a publisher for that book than for *In Contradiction*. Cambridge University Press accepted it, though when the initial print run had sold out they refused to do another. They said that it had 'sold about the number of copies we expect for a book of this kind'. The copyright then reverted to me. Peter Momtichiloff at Oxford University Press agreed to bring out a second edition, and that began the strong and happy relationship with Peter and Oxford University Press I have had since then.

28.4 From Explosion to Paraconsistency

If one asks a contemporary philosopher why dialetheism cannot be true, the first thing they are likely to say is that contradictions entail everything, and since it is clear that not everything is true, contradictions cannot be true. Appealing to the principle of inference that a contradiction implies everything, *ex contradictione sequitur quodlibet*—or Explosion, to give it its modern and more colourful name—is clearly question-begging here. A dialetheist holds that some contradictions are true; if they hold that not everything is true, they will hold Explosion to be invalid. However, the matter of Explosion takes us to another way in which modern philosophy, or logic anyway, has crossed the bridge into the land where inconsistencies may be tolerated.

Explosion is not to be found in Ancient Greek logic. Indeed, Aristotle himself tells us that syllogisms with contradictory premises may be invalid. The discovery/invention of the principle seems to arise in Western logic about the twelfth century. And the appropriateness of its use is discussed at length in Medieval logic after that. The principle becomes baked in to logic only around the turn of the twentieth century, with the rise of classical logic.

Its validity is clearly highly counter-intuitive, however, as teaching any first-year logic class suffices to establish. It is therefore unsurprising that in the rise of non-classical logics some appeared in which Explosion is invalid. Such logics are, by definition, paraconsistent. Paraconsistent logics, based on very different principles, were developed within a period of about 20 years by logicians working quite independently of each other in different countries—indeed continents: Jaśkowsi (Poland), Halldén (Sweden), da Costa (Brazil), Anderson and Belnap (USA).

The name itself was coined by the Peruvian philosopher Miró Quesada. The prefix 'para' in Ancient Greek is ambiguous. It can mean something like 'quasi', as in 'para-military', parachute (sort of falling); or it can mean 'beyond', as in 'paradox' (beyond belief). I had always thought that in 'paraconsistent logic' it had the latter meaning. But some years ago Newton da Costa told me that Quesada had the former meaning in mind. I still think that 'beyond the consistent' is better than 'sort of consistent'. For an exploration of things beyond the consistent is exactly what paraconsistent logic allows.

The use of a paraconsistent logic is clearly necessary for dialetheism unless one wishes it to lapse into trivialism—the view that everything is true. Most paraconsistent logicians are not dialetheists, however. They simply feel that inconsistent information/theories/scenarios should not blow up in one's face. Even so, the mere thought that contradictions might be tolerable in any sense was anathema to most people in the early years of paraconsistent logic, so the logic had a very hostile reception. That changed as the mathematical development of such logics put runs on the board that one could not deny—whether one approved of such logics or not. Paraconsistent logic is now a very well established branch of non-classical logic.[4]

LP is a paraconsistent logic, and I put it to the service of dialetheism. Indeed, I do not think that dialetheism would have been taken seriously at all without developments in paraconsistent logic. When I developed *LP* in the UK, I knew nothing about earlier developments of paraconsistent logic in other parts of the world. Things changed when I moved to Australia, where I was offered my first permanent academic position at the University of Western Australia. I was, in fact, emigrating, though at the time I did not know this. I now regard the move to Australia as an exceptionally serendipitous event in my academic life. There, I became part of a community of philosophers who were both open-minded and tough-minded, just the atmosphere in which new ideas can flourish if they have value.

In particular, I became part of the group of logicians centred around Richard Routley (later, Sylvan) and Bob Meyer (later Meyer) at the Australian National University in Canberra (though, like many of the group, my job was elsewhere). The group was very much concerned with relevant logic, and throughout those years, I worked a great deal on technical issues in this and related areas, exploring or discarding many logical avenues. For the most part, the logicians there were not

---

[4] Again, for more on the history of paraconsistent logic, see Priest (2007).

dialetheists, but since relevant logic is one kind of paraconsistent logic, I found myself very much with a bunch of fellow travellers.

I still remember my first meeting with Richard Sylvan. I read a version of 'The Logic of Paradox' at the first conference I ever attended in Australia, a meeting of the Australasian Association for Logic in Canberra. After the paper, we were walking up the stairs to the tea room. Richard turned round to me and said 'So you're a dialectician then?' So began many happy years of fruitful collaboration, which ended only with Richard's untimely death. 'Dialectician' was the word that Richard was using for what we would now call 'dialetheist'. Richard had been toying with the idea that the world (all that is the case) was inconsistent before I arrived. It was my arrival, I think, that pushed him over the edge into dialetheism.

Richard knew about developments in paraconsistent logic in other countries; and so it was that I came to know about them. It was clear that paraconsistent ideas were being worked on in several different countries, by relatively isolated groups of logicians. We thought that it was about time to make the people involved more aware of their common interests, so we collected a bunch of papers from all those we knew to be actively working in the area and published *Paraconsistent Logics* ("the Big Black Book"). Because of oversights by the publisher, the book's publication was delayed for nearly a decade, and when it appeared it sold for about US$200. Perhaps, we are now accustomed to logic books costing this much; but at the time this was a small king's-ransom. I think the only people who ended up with the book were the contributors and a few libraries. We also decided that there should be a conference bringing all those working on paraconsistent logic together. We figured that Australia was not the place to hold it: it was too far and expensive for most people to come. Diderik Batens generously offered to organise it in Gent, and so the first World Conference on paraconsistency took place there. Sadly, Richard died suddenly and unexpectedly a few months before the conference.

When we were producing the Big Black Book, Richard and I agreed that we needed a distinctive name for the view that some contradictions are true. At that time, the word 'paraconsistency' was being used for both the failure of Explosion and for dialetheism. (Sometimes 'strong paraconsistency' was used for the latter.) This was leading to many unfortunate confusions. Richard soon agreed that 'dialectics' was a poor name: it came with too much baggage. We couldn't think of anything appropriate, so we decided to search for some foreign dictionaries. We went to the reference library at the ANU and looked up things like 'contradiction', and 'inconsistent' in the Ancient Greek and Latin dictionaries. No luck. We then tried every other dictionary the library possessed—including the Gaelic and Hebrew dictionaries. Still no luck. So, drawing on a remark by Wittgenstein in his *Remarks on the Foundations of Mathematics*, where he likens the liar sentence to a Janus-headed figure facing both truth and falsity, we coined the rather ugly neologism 'dialetheism' (two-way truth). Unfortunately we forgot to agree on how to spell it. I spelled it with the 'e'; Richard without. So, both appear in the literature at that time. The spelling with the 'e' has now become standard, probably because of *In Contradiction*.

We did have a very narrow escape in this process, however. When looking up words in English to Greek, we found one that we liked. (I forget now what the Greek word was.) So we fixed on that. As we were leaving the library, it occurred to us that it might be a good idea to look up the word in the reverse direction, the Greek to English. It was translated as something like 'contradictory, stupid, absurd'. Needless to say, we then junked our choice; but we came within a hair's breadth of providing our critics with a rhetorical own goal.

28.5 From Existence to the Non-existence

The view that some objects do not exist is usually, now, associated with the Austrian philosopher Alexius Meinong. Meinong's view is bizarre—just a touch short of insanity, so the story usually goes. Fortunately, drawing on Frege's view of quantification, it was disposed of by Russell—albeit the case that he had held a similar view earlier in his life. Russell's view was hammered home by Quine, and common sense once more prevailed.

The view is distorted on many fronts. First, Meinong was an important philosopher, as Russell himself realised. He was concerned with intentionality; and it is clear that one can admire something, desire something, fear something, where that something may or may not exist. One admires/desires/fears *something* none the less. So some things do not exist. All perfectly straightforward.

Next, Meinong had history on his side. Nearly, all logicians in the history of Western logic—from Aristotle onwards—held some things not to exist. Medieval logicians had sophisticated theories of merely possible objects, in their doctrine of ampliation. Some even held that this doctrine extended to impossible objects in the context of intentionality. It was the Russell/Quine view which was historically aberrant.

Third, attributing this view to Frege is incorrect. Frege points out that the German phrase that is normally translated into English as 'there is', *es gibt*—the verb in which, note, is part of the verb 'to give' (*geben*) not a part of the verb 'to be' (*sein*)—can be used to mean much the same thing as the quantifier *some*. However, he, himself, points out that this use has nothing to do with existence or reality in any metaphysically loaded sense.

Finally, the arguments used by Russell and Quine in support of the view that everything exists are frightful. Their effect was, arguably, more the result of Quine's silver rhetoric than their rational content. (Again, I leave the reader to ponder what this says about rationality and philosophy.)

Be all this as it may, the Quine/Russell view was high orthodoxy in the second half of the twentieth century. In the 1980s, a number of philosophers started to challenge the orthodoxy; most relentlessly, to my mind, Richard Sylvan, who coined the neologism *noneism* to describe the view that some things do not exist. (It's pronounced by saying *none*, and just sticking *ism* on the end.) Meinong had noneist common sense on his side, not Russell and Quine; but I still hear the view that it is just plain obvious that *some* means *some existent* expressed by older member of the Anglo philosophical profession. I sense that something of a sea change is underway, though. Many of the younger philosophers I meet are not scared to take on board the view that some objects do not exist, and even to endorse it. Thus, it seems to me that philosophy crossed the boundary into the dark side in the first part of the twentieth century, and is now crossing the boundary back.[5]

When I met Richard Sylvan I was an orthodox Quinean about these matters. Richard's noneist views struck me as outrageous. However, over the years in which we argued about these things, I came to agree that all my arguments were pretty hopeless. In due course, I became a noneist.

It was only later that I discovered that the orthodox history was all wrong. In my early sojourn in St. Andrews, Steve Read ran a reading group on Desmond Henry's *Medieval Logic and Metaphysics*. This opened my eyes to the richness of Medieval logic. I knew nothing about it before that. Since then, I have had a lively interest in the topic. Indeed, Steve and I have frequently written together on aspects of it. It was he who, many years later, when I was back at St. Andrews as an Arché Professorial Fellow, taught me what medieval logicians had to say about existence. I also went back and read Frege and Russell more carefully, to fill in that part of the picture. I was in for still another shock later. For a logician 'there is' is a paradigm quantifier. I learned only a few years ago from linguists that it is not a quantifier—quantifiers are things like, *all, some, many, most*—and that there is currently no agreement amongst linguists about how, exactly, to understand constructions with the dummy subject *there*.

Even when I had come to accept that noneism is a perfectly coherent and common sense view, I did not immediately accept it. One still needed an account of what properties non-existent objects have. An answer to this question is provided by some version of the Characterisation Principle (CP): an object which is characterised as being so and so is, indeed, so and so. Natural as this principle seems, no one (other than a trivialist) can accept it, since it delivers a two line argument to any conclusion. The CP has to be restricted in some way, and I found all the suggestions as to how to do this (including Richard's) unsatisfactory.

I finally became a noneist when I found a solution that satisfied me. A characterised object has the properties it is characterised as having, not necessarily at this world (though it may), but at some possible or impossible world. The view has now come to be called *Modal Meinongianism*, for obvious reasons. This has a pleasing alliteration, but I don't entirely care for it, since it ignores the entire history of logic leading up to Meinong. Better, I think, would be something like *worldly noneism*. Whatever one calls it, Richard had unfortunately died before I came up with the idea, so I never had the pleasure of discussing it with him. But it resulted in the book *Towards Non-Being*. This book is part of the move that philosophy is making to cross the boundary back to its healthy earlier view.

---

[5]For more reflections on the matter, see Priest (2008).

28.6 From Analytic Philosophy to Continental Philosophy

The next boundary I want to discuss is that between so called 'analytic philosophy' and 'continental philosophy'. Characterising each side of this boundary is fraught. The standard labelling is hopeless. Philosophical analysis was a methodology deployed by some philosophers in England, and maybe Austro-Germany, in the first part of the twentieth century; but it hardly characterises most of what goes on on the analytic side of the divide nowadays. 'Continental philosophy' is even worse. Even if we understand ourselves to be talking about the continent of Europe, Britain is part of this (and will be even if/when the UK cuts its throat and leaves the EU). Worse again, many of the founders of this side of the divide were German or Austrian (Frege, Wittgenstein, Carnap). However, I don't have a terminology to suggest that is not equally misleading, so I will stick with this.

Whatever, one calls the two sides, it is clear that there are various differences which mark them roughly. There are characteristic differences in style of writing and expression. The two sides tend to publish in different journals. There are differences between the philosophers that each side tends to refer to—and even talk to: there is not much communication across the divide. Indeed, philosophers on each side of the divide are often somewhat rude about the philosophers on the other side.

The genesis of the divide may be located around the turn of the twentieth century with the works of Frege and Husserl. Not that these two philosophers would have seen each other as belonging to different traditions. The initial concern of both was the philosophy of mathematics; both were driven to problems in the philosopher of language; both engaged with the thought of the other and criticised it (not a sufficient condition to locate two philosophers in different traditions!). The difference between the two was in the tools that each forged to attack their problems. Frege invented modern formal logic; Husserl invented phenomenology. It was drawing on these two tools that, at last initially, characterised the differences between the two sides of the divide, though other factors were soon added. Positivism was added to the analytic side; existentialism was added to the continental side. Later, differences also emerged. Thus, we had naturalism on the analytic side, and structuralism and post-structuralism on the continental side.

The closer one looks at the divide, the less substantial it becomes, however. It is not just that they have a shared beginning. The problems that each side attacks are often similar: there is a common interest in questions of epistemology, the philosophy of language, political philosophy. Indeed, there are even philosophers who play similar roles on each side of the divide: Kuhn and Foucault argue that science is characterised by ruptures in traditions; Quine and Derrida argue meaning is indeterminate; and so on. Doubtless, the philosophers on each side of the divide tend to express themselves in different ways. But philosophy can be written in many ways. Compare Plato, Aristotle, Kant—philosophers who would be claimed by both sides. Not to mention the earlier and later Wittgenstein—whichever side of the divide one locates him on. Doubtless, there are also turf wars and institutional struggles; but these are features of most university philosophy departments everywhere.

It is a happy fact, then, that a number of philosophers on each side of the boundary are starting to engage with the work of the other side. Thus, philosophers such as Bob Brandom, Adrian Moore and Markus Gabriel draws happily from both sides of the tradition. I have no doubt that there is good philosophy and bad philosophy on each side of the divide. And doubtless, there are characteristic failings on both sides. Analytic writing can be nit-picking and boring. Continental writing can be rambling and pretentious. But there are great philosophers on both sides of the divide; and we have something to learn from great philosophers of any tradition.[6]

Being a logician, my early years of philosophy were very much influenced by Frege, Russell, Carnap, and Quine. And my work often makes use of the tools of formal logic. So I suppose that it would be natural to think of me as an analytic philosopher. But I read a number of 'continental' writers early in my professional life: Sartre (my wife was studying French literature), Foucault (I was teaching the philosophy of science), Hegel (in connection with dialetheism). Others came later. When I was writing *Beyond the Limits of Thought*, John Frow (then Professor of English at the University of Queensland) suggested that I should read Derrida. I struggled with his writing, but finally made some sense of it. Later, I discovered Heidegger, and I came to the conclusion that so much of Derrida is simply a post-structuralist embroidering of Heidegger, expressed with a written style that is willfully obscure. Later again, I discovered Nietzsche, one of the few philosophers one can read just for the pleasure of his style.

Of course, I have continued to read on the other side of the divide too. I really don't set much store by the divide at all; and in my work, I have drawn on insights from both sides, as in *Beyond the Limits of Thought*—as well, of course, on insights from the many great philosophers in history who predate the analytic/continental divide.

I certainly hope that in doing so, my work is helping to bridge the boundary between the two traditions, and, ultimately, render it irrelevant to philosophy.

28.7 From Western Philosophy to Eastern Philosophy

The distinction between analytic and continental philosophy pales into insignificance compared with that between Eastern and Western philosophy—or better, between Eastern and Western philosophies: there are many different kinds of each. From the perspective of the East, the distinction between analytic and continental philosophy appears as just a family tiff.

The Eastern philosophical traditions are as venerable as any in the West. They are just as sophisticated, and just as deep. Whether there was any communication across the Silk Route in the period of Classical Greek, Indian, and Chinese philosophy, we will probably never know; but it is fair to say that the Eastern and Western traditions have developed largely independently of each other. (Arabic philosophy, note, is a

---

[6]For further reflections on these matters, see Priest (2003).

Western philosophy. Religiously, it comes out of the same matrix as Judaism and Christianity; and its philosophical heritage is that of Ancient Greece.)

To the extent that the Asian traditions have registered much in the West, they have been the property of people working in departments of philology and comparative religion. Needless to say, the people in these traditions have not tended to engage in the Eastern texts as would a philosopher. Moreover, the translations and commentaries made have not been as philosophically acute as one might have wished. To translate is to interpret, and it takes someone with philosophical skills to make a good translation of a philosophical text.

There have been a few Western philosophers who had some limited—and pretty inaccurate—knowledge of some of the Asian traditions. Hegel and Schopenhauer come to mind. But it is fair to say that most Western philosophers have known little about the Eastern traditions—and cared even less. In fact, as recently as a few decades ago, it was common to hear Western philosophers say that these traditions were not philosophy at all, but mere religion, mysticism, wise-man pronouncements. Clearly, people who held such views had never looked carefully at the texts. If one does so, it is clear that they engage in philosophical issues and debates. Indeed, many (though not all) of the issues engaged with are ones very familiar to Western philosophers. Where this is so, what is said about them is sometimes similar to Western views; sometimes not.

Happily, one rarely hears this view expressed nowadays, though it is still common enough to hear it said that the Asian philosophical traditions are fringe philosophy, like feminist philosophy and aesthetics, not part of 'core philosophy' at all. (Let me hasten to add that I am most certainly not endorsing this view of feminist philosophy and aesthetics. I am merely reporting what I hear.) It beggars belief that one can write off half of the world's philosophy in this way.

But there is clearly a sea change underway. Western philosophers are coming to engage with these traditions, writing and thinking about them, and teaching them. It will be a gradual process. As more Western philosophers know about these matters, the more they will be taught, the more Western philosophers will know about these matters, and so on. I expect to see these traditions as an integral part of the Western philosophical curriculum a few decades hence.

I note that philosophers from Asian countries have been engaging in the Western traditions for at least a century now: Nishida and Nishitani in Japan; Aurobindo and Bhattacharya in India; to say nothing of the influence of Marxist philosophy on thinkers is all the Asian countries.

Of course, to understand the texts from these traditions in a sophisticated way, one needs a pertinent knowledge of the languages in which they are written, the societies in which they are embedded, and so on. Such is equally true, of course of Ancient Greek, Arabic or Medieval Christian philosophy. Fortunately, translations of the Asian texts are now being made by good philosophers with the appropriate linguistic skills. And of course, the fact that one does not have these skills does not stop one from engaging in the philosophical content of the texts, any more than one has to have Ancient Greek to discuss the philosophical content of Ancient Greek philosophical texts. One just has to be aware of one's limitations, and respect the skills of the scholars who do have these skills. (Think how many languages philosophy is written in: English, French, German, Latin, Ancient Greek, Arabic, Classical Chinese, Sanskrit—to name but a few. If one could engage only in those texts for which one could speak the language, one's philosophical compass would be deeply impoverished.)[7]

In my first couple of decades as a professional philosopher, I did not have an antagonistic attitude to the Asian philosophical traditions. I had no attitude at all: they were just not on my radar. But in the 1990s, I met Jay Garfield. I had just finished *Beyond the Limits of Thought*, and he had just finished his translation of Nāgārjuna's *Mūlamadhyamakakārikā*. We discovered a number of mutual interests. Indeed, Jay has since become a third philosopher with whom I have co-authored many things.

When I met Jay, I was shocked to learn of my ignorance in the Asian philosophical traditions; and since then, I have made a point of trying to educate myself: reading, teaching, travelling to India and Japan to study. And I have come to draw on Asian traditions—especially the Buddhist ones—more and more in my work, as, for example, in *One* (a book which also provided a new opportunity to deploy the techniques of non-classical logic: non-transitive identity). As with the analytic/continental divide, I have certainly not stopped drawing on the Western traditions. And though I am by no means a scholar of either Eastern or Western philosophy, I hope that my work now draws on some of the best of both. I also hope that my philosophical writings which do so are helping Western philosophy to cross the East/West boundary.

28.8 From Logic and Metaphysics to Political Philosophy

There is, of course, much more to be said about all the developments in philosophy which I have briefly discussed. There are also many other significant developments in Western philosophy in the past 100 years which I have not discussed at all. That, however, will do for the present context.

Let me end by returning to the issue of personal boundaries. I started by crossing the divide between mathematics and philosophy. And I think it fair to say that most of my philosophical work has been in logic and metaphysics, with bits of the history of philosophy (East and West) thrown in. In the last couple of chapters of *One*, I ventured into the realm of Buddhist ethics, though; and as the very end of the book shows, issues of political philosophy surfaced. I have always had an interest in politics and its philosophy. When I was at the University of Western Australia, Val Kerruish (a lecturer in the Law Faculty), set up a reading group on Marx, and over the next several years, we read a great deal of his work (all three volumes of *Capital*, *Grundrisse*, and many of the earlier works). I was struck by the acuity of Marx' analysis of capitalism. I have never written anything much about political philosophy, however. That is the aim of the book I am now working on—though whether I will be able to say anything satisfactory about the topic, remains to be

---

[7]For further reflections on these matters, see Priest (2011).

seen. However, crossing into political philosophy is the next personal philosophical boundary I aim to cross.

References

Priest, G. (2003). Where is philosophy at the start of the twenty-first century? *Proceedings of the Aristotelian Society, 103*, 85–96.
Priest, G. (2007). Paraconsistency and dialetheism. In D. Gabbay & J. Woods (Eds.), *Handbook of the history of logic* (Vol. 8, pp. 129–204). Amsterdam: North Holland.
Priest, G. (2008). The closing of the mind: How the particular quantifier became existentially loaded behind our backs. *Review of Symbolic Logic 1*, 42–55. Reprinted as Chap. 16 of the 2nd edn of *Towards Non-Being*, Oxford: Oxford University Press, 2016.
Priest, G. (2011). Why Asian philosophy? In G. Oppy & N. Trakakis (Eds.), *The antipodean philosopher*, Chap. 18. Lanham, MD: Lexington Books.
Priest, G. (2014). Revising logic. In P. Rush (Ed.), *The metaphysics of logic*, Chap. 12. Cambridge: Cambridge University Press.

Graham Priest: Publications

As of the end of 2018. All the papers are available on grahampriest.net.

a) Papers

'The Conventionalist Philosophy of Mathematics', *Proc. Bertrand Russell Memorial Logic Conference*, Denmark 1971, eds. J. Bell *et al*, Leeds 1973, pp. 115-132.
'Gruesome Simplicity', *Philosophy of Science* 1976, 43, 432-7.
'Modality as a Metaconcept', *Notre Dame Journal of Formal Logic* 1976, XVII, 401-414.
'The Formalization of Ockham's Theory of Supposition' (with S. Read), *Mind* 1977, LXXXVI, 109-113.
'A Refoundation of Modal Logic', *Notre Dame Journal of Formal Logic* 1977, XVIII, 340-354.
'Logic of Paradox', *Journal of Philosophical Logic* 1979, 8, 219-241. Translated into French as 'La Logique du Paradoxe', *Philosophie* 94 (2007), 72-94.
'Indefinite Descriptions', *Logique et Analyse* 1979, 22, 5-21.
'A Note on the Sorites Paradox', *Australasian Journal of Philosophy* 1979, 57, 74-75.
'Two Dogmas of Quineanism', *Philosophical Quarterly* 1979, 29, 289-301.
'Sense, Entailment and Modus Ponens', *Journal of Philosophical Logic* 1980, 9, 415-35.
'Merely Confused Supposition' (with S. Read), *Franciscan Studies* 1980, 40, XVIII, 265-297.
'Ockham's Rejection of Ampliation' (with S. Read), *Mind* 1981, 90, 274-9.
'The Argument from Design', *Australasian Journal of Philosophy* 1981, 59, 422-43.
'The Logical Paradoxes and the Law of Excluded Middle', *Philosophical Quarterly* 1983, 33, 160-65.
'To be and not to be: Dialectical Tense Logic', *Studia Logica* 1982, 41, 249-686; translated into Bulgarian and reprinted in *Filosofska Missal* XL(8), 1984, 63-76.
'Lessons from Pseudo-Scotus' (with R. Routley), *Philosophical Studies* 1982, 42, 189-99.
'The Truth Teller Paradox' (with C. Mortensen), *Logique et Analyse* 1981, 95-6, 381-8.
'An Anti-Realist Account of Mathematical Truth', *Synthese* 1983, 57, 49-65. Reprinted as ch. 8 of D. Jacquette (ed.) *Philosophy of Mathematics*, Blackwell, 2002.
'Logic of Paradox Revisited', *Journal of Philosophical Logic*, 1984, 12, 153-179.
'Semantic Closure', *Studia Logica* 1984, 43, 117-29.
'Introduction to Paraconsistent Logic' (with R. Routley), *Studia Logica* 1983, 44, 3-16.
'Hypercontradictions', *Logique et Analyse* 1984, 107, 237-43.
'Hume's Final Argument', *History of Philosophy Quarterly* 1985, 2, 349-352.

'Inconsistencies in Motion', *American Philosophical Quarterly* 1985, 22, 339-46.
'Contradiction, Belief and Rationality', *Proc. Aristotelian Society* 1985/6, LXXXVI, 99-116.
'Tense and Truth Conditions', *Analysis* 1986, 46, 162-6.
'The Logic of Nuclear Armaments', *Critical Philosophy* 1986, 3, 107-113.
'Unstable Solutions to the Liar Paradox' in *Self Reference: Reflections and Reflexivity*, S.J. Bartlett and P. Suber (eds.), Nijhoff, 1987.
'Tense, *Tense* and TENSE', *Analysis* 47, 1987, 177-9.
'Reasoning about Truth', *Technical Report TR-ARP-2/88*, Automated Reasoning Project, Australian National University, 1988.
'Consistency by Default', *Technical Report TR-ARP-3/88*, Automated Reasoning Project, Australian National University, 1988.
'When Inconsistency is Inescapable', *South African Journal of Philosophy*, 7, 1988, 83-89.
'Primary Qualities are Secondary Qualities Too', *British Journal for the Philosophy of Science*, 1989, 40, 29-37.
'Contradiction, Assertion, and "Frege's Point"' (with R. Sylvan), *Analysis*, 49, 1989, 23-6.
'Reasoning About Truth', *Artificial Intelligence*, 39, 1989, 231-44.
'Denyer's $ not Backed by Sterling Arguments', *Mind*, 98, (1989), 265-8.
'Classical Logic *Aufgehoben*' in *Paraconsistent Logic*, G. Priest, R. Routley and J. Norman (eds.), Philosophia Verlag, 1989.
'Reductio ad Absurdum et Modus Tollendo Ponens' in *Paraconsistent Logic*, G. Priest, R. Routley and J. Norman (eds.), Philosophia Verlag, 1989. Reprinted in Rumanian in I. Lucica (ed.), *Ex Falso Quodlibet: studii de logica paraconsistenta* (in Romanian), Editura Technica, 2004.
'Relevance, Truth and Meaning' (with J. Crosthwaite) in *Directions of Relevant Logic*, R. Sylvan and J. Norman (eds.), Nijhoff, 1989.
'Gegen Wessel', *Philosophische Logik* 1989, 2, 109-20.
'Dialectic and Dialetheic', *Science and Society* 1990, 53, 388-415.
'Boolean Negation, and All That', *Journal of Philosophical Logic* 1990, 19, 201-15.
'Was Marx a Dialetheist?', *Science and Society*, 1991, 54, 468-75.
The Nature of Philosophy and its Place in the University, University of Queensland Press, 1991.
'Minimally Inconsistent LP', *Studia Logica*, 1991, 50, 321-331.
'Intensional Paradoxes', *Notre Dame Journal of Formal Logic* 32 (1991), 193-211.
'The Limits of Thought - and Beyond', *Mind* 100 (1991), 361-70.
'Simplified Semantics for Basic Relevant Logics' (with R. Sylvan*), Journal of Philosophical Logic*, 1992, 21, 217-32.
'Goedel's Theorem and Creativity', in *Creativity*, ed. T. Dartnall, Kluwer Academic Publishers, 1994. Reprinted with a different introduction as 'Goedel's Theorem and the Mind... Again', in *Philosophy of Mind: the place of philosophy in the study of mind*, eds. M. Michaelis and J. O'Leary-Hawthorne, Kluwer, 1994.
'Another Disguise of the same Fundamental Problems: Barwise and Etchemendy on the Liar', *Australasian Journal of Philosophy*, 71 (1993), 60-9.
'Yu and Your Mind', *Synthese*, 95 (1993), 459-60.
'On Time', *Philosophia*, 50 (1992), 9-18.
'Can Contradictions be True?, II', *Proc. Aristotelian Society*, Supplementary Volume 67 (1993), 35-54.
'Derrida and Self-Reference', *Australasian Journal of Philosophy* 72 (1994), 103-11.
'Paraconsistent Dialogues' (with J.McKenzie), *Logique et Analyse*, 131-2 (1990), 339-57.
'The Structure of the Paradoxes of Self-Reference', *Mind* 103 (1994), 25-34.
'Etchemendy and Logical Consequence', *Canadian Journal of Philosophy* 25 (1995), 283-92.
'Gaps and Gluts: Reply to Parsons', *Canadian Journal of Philosophy*, 25 (1995), 57-66.
'What is a Non-Normal World?', *Logique et Analyse* 35 (1992), 291-302.
'Sorites and Identity', *Logique et Analyse*, 135-6 (1991), 293-6.
'Is Arithmetic Consistent?', *Mind*, 103 (1994), 337-49.

'Some Priorities of Berkeley', *Logic and Reality: Essays on the Legacy of Arthur Prior*, ed. B. J. Copeland, Oxford University Press, 1996.

'The Definition of Sexual Harassment' (with J. Crosthwaite*), Australasian Journal of Philosophy* 74 (1996), 66-82. Reprinted in G. Lee Bowie and Meredith Michaels (eds.), *13 Questions in Ethics and Social Philosophy*, Harcourt, Brace, Jovanovich, 2nd ed., 1997.

'Paraconsistent Logic', *Encyclopedia of Philosophy*, Vol.7, 208-11, ed. E. Craig, Routledge, 1998.

'Number', *Encyclopedia of Philosophy*, Vol.7, 47-54, ed. E.Craig, Routledge, 1998.

'Logic, Nonstandard', pp. 307-10, *Encyclopedia of Philosophy; Supplement*, ed. D. Borshert, MacMillan, 1996.

'On a Paradox of Hilbert and Bernays', *Journal of Philosophical Logic* 26 (1997), 45-56.

'Everett's Trilogy', *Mind*, 105 (1996), 631-47.

'On Inconsistent Arithmetics: Reply to Denyer', *Mind*, 105 (1996), 649-59.

'Paraconsistent Logic', *Handbook of Philosophical Logic*, Vol. 6, pp. 287 – 393, eds. D.Gabbay and F. Guenthner, 2nd edition, Kluwer Academic Publishers, 2002.

'Paraconsistent Logic', *Encyclopaedia of Mathematics; Supplement*, ed. M. Hazenwinkle, Kluwer Academic Publishers, 1997, 400-1.

'The Trivial Object and the Non-Triviality of a Semantically Closed Theory with Descriptions', *Journal of Applied and Non-Classical Logic*, 8 (1998), 171-83.

'The Linguistic Construction of Reality', *Exordium* 6 (1997), 1-7.

'Inconsistent Models of Arithmetic; I Finite Models', *Journal of Philosophical Logic*, 26 (1997), 223-35.

'What not? A Defence of a Dialetheic Account of Negation', in D. Gabbay and H. Wansing (eds.), *What is Negation?*, Kluwer Academic Publishers, 1999.

'What Could the Least Inconsistent Number be?', *Logique et Analyse*, 37 (1994), 3-12.

'Sexual Perversion', *Australasian Journal of Philosophy* 75 (1997), 360-72. Translated into Croatian, ch. 9 of I. Primoratz (ed.), *Suvremena filozofija seksualnosti*, Zagreb: KruZak, 2003.

'Paraconsistent Logic' (with K. Tanaka*), Stanford Internet Encyclopedia of Philosophy*, created 1996.

'Multiple Denotation, Ambiguity and the Strange Case of the Missing Amoeba', *Logique et Analyse*, 38 (1995), 361-73.

'On Alternative Geometries, Arithmetics and Logics, a Tribute to Łukasiewicz', *Studia Logica* 74 (2003), 441-468.

'Yablo's Paradox', *Analysis* 57 (1997), 236-42.

'Fuzzy Identity and Local Validity', *Monist* 81 (1998), 331-42.

'Sylvan's Box', *Notre Dame Journal of Formal Logic* 38 (1997), 573-82.

'To be *and* Not to Be - That is the Answer. On Aristotle on the Law of Non-Contradiction', *Philosophiegeschichte und Logische Analyse* 1 (1998), 91-130.

'Validity', pp183-206 of A. Varzi (ed.) *The Nature of Logic*, CSLI Publications, 1999. (*European Review of Philosophy*, vol. 4). An abbreviated version under the same title appears as pp.18-25 of *The Logica Yearbook, 1997*, ed. T. Childers, Institute of Philosophy, Czech Republic.

'Language, its Possibility, and Ineffability', pp. 790-794 of P. Weingartner, G. Schurz and G. Dorn (eds.), *Proceedings of the 20th International Wittgenstein Symposium*, The Austrian Ludwig Wittgenstein Society, 1997.

'Negation as Cancellation, and Connexivism', *Topoi* 18 (1999), 141-8.

'On a Version of one of Zeno's Paradoxes', *Analysis* 59 (1999), 1-2.

'Inconsistency and the Empirical Sciences', in J. Meheus (ed.), *Inconsistency in Science*, Kluwer Academic Publishers, 2002.

'Editor's Introduction', *Notre Dame Journal of Formal Logic* 38 (1997), 481-47.

'Motivations for Paraconsistency: the Slippery Slope from Classical Logic to Dialetheism', in D. Batens *et al* (eds.), *Frontiers of Paraconsistent Logic*, Research Studies Press, 2000.

'On an Error in Grove's Proof' (with K. Tanaka), *Logique et Analyse* 158 (1997), 215-7.

'What's so Bad about Contradictions?', *Journal of Philosophy* 95 (1998), pp. 410-26

'Inconsistent Models of Arithmetic II; the General Case', *Journal of Symbolic Logic* 65 (2000), 1519-29.
'The Import of Inclosure; some Comments on Grattan-Guinness', *Mind* 107 (1998), 835-40.
'Dialetheism', *Stanford Internet Encyclopedia of Philosophy*, created 1998.
'Worlds Apart', *Mind! 2000* (a supplement to *Mind* 109 (2000)), 25-31.
'Perceiving Contradictions', *Australasian Journal of Philosophy* 77 (1999), 439-46.
'Semantic Closure, Descriptions and Triviality', the *Journal of Philosophical Logic* 28 (1999), 549-58.
'On the Principle of Uniform Solution: a Reply to Smith', *Mind* 109 (2000), 123-6.
'Logic: One or Many' in J. Woods, and B. Brown (eds.), pp. 23-28 of *Logical Consequence: Rival Approaches Proceedings of the 1999 Conference of the Society of Exact Philosophy*, Stanmore: Hermes, 2001.
'Truth and Contradiction', *Philosophical Quarterly* 50 (2000), 305-19.
'Paraconsistent Belief Revision', *Theoria* 68 (2001), 214-28.
'Could everything be True?', *Australasian Journal of Philosophy* 78 (2000), 189-95.
'The Logic of Backwards Inductions', *Economics and Philosophy* 16 (2000), 267-85.
'Vasil'év and Imaginary Logic', *History and Philosophy of Logic* 21 (2000), 135-46.
'Geometries and Arithmetics', pp. 65-78 of P. Weingartner (ed.), *Alternative Logics; Do Sciences Need Them?*, Springer Verlag, 2003.
'Logicians Setting Together Contradictories. A Perspective on Relevance, Paraconsistency, and Dialetheism', ch. 14 of D. Jacquette (ed.), *A Companion to Philosophical Logic*, Blackwell, 2002.
'Fuzzy Relevant Logic', *Paraconsistency: the Logical Way to the Inconsistent*, ed. W. Carnielli et al., Marcel Dekker, 2002.
'Why it's Irrational to Believe in Consistency', pp. 284-93 of *Rationality and Irrationality; Proc. 23rd International Wittgenstein Symposium*, eds., B. Brogaard and B. Smith, 2001.
'Objects of Thought', *Australasian Journal of Philosophy* 78 (2000), 494-502.
'Heidegger and the Grammar of Being', ch. 10 of R. Gaskin (ed.), *Grammar in Early 20th Century Philosophy*, Routledge, 2001.
'Nagarjuna and the Limits of Thought' (with Jay Garfield), *Philosophy East West 53 (2003), 1-21*. Reprinted as ch. 5 of J. Garfield, *Empty Words*, Oxford University Press, 2002.
'Paraconsistency and Dialetheism', pp. 129-204 of *Handbook of the History of Logic*, Vol. 8, eds. D. Gabbay and J. Woods, North Holland, 2007.
'Rational Dilemmas', *Analysis* 62 (2002), 11-16.
'Meinongianism and the Philosophy of Mathematics', *Philosophica Mathematica* 11 (2003), 3-15.
'The Hooded Man', *Journal of Philosophical Logic* 31 (2002), 445-67.
'Where is Philosophy at the Start of the Twenty First century?', *Proceedings of the Aristotelian Society* 103 (2003), 85-96.
'Consistency, Paraconsistency and the Logical Limitative Theorems', in *Grenzen und Grenzüberschreitungen (XIX Deutscher Kongress für Philosophie)*, ed. W. Hogrebe and J. Bromand, Akademie Verlag, 2004.
'The Paradoxes of Denotation', ch. 7 of *Self-Reference*, eds. T. Bolander, V. F. Hendricks and S. A Pedersen, CLSI Lecture Notes, Stanford University, 2006.
'A Site for Sorites', pp. 9-23 of J. C. Beall (ed.), *Liars and Heaps: New Essays on Paradox*, Oxford University Press, 2003.
'Wittgenstein's Remarks on Goedel's Theorem', ch. 8 of M. Kölbel and B. Weiss (ed.) *Wittgenstein's Lasting Significance*, Routledge, 2004.
'Inconsistent Arithmetic: Issues Technical and Philosophical', pp. 273-99 of V. F. Hendricks and J. Malinowski (eds.), *Trends in Logic: 50 Years of Studia Logica (Studia Logica Library*, Vol. 21), Kluwer Academic Publishers, 2003.
'Conditionals: a Debate with Jackson', ch. 13 of I. Ravenscroft (ed.), *Minds, Worlds and Conditionals: Themes from the Philosophy of Frank Jackson*. Oxford University Press, 2009.

Spiking the Field Artillery', in J. C. Beall and B. Armour-Garb (eds.), ch. 3 of *Truth and Deflationism*, Oxford University Press, 2005.
'Problems with the Argument for Fine Tuning' (with Mark Colyvan and Jay Garfield), *Synthese* 145 (2005), 325-38.
'Words Without Knowledge', *Philosophy and Phenomenological Research* 71 (2005), 686-94.
'Foreword (Cuivant Inainte)' to I. Lucica (ed.), *Ex Falso Quodlibet: studii de logica paraconsistenta* (in Romanian), Editura Technica, 2004.
'Chunk and Permeate I: the Infinitesimal Calculus' (with Bryson Brown), *Journal of Philosophical Logic* 33 (2004), 379-88.
'Intentionality - Meinongianisn and the Medievals', (with Stephen Read*)*, *Australasian Journal of Philosophy*, 82 (2004), 421-442.
'Logic, Paraconsistent', in D. Borchert (ed.), *Encyclopedia of Philosophy* (second edition), Macmillan, 2006, Vol. 7, 105-6.
'Logic, Relevant (Relevance)', in D. Borchert (ed.), *Encyclopedia of Philosophy* (second edition), Macmillan, 2006, Vol. 8, 358-9.
'Motion', in D. Borchert (ed.), *Encyclopedia of Philosophy* (second edition), Macmillan, 2006, Vol. 6, 409-11.
'The Limits of Language' in K. Brown (ed.), *Encyclopedia of Language and Linguistics*, (second edition) Vol.7, 156-9, Elsevier, 2005.
'Analetheism: a Phyrric Victory' (with Brad Armour-Garb), *Analysis* 65 (2005), 167-73.
'Logic, Many-Valued', in D. Borchert (ed.), *Encyclopedia of Philosophy* (second edition), Macmillan, 2006, Vol. 5, 688-95.
'Logic, Non-Classical', in D. Borchert (ed.), *Encyclopedia of Philosophy* (second edition), Macmillan, 2006, Vol. 5, 485-93.
'The Proliferation of Non-Classical Logics', in D. Borchert (ed.), *Encyclopedia of Philosophy* (second edition), Macmillan, 2006, Vol. 5, 482-4.
'Reply to Slater', pp. 467-74 of J-Y Beziau, W. Carnielli and D. Gabbay (eds.), *Handbook of Paraconsistency*, College Publications, 2007.
'Restricted Quantification in Relevant Logic', (with JC Beall, R. Brady, A. Hazen, and G. Restall) *Journal of Philosophical Logic* 35/6 (2006), 587-98.
'What is Philosophy?', *Philosophy* 81 (2006), 189-207.
'Not so Deep Inconsistency: a Reply to Eklund' (with JC Beall), *Australasian Journal of Logic* 5 (2007) 74-84.
'The Limits of Knowledge', *Logica Yearbook 2005*, M. Bilkova and Ondrej Tomala, Prague, 2006, pp. 165-76, and *Philosophy of Uncertainty and Medical Decisions* (*Bulletin of Death and Life Studies*, Vol. 2), Graduate School of Humanities and Sociology, The University of Tokyo, 2006, pp. 53-63.
'Beyond the Limits of Knowledge', ch. 7 (pp. 93-104) of J. Salerno (ed.), *New Essays on the Knowability Paradox*, Oxford University Press, 2009.
Foreword to the Romanian translation of the second edition of *Beyond the Limits of Thought*. Pp. 23-25 of *Dincolo de limitele gândirri*, (trs. Dimitru Gheorghiu), Paralela 45, 2007.
Foreword to F. Berto, *Teorie dell'assurdo. I rivali del Principio di Non-Contraddizione*, Carocci, 2006.
'Analysing of the Iraqi Adventure', *Ormond Papers* 22 (2005), 147-50.
'A Hundred Flowers', *Topoi* 25 (2006), 91-5.
'The Way of the Dialetheist: Contradictions in Buddhism' (with J. Garfield and Y. Deguchi), *Philosophy East and West* 58 (2008), 395-402.
'Logic', $\sqrt{News}$ 10 (2007), 5-8. Reprinted with minor amendments in *The Dictionary*, ed. Giandomenico Sica, Polymetrica, http://www.polimetrica.eu/site/?p=115.
'60% Proof: Proof, Lakatos and Paraconsistency', (with Neil Thomason), *Australasian Journal of Logic* 5 (2007), 89-100.
'Jaina Logic: a Contemporary Perspective', *History and Philosophy of Logic* 29 (2008), 263-278.

'How the Particular Quantifier became Existentially Loaded Behind our Backs', *Soochow Journal of Philosophical Studies* 16 (2007), 197-213.
'Precis of Towards Non-Being', *Review of Metaphysics* 74 (2008), 185-90.
'Replies to Nolan and Kroon', *Philosophy and Phenomenological Research* 74 (2008), 208-14.
'Revenge, Field, and ZF', ch. 9 of JC Beall (ed.), *Revenge of the Liar: New Essays on the Paradox*, Oxford University Press, 2007.
'The Structure of Emptiness', *Philosophy East and West* 59 (2009), 467-480.
'Many-Valued Modal Logic: a Simple Approach', *Review of Symbolic Logic* 1 (2008), 190-203.
'The Closing of the Mind: How the Particular Quantifier Became Existentially Loaded Behind our Backs', *Review of Symbolic Logic* 1 (2008), 42-55.
'The Australasian Association of Philosophy' (with Eliza Goddard), in G. Oppy and N. Trakakis, *A Companion to Philosophy in Australia and New Zealand*, Monash University Publishing, 2010.
'Mountains are Just Mountains' (with Jay Garfield), pp. 71-82 of J. Garfield and M. D'Amato (eds.), *Pointing at the Moon: Buddhism. Logic and Analytic Philosophy*, Oxford University Press, 2009.
Translation of 'Logic of Paradox', 'La Logique du paraqdoxe', *Philosophie* 94 (2007), 72-94.
'Graham Priest and Diderik Batens Interview Each Other', *The Reasoner* 2 (No. 8) (2008), 2-4.
'Creating Non-Existents: Some Initial Thoughts', *Studies in Logic*, 1 (2008), 18-24.
'Neighbourhood Semantics for Intentional Operators', *Review of Symbolic Logic*, 2 (2009), 360-373.
'Badici on Inclosures and the Liar Paradox', *Australasian Journal of Philosophy* 88 (2010), 359-366.
'Two Truths: Two Models', ch. 13 of the Cowherds (eds.), *Moonshadows*, Oxford University Press, 2010.
'The Truth(s) about the Two Truths', (with T. Tillemans and M. Siderits) ch. 8 the Cowherds (eds.), *Moonshadows*, Oxford University Press, 2010.
'A Case of Mistaken Identity', ch. 11, pp. 205-222 of J. Lear and A. Oliver, *The Force of Argument*, Routledge, 2010.
'Not to Be', ch. 23 of R. Le Poidevin, P. Simons, A. McGonigal, and R. Cameron (eds.), *The Routledge Companion to Metaphysics*, Routledge 2009.
Translation of 'Objects of Thought' into Japanese, in *Human Ontology* 15 (2009), 1-12. (Trans. S. Yamaguchi.)
'Logical Pluralism *Hollandaise*', *The Australasian Journal of Logic*, 6 (2008), 210-14.
'Inclosures, Vagueness and Self-Reference', *Notre Dame Journal of Formal Logic* 51 (2010), pp. 69-84.
'Obituary for Leonard Goddard', *The Australasian Journal of Philosophy* 87 (2009) 693-4.
'Hopes Fade for Saving Truth', *Philosophy* 85 (2010), 109-140.
'Non-Transitive Identity', ch. 23 of R. Dietz and S. Moruzzi (eds.), *Cuts and Clouds: Vagueness, its nature and its Logic*, Oxford University Press, 2010.
'Foreword' to P. Kabay, *On the Plenitude of Truth: a Defense of Trivialism*, LAP-LAMBERT Academic Publishing, 2010.
'Dualising Intuitionistic Negation', *Principia* 13 (2009), 165-89.
'The Logic of the Catuskoti', *Comparative Philosophy* 1 (2010), 32-54.
'Contradiction and the Structure of Unity', pp. 35-42 of *Analytic Philosophy in China, 2009*, ed. Yi, Jiang, 2010.
'Paradoxical Truth', *New York Times*, 28 November 2010. http://opinionator.blogs.nytimes.com/2010/11/28/paradoxical-truth/
'Quine: Naturalism Unravelled', pp. 19-30 of M. Dumitru and C. Stoenescu (eds.), *Cuvinte, teorii si lucruri: Quine in perspectiva*, editura Pelican (Bucharest), 2010.
'An Interview with Bodhidharma', ch. 2 of G. Priest and D. Young (eds.), *Beating and Nothingness: Philosophy and the Martial Arts*, Open Court, 2010.
'Creating Non-Existents', pp. 107-118 of F. Lihoreau (ed.), *Truth in Fiction*, Ontos Verlag, 2011.
'First-Order da Costa Logic', *Studia Logica* 97, (2011), pp.183-198.
'Four Corners – East and West', pp. 12-18 of M. Banerjee and A. Seth (eds.), *Logic and Its Applications*, Springer, 2011.

'Can u do that?', (with JC Beall and Z. Weber), *Analysis* 71 (2011), 280-5.
'Paraconsistent Set Theory', ch. 8 of D. DeVidi, M. Hallett, and P. Clarke (eds.), *Logic, Mathematics, Philosophy: Vintage Enthusiasms. Essays in Honour of John L. Bell*, Springer, 2011.
'Against Against Non-Being', *Review of Symbolic Logic* 4 (2011), 237-53.
'Envelopes and Indifference' (with Greg Restall), pp. 283-290 in *Dialogues, Logics and Other Strange Things*, essays in honour of Shahid Rahman, edited by Cédric Dégremont, Laurent Keiff and Helge Rückert, College Publications, 2008.
'Realism, Anti-realism and Paraconsistency', ch. 10 of S. Rahman, G. Primeiero, and M. Marion (eds.), *The Realism and Antirealism Debate in the Age of Alternative Logics*, Springer, 2012.
'A brief remembrance of Michael Dummett', The Opinionator, *New York Times*, http://opinionator.blogs.nytimes.com/2012/01/04/remembering-michael-dummett/. 4th January, 2012.
'A Note on the Axiom of Countability', *Al-Mukhatabat* 1 (2012), 23-32; http://mukhatabat.unblog.fr/
'An Interview with Graham Priest', pp. 183-198 of G. Oppy and N. Trakakis (eds.), *The Antipodean Philosopher, Vol. 2: Interviews with Australian and New Zealand Philosophers*, Lexington Books, 2012.
'Why Asian Philosophy?', ch. 18 of G. Oppy and N. Trakakis (eds.), *The Antipodean Philosopher*, Lexington Books, 2011.
'Logically Speaking', interview with 3AM Magazine, http://www.3ammagazine.com/3am/logically-speaking/.
'The Sun may, Indeed, not Rise Tomorrow: a Reply to Beall', *Analysis* 72 (2012), 739-41.
'Definition Inclosed: a Reply to Zhong', *Australasian Journal of Philosophy*, 90 (2012), pp. 1-7.
'On the Ternary Relation and Conditionality' (with Jc Beall, Ross Brady, J. Michael Dunn, A. P. Hazen, Edwin Mares, Robert K. Meyer, Greg Restall, David Ripley, John Slaney, Richard Sylvan), *Journal of Philosophical Logic*, 41 (2012), pp. 595–612.
'The Axiom of Countability', pp. 164-9 of *Philosophy, Mathematics, Linguistics: Aspects of Interaction*, Euler International Mathematical Institute, St Petersburg, 2012.
'What does the Ternary R Mean?' pp. 19-24 of *Philosophy, Mathematics, Linguistics: Aspects of Interaction*, Euler International Mathematical Institute, St Petersburg, 2009.
'Between the Horns of Idealism and realism: the Middle Way of Madhyamaka', in Steven Emmanuel (ed.), *A Companion to Buddhist Philosophy*, Wiley-Blackwell, 2013.
'Mathematical Pluralism', *Logic Journal of IGPL* 21 (2013), pp. 4-14.
'Jaina Logic: a Contemporary Perspective'. Pp. 142-93 of M. N. Mitra, M. K. Chakraborty, and S. Sarukkai (eds.), *Studies in Logic: A Dialogue Between East and West*, Sanctum Books, 2012.
'Read on Bradwardine on the Liar', pp. 155-161 of C. Dutilh Novaes and O. Hjorland (eds.), *Insolubles and Consequences: Essays in Honour of Stephen Read*, College Publications, 2012.
'Vague Inclosures', ch. 20 (pp. 367-77) of K. Tanaka, F. Berto, E. Mares, and F. Paoli (eds.), *Paraconsistency: Logic and Applications*, Springer Verlag, 2012.
'Lost in Translation: a Reply to Woodward', *Philosophy and Phenomenological Research* 86 (2013), pp. 194-9.
'A Prolegomenon to any Planning for the Future', *Ormond Papers* 29 (2012), pp. 136.43.
'Nāgārjuna's *Mūlamadhyamakakārika*', *Topoi* 32 (2013), 129-34.
'Philosophy Sans Frontieres: Analytic and Continental Philosophy – a View from the East', ch. 8 of B. Mou and R. Tiezen (eds.), *Constructive Engagement of Analytic and Continental Approaches to Philosophy*, Brill, 2013.
Czech translation of Garfield and Priest (2003) [113]. Pp. 305-330 of Jiri Holba (ed. and trans.), *Nagardzuna: Filosofie Stredni Cesty*, Oikoymenh, 2012.
'The Martial Arts and Buddhist Philosophy', pp. 17-28, *Royal Institute of Philosophy, Supp. Volume*, 73, 2013.
'The Parmenides: a Dialetheic Interpretation', in *Plato: the Journal of the International Plato Society* 12 (2012). Appeared August, 2013, http://gramata.univ-paris1.fr/Plato/article120.html.
'A Mountain by any Other Name: a Response to Koji Tanaka' (with Jay Garfield and Yasuo Deguchi), *Philosophy East & West*, 63 (2013), pp. 335-43.

'Two Plus Two Equals One: a Response to Brook Ziporyn' (with Jay Garfield and Yasuo Deguchi), *Philosophy East & West*, 63 (2013), pp. 353-58.

'The Contradictions are True – and it's not Out of This World!: a Response to Takashi Yagisawa' (with Jay Garfield and Yasuo Deguchi), *Philosophy East & West*, 63 (2013), pp. 370-72.

'Does a Table Have Buddha Nature? A Moment of Yes and No. Answer! But not in Words of Signs: a Response to Mark Siderits', (with Jay Garfield and Yasuo Deguchi), *Philosophy East & West*, 63 (2013), pp. 387-98.

"Those Concepts Proliferate Everywhere: a Response to Constance Kassor', (with Jay Garfield and Yasuo Deguchi), *Philosophy East & West*, 63 (2013), pp. 411-16.

'How We Think Madhyamikas Think: a Response to Tom Tillemans' (with Jay Garfield and Yasuo Deguchi), *Philosophy East & West*, 63 (2013), pp. 462-35.

'Zen and the Art of Harley Riding', pp. 2-12 of B. E. Rollin, C. M. Gray, K. Mommer, and C. Pineo (eds.), *Harley Davidson and Philosophy* (Popular Culture and Philosophy, Vol. 18), Open Court, 2006.

'The Prime Minister' ('When Tony Abbot met Socrates'), *The Conversation*, Nov 27, 2013, http://theconversation.com/when-tony-abbott-met-socrates-20360.

'Logic and Buddhist Metaphysics', Oxford VSI Blog, http://blog.oup.com/2013/12/logic-and-buddhist-metaphysics/.

'Indefinite Extensibility – Dialetheic Style', *Studia Logica* 101 (2013), pp. 1263-1275.

'Foreword', *Kereknyomok* 2013/7, p. 6.

'Entravista – Graham Priest', *Polemos* 2 (2013), 167-194.

'Logical Pluralism: Another Application of Chunk and Permeate', *Erkenntnis* 29 (2014), 331-8.

'Plurivalent Logic', *Australasian Journal of Logic* 11 (2014); article 1. http://ojs.victoria.ac.nz/ajl/article/view/1830.

'Philosophy and its History', *Oxford University Press VSI Blog*, http://blog.oup.com/2014/04/philosophy-history/

'Lost Platonic Dialogue Found', pp. 105-23 of *Four Lives: a Celebration of Raymond Smullyan*, ed. J. Rosenhouse, Dover Publications, 2014.

'Beyond True and False', *Aeon*, 5 May 2014, http://aeon.co/magazine/world-views/logic-of-buddhist-philosophy/

'Chunk and Permeate III: the Dirac Delta Function' (with R. Benham and C. Mortensen), *Synthese* 191 (2014), pp. 3057-62.

'Contradictory Concepts', ch. 10 of *Logic, Reasoning and Rationality*, eds. E. Weber, D. Wouters, and J. Meheus, Springer, 2014.

'Contradictory Concepts' (short version), ch. 1 of *Contradictions: Logic, History, and Actuality*, ed. E. Ficara, De Gruyter, 2012.

'Speaking of the Ineffable…', ch. 7 of *Nothingness in Asian Philosophy*, eds. J. Lee and D. Berger, Routledge, 2014.

'Buddhism, Logic, and All That', *Scientia Salon*, September 8, 2014, http://scientiasalon.wordpress.com/2014/09/08/logic-buddhism-and-all-that/.

'The Martial Arts and Buddhist Philosophy', ch. 11 of *Philosophy and the Martial Arts: Engagement*, ed. G. Priest and D. Young, Routledge, 2014.

'Revising Logic', ch. 12 of P. Rush (ed.), *The Metaphysics of Logic*, Cambridge: Cambridge University Press, 2014.

'Much Ado about Nothing', *Australasian Journal of Logic* 11 (2014), Article 4, http://ojs.victoria.ac.nz/ajl/issue/view/209.

'*Sein* Language', *Monist* 97 (2014), 430-42.

'External Curries' (with Heinrich Wansing), *Journal of Philosophical Logic*, 44 (2015), pp. 453-71.

'Introduction' (to a special issue of the *AJP* on David Lewis) (with Frank Jackson), *Australasian Journal of Philosophy* 82 (2004), 1-2.

'Introduction: Contemporary Australian Research in Logic', (with Chris Mortensen), *Logique et Analyse*, 137-8 (1992), 3-4.

'Three Heresies in Logic and Metaphysics', *Polish Journal of Philosophy* 7 (2013), 9-20.

'Replies', *Polish Journal of Philosophy* 7 (2013), 93-108.
'Chunk and Permeate II: Bohr's Hydrogen Atom' (with Bryson Brown), *European Journal for the Philosophy of Science* 5, 297-314.
'Fusion and Confusion', *Topoi* 34 (2015), 55–61.
'In the Same Way that This is: a Reply to Dotson', *Comparative Philosophy* 3 (2012), 3-9. http://scholarworks.sjsu.edu/comparativephilosophy/vol3/iss2/5/.
Introduction to Japanese translation of *Towards Non-Being*. (Tokyo: Keiso Shobo, 2011).
'The Net of Indra', pp. 113-127 of K. Tanaka, et al, *The Moon Points Back*, Oxford University Press, 2015.
'Philosophie Sans Frontières', OUP Blog, http://blog.oup.com/2015/05/eastern-western-philosophy-tradition/, May 23, 2015.
'Introduction' (with K. Tanaka et al), pp. xi-xvi of *The Moon Points Back*, Oxford University Press, 2015, eds. K. Tanaka, Y. Deguchi, J. Garfield, and G. Priest.
'Kripke's Thought Paradoxes and the 5th Antinomy' ch. 24 of T. Achourioti, H. Galinon, J. Fernandez, and K. Fujimoto (eds.), *Unifying the Philosophy of Truth*, Springer, 2015.
'Is the Ternary R Depraved?', ch. 4 of C. Caret and O. Hjortland (eds.), *Foundations of Logical Consequence*, Oxford University Press, 2015.
'Modal Meinongianism and Characterization. Reply to Kroon', *Grazer Philosophische Studien* 90 (2014), pp. 183-200.
'Alethic Values', *Newsletter of the APA, Asian and Asian-American Philosophers and Philosophies*, 14(2), (2015), pp. 2-4.
'Nineteenth Century German Logic', ch. 20 of M. Forster and K. Gjesdal (eds.), *The Oxford Handbook of German Philosophy in the Nineteenth Century*, Oxford University Press 2015.
'What I'm Reading', *Meanjin* 74 (2015), http://meanjin.com.au/blog/what-im-reading-graham-priest/.
'Graham Priest on Philosophy and Buddhism', *Philosophy Bites*, http://philosophybites.com/2015/10/graham-priest-on-buddhism-and-philosophy.html
'The Answer to the Question of Being', pp. 249-58 of J. Bell, A. Cutrofello, and P. Livingston (eds.), *Beyond the Analytic-Continental Divide: Pluralist Philosophy in the Twenty-First Century*, London: Routledge, 2015.
'Introduction: Why ask About Madhyamaka and Ethics?' (with Jay Garfield), pp. 1-6 of the Cowherds (eds.), *Moonpaths: Ethics and Emptiness*, Oxford University Press, 2015.
'The Santideva Passage: *Bodhicaryvatara* VIII 90-103' (with Stephen Jenkins and Jay Garfield), pp. 55-76 of the Cowherds (eds.), *Moonpaths: Ethics and Emptiness*, Oxford University Press, 2015.
'Compassion and the Net of Indra', pp. 221-39 of the Cowherds (eds.), *Moonpaths: Ethics and Emptiness*, Oxford University Press, 2015.
'Paradoxical Truth', pp. 166-71 of *The Stone Reader: Modern Philosophy in 133 Arguments*, ed. P. Catapano and S. Critchley, Liveright Publishing Co, 2016.
'Torn by Reason: Lukasiewicz on the Principle of Non-Contradiction', ch. 18 of S. Costreie (ed.), *Early Analytic Philosophy: Some New Perspectives on the Tradition*, Springer, 2016.
'None of the Above: The Catuskoti in Indian Buddhist Logic', ch. 24 of J.-Y. Beziau, M. Chakraborty, and S. Dutta (eds.), *New Directions in Paraconsistent Logic*, Springer, 2015.
'Tolerating Gluts' (with Z. Weber, D. Ripley, D. Hyde, and M. Colyvan), *Mind* 123 (2014): 813-28.
'Comment on Restall', *Thought* 5 (2016), p. 125.
'Is Moonshadows Lunacy? The Cowherds Respond', (as one of the Cowherds) *Philosophy East and West* 66 (2016), pp. 617-21.
'Reflections on Zen and the Art of Motorcycle Maintenance', *International Journal of Motorcycle Studies*, 10 (2014), http://ijms.nova.edu/Fall2014/IJMS_Rndtble.Priest.html.
'Logical Disputes and the *a Priori*', *Principios: Rivista de Filosofia* 23 (2016), pp. 29-57.
A Dictionary of Logic (with Thomas Ferguson) (2016)Oxford University Press, http://www.oxfordreference.com/view/10.1093/acref/9780191816802.001.0001/acref-9780191816802

'The Enduring Evolution of Logic', with Thomas Ferguson, *OUP Blog*, July 2016, http://blog.oup.com/2016/07/history-of-logic/.

'Old Wine in (Slightly Leaky) New Bottles: Some Comments on Beall', *Australasian Journal of Logic* 13(5) (2016), Article 1, https://ojs.victoria.ac.nz/ajl/article/view/3934/3544.

'Thinking the Impossible', *Philosophical Studies*, 173 (2016), pp. 2649–2662.

'It is and it Isn't' (with Damon Young), *Aeon* (2016), https://aeon.co/essays/how-can-duchamp-s-fountain-be-both-art-and-not-art.

'The Strange Case of the Missing Object', *OUP Blog*, September 2016, http://blog.oup.com/2016/09/non-existent-objects-philosophy/

'Foreword: Edward Conze and the Law of Non-Contradiction', pp. vii-x of E. Conze, *The Principle of Contradiction: On the Theory of Dialectical Materialism* (tr. Holger Heine), Lanham, MD: Lexington Books, 2016.

'Answers to Five Questions', pp. 153-160 of T. Adajian and T. Lupher (eds.), *Philosophy of Logic: Five Questions*, Automatic Press, 2016.

'Logical Disputes and the *a Priori*', *Logique et Analyse* 236 (2016), pp. 347-66.

'Nasz Świat Nie Jest Najlepszym z Możliwych' ('Our World is not the Best Possible'), *Filosofuj* 2016 #6 (12), pp. 26-27.

'Where Laws Conflict: an Application of Chunk and Permeate', ch. 8 of H. P. Glenn and L. D. Smith (eds.), *Law and the New Logics*, Cambridge University Press, 2017.

'Stop Making Sense', *Philosophical Topics* 43 (2015), pp. 285-99. [Note that this appeared in 2017, notwithstanding the journal date.]

'Three Questions to Graham Priest', *Paraconsistent Logic Newsletter*, Spring 2017, http://www.paraconsistency.org/spring2017.html.

'What If? The Exploration of an Idea', *Australasian Journal of Logic* 14 (2017), https://ojs.victoria.ac.nz/ajl/article/view/4028/3574.

'Speaking of the Ineffable, East and West', *European Journal of Analytic Philosophy* 11 (2015), pp. 6-21. [Note: despite the date of the volume, this issue actually appeared in 2017.]

'What is it like to be a philosopher? Interview with Graham Priest', http://www.whatisitliketobeaphilosopher.com.

'Thinking the Impossible', *Argumenta*, 2, (2017), pp. 181-194.

'A Note on the Axioms of Countability', *IfCoLog Journal of Logics and their Applications* 4 (2017), pp, 1351-6. Republished as pp 177-81 of O. Proserov (ed), *Proceedings of the International Conference Philosophy, Mathematics, Linguistics: Aspects of Interaction, 2012 (PhML-2012)*, College Publications 2017.

'Contradiction and the Instant of Change Revisited', *Vivarium* 55 (2017), pp. 217-26.

'Eubulides and his Paradoxes', *Oxford University Blog*, https://blog.oup.com/2017/08/eubulides-paradoxes-philosophy/.

'Précis of *One*', *International Journal of Philosophical Studies* 25 (2017), pp. 532-5.

'Entangled Gluons: Replies to Casati, Han, Kim, and Yagisawa', *International Journal of Philosophical Studies* 25 (2017), pp. 560-8.

'Metaphysical Grounding, East and West' (with Ricki Bliss), pp. 63-85 of S. Emmanuel (ed.), *Buddhist Philosophy: a Comparative Approach*, Wiley Blackwell, 2017.

'What is the Specificity of Classical Mathematics?', *Thought* 6 (2017), pp. 115-21.

'Things are Not What they Seem', pp. 225-236 of M. Silva (ed.), *How Colours Matter to Philosophy*, Springer, 2017.

'Plurivant Logics', pp. 169-79 of V. Markin and D. Zaitsev (eds.), *The Logical Legacy of Nokolai Vasiliev and Modern Logic*, Springer 2017.

'Interview with Graham Priest', *Figure/Ground*, January 1, 2018, http://figureground.org/interview-graham-priest/.

'Buddhist Ethics: a Perspective', ch. 5 of J. H. Davis (ed.), *A Mirror is for Reflection: Understanding Buddhist Ethics,* New York: Oxford University Press, 2017.

Preface (pp. 13-18) of *The Vindication of Nothingness*, by Marco Simionato, Editiones Scholasticae (Taylor and Francis), 2017.

'Inside *Aussersein*' (with Filippo Casati), *The IfCoLog Journal of Logics and their Applications* 4 (2017), pp. 3583-96.
'Contradiction and the Instant of Change Revisited', pp 217-226 of F. Goubier and M. Roques (eds.), *The Instant of Change in Medieval Philosophy and Beyond*, Brill, 2018. [A reprint of the 2017 *Vivarium* article.)
'The Geography of Fundamentality' (with R. Bliss), Ch. 0 of R. Bliss and G. Priest (eds.), *Reality and its Structure: Essays in Fundamentality*, Oxford University Press, 2018.
'Buddhist Dependence', Ch. 6 of R. Bliss and G. Priest (eds.), *Reality and its Structure: Essays in Fundamentality*, Oxford University Press, 2018.
'Introduction: Approaching Sylvan and this Collection of Essays' (with Filippo Casati and Chris Mortensen), *Australasian Journal of Logic* 15(2), 2018, Article 1. https://ojs.victoria.ac.nz/ajl/article/view/4854.
'Some New Thoughts on Conditionals', *Topoi* 27 (2018), pp. 369-77.
'Heidegger and Dōgen on the Ineffable' (with Filippo Casati), ch. 14 of R. H. Scott ad G. S. Moss (eds.), *The Significance of Indeterminacy*, Routledge, 2019. (Despite the date, this appeared in 2018.)
'Wittgensteins Bemarkungen zu Goedel's Theorem', pp. 143-69 of J. Bromand (ed.), *Wittgenstein und die Philosophie der Mathematik*, Mentis. (A translation of 122.)
'Paradoxical Propositions', *Philosophical Issues* 28 (2018), pp. 300-307.
'Marxism and Buddhism: Not such Strange Bedfellows', *Journal of the American Philosophical Association*, 4 (2018), pp. 2-13.
'Characterisation, Existence and Necessity', pp. 250-269 of G. Oppy (ed), *Ontological Arguments*, Cambridge University Press, 2018.
Farsi translation of 'The Prime Minister' (221 above), *Logos* (*An Iranian Quarterly Journal of Philosophy*) 1 (2018), pp. 16-26.
'Williamson on Counterpossibles' (with Franz Berto, Rohan French, and Dave Ripley), *Journal of Philosophical Logic* 47 (2018), pp. 693-713.
'Berry's Paradox… Again', *Australasian Journal of Logic* 16 (2019), pp. 41-48.

b) Monographs and Books

On Paraconsistency (with R. Routley). Research Report #13, Research School of Social Sciences, Australian National University 1983. Reprinted as the introductory chapters of *Paraconsistent Logic*, G. Priest, R. Routley and J. Norman (eds.), Philosophia Verlag, 1989. Translated into Romanian as chapters in I. Lucica (ed.), *Ex Falso Quodlibet: studii de logica paraconsistenta* (in Romanian), Editura Technica, 2004.
In Contradiction: a study of the transconsistent, Martinus Nijhoff, 1987. Second edition Oxford University Press, 2006.
Beyond the Limits of Thought, Cambridge University Press, 1995. Second edition, Oxford University Press, 2002.
Logic: a Very Short Introduction, Oxford University Press, 2000. Translated into Portuguese as *Lógica para Começar*, Temas & Debates, 2002. Translated into Spanish as *Una Brevísima Introducción a la Lógica*, Oceano, 2006. Translated into Czech, as *Logica*, Dokorán, 2007. Translated into Japanese, Iwanami Shoten, 2008. Translated into Chinese by Yilan Press, 2011. Translated in Farsi, Press?, 2007. Translated into Italian, Condice Edizioni, 2012. Translated into Turkish as *Mantik*, Dost Kitabevi, 2017. Second edition, 2017. Translated into Finnish, as *Logiikka*, Niin & Näin, 2017.
Introduction to Non-Classical Logic, Cambridge University, 2001.
Towards Non-Being: the Semantics and Metaphysics of Intentionality, Oxford University Press, 2005. Second edition, 2016.

Doubt Truth to be a Liar, Oxford University Press, 2006.
Dincolo de limitele gândirri, (trs. Dimitru Gheorghiu), Paralela 45, 2007. (Romanian translation of the second edition of 3.)
Introduction to Non-Classical Logic: from If to Is, Cambridge University Press, 2008.
German translation of Part 1 of *Introduction to Non-Classical Logic: From If to Is*: *Einführung in die nicht-klassische logic*, Mentis 2008.
Logic: A Brief Insight, Sterling 2010.
Japanese translation of *Towards Non-Being* (Tokyo: Keiso Shobo, 2011). (Translators Naoya Fujikawa and Minao Kukita.)
One: Being an Investigation of the Unity of reality and of it Parts, including the Singular Object which is Nothingness, Oxford University Press, 2014.
Uno (Spanish Translation of 13), Alpha Decay, 2016.
The Fifth Corner of Four: and Essay in Buddhist Metaphysics and the Catuṣkoṭi, Oxford University Press, 2018.

c) Works Edited

Proc. Bertrand Russell Memorial Logic Conference, eds. J. Bell *et al*, Leeds 1973.
Studia Logica 1983, 43. [A double issue on paraconsistent logic, guest-edited by myself and R. Routley.]
Paraconsistent Logic: essays on the inconsistent, eds. G. Priest and R. Routley, J. Norman, Philosophia Verlag, 1989.
Logique et Analyse 1992, 137-8 and 139-40. [Two issues entitled 'Contemporary Australian Research in Logic', guest-edited by myself and C.Mortensen.]
Notre Dame Journal of Formal Logic, 1997, #4. [An issue on impossible worlds, guest edited.]
Frontiers of Paraconsistency, (with Diderik Batens, Chris Mortensen and Jean-Paul van Bendegem), Research Studies Press, 2000.
Sociative Logics and their Applications: essays by the late Richard Sylvan' (with Dominic Hyde), Ashgate Publishing Co., 2000
The Monist 86(1), (2003). (with Roger Lamb). [An issue on the topic of perversion.]
New Essays on the Law of Non-Contradiction, eds. G. Priest, J. C. Beall and B. Armour-Garb, Oxford University Press, 2004.
Australasian Journal of Philosophy 82(1), (2004), (with Frank Jackson). [A special issue on David Lewis and his philosophy.]
Lewisian Themes: the Philosophy of David K Lewis, with Frank Jackson, Oxford University Press, 2004.
Moonshadows: Conventional Truth in Buddhist Philosophy, as one of The Cowherds (eds.), Oxford University Press, 2010.
Beating and Nothingness: Philosophy and the Martial Arts (with Damon Young), Open Court, 2010.
Philosophy and the Martial Arts: Engagement (with Damon Young), Routledge, 2014.
The Moon Points Back (with K. Tanaka, Y. Deguchi, and J. Garfield), Oxford University Press, 2015.
Moonpaths: Ethics and Emptiness, as one of the Cowherds (eds.), Oxford University Press, 2015.
Reality and its Structure: Essays in Fundamentality (with Ricki Bliss), Oxford University Press, 2018.
The Australasian Journal of Logic, Special Issue on Richard Sylvan, 15(2), 2018 (with Filippo Casati and Chris Mortensen). https://ojs.victoria.ac.nz/ajl/issue/view/568

d) Reviews, Critical Notices, etc.

Logic: depth grammar of rationality, P.K. Bastable, *Times Higher Educational Supplement* 210, 31 October 1975, p.20.
Deviant Logic, S. Haack, *Philosophical Quarterly* 1975, 25, 371-3.
Mathematical Knowledge, M. Steiner, *Philosophical Quarterly* 1976, 26, 281-2.
Can Theories be Refuted? ed. S. Harding, *Philosophical Quarterly* 1977, 27, 73-4.
From Belief to Understanding, R. Campbell, *Philosophical Quarterly* 1978, 28, 92-4.
Willard Van Orman Quine, A. Orenstein, *Philosophical Quarterly* 1979, 29, 173-5.
Knowledge and Science, H. Kannegeiser, *Philosophical Quarterly* 1979, 29, 366-7.
Ways of Meaning, M. Platts, *Australasian Journal of Philosophy* 1980, 58,74-6.
Theory and Meaning, D. Papineau, *Philosophical Quarterly* 1981, 31, 77-79.
Russell, R.M. Sainsbury, *Australasian Journal of Philosophy* 1981, 59, 346-7.
Cognitive Systematization, N. Rescher, *Australasian Journal of Philosophy* 1982, 60, 185-6.
On Philosophical Method, H.N. Castañeda, *Australasian Journal of Philosophy* 1982, 60, 296-8.
A Companion to Modal Logic, G. Hughes and M.J. Cresswell, *Australasian Journal of Philosophy* 1986, 64, 220-1.
Truthlikeness, I. Niiniluoto, *Metascience* 6, 1988, 39-40.
Conditionals, F. Jackson, *Australasian Journal of Philosophy*, 1989, 67, 236-9.
Paradoxes, R. M. Sainsbury, *History and Philosophy of Logic* 1990, 11, 245-7.
Computers, Brains and Minds: Essays in Cognitive Science, eds. P. Slezak and W. Albury, *Metascience* 1990, 8, 137-8.
Truth, Vagueness and Paradox, V. McGee, *Mind*, 103 (1994), 387-91.
Entities and Indices, M. J. Cresswell, *Australasian Journal of Philosophy*, 71 (1993), 336-7.
Entailment, Vol. II, A. Anderson, N. Belnap & J. M. Dunn, *Australasian Journal of Philosophy* 72 (1994), 112-4.
Dialectical Investigations, B. Ollman, *Science and Society*, 58 (1994-5), 496-8.
From Dedekind to Goedel, J. Hintikka (ed.), *Studia Logica* 60 (1998), 333-6.
Mathematical Objects and Mathematical Knowledge, M. Resnik (ed.). *Studia Logica* 61 (1998), 293-6.
George Boole: Selected Manuscripts on Logic and Philosophy, I. Grattan-Guinness and G. Bornet (eds.), *Studia Logica* 63 (1999), 143-146.
Logic with Trees, C.Howson, *Studia Logica* 63 (1999), 140-143.
Logiques Classiques et Non-Classique, Newton da Costa, *Studia Logica* 64 (2000), 435-43.
Understanding Truth, Scott Soames, *British Journal for the Philosophy of Science* 52 (2001), 211-5.
Questions of Time and Tense, R. Le Poidevin, *Mind* 110 (2001), 218-22.
Papers in Philosophical Logic, Papers in Metaphysics and Epistemology, Papers in Ethics and Social Theory, David Lewis, *Nous* 36 (2002), 351-58.
The Later Heidegger, George Pattison, *Philosophical Quarterly* 52 (2002), 401-3.
Hegel's Dialectical Logic, E. Bencivenga, *Mind* 111 (2002), 643-46.
Knowledge and its Limits, T. Williamson, *Journal of Philosophy* 100 (2003), 268-71.
Possibilities and Paradox, J. C. Beall and B. van Fraassen, *Studia Logica* 79 (2005), 310-313.
Relevant Logic, Ed Mares, *Australasian Journal of Philosophy* 83 (2005), 431-3.
Free Logic, Karel Lambert, *Philosophia Mathematica* 13 (2005), 326-328.
Truth and Paradox, Tim Maudlin, *Journal of Philosophy*, 102 (2005), 483-6.
Fear of Knowledge: against Relativism and Constructivism, Paul Boghossian, *Review of Metaphysics* 61 (2007), 120-2.
Mainstream and Formal Epistemology, Vincent F. Hendricks, *British Journal for the Philosophy of Science*, 60 (2009) 433-437.
Absolute Generality, Agustín Rayo and Gabriel Uzquiano (eds.), *Notre Dame Philosophical Reviews*, 2007.09.17, http://ndpr.nd.edu/review.cfm?id=11144.
Vagueness, Logic and Ontology, Dominic Hyde, and *Vagueness and Degrees of Truth*, Nicholas, J. J. Smith, *History and Philosophy of Logic* 31 (2010), 177-84.

Talking about Nothing, Jody Azzouni, *Philosophia Mathamatica* 19 (2011), 359-63.
The Bodhisattva's Brain: Buddhism Naturalised, Owen Flannagan, Philosophical Quarterly 62 (2012), 862-4.
Indian Buddhist Philosophy, Amber Carpenter, *Philosophical Quarterly* 65 (2015), 585-7.
The Boundary Stones of Thought: an Essay in the Philosophy of Logic, Ian Rumfitt, *Journal of Philosophy* 112 (2016), 570-6.
Replacing Truth, Kevin Scharp, *Mind* 125 (2016), 553-8.
Logical Studies of Paraconsistent Reasoning in Science and Mathematics, eds. H. Andreas and P. Verdée, *Studia Logica* 106 (2018), 1213-18.

Lightning Source UK Ltd.
Milton Keynes UK
UKHW020649080120
356578UK00001B/34/P